Smart Innovation, Systems and Technologies

Volume 55

Series editors

Robert James Howlett, KES International, Shoreham-by-sea, UK
e-mail: rjhowlett@kesinternational.org

Lakhmi C. Jain, University of Canberra, Canberra, Australia;
Bournemouth University, UK;
KES International, UK
e-mails: jainlc2002@yahoo.co.uk; Lakhmi.Jain@canberra.edu.au

About this Series

The Smart Innovation, Systems and Technologies book series encompasses the topics of knowledge, intelligence, innovation and sustainability. The aim of the series is to make available a platform for the publication of books on all aspects of single and multi-disciplinary research on these themes in order to make the latest results available in a readily-accessible form. Volumes on interdisciplinary research combining two or more of these areas is particularly sought.

The series covers systems and paradigms that employ knowledge and intelligence in a broad sense. Its scope is systems having embedded knowledge and intelligence, which may be applied to the solution of world problems in industry, the environment and the community. It also focusses on the knowledge-transfer methodologies and innovation strategies employed to make this happen effectively. The combination of intelligent systems tools and a broad range of applications introduces a need for a synergy of disciplines from science, technology, business and the humanities. The series will include conference proceedings, edited collections, monographs, handbooks, reference books, and other relevant types of book in areas of science and technology where smart systems and technologies can offer innovative solutions.

High quality content is an essential feature for all book proposals accepted for the series. It is expected that editors of all accepted volumes will ensure that contributions are subjected to an appropriate level of reviewing process and adhere to KES quality principles.

More information about this series at http://www.springer.com/series/8767

Giuseppe De Pietro · Luigi Gallo
Robert J. Howlett · Lakhmi C. Jain
Editors

Intelligent Interactive Multimedia Systems and Services 2016

Springer

Editors
Giuseppe De Pietro
Institute of High Performance Computing and Networking
National Research Council
Naples
Italy

Luigi Gallo
Institute of High Performance Computing and Networking
National Research Council
Naples
Italy

Robert J. Howlett
KES International
Shoreham-by-sea
UK

Lakhmi C. Jain
University of Canberra
Canberra
Australia

and

Bournemouth University
Poole
UK

and

KES International
Shoreham-by-sea
UK

ISSN 2190-3018 ISSN 2190-3026 (electronic)
Smart Innovation, Systems and Technologies
ISBN 978-3-319-81871-9 ISBN 978-3-319-39345-2 (eBook)
DOI 10.1007/978-3-319-39345-2

Printed on acid-free paper

This Springer imprint is published by Springer Nature
The registered company is Springer International Publishing AG Switzerland

Preface

Dear Readers,

We introduce to you a series of carefully selected papers presented during the 9th KES International Conference on Intelligent Interactive Multimedia Systems and Services (IIMSS-16).

At a time when computers are more widespread than ever and computer users range from highly qualified scientists to non-computer expert professionals, Intelligent Interactive Systems are becoming a necessity in modern computer systems. The solution of "one-fits-all" is no longer applicable to wide ranges of users of various backgrounds and needs. Therefore, one important goal of many intelligent interactive systems is dynamic personalization and adaptivity to users. Multimedia Systems refer to the coordinated storage, processing, transmission, and retrieval of multiple forms of information, such as audio, image, video, animation, graphics, and text. The growth rate of multimedia services has become explosive, as technological progress matches consumer needs for content.

The IIMSS-16 conference took place as part of the Smart Digital Futures 2016 multi-theme conference, which groups AMSTA-16, IDT-16, InMed-16, and SEEL-16 with IIMSS-16 in one venue. It was a forum for researchers and scientists to share work and experiences on intelligent interactive systems and on multimedia systems and services. It included a general track and nine invited sessions.

The general track (Chaps. "Analysis of Similarity Measurements in CBIR Using Clustered Tamura Features for Biomedical Images"–"How to Manage Keys and Reconfiguration in WSNs Exploiting SRAM Based PUFs") focused on intelligent image or video storage, retrieval, transmission, and analysis. The invited session "Intelligent Video Processing and Transmission Systems" (Chaps. "Fast Salient Object Detection in Non-stationary Video Sequences Based on Spatial Saliency Maps"–"Development Prospects of the Visual Data Compression Technologies and Advantages of New Approaches") specifically focused on functionalities and architectures of systems for video processing and transmission. The invited session "Innovative Information Services for Advanced Knowledge Activity" (Chaps. "A Near-far Resistant Preambleless Blind Receiver with Eigenbeams Applicable to

Sensor Networks"–"Trends in Teaching/Learning Research Through Analysis of Conference Presentation Articles") focused on novel functionalities for information services. The invited session "Autonomous System" (Chaps. "Motion Prediction for Ship-Based Autonomous Air Vehicle Operations"–"Active Suspension Investigation Using Physical Networks") considered issues such as motion prediction, operating systems, and networks for what concerns autonomous systems. The invited session "Mobility Data Analysis and Mining" (Chaps. "Automatic Generation of Trajectory Data Warehouse Schemas"–"A Survey on Web Service Mining Using QoS and Recommendation Based on Multidimensional Approach") focused on novel modelling and analysis approaches for mobility data. The invited session "Intelligent Computer Systems Enhancing Creativity" (Chaps. "Mapping and Pocketing Techniques for Laser Marking of 2D Shapes on 3D Curved Surfaces"–"Experience-driven Framework for Technologically-enhanced Environments: Key Challenges and Potential Solutions") provided insight into the most recent efforts, challenges, and best practices across the fields of computer-aided creativity and innovation. The invited session "Internet of Things: Architecture, Technologies and Applications" (Chaps. "Touchless Disambiguation Techniques for Wearable Augmented Reality Systems"–"Opinions Analysis in Social Networks for Cultural Heritage Applications") focused on IoT approaches, especially considering cultural heritage scenarios. The invited session "Interactive Cognitive Systems" (Chaps. "A Forward-Selection Algorithm for SVM-Based Question Classification in Cognitive Systems"–"A Model of a Social Chatbot") focused on adaptive and human-like cognitive systems and on artificial intelligence systems and robotics. The invited session "Smart Environments and Information Systems" (Chaps. "An Experience of Engineering of MAS for Smart Environments: Extension of ASPECS"–"Soft Sensor Network For Environmental Monitoring") discussed the requirements of the information systems supporting smart environments, as well as the methods and techniques that are currently being explored. Finally, the invited session "New Technologies and Virtual Reality in Health Systems" (Chaps. "The Use of Eye Tracking (ET) in Targeting Sports: A Review of the Studies on Quiet Eye (QE)"–"The Elapsed Time During a Virtual Reality Treatment for Stressful Procedures. A Pool Analysis on Breast Cancer Patients During Chemotherapy") focused on advanced functionalities for VR-based applications in health care.

Our gratitude goes to many people who have greatly contributed to putting together a fine scientific programme and exciting social events for IIMSS 2016. We acknowledge the commitment and hard work of the programme chairs and the invited session organizers. They have kept the scientific programme in focus and made the discussions interesting and valuable. We recognize the excellent job done by the programme committee members and the extra reviewers. They evaluated all the papers on a very tight time schedule. We are grateful for their dedication and contributions. We could not have done it without them. More importantly, we thank the authors for submitting and trusting their work to the IIMSS conference.

We hope that readers will find in this book an interesting source of knowledge in fundamental and applied facets of intelligent interactive multimedia and, maybe, even some motivation for further research.

Giuseppe De Pietro
Luigi Gallo
Robert J. Howlett
Lakhmi C. Jain

Organization

Honorary Chairs

Toyohide Watanabe, Nagoya University, Japan
Lakhmi C. Jain, University of Canberra, Canberra, Australia;
Bournemouth University, UK

Co-General Chairs

Giuseppe De Pietro, National Research Council, Italy
Luigi Gallo, National Research Council, Italy

Executive Chair

Robert J. Howlett, University of Bournemouth, UK

Programme Chair

Antonino Mazzeo, University of Naples Federico II, Italy

Publicity Chair

Giuseppe Caggianese, National Research Council, Italy

Invited Session Chairs

Intelligent Video Processing and Transmission Systems

Margarita N. Favorskaya, Siberian State Aerospace University, Russia
Lakhmi C. Jain, University of Canberra, Australia and Bournemouth University, UK
Mikhail Sergeev, Saint-Petersburg State University of Aerospace Instrumentation, ITMO University, Russia

Innovative Information Services for Advanced Knowledge Activity

Koichi Asakura, Daido University, Japan
Toyohide Watanabe, Nagoya Industrial Research Institute, Japan

Autonomous System

Milan Simic, RMIT University, Australia
Reza Nakhaie Jazar, RMIT University, Australia

Mobility Data Analysis and Mining

Jalel Akaichi, King Khalid University, Saudi Arabia

Intelligent Computer Systems Enhancing Creativity

Raffaele De Amicis, Graphitech, Italy
David Oyarzun, Vicomtech-IK4, Spain

Internet of Things: Architecture, Technologies and Applications

Francesco Piccialli, University of Naples Federico II, Italy
Angelo Chianese, University of Naples Federico II, Italy

Interactive Cognitive Systems

Ignazio Infantino, National Research Council, Italy
Massimo Esposito, National Research Council, Italy

Smart Environments and Information Systems

Massimo Cossentino, National Research Council of Italy, Italy
Vincent Hilaire, Université de Belfort-Montbeliard, France
Juan Pavon, Universidad Complutense de Madrid, Spain

New Technologies and Virtual Reality in Health Systems

Antonio Giordano, College of Science and Technology, Temple University, USA

International Programme Committee

Jalel Akaichi, King Khalid University, Saudi Arabia
Raffaele De Amicis, Graphitech, Italy
Koichi Asakura, Daido University, Japan
Monica Bianchini, Università degli Studi di Siena, Italy
El Fazziki Aziz, Cadi Ayyad University of Marrakesh, Morocco
Angelo Chianese, University of Naples Federico II, Italy
Massimo Cossentino, National Research Council of Italy, Italy
Mario Döller, University of applied science Kufstein Tirol, Austria
Dinu Dragan, Faculty of Technical Sciences, University of Novi Sad, Serbia
Massimo Esposito, National Research Council of Italy, Italy
Margarita Favorskaya, Siberian State Aerospace University, Russia
Colette Faucher, LSIS-Polytech Marseille, France
Christos Grecos, Sohar University, Oman
Vincent Hilaire, Université de Belfort-Montbeliard, France
Katsuhiro Honda, Osaka Prefecture University, Japan
Hsiang-Cheh Huang, National University of Kaohsiung, Taiwan
Ignazio Infantino, National Research Council of Italy, Italy
Lakhmi C. Jain, University of Canberra, Australia and Bournemouth University, UK
Reza Nakhaie Jazar, SAMME, RMIT University, Australia
Dimitris Kanellopoulos, University of Patras, Greece
Chengjun Liu, New Jersey Institute of Technology, USA
Cristian Mihăescu, University of Craiova, Romania
Vincent Oria, New Jersey Institute of Technology, USA
David Oyarzun, Vicomtech-IK4, Spain
Juan Pavon, Universidad Complutense de Madrid, Spain
Francesco Piccialli, University of Naples Federico II, Italy
Radu-Emil Precup, Politehnica University of Timisoara, Romania
Luca Sabatucci, National Research Council of Italy, Italy
Mohammed Sadgal, Cadi Ayyad University of Marrakesh, Morocco
A. Sadiq, Cadi Ayyad University of Marrakesh, Morocco
Mikhail Sergeev, ITMO University, Russia
Milan Simic, SAMME, RMIT University, Melbourne, Australia
Taketoshi Ushiama, Kyushu University, Japan
Toyohide Watanabe, Nagoya Industrial Science Research Institute, Japan

Contents

Analysis of Similarity Measurements in CBIR Using Clustered Tamura Features for Biomedical Images

Nadia Brancati and Francesco Camastra

Abstract Content based image retrieval (CBIR) is an important research topic in many applications, in particular in the biomedical field. In this domain, the CBIR has the aim of helping to improve the diagnosis, retrieving images of patients for which a diagnosis has already been made, similar to the current image. The main issue of CBIR is the selection of the visual contents (feature descriptors) of the images to be extracted for a correct image retrieval. The second issue is the choice of the similarity measurement to use to compare the feature descriptors of the query image to ones of the other images of the database. This paper focuses on a comparison among different similarity measurements in CBIR, with particular interest to a biomedical images database. The adopted technique for CBIR is based on clustered Tamura features. The selected similarity measurements are used both to evaluate the adopted technique for CBIR and to estimate the stability of the results. A comparison with some methods in literature has been carried out, showing the best results for the proposed technique.

Keywords Content based image retrieval · Tamura features · Similarity measurement · CBIR by clustering

1 Introduction and Background

In the biomedical field, information systems help to improve the efficiency and the quality of a diagnosis. In particular, for the clinical decision-making process it can be very useful to find images with characteristics similar (same anatomic region, same

N. Brancati (✉)
Institute for High Performance Computing and Networking,
National Research Council of Italy (ICAR-CNR), Naples, Italy
e-mail: nadia.brancati@cnr.it; nadia.brancati@nai.icar.cnr.it

F. Camastra
Department of Science and Technology,
University of Naples Parthenope, Naples, Italy
e-mail: camastra@ieee.org

G. De Pietro et al. (eds.), *Intelligent Interactive Multimedia Systems and Services 2016*, Smart Innovation, Systems and Technologies 55,
DOI 10.1007/978-3-319-39345-2_1

disease, ...) to a given image. For this purpose a content based image retrieval (CBIR) system could be used [11, 14, 19], both to benefit the management of increasingly large image collections, and to support clinical care, biomedical research, and education. However, although the number of experimental algorithms comprehending specific problems and databases is growing, few systems exist with relative success [3, 7, 29]. So, biomedical applications are one of the priority areas where CBIR can meet more success outside the experimental sphere, due to population aging in developed countries.

The CBIR is the technique that allows retrieving images similar to a query image, in a large unannotated database. The retrieval of the similar images is based on the extraction of some visual contents of the images, called feature descriptors.

Many different feature descriptors have been proposed and used in the past years [19, 25, 26]. These feature descriptors are usually low level features, easy to extract and they are mainly of two types: *global* as shape, color or texture [1, 5, 6, 12, 24] and *local*, that focus mainly on key points or salient patches [10, 15, 21, 22, 28].

After the choice of the most appropriate feature descriptors, these are extracted both from all the images in the database and from the query image. At this point, a similarity measurement should be chosen to compute the distance between the feature descriptors of the query image and the feature descriptors of all the images in the database. The choice of the appropriate similarity measurement could be another crucial element for the correct design of a CBIR system [16].

In some cases, it results necessary to make CBIR techniques more efficient and accurate, above all when the databases are very large. In this case, the aim is to decrease the number of the images for which to compute the distance from the query image. Many clustering techniques with some feature descriptors of the images can be used [2, 8, 9, 18].

In this paper, the retrieval of images more similar to a query image is performed using textural features, in particular Tamura features, that correspond to human visual perception [27]. Then, a clustering using K-Means Algorithm [13] is carried out to obtain homogeneous groups, based on Tamura features [23].

The main contribution of the paper is the validation of the use of Tamura features, for content based image retrieval, in particular for biomedical databases, compared with the use of local descriptors. Moreover, a comparison among different similarity measurements is performed both to evaluate the adopted technique for CBIR and to estimate the stability of the results.

2 CBIR Steps

CBIR technique proposed in this paper is composed by the following steps:

- the extraction of Tamura features from all the images in the database and from the query image;

- the clustering of the Tamura features, extracted from all the images in the database, using K-Means algorithm;
- the computation of five distance metrics both to evaluate the adopted technique for CBIR and to estimate the stability of the results.

2.1 Tamura Features

Tamura features correspond to human visual perception. They were designed in accord to psychological studies on the human perception and they capture the high-level perceptual attributes of a texture. They define six textural features: coarseness, directionality, contrast, roughness, line-likeness and regularity. The first three features are the most similar to human visual perception, and they are considered in the present work; they are extracted both from all images in the database and from the query image. More details of these features can be found in [27].

Coarseness The aim of this feature is to find a repetitive pattern in the texture, which can have several orders of magnitude, depending on whether you are in front of a coarse or fine texture. So, operators to several orders of magnitude are computed. If the texture is fine, the highest response will be given by the operator of magnitude lower, vice versa, if the texture is coarse, the highest response will be given by the operator of magnitude greater.

The computation of the coarseness is given from:

$$F_{crs} = \frac{1}{m \times n} \sum_{i=0}^{m} \sum_{j=0}^{n} S_{best}(i,j)$$

where $m \times n$ is the resolution of the image and $S_{best}(i,j)$ is computed for each pixel and it provides the information about the magnitude of the pattern. It is important to underline that the coarseness feature is influenced both by the size of the pattern to find and by its repetitiveness.

Contrast The contrast of Tamura features takes into account both the variation range of the gray levels and the polarization of white and black pixels. A measurement for the variation range is the variance σ^2 of the pixels of the image. In fact, they measure the dispersion present in the distribution of the gray levels. However this single measurement does not appear to be very significant when the image histogram shows a prominent peak towards white or towards black. A measurement for the polarization of white and black pixels is given by the *kurtosis*, defined as $\alpha_4 = \mu_4/\sigma_4$, where μ_4 is the moment of fourth order. At this point, the two measurements are combined, obtaining the feature of the contrast:

$$F_{con} = \frac{\sigma}{(\alpha_4)^n}$$

where n can be equal to $8, 4, 2, 1, 1/2, 1/4, 1/8$. In this paper, for the experiments, n is set to 1.

Directionality The directionality is a feature that can be calculated from the analysis of the Fourier spectrum. However, the features obtained by the Fourier spectrum do not behave in the same way as those calculable in the spatial domain. Tamura preferred to get a global feature of the image, analysing the histogram of the directions of the edges, the form of the gradient image and the peaks of the histogram. In particular, the sum of the moments of second order around each peak, from to a valley to another valley is computed, and this measurement is defined in the following way:

$$F_{dir} = 1 - rn_p \sum_{p}^{n_p} \sum_{\phi \in w_p} (\phi - \phi_p)^2 H_D(\phi)$$

where n_p is the number of the peaks, ϕ_p is the pth peak of H_D, w_p is the range of the pth peak between two valleys, ϕ is the quantized directionality, r is a normalization factor, related to the quantized levels of ϕ.

2.2 Clustering

After the extraction of the Tamura features from all images in the database, these features are clustered, using K-Means algorithm. For the experiments in this paper, the number of clusters is set to 2. These information are saved in a data structure in order to use them for the following experiments.

2.3 Distance Metrics

The choice of the similarity measurement is the second issue in CBIR. For the proposed technique, first the distances between the feature descriptors of the query image and the centroids of the clusters are computed. The cluster at minimum distance is selected. Then, the distances between the feature descriptors of the query image and the feature descriptors of all the images of the selected cluster are computed. The images with feature descriptors with small distance from the feature descriptors of the query image are considered as the images more similar to the query image. In this work, some distance metrics are used as similarity measurements. In particular, well-known distance metrics are used [4]:

- **Euclidean distance**;
- **City block distance**;
- **Minkowski distance**, with order $p = 3$.

Moreover, let the vector of the feature descriptors of the query image be represented by Q, and the vector of the feature descriptors of an image of the database be represented by I, two additional distance metrics are calculated:

Canberra distance, that normalizes each feature pair difference by dividing it by the sum of a pair of feature descriptors:

$$D = \sum_{i=1}^{n} \frac{(|Q_i - I_i|)}{|Q_i| + |I_i|}$$

d1 distance, where the distance between two vectors of feature descriptors is calculated based on the formula described in [21]:

$$D = \sum_{i=1}^{n} \frac{|Q_i - I_i|}{|1 + Q_i + I_i|}$$

3 Experimental Results and Discussion

In order to analyse the performance of the proposed technique, some experiments have been performed. The experiments have been conducted on the Open Access Series of Imaging Studies (OASIS) [17]. It is a series of magnetic resonance imaging (MRI) database that is publicly available for study and analysis. This dataset consists of a cross-sectional collection of 421 subjects aged between 18 to 96 years, including individuals with early-stage Alzheimer's Disease (AD). For image retrieval purpose, these 421 images are grouped into four categories (124, 102, 89, and 106 images) based on the ventricular shape in the images. Sample images for each category are displayed in the Fig. 1.

For the proposed CBIR technique, each image of the database is used as a query image and the distances from the clusters are calculated, based on the distance metrics of the Sect. 2.3. When the cluster containing images more similar to the query

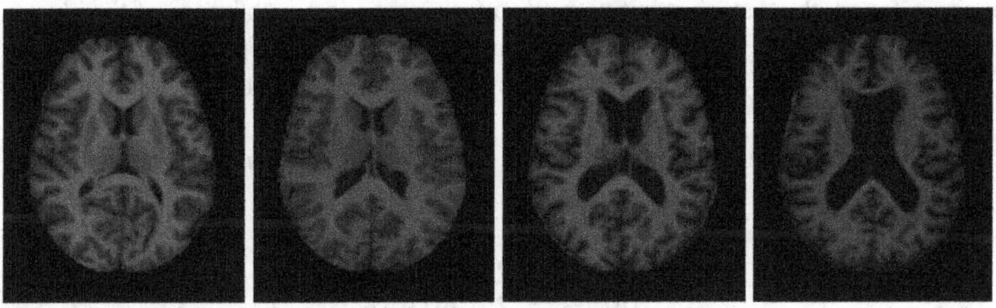

Fig. 1 Sample images from OASIS database (one image per category)

Query

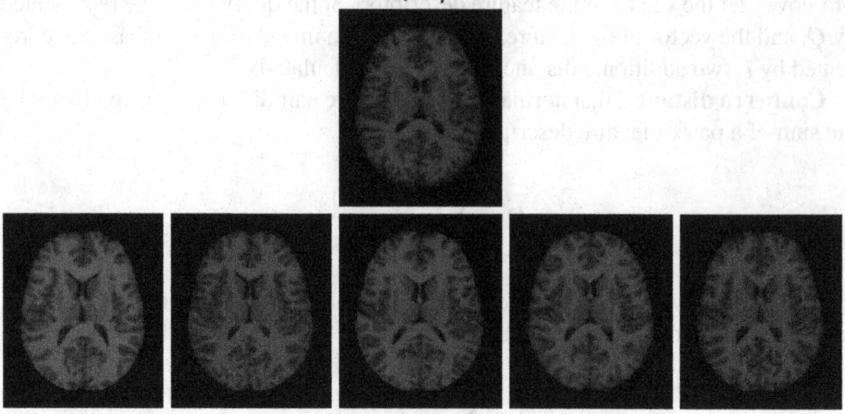

Fig. 2 The results for a query image of the group 1

Query

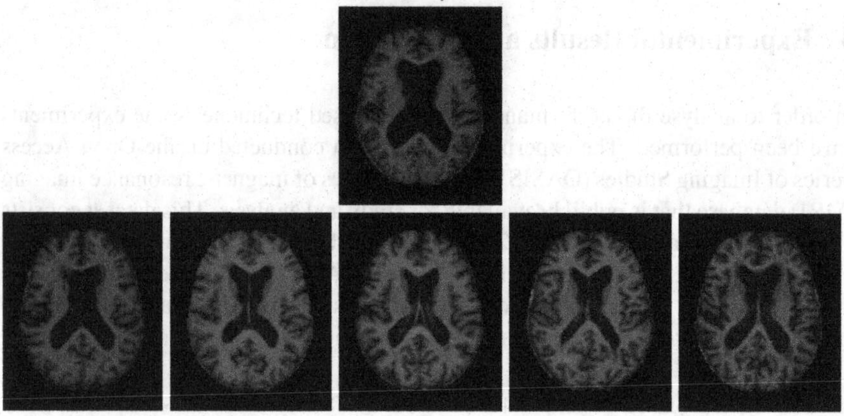

Fig. 3 The results for a query image of the group 4

image is found, the distances between the query image and each image of the cluster are computed, and the images more similar to the query are displayed.

Some results for two query image examples, for groups 1 and 4, are shown in the Figs. 2 and 3.

The *average retrieval precision* (ARP) and the *average retrieval rate* (ARR) are calculated, to evaluate the performance:

$$ARP = \frac{1}{N} \sum_{i=1}^{N} PR(Q_i)\Big|_m$$

$$ARR = \frac{1}{N}\sum_{i=1}^{N} RE(Q_i)\Big|_p$$

where Q is the query image and N is the total number of images in the database and where *precision* (PR) and *recall* (RE) are:

$$PR(Q) = \frac{Number\ of\ Relevant\ Images\ Retrieved}{Total\ Number\ of\ Images\ Retrieved}$$

$$RE(Q) = \frac{Number\ of\ Relevant\ Images\ Retrieved}{Total\ Number\ of\ Relevant\ Images\ in\ the\ Database}$$

As specified in [20], to calculate *precision* and *recall* the number of images retrieved should be specified, (e.g. *precision* with $m = 20$ images and *recall* with $p = 100$ images are retrieved). So, for the current experiment, the number m of images retrieved for *precision* is 100 and for *recall* is specific for each group, i.e. $p = 124$ for the group 1, $p = 102$ for the group 2, $p = 89$ for the group 3, $p = 106$ for the group 4.

In Tables 1 and 2 the results of ARP and ARR for all similarity measurements and for all groups are reported. The performance of the proposed technique is better for group 1 and 4, independently from the chosen similarity measurement. Moreover, all similarity measurements for each group, provide values very similar among them, showing a stability and a robustness of the technique. However, ARP and ARR show a consistent behaviour for all groups: for the groups 1 and 2, they show the best values for the City Block distance, for the group 3 the best values are given by the Minkowski distance, for the group 4 are given by the d1 distance and for all groups the best values of ARP are given by the City Block and Minkowski distance and the best values of ARR are given by the City Block, Minkowski and Euclidean distance.

Finally, a comparison with some methods in literature are reported in Table 3 [28]. In order to compare with these methods, the number of retrieved images is 10 ($m = 10$), as in [28]. As similarity measurement the Euclidean distance has been chosen.

Table 1 Average retrieval precision (ARP) of all similarity measurements for each category

ARP (%)					
	Group 1	Group 2	Group 3	Group 4	**Total**
Euclidean	0.500	0.378	0.334	0.710	0.480
City Block	**0.505**	**0.381**	0.329	0.711	**0.482**
Canberra	0.461	0.375	0.276	0.735	0.462
Minkowski	0.497	0.379	**0.338**	0.713	**0.482**
d1	0.465	0.377	0.279	**0.736**	0.465

Table 2 Average retrieval rate (ARR) of all similarity measurements for each category

ARR (%)	Group 1	Group 2	Group 3	Group 4	Total
Euclidean	0.464	0.392	0.357	0.817	**0.507**
City Block	**0.465**	**0.394**	0.351	0.817	**0.507**
Canberra	0.425	0.388	0.297	0.843	0.488
Minkowski	0.463	0.388	**0.358**	0.819	**0.507**
d1	0.428	0.391	0.301	**0.844**	0.491

Table 3 Comparison of the CBIR proposed technique with other methods in literature

ARP ($m = 10$) (%)	Group 1	Group 2	Group 3	Group 4	Total
CSLBP	0.46	0.36	0.29	0.4	0.38
LEPINV	0.48	0.34	0.29	0.41	0.38
LEPSEG	0.51	0.34	0.29	0.43	0.39
LBP	0.56	0.34	0.34	0.45	0.42
LMEBP	0.55	0.35	**0.39**	0.54	0.46
DLEP	0.51	0.37	0.38	0.53	0.45
CSLBCoP	0.55	**0.39**	0.38	0.64	0.49
CBIR proposed	**0.58**	0.34	0.37	**0.78**	**0.52**

These methods are all based on local information of the pixels [28]:

- **CSLBP**: center simmetric local binary pattern;
- **LEPINV**: local edge pattern for image retrieval;
- **LEPSEG**: local edge pattern for segmentation;
- **LBP**: local binary pattern;
- **LMEBP**: local maximum edge binary pattern;
- **DLEP**: directional local extrema pattern;
- **CSLBcoP**: CSLBP + gray level co-occurrence matrix (GCLM).

The results show that the proposed CBIR technique outperforms the other methods for groups 1 and 4 and in terms of total ARP.

The promising results prove that the Tamura features represent good global visual descriptors and that the local visual descriptors are less representative for the kind of examined images. In particular, the difference between ventricular shape of healthy subjects and subjects with early-stage Alzheimer's Disease is well highlighted by Tamura features, above all for the groups 1 and 4, for which the best results are obtained.

4 Conclusion

In this paper a CBIR technique based on the clustering of the Tamura features has been proposed. A biomedical database, containing MRI images (OASIS) has been used for the experiments and different similarity measurements have been used both to evaluate the proposed technique and to verify the stability of the technique. The results show that the proposed technique is stable and robust, independently from the selected distance metric; in fact both the Average Retrieval Precision (ARP) and the Average Retrieval Rate (ARR) show similar results for all distance metrics. Moreover, the comparison with other methods in literature, that use local information of the pixels, show that the proposed technique, based on global information, outperforms these methods in terms of ARP, with a number of retrieved images equal to 10.

References

1. Baraldi, A., Parmiggiani, F.: An investigation of the textural characteristics associated with gray level cooccurrence matrix statistical parameters. IEEE Trans. Geosci. Remote Sensing **33**(2), 293–304 (1995)
2. Chen, Y., Wang, J.Z., Krovetz, R.: Content-based image retrieval by clustering. In: Proceedings of the 5th ACM SIGMM international workshop on Multimedia information retrieval. pp. 193–200. ACM (2003)
3. Deserno, T.M., Güld, M.O., Plodowski, B., Spitzer, K., Wein, B.B., Schubert, H., Ney, H., Seidl, T.: Extended query refinement for medical image retrieval. J. Digital Imaging **21**(3), 280–289 (2008)
4. Deza, M.M., Deza, E.: Encyclopedia of Distances. Springer (2009)
5. Ergen, B., Baykara, M.: Texture based feature extraction methods for content based medical image retrieval systems. Bio-Med. Mater. Eng. **24**(6), 3055–3062 (2014)
6. Haridas, K., Thanamani, A.S.: Well-organized content based image retrieval system in RGB Color Histogram, Tamura Texture and Gabor Feature. Int. J. Adv. Res. Comput. Commun. Engineering **3**(10) (2014)
7. Hsu, W., Long, L.R., Antani, S., et al.: SPIRS: a framework for content-based image retrieval from large biomedical databases. MedInfo **12**, 188–192 (2007)
8. Jain, M., Singh, S.: A survey on: content based image retrieval systems using clustering techniques for large data sets. Int. J. Managing Inf. Technol. (IJMIT) **3**(4), 23–39 (2011)
9. Jain, M., Singh, S.: An experimental study on content based image retrieval based on number of clusters using hierarchical clustering algorithm. Int. J. Sig. Process. Image Process. Pattern Recogn. **7**(4) (2014)
10. Jasmine, K.P., Kumar, P.R.: Multi-resolution joint LBP Histograms for biomedical image retrieval. Int. J. Comput. Appl. **95**(3) (2014)
11. Lehmann, T.M., Gold, M., Thies, C., Fischer, B., Spitzer, K., Keysers, D., Ney, H., Kohnen, M., Schubert, H., Wein, B.B.: Content-based image retrieval in medical applications. Methods Inf. Med. **43**(4), 354–361 (2004)
12. Liu, G.H., Yang, J.Y.: Content-based image retrieval using color difference histogram. Pattern Recogn. **46**(1), 188–198 (2013)
13. Lloyd, S.P.: Least squares quantization in PCM. IEEE Trans. Inf. Theory **28**(2), 129–137 (1982)
14. Long, L.R., Antani, S., Deserno, T.M., Thoma, G.R.: Content-based image retrieval in medicine: retrospective assessment, state of the art, and future directions. Int. J. Healthc. Inf. Syst. Inf. Official Publ. Inf. Resour. Manag. Assoc. **4**(1), 1 (2009)

15. Lowe, D.G.: Distinctive image features from scale-invariant keypoints. Int. J. Comput. Vis. **60**(2), 91–110 (2004)
16. Malik, F., Baharudin, B.: Analysis of distance metrics in content-based image retrieval using statistical quantized histogram texture features in the DCT domain. J. King Saud Univ.-Comput. Inf. Sci. **25**(2), 207–218 (2013)
17. Marcus, D.S., Wang, T.H., Parker, J., Csernansky, J.G., Morris, J.C., Buckner, R.L.: Open access series of imaging studies (OASIS): cross-sectional MRI data in young, middle aged, nondemented, and demented older adults. J. Cogn. Neurosci. **19**(9), 1498–1507 (2007)
18. Mishra, P., Sonam, M., Vijayalakshmi, M.S.: Content based image retrieval using clustering technique: a survey. Int. J. Res. Comput. Eng. Electron. **3**(2) (2014)
19. Müller, H., Michoux, N., Bandon, D., Geissbuhler, A.: A review of content-based image retrieval systems in medical applications—clinical benefits and future directions. Int. J. Med. Inf. **73**(1), 1–23 (2004)
20. Müller, H., Müller, W., Squire, D.M., Marchand-Maillet, S., Pun, T.: Performance evaluation in content-based image retrieval: overview and proposals. Pattern Recogn. Lett. **22**(5), 593–601 (2001)
21. Murala, S., Maheshwari, R., Balasubramanian, R.: Directional binary wavelet patterns for biomedical image indexing and retrieval. J. Med. Syst. **36**(5), 2865–2879 (2012)
22. Murala, S., Wu, Q.: Local mesh patterns versus local binary patterns: biomedical image indexing and retrieval. IEEE J. Biomed. Health Inf. **18**(3), 929–938 (2014)
23. Murthy, V., Vamsidhar, E., Kumar, J.S., Rao, P.S.: Content based image retrieval using hierarchical and k-means clustering techniques. Int. J. Eng. Sci. Technol. **2**(3), 209–212 (2010)
24. Puviarasan, N., Bhavani, R., Vasanthi, A.: Image retrieval using combination of texture and shape features. Image **3**(3) (2014)
25. Rui, Y., Huang, T.S., Chang, S.F.: Image retrieval: current techniques, promising directions, and open issues. J. Vis. Commun. Image Represent. **10**(1), 39–62 (1999)
26. Smeulders, A.W., Worring, M., Santini, S., Gupta, A., Jain, R.: Content-based image retrieval at the end of the early years. IEEE Trans. Pattern Anal. Mach. Intell. **22**(12), 1349–1380 (2000)
27. Tamura, H., Mori, S., Yamawaki, T.: Textural features corresponding to visual perception. IEEE Trans. Syst. Man Cybern. **8**(6), 460–473 (1978)
28. Verma, M., Raman, B.: Center symmetric local binary co-occurrence pattern for texture, face and bio-medical image retrieval. J. Vis. Commun. Image Represent. **32**, 224–236 (2015)
29. Xue, Z., Long, L.R., Antani, S., Jeronimo, J., Thoma, G.R.: A web-accessible content-based cervicographic image retrieval system. In: Medical imaging. International Society for Optics and Photonics, pp. 691907–691907 (2008)

2-Stripes Block-Circulant LDPC Codes for Single Bursts Correction

Evgenii Krouk and Andrei Ovchinnikov

Abstract In this paper the low-density parity-check (LDPC) codes are considered applied to correction of error bursts. Errors grouping and forming of so-called bursts are typical effect in real communication and data storage systems, however, this effect is typically ignored, and the coding task is reduced to correction of independent errors, which makes the practical characteristics of coding systems worse comparing to possibly reachable. Nevertheless, LDPC codes are able to protect from burst errors as well as independent ones. The main result of the paper is dedicated to evaluation of maximum correctable burst length of Gilbert codes, which are the 2-stripes special case of LDPC block-permutation codes, the construction which is often used in modern practical applications and research.

Keywords LDPC codes · Bursts-correcting codes · Gilbert codes

1 Introduction

During the development of modern practical communication systems the channel models which are commonly used (binary-symmetric channel or Gaussian channel) are often inadequate since they consider independent errors. At the same time in real communication channels the effect of "memory" occurs (for example due to fading [1]) leading to the dependencies between erroneous symbols. To fight with such errors grouping the interleaving procedure is often used [1].

However, usage of interleaver leads to typical channel behaviour loss, the channel is transformed to memoryless, this decreases the possible transmission rates, and increases the complexity and delay of transmitter and receiver [2, 3]. This is

E. Krouk · A. Ovchinnikov (✉)
Saint-Petersburg State University of Aerospace Instrumentation,
B.Morskaya 67, 190000 Saint Petersburg, Russia
e-mail: mldoc@ieee.org

E. Krouk
e-mail: ekrouk@vu.spb.ru

G. De Pietro et al. (eds.), *Intelligent Interactive Multimedia Systems and Services 2016*, Smart Innovation, Systems and Technologies 55,
DOI 10.1007/978-3-319-39345-2_2

because the classical coding theory usually proposes code constructions for independent errors which are simpler to analyse. So the important task is to construct coding schemes oriented on typical channel errors, in particular, on correcting the error bursts, that is, the error patterns when first and last erroneous symbols are no far than some value b from each other (and which is called the burst length). Besides, the effect of errors grouping is typical for data storage systems.

In the coding theory the classes of burst-correcting codes are known. For example, these are Fire codes or Reed-Solomon codes [4]. During the last decades a lot attention was given to low-density parity-check (LDPC) codes, particularly block-permutation constructions [5]. Gilbert codes which are considered in this paper are the simple special case of such construction and were proposed initially for burst-correction. However, the exact burst-correction capability of these codes was unknown.

The paper is organized as follows. Section 2 describes Gilbert codes and known estimations of its burst-correction capability. In Sect. 3 the procedure is derived allowing computation of exact value of maximum correctable burst length. The Sect. 4 concludes the paper.

2 Gilbert Codes

LDPC-codes were invented by Gallager [6, 7] and later investigated in many works [8–11]. While possessing comparatively poor minimal distance, these codes, however, provide high error-correction capability with very low decoding complexity. It was shown that LDPC codes may overcome turbo-codes and approach to channel capacity [12]. Additionally, some LDPC constructions (and block-permutation constructions as well) are cyclic or quasi-cyclic, allowing effective coder and decoder implementation.

Block-permutation codes are one of the most prominent and widely used class of LDPC codes [3, 5, 13]. The simple special case of this class are Gilbert codes, which were proposed in [14] as burst-correction codes. Gilbert codes may be defined by the parity-check matrix H_ℓ,

$$H_\ell = \begin{bmatrix} I_m & I_m & I_m & \cdots & I_m \\ I_m & C & C^2 & \dots & C^{\ell-1} \end{bmatrix}, \quad (1)$$

where I_m is $(m \times m)$-unity matrix, C is $(m \times m)$-matrix of cyclic permutation:

$$C = \begin{bmatrix} 0 & 0 & 0 & \dots & 0 & 1 \\ 1 & 0 & 0 & \dots & 0 & 0 \\ 0 & 1 & 0 & \dots & 0 & 0 \\ \dots & \dots & \dots & \dots & \dots & \dots \\ 0 & 0 & 0 & \dots & 1 & 0 \end{bmatrix}, \quad (2)$$

and $\ell \leq m$.

Many works were dedicated to estimation of burst-correcting capability of these codes, as well as their modifications and extensions [15–19]. In [20] the estimation of maximum correctable burst length b for the codes defined by matrix (1) is given by inequality

$$b \leq \min_{\gamma \in \{0, \ell-2\}} \max\{\gamma - 1, m - \gamma - 1\}. \tag{3}$$

However, estimation (3) is not correct, giving only the lower bound of the maximum correctable burst length. The exactness of this estimation decreases with growth of ℓ.

In the next section we will give the method of exact evaluation of b.

3 Burst-Correction Capability of Gilbert Codes

The main result of this section and paper is the following theorem.

Theorem 1 *Code with parity-check matrix H_ℓ defined by (1) can correct single bursts of maximal length b_ℓ, where b_ℓ is calculated by the first satisfied condition:*

1. $b_3 = m - 1$, *m is odd.*
2. *If $\ell > \lceil m/2 \rceil + 1$, then*

$$\begin{cases} b_\ell = m - \lceil m/2 \rceil + 1, & m \text{ odd}, \\ b_\ell = m/2 - 1, & m \text{ even}. \end{cases}$$

3. *If $\ell \leq \lceil m/2 \rceil + 1$, then*

$$\begin{array}{ll} b_\ell = m - \ell + 1, & \text{if } m \vdots (\ell - 1), \\ b_\ell = m - \ell + 1, & \text{if } \exists k > 0 : (m - \ell + 3 - k \cdot (\ell - 1)) \vdots (\ell - 2), \\ b_\ell = m - \ell + 2, & \text{if } \exists k > 0 : (m - k \cdot (\ell - 3)) \vdots (\ell - 2), \\ b_\ell = m - \ell + 2, & \text{if } \exists k > 0 : (m - k \cdot (\ell - 2)) \vdots (\ell - 1), \\ b_\ell = m - \ell + 2, & \text{if } \exists k > 0 : (m - k \cdot (\ell - 1)) \vdots (\ell - 2). \end{array}$$

4. *If all preceding conditions are unsatisfied, then:*

$$b_\ell = b_{\ell-1}.$$

Proof To prove the statement of the theorem we will introduce some notations and prove some lemmas. Represent the matrix (1) as $H_\ell = [h_0, h_1, \ldots, h_{\ell-1}]$, where h_γ—$(2m \times m)$-block-column, which we will call as block.

The code can correct single error burst of length b, if and only if all packets of length b are in different cosets, i.e. there are no two error vectors e_1 and e_2 (forming the bursts of length no more than b), such that

$$e_1 \cdot H_\ell^T = e_2 \cdot H_\ell^T. \tag{4}$$

Since e_1 and e_2 are error bursts, the analysis of (4) may be reduced to consideration of submatrices of H_ℓ consisting of $2m$ rows and no more than b columns.

Clearly, code with parity-check matrix (1) cannot correct bursts of length m, since the sum of all m columns from any block of H_ℓ gives all-one column.

Each burst B of length $b < m$ affects no more than two adjacent blocks from H_ℓ, let these blocks be γ and $\gamma + 1$. Let us replace in h_γ and $h_{\gamma+1}$ all columns with numbers not from B by zeros and sum obtained matrices h'_γ and $h'_{\gamma+1}$, obtaining

$$Q(B) = (h'_\gamma + h'_{\gamma+1})^T. \tag{5}$$

This matrix has the form

$$Q^T(B) = \left[\begin{array}{c|cccc|c|ccc}
 & & & & & i & i+1 & & \\
 & 1 & 0 & \dots & 0 & 0 & 0 & \dots & 0 \\
 & 0 & 1 & \dots & 0 & 0 & 0 & \dots & 0 \\
 & \dots & \dots & \dots & \dots & \dots & \dots & \dots & \dots \\
 & 0 & 0 & \dots & 1 & 0 & 0 & \dots & 0 \\
\hline
i & 0 & 0 & \dots & 0 & 0 & 0 & \dots & 0 \\
\hline
 & 0 & 0 & \dots & 0 & 0 & 1 & \dots & 0 \\
 & \dots & \dots & \dots & \dots & \dots & \dots & \dots & \dots \\
 & 0 & 0 & \dots & 0 & 0 & 0 & \dots & 1 \\
\hline
 & & & & & 0 & & & \\
 & & & \tilde{C}^{\gamma+1} & & \dots & & \tilde{C}^{\gamma} & \\
 & & & & & 0 & & &
\end{array}\right], \tag{6}$$

where $\tilde{C}^\gamma$ and $\tilde{C}^{\gamma+1}$ are parts of matrices C^γ and $C^{\gamma+1}$, and $i + 1$ is the beginning position modulo m of burst B, which we will call as relative beginning of B.

From (4) and (5) it follows that the code with parity-check matrix H_ℓ cannot correct bursts of length $b < m$ if and only if for at least one pair of bursts B_1 and B_2 (of length b) there exists the pair of vectors $\bar{x}_1$ and $\bar{x}_2$ of length m such that

$$\bar{x}_1 \cdot Q(B_1) = \bar{x}_2 \cdot Q(B_2). \tag{7}$$

For any burst B the matrix $Q(B)$ may be represented as

$$Q(B) = [Q_1(B), Q_2(B)],$$

where $Q_1(B)$ is $(m \times m)$-unity matrix with ith row replaced by zeros.

Then the condition (7) may be written as

$$\bar{x}_1 \cdot Q_1(B_1) = \bar{x}_2 \cdot Q_1(B_2), \tag{8}$$

$$\bar{x}_1 \cdot Q_2(B_1) = \bar{x}_2 \cdot Q_2(B_2). \tag{9}$$

Removing the zero row in $Q_1(B)$, we obtain $((m-1)\times(2m))$-matrix $Q'(B) = [Q'_1(B), Q'_2(B)]$.

Denote as $(\bar{y}\backslash s)$ the vector of length $m-1$ obtained from $\bar{y}$ (of length m) by removing sth position (with correspondent shift of remained digits by one position to the right). Then the following lemma may be formulated.

Lemma 1 *The condition (7) is satisfied if and only if the vector $\bar{y}$ of length m exists such that*

$$\begin{cases} (\bar{y}\backslash i)\cdot Q'_2(B_1) = (\bar{y}\backslash j)\cdot Q'_2(B_2), \\ y_i = y_j = 0. \end{cases} \tag{10}$$

where $i+1, j+1$ are relative beginnings of bursts B_1 and B_2 correspondingly.

Proof Consider the expression (8). Matrices $Q_1(B_1)$ and $Q_1(B_2)$ are unity $(m\times m)$-matrices with ith and jth zero rows correspondingly. This means that multiplication of $\bar{x}_1$ by $Q_1(B_1)$ gives the vector $\bar{x}_1$ with ith position equal to zero. From (8) it follows that $\bar{x}_1$ and $\bar{x}_2$ are coincide and contain zeros on positions i and j. That is, the vector $\bar{y} = \bar{x}_1 = \bar{x}_2$ may always be defined with $y_i = y_j = 0$.

Rewrite (8) and (9) as

$$\begin{cases} (\bar{y}\backslash i)\cdot Q'_1(B_1) = (\bar{y}\backslash j)\cdot Q'_1(B_2) \\ (\bar{y}\backslash i)\cdot Q'_2(B_1) = (\bar{y}\backslash j)\cdot Q'_2(B_2) \end{cases}$$

Evidently, first equation is satisfied for any $\bar{y}$, if $y_i = y_j = 0$, this gives the lemma's statement.

Lemma 2 *For any burst B:*

$$Q'_2(B) = [I_{m-1}, \bar{0}]\cdot C^{\gamma}, \tag{11}$$

where I_{m-1} is unity $((m-1)\times(m-1))$-matrix, $\bar{0}$ is zero vector-column of length $m-1$ and γ is integer between 0 and $\ell-1$.

Proof To prove the lemma it is enough to show that for any burst B the correspondent matrix $Q'_2(B)$ is circulant $((m-1)\times m)$-matrix.

Let B is burst of length $m-1$ containing $s \geq 0$ last columns of block h_i and $m-s-1$ first columns of block h_{i+1}. Then $Q'_2(B) = [q_1(B), q_2(B)]^T$, where $q_1(B)$ is the matrix from first $m-s-1$ columns of C^i and $q_2(B)$ is the matrix from last s columns of C^{i-1}. Matrices $q_1(B)$ and $q_2(B)$ are circulant by construction and we need to show that $(m-s-1)$th column of C^{i-1} (i.e. first column of $q_1(B)$) is cyclic shift of the $(m-s-1)$th column of C^i (i.e. the last column of $q_2(B)$) to conclude the proof. This follows from the fact that $(m-s)$th column of C^{i-1} is equal to $(m-s-1)$th column of C^i.

From Lemmas 1, 2 and condition (7) it follows that two bursts of length $b < m$ have the same syndrome if and only if there exists $\bar{y}$ of length m, such that for some integer γ the following condition holds:

$$\begin{cases} (\bar{y}\backslash i, 0) = (\bar{y}\backslash j, 0) \cdot C^{\gamma}, \\ y_i = y_j = 0, \end{cases} \tag{12}$$

where $(\bar{y}\backslash i, 0)$ is vector of length m obtained by adding zero position to $(\bar{y}\backslash i)$.

Let us put in correspondence to vector $\bar{y}$ the polynomial $y(x) = \sum_{k=0}^{m-1} y_k x^k$. Then (12) in polynomial representation will be

$$\begin{cases} (y(x)\backslash i) = (y(x)\backslash j) \cdot x^{\gamma} \bmod x^m - 1, \\ y_i = y_j = 0 \end{cases} \tag{13}$$

(coefficients of $(y(x)\backslash i)$ and $(y(x)\backslash j)$ are defined by vectors $(\bar{y}\backslash i, 0)$ and $(\bar{y}\backslash j, 0)$ correspondingly).

Lemma 3 *If* $y(x) = \sum_{k=0}^{m-1} y_k x^k$*—polynomial, satisfying (13), then for any non-zero* y_k *one of the following holds:*

$$y_k = y_{(k-\gamma) \bmod m}, \quad y_k = y_{(k-\gamma+1) \bmod m}, \quad y_k = y_{(k-\gamma-1) \bmod m}.$$

Proof Polynomial $(y(x)\backslash i)$ may be represented as

$$(y(x)\backslash i) = \sum_{k=0}^{i-1} y_k x^k + \sum_{k=i}^{m-2} y_{k+1} x^k = A^i(x) + B^i(x). \tag{14}$$

Then $((y(x)\backslash j) \cdot x^{\gamma}) \bmod x^m - 1$ may be written as

$$\begin{aligned} ((y(x)\backslash j) \cdot x^{\gamma}) \bmod x^m - 1 &= \left(\sum_{k=0}^{j-1} y_k x^{k+\gamma} + \sum_{k=j}^{m-2} y_{k+1} x^{k+\gamma} \right) \bmod x^m - 1 = \\ &= (A^{\gamma}(x) + B^{\gamma}(x)) \bmod x^m - 1. \end{aligned} \tag{15}$$

From (13) the equality of coefficients correspondent to the same degrees of polynomials (14) and (15) is follows.

Consider the coefficients for degrees $k = \overline{0, i-1}$ (i.e. from $A^i(x)$). They are correspondent to coefficients of the same degrees either from $A^{\gamma}(x)$ or from $B^{\gamma}(x)$. However from (15) it follows that any degree k of $((y(x)\backslash j) \cdot x^{\gamma}) \bmod x^m - 1$ may be represented as $(s + \gamma) \bmod m$. Then

$$\begin{aligned} y_k \in A^i(x) = y_k \in A^{\gamma}(x) = y_{(k-\gamma) \bmod m}, \text{ or} \\ y_k \in A^i(x) = y_k \in B^{\gamma}(x) = y_{(k-\gamma+1) \bmod m}. \end{aligned} \tag{16}$$

Similarly for degrees $k = \overline{i, m-2}$

$$\begin{aligned} y_k \in B^i(x) = y_{k-1} \in A^{\gamma}(x) = y_{(k-\gamma) \bmod m}, \text{ or} \\ y_k \in B^i(x) = y_{k-1} \in B^{\gamma}(x) = y_{(k-\gamma-1) \bmod m}. \end{aligned} \tag{17}$$

From (16) and (17) we get the lemma's statement.

Corollary 1 *For any $y(x)$ satisfying (13) the non-negative integers α_1, α_2, α_3 exists such that*

$$\alpha_1(\gamma + 1) + \alpha_2\gamma + \alpha_3(\gamma - 1) = m. \tag{18}$$

Proof Let y_k be non-zero element of $\bar{y}$. Then, according to Lemma 2, one of the elements $y_{(k-\gamma) \bmod m}$, $y_{(k-\gamma+1) \bmod m}$ or $y_{(k-\gamma-1) \bmod m}$ is also non-zero. Continuing these steps the non-zero y_s can be found for which one of the following statements holds:

$$y_s = y_{(s-\gamma) \bmod m} = y_k, y_s = y_{(s-\gamma+1) \bmod m} = y_k, y_s = y_{(s-\gamma-1) \bmod m} = y_k. \tag{19}$$

From this there exists such nonnegative α_1, α_2, α_3 that:

$$\alpha_1(\gamma + 1) + \alpha_2\gamma + \alpha_3(\gamma - 1) = 0 \bmod m. \tag{20}$$

To prove the Corollary 1 it is enough to show that there are no other non-zero elements between two non-zero elements of $\bar{y}$.

If this is not true, then one of the Eqs. (19) is satisfied after more that one passing through the vector $\bar{y}$ (in other words, the value of $k - \gamma$, $k - \gamma - 1$ or $k - \gamma + 1$ becomes negative before satisfying (19), and therefore is taken modulo m more than once). In this case the vector $\bar{y}$ is either all-one vector and the minimal length of uncorrectable burst is m, which is impossible, or it contains zero elements. Then the γ' exists, perhaps less than γ from (20), equal to the number of positions between non-zero elements or differs from it by one, for this γ' the condition (13) also holds, and before satisfying (19) the position number is taken modulo m only once. From this the statement of Corollary 1 follows.

Corollary 1 also means that from all γ, for which (13) holds, the least non-zero value should be chosen.

Lets call as section of $\bar{y}$ the sequence consisting of one and consequent zeros. For example, $\bar{y} = 100010000$ contains two sections of length 4 and 5.

From Corollary 1 it follows that $\bar{y}$ is concatenation of sections of lengths γ, $\gamma + 1$ and $\gamma - 1$. Lemma 4 and Corollary 2 specify the locations of these sections.

Lemma 4 *For vector, satisfying (13), the following holds:*

$$j + \gamma - 1 \leq m - 1. \tag{21}$$

Proof Indeed, let y_{j+1} be the last non-zero element of $\bar{y}$ (otherwise (21) evidently holds). Recall that the element y_{j+1} is always equal to one, since j is relative burst beginning. We will assume that the coefficient y_0 in $y(x)$ is always equal to one. If this is not true, the correspondent $\bar{y}$ may be cyclically shifted, preserving all results of (13), Lemma 3 and Corollary 1. Then from Lemma 3 and Corollary 1, the following cases are possible:

$$y_0 = y_{m \bmod m} = y_{(m-\gamma-1) \bmod m} = y_{j+1},$$
$$y_0 = y_{m \bmod m} = y_{(m-\gamma+1) \bmod m} = y_{j+1},$$
$$y_0 = y_{m \bmod m} = y_{(m-\gamma) \bmod m} = y_{j+1},$$

from this we get $j+\gamma-1 \leq m-1$.

Corollary 2 *If $i \leq j+\gamma$, then in sections, forming $\bar{y}$, the section of length $\gamma-1$ is either absent, or appeared exactly once and located last.*

Proof From Lemma 4 and inequality (21) the polynomial (15) may be represented as

$$\begin{aligned} &((y(x)\backslash j)\cdot x^{\gamma}) \bmod x^m - 1 = \\ &= \sum_{k=0}^{j-1} y_k x^{k+\gamma} + \sum_{k=j}^{m-\gamma-2} y_{k+1}x^{k+\gamma} + \sum_{k=m-\gamma-1}^{m-2} y_{k+1}x^{(k+\gamma) \bmod m} = \\ &= \sum_{k=0}^{\gamma-2} y_{(k-\gamma+1) \bmod m}x^k + y_j x^{\gamma-1} + \sum_{k=j+\gamma}^{m-2} y_{k-\gamma+1}x^k = \\ &= A^m(x) + y_j x^{\gamma-1} + B^m(x) + C^m(x). \end{aligned} \qquad (22)$$

As we assumed before, let $y_0 = 1$. At first we will not consider cases when relative beginning of one or both bursts coincides with the beginning of correspondent block h_γ of matrix (1), these cases will be considered separately. Then $i+1 \geq \gamma-1$ or $i \geq \gamma-2$. Consider $i = \gamma-2$ and $i = \gamma-1$. In the first case from (14) and (22) we get $y_{i+1} = y_{-1 \bmod m} = y_{m-1}$, which is possible only if $\bar{y}$ contains only ones. If there are no other solutions of (13), then the maximal length of correctable burst is $m-1$. We will exclude such vectors from consideration.

In the second case, for $i = \gamma-1$ we get $y_{i+1} = y_j$, which is impossible since the element y_{i+1} is always non-zero, while y_j is always zero. From this we get $i \geq \gamma$, i.e. degrees numbers from 0 to $\gamma-1$ are completely belong to $A^i(x)$, and y_0 is the only non-zero coefficient from coefficients y_k from $A^m(x)$ (this follows from the fact that the minimal number of positions between non-zero elements is $\gamma-1$, and coefficient of $x^{\gamma-1}$ is zero. Degrees higher than $\gamma-1$ belong to $B^m(x)$ and $C^m(x)$).

Let $i \leq j+\gamma$. Then from (14) and (22), $C^m(x) \subset B^i(x)$, and the following expressions hold:

$$\begin{aligned} y_k &= y_{(k-\gamma+1) \bmod m}, && 0 \leqslant k < \gamma; \\ y_k &= y_{k-\gamma}, && \gamma \leqslant k < i; \\ y_k &= y_{k-\gamma-1}, && i \leqslant k < (j+\gamma); \\ y_k &= y_{k-\gamma}, && (j+\gamma) \leqslant k \leqslant m-2. \end{aligned}$$

So, the expression $y_k = y_{(k-\gamma+1) \bmod m}$ may holds only for one non-zero element of $\bar{y}$, namely y_0.

From this also follows that for $i < j+\gamma$ the equation $\alpha_3 = 1$ holds and the only section of length $\gamma-1$, appeared in $\bar{y}$, is the last section.

Consider the special case $i = j+\gamma$. Here $C^m(x) = B^i(x)$, and the vector $\bar{y}$ does not contain the sections of length $\gamma+1$ at all, it has one (last) section of length $\gamma-1$, and other sections of length γ (in other words, $\alpha_1 = 0$, $\alpha_3 = 1$).

Let us analyse the results of Lemma 3 and its corollaries. Let we have the polynomial $y(x)$, satisfying (13), and correspondent vector $\bar{y}$. If we denote as i^0 and j^0 the number of zero positions of $\bar{y}$ before $(i+1)$ and $(j+1)$th positions, then the length of incorrectable burst is estimated as $b = \max(m - i^0, m - j^0)$.

However, it is unknown how to solve (13) having only m and ℓ. From the other hand, the properties of polynomial $y(x)$ are known, reflected in Lemma 3 and corollaries. It is clear that, defining the coefficients α_1, α_2 and α_3 of (18), we will define the polynomial $y(x)$ satisfying (13). At the same time for given m there may be several solutions of (18). In this case the worst scenario should be chosen, i.e. polynomials giving the minimal length of incorrectable burst. In the following we will show that coefficients α_1, α_2, α_3 of (18), as well as the minimal length of incorrectable burst, are dependent on mutual placement of bursts' relative beginnings i and j.

Consider the case when the beginnings of one or both bursts are coincide with the beginning of the correspondent block of H_ℓ. Then the values $i+1$ or (and) $j+1$ should be equal to zero, and from this $i = -1 \mod m$ or (and) $j = -1 \mod m$ (since from $i = -1 \mod m$ follows $(\bar{y}\backslash i, 0) = \bar{y}$).

The following lemma connects the parameters m, γ, coefficients α_1, α_2, and α_3 from (18) and relative beginnings i and j of error bursts.

Lemma 5 *For the vector $\bar{y}$ of length m and parameters i, j and γ, satisfying (12), the values of coefficients α_1, α_2, and α_3 of (18) are defined by the values i and j according to the Table 1.*

Proof We will subsequently consider all possible locations of bursts' relative beginnings i and j, using polynomial representations (14) and (22).

1. Let $i \leq j$ and $i = j = -1$. This case corresponds to the fist row of Table 1. This means that the equality $y(x) = (y(x)\backslash i) = (y(x)\backslash j)$ holds, that is

$$(y(x)\backslash i) = \sum_{k=0}^{m-1} y_k x^k,$$

$$((y(x)\backslash j) \cdot x^\gamma) \bmod x^m - 1 = \sum_{k=0}^{m-1} y_k x^{(k+\gamma) \bmod m},$$

Table 1 Relative beginnings of bursts, coefficients of (18) and minimal lengths of incorrectable bursts

#	Condition	Incorr. burst length	Coeff. of (18)	Value of γ
1	$i = -1, j = -1$	$m - \gamma - 1$	$\alpha_1 = 0, \alpha_3 = 0$	$\ell - 1$
2	$-1 < i < (j + \gamma), j \neq -1$	$m - \gamma$	$\alpha_1 \neq 0, \alpha_3 = 1$	$\ell - 2$
3	$i \geq (j + \gamma), j \neq -1$	$m - \gamma + 1$	$\alpha_1 = 0, \alpha_3 \neq 0$	$\ell - 2$
4	$i = -1, j \neq -1$	$m - \gamma + 2$	$\alpha_1 = 0, \alpha_3 \neq 0$	$\ell - 1$
5	$i \neq -1, j = -1$	$m - \gamma + 1$	$\alpha_1 \neq 0, \alpha_3 = 0$	$\ell - 2$

from this $y_k = y_{(k-\gamma) \bmod m}$ for all k. Taking into account (18) we get $\alpha_1 = 0$, $\alpha_3 = 0$. The length b of incorrectable burst in this case is $b = m - \gamma - 1$.

2. Now let $i \le j, i \ne -1, j \ne -1$.

   In the proof of Corollary 2 it was shown that if $i \ne -1$, then $i \ge \gamma$. Consider different possible values of i and j.

   (a) $i = \gamma$. In this case from (14) and (22) we get $B^m(x) \subset B^i(x)$, $C^m(x) \subset B^i(x)$. In polynomial (14) the coefficient y_{j+1} is always correspondent to degree x^j, and degree x^j in polynomial (22) is always belong to $B^m(x)$, so $y_{i+1} = y_0$, $y_{j+1} = y_{j-\gamma}$. That is, before $(i+1)$ and $(j+1)$th positions of $\bar{y}$ there is the section of length $\gamma + 1$, then the length of incorrectable burst is $m - \gamma$.
   (b) $(\gamma + 1) \le i < (j + \gamma)$. In this case the degree x^i belongs to $B^m(x)$, $y_{i+1} = y_{i-\gamma}$, $y_{j+1} = y_{j-\gamma}$, and this is the generalization of the previous case 2a, the length of incorrectable burst is also $m - \gamma$.

   These cases are correspondent to the second row of Table 1.

3. Let $i \ge j + \gamma$. As it was shown in the proof of Corollary 2, if $i = j + \gamma$, then $\bar{y}$ consists of one section of length $\gamma - 1$ and other sections of length γ. However, since neither i nor j in considered case are not equal to -1, there are sections of length γ before relative beginnings of bursts, so $b = m - \gamma + 1$.
   If $i > j + \gamma$, degree x^i belongs to $C^m(x)$, and $B^m(x) \subset A^i(x)$, so $y_{i+1} = y_{i-\gamma+1}, y_{j+1} = y_{j-\gamma+1}$. In this case in $\bar{y}$ there are sections of length γ before positions y_{i+1} and y_{j+1}, so $b = m - \gamma + 1$.
   Now we need to consider cases when i or j equal to -1.
4. Let $i = -1 = m - 1, j \ne -1$. Then (14) and (22) may be written as

$$(y(x)\backslash i) = \sum_{k=0}^{m-1} y_k x^k = A^i(x),$$
$$((y(x)\backslash j) \cdot x^\gamma) \bmod x^m - 1 = A^m(x) + y_j x^{\gamma-1} + B^m(x) + C^m(x).$$

   From this for any k one of the equations is hold: $y_k = y_{(k-\gamma) \bmod m}$, $y_k = y_{(k-\gamma+1) \bmod m}$. In other words, coefficient α_1 in (18) is equal to zero. This corresponds to the previous case, when $B^m(x) \subset A^i(x)$, and there are $\gamma - 1$ zeros before position $(j+1)$. However, there are $\gamma - 2$ zeros before position $i+1$, so we should select $b = \max(m - \gamma + 1, m - \gamma + 2) = m - \gamma + 2$. That is, in this case $b = m - \gamma + 2$, $\alpha_3 \ne 0$, $\alpha_1 = 0$.
5. Let $i \ne -1, j = -1 = m - 1$. Then

$$(y(x)\backslash i) = \textstyle\sum_{k=0}^{i-1} y_k x^k + \sum_{k=i}^{m-2} y_{k+1} x^k = A^i(x) + B^i(x),$$
$$((y(x)\backslash j) \cdot x^\gamma) \bmod x^m - 1 = \sum_{k=0}^{m-1} y_{(k-\gamma) \bmod m} x^k. \tag{23}$$

We get $y_{i+1} = y_{i-\gamma}$, $y_{j+1} = y_0 = y_{(-\gamma) \bmod m}$. This gives $b = m - \gamma + 1$, $\alpha_1 \neq 0$, $\alpha_3 = 0$.

Considered cases give the statement of Lemma 5.

If we know the values of coefficients in (18), we may calculate the minimal length of incorrectable burst. However, in third and fourth rows of Table 1 we have the same coefficients but different burst lengths. But taking into account the value of γ we may see that in fact these values coincide. Let us analyse the relationship between the value of γ (which is unknown during code construction) and ℓ (which is the code's parameter).

First consider $i \neq -1, j = -1 = m - 1$ (fifth row of Table 1). As it was shown, $\bar{y}$ for such values i and j consists of sections of lengths γ and $\gamma + 1$, and the last section has the length γ. However, since $i \neq -1$ and $y_0 = 1$, this means that the first error burst occupies blocks h_0 and h_1, and in the second burst the position correspondent to the column of $Q_2(B_2)$ with one in the first row (counting from zero), or equivalently to the $(m + 1)$th row of block h_γ, should contain non-zero value. From this it follows that the number of this position is $(0 - \gamma) \bmod m = m - \gamma$. But this is possible only if second burst occupies the block $h_{\gamma+1}$. Then the upper diagonal of $Q_2(B_2)$ contains $\gamma + 1$ ones, and the number of the column containing second non-zero element in this diagonal (i.e. the position of one in the $(m + 1)$th row of $h_{\gamma+1}$) is $m - \gamma$. So, since the burst B_2 occupies one block, the number of this block is $\gamma + 1$ instead of γ, so the value of γ cannot exceed $\ell - 2$.

Consider in more detail the connection between γ and ℓ. It is clear that $\gamma \leq \ell - 1$, since the burst begins in block h_γ. However, the burst may occupy two blocks and we need to check whether the burst affects the block with number exceeding $\ell - 1$. Let for some ℓ the code can correct bursts of length no more than b_ℓ. Then there are two uncorrectable bursts of length $b_\ell + 1$ occupying the first and last blocks of H_ℓ (it is supposed that the second burst ends in the last block but it may begins in the preceding one). If the second block does not affect the last block, this means that these bursts cannot be corrected by the code with number of blocks less than ℓ, then we may consider this shorter code as the code with bursts in the first and last blocks.

Next, each vector $\bar{y}$, which is defined by the incorrectable bursts (see (4) and (12)) and is solution of (12) and (18), has the correspondent value of γ equal either to $\ell - 1$ or to $\ell - 2$. Consider the code with the number of blocks increased by one. The length of incorrectable burst may either reduced or remain the same. In the first case we have $\gamma' = \gamma + 1$, in the second case $\gamma' = \gamma$. If for $\gamma' = \gamma + 1$ there are no solutions of equation (18) in terms of α_1, α_2 and α_3, then $b_{\ell+1} = b_\ell$ (b_ℓ means the maximum length of correctable burst for the code with parameter ℓ).

Continuing the analysis of γ in similar way, we obtain the values given in the last column of Table 1.

Summarizing the analysis, we may formulate the following results.

1. If the condition in the first row of Table 1 is satisfied, this means that $m \vdots \gamma$, or $m \vdots (\ell - 1)$ for given m and ℓ, and the code cannot correct bursts of length $b = m - \ell + 2$.

2. $\exists k > 0 : (m - (\gamma - 1) - k \cdot (\gamma + 1)) \vdots \gamma$, or $\exists k > 0 : (m - \ell + 3 - k \cdot (\ell - 1)) \vdots (\ell - 2)$, so $b = m - \ell + 2$.
3. $\exists k > 0 : (m - k \cdot (\gamma - 1)) \vdots \gamma$, or $\exists k > 0 : (m - k \cdot (\ell - 3)) \vdots (\ell - 2)$, so $b = m - \ell + 3$.
4. $\exists k > 0 : (m - k \cdot (\gamma - 1)) \vdots \gamma$, or $\exists k > 0 : (m - k \cdot (\ell - 2)) \vdots (\ell - 1)$, so $b = m - \ell + 3$.
5. $\exists k > 0 : (m - k \cdot (\gamma + 1)) \vdots \gamma$, or $\exists k > 0 : (m - k \cdot (\ell - 1)) \vdots (\ell - 2)$, so $b = m - \ell + 3$.

Given the code parameters m and ℓ we should first consider the conditions giving the least length of uncorrectable burst (i.e. 1 and 2). Then the conditions 3, 4 and 5 should be considered. If there are no satisfied conditions, then the minimal incorrectable burst is not in the last block of H_ℓ, then we should decrease ℓ and repeat the procedure.

The following two lemmas conclude the proof of Theorem 1.

Lemma 6 *If $\ell > \lceil m/2 \rceil + 1$, then the code defined by (1) can correct single bursts of length $b \le b_\ell$, where*

$$\begin{cases} b_\ell = m - \lceil m/2 \rceil + 1, & m \text{ odd}, \\ b_\ell = m/2 - 1, & m \text{ even}. \end{cases} \tag{24}$$

Proof According to Lemma 5 and its corollaries, for each non-zero element of $\bar{y}$ there are two other non-zero elements at distance γ, $\gamma + 1$ or $\gamma - 1$ positions to the left and to the right (that is, non-zero elements of neighbour sections). If $\ell > \lceil m/2 \rceil + 1$, then the maximum value of γ may exceed $m/2$, this means that the number of sections in $\bar{y}$ is minimal in this case and is equal to 2, so the error burst contains only two ones, which are the beginning and the end of the burst, "moving" to each other by the cyclic shift by γ positions.

Further increasing of ℓ (exceeding $\lceil m/2 \rceil + 1$) will not lead to the decreasing of uncorrectable burst length. It is clear that in this case the maximal length of correctable burst will be defined by (24).

Lemma 7 *For $\ell = 3$ and m odd, the code with parity-check matrix (1) can correct single bursts of length $b = m - 1$.*

Proof It is clear that for $\ell = 3$ the error bursts occupy the first and last blocks of H_3. Otherwise, γ would be equal to one and $\bar{y}$ would consists of all-ones. Then $\gamma = 2$, which is possible only for $i = -1$, $j = -1$, and from Table 1 the burst length is $m - \gamma + 1$, which is $m - 1$ for $\gamma = 2$.

It is easy to check that for even m and $\ell = 3$ one may always define the vector $\bar{y}$ satisfying (13) and containing ones on even positions and zeros on others.

From the results given in Table 1, Lemmas 6 and 7 the statement of Theorem 1 follows.

4 Conclusion

In this paper the burst-correcting capability of Gilbert codes is considered, when correcting single error bursts. The procedure is formulated, allowing to calculate the exact value of maximal correctable burst length depending on the parameters of the code. Its worth mentioning that extending Gilbert codes by adding extra parity-checks (or block-rows) to the parity-check matrix may improve the burst-correcting capabilities, however, for any block-permutation LDPC-code this value will be less than the block size.

References

1. Proakis, J., Salehi, M.: Digital Communications. McGraw-Hill (2007)
2. Krouk, E.A., Ovchinnikov, A.A.: Metrics for distributed systems. In: 2014 XIV International Symposium on Problems of Redundancy in Information and Control Systems (REDUNDANCY), pp. 66–70, 1–5 June 2014
3. Krouk, E., Semenov, S., authors.: Krouk, E., Semenov, S. (eds.) Modulation and Coding Techniques in Wireless Communications. Wiley (2011)
4. MacWilliams, F., Sloane, N.: The Theory of Error-Correcting Codes. North-Holland Publishing Company (1977)
5. Lin, S., Ryan, W.: Channel Codes: Classical and Modern. Cambridge University Press (2009)
6. Gallager, R.G.: Low density parity check codes. IRE Trans. Inf. Theory (1962)
7. Gallager, R.G.: Low Density Parity Check Codes. MIT Press, Cambridge, MA (1963)
8. MacKay, D.: Good error correcting codes based on very sparse matrices. IEEE Trans. Inf. Theory **45** (1999)
9. MacKay, D., Neal, R.: Near shannon limit performance of low-density parity-check codes. IEEE Trans. Inf. Theory **47**(2) (2001)
10. Richardson, T.J., Urbanke, R.L.: The capacity of low-density parity-check codes under message-passing decoding. IEEE Trans. Inf. Theory **47**(2) (2001)
11. Zyablov, V., Pinsker, M.: Estimation of the error-correction complexity for Gallager low-density codes. Probl. Inf. Trans. **XI**(1), 18–28 (1975)
12. Forney, G.D., Richardson, T.J., Urbanke, R.L., Chung, S.-Y.: On the design of low-density parity-check codes within 0.0045 db of the Shannon Limit. IEEE Commun. Lett. **5**(2) (2001)
13. Kozlov, A., Krouk, E., Ovchinnikov, A.: An approach to development of block-commutative codes with low density of parity check. Izvestiya vuzov. Priborostroenie. **8**, 9–14 (2013). (In Russian)
14. Gilbert, E.N.: A problem in binary encoding. Proc. Symp. Appl. Math. **10**, 291–297 (1960)
15. Arazi, B.: The optimal Burst error-correcting capability of the codes generated by $f(x) = (x^p + 1)(x^q + 1)/(x + 1)$. Inf. Contr. **39**(3), 303–314 (1978)
16. Bahl, L.R., Chien, R.T.: On Gilbert Burst-error-correcting codes. IEEE Trans. Inf. Theory **15**(3) (1969)
17. Neumann, P.G.: A note on Gilbert Burst-correcting codes. IEEE Trans. Inf. Theory, IT-11:377 (1965)
18. Zhang, W., Wolf, J.: A class of Binary Burst error-correcting quasi-cyclic codes. IEEE Trans. Inf. Theory, IT-34:463–479 (1988)
19. Krouk, E., Ovchinnikov, A.: 3-Stripes Gilbert low density parity-check codes. US Patent 7,882,415
20. Krouk, E.A., Semenov, S.V.: Low-density parity-check Burst error-correcting codes. In: International Workshop Algebraic and Combinatorial Coding Theory, Leningrad, pp. 121–124 (1990)

Data Dictionary Extraction for Robust Emergency Detection

Emanuele Cipolla and Filippo Vella

Abstract In this work we aim at generating association rules starting from meteorological measurements from a set of heterogeneous sensors displaced in a region. To create rules starting from the statistical distribution of the data we adaptively extract dictionaries of values. We use these dictionaries to reduce the data dimensionality and represent the values in a symbolic form. This representation is driven by the set of values in the training set and is suitable for the extraction of rules with traditional methods. Furthermore we adopt the boosting technique to build strong classifiers out of simpler association rules: their use shows promising results with respect to their accuracy a sensible increase in performance.

1 Introduction

Exploiting sensor networks for environmental monitoring enables the study of a wide variety of geophysical phenomena based on time-delayed measurements that can be furtherly processed to seek new, previously unknown connection among events. On the other hand, the development of knowledge discovery in databases has required scientists and engineers to focus more and more on data-driven discovery while modeling their domains of interest. More recently, the Big Data paradigm has been adopted to deal with huge quantities of data differing in range and representation, with a variable trustworthiness that the availability of powerful hardware commodities have made operable. Discretized time series are useful to reduce the cardinality of the set of symbols without a significant loss of information. A statistical approach, such as [1], may detect current anomalies in those data measuring entropy by means of mutual information or other suitable representations; discretized data may also be

E. Cipolla (✉) · F. Vella
Institute for High Performance Computing and Networking - ICAR,
National Research Council of Italy, Palermo, Italy
e-mail: cipolla@pa.icar.cnr.it

F. Vella
e-mail: vella@pa.icar.cnr.it

G. De Pietro et al. (eds.), *Intelligent Interactive Multimedia Systems and Services 2016*, Smart Innovation, Systems and Technologies 55,
DOI 10.1007/978-3-319-39345-2_3

used to extract rules based on frequent items. A first, straightforward, choice is to divide the range of all the values in a limited set of intervals and associate to each interval a different symbol. In this case the data levels are fixed and the thresholds for all the levels are generated by the previous knowledge about the experimental domain. A different choice is to create a set of symbols according the distribution of the data on a given set. The symbols are chosen according to the distribution of the data in the vector space. A reasonable choice is to use a clusterization process and extract the centroids from a given set of values (e.g. the training set). Each value of the set can therefore be replaced with a symbol that corresponds to the centroid associated to it. *k*-means is a well known algorithm for the clusterization of data.The values obtained with this method are compared with a set of centroids obtained with Vector Quantization algorithm evaluating advantages and shortcomings. In this work, we use a parallelized Parallel FP-Growth (PFP) [2], based on the seminal FP-Growth algorithm described in [3]. We used the Apache Mahout™ implementation of PFP to generate rules as it leverages the open source Hadoop environment that has become the *de facto* standard for storing, processing and analyzing Big Data. The whole set of rules found is the input to the AdaBoost algorithm, first proposed by [4]. Our main contribution to the topic lies in the use of pattern mining techniques to find co-occurrence relationships leading to risk situations, enhancing historic datasets gathered by sensor networks with emergency notifications commonly found online newspapers and weblogs; acknowledging that association rules can be heavily dependent on the training set, we strive to provide stronger classifiers built using boosting. We tested the proposed technique on a dataset of Tuscanian meteorological data ranging from 2000 to 2010, and we have compared these values with the emergency detection in the same region along the same years, with promising results.

2 Frequent Pattern Mining

Frequent pattern mining deals with finding relationships among the items in a database. This problem was originally proposed by Agrawal in [5]: given a database D with transactions $T_1 \dots T_N$, determine all patterns P that are present in at least a fraction of the transactions. The set T_i of identifiers related to attributes having a boolean `TRUE` value is called *transaction*.

An example domain of interest is composed of market baskets: each attribute corresponds to an item available in a superstore, and the binary value represents whether or not it is present in the transaction: an interesting pattern is thus present if two or more items are frequently bought together. The aforementioned approach has successfully been applied to several other applications in the context of data mining since then.

2.1 Association Rule Mining

Agrawal et al. [5] presented association rule mining as a way to identify strong rules using different measures of interestingness, such as high frequency and strong correlation.

Let $D = \{T_1, T_2, \ldots, T_n\}$ be a transaction database. A set $P \subseteq T_i$ is called an l-sized *itemset* if the number of items it contains is l, and has a support $supp\left(P_i\right) = \frac{|P_i(t)|}{|D|}$ that is the ratio of transactions in D containing X. X will be deemed *frequent* if its support is equal to, or greater than, a given threshold minimal support. An association rule R is the implication $X \implies Y$, where itemsets X and Y do not intersect. An evaluation on the validity of each rule can be performed using several quality measurements:

- the support of R is the support of $X \cup Y$, and states the frequencies of occurring patterns;
- the confidence of R $conf\,(X \implies Y)$, defined as the ratio $\frac{supp(X \cup Y)}{supp(X)}$, states the strength of implication.

Given a minimal support s_{MIN} and minimal confidence c_{MIN} by users or experts, $X \implies Y$ is considered a valid rule if both $supp\,(X \implies Y) \geq s_{MIN}$ and $conf\,(X \implies Y) \geq c_{MIN}$.

2.2 PFP: The FP-Growth Algorithm in a Parallelized Environment

In 2008, Dean and Ghemawat [6] presented MapReduce, a framework for processing parallelizable problems across datasets using a large number of inter-connected computer systems, called *worker nodes*, taking advantage of locality of data in order to reduce transmission distances. The FP-Growth Algorithm, a divide et impera algorithm that extracts frequent patterns by pattern fragment growth proposed by Han in [3], has been ported to the MapReduce framework by Li et al. [2].

Given a transaction database D, the three MapReduce phases used to parallelize FP-Growth can be outlined as follows:

1. **Sharding**: D is divided into several parts, called *shards*, stored on P different computers.
2. **Parallel Counting**: The support values of all items that appear in D is counted, one shard per mapper. This step implicitly discovers the items vocabulary I, which is usually unknown for a huge D. The result is stored in a frequency list.
3. **Grouping Items**: Dividing all the $|I|$ items on the frequency list into Q groups. The list of groups is called group list (*G-list*), where each group is given a unique group identifier (gid).

4. **Parallel FP-Growth**: During the map stage, transactions are rearranged group-wise: when all mapper instances have finished their work, for each group-id, the MapReduce infrastructure automatically gathers every group-dependent transaction into a shard. Each reducer builds a local FP-tree and recursively grows its conditional FP-trees, returning discovered patterns.
5. **Aggregating**: The results generated in Step 4 are coalesced into the final FP-Tree.

3 Construction of a Robust Classifier Through Boosting

The word *boosting* refers to a general method of rule production that combines less accurate rules to form more accurate ones. A so-called "weak learning algorithm", given labeled training examples, produces several basic classifiers: the goal of boosting is to improve their global performance combining their calls, assuming that they fare better than a classifier whose every prediction is a random guess.

We have chosen to improve the performance of our association rules using the AdaBoost meta-algorithm, first proposed by Freund and Schapire in [4]. AdaBoost takes as input a set of training examples $(x_1, y_1), \ldots, (x_m, y_m)$ where each x_i is an instance from X and each y_i is the associated label or class: in this work $y_i = 0$ for negative examples, $y_i = 1$ otherwise. We repeat the weak classifier training process exactly T times.

At each iteration $t = 1, \ldots, T$ a base classifier $h_t : X \in 0.1$ having low weighted error $\epsilon_t = \sum_i w_i \left| h_j(x_i) - y_i \right|$ is chosen. A parameter α_t, with $\alpha_t > 0 \iff \epsilon_t < 1$, is chosen, so that the more accurate the base classifier h_t is, the more importance we assign to it. To give prominence to hard-to-classify items, weights for the next iteration are defined as $w_{t+1}, i = w_t \beta_t^{1-e_i}$, where $e_i = 0$ if example x_i has been correctly classified, $e_i = 1$ otherwise, and $\beta_t = \frac{\epsilon_t}{1-\epsilon_t}$. The final strong classifier is:

$$H(x) = \begin{cases} 1 \text{ if } \sum_{t=1}^{T} \alpha_t h_t(x) \geq \frac{1}{2} \sum_{t=1}^{T} \alpha_t; \\ 0 \text{ otherwise.} \end{cases} \tag{1}$$

where $a_t = \log \frac{1}{\beta_t}$.

Association rule extraction algorithms produce, on average, many more rules than classification algorithms, because they do not repeatedly partition record space in smaller subsets—on the other hand, this means that association rules are much more granular, and their extraction algorithms are generally slower. Anyway, a balance between granularity and performance may be found imposing support and confidence thresholds on itemsets. It turns out that association rules can be used as classifiers if a discretization (as shown in Sect. 4) of the attribute space is performed, so that the established bins can serve as feature sets. Association rule mining can be thus applied to find patterns of the form < *featuresets* > $\Longrightarrow$ *ClassLabels*, ranking rules first by confidence, and then by support, as shown in Yoon and Lee's [7]. While we follow a similar approach, we diverge in several aspects:

- we perform boosting on the whole set of generated rules;
- a correct weak classification occurs if both or neither of antecedent and consequent are present;
- we penalize the weight only if the error rate of a given weak classifier is lower than 50 %.

4 Information Representation

The datasets used in this work has been made available by Servizio Idrogeologico Regionale della Toscana (SIR).[1] Their sensors and surveillance network, spanning the entire surface of Tuscany, can provide both real-time and historic samples from hydrometric, pluviometric, thermometric, hygrometric, freatimetric and mareographic sensors, allowing a general characterization of hydroclimatic phenomena.

Generally, stations in a sensor network are placed in a way that ensures optimum coverage of a given region: different restrictions due to the domain of interest and regulations already in force when considering the placement need to be taken into account, so any two given networks may have very different topologies. Given a station, relevant neighbors belonging to the other networks must be found. In this work, we group values using concentric circles having radiuses $r_1 = 25\,\text{km}$, $r_2 = 50\,\text{km}$, $r_3 = 75\,\text{km}$ centered on basin stations, as they constitute the sparsest network among those managed by SIR.

An outline of the data transformation steps we perform follows:

1. *Per-network grouping*: As every station stores a small subset of data, each station is polled by a central facility at regular intervals. SIR provides a single file for each station in a given network. For our convenience, a single table is created for gathering data coming from all the stations in the same network;
2. *Discretizations*: Each sensor measure is replaced with a discretized value. This quantized representation is needed since the rule extraction algorithms extract connections among recurring symbols.
3. *Basket arrangement and emergency binding*: The output of the discretization process must be converted to a transactional format for use with the association rules extraction algorithm. There will be a transaction row vector r per day per station. In each of them the column (B_k) will have a `TRUE` value if the discretized value is k, `FALSE` otherwise. An emergency flag is set if for a given date the basin station was near enough to dangerous phenomena.

[1] http://www.sir.toscana.it/.

4. *Inverse mapping*: Apache Mahout™ requires transactions items to be expressed using integer keys, so we map the column names in the basket arrangement made in the previous step, keeping trace of the mappings to the original items to properly present results in the output study phase.

4.1 Discretization

The continuous nature of meteorological measurements is not suitable for association rules extraction: a reduced set of values is thus considered using a discretization process. We used both the *k*-means and Linde-Buzo-Gray algorithms to extract the discretization interval bounds.

k-means, first proposed by Lloyd in [8], takes as input the number *k* of clusters to generate and a set of observation vectors to cluster, returning exactly *k* centroids, initially chosen at random and converging to a stable position that minimizes the sum of the quadratic distance between the observations and the centroids.

A vector v belongs to cluster i if it is closer to centroid i than any other centroid; c_i is said to be the *dominating centroid* of v. Since vector quantization is a natural application for *k*-means, information theory terminology is often used: the centroid index or cluster index is also referred to as a *code* and the table mapping codes to centroids is often referred as a *codebook*, so *k*-means can be used to quantize vectors. Quantization aims at finding an encoding of vectors that minimizes the sum of the squared distances (SS) between each observation vector x_i^j and its dominating centroid c_j, called *distortion*:

$$J = \sum_{j=1}^{k} \sum_{i=1}^{n} \left\| x_i^{(j)} - c_j \right\|^2 \tag{2}$$

k-means terminates either when the change in distortion is lower than a given threshold or a maximum number of iterations is reached.

Linde, Buzo and Gray algorithm, first proposed in [9], is another iterative algorithm which assures both proximity to centroids and distorsion minimization. The initial codebook is obtained by splitting into two vectors the average of the entire training sequence. At the end of the first iteration, the two codevectors are splitted into four and so on. The process is repeated until the desired number of codevectors is obtained.

A simple test, probably first proposed by R. Thorndike in 1953 and called *elbow method*, can be used to choose the right *k* with respect to the percentage of variance: if you plot the SS against the value of *k*, you will see that the error decreases as *k* gets larger; this is because when the number of clusters increases, they should be smaller, so distortion is also smaller. The idea of the elbow method is to choose the *k* at which the SSE decreases abruptly. This produces an "elbow effect" in the graph.

4.2 Emergency Information

While sensor networks provide quantitative figures with a certain degree of reliability, they do not convey any information about emergencies: we need this information in order to train a classifier that can identify potential unforeseen climaxes. We assume that if an emergency situation has been reported in the past, traces can be found in the World Wide Web by means of online newspaper articles or even posts on a personal blog: relevant content may hopefully contain a word in the set A of words describing the phenomenon, and another one in the geocoded set B of Tuscanian cities. The set A is formed by Italian keywords about hydrogeological emergencies such as: *esondazione*, *violento temporale*, *diluvio*, *allagamento*, *inondazione*, *rovinosa tempesta*, *violento acquazzone*. The set B is formed by the names of the cities in the Tuscany region such as: *Firenze*, *Pisa*, *Livorno*, *Grosseto*, *Lucca*, *Siena*, *Massa*, *Carrara*, *Pistoia*. As WWW pages cannot be easily dated, we used the subset of search results having day, month and year information in their URLs, usually found in weblogs and digital magazines, as they require specific expertise to get altered after publication. This subset has been filtered by visual check to remove spurious and incomplete data, but we are unable to exclude that some of the remaining information has not been altered by the content authors, either willingly or because of an error. The emergency flag is then set to `TRUE` for every basin station placed at a distance less than 75 Km from each interesting location that has been found.

5 Experimental Setup and Results

A subset of SIR basin levels, rain measures and phreatic zone data for the years 2000–2010 has been used. The Elbow test has been performed after having repeatedly run the k-means algorithm on each set of measures. We chose to use 3 bins for rain data, 4 bins for basin level data, and 6 bins for phreatic data. In the latter case, we actually have two elbow conditions, and the second one happens just before a slight increase in the sum of squares: we arbitrarily chose the number of bins that gives the absolute minimum sum of squares (see Fig. 1).

A similar profiling has been performed using Linde-Buzo-Gray quantization on our discrete data; the minimum value of distorsion is achieved using 8 codevectors (see Fig. 2).

A number of software tools have been developed specifically to extract and aggregate data provided by SIR, and parse Apache Mahout™ output. After the creation of the basket connecting all the basin level station with the nearest rain of phreatic values, the data have been divided in two subsets: a **training** subset, containing a 60 % of the items in the original set, to be used as PFP input for association rules extraction and a **test** subset, containing the remaining 40 %, over which the extracted rules have been tested. Candidates for both sets are chosen using a random sampling,

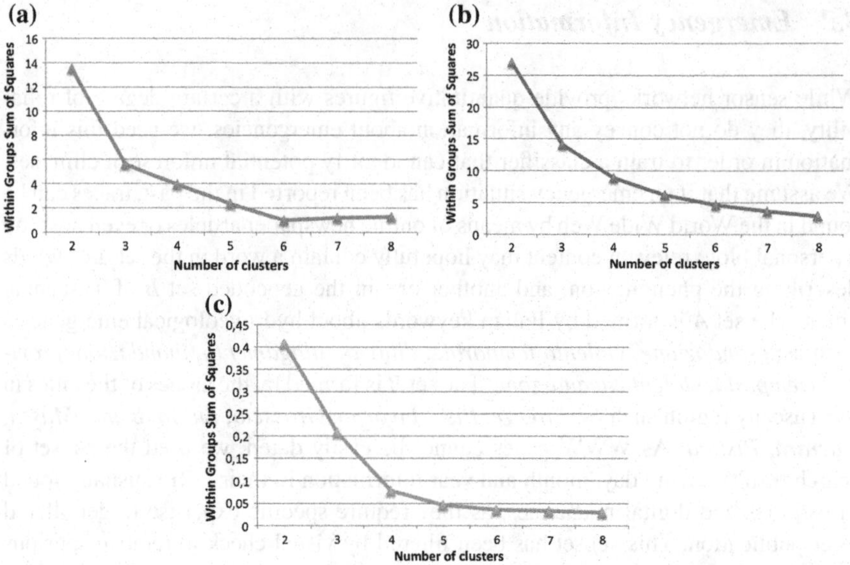

Fig. 1 Average distortion for *k*-means algorithm. **a** Phreatic levels. **b** Rain levels. **c** Basin levels

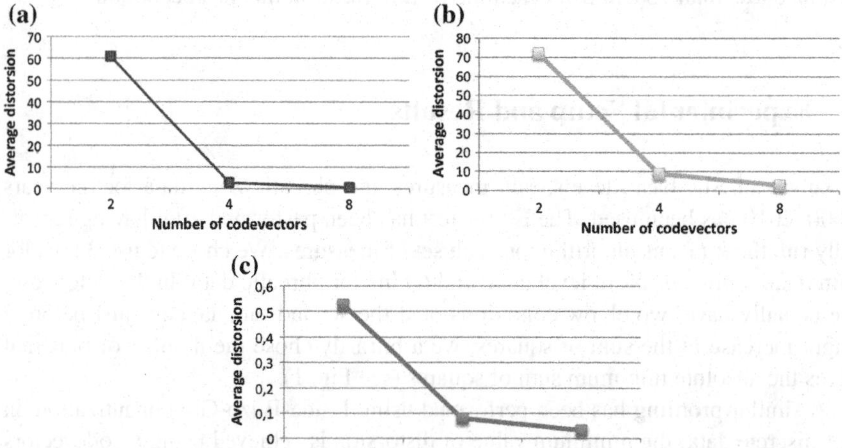

Fig. 2 Average distortion for LBG algorithm **a** Phreatic levels. **b** Rain levels. **c** Basin levels

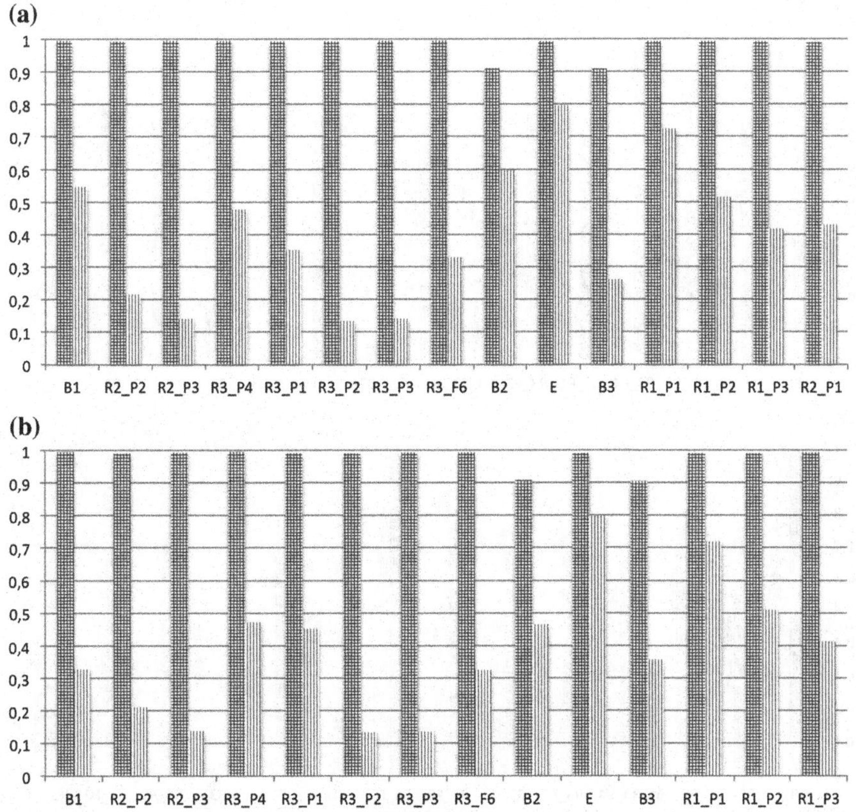

Fig. 3 Accuracy of boosted classifiers (*in blue*) versus accuracy values of association rules for the same consequent (*in red*). **a** LBG quantization. **b** *k*-means quantization (Color figure online)

having discarded a small subset of rules having either a confidence ratio inferior to 25 % or a missing value as consequent.

The extracted rules have been evaluated considering their performance over the test set. A **True Positive** (TP) classification takes place when both the antecedent and the consequent of the rule are satisfied, while in a **True Negative** (TN) one neither of them is. A **False Positive** (FP) classification satisfies the antecedent, but not the consequent: for **False Negatives**(FN), the reverse applies.

The accuracy Acc is defined as $\frac{TP+TN}{TP+TN+FP+FN}$ and presented in Fig. 3.

Precision $prec = \frac{TP}{TP+FP}$ and Recall $rec = \frac{TP}{TP+FN}$ are shown in Figs. 4 and 5, respectively.

The harmonic mean of Precision and Recall, called F_1-score, is $F_1 = 2 \times \frac{prec \times rec}{prec+rec}$. It is shown in Fig. 6.

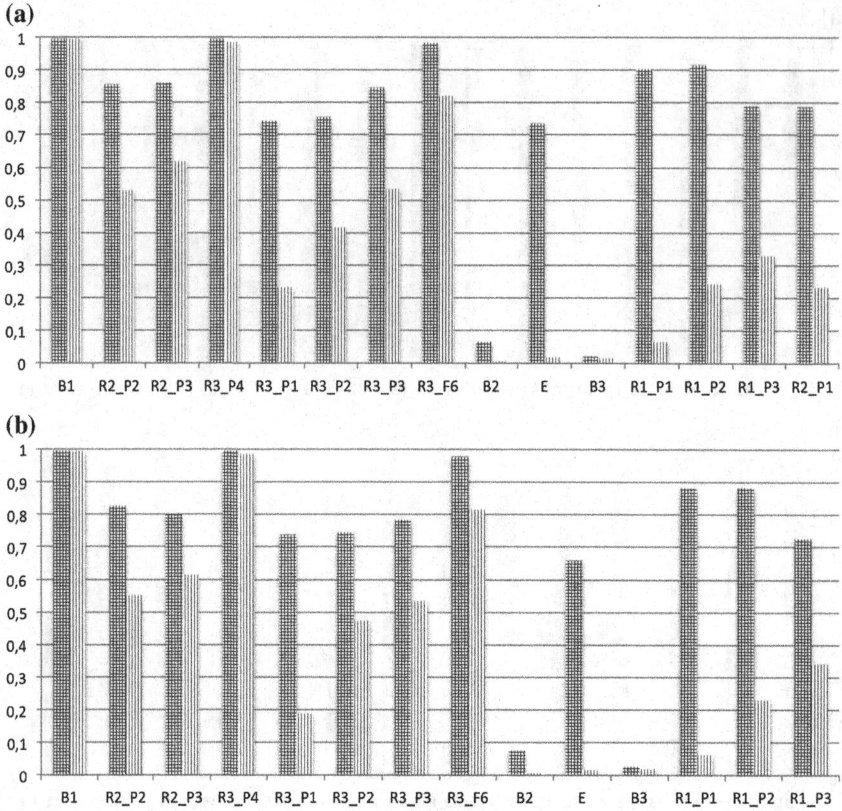

Fig. 4 Precision of boosted classifiers (*in blue*) versus average of precision values of association rules for the same consequent (*in red*). **a** LBG quantization. **b** *k*-means quantization (Color figure online)

Finally, we compared the Precision of those boosted classifiers yielded after using both techinques for vector quantizations on our source data with the Precision of similarly boosted classifiers generated from the SIR-inspired 7-bin discretization we presented in our previous work [10].

While the use of different quantization techniques has nearly halved the number of produced classifiers, those ones remaining are much more accurate. As you can see in Fig. 7, while LBG and *k*-means performance is nearly on par, with LBG being marginally better, they both outperform our old classifiers with fixed symbols. This is also true for the classifier for the emergency symbol E, whose Precision has grown by 68 %.

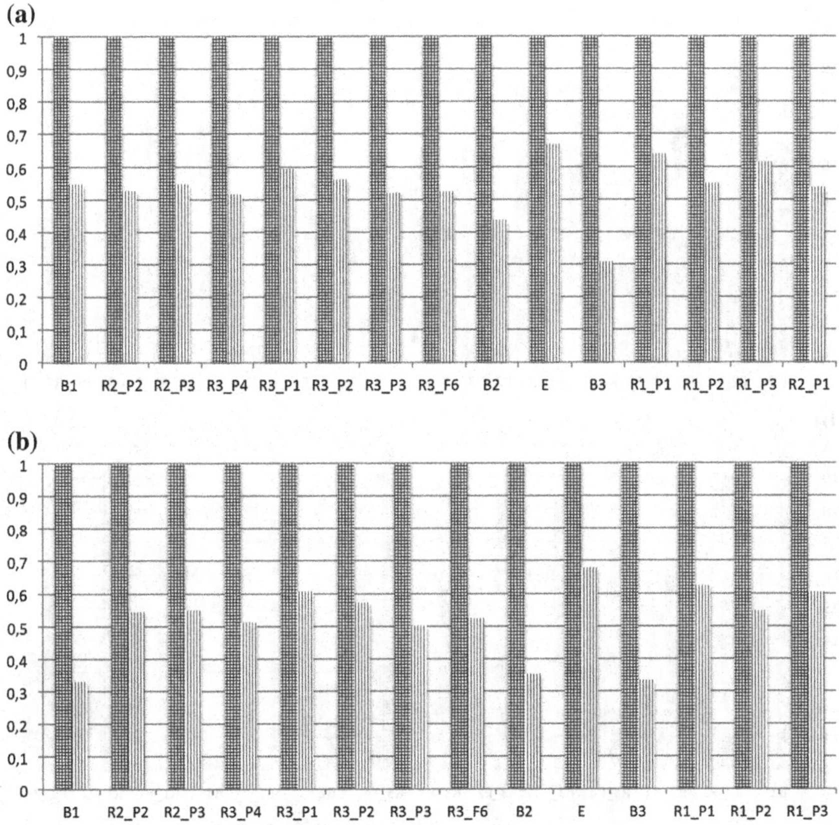

Fig. 5 Recall of boosted classifiers (*in blue*) versus average of Recall values of association rules for the same consequent (*in red*). **a** LBG quantization. **b** *k*-means quantization (Color figure online)

6 Conclusions

In this paper, a method to detect relationships between different measure types in a sensor network has been devised for analysis and emergency detection purposes. A set of association rules has been extracted using a subset of 10 years of Tuscanian Open Data by SIR containing geophysical measures. Emergency information have been extracted through queries to the Bing™ Search Engine. Stronger classifiers have been generated using the AdaBoost meta-algorithm on association rules extracted by the PFP algorithm. Having generated a strong classifier per symbol, we have grouped each weak classifier and took their average performance as reference values to evaluate possible improvements. The use of vector quantization has significantly improved the accuracy of our boosted classifiers over the same dataset, especially that of our classifier for emergency situations.

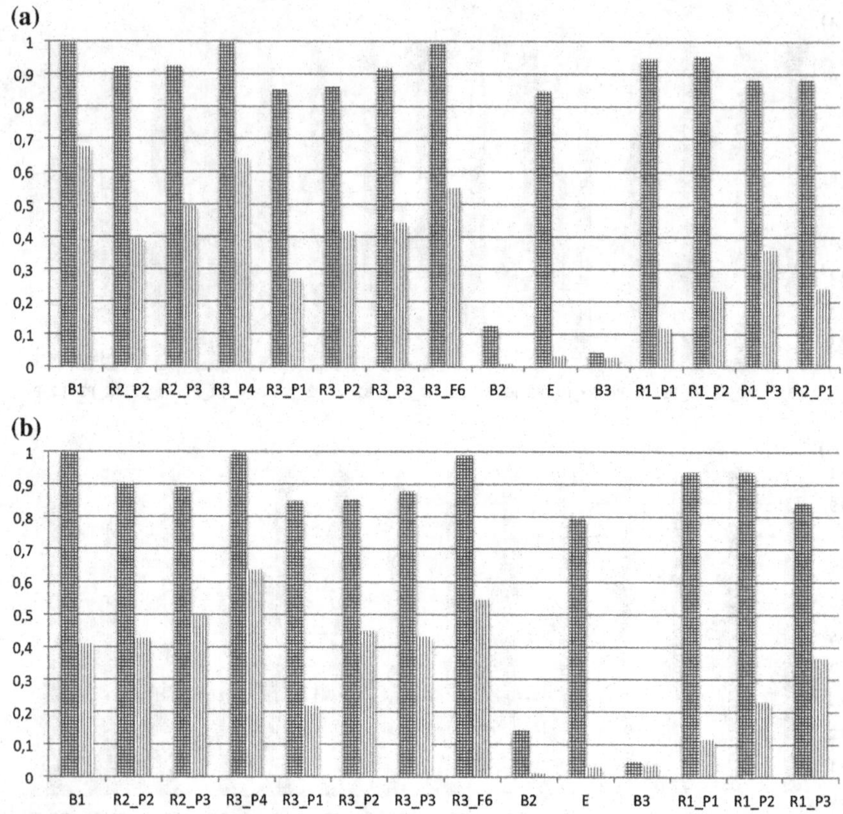

Fig. 6 F1-score of boosted classifiers (*in blue*) versus average of F1-score values of association rules for the same consequent (*in red*). **a** LBG quantization. **b** *k*-means quantization (Color figure online)

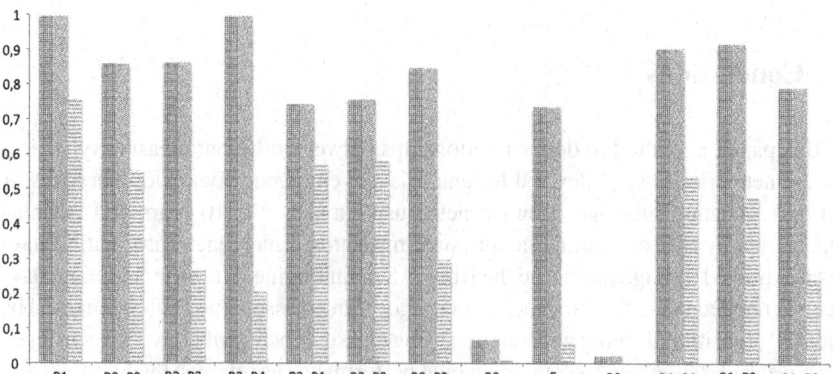

Fig. 7 Comparison of precision for *k*-means, LBG, 7-bin quantization

Acknowledgments This work has been partially funded by project SIGMA PONOI_00683 Sistema Integrato di sensori in ambiente cloud per la Gestione Multirischio Avanzata PON MIUR R&C 2007–2013.

References

1. Cipolla, E., Maniscalco, U., Rizzo, R., Stabile, D., Vella, F.: Analysis and visualization of meteorological emergencies. J. Ambient Intell. Hum. Comput. 1–12 (2016)
2. Li, H., Wang, Y., Zhang, D., Zhang, M., Chang, E.Y.: Pfp: parallel fp-growth for query recommendation. In: Proceedings of the 2008 ACM Conference on Recommender Systems, ser. RecSys '08, pp. 107–114. ACM, New York, NY, USA (2008)
3. Han, J., Pei, J., Yin, Y.: Mining frequent patterns without candidate generation. SIGMOD Rec. **29**(2), 1–12 (2000)
4. Freund, Y., Schapire, R.E.: A decision-theoretic generalization of on-line learning and an application to boosting. J. Comput. Syst. Sci. **55**(1), 119–139 (1997)
5. Agrawal, R., Imieliński, T., Swami, A.: Mining association rules between sets of items in large databases. SIGMOD Rec. **22**(2), 207–216 (1993)
6. Dean, J., Ghemawat, S.: Mapreduce: simplified data processing on large clusters. Commun. ACM **51**(1), 107–113 (2008)
7. Yoon, Y., Lee, G.G.: Text categorization based on boosting association rules. In: IEEE International Conference on Semantic Computing, vol. 2008, pp. 136–143 (2008)
8. Lloyd, S.: Least squares quantization in pcm. IEEE Trans. Inf. Theory **28**(2), 129–137 (1982)
9. Linde, Y., Buzo, A., Gray, R.M.: An algorithm for vector quantizer design. IEEE Trans. Commun. **28**, 84–95 (1980)
10. Cipolla, E., Vella, F., Boosting of association rules for robust emergency detection. In: 11th International Conference on Signal-Image Technology & Internet-Based Systems, SITIS 2015, pp. 185–191. Bangkok, Thailand, 23-27 Nov 2015

SmartCARE—An ICT Platform in the Domain of Stroke Pathology to Manage Rehabilitation Treatment and Telemonitoring at Home

Francesco Adinolfi, Giuseppe Caggianese, Luigi Gallo, Juan Grosso, Francesco Infarinato, Nazzareno Marchese, Patrizio Sale and Emiliano Spaltro

Abstract This paper describes the SmartCARE ICT eco-system, which goal is to deliver advanced health collaboration services in the rehabilitation domain. The system provides a set of tools that enable the continuity of care at home to patients affected by stroke diseases. Moreover, by taking advantage of motion sensing-based serious games and virtual companions, the system can stimulate the patient at being more reactive both at neuro-motorial and neuro-cognitive levels.

F. Adinolfi · N. Marchese
ITSLab s.r.l., Torre Annunziata, Naples, Italy
e-mail: francesco.adinolfi@itslab.it

N. Marchese
e-mail: nazzareno.marchese@itslab.it

G. Caggianese (✉) · L. Gallo
Institute for High Performance Computing and Networking,
National Research Council of Italy (ICAR-CNR), Naples, Italy
e-mail: giuseppe.caggianese@na.icar.cnr.it

L. Gallo
e-mail: luigi.gallo@na.icar.cnr.it

F. Infarinato
Rehabilitation Bioengineering Laboratory, IRCCS San Raffaele Pisana, Rome, Italy
e-mail: francesco.infarinato@sanraffaele.it

P. Sale
Fondazione Ospedale San Camillo IRCCS, Venezia Lido, Italy
e-mail: patrizio.sale@gmail.com

J. Grosso · E. Spaltro
ALPHA Consult, Milano, Italy
e-mail: jg@alphacons.eu

E. Spaltro
e-mail: es@alphacons.eu

G. De Pietro et al. (eds.), *Intelligent Interactive Multimedia Systems and Services 2016*, Smart Innovation, Systems and Technologies 55,
DOI 10.1007/978-3-319-39345-2_4

Keywords Stroke · Rehabilitation · Telemonitoring · Microsoft kinect · Unity 3D · Microsoft band · Serious game · Wearable devices

1 Introduction

Demographic studies show a continuous increase in the European population and aging: the number of people older than 50 years is above 150 million [5]. Moreover, recent analysis show a correlation between two increasing trend for next years: average age of the population and demand for health services [22]. However, the increase in demand for health services is linked to the non-sustainable increase in public spending on health care. This causes serious problems to administrations and existing clinical facilities as it becomes no longer possible to meet the needs of all patients without resorting to a significant increase of economic and structural resources used. In fact, in the framework of clinical neurology, over the past 15 years, a gradual reduction in the average length of the hospitalization in patients with stroke outcome has been witnessed, both in the acute care than in rehabilitation hospitals. Patients are often required to return to their home without having completed their rehabilitation process contrary to what is stated in the most modern theories on the modulation of motor, language and cognitive relearning, after brain injury, which confirm the importance of continuous monitoring of performance and feedback, to be returned to the patient while performing the exercise [3, 21]. Obviously, this situation can result in a particularly unfavourable impact on the recovery of functional ability, in particular, upper limb motor recovery and cognitive ability need of long training times no longer achieved at the hospital centres.

In this scenario, the application of Information and Communication Technology (ICT) together with an increasingly interoperability among different e-Health applications [1, 9] can be the right instrument to cover the demand for the continuation of the rehabilitation process, even with the patient discharged from the hospital. Therefore, the just mentioned theories can become practical through ICT where "smart" systems for instrumental, accurate, and objective detection of the motion parameters and the characteristics of the rehabilitative gesture should represent also a possibility to limit the costs. However this approach still present some issue to be faced. For example, in Italy there is an infrastructural digital divide due to the number of people devoid of broadband communication coverage (about to 4.8 % of the entire population). To afford this problem, the use of satellite communications allows patients to access healthcare services without hospitalization, wherever their home is located. Moreover, is important notice that the digital divide refers also to social, economic and geographical reasons. This means the population has also difficulties in accessing or using post-clinical practices and, in this light, the above-mentioned 4.8 % of the entire population becomes a more significant target of innovative rehabilitation strategies.

Nowadays there are many technical platforms [4] candidate to afford e-Health challenges in the future. In particular, in the framework of clinical neurology, ICT can cover the demand for the continuation of the rehabilitation process, even with the patient discharged from the hospital. ICT solutions can: (i) allow remote interaction between staff and patients; (ii) nearly set to zero the times (and the need) of transport; iii) ensure the continuity of patient care between hospital and community; (iv) protecting the equity of access to local healthcare services [6, 7].

SmartCARE [20] project aims at realizing an ICT eco-system of advanced e-Health services. SmartCARE enables integrated home-care services assuring a constant verification of the therapeutic process, avoiding the presence on-site of the medical staff. Furthermore, SmartCARE: (i) allows a punctual assessment of the efficacy of the treatment via multimodal measurements; (ii) enables the communication between medical staff and caregivers; (iii) provides tools for the safety control of patients from the medical staff and family components. Accordingly, health operational costs are reduced and the clinical effectiveness of care and assistance is assured, while preserving the quality of patient's treatment. The SmartCARE platform is structured to offer rehabilitation and monitoring services by integrating measurement devices, innovative interaction paradigms, customized motorial/cognitive neuro-rehabilitation treatments, continuous health status monitoring, cloud and interoperable information systems. In the domain of stroke pathology, within a therapeutic process, physicians and patients can meet and manage daily activities in a collaborative way, enabling an optimization of healthcare rehabilitation at home.

The paper is organized as follows. The following sections describe the main system characteristics in terms of treatment (Sect. 2), technological equipment are reported (Sect. 3) and human-machine interface (Sect. 4). Finally, in Sect. 5, a brief analysis of integrated home-care platforms from a business perspective, is reported.

2 The Rehabilitation Treatment

People with disabilities or illnesses that cause lack of movement may be limited and restricted in their participation in the active life, through the reduction of motor function. The main goal of rehabilitation is to restore the functional capacity so as to enable a full integration of the patient into the community and, if possible, in the working environment [17]. Rehabilitation should start immediately after the incident occurs: the sooner the rehabilitation starts, the better results can be expected. Although higher intensity means better results, therapists cannot accommodate such high intensities since people need several hours of daily supervision [18]. Continuous and intensive rehabilitation efforts carried out even 6 months after the incident have an effect on improving motor functionality. However, the number of exercises in a therapy session is typically limited. Due to this limitation in therapy, therapists often prescribe exercises to be carried out at home, but such exercises are sometimes boring [12].

Research has also shown that rehabilitation of upper body limbs can have an effect in rehabilitating neglect, a condition in which the person is unable or has a hard time attending to one part of the visual field. Virtual reality devices are generally good because if immersed they allow the patient to do tasks that are normally impossible thus further motivating them [11]. The heterogeneity of impairments that patients suffer from is a relevant factor for planning, developing and evaluating treatments. Usually, an individual rehabilitation tasks plan is required for each patient, aspect that leads to a substantial increase in cost and constitutes one of the main problem for the therapists. In fact, the therapist has to find a way to plan a set of tasks always exciting and challenging for the patient to constantly motivate him, since the motivation is considered an important factor for the rehabilitation success.

Recent research in the field revealed the possibility to obtain a gradually recover of the cognitive functionality through repetitive, intensive and task oriented training [12]. However, traditional rehabilitation exercises, due to their repetitive nature, result to be boring leading to neglect the exercises required for recovery. SmartCARE proposes a patient-centred design of new approach to functional recovery trough the development of a serious game environment aimed at merging both motor rehabilitation and cognitive rehabilitation. This treatment focuses on:

- Improving functional independence and physical performance,
- Prevent and managing pain, physical disabilities that cause limits to participation,
- Promoting fitness, health and wellness.

Physiotherapy is an interaction process between the therapist and the patient, where the therapist assists the patient's level of mobility, such as strength, endurance and other physical abilities to determine the cause of their limitations in their physical functions. During the normal process of therapy, the therapist is guiding the patient to perform the exercises correctly. Conversely, in the serious game scenario, the system should plays the role of the therapist, giving a constant feedback to the patient. In this direction, the proposed system allows to meet the needs of patients with those of therapists, so allowing a fine tailoring of the game for each patient and a precise mapping of the patient's abilities in to the serious game dimensionalities. The system is based entirely on the use of touchless interaction exploiting low cost motion sensing technologies. The user/patient placed in front of the station does not need to touch with any kind of device, neither worn gloves nor use markers. The sensor captures the patient movements in his/her motor space mapping them into the virtual space in order to allow the patient to interact with the synthetic elements of the scene.

Cognitive stimulation is configured as an intervention directed to the overall well-being of the persons to increase their involvement in tasks aimed at the reactivation of the residual cognitive ability and the slowing of functional loss due to the neurological impairment. Therefore, in consequence to a loss of physiological neuronal matter, it would be possible the recovery of some connections inhibited, through the experiences of systematic stimulation. This is one of more target of the SmartCARE system.

3 Main Technological Characteristics

The SmartCARE system is based on sensor devices, integrated in a home station pc-based to support physiological measurements gathered during rehabilitation exercise executions. It is based also on a health information system, used remotely to feed a collaborative network among patients, therapists, physicians and to manage customized treatment plan (Fig. 2). The platform offers full support to user interfaces, combined with biological sensor in order to enable gaming experience for rehabilitation paths. The adoption of a graphical metaphor (e.g., an avatar as in Fig. 1) for rehabilitation of post-stroke patients helps in recovery of the motor skills, limb-eye coordination, orientation in space and everyday tasks. An exercise range from simple limb movements (goal directed) or aimed at achieving a given goal (e.g. putting a coffee cup on a table) to the improvement of lost motor skills (e.g. virtual driving). The detection of the main vital activities is executed through a wearable device support that detects physiological parameters and sends them to control unit. The rehabilitation exercises become more intuitive in their approach by using exercise templates with feedback showing correctness of performed exercises.

The home station is developed integrating different technologies to support the applicative needs of the rehabilitation process. The Microsoft Kinect V2 sensor is used to trace and measure the patient's motions and positioning. This choice was supported by evidence on usage of this device in many similar clinical studies [10]. The system integrates also the Microsoft Band bracelet used to correlate the dynamic movement measurements with physiological human parameters (e.g. pulse rate, calories, skin temperature), during the rehabilitation exercise. The integration is

Fig. 1 The avatar of the patient and the representation of the measurements during a motor rehabilitation exercise

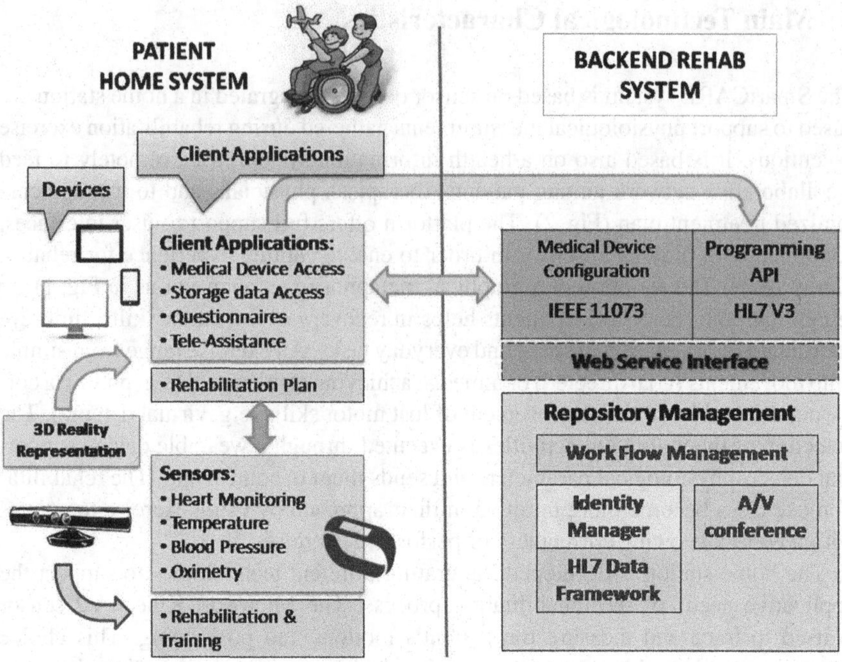

Fig. 2 SmartCARE—Main conceptual view

accomplished using the Unity 3D software platform, which handles the presentation logic containing patient's virtual representation and measurements.

4 The Natural User Interface

In order to meet post stroke patient needs, researchers have been trying to break away from tradition therapist-based rehabilitation focusing on patients' personalized rehabilitation tasks. All this has been made possible thanks to new low cost available technologies allowing to turn tele-health, tele-medicine and tele-rehabilitation from concept to reality.

However, there are two types of common approaches to the post stroke tele-rehabilitation. The first is based on exploiting the serious game purposefully for cognitive rehabilitation. In fact, in [15] is presented a possible classification based on their fundamental characteristics and in [13] are illustrated the necessary components of a comprehensive software platform with a general discussion of game design and use in rehabilitation, including motivational aspects from patient and therapist perspectives. In [14], is explored how Natural User Interface (NUI) can benefit the process of rehabilitation proposing a valid interface prototype for cognitive serious game, which allows alternative interaction modalities. Moreover, in [19] are presented cognitive rehabilitation game system based on serious games which, aim to avoid boring exercises and patients to neglect the prescribed exercises.

Secondly, other approaches exploit the applicability of the serious game for the motor rehabilitation by using low-cost video-capture technology suitable for deployment at home. For instance, in [2] is presented an approach for the upper limbs whereas in [8] also the patient's legs are used into the serious game. In this context, we propose a patient-centred system that exploits a serious game environment to merge both motor and cognitive rehabilitation. A similar approach was presented in [16] for older adult mental and physical exercises where the patient's movements are used to animate an avatar puppet in the game virtual world. However, in the proposed system, the user physical exercises tracked by the motion sensor are not used to simply control the movement of an avatar neither are statically associated to a specific game action. We propose an adaptive patient-game interaction in which each physical exercise that the user can carry out can be exploited to control an opportunely tailored aspect of the game interface. More in detail, each serious game in the system is decoupled by the interaction modality that becomes specific for each patient. Each game exhibits a set of dimensionalities in the form of degree of freedom (DOF) needed, for instance, to move a pointer in the virtual scene or to confirm a selection; the therapist is allowed to associate a specific motor exercise to each one of these game dimensionalities. In this way patients with different inabilities can play exactly the same game. The proposed system meets the needs of patients aiming that an appropriate mapping of the motor rehabilitation goals with the game dimensionalities will motivates the patient in both the rehabilitation aspects.

The conceptual design of the Patient Specific Rehabilitation Gaming System (PSRGS), illustrated in Fig. 3, depicts the main actors and activities involved in the rehabilitation process and the flow of data. It comprises four different phases: Diagnosis, Game Customization, Game Activity, Game Feedback. Each of these phases plays an important role in the rehabilitation process of the patient. In the following, we describe such phases in more details.

Diagnosis: The first step is the initial assessment of the patient. The patient's level of impairments is evaluated from a Doctor who compiles his/her clinical assessment. Such assessment includes information useful to rehabilitation professionals to assess the patient's ability, needs, preferences and expectations.

Game Customization: The customization of the game is the main focus of the proposed approach, the therapist exploits the assessment of the Doctor to tailor the game to the specific patient, so enhancing the quality of training, the self-esteem and the motivation of the patient herself. The Therapist, by using the Cognitive Tailoring Tool, associates the intended motor rehabilitation objectives to the game purpose allowing the patient to play executing the required exercise. Contextually, the therapist can modulate the cognitive difficulty level of the game in order to stimulate the cognitive rehabilitation and trigger individual's motivational forces towards achieving the intended outcome.

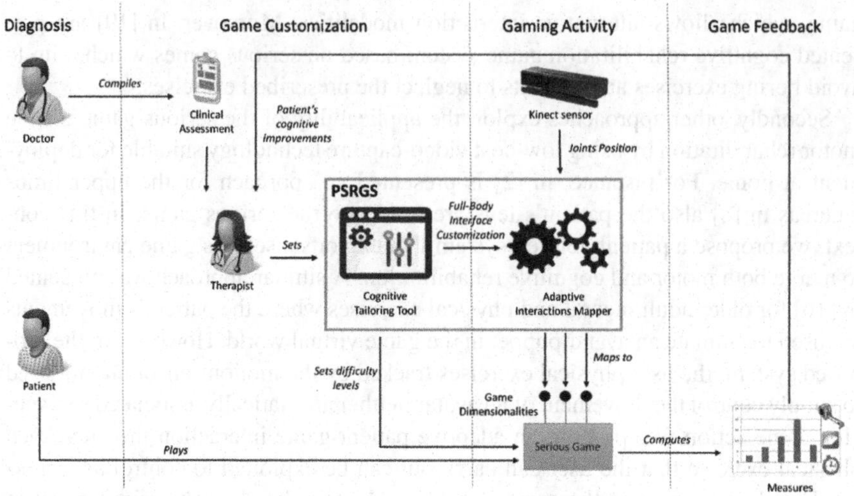

Fig. 3 The conceptual design of the Patient Specific Rehabilitation Gaming System (PSRGS)

Game Activity: During a game section, the patients movements collected by the motion sensors become the input for the real-time module of the platform, the Adaptive Interactions Mapper. This module maps in real-time the patient's movements to the serious game dimensionalities, by taking into account the guidelines of the therapist expressed using the cognitive tailoring tool. For instance, the patient's right arm abduction can be mapped to the vertical movement of a pointer in the game scene.

Game Feedback: The game play activity generates a set of measures regarding both the patient's movements and the cognitive game score as for instance the time taken to achieve the goal. Finally, the collected measures become a new input for the Game Customization phase, where can be used from the therapist to evaluate the progress of the patient and, eventually, change the therapy.

5 Business Implications

A process of rehabilitation is provided by a multidisciplinary team with several professionals from different areas. More so, reaching the rehabilitation centres is difficult for the majority of patients and often, this is done several times a month for a correct treatment. These facts make evident the need for more human and material resources to ensure the quality of the treatment and the quality of life of the patients and their families (e.g. more granularity of medical centres); however, an increment in the costs for the national health system is unavoidable in this case.

SmartCARE proposes to extend the rehabilitative care beyond the hospital setting and into the familiar environment, whilst helping to keep unchanged the effectiveness of the intervention and contained and sustainable its costs. The project partners have identified opportunities for both main stakeholders: medical centres and patients with their respective carers/family members. In the first case, in Italy it is estimated that to correctly address an emergency from an ictus, there should be at least 40 % more medical centres with stroke units and better distributed throughout the entire country. This lack of specialised medical centres is also similar for the post-ictus rehabilitation to which it should be also added the assistance from professionals at the patient's premises once he/she gets the discharge. Both deficiencies suppose high investments from the national health system if a normal approach is followed. However, SmartCARE could significantly reduce the number of professionals required to tackle this situation by efficiently utilising the human resources through a high-level tele-health application. On the other hand, it is studied that c. 78 % of families with a member that has suffered from an ictus, have worsen their quality of life significantly and in 56 % of the cases, there is a loss of spare time or even work time in order to help the patient (e.g. accompanying him/her to a medical centre for rehabilitation). By using the SmartCARE system, patients and caregivers could monetarily benefit from savings on transportation, missed work income and normal treatment and could increase their quality of life significantly when compared to a standard rehabilitation process.

6 Conclusions

In this paper, we have described an e-Health platform that supports the rehabilitation of patients suffering of stroke disease. The platform is enabled to deliver rehabilitation at home, under the control of physicians, assessing remotely, in a service centre, the progress of the therapy, and collaborating with caregivers to assure the continuity of care. The platform is structured to gather physical and physiological parameters while a patient executes neuro-motorial and neuro-cognitive exercises, customized on specific pathologies. The innovation of the platform is the integration of a markerless motion tracking for an adaptive touchless interaction mechanisms with a set of physiological wearable devices. Moreover, the virtual assistant and the serious games stimulate the patient during the therapy improving his/her neurological response. Finally, the main goal is to allow the accessing e-Health rehabilitation services at home, providing a collaborative eco-system that is centred around the patient, based on measurements, workflow, interaction, and sharing of relevant information among the stakeholders.

Acknowledgments The proposed ICT eco-system has been developed within the project Smart CARE—Satellite enhanced Multi-channel e-health Assistance for Remote Tele-rehabilitation and CAREgiving, funded by the European Space Agency (ESA) and sponsored by Lazio Region—Italy.

References

1. Amato, F., Fasolino, A.R., Mazzeo, A., Moscato, V., Picariello, A., Romano, S., Tramontana, P.: Ensuring semantic interoperability for e-health applications. In: 2011 International Conference on Complex, Intelligent and Software Intensive Systems (CISIS), pp. 315–320, June 2011
2. Burke, J.W., McNeill, M., Charles, D.K., Morrow, P.J., Crosbie, J.H., McDonough, S.M.: Optimising engagement for stroke rehabilitation using serious games. Vis. Comput. **25**(12), 1085–1099 (2009)
3. Dobkin, B.H.: Wearable motion sensors to continuously measure real-world physical activities. Curr. Opin. Neurol. **26**(6), 602 (2013)
4. Healthcare it: Top 5 ehealth trends reshaping the industry in 2015. https://www.accenture.com/us-en/insight-healthcare-technology-vision-2015.aspx
5. Key figures on europe—2015 edition. http://ec.europa.eu/eurostat/web/products-statistical-books/-/KS-EI-15-001
6. Kwakkel, G., Kollen, B.J., van der Grond, J., Prevo, A.J.: Probability of regaining dexterity in the flaccid upper limb impact of severity of paresis and time since onset in acute stroke. Stroke **34**(9), 2181–2186 (2003)
7. Lo, A.C., Guarino, P.D., Richards, L.G., Haselkorn, J.K., Wittenberg, G.F., Federman, D.G., Ringer, R.J., Wagner, T.H., Krebs, H.I., Volpe, B.T., et al.: Robot-assisted therapy for long-term upper-limb impairment after stroke. N. Engl. J. Med. **362**(19), 1772–1783 (2010)
8. Martins, T., Carvalho, V., Soares, F.: A serious game for rehabilitation of neurological disabilities: Premilinary study. In: 2015 IEEE 4th Portuguese Meeting on Bioengineering (ENBENG), pp. 1–5. IEEE (2015)
9. Moscato, F., Amato, F., Amato, A., Aversa, R.: Model-driven engineering of cloud components in metamorp(h)osy. Int. J. Grid and Util. Comput. **5**(2), 107–122 (2014)
10. Mousavi Hondori, H., Khademi, M.: A review on technical and clinical impact of microsoft kinect on physical therapy and rehabilitation. J. Med. Eng. **2014** (2014)
11. Nasr, N., Leon, B., Mountain, G., Nijenhuis, S.M., Prange, G., Sale, P., Amirabdollahian, F.: The experience of living with stroke and using technology: opportunities to engage and co-design with end users. Disabil. Rehabil. Assist. Technol. 1–8 (2015)
12. Nijenhuis, S.M., Prange, G.B., Amirabdollahian, F., Sale, P., Infarinato, F., Nasr, N., Mountain, G., Hermens, H.J., Stienen, A.H., Buurke, J.H., et al.: Feasibility study into self-administered training at home using an arm and hand device with motivational gaming environment in chronic stroke. J. Neuroeng. Rehabil. **12**(1), 1 (2015)
13. Perry, J.C., Andureu, J., Cavallaro, F.I., Veneman, J., Carmien, S., Keller, T.: Effective game use in neurorehabilitation: user-centered perspectives. In: Handbook of Research on Improving Learning and Motivation through Educational Games, IGI Global (2010)
14. Rego, P.A., Moreira, P.M., Reis, L.P.: Natural user interfaces in serious games for rehabilitation. In: 2011 6th Iberian Conference on Information Systems and Technologies (CISTI), pp. 1–4. IEEE (2011)
15. Rego, P., Moreira, P.M., Reis, L.P.: Serious games for rehabilitation: a survey and a classification towards a taxonomy. In: 2010 5th Iberian Conference on Information Systems and Technologies (CISTI), pp. 1–6. IEEE (2010)
16. Sáenz-de Urturi, Z., Zapirain, B.G., Zorrilla, A.M.: Kinect-based virtual game for motor and cognitive rehabilitation: a pilot study for older adults. In: Proceedings of the 8th International Conference on Pervasive Computing Technologies for Healthcare, pp. 262–265. ICST (Institute for Computer Sciences, Social-Informatics and Telecommunications Engineering) (2014)
17. Sale, P., Ceravolo, M.G., Franceschini, M.: Action observation therapy in the subacute phase promotes dexterity recovery in right-hemisphere stroke patients. Biomed. Res. Int. **2014** (2014)
18. Sale, P., Franceschini, M., Mazzoleni, S., Palma, E., Agosti, M., Posteraro, F.: Effects of upper limb robot-assisted therapy on motor recovery in subacute stroke patients. J. Neuroeng. Rehabil. **11**(1), 1 (2014)

19. Shapii, A., Mat Zin, N.A., Elaklouk, A.M.: A game system for cognitive rehabilitation. Biomed. Res. Int. **2015** (2015)
20. Smartcare-satellite enhanced multi-channel e-health assistance for remote tele-rehabilitation and caregiving. https://artes-apps.esa.int/projects/smartcare
21. Solana, J., Cáceres, C., García-Molina, A., Opisso, E., Roig, T., Tormos, J.M., Gómez, E.J.: Improving brain injury cognitive rehabilitation by personalized telerehabilitation services: guttmann neuropersonal trainer. IEEE J. Biomed. Health Inf. **19**(1), 124–131 (2015)
22. The 2015 ageing report. http://europa.eu/epc/pdf/ageing_report_2015_en.pdf

Optimal Design of IPsec-Based Mobile Virtual Private Networks for Secure Transfer of Multimedia Data

Alexander V. Uskov, Natalia A. Serdyukova, Vladimir I. Serdyukov, Adam Byerly and Colleen Heinemann

Abstract Optimal design of IPsec-based mobile virtual private networks (MVPN) for a secure transfer of multimedia data, in general case, depends on multiple factors and parameters such as to-be-selected MVPN's architectural model, hardware and software setups and technical platform's solutions, network topology models, modes of tunnel's operation, levels of the Open Systems Interconnection (OSI) model, encryption/decryption algorithms, modes of cipher operation, security protocols, security associations and key management techniques, connectivity modes, parameters of security algorithms, computer architectures, number of tunnels in MVPN, and other factors. This paper presents the outcomes of research project on multi-objective optimization of IPsec mobile virtual private network design based on non-isomorphic groups of order 4—Cayley Tables—for three major MVPN factors: (1) level of data transfer security provided by MVPN, (2) MVPN performance and (3) cost of designed virtual private network.

Keywords Multimedia system · Data security · Virtual private network · Multi-objective optimization

A.V. Uskov (✉) · A. Byerly · C. Heinemann
Bradley University, Peoria, IL, USA
e-mail: auskov@bradley.edu

N.A. Serdyukova
Plekhanov Russian University of Economics, Moscow, Russia

V.I. Serdyukov
Institute of Management of Education of Russian Academy of Education, and Bauman Moscow State Technical University, Moscow, Russia

G. De Pietro et al. (eds.), *Intelligent Interactive Multimedia Systems and Services 2016*, Smart Innovation, Systems and Technologies 55,
DOI 10.1007/978-3-319-39345-2_5

1 Introduction

1.1 Web-Based Rich Multimedia Systems

Modern advanced Web-based rich multimedia (RMM) systems (WBRMM) systems incorporate mobile computing (MC), streaming and rich multimedia technologies, and, as a result, provide mobile RMM content developers with an opportunity to quickly create a complex webified RMM data (i.e. data that integrates text, PPT slides, charts, video, audio, computer animation/simulation) and deliver (stream) data to the end user [1–5]. For example, one of the first advanced WBRMM systems—the *InterLabs* system—was designed and developed as a part of the National Science Foundation (NSF) grants ## 9950029 and 0409615 [5]. Using the developed-by-co-author *InterLabs* WBRMM system in *Developer* mode (Fig. 1), the RMM content developer can (1) create various RMM components (such as audio, video, narrated computer exercises based on computer screen capturing technology, computer simulation and/or animation, etc.), (2) synchronize RMM components, (3) synchronize various RMM files, (4) webify the developed RMM content, (5) compress and code RMM content into ready-to-be-streamed files, and (6) upload the developed WBRMM files onto a streaming server. As a result, the end users from all over the world are able to play developed and streamed RMM content.

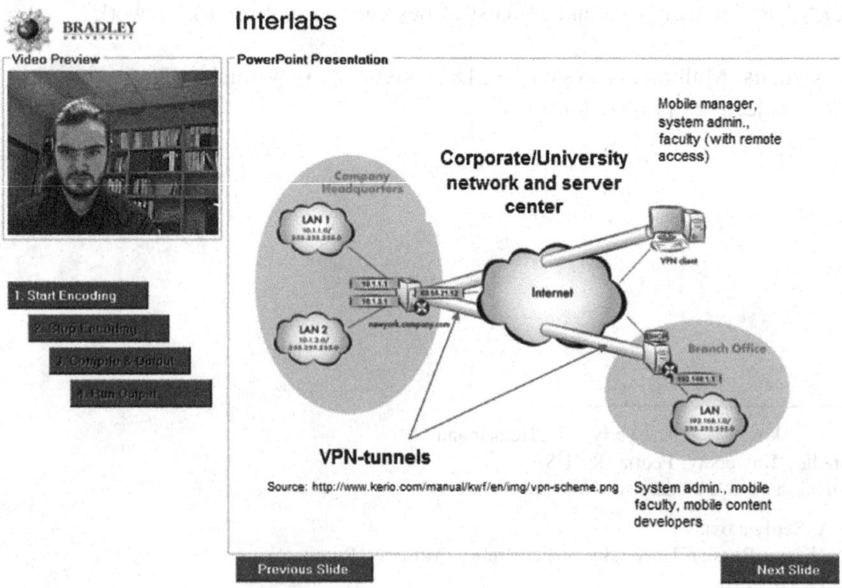

Fig. 1 The *InterLabs* Web-Lecturing system with rich multimedia content (a structure of mobile virtual private network with secure VPN channels inside public Internet is presented on the *right* side of this figure)

In order to significantly increase security of RMM files transfer in WBRMM systems over the Internet in general, and, mobile networks, in particular, various network solutions can be used. However, one of the most effective ones is based on active utilization of Virtual Private Network (VPN) technology—one of the most reliable technologies to provide data protection, confidentiality, integrity, data origin authentication, replay protection and access control [6, 7].

1.2 Architectural Model of IPsec VPN-Based WBRMM System

VPNs are created by using tunneling, authentication, and encryption to provide a virtual secure line—a VPN tunnel—inside a network. VPNs allow the provisioning of private network services for an organization or organizations over public or shared infrastructure such as the Internet or intranet, or service provider backbone network.

IPsec is a framework of open standards for ensuring private communications over public Internet Protocol (IP) based networks.

The developed architectural model of IPsec VPN-based WBRMM system is given on Fig. 2 [3]. It contains the following major components: (1) RMM content developers with graphics stations; (2) RMM files of different types, including: 2–1—webified versions of static multimedia files (text, static graphics, etc.); 2–2—audio files; 2–3—video and dynamic graphics files; (3) encoding of source RMM files; (4) compression and mutual synchronization of RMM files; integration of those files with various communications and collaborative technologies and WBRMM systems' supporting tools; (5) structurization of integrated data and files and their upload of streaming/Web/data/application/communication servers: 5–1—through VPN or MVPN tunnels in commodity Internet; 5–2—through corporate intranet; (6) streaming servers with 6–1—storage units, 6–2—quality of service (QoS) on server side, 6–3—security and transfer of encoded RMM files (using IP, RTP, RTCP, UDP, TCP, RTSP, SIP, and other protocols); (7) secure transfer of streaming RMM files over VPNs in commodity Internet from servers to users; (8) user's personal desktop or mobile computer that provides 8–1—downloading of RMM files on client side using various data protocols, 8–2—QoS on client side, 8–3 RMM files decoding, 8–4—synchronization of decoded RMM files, and 8–5—visualization of RMM files; (9) users of RMM files.

VPN or MVPN tunnels are used in processes 5–2 and 7 on Fig. 2. In order to provide WBRMM system's users with required (optimal) levels of security and network performance, a network system/security administrator must design and build effective virtual MVPN tunnels by selecting various MVPN's specifications, including: (1) MVPN's architectural model, (2) a set of software/hardware solutions in MVPN, (3) MVPN's topology model, (4) a set of technologies to be used in MVPN, (5) secure or trusted MVPN's model; (6) MVPN's tunnel modes, (7) level

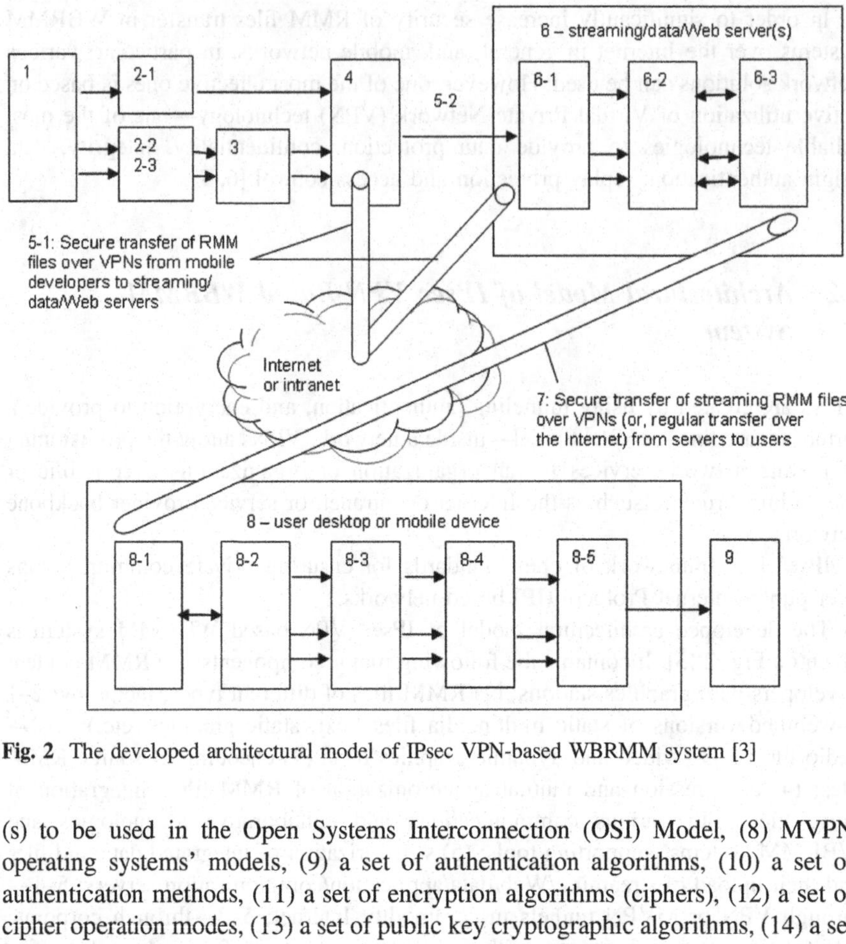

Fig. 2 The developed architectural model of IPsec VPN-based WBRMM system [3]

(s) to be used in the Open Systems Interconnection (OSI) Model, (8) MVPN operating systems' models, (9) a set of authentication algorithms, (10) a set of authentication methods, (11) a set of encryption algorithms (ciphers), (12) a set of cipher operation modes, (13) a set of public key cryptographic algorithms, (14) a set of security protocols, (15) a set of MVPN's operation modes, (16) a set of security associations and key management techniques, (17) a set of MVPN's connectivity modes, (18) a set of MVPN's availability levels, and (19) a set of multiple parameters for various security algorithms to be used in MVPN.

Unfortunately, a conceptual design and development of a particular optimal MVPN for an arbitrary customer is a very difficult problem due to MVPN's (1) dozens of available possible technologies, algorithms, models, and protocols [8], (2) hundreds of possible numeric parameters to be selected [8, 9], (3) absence of clear multi-objective optimization procedures for a specific applied area [10, 11]. As a result, MVPN's optimal design below is based on three most important optimization objectives: *SEC*—a level data security provided by MVPN, *PERF*—a level of MVPN performance, and *COST*—a total design and development cost of MVPN.

The ongoing multi-aspect research project, in general, is aimed at the development of a new methodology for mathematical modeling of an actual system (in this case—VPN or MVPN network); this methodology is based on an integrative application of (a) general system theory [12], (b) theory of algebraic systems [13], (c) theory of groups [14], and (d) generalization of purities [15] to various qualitative characteristics and quantitative indicators that adequately describe a system, object, process or phenomenon [16–20]. This paper presents the proposed formal approach, models, and algorithms for a multi-objective optimization of IPsec VPN system's conceptual design.

2 Conceptual Modeling of MVPNs

The conceptual model of MVPN (CM-MVPN) is described below based on notation introduced in [3].

Definition 1: Mobile Virtual Private Network *Mobile Virtual Private Network is described as n-tuple of n elements that can be chosen from following sets:*

$$\begin{aligned} CM-MVPN = \ < &\{O\}, \{S\}, \{R\}, \{AP\}, \{SH\}, \{TP\}, \{TECH\}, \{TS\}, \\ &\{TUN\}, \{YP\}, \{OS\}, \{AA\}, \{AM\}, \{AC\}, \{OPER\}, \\ &\{PKA\}, \{SP\}, \{MOD\}, \{AK\}, \{CT\}, \{AV\}, \{PAR\} > \end{aligned} \tag{1}$$

where:

- O *a set of all objects in organization's mobile network (desktop computers, mobile devices, servers, routers, switches,* etc.*),*
- S *a set of all subjects (users) in organization' mobile network, including a subset of users-human beings (U), and a subset of users-software applications (SWA),*
- R *a set of all legal access functions of subjects S to objects O in organization's mobile network, for example, read, copy, modify functions,*
- AP *a set of MVPN architectural models,*
- SH *a set of software/hardware solutions in MVPN,*
- TP *a set of MVPN topology models,*
- $TECH$ *a set of technologies used in MVPN,*
- TS *a set of secure and trusted MVPNs,*
- TUN *a set of MPVN's tunnel modes,*
- YP *a set of seven levels in the Open Systems Interconnection (OSI) model,*
- OS *a set of MVPN operating systems' models,*
- AA *a set of authentication algorithms used in MVPN,*
- AM *a set of authentication methods used in MVPN,*
- AC *a set of encryption algorithms (ciphers) used in MVPN,*
- $OPER$ *a set of cipher operation modes used in MVPN,*
- PKA *a set of public key cryptographic algorithms used in MVPN,*
- SP *a set of security protocols used in MVPN,*
- MOD *a set of MVPN's operation modes,*

AK	*a set of security associations and key management techniques used in MVPN,*
CT	*a set of MVPN's connectivity modes,*
AV	*a set of MVPN's availability levels,*
PAR	*a set of parameters for various security algorithms to be used in MVPN.*

Multiple details on elements of all designated sets are available in [3, 4]. The MVPN's conceptual model *CM-MVPN* is designated as F_0 below; it can be presented as f_{00} function of the following arguments:

$$F_0 = f_{00}\,(O, S, R, AP, SH, TP, TECH, TS, TUN, YP, OS, AA, AM, AC, OPER, PKA, SP, MOD, AK, CT, AV, PAR) \tag{2}$$

Definition 2: Information Security Space of MVPN (IS-MVPN Model) *Information security space of MVPN is described as the Cartesian product over MVPN's sets of elements:*

$$IS-MVPN = O \times S \times R \times AP \times SH \times TP \times TECH \times TS \times TUN \times YP \times OS \times AA \times AM \times AC \times OPER \times PKA \times SP \times MOD \times AK \times CT \times AV \times PAR \tag{3}$$

3 Groups-Based Multi-objective Optimization

3.1 Index of Effectiveness of Multi-objective Optimization for Closed System of Goals

Below we consider *SEC* (i.e. a level of information security provided by MVPN), *PERF* (a level of MVPN performance) and *COST* (a total design and development cost of MVPN) as a closed system of goals. As a result, we propose an algorithm to identify the index of effectiveness of multi-objective optimization function for a closed system of goals.

Algorithm 1: Index of Effectiveness of Multi-Objective Optimization Function Let $\{A_1, \ldots, A_n\}$ be a closed system of goals, and $\{\alpha_1, \ldots, \alpha_n\}$ be a set of goals. The result of the application of method $\propto_i$ to achieve goal A_j is $B_{ij}, j = 1, \ldots, n(i)$.

Step 1: A relation $\lesssim$ of partial order is put on the set of all goals. In this case, $A_i \lesssim A_j$ means that achievement of A_i is more preferable than achievement of goal A_j. After that, all ways to assign partial order on $\{A_1, \ldots, A_n\}$ are considered, or, in other words, all lattices $R_\propto = \langle\{A_1, \ldots, A_n\}, \lesssim\rangle$ of the priorities of the goals are considered.

Step 2: A graph of achievements of all goals is constructed.

Step 3: A selection of a measurement scale takes place, i.e. a correspondence of obtained results B_{ij} to goals A_i, $i=1, \ldots, n, j=1, \ldots, n(i)$. Then the correspondence matrix $M=\|\propto_{ij}\|$ is constructed using the following rules:

3.1. Let $M=max\{n, n(i)|i=1, \ldots, n\}$, μ_{ij} be a measure of correspondence $\mu_{ij}(\propto_{ij})$ of a result B_{ij} (that is received by using method $\propto_{ij}$) to goal $A_i, i=1, \ldots, n, j=1, \ldots, n(i)$.

3.2. Let $\mu_{1n+1}=0=\cdots=\mu_{im}$ in order to make M square.

3.3. If $n<m$, then let $\mu_{n+1j}=0=\cdots=\mu_{mj}; j=1, \ldots, m$. (A note: a scale to measure μ_{ij} may be either discrete or continuous one).

Step 4: Matrix $G=\|\gamma_{ij}\|$ of interference of the goals is constructed, using the following rules:

4.1. If $A_i \precsim A_j$ (i.e. if goal A_i should be achieved first and goal A_j after that), then $\gamma_{ij}=-1$ in case goal $A_i$impedes goal A_j. If goal A_i has no influence on goal A_j, then $\gamma_{ij}=0$. If an achievement of goal A_i contributes to the achievement of goal A_j, then $\gamma_{ij}=1$.

4.2. If goal A_i is not comparable with goal A_j, then $\gamma_{ij}=0$. (A note: a scale for matrix G may be either discrete $\{-1;0;1\}$ or continuous one $[-1;1]$.

Step 5: Matrix of effectiveness of the used methods *GM* is computed.

The absolute index of effectiveness is the determinant |*GM*| of *GM*.

Let $|\lambda E-GM|=0$ be a characteristic equation of *GM*, and $\lambda_1, \ldots, \lambda_m$—it's different roots. Then a product $\lambda_1 \times \cdots \times \lambda_m=|GM|$ and the sum $\lambda_1+\cdots+\lambda_m=tr(GM)=\sum_{i=1}^{m} G_i M^i$. As a result, a relative index of effectiveness of means $\{\alpha_1, \ldots, \alpha_n\}$ to achieve goal A_i is equal to $\lambda_1^{'}=\frac{\lambda_1 \times \ldots \lambda_{i-1} \times \lambda_{i+1} \times \cdots \times \lambda_m}{|GM|}$

Let $(GM)X=b$ be a system of linear equations where X defines a desired outcome. If $|GM| \neq 0$, then this system has a unique solution, i.e. $(GM)^{-1}b$. If X is a proper vector of matrix *GM* and it belongs to a proper meaning λ, then $(GM)X=\lambda X$ and $\lambda X=b$. It is a well-known fact that proper meanings $\lambda_1, \ldots, \lambda_m$ are speeds of changing of proper vectors $X_1, \ldots, X_m$, and, thus, a maximum value of a proper meaning corresponds to the desired outcome X. As a result, we can formulate the following theorem.

Theorem 1: Index of Effectiveness of Multi-Objective Optimization *A system of goals is accessible if and only if a determinant of matrix of effectiveness of methods used to achieve goals is non-zero.*

3.2 Convolution of Complex Hierarchical Function and Optimization Algorithm

We need to show that the *CM-MVPN* model (2), i.e. the top level function $F_0 = f_{00}$, is a subset of the Cartesian degree $(R^{n_l+1})^{n_l \ldots n_1}$ or $((R^{n_l+1})^{n_l})\ldots)^{n_1}$ of the set of real numbers R, and, therefore, $F_0 = f_{00}$ function is a fractal.

Let us consider a complex function of zero level $F_0 = f_{00} = f_{00}(f_{01}, \ldots, f_{0i_0}, \ldots, f_{0n_0})$, $n_0 \le k_0, k_0 \in N; 1 \le i_0 \le n_0$. Each argument of this function, in its own turn, is a complex function of the first level which depends on n_1 arguments $f_{0i_0} = f_{0i_0}(f_{11}, \ldots, f_{1i_1}, \ldots, f_{1n_1})$, $n_1 \le k_1, k_1 \in N; 1 \le i_1 \le n_1$.

Let us assume that a complex function of the level m–1 is constructed. Then a complex function of the level m is constructed as follows: $f_{m-1i_{m-1}} = f_{m-1i_{m-1}}(f_{m1}, \ldots, f_{mi_m}, \ldots, f_{mn_m})$, $n_m \le k_m, k_m \in N; 1 \le i_m \le n_m$. Every argument of this function is a complex function of the level $m + 1$ and it depends on n_{m+1} arguments.

A complex function of the level $m + 1$ is constructed as follows: $f_{mi_m} = f_{mi_m}(f_{m+1.1}, \ldots, f_{m+1.i_{m+1}}, \ldots, f_{m+1.n_{m+1}})$, $n_{m+1} \le k_{m+1}, k_{m+1} \in N; 1 \le i_{m+1} \le n_{m+1}$. Every argument of this function is a complex function of the level $m + 2$ and it depends on n_{m+2} arguments.

Let S_m be a set of all functions of the level m, specifically $S_m = \{f_{mi_m} | n_m \le k_m, k_m \in N; 1 \le i_m \le n_m\}$. Then there is one function of zero level (i.e. a mega—function) $F_0 = f_{00}$, and there are k_m functions of level m, specifically $f_{m1}, \ldots, f_{mi_m}, \ldots, f_{mn_m}$.

In general case, we can assume that $f_{mi_m} : S_m^{n_{m+1}} \to R$, where R is the set of all real numbers. Also, it is possible to identify every function that is the factor in the description of the process with its numerical value. Additionally, without loss of generality, we can assume that this numerical index is the index of effectiveness of the corresponding factor; below it is denoted as e_{mi_m}. If we denote the lowest obtained level of functions' hierarchy by l, then $f_{li_l} = f_{li_l}(f_{l+1.1}, \ldots, f_{l+1.i_{l+1}}, \ldots, f_{l+1.n_{l+1}})$, $n_{l+1} \le k_{l+1}, k_{l+1} \in N; 1 \le i_{l+1} \le n_{l+1}$,

where $f_{l+1.1}, \ldots, f_{l+1.i_{l+1}}, \ldots, f_{l+1.n_{l+1}} \in R$, $f_{l-1i_{l-1}} = f_{l-1.i_{l-1}}(f_{l1}, \ldots, f_{li_l}, \ldots, f_{ln_l})$, $n_l \le k_l, k_l \in N; 1 \le i_l \le n_l$, and $f_{l-1i_{l-1}} = \prod_{i_l=1}^{n_l} f_{li_l}$.

The $f(x_1, \ldots, x_n) = y$ function depends on n arguments; thus, it could be represented as the sequence $(x_1, \ldots, x_n, y)$ of length $n + 1$. As a result, the top level function $F_0 = f_{00}$ is a subset of the Cartesian degree $(R^{n_l+1})^{n_l \ldots n_1}$ or $((R^{n_l+1})^{n_l})\ldots)^{n_1}$ of the set of real numbers R, and, therefore, $F_0 = f_{00}$ function is a fractal.

Now it is necessary to develop an algorithm for a construction of preferences matrix for a complex hierarchical function F_0.

Algorithm 2: Algorithm for a Construction of Preferences Matrix

Step 1. Construct a matrix of preferences and the indexes of effectiveness e_{mi_m} for each particular function f_{mi_m} in accordance with developed Algorithm 1 for every level m, and, then, for mutual index of effectiveness e_m of each level m.

Step 2. Construct a matrix of preferences of the levels and the indexes of effectiveness of mega function F_0.

This algorithm will enable us to construct an optimization algorithm for complex hierarchical function F_0.

Algorithm 3: Algorithm for Multi-Objective Optimization of Complex Hierarchical Function F_0

Step 1. Apply Algorithm 1 to each of set as designated in (2): *O, S, R, AP, SH, TP, TECH, TS, TUN, YP, OS, AA, AM, AC, OPER, PKA, SP, MOD, AK, CT, AV, PAR.* As a result, we will receive: $O_1 \in \mathrm{O}$, $S_1 \in S$, $R_1 \in R$, $AP_1 \in \mathrm{AP}$, $SH_1 \in \mathrm{SH}$, $TP_1 \in \mathrm{TP}, \ldots, YP_1 \in YP$, $MOD_1 \in MOD$, $AK_1 \in \mathrm{AK}$, $CT_1 \in \mathrm{CT}$, $AV_1 \in \mathrm{AV}$, $PAR_1 \in PAR$. In this case, the left side of each inclusion can consists of a single element or several elements. If left side of any inclusion consists of several elements, then it is necessary to use Algorithm 1 again (in this case, an optimization of the second level will be executed), and so on. As a result, horizontal or tier optimization will be completed.

Step 2. Apply Algorithm 2 to each outcomes of Step 1, specifically O_1, S_1, R_1, AP_1, SH_1, $TP_1, \ldots, YP_1$, $MOD_1, AK_1, CT_1, AV_1, PAR_1$. As a result, we will get the optimal solution for the complex hierarchical function F_0, i.e. VPN or MVPN design

4 Four Factor Groups-Based Model of VPN's Multi-objective Optimization

Based on introduced notation, we will use the following most important particular objective in multi-objective optimization of VPN design: a_1—data security provided by VPN components (*SEC* objective), a_2—VPN's performance (*PERF* objective), a_3—VPN's total cost (*COST* objective). Let e be a neutral element (or, singular objective). (A note: In general case, there may be more than three particular optimization objectives in VPN's conceptual design. However, for the purpose of this research project, we assume that we can perform adequate multi-objective optimization of VPN or MPVN design based on those three designated most important optimization objectives).

Let's assume that system objectives are closed relatively to the operation of composition $*$ that is under the consistent application of objectives; additionally, assume that $*$ is an associative binary operation.

Let us consider an operation of VPN by describing four designated factors $\{a_1, a_2, a_3, e\}$. There exist two different non—isomorphic groups of the order 4, so that there are only the following two different ways of functioning of the described system which determine the impact of these factors, namely options defined by the developed *Cayley Tables*. Cyclic group of the order 4: $Z_4 = \langle F_4 \| a_1^4 = e, a_1^2 = a_2, a_1^3 = a_3 \rangle$ is presented on Fig. 3. The Cartesian product of two cyclic groups of the 2nd order (i.e. a quad group) is presented on Fig. 4; it can be described as follows: $Z_2^2 = \langle F_4 \| a_1^2 = \mathrm{e}_4,\ a_2^2 = \mathrm{e},\ a_3^2 = \mathrm{e},\ a_4^2 = \mathrm{e} \rangle$.

A qualitative interpretation of cyclic group (Fig. 3) is as follows (see 2nd row in the dimmed area on Fig. 3): (a) a change of VPN's security level *SEC (*factor a_1) causes an adequate change of VPN performance *PERF* (factor a_2), (b) a change of VPN performance *PERF* (factor a_2) causes an adequate change of VPN's overall cost *COST* (factor a_3), etc. A qualitative interpretation of a quad group (Fig. 4) is as follows (see 2nd row in the dimmed area on Fig. 4): a change of VPN's security level *SEC* (factor a_1) and VPN performance *COST* (factor a_3) will ultimately lead to a change of VPN performance *PERF* (factor a_2).

The developed 4-factor group model is the simplest example of an application of the algebraic formalization [13, 14] to manage risks in VPN networks. In fact, the possible risks are determined by the points of distinction in the Cayley Tables (Figs. 3 and 4). Let Fig. 3 $a_i^\circ a_j = a_k$ (Fig. 3) and $a_i^\circ a_j = a_m$ (Fig. 4). If at time $t = t_\propto$ a numeric indicator of factor a_k changes a trend of $y = f(a, t)$, and the indicator of factor a_{m} does not change that trend, then an operation of a system is described by a cyclic group on Fig. 3. If this scenario is undesirable, then it is necessary to adjust the optimization process of a system. The proposed methodology makes it possible to build corresponding models that depend on any finite number of factors and options—this enables designers to regulate (control) undesirable developments in optimization process.

Fig. 3 Cyclic group of order 4

*	e	a_1	a_2	a_3
e	e	a_1	a_2	a_3
a_1	a_1	a_2	a_3	e
a_2	a_2	a_3	e	a_1
a_3	a_3	e	a_1	a_2

Fig. 4 The Cartesian product of two cyclic groups of the 2nd order (i.e. a quad group)

*	e	a_1	a_2	a_3
e	e	a_1	a_2	a_3
a_1	a_1	e	a_3	a_2
a_2	a_2	a_3	e	a_1
a_3	a_3	a_2	a_1	e

5 Conclusions. Next Step

Conclusions. The performed research enables us to make the following conclusions.

(1) A new methodology for mathematical modeling and multi-objective optimization of a system has been developed and presented with an application to IPsec VPN or MVPN in WBRMM systems. This methodology is based on an integrative application of (a) general system theory [12], (b) theory of algebraic systems [13], (c) theory of groups [14], and (d) generalization of purities [15] to various sets of components and characteristics of modern IPsec virtual private networks [3, 8, 9].

(2) The developed (a) conceptual design models of MVPNs, (b) a description and convolution of a complex hierarchical function F_0, (c) algorithm of multi-objective optimization, (d) theorem 1, (e) Algorithms 1, 2 and 3, and (f) developed cyclic group (Fig. 3), quad group (Fig. 4) and dynamics of factors a_1, a_2, a_3 enable security system administrators to manage various risks in IPsec VPN networks of WBRMM systems.

(3) The developed 4-factor model is the simplest example of an application of the algebraic formalization [13, 14] to manage risks in VPN networks. In fact, the possible risks are determined by the points of distinction in the *Cayley Tables* (Figs. 3 and 4). The proposed methodology makes it possible to build corresponding optimization models that depend on any finite number of factors and options.

(4) Any 4-factor model of the closed associative system with feedback elements would function similarly to two above-mentioned groups, specifically cyclic group of the 4th order (Fig. 3), and quad group (Fig. 4). In addition, if a developed 4-factor model does not work, then the reason is that a qualitative analysis did not reveal all important quality factors of that system or object (or, in the above presented mathematical notation—arguments of a complex hierarchical function F_0).

(5) The automorphic group of quad group (i.e. non-cyclic groups of order (4) is isomorphic to a symmetric group S_3 of order 3 [13]. This means that if a design and development of a system follows an option with a quad group (Fig. 4), then there will be a total of six diagrams of factors' interaction—those diagrams will be close to the developed original quad group (Fig. 4).

Next step. The next planned step in this project is to develop a software recommender system that will incorporate the aforementioned (a) sets and parameters of IPsec MVPNs, and (b) developed mathematical foundation for groups-based multi-objective optimization.

References

1. Karmakar, G., Dooley, L. (Eds.) Mobile multimedia communications: concepts, applications, and challenges, IGI Global, p. 420 (2008). ISBN: 978-1-59140-766-9
2. Sharma, D., Favorskaya, M., Jain, L., Howlett, R., (eds.).: Fusion of Smart, Multimedia and Computer Gaming Technology: Research, Systems and Perspectives, Springer International Publishing (2015). ISBN: 978-3-319-14644-7
3. Uskov, A.V., Avagyan, H.: Fusion of secure IPsec-based virtual private network, mobile computing and rich multimedia technology, In: Sharma, D., et al. (eds.) Fusion of Smart, Multimedia and Computer Gaming Technology: Research, Systems and Perspectives, Springer, pp. 37–71 (2015) ISBN: 978-3-319-14644-7
4. Uskov, A.: IPsec VPN-based security of web-based rich multimedia systems. In: Proceedings of 6th International Conference on Intelligent Interactive Multimedia Systems and Services (IIMSS-2013), pp. 31–40. IOS Press, Sesimbra, Portugal (2013)
5. Uskov, V.L., Uskov, A.V.: Blending streaming multimedia and communication technology in advanced web-based education. Int. J. Adv. Technol. Learn. **1**(1), 54–66 (2004)
6. Al-Khayatt, S. et al. Performance of multimedia applications with IPsec tunneling. In: Proceedings of IEEE International conference on Coding and Computing, pp. 134–138. IEEE, 8–10 Apr 2002
7. Adeyanka, O.: Analysis of IPsec VPNs performance in a multimedia environment, In: Proceedings of 4th International Conference on Intelligent Environments, pp. 1–5, IEEE, EIT, Seattle WA, 21–22 July 2008. ISBN: 978-0-86341-894-5, doi:10.1049/cp:20081131
8. Lewis, M.: Comparing, Designing, and Deploying VPNs. Cisco Press, Indianapolis, IN (2006)
9. Bollapragada, V., et al.: IPSec VPN Design. Cisco Press, Indianapolis, IN (2005)
10. Donoso, Y., Fabregat, R.: Multi-Objective Optimization in Computer Networks Using Metaheuristics, p. 472. Auerbach Publications, Boca-Raton, FL (2007)
11. Marler, T. Multi-Objective Optimization: Concepts and Methods for Engineering. VDM Verlag (2009)
12. Mesarovich, M., Takahara, Y.: General system theory: mathematical foundations. Mathematics in Science and Engineering, vol. 113. Academic Press, New York (1975)
13. Maltcev, A.I.: Algebraic Systems, p. 392. Nauka, Moscow (1970). (in Russian)
14. Kurosh, A.G.: Theory of Groups, p. 648c. Nauka, Moscow (1967). (in Russian)
15. Serdyukova, N.A.: On Generalizations of Purities, Algebra & Logic, vol. 30, № 4, pp. 282–296 (1991). http://link.springer.com/article/10.1007%2FBF01985063
16. Serdyukova, N.A.: Optimization of Tax System of Russia, Parts I and II, Budget and Treasury Academy, Rostov State Economic University (2002). (in Russian)
17. Serdyukova, N.A., Serdyukov, V.I.: The new scheme of a formalization of an expert system in teaching. In: Proceedings of ICEE/ICIT-2014 International Conference, Riga, Latvia (2014)
18. Barajas, M.: Monitoring and Evaluation of Research in Learning Innovations (MERLIN). http://www.ub.edu/euelearning/merlin/docs/finalreprt.pdf
19. Serdyukova, N.A., Serdyukov, V.I., Slepov, V.A.: Formalization of knowledge systems on the basis of system approach, In: Proceedings of International conference on Smart Education and Smart e—Learning SEEL-2015, vol. 41, p.p. 371—380. Springer (2015)
20. Serdyukova, N.A., Serdyukov, V.I.: Modeling, simulations and optimization based on algebraic formalization of the system, In: Proceedings of 19th International Conference on Engineering Education, Zagreb, Zadar (Croatia), pp. 576–582 (2015). http://icee2015.zsem.hr/images/ICEE2015_Proceedings.pdf Accessed 20—24 July 2015

Malicious Event Detecting in Twitter Communities

Flora Amato, Giovanni Cozzolino, Antonino Mazzeo and Sara Romano

Abstract Social networking services gain more often interest for research goals in several fields and applications. The number of active users of social networking services like Twitter raised up to 320 million per month in 2015. The rich knowledge that has accumulated in the social sites enables to catch the reflection of real world events. In this work we present a general framework for event detection from Twitter. The framework implements techniques that can be exploited for malicious event detection in Twitter communities.

1 Introduction

Today social media networks are considered as powerful source of knowledge where people share and exchange information reflecting real-world events. The events reflected in social media range from popular, as for example concerts of a very popular music band, to local events as a protest or an accident. Millions of people daily generate an enormous size of data that contains different kind of informations. It is estimated that the number of active users of social networking services like Twitter raised up to 320 million per month in 2015, with more than 400 million of tweets

F. Amato (✉) · G. Cozzolino · A. Mazzeo · S. Romano
DIETI - Department of Electrical Engineering and Information Technology,
University of Naples "Federico II", Naples, Italy
e-mail: flora.amato@unina.it

G. Cozzolino
e-mail: giovanni.cozzolino@unina.it

A. Mazzeo
e-mail: antonino.mazzeo@unina.it

S. Romano
e-mail: sara.romano@unina.it

S. Romano
Centro Regionale Information Communication Technology, CeRICT scrl Complesso
Universitario di Monte Sant'Angelo, Naples, Italy

G. De Pietro et al. (eds.), *Intelligent Interactive Multimedia Systems and Services 2016*, Smart Innovation, Systems and Technologies 55,
DOI 10.1007/978-3-319-39345-2_6

daily generated. Thus, social networks have become a virtually unlimited source of knowledge, that can be used for scientific as well as commercial purposes. In fact, the analysis of on-line public information collected by social networks is becoming increasingly popular, not only for the detection of particular trend, close to the classic market research problem, but also to solve problems of different nature, such as the identification of fraudulent behaviour, optimization of web sites, tracking the geographical location of particular users or, more generally, to find meaningful patterns in a certain set of data. In particular, the rich knowledge that has accumulated in Twitter enables to catch the happening of real world events in real-time. These event messages can provide a set of unique perspectives, regardless of the event type [1, 2], reflecting the points of view of users who are interested or even participate in an event. In particular, for unplanned events (e.g., the Iran election protests, earthquakes), Twitter users sometimes spread news prior to the traditional news media [3]. Even for planned events, Twitter users often post messages in anticipation of the event, which can lead to early identification of interest in these events. Additionally, Twitter users often post information on local, community-specific events where traditional news coverage is low or nonexistent. Thus Twitter can be considered as a collector of real-time information that could be used by public authorities as an additional information source for obtaining warnings on event occurrence. In the last few years, particular interest has given to the extraction and analysis of information from social media by authorities and structures responsible for the protection of public order and safety. Increasingly, through the information posted on social media, public authorities are able to conduct a variety of activities, such as prevention of terrorism and bio-terrorism, prevention of public order problems and safety guarantee during demonstrations with large participation of people.

The heterogeneity of information and the huge scale of data makes the identification of events from Twitter a challenging problem. In fact, Twitter messages, or tweets, has a variety of content types, including personal updates, not related to any particular real-world event, information about event happenings, retweets of messages which are of interest for a user and so on [4]. As an additional challenge, Twitter messages contain little textual information, having by design the limit of 140 characters and often exhibit low quality [5]. Several research efforts have focused on identifying events in Twitter [6, 7]. For example, in recent years there has been a lot of research efforts in analyzing Tweets to enhance health related alerts [8], following a natural calamity or a bio-terrorist attack, which can urge a rapid response from health authority, as well as disease monitoring for prevention. The previous work in this area includes validating the timeliness of Twitter by correlating Tweets with the real-world statistics of disease activities, e.g., Influenza-like-Illness (ILI) rates [9–11], E-coli [12], cholera [13] or officially notified cases of dengue [14].

In this work we present a general framework to be adopted for event detection from Twitter. The proposed system is aimed at detecting anomalies in Twitter stream. Event related anomalies are patterns in data that do not fit the pattern of the expected normal behavior [15]. Those anomalies might be induced in the data for malicious activity, as for example cyber-intrusion or terrorist activity. The proposed framework aims at analyzing tweets and automatically extract relevant event related information in order to raise a signal as soon as a malicious event activity is detected.

The reminder of the paper is structured as follows: in Sect. 2 we describe two examples that motivate our work; in Sect. 3 we present an overview of the framework along with the goals and requirements that are of importance for event detection from Twitter. Moreover in this Section we describe the main processing stages of the system. In the end, in Sect. 4 we present some conclusions and future work.

2 Motivating Example

Social Network, as Twitter, have become a virtually unlimited source of knowledge, that can be used for scientific as well as commercial purposes. Twitter users spreads a huge amount of information ranging on a several "facts". The analysis of the information collected by social networks is becoming increasingly popular for the detection of trends, fraudulent behavior or, more generally, for finding meaningful patterns in a certain set of data. This process of extraction and analysis of large amounts of data is commonly defined data mining.

Particular attention has been paid to the analysis of the User-Generated Content (UGC) coming from Twitter, which is one of the most popular microblogging websites. As described in the Sect. 1, Twitter textual data (i.e., tweets) can be analyzed to discover user thoughts associated with specific events, as well as aspects characterizing events according to user perception. Tweets are short, user-generated, textual messages of at most 140 characters long and publicly visible by default. For each tweet a list of additional features (e.g., GPS coordinates, timestamp) on the context in which tweets have been posted is also available.

In this Section we show two examples of text mining on live tweets to monitor possible public security issues. In the first one we describe the topic relevance, in the second one we analyze the results, using some realistic data taken during a sport event.

In the last years, especially from the episodes following Sept. 11, the fear of bioterrorism is in the list of most urgent healthcare concerns, together with previous health issues—cancer, AIDS, smoking or alcohol/drug abuse, heart disease and the cost of healthcare and insurance. Gallup[1] has asked Americans since 2001 to name the most urgent health problem facing the U.S.. Cost of and access to healthcare have generally topped the list in last few years, while Americans most frequently mentioned diseases such as cancer and AIDS in the 1990s. Other health issues have appeared near the top of the list over the past 15 years, including 2014, when Ebola was listed among the top three health concerns. This was most likely due to multiple Ebola outbreaks in West Africa and a few confirmed cases all around the world, which prompted widespread media coverage of the disease. In 2015, less than 0.5 of Americans listed Ebola as the most urgent issue, as the threat of the virus has subsided.

[1] http://www.gallup.com/.

Similar short-lived spikes in the responses to this question have occurred regarding other recent health threats, including the H1N1/swine flu outbreak in 2009 and anthrax and bioterrorism attacks in 2001.

In our first scenario, we consider as application domain a global health treat like bioterrorism, so the system aim is to monitor the Twitter Stream detecting tweets containing given keywords related to bioterrorism attacks, like "attack", "terror", "anthrax, "terrorism", "jihadi", etc. In this scenario we focus our attention on gathering all the upcoming tweets about a particular event using Twitter' streaming API. Depending on the search term, we can gather tons of tweets within a few minutes. This is especially true for live events with a world-wide coverage (World Cups, Academy Awards, Election Day, you name it). A working example that gathers all the new tweets with the #anthrax hashtag, or containing the word "anthrax", shows that a large number of collected tweets is not related to an upcoming real treat. In our case, for example, the word "anthrax" is also related to a metal music band, so we need a technique to distinguish relevant tweets from not relevant ones.

As for the second example, we consider a given event as search domain to make a first selection of relevant tweets on which we adopt deeper textual analysis to detect the positive ones. A post hoc analysis of the information posted on Twitter, shows that, after the occurrence of a given event, the number of tweets related to that event is influenced by two main factors:

1. density of users in different geographic areas;
2. popularity of the event in the different geographic areas.

To argue this consideration we consider, as a simplified example, a sport event as given event; more specifically, we focus our attention to the soccer match between SSC Napoli and Legia Varsavia teams, played on 10 December 2015 during qualifying stages of Europa League competition. Before the match, tension, acts of vandalism and violence breaks out as opposite hooligans fight in city centre. The news was reported by main national newspapers, as well as main on-line information sites. Obviously this news was also reported by people and mass media on their Twitter profiles.

We analyse relevant tweet concerning this given event, to estimate user perception of the episodes of violence. Searching the string "Napoli Legia Varsavia" and filtering tweets by date between 9 December and 11 December, we collect the total amount of tweets concerning the given event. Refining the result with the words "arresti", "scontri", "polizia", ect., we obtain the number of tweets related with the public security disorders, instead of the soccer match. In Table 1 we summary the results.

From this analysis we note that a considerable number of tweets is not related with the event itself, but rather with the occurrence of acts of vandalism and violence during the social event. Moreover, the proportion between total number of tweets and the number of tweets concerning with public security issue depends by the severity of the episodes.

Table 1 Search results

	Event related	Public security concern
Date of interest:	9 December 11 December	
Query string:	Napoli Legia Varsavia	Napoli Legia Varsavia scontri
		Napoli Legia Varsavia arresti
		Napoli Legia Varsavia polizia
Number of tweets:	709	96

3 Framework Description

In this work we present a general framework to be adopted for event detection from Twitter. In particular, we aim at detecting anomalies in Twitter stream related to malicious activity, as for example cyber-intrusion or terrorist activity. In this section we present an overview of the main processing stages of the framework along with the goals and requirements that are of importance for event detection from Twitter. Event detection from Twitter messages must efficiently and accurately filter relevant information about events of specific interest, which is hidden within a large amount of insignificant information. The proposed framework aims at analyzing tweets and automatically extract relevant event related information in order to raise a signal as soon as a malicious event activity is detected. It is worth noting that in this work we consider an event occurrence in Twitter stream whenever an anomaly bursts in the analyzed data stream. The Twitter stream processing pipeline consists of three main stages, as illustrated in Fig. 1. The system aims at detecting user defined phenomenon form the twitter stream in order to give alerts in case of event occurrence. In the following we describe the main stages of the proposed system.

3.1 Collection and Filtering

The first stage of the system is devoted to collect and filter tweets related to an event. It is based on Information extraction task. Information extraction is the process of automatically scanning text for information relevant to some interest, including extracting entities, relations, and, most challenging, events (something happened in particular place at particular time) [16]. It makes the information in the text more accessible for further processing. The increasing availability of on-line sources of information in the form of natural-language texts increased accessibility of textual information. The overwhelming quantity of available information has led to a strong interest in technology for processing text automatically in order to extract task-relevant information. Information extraction main task is to automatically extract structured information from unstructured and/or semi-structured documents exploiting different kinds of text analysis. Those are mostly related to techniques of Natural

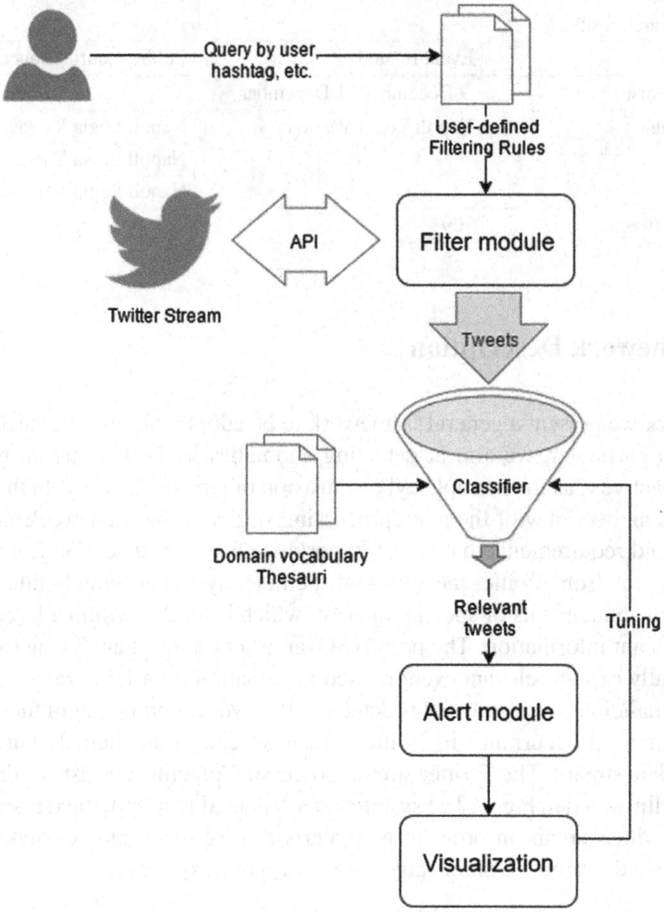

Fig. 1 Overview of the twitter stream processing framework for event detection and alerting

Language Processing (NLP) and to cross-disciplinary perspectives including Statistical and Computational Linguistics, whose objective is to study and analyze natural language and its functioning through computational tools and models. Moreover techniques of information extraction can be associated with text mining and semantic technologies activities in order to detect relevant concepts from textual data aiming at detecting events, indexing and retrieval of information as well as long term preservation issues [17–19]. Standard approaches used for implementing IE systems rely mostly on:

- Hand-written regular expressions. Hand-coded systems often rely on extensive lists of people, organizations, locations, and other entity types.
- Machine Learning (ML) based Systems. Hand annotated corpus is costly thus ML methods are used to automatically train an IE system to produce text annotation.

Those systems are mostly based on supervised techniques to learn extraction patterns from plain or semi-structured texts. It is possible to distinguish two types of ML systems:

- Classifier based. A part of manually annotated corpus is used to train the IE system in order to produce text annotation [20].
- Active learning (or bootstrapping). In preparing for conventional supervised learning, one selects a corpus and annotates the entire corpus from beginning to end. The idea of active learning involves having the system select examples for the user to annotate which are likely to be informative which are likely to improve the accuracy of the model [21]. Some examples of IE systems are [22].

In our work we are interested in detecting specified event which relies on specific information and features that are known about the event such as type and description, which are provided by the a domain expert of the event context. These features are exploited by adapting traditional information extraction techniques to the Twitter messages characteristics. In the first stage, the tweets are collected exploiting the REST API provided from Twitter that allow programmatic access to read data from the stream. Those API are costumed in order to crawl data exploiting user defined rules (keywords, hashtag, user profile, etc.). The tweets are annotated, in the filter module, with locations and temporal expressions using a series of language processing tools for tokenization, part-of-speech tagging, temporal expression extraction, and tools for named entity recognition.

In order to access to Twitter data programmatically, it's necessary to register a trusted application, associated with an user account, that interacts with Twitter APIs. Registering a trusted application, Twitter provides the required credentials (consumer keys) that can be used to authenticate the REST calls via the OAuth protocol. To create and manage REST calls we integrate in the Filter Module a Python wrapper for the Twitter API, called Tweepy, that includes a set of class enabling the interaction between the system and Twitter Stream. For example, the API class provides access to the entire twitter RESTful API methods. Each method can accept various parameters and return responses.

To gather all the upcoming tweets about a particular event, we call the Streaming API extending Tweepy's StreamListener() class in order to customise the way we process the incoming data. In the Listing 1.1 we simply gather all the new tweets with the #anthrax hashtag:

Listing 1.1 Custom Listener to gather all the new tweets with the #python hashtag and save them in a file

```
class TwitterListener(StreamListener):

    def on_data(self, data):
        try:
            with open('anthrax.json', 'a') as f:
                f.write(data)
                return True
```

```
        except BaseException as e:
            print("Error_on_data:_%s" % str(e))
        return True

    def on_error(self, status):
        print(status)
        return True

twitter_stream = Stream(auth, TwitterListener())
twitter_stream.filter(track=['#anthrax'])
```

All gathered tweets will be furthermore filtered and then passed to the Classification module for deeper textual processing.

3.2 *Classifying Relevant Information*

The collected and annotated tweets are then classified, in the second stage, with respect to their relevance. As stated before in the paper, event detection from Twitter messages is a challenging task since the relevant information about a specific event is hidden within a large amount of insignificant messages. Thus a classifier must efficiently and accurately filter relevant information. The classifier is trained exploiting a domain dependent thesaurus that include a list of relevant terms for event detection. It is responsible for filtering out irrelevant messages.

3.3 *Generating Alerts*

Once the set of irrelevant tweets are discarded, those remaining must be conveniently aggregated with respect to an event of interest. The third stage of the proposed system is aimed at detecting anomalies in the event related set of tweets. Anomalies are patterns in data that do not fit the pattern of the expected normal behavior. Anomalies might be induced in the data for a variety of reasons, such as malicious activity, as for example cyber-intrusion or terrorist activity, but all of the reasons have a common characteristic that they are interesting to the analyst. In literature there are several techniques exploited for the anomaly detection. Those range between several disciplines and approaches such as statistics, machine learning, data mining, information theory, spectral theory.

In our proposed system, the set of event related tweets are processed by the alert module that is responsible to raise an alert in case of anomaly within the collected tweets. The alert module implements a set of burst detection algorithms in order to generate alerts whenever an event occurs that means an anomaly in the analyzed data stream is detected. The detected events are then displayed to the end user exploiting the visualization module. An important aspect is that the analyzed data can be tuned by the end user for novelty detection which aims at detecting previously unobserved (emergent, novel) patterns in the data, such as a new topic. The novel patterns are novel patterns are typically incorporated into the normal model after being detected. Moreover with the tuning phase the parameters for the tweets classification module can be modified in case the messages that raised an alert are considered not relevant for the event occurrence.

4 Conclusions and Future Work

Social networking services are matter of interest for research activities in several fields and applications. The number of active users of social networking services like Twitter raised up to 320 million per month in 2015. The rich knowledge that has accumulated in the social sites enables to catch the reflection of real world events [23].

In this work we proposed a general framework for event detection from Twitter. The framework aims at detecting malicious events in Twitter communities, exploiting techniques related to text analysis and information classification. Once the set of tweets are classified as related to criminal activity, they are processed by the alert module that is responsible to raise an alert in case of anomaly within the collected tweets. The alert module implements a set of burst detection algorithms in order to generate alerts whenever an event occurs that means an anomaly in the analyzed data stream is detected.

As future work we intend to provide the implementation of our system tuned with a domain related to the bio-terrorism.

References

1. Diakopoulos, N., Naaman, M., Kivran-Swaine, F.: Diamonds in the rough: social media visual analytics for journalistic inquiry. In: 2010 IEEE Symposium on Visual Analytics Science and Technology (VAST), pp. 115–122, Oct 2010
2. Yardi, S., Boyd, D.: Tweeting from the town square: measuring geographic local networks. In: International Conference on Weblogs and Social Media. American Association for Artificial Intelligence. http://research.microsoft.com/apps/pubs/default.aspx?id=122433 (2010). Accessed May 2010
3. Sakaki, T., Okazaki, M., Matsuo, Y.: Earthquake shakes twitter users: Real-time event detection by social sensors. In: Proceedings of the 19th International Conference on World Wide Web, ser. WWW '10, pp. 851–860. ACM, New York, NY, USA (2010). http://doi.acm.org/10.1145/1772690.1772777
4. Naaman, M., Boase, J., Lai, C.-H.: Is it really about me?: message content in social awareness streams. In: Proceedings of the 2010 ACM Conference on Computer Supported Cooperative Work, ser. CSCW '10, pp. 189–192. ACM, New York, NY, USA (2010). http://doi.acm.org/10.1145/1718918.1718953
5. Becker, H., Naaman, M., Gravano, L.: Beyond trending topics: real-world event identification on twitter (2011)

6. Becker, H., Naaman, M., Gravano, L.: Learning similarity metrics for event identification in social media. In: Proceedings of the Third ACM International Conference on Web Search and Data Mining, ser. WSDM '10, pp. 291–300. ACM, New York, NY, USA (2010). http://doi.acm.org/10.1145/1718487.1718524
7. Atefeh, F., Khreich, W.: A survey of techniques for event detection in twitter. Comput. Intell. **3**(1)
8. Essmaeel, K., Gallo, L., Damiani, E., De Pietro, G., Dipanda, A.: Comparative evaluation of methods for filtering kinect depth data. Multimedia Tools Appl. **74**(17), 7331–7354 (2015)
9. Aramaki, E., Maskawa, S., Morita, M.: Twitter catches the flu: detecting influenza epidemics using twitter. In: Proceedings of Conference on Empirical Methods in Natural Language Processing (2011)
10. Lamb, A., Paul, M.J., Dredze, M., Separating fact from fear: tracking flu infections on twitter. In: HLT-NAACL, pp. 789–795 (2013)
11. Chon, J., Raymond, R., Wang, H., Wang, F.: Modeling flu trends with real-time geo-tagged twitter data streams. In: Wireless Algorithms, Systems, and Applications, pp. 60–69. Springer (2015)
12. Diaz-Aviles, E., Stewart, A.: Tracking twitter for epidemic intelligence: case study: Ehec/hus outbreak in germany, 2011. In: Proceedings of the 4th Annual ACM Web Science Conference, ser. WebSci '12, pp. 82–85. ACM, New York, NY, USA (2012). http://doi.acm.org/10.1145/2380718.2380730
13. Chunara, R., Andrews, J.R., Brownstein, J.S.: Social and news media enable estimation of epidemiological patterns early in the 2010 haitian cholera outbreak. Am. J. Trop. Med. Hyg. **86**(1), 39–45 (2012). http://www.ajtmh.org/content/86/1/39.abstract
14. Gomide, J., Veloso, A., Meira, W., Almeida, V., Benevenuto, F. Ferraz, F., Teixeira, M.: Dengue surveillance based on a computational model of spatio-temporal locality of twitter. In: Proceedings of ACM WebSci'2011 (2011)
15. Chandola, V., Banerjee, A., Kumar, V.: Anomaly detection: a survey. ACM Comput. Surv. **41**(3), 15:1–15:58 (2009). http://doi.acm.org/10.1145/1541880.1541882
16. Hobbs, J.R., Riloff, E., Information extraction. In: Indurkhya, N., Damerau, F.J. (eds.) Handbook of Natural Language Processing, Second Edition, Boca Raton. FL: CRC Press, Taylor and Francis Group (2010)
17. Amato, F., Mazzeo, A., Penta, A., Picariello, A.: Using NLP and ontologies for notary document management systems. In: Proceedings of the 19th International Conference on Database and Expert Systems Application (2008)
18. Amato, F., Casola, V., Mazzocca, N., Romano, S.: A semantic approach for fine-grain access control of e-health documents. Logic J. IGPL **21**(4), 692–701 (2013)
19. Amato, F., Fasolino, A., Mazzeo, A., Moscato, V., Picariello, A., Romano, S., Tramontana, P.: Ensuring Semantic Interoperability for E-health Applications, pp. 315–320 (2011)
20. Riloff, E.: Automatically constructing a dictionary for information extraction tasks. In: Proceedings of the Eleventh National Conference on Artificial Intelligence (1993)
21. Grishman, R.: Information Extraction: Capabilities and Challenges (2012)
22. Pantel, P., Pennacchiotti, M.: Espresso: leveraging generic patterns for automatically harvesting semantic relations. In: Proceedings of the 21st International Conference on Computational Linguistics and the 44th annual meeting of the Association for Computational Linguistics (2006)
23. Caggianese, G., Neroni, P., Gallo, L.: Natural interaction and wearable augmented reality for the enjoyment of the cultural heritage in outdoor conditions. In: Augmented and Virtual Reality, pp. 267–282. Springer (2014)

Adopting Decision Tree Based Policy Enforcement Mechanism to Protect Reconfigurable Devices

Mario Barbareschi, Antonino Mazzeo and Salvatore Miranda

Abstract The Field Programmable Gate Array technology invaded the electronic market by offering economic advantages and many attractive features, such as the possibility to dynamically reprogram the hardware configuration in field. However, FPGA devices are not free of secure drawbacks, which include the possibility of install third-party components which may damage the system on which they are hosted. In this paper, we devise a policy enforcement mechanism to monitor and control the access of a dynamically installed component and we design it by employing Decision Trees. We demonstrate, with a significant experimental setup conducted on a commercial device, namely the Xilinx Zynq-7020, the efficacy of the DT based policy enforcer.

1 Introduction

Reconfigurable systems offer a programmable layer onto hardware circuits and they can be dynamically configured and executed to offer new functionalities and extend the system capabilities. Many research efforts have been accomplished in direction of Reconfigurable Computing, mainly thanks to the greatly advances made by the Field Programmable Gate Array (FPGA) technology. FPGA devices are able to provide design flexibility, very similar to the software, and balance between performance

M. Barbareschi (✉) · A. Mazzeo · S. Miranda
DIETI - Department of Electrical Engineering and Information Technologies,
University of Naples Federico II, Via Claudio 21, 80125 Naples, Italy
e-mail: mario.barbareschi@unina.it

A. Mazzeo
e-mail: mazzeo@unina.it

S. Miranda
e-mail: miranda.salvatore1@gmail.com

M. Barbareschi · A. Mazzeo
CeRICT scrl - Centro Regionale Information Communication Technology,
Naples, Italy

G. De Pietro et al. (eds.), *Intelligent Interactive Multimedia Systems and Services 2016*, Smart Innovation, Systems and Technologies 55,
DOI 10.1007/978-3-319-39345-2_7

and cost. They can be programmed with an arbitrary configuration by means of a bitstream file in order to implement designed hardware circuits. Considering their benefits, FPGAs are suitable for innovative application contexts (e.g. the SNOPS project [1]), in which the security is a fundamental concern. For instance, in e-Health applications the integrity of all sensitive information and then the Semantic Interoperability must be guaranteed [2].

Modern FPGAs open new application scenarios by enabling dynamic partial reconfiguration (DPR) in-field: a device deployed on the FPGA technology can be partially updated and the part of the circuit which is not involved into the reconfiguration continues to work without interruption. The DPR feature proved to be very useful in applications which are characterized by high processing requirements and need for adaptive remote updates [4]. In such context, the Reconfigurable Computing introduces also the possibility to distribute hardware contents to end-user devices pretty much like software components. We can easily envision a scenario in which a market system dispenses applications which integrate in themselves hardware accelerators to install on a reconfigurable device, aiming at speed-up a specific high-computational demanding task.

The FPGA reconfigurability arose a significant number of challenges, most of them related to security. First of all, the bitstream whereby the FPGA partially updates its configuration (namely the partial bitstream), as well as the full bitstream, have to be protected against violations of its intellectual properties (IPs). Previous work proposed actual solutions to tackle the piracy of IP cores and to protect reconfigurable devices against attacks. Moreover the FPGA technology introduced the cryptography on the bitstreams [11, 13, 16, 18]. Conversely, the protection of static configurations of FPGAs against new installed IP cores is still an open issue. Indeed, as the bitstream encoding is kept secret by FPGA vendors and cannot be queried or analyzed in someway, none is able to guarantee if a partial bitstream is free of malicious or faulty IP cores which may be potentially harmful for the system. As stated in [17], several security problems might occur during the reconfigurable hardware life cycle.

Huffmire et al. [15] introduce the risks associated with the adoption of third-parties IP cores in a complex design and devised a system level technique to forbid unwanted accesses to such IPs by arbitrating memory accesses through a static enforcement module. Their approach uses permissions which are declared at design time and cannot be modified at runtime.

Conversely, in [14] the authors provided physical and logical isolation of hardware components onto the FPGA by wrapping each of them within a *moat*, which is defined as a channel around an IP core in which the routing is not enabled. However, the solely proposed isolation method is not completely effective against malicious IP cores that are programmed to access sensitive parts of the system.

Being mindful of previous considerations, in this paper we briefly introduce a model of reconfigurable system based on the SoC-FPGA technology and define a policy enforcement mechanism based on Decision Trees (DTs), which is able to control the memory accesses of third-party components. Furthermore, we also provide a real implementation of the proposed mechanism employing a commercial medium-end FPGA, namely the Xilinx Zynq-7020, embedded in the ZedBoard development kit and we show the experimental result of implemented DTs.

2 System Model and Policy Enforcement

From the previous introduced scenario, we model the SoC-FPGA logically partitioned in two regions. The first one is the static partition, which represents the region that will never be overwritten. Such region is initially configured by the full bitstream. The second one is the reconfigurable partition, which includes one or more reconfigurable slots that can be programmed with different partial bitstreams over time. Every reconfigurable slot has to be characterized by a well-defined interface to let the hosted core communicate with other parts of the system.

Partial bitstreams are issued by IP core vendors, which designs the intellectual property core which has to be installed on a reconfigurable slot by a end-user. Using partial bitstreams, the end-user is able to expand system capabilities by install additional functionalities on his own device that include a hardware-configurable subsystem to be installed onto reconfigurable slots. Vendors can protect their own IPs by means of cryptography, which is available on modern FPGAs [11]. Apart the correctness of the installed third-party IP core, the end-user requires to keep the whole system properly working and the device to be protected against unauthorized accesses to personal sensitive data.

Despite previously described solutions, to protect the system from third-party IP cores, we propose a policy enforcement mechanism embedded in a bridge-based component. Each bridge embeds a controller to enforce access policies on every transaction initiated by the wrapped slot. In particular, each controller is programmed using policies which represent the set of legal transactions that can be performed by the IP core within the slot. Therefore, if a transaction is considered legit, the bridge propagates it towards the destination, otherwise it is blocked by means of using a dummy slave which completes the transaction returning back an appropriate error response, according to the communication protocol. From the IP core point of view, the dummy slave acts as the real destination, hence the solution is transparently implemented.

Considering the hypothesis in which each system resource is identifiable through addresses, a generic policy can be formalized as (address, permission) or (address range, permission). The permission field strictly depends on the available transactions for the target bus and on the considered access model. Hence, a set of policies classifies different memory ranges with a corresponding permission.

3 Implementing the Decision Tree Based Policy Enforcer

DT models are very suitable to enforce formally defined rules in many application areas. Moreover, they can be automatically defined as data model by means of data mining algorithm, such as the C4.5 [21]. Policies which specify permission associated to each memory range can be inherently modeled as a DT. Indeed, the search operation for a permission of a given address can be accomplished by executing an algorithm which resembles the binary search. We can figure out that it splits the memory in ranges and looking for the range in which the given address belongs to. Such procedure perfectly matches the DT structure, as long as the leaves are properly labeled with associated permissions.

To be suitable in the context of memory protection, DT visiting algorithm has to be realized as a hardware architecture and it has to introduce an overhead as low as possible to avoid a bottleneck [9, 10, 12]. Several research papers discussed about hardware DT architectures, aiming at speed-up performance related to their visiting algorithms. Static hardware DTs were adopted in [6, 7] and demonstrated to be scalable in [3], while in [5] they were realized with the programmability feature to classify physical data collected within wireless sensor network. In [20] authors provided high performance architecture to implement DTs by compounding processing element in a pipeline structure.

These implementation do not take advantage of the case in which the DT has characterized by a single feature input. In Fig. 1a we report an example of memory permissions associated to address ranges and a corresponding DT. To retrieve a permission associated to a specific address value, the visiting algorithm proceeds by comparing it with the value reported by each node and navigating the tree according to the comparison outcome (Fig. 1b).

Integrating the DT visiting algorithm within the reconfigurable slot involves the design for a hardware unit which: (i) performs the search for the permission associated with an input value, and (ii) can be dynamically programmed with required policies. Hence, we proposed a novel scheme reported in Fig. 2, which consists of 2 main stages:

1. the address value is compared with every value reported in the DT nodes and *xor*-gates detect the exact range in which the value falls into;
2. a policy memory (RAM) outputs the associated permission.

Obviously, the dynamic configurability is assured by having configurable both the comparators and policy memory. In addition, such architecture is suitable for being implemented in a pipelined fashion.

More in detail, the 0 and 1 constant values allow to take in account address input values among $[0, A]$ and $(D, 2^n - 1]$, where 0 and $2^n - 1$ are the limits of the representation interval and n is the width of the addresses, which depends on the architecture of the system in which the policy enforcer will be placed. As regards the policy memory, it contains only the permissions associated to the stored ranges. Such memory is addressed using the output of an encoder. Thanks to the xor-gates structure, the

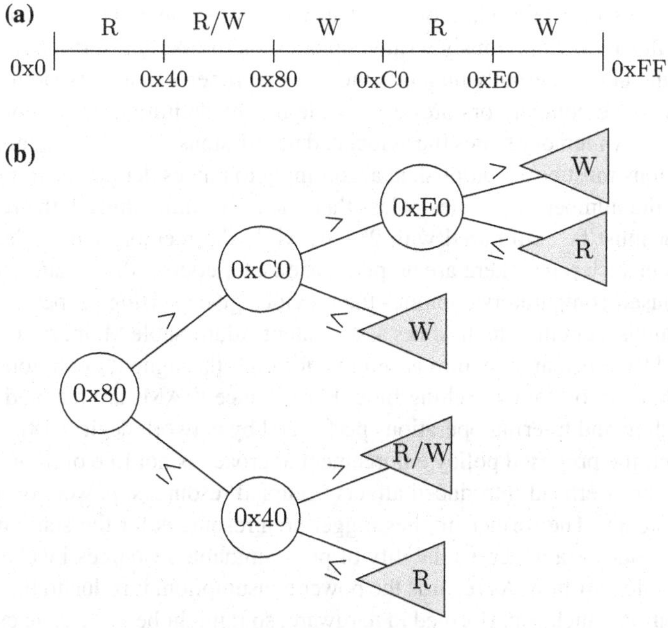

Fig. 1 Decision tree model of memory permissions. **a** Example of permissions associated to memory address ranges. **b** Decision tree model of memory permissions

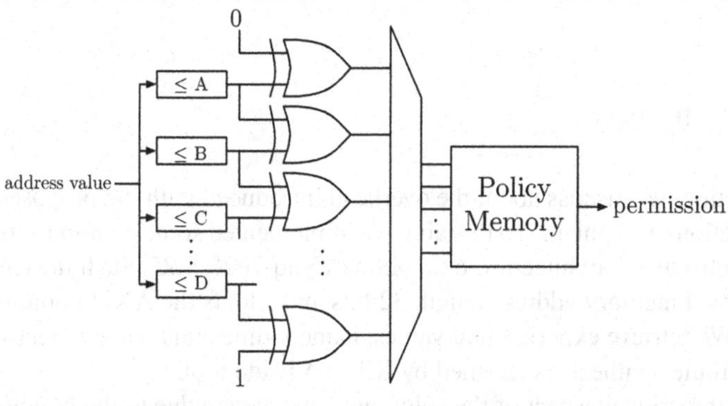

Fig. 2 A schematic programmable DT-based search architecture

encoder inputs are characterized by only one high bit. Therefore, during enforcement operations, the encoder addresses the policy memory such that the output permission is granted for the corresponding range.

From the previous scheme, it is worth to notice that the values reported in the comparators (namely A, B, C and D) must be sorted such that A ≤ B ≤ C ≤ D. To handle and keep this property after every policy update, the DT visiting architecture must be configured by means of a dedicated insertion procedure, which can be provided by a software module or by a hardware component. The former solution implies to flush away previously written policies and properly insert the new policy set, while the latter internally manages the sorting property in a transparent manner.

Moreover, the comparators are responsible for the definition of memory ranges while the policy memory stores the associated permissions. Hence, the configuration of permissions must be accomplished accordingly to ranges defined in comparators. Whenever the number of policies is less than the maximum allowed, the remaining comparator must be configured with $2^n - 1$, while the memory can be left blank, meant that in such range there are no permissions. Of course, the default configurations of unused comparators complies the previous given sorting property.

The result is a circuit which works as a Content Addressable Memory (CAM). As devised in [19], such architecture is suitable for high-throughput operations since it allows to have a constant searching time. For instance, CAMs are adopted for both the forwarding and filtering operations performed by network devices [8].

However, the proposed policy enforcement approach is not free of drawbacks. In particular, the overhead introduced affects occupied resources, power consumption and time latency. The former implies bigger area resources for the static partition, which turns out for a lower availability of programmable resources involved in the reconfigurable partition. As regards the power consumption, it is due to the enforcement operation which is performed in hardware, so it might be unsuitable especially for low-power devices. Instead, the latter affects the communication throughput, due to the delay inherently added by the controller during the policy checking. In the next Section we evaluate the introduced overhead with respect to some design parameters.

4 Evaluations

In this Section, we discuss about the overhead introduced with the proposed bridge-based isolation mechanism. To this aim, we implemented some configurations of the previous introduced architecture on a Xilinx Zynq-7020 SoC. Such device is characterized by a memory address length 32 bits and adopts the AXI4 communication protocol. We retrieve experimental values, namely time and area, by means of post-place and route synthesis performed by Xilinx Vivado tool.

One of the critical aspect of the solution is the latency due to the policy enforcement operation. Indeed, as previously stated, the policy enforcement has to introduce a negligible overhead in order to avoid the insertion of a bottleneck.

The DT based policy enforcement architecture depends only on the maximum number of policies that can be installed. Indeed, a greater number of policies involves more comparators and deeper permission memory. With respect to an implementation, we can state that the searching time is constant with the number of installed policies, i.e. $O(1)$. Obliviously, the speed of circuit strictly varies on the number of maximum supported policies. Figure 3 illustrates the maximum frequency of the DT-based policy enforcer varying the number of policies. As one can notice, the maximum frequency decreases. This is due to the additional time required by both

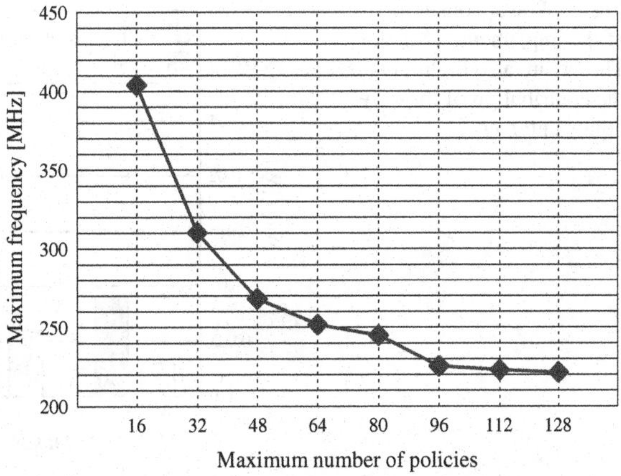

Fig. 3 Maximum frequency of the DT architecture as the policy memory depth changes

the encoder and permission memory when policies grow. However, the maximum frequency also depends on synthesis and implementation strategies (e.g. maximum fanout, optimization goals and efforts) and on the number of configurable logic blocks (CLBs) involved, which affects the critical path delay.

Similarly, the occupied area and energy consumption increases with the number of maximum policies and, as the frequency, there is a dependency with the implementation strategies. Figure 4 shows the resources utilization as the policies memory changes, while the Fig. 5 illustrates the power consumption, comparing the contribution of the only policy enforcer with the overall power consumption, so including the logic to configure the DT architecture and the logic which monitors AXI transactions issued by third-party IP cores. In both Figures there is a linear relation between the number of policies and the resource/power overheads.

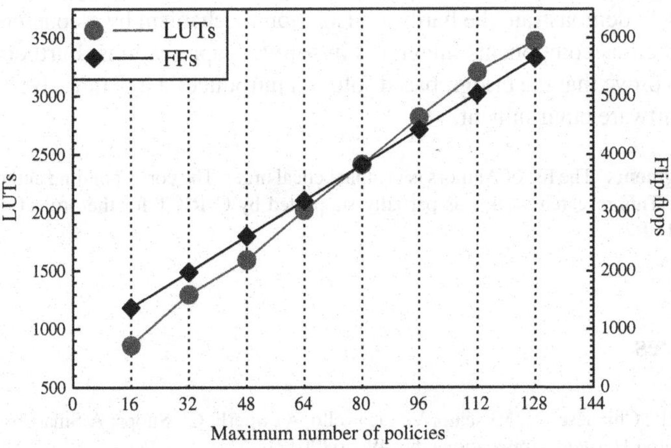

Fig. 4 Resource utilization of the implemented prototype as the policy memory depth changes

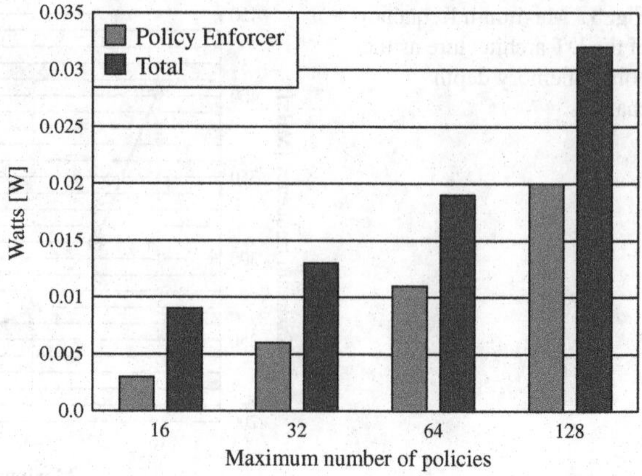

Fig. 5 Power consumption of the implemented bridge. The Figure also highlights the contribution of the only policy enforcer

5 Conclusion and Future Work

Advances of the FPGA technology made them suitable for a wide range of applications characterized by a high adaptability demand. FPGA based devices are actually able to dynamically change their circuits in-field, providing ease for the design of partial bitstreams and guaranteeing inherently security mechanisms to protect them. However, the end-user has no tools to check any properties on bitstreams and there are no possibilities to trust the IP cores. In this paper we presented a policy enforcement mechanism which can be adopted to manage the transactions performed on an interconnection system by IP cores. In particular, we exploited a solution based on a programmable decision tree architecture and we demonstrated that it is suitable to accomplish the enforcement task as introduces negligible overhead.

As part of future work, we plan to integrate our approach within a real embedded SoC-FPGA device which runs a complete software stack. In this way, we will be able to completely demonstrate the hardware isolation mechanism by a concrete realization of malicious components integrated in software applications. Furthermore, we aim also to prove that the bridge-based solution introduces a low intrusiveness grade onto the software environment.

Acknowledgments The list of Authors is in alphabetical order. The corresponding author is Mario Barbareschi. This research work was partially supported by CeRICT for the project NEMBO—PONPE_00159.

References

1. Amato, F., Chianese, A., Moscato, V., Picariello, A., Sperli, G.: Snops: A Smart Environment for Cultural Heritage Applications, pp. 49–56 (2012)
2. Amato, F., Fasolino, A., Mazzeo, A., Moscato, V., Picariello, A., Romano, S., Tramontana, P.: Ensuring Semantic Interoperability for E-health Applications, pp. 315–320 (2011)

3. Barbareschi, M.: Implementing hardware decision tree prediction: a scalable approach. In: IEEE International Conference on Advanced Information Networking and Applications (AINA-2016). IEEE (2016)
4. Barbareschi, M., Battista, E., Casola, V., Mazzocca, A.M.E.N.: On the adoption of fpga for protecting cyber physical infrastructures. In: 2013 Eighth International Conference on P2P, Parallel, Grid, Cloud and Internet Computing (3PGCIC), pp. 430–435. IEEE (2013)
5. Barbareschi, M., Battista, E., Mazzocca, N., Venkatesan, S.: A hardware accelerator for data classification within the sensing infrastructure. In: 2014 IEEE 15th International Conference on Information Reuse and Integration (IRI), pp. 400–405. IEEE (2014)
6. Barbareschi, M., Mazzeo, A., Vespoli, A.: Network traffic analysis using android on a hybrid computing architecture. In: Algorithms and Architectures for Parallel Processing, pp. 141–148. Springer (2013)
7. Barbareschi, M., Mazzeo, A., Vespoli, A.: Malicious traffic analysis on mobile devices: a hardware solution. Int. J. Big Data Intell. **2**(2), 117–126 (2015)
8. Chen, H., Chen, Y., Summerville, D.H.: A survey on the application of fpgas for network infrastructure security. Commun. Surv. Tutorials IEEE **13**(4), 541–561 (2011)
9. Cilardo, A.: New techniques and tools for application-dependent testing of fpga-based components. IEEE Trans. Ind. Inf. **11**(1), 94–103 (2015)
10. Cilardo, A., Mazzocca, N.: Exploiting vulnerabilities in cryptographic hash functions based on reconfigurable hardware. IEEE Trans. Inf. Forensics Secur. **8**(5), 810–820 (2013)
11. Cilardo, A., Barbareschi, M., Mazzeo, A.: Secure distribution infrastructure for hardware digital contents. CDT, IET **8**(6), 300–310 (2014)
12. Cilardo, A., Gallo, L., Mazzocca, N.: Design space exploration for high-level synthesis of multi-threaded applications. J. Syst. Archit. **59**(10, Part D), 1171–1183 (2013). http://www.sciencedirect.com/science/article/pii/S1383762113001537
13. Drimer, S., Güneysu, T., Kuhn, M.G., Paar, C.: Protecting multiple cores in a single fpga design. http://www.cl.cam.ac.uk/sd410/ (2008). Accessed May 2008
14. Huffmire, T., Brotherton, B., Wang, G., Sherwood, T., Kastner, R., Levin, T., Nguyen, T., Irvine, C.: Moats and drawbridges: an isolation primitive for reconfigurable hardware based systems. In: IEEE Symposium on Security and Privacy, 2007. SP'07, pp. 281–295. IEEE (2007)
15. Huffmire, T., Prasad, S., Sherwood, T., Kastner, R.: Policy-driven memory protection for reconfigurable hardware. In: Computer Security–ESORICS 2006, pp. 461–478. Springer (2006)
16. Kashyap, H., Chaves, R.: Compact and on-the-fly secure dynamic reconfiguration for volatile fpgas. ACM Trans. Reconfigurable Technol. Syst. (TRETS) **9**(2), 11 (2016)
17. Kastner, R., Huffmire, T.: Threats and challenges in reconfigurable hardware security. Technical report, DTIC Document (2008)
18. Maes, R., Schellekens, D., Verbauwhede, I.: A pay-per-use licensing scheme for hardware ip cores in recent sram-based fpgas. IEEE Trans. Inf. Forensics Secur. **7**(1), 98–108 (2012)
19. Pagiamtzis, K., Sheikholeslami, A.: Content-addressable memory (cam) circuits and architectures: a tutorial and survey. IEEE J. Solid-State Circuits **41**(3), 712–727 (2006)
20. Qu, Y.R., Zhou, S., Prasanna, V.K.: High-performance architecture for dynamically updatable packet classification on fpga. In: Proceedings of the Ninth ACM/IEEE Symposium on Architectures for Networking and Communications Systems. pp. 125–136. IEEE Press (2013)
21. Quinlan, J.R.: C4. 5: Programs for Machine Learning. Elsevier (2014)

Arabic Named Entity Recognition—A Survey and Analysis

Amal Dandashi, Jihad Al Jaam and Sebti Foufou

Abstract As Arabic digital data has been increasing in abundance; the need for processing this information is growing. Named entity recognition (NER) is an information extraction technique that is vital to the processes of natural language processing (NLP). The ambiguous characteristics of the Arabic language make tasks related to NER and NLP very challenging. In addition to that, work related to Arabic NER is rather limited and under-studied. In this study, we survey previous works and methodologies and provide an analysis and discussion on the feature sets used, evaluation tools and advantages and disadvantages of each technique.

1 Introduction

Named Entity Recognition (NER) was initially introduced as an information extraction technique. NER is a task that locates, extracts and automatically classifies named entities into predefined classes in unstructured texts (Nadeau and Sekine [7]. It covers proper names, temporal expressions and numerical expressions. Proper names are classified into three main groups: persons, locations and organizations. A class can be divided into sub-classes to form an entire hierarchy, i.e., location can be classified into city, state and country. The majority of NER studies have been focused on the English language, as it is the internationally dominant language, while research on other languages for the NER task has been limited.

Arabic is a richly morphological language of complex syntax. The lack of simplicity in the characteristics and specifications of the Arabic language make it a challenging task for NER techniques. Arabic can be classified into three types:

A. Dandashi (✉) · J. Al Jaam · S. Foufou
Department of Computer Science, Qatar University, Doha, Qatar
e-mail: amal.dandashi@qu.edu.qa; amaldandashi@gmail.com

J. Al Jaam
e-mail: jaam@qu.edu.qa

S. Foufou
e-mail: a.karkar@qu.edu.qa

G. De Pietro et al. (eds.), *Intelligent Interactive Multimedia Systems and Services 2016*, Smart Innovation, Systems and Technologies 55,
DOI 10.1007/978-3-319-39345-2_8

Classical Arabic, Modern Standard Arabic and Colloquial Arabic. It is imperative for the task of NER to be able to distinguish between those three types. Classical Arabic is the formal version of Arabic used for over 1,500 years in religious scripts. Modern Standard Arabic is that used in today's newspapers, magazines, books, etc. Colloquial Arabic is the spoken Arabic used by Arabs in their informal day to day speech and differs in dialect for each country and city. There are several specifications of the Arabic language that do not make NER an easy task; lack of capitalization, agglutination, optional short vowels, ambiguity inherent in named entities and lack of uniformity in writing styles. Add to that common spelling mistakes and shortage of technological resources such as tagged corporas and gazetteers, we have several issues to tackle for tasks associated to natural language processing (NLP).

In this study, we aim to analyze and report on the performance of recent NER studies dedicated to the Arabic language. In Sect. 2, we include analysis on the features and tag sets used, algorithm and methodology, performance evaluation and pros and cons of each technique. In Sect. 3 we include a descriptive listing of Arabic NER tools currently available. Finally, Sect. 4 concludes the study.

2 Analysis on Recent Studies

2.1 Study 1

The impact of using different sets of features in three different machine learning frameworks is investigated by Benajiba et al. [3], for the Arabic NER task. The machine learning frameworks tested are support vector machines (SVM), maximum entropy (ME), and conditional random fields (CRM). Nine different data sets of genres and annotations are explored along with lexical, contextual and morphological features. In order to evaluate robustness to noise of each approach, different feature impacts are measured in isolation and incremental combination.

Methodology: The three comparable approaches which were used for named entity recognition (NER) are:

1. SVM: known to be robust to noise and have strong generalization capabilities for large feature sets. The *Yamcha* toolkit is used to apply SVM to the NER task.
2. ME: known to provide model with the least biases possible. A customized ME approach is implemented to carry out the experiments, and the *Yasmet* tool is used for weight estimation.
3. CRF: oriented towards segmenting and labeling sequence data, and can represent probability distributions. Conditional probabilities of classes are maximized during the training phase. *CRF++* is used for experimentation.

The optimal feature sets investigated are: Contextual (CXT), lexical, gazetteers, morphological features, part-of-speech tags, nationality and corresponding English capitalization.

Performance Evaluation:

1. The NLE-corpus is used. It comprises of text collected from several newswire web resources, and the text is manually annotated. The tag set used is CoNLL set which includes four classes: Person, Location, Organization and Miscellaneous.
2. The ACE 2003, 2004 and 2005 corpora are used. The data in ACE is annotated for the following tasks: entity detection and tracking, relation detection and recognition, event detection and recognition. ACE 2003 consists of two data genres: Broadcast News (BN) and Newswire (NW). The ACE 2004 also includes the Arabic Treebank (ATB). The ACE 2005 includes the Weblogs (WL) but not the ATB genre.

Experimental setup:

1. Metrics: CoNLL evaluation metrics are used for precision, recall and F1-measure (harmonic mean between precision and recall).
2. Experiments: Three sets of experiments take place; a baseline, a parameter setting of experiments and feature-engineering experiments. The latter involves exploring individual features, ranking features according to impact and evaluating the SVM, ME and CRF approaches. In order to find the optimal number of features, evaluation of SVM, ME and CRF approaches involves combining the top N-elements of the ranked features each increment, starting from N = 1 up to N = 22.

The performance evaluations are done using fivefold cross validation on each corpus independently. For the NLE-corpus, the same ratio of test data size to training data size that has been used in CoNLL competitions has been also used for this study. For the ACE data, the authors used the same ratio of test data size to training data size which has been used for the ACE evaluation.

Almost all the corpora tested achieved high performance and improvement over the baseline. For ACE 2003, an F1 score of 83.34 for the BN genre. The WL genre yielded the worst results, which may be due to the inclusion of dialectical language and randomness of the WL data compared to the other genres. The NW genre of the ACE 2005 data yielded an F-measure of 77.06.

The features ranked according to impact are as follows, first place is the POS tagger, second is the CAP feature, and that is confirmation that the lack of capitalization in languages such as Arabic complicates the NER task considerably. Several of the MFs are ranked in a scattered manner ranging from third to twenty-second. LEX features ranks range from fifth till nineteenth place, and show that marking the first and last three characters of a word can be useful for a NER approach. Incremental features selection resulted in better performance, since using a selected feature set leads to avoidance of providing the classifier with noisy information.

The ME approach is the most sensitive to noise and obtained significantly lower results than that of the CRF and SVM, specifically when the number of features exceeded six. When the top seven features were used, the SVM approach depicted better performance than that of the CRF approach. Otherwise, the performance of SVM and CRF obtained similar results. And while the latter two approaches give different false alarms, they tend to miss the same named entities and thus cannot be used to complement each other.

2.2 *Study 2*

Zitouni et al. [11] present a statistical approach to Arabic mention detection and chaining (MDC) systems. The approach is based on the Maximum Entropy (ME) principle. The system first detects mentions in an input document and then chains the identified mentions into entities. The ME framework allows for a large range of feature types, including lexical, morphological, syntactic and semantic features. The authors consider the additional challenges of the Arabic language, as one needs to correctly identify and correct enclitic pronouns by adding segmentation as a processing step. In Arabic text, nominals and pronouns are attached to words, hence special attention needs to be given when processing Arabic text in order to be able to detect partial parts of a word as a mention.

Methodology: The MDC system in this study proceeds in two main phases: detecting mentions and then partitioning the detected mentions into entities. The latter phases are based on the ME technique. This approach is language independent and must be modified to accommodate the Arabic Language specifications. The methodology is detailed as follows:

1. Arabic word segmentation: The weighted finite state transducer (WFST) is used for this section, which was initially a manually trained corpus and later refined using unsupervised learning on a large corpus of 155 million words. The segmentation process is as follows; partition Arabic text into a sequence of segments (tokens), and separation of delimited words into prefixes, stems and suffixes.
2. Mention detection: identify and characterize the main actors in an Arabic text; people, locations, organizations, geopolitical entities. It is formulated as a classification problem, where the labels are assigned to tokens in the text, indicating whether it starts a specific mention, is inside a given mention or is outside any mention. The ME classifier is selected for this process, as it integrates arbitrary types of information and makes classification decisions by aggregating all information for a specific classification. ME associates a set of weights with features, and computes the probability distribution. Weights are estimated during the training phase of the data set to maximize the likelihood. In this study ME

model is trained using the sequential conditional generalized iterative scaling technique (SCGIS), and used a Gaussian prior for regularization.
3. Features: the features used in this mention detection system are: Lexical, Stem n-gram gazetteer-based, syntactic and features from other named-entity classifiers.
4. Mention chaining: uses a machine learning based approach:

- Entities in a document are created from mentions incrementally and in a synchronous fashion. This is done by constructing a data structure called a Bell tree.
- First mention is used to create the root of the Bell tree.
- Subsequent mentions may start a new entity or link with an existing entity.
- End result is a Bell tree structure with each leaf node representing a coreference outcome.
- Coreference resolution problem is converted into scoring the competing paths in Bell tree.

5. Automatic Content Extraction (ACE) entity detection and recognition: detecting certain specified types of entities. All mentions pertaining to each entity are to be detected and four pieces of information are to be defined as follows:

- Mention type can classify as: person, organization, location, geopolitical entity, facility, vehicle or weapon.
- Mention subtype or subcategory of a mention type
- Mention class, whether generic or specific
- Mention type, whether named, nominal or pronominal.

Performance Evaluation: The Arabic ACE 2007 is used for experimentation. 323 documents (80000 words) are used for training and 56 documents (18000) for testing. The result is 17,634 mentions (7816 named, 8831 nominal and 987 pronominal) for training, and 3566 for testing (1673 named, 1682 nominal, and 211 pronominal).

Performance metrics for mention detection use F-measure to report evaluation results. Measuring performance for mention chaining involves the use of two performance metrics; the first is a Constrained-Entity Alignment F-measure (CEAF), which measures the percentage of mentions that are in the right entities. The second metric is the ACE value, the metric used for the ACE task. The ACE value is computed by summing values of aligned system entities, subtracting values of false alarm entities, and normalizing over the values of reference entities. A perfect coreference system would get a 100 % ACE-value.

Discussion: The Arabic main-type mention detection system obtained an F-measure of 80.0 in the ACE evaluation. The authors started the system evaluation by only allowing access to the lexical features and gradually increasing features with each increment. A system using only lexical, stem and gazetteer features achieves a measure of 76.5. Syntactic and other classifier feature outputs add more than 3 F-measure points to overall performance (80 vs. 76.5).

Results of the system evaluation depict a reasonable performance of the mention types: geopolitical entities, person and vehicles, with an F-measure of 89.2, 81.5 and 75.0, respectively. The remaining mention types received a low performance with the F-measure ranging from 68.5–55.1.

Lexical attributes and distance features were found to be the most essential for coreferencing. Lexical attributes features lead to 76.3 F-measure and 68.8 ACE-value, and addition of the distance feature improves F-measure by 4.1 and ACE-Value by 6.1 points. While Stem Match and syntactic features do not help as much as the latter features, they do improve performance in small increments by acting as extra optimizing features, which is a typical component of a statistical system that has reached a good performance level. While syntactic feature improves ACE-value by 0.4 %, it slightly delays the CEAF by 0.3 %, which is an indication that different configurations may lead to better results.

2.3 Study 3

Zitouni and Benajiba [10] proposed a semi-supervised approach that utilizes multilingual parallel data (English-Arabic) in an effort to enhance the mention detection (MD) task. The challenge with MD is directly related to the complexity of the morphology of a given language. For this reason, the authors explore the idea of using a MD system designed to suit the needs of a resource rich language (RRL), namely English, to improve the performance of a MD system for a target language (TL), namely Arabic. To maximize the scope of the study to be potentially applicable for several languages, the authors have chosen to experiment with a limited set of annotation types for the TL, as well as a larger set. The proposed system uses the maximum entropy (ME) principle, which combines arbitrary types of information to make classification decisions.

The hypothesis of this study is that an MD system with a rich feature set such as English may be used to boost performance for TL such as Arabic, provided that the donor language system's resources are capable of surpassing its TL counterpart's. To test this theory, MD tags are projected from RRL to TL via a parallel corpus, and several linguistic features about the automatically tagged words are extracted. Then several experiments have been conducted, involving adding these new features to the TL baseline MD system. The benefit of the ME technique lies with its ability to integrate multiple features seamlessly. However, in some cases it may lead to an overestimation of confidence, particularly in low-frequency features. This problem arises when a hard constraint is reinforced on a feature whose estimate is not reliable. The adjustment used in this experiment involves using the sequential conditional generalized iterative scaling (SCGI) technique.

Methodology: In order to validate the hypothesis of this study, English and Arabic corpuses are used, part of the ACE 2007 evaluation. In Arabic, words are composed of zero or more prefixes, a stem, and a zero or more suffixes, which are all considered tokens. Any contiguous sequence of tokens may present a mention.

The first step is segmentation of text, and then classification is performed on tokens. MD systems across these English and Arabic languages use a large range of features, which may be classified as: lexical, syntactic and information obtained from other named-entity classifiers with customized semantic tags. As additional features, the assigned classifier tags are utilized: lexical, syntactic, semantic and stem-n-gram features.

Cross language mention projection: In order to use a RRL like English, to enhance MD in a TL, like Arabic, a parallel corpus with word alignment must be utilized, and a MD system in the RRL must be available. The steps are as follows:

1. Extracting necessary features
2. Running MD on the RRL text of the parallel corpus, resulting with tagged text
3. Utilizing word alignment file to project mentions from RRL to TL
4. Alignment file has many-to-many structure, which describe which words from the RRL side correspond to the words on the TL side.
5. Words with no alignment will not be tagged
6. Result is a text in the TL annotated with mentions obtained by propagation from RRL.
7. Once Arabic corpus is tagged, features are extracted: Gazetteers, model-based features, lexical content, surrounding phrase head-words, and parser-based features.

Performance Evaluation: The main goal is to investigate the effect of using a semi-supervised multilingual approach to enhance the Arabic MD system. First the use of an unlabeled data corpus and a MD system based on a RRL (English) is explored to show how Arabic MD system benefits. Then the monolingual technique is explored in order to compare results. Experiments are conducted on specified partitions of the ACE 2007 data set.

Corpus Alignment techniques used: (1) hand aligned data, (2) automatically aligned data, (3) Combined hand aligned and automatically aligned data, and (4) monolingual data.

The following features are used for the cross-lingual (English-Arabic system): baseline, En-nLexCon, En-nPhHead, En-nSynEnv, En-Gaz, En-Model and Comb.

Discussion: The impact of features obtained through cross-lingual system proved to be far more effective in enhancing system performance.

Hand-aligned data results: This model achieved a 57.7 F-measure. The low performance is an indication of the level of noise, and we may understand that better performance can be achieved when there is no human annotated data. However, when the Arabic MD system is poor in resources, significant improvements are obtained. When only Lexical features are used, there is an improvement of approximately1.9 points. When the Arabic MD system uses a rich feature-set Syntac, a 1.5 point improvement is obtained. The model based feature (En-Model) depicted the greatest performance (76.22) when the Arabic MD system uses a feature-set that includes more than just the Lexical features. Comb features achieved higher performance than Baseline features (77.18 and 75. 68, respectively).

Automatically-aligned data results: the 22 million word corpus is split into four subsets, each with 5, 10, 17 and 22 million words, respectively. From each of these subsets, the impact of the features extracted from each subset is analyzed separately. The most optimal results were found from the 17 million word subset. Best results have been obtained with the use of the En-nSynEnv feature for the 17 million word subset, with an F-measure of 76.02.

Combined hand-aligned and automatically-aligned data: automatically aligned data helped capture more of the unseen mentions, whereas the hand-aligned data helped decrease the number of false alarms. When the Comb feature is used with the Stem baseline model, the F-measure obtained (76.85) is 1.2 points higher than the baseline model that uses Lexical, Stem and Syntactic features (75.68). Using a propagation approach like this one, definitely helps bootstrap the process and achieves better performance.

Monolingual data: the same 17 million words subset features are used to demonstrate performance for monolingual data experimentation. The only difference is that the Arabic data is tagged with the baseline Arabic MD system. The En-nSyncEnv system using only Lexical features with project from the English MD system has very similar performance with the Baseline system used with the monolingual experiment (75.62, 75.68, respectively). When more resources are used with Arabic, such as Syntac feature, together with the monolingual features, Ar-nSynEnv, results are very comparable to those achieved when the system uses Lexical features with cross-lingual En-nSyncEnv features (75.61, 75.62, respectively).

The proposed approach of cross-lingual propagation MD systems is efficient and practical due to the fact that various resources are already available in RRL such as English. There are many parallel corpora that cover many language pairs that can be used. The proposed approach uses parallel aligned corpora and MD system designed to enrich a TL system by using information in a RRL system. Experimentation led to the conclusion that this approach is more effective than dealing with languages with limited resources with the use of unsupervised data. Results have shown that performance decreases when too many resources are used, but even when all resources are used, significant gains are observed. When the TL has access to limited resources, a 2.2 point improvement is made, whereas when all resources are used for the TL, only a 0.5 point improvement is observed. With absence of human-annotated Arabic data, performance F-measure reaches to 57.6 using only mention propagation from RRL system. In conclusion, the approach using cross-lingual information outperforms the approach using monolingual information.

2.4 Study 4

Chen and Ng [4] propose a hybrid approach to coreference resolution aimed for the CoNLL-2012 shared task is proposed. The hybrid approach consists of combining specific features of rule-based and learning-based techniques. The system models coreference in OntoNotes for three challenging languages that come from very different language families; English, Chinese and Arabic. The system is designed to resolve references in all three languages. There are four main tracks of the proposed system; closed track for all three languages, and an open track for the Chinese language. The system also exploits genre specific information, and optimizes parameters with respect to each genre. The system performs mention detection as an initial step, which is followed by coreference resolution. The parameters of those two components are jointly optimized with respect to the desired evaluation measure.

Methodology:

1. Mention detection component: two step approach is employed;
   a. Extraction step: identifying named entities (NE) and employing language-specific heuristics to extract mentions. In order to increase upper bound on recall, mention extraction is done by utilizing syntactic parse trees.
   b. Pruning step: improve precision by employment of language-specific heuristic pruning and language-independent learning-based pruning.
2. Coreference Resolution: Employs heuristics and machine learning, via the Stanford multi-pass sieve approach. As most of these sieves are unlexicalized, the multi-pass sieve (heuristic rules) approach is optimized by incorporating lexical information via machine learning techniques. Each sieve heuristic rule extracts a conditional-based coreference relation between two mentions. Sieves are ordered by precision, and in order to resolve a set of mentions in a document, the resolver makes multiple passes over them, finally using the rules only in the final sieve to find an antecedent for each mention.
   a. Sieves for English: Composed of 12 sieves, modeled after those employed by the Stanford Resolver, with some optimizations.
   b. Sieves for Chinese: For Chinese, the authors participate in close and open track. The sieves used for both tracks are the same. The Chinese resolver is composed of 9 sieves: Chinese Head Match, Precise Construct, Pronouns, Discourse Processing, Exact String Match, Strict Head Match (A,B,C), and Proper Head Match.
   c. Sieves for Arabic: Only one sieve is employed for Arabic; the Exact Match sieve. Other sieves such as Head Match and Pronouns sieve were also implemented but excluded from the study as they did not yield good results.

Performance Evaluation: For each language genre, the following steps are implemented: (1) learn lexical probabilities from training set, (2) obtain most optimal parameter values, using two parameter estimation algorithms, and (3) chose the parameter with that obtained the better performance on the development set to be the final set of parameter estimates for the resolver.

Discussion: Mention detection results and coreference results are both depicted in terms of recall, precision and F-measure. Ablation experiments are performed, showing for each language track combination the results obtained without editing: the rule relaxation parameters, the probability thresholds, and all parameters tested. However, there were no rule relaxation parameters for Arabic.

The average F-measure for the English closed track is 60.3, while the average for the Chinese closed track is 68.6, and the closed Arabic track 47.3. The Arabic track obtained the lowest results due to the fact that the Arabic language is a highly inflectional language and the authors claim to have little linguistic knowledge of the language to design effective sieves. As for the open track, which was implemented solely for the Chinese language, the average F-measure obtained is 79.0.

2.5 *Study 5*

Arabic knowledge bases are valuable lexical semantic resources with high influence on several Natural Language Processing tasks. There are two main types of knowledge bases: collaborative knowledge bases (CKB) and linguistic knowledge bases (LKB). The main difference between the two is that CKBs are developed by voluntary nonprofessionals on the web that follow nonbinding guidelines, while LKBs are developed by linguistic professionals and are guided by theoretical linguistic models. CKBs are ever-growing and available to everyone for free use (such as Wikipedia or Wiktionary), while LKBs are rigid and unavailable for free use, apart from WordNet. Aljazeera in itself, is considered to be among the richest Arabic knowledge bases on the web. While it is not considered a CKB nor an LKB, it shares characteristics from both; it is developed and edited by linguistic professionals, it is ever-growing and freely available online, and its content is semantically interlinked.

All three Arabic knowledge bases (CKBs, LKBs and Aljazeera.net) have several applications in Natural Language Processing. However, the following challenges must be noted: (3) Arabic CKBs are less structured than LKBs and Aljazeera.net, and contain more noisy information, (2) Arabic CKBs rely on social control for accuracy and precision, whereas LKBs and Aljazeera.net rely on professional editorial provision, and (3) While CKBs and LKBs have encyclopedic and linguistic orientation, Aljazeera.net is formed on events-based orientation. Al-Kouz et al. [1] form the following hypothesis for this study: Arabic CKBs, LKBs and Aljazeera.net could form a complementary resource, and that an Arabic Semantic Graph (ASG) could be constructed based on these complementary knowledge bases. The authors hence present a framework design for the development of a semantic graph extractor for Aljazeera.net, a high performance Java-based API that has capabilities to extract implicit and explicit information.

Methodology: The Aljazeera.net content is rich in quantity and quality. An efficient framework design is needed in order to use CKBs and Aljazeera as a complementary knowledge resource for large scale NLP tasks. The authors have developed a general purpose framework designed to extract and build the ASG from these knowledge bases. The Arabic Semantic Graph Extraction Framework (ASGEF) comes with high performance Java API(JAKL) packages that are to be available freely for the research communities. The APIs are designed to work on an object oriented programming paradigm with the following objects in consideration; ALJAZEERA, NEWSARTICLE, SEMANTICENTITY, NAMEDENTITY, SEMANTICENRICHMENT, and CATEGORY.

1. The ASGEF is to be built based on Aljazeera.net platform as the main knowledge resource, and CLBs used as complementary semantic enrichment sources.
2. Java based Data Machine and Time Machine APIs are developed in order to utilize Aljazeera.net platform.
3. Time Machine can be used to crawl Aljazeera.net from a specific point in time in offline mode, only after first use of Data Machine.
4. Time Machine can also be used to crawl new content on Aljazeera.net in online mode.
5. Parsing stage starts: two parsers are used, the File Parser and the Web Parser.
6. File Parser is capable of parsing hierarchical directory structure, and translating it into Arabic hierarchical related entities.
7. Ontology Builder API then transforms hierarchical related entities into Arabic Ontology.
8. Data Set Builder API transforms hierarchical related entities into a manually annotated data set.
9. The Web Parser parses HTML pages within the hierarchical directory structure to extract text.
10. The Content Extractor API employs the File and Web Parser results to extract a semi-structured data set.
11. The result is extracted semi-structured data with explicit semantic data represented as interlinked articles and hyperlinked named entities. This data is applicable to named entity detection tasks.
12. Java APIs are utilized to access local dumps of Arabic CKB resources.
13. The SGB provides the JAKL, which operates on semantically related and enriched entities that created from Aljazeera.net, Wikipedia, and Wiktionary.
14. The output of the SGB is the Arabic Semantic Graph, which can be published to Linked Open Data (LOD).

Discussion: The system proposed in this study could be used in several NLP tasks and large-scale research projects that involve analysis, computation of semantic relations between entities, and access of Arabic knowledge base semantic graph structure, among other information retrieval applications. The ASGEF architecture has the following functionalities: (1) enable large-scale Arabic NLP tasks with computational efficiency, (2) enable reproducible experimental results based on Aljazeera.net along with CKBs such as Wikipedia and Wiktionary, (3) enable reliable mathematical representation of knowledge, (4) free and easy to use.

3 Arabic NER Tools

3.1 *Fassieh*

Attia et al. [2] introduce an Arabic annotation tool called Fassieh. It enables the production of large annotated Arabic text corpora, classified according to the following Arabic linguistic models: Part of speech tagging, morphological analysis, phonetic transcription and discretization, and lexical semantic analysis. The inherent ambiguity of these analysis models is statistically resolved with Fassieh. The system also presents various other auxiliary features which enable a normalized, guided and efficient proof-reading for the factorized corpus. These features include morpheme-based dictionaries, short-contect statistical ranking, illustrative GUI tools such as character and word status coloring. Fassieh is not only an annotation tool, but also incorporates evaluation, demonstrative and tutorial functions for Arabic Natural Language Processing.

3.2 *MATAR*

Zaraket and Jaber [9] introduce and open source tagging tool with a visual interface. This tool enables the annotation of Arabic text corpora with the automated utilization of morphological tags. MATAR allows for a Boolean-based specification of tags, considering predicates and relations between morphological parts of text and values of a given morphological feature. Users can enter manual tags, edit existing ones, compare tag sets and compute accuracy results, all through a user-friendly interface.

3.3 *MADAMIRA*

MADARMIRA is a system designed by Pasha et al. [8] that can be utilized for morphological analysis and disambiguation of Arabic text. It combines aspects of two previously used systems for NLP; MADA [6] and AMIRA (Diab et al. [5]. MADAMIRA optimizes both previously mentioned systems with a more streamlined, robust, portable, extensible and faster Java-based implementation. It includes several tasks useful for NLP processes: part-of-speech tagging, tokenized forms of words, diacritization, lemma stemming, base phrases, and NER.

4 Conclusion

NER is among the most vital processes for the development of NLP systems. Accurate NER mechanisms ensure the success of a range of NLP systems like machine translation and information retrieval. Arabic textual information resources are increasing all over the internet, which makes the task of automated NER a necessary one, for the sake of classifying online data such as web pages, articles, informative texts, emails, blogs, etc. This study provides a survey and analysis of progress done towards Arabic NER. As the presence of named entities in one language leads to a correspondence in other languages, studies of NER in a specific language allots valuable insight and research for developing multi-lingual NLP systems. We hope this analytical study provides fruitful guidance for researchers dealing with Arabic NER systems.

Acknowledgements This publication was made possible by GSRA grant # 1-1-1202-13026 from the Qatar National Research Fund (a member of Qatar Foundation). The findings achieved herein are solely the responsibility of the author(s).

References

1. Al-Kouz, A., Awajan, A., Jeet, M., Al-Zaqqa, A.: Extracting Arabic semantic graph from Aljazeera. net. In: 2013 IEEE Jordan Conference on Applied Electrical Engineering and Computing Technologies (AEECT), (pp. 1–6). IEEE, Dec 2013
2. Attia, M., Rashwan, M.A., Al-Badrashiny, M.A.S.A.A.: Fassieh, a semi-automatic visual interactive tool for morphological, PoS-Tags, phonetic, and semantic annotation of Arabic Text Corpora. IEEE Trans. Audio Speech Lang. Process. **17**(5), 916–925 (2009)
3. Benajiba, Y., Diab, M., Rosso, P.: Arabic named entity recognition: A feature-driven study. IEEE Trans. Audio Speech Lang. Process. **17**(5), 926–934 (2009)
4. Chen, C., Ng, V.: Combining the best of two worlds: a hybrid approach to multilingual coreference resolution. In: Joint Conference on EMNLP and CoNLL-Shared Task, pp. 56–63. Association for Computational Linguistics, July 2012
5. Diab, M.: Second generation AMIRA tools for Arabic processing: fast and robust tokenization, POS tagging, and base phrase chunking. In: 2nd International Conference on Arabic Language Resources and Tools (2009)
6. Habash, N., Roth, R., Rambow, O., Eskander, R., Tomeh, N.: Morphological analysis and disambiguation for dialectal Arabic. In: HLT-NAACL, pp. 426–432 (2013)
7. Nadeau, D., Sekine, S.: A survey of named entity recognition and classification. Lingvisticae Investigationes **30**(1), 3–26 (2007)
8. Pasha, A., Al-Badrashiny, M., Diab, M., El Kholy, A., Eskander, R., Habash, N., Roth, R.M.: (2014). Madamira: a fast, comprehensive tool for morphological analysis and disambiguation of Arabic. In: Proceedings of the Language Resources and Evaluation Conference (LREC). Reykjavik, Iceland
9. Zaraket, F.A., Jaber, A.: MATAr: Morphology-based Tagger for Arabic. In Computer Systems and Applications (AICCSA), 2013 ACS International Conference on, pp. 1–4. IEEE, May 2013

10. Zitouni, I., Benajiba, Y.: Aligned-parallel-corpora based semi-supervised learning for Arabic mention detection. IEEE/ACM Trans. Audio Speech Lang. Process. **22**(2), 314–324 (2014)
11. Zitouni, I., Luo, X., Florian, R.: A cascaded approach to mention detection and chaining in arabic. IEEE Trans. Audio Speech Lang. Process. **17**(5), 935–944 (2009)

Exploitation of Web Resources Towards Increased Conversions and Effectiveness

Jarosław Jankowski, Jarosław Wątróbski, Paweł Ziemba and Wojciech Sałabun

Abstract The development of Internet technologies is impacting the need to increase their effectiveness in achieving commercial goals. Selecting the most appropriate parameters for their functioning is highly important particularly in the area of online marketing and e-commerce platforms. Moreover, the focus on system performance is observed in terms of conversions and direct responses leading to maximum results. The proposed approach targets the reduction of complexity in the decision-making process with a multistage factorial analysis that is applied to advance the performance of internet websites towards better conversions.

Keywords Website conversion · Online marketing · User experience

1 Introduction

A company's success in online environments is affected greatly by a series of factors which must be taken into consideration when creating any plan of action regarding policy. The trend of developing internet platforms together with delivering attractive content in order to appeal to audiences and build long term relationships with consumers plays a crucial role. Increasing website effectiveness requires techniques related to effective personalization [19] or proper usage of

---

J. Jankowski (✉) · J. Wątróbski · W. Sałabun
West Pomeranian University of Technology, Żołnierska 49, 71-210 Szczecin, Poland
e-mail: jjankowski@wi.zut.edu.pl

J. Wątróbski
e-mail: jwatrobski@wi.zut.edu.pl

W. Sałabun
e-mail: wsalabun@wi.zut.edu.pl

P. Ziemba
The Jacob of Paradyż University of Applied Sciences in Gorzów Wielkopolski, Teatralna 25, 66-400 Gorzów Wielkopolski, Poland
e-mail: pziemba@pwsz.pl

G. De Pietro et al. (eds.), *Intelligent Interactive Multimedia Systems and Services 2016*, Smart Innovation, Systems and Technologies 55,
DOI 10.1007/978-3-319-39345-2_9

recommendations [15] and persuasive marketing content [7]. Systems are based on personalization and they are adjusting content to a user's needs with the characteristics of social networks [9], adaptive websites, elements of interfaces [5] and techniques like website morphing [6]. Website effectiveness is key area for improvement in e-commerce systems or social networking platforms. The approach proposed in this paper focuses on the key elements which affect the effectiveness of internet websites and selective areas by using a factor analysis. The methods discussed in the article are meant for the advancement of the efficacy of plans, both complete and partial, and the development of a multistage optimization process. The article is organized as follows: Sect. 2 discusses the main areas of website optimization towards better conversions; Sect. 3 presents a methodological background for increasing website performance; Sect. 4 discusses the proposed multistage approach; and Sect. 5 discusses the results of the experiments and presents a summary.

2 Selected Area of Website Effectiveness Improvement

The scheme used for exploring the effects of online platforms and the scope of their contribution is most commonly associated with the fact that the websites want to generate interfaces that are more user friendly. The scheme is concerned mainly with the engaging the audience by utilizing multiple options, such as persuasion to attract the audience's attention. However, in position papers, the parameters for examining the effectiveness of a website are different. K. Kaplanidou states that conventional measurement methods, for instance, sets of acquired websites, the number of external links, and a user's time interval of stay on a website, have been examined properly. Such assessment provides only a distant view of the website's user activity [10]. Whereas, E.W Welch and S. Pandey have gone through an extensive investigation on a large scale to inquire about multifaceted issues of the websites covering: content quality, interwoven terms of managerial mechanisms, technology, and customer response [20]. Moreover, S. Kelly specifies that the prospects of behavioural study in the system and the multifaceted mode of navigational paths will provide precise measurements interconnected with decisions and actions to be expressed by users [11]. On the another hand, M. Sicilia and S. Ruiz testify regarding the effectiveness of the idea that strengthening the association of behavioural patterns and the link of advertising messages within the website is useful [18]. Research on this same project has also indicated that furthering messages in the system is a definite symbol of the success rate of achieving the targets set for the website. A. Scharl has worked significantly on measuring the effectiveness of the system for the tourist industry [17]. Whereas, M. Birgelen and other associates have conducted thorough research on the volatility of commercial websites and their assessment [2].

Earlier work is associated with the field points out issues regarding the process and multiple methods related to the assessing a website [21, 23, 22]. Fundamental elements include the process of consumer acquisition through the website, which is generally called lead generation, and certain actions connected to the activity within websites. The main factors include: the amount of acquired information, number of users coming back to the website, the minimum page numbers visited by the user during his or her single stay at the website, and the time that a user usually spends on the website. The bounce rate indicates the number of users visiting only the main site, which is also very crucial for goal attainment and depicts the interest of a visitor to the content of a website. B.J. Jansen reveals that the objective of assessment can possibly be an attempt to explore more advertising associates in the field with brighter options and a comparative study in the market to know a website's standing [8]. Matters related to website effectiveness have a wider zone of correlation, and therefore, mingle information from multiple sources and areas.

As there has been a consistent rate of growth in the development of website effectiveness inducing measures and techniques associated with maximizing the usage of decision-making tools, conversion oriented websites have come into the business arena. Ideas based on enhancing the usability of a website and user-centric designing have been incorporated. The utility of tools to enhance the usability of a website has been used on multiple stages, which are very important for the smooth working of internet systems. J. Nielsen has defined term usability as a qualitative quality which develops from the ease of use in interactive campaigns [16]. However, Chairman-Hewett states that the Human-Computer Interaction (HCI) is as a decisive matter in the process [3]. This is an ever growing concept interlinked with the fields of design, design testing, and operational requirements of computer systems, which all work to build communication among users. HCI includes multiple disciplines ranging from IT, behavioural sociology, and the consciousness of human factors. In order to acquire an expected number of interactions, it is important to pay attention to the method of interface and structural information designs.

P. Morville has studied the concept of information architecture (IA) [14] and has shown that it relates to the development of the structures of data and navigational paths. This includes the organizational processes, navigational designs, and efficacy of search systems to acquire data and organize it. IA, is closely linked with user-centered design (UCD). Both of the terms include mutual features, for example, the objective of designing in IA and UCD provides simple access to information. IA is most closely connected with the information processing paradigm whereas UCD connects more closely with the demands of the user. W. O. Galitz explains the most significant part of IA is its ability to divide the website into multiple operational blocks and remove the website material which is not very handy for utilization in the communication process [4]. M. Parrow has highlighted the broader application of a websites relevant physiological humanistic matters, including: the impact of multiple methods on the matters of designing, assessment, and the final solicitation of products. The likelihood model developed by R.E. Petty and J.T. Cacioppo [x18x] allows for elaboration on the way in which internet website development has affected perception, coined as an explanation of the physical being to understand the contagious reasoning [12]. The research conducted in this regard explains several factors contributing to stimulus effects and can be studied in regards to the optimization of real online systems. The author also stressed the point that efficient call to action provides a skimming and scanning of the website content, particularly handy for users who are not able to view all the website content. In e-commerce websites, multiple factors play a role in the ultimate success, such as the purchasing process, stages of transactions being understood, and producing and managing the shopping cart in the most user-friendly manner. Moreover, dynamic events can also occur in the development of the pricing of the products and the application of multiple price discriminatory methods. If the content of a website and space provided is balanced in an adequate manner, then a website's effectiveness can be enhanced and the proper allocation of resources can be ensured.

3 Conceptual Framework and Design of Experiments

This study focuses on the research in chosen fields of assessment and the interactive measurements, which reveal the success in the attainment of set goals and the significance of user interface during the entire process as related to the online system and the particular environment surrounding them. The factor of effectiveness (EF) is determined as a consequence of the actions taken under the proposition that is identified by terms between the input and the results that are generated within the course of time t, in accordance with the following formula:

$$EF_t = \frac{E_t}{I_t} \tag{1}$$

in which E_t—effect is registered as a consequence of the undertaken action within time t and I_t—input is represented by the costs for given time t. The results being generated depend largely on the level of sales being maintained and for internet services provided. Interaction and the actions taken by the users are also taken into account. Interweave relations between input and effects can be utilized to determine the rates of economic effectiveness. For instance, information about input and efforts being generated are developed through the efficacy equation, such as E − I represents the disparity of values between effects and input, I/E is the relationship between input and effects, and (E − I)/I shows the difference between the effects and input related to the input. In regards to the question of internet websites, the effects of the process can be examined by a few aspects and are not always connected directly with profitable earnings. The major motivation behind this article is to make the most of the factors that explain the fractional share of the users and users' attraction towards some of the goods and services with respect to the total number of users. The assessment tools used for such calculations are the conversion factors, C_t, which ultimately explain the relationship between the desired interaction, I_t, to the number of the users on the website, U_t, in a specific time period, t.

While resolving the multi-faceted monitoring of the system and the understanding of the flowing advertising campaigns, a sharp distinction can be made between partial conversions for the selected audience. Partial conversion, $\mathrm{CR}^{\mathrm{c}}_{\mathrm{i,j,t}}$, is determined by a comparison to the advertising campaign, *c* (*c = 1, …, k*), for a website, *I* (*i = 1, …, w*), and a type of action, *j* (*j = 1, …, k*), in the period of time, *t*. This can be determined using the following formula:

$$\mathrm{CR}^{\mathrm{c}}_{\mathrm{i,j,t}} = \frac{I^{c}_{i,j,t}}{U^{c}_{i,t}} \quad (2)$$

In which $I^{c}_{i,j,t}$, the number of interactions of *j* type, produced inside a website, *I*, by users in response to an advertising campaign, *c*, in the period of time, *t*. $U^{c}_{i,t}$, or the total number of website users in time, *t*, is obtained as an effect of an advertising campaign, *c*. With such a resolve of effectiveness and the measurement factors at hand, the most important function of the optimized process application is the increase of the levels to the highest point, which relates to the strength of full actions within the internet facility being provided. The process of increasing the optimization capacity of a website in an attempt to increase its effectiveness is complicated. The variety of elements, which affect the overall results, need to be identified. It is difficult to explain each and every element, and the role they have in smoothing out the communication process between the user and the number of commands he wants to adhere to. However, this process of optimizing the website can be accounted for in multiple stages using a variety of methods. Firstly, we can minimize the input segments and explain the number of variables assigned to the process. It will decrease the time taken and apparent costs of the process. On the basis of the initial analysis, 11 variables have been chosen, which are interconnected through textual content, interactive elements, and graphical representation. These elements are presented in Table 1.

4 Empirical Results

The main objective of this experiment was to assess the interconnectivity between influencers and the impact on the user actions landing on the website. With respect to the interactive elements and their mutual effect on the user, A/B testing was replaced by multi-dimensional testing to account for the unique mechanisms. These diverse phenomena will lead to a diverse website mechanism, such as the final plan constructed by such a process will accommodate around 4,096 combinations in total. In order to check the validity of acquired results, they can be eluded to almost half a million visits on the website at a minimum with the assumed one hundred visits per combination. This large scale expectation makes it difficult to conduct an analysis practically. However, this experiment is only in its initial stage, and the changes will not be permanently applied to the main website's color and content.

Table 1 Set of variables and components for first level optimization

Id	Variable	Element	OFF	ON	Remarks
1	H1	Header 1	−1	1	First main header on the test page with the biggest font
2	H2	Header 2	−1	1	Second header on the page with middle-sized font
3	H3	Header 3	−1	1	Third header on the test page with small-sized font
4	S	Secure logo	−1	1	Additional image with secure logo
5	C	Number of users	−1	1	Counter for the number of users logged into the system
6	B	Prize related element	−1	1	Information about additional bonuses
7	F	Free logo	−1	1	Special graphical element showing that system usage, which is free
8	I2	Additional image	−1	1	Additional image near to header showing some system characteristics
9	V	Buttons localisation	−1	1	Vertical or horizontal locations of registration buttons.
10	I1	Image of text	−1	1	Switch 1 for an image or −1 for a text
11	RR	Rectangular or round shape	−1	1	Switch 1 for round elements −1 for rectangular

Figure 1 exemplifies the edgy variants that will be generated using the foundation of such plan. The main purpose of it is to identify the most detailed ways of selecting the appropriate text and content for a website.

In an attempt to reduce the possibility of delinquency in the matter, a methodology of object structure was generated for developing unique stages on the grounds of an elimination model being utilized in factor analysis. These methods are based on the notion to produce practically applicable plans. The research work of C.K. Bayen, and I.B. Rubin developed a plan along with a fractional elements [1] and S.N. Morgan proposed other approaches [13], which were used to support the planning of this experiment in order to determine the extremes of many variables within the formula:

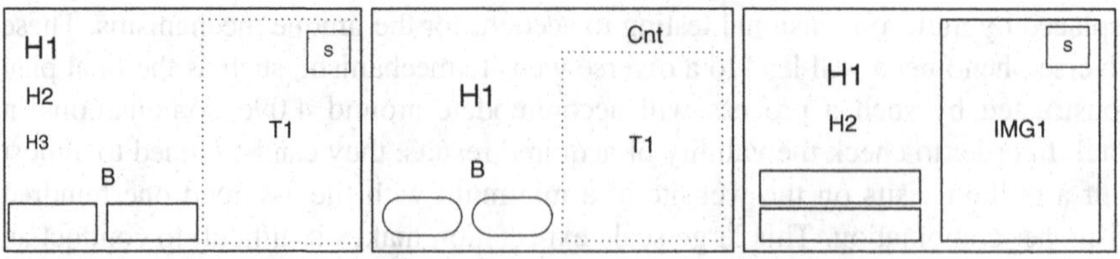

Fig. 1 Sample variants of website layout based on a set of devised variables

$$x = f(u_1, u_2, \ldots, u_n) \tag{3}$$

The methodology used was based on the assumption that f(u) is an unidentified constant and has only one extreme value. Variables u_k have values on two layers, such as us0 + Δus, us0 − Δus and unknown values have generated the unique points, for example, $u_{1,0}$, $u_{2,0}$, ..., $u_{s,0,}$ which take on the following form:

$$\hat{x} = b_0 + b_1u_1 + \cdots + b_su_s + b_{11}u_1^2 + \cdots + b_{ss}u_s^2 + + b_{12}u_1u_2 + \cdots + b_{s-1,s}u_{s-1}u_s \tag{4}$$

In which b_0, b_1,...,b_k, b_{11},...,b_{ss}, b_{12},...,b_{s-1}, b_s are taken up as factors to be examined. Whereas, the methodology adopted by Plackett and Burman involves the use of an entire lot of elements in order to determine the actual influence [x7x]. Table 2 shows the diversity in the experiment plan, which was reduced to 12 phases.

The website developed during this process was mainly used during the advertising campaigns. Users selected any of the 12 prospective options, and the column of CR represents the proportional factor of the user's conversion. The data generated was processed further. Data represented in Fig. 2 was generated by applying Pareto testing, and the data obtained was further organized in an orderly manner from the biggest to the smallest value. Bigger numbers reveal the inauspicious actions that were taken. Positive influence was added by the incorporation of graphical values in the data. Most stooling of the headers had the variant presentation of buttons, which were calls to action and the round buttons variable, RR. On the basis of the calculations gathered, a negative effect was noticed with respect to the security of the elements. This was due to the apparent decrease in registrations. However, a positive impact was conceived later by the application of text concerning the charge for service, Variable F. The addition of 12 graphical elements added a negative influence on the generated results, but not on a broader perspective understanding of the phenomenon. The main elements having the most positive

Table 2 Two-value reduced plan of experiment

Id	H_1	H_2	H_3	S	C	B	F	I_2	V	I_1	RR	CR
1	−1.00	1.00	1.00	−1.00	1.00	−1.00	−1.00	−1.00	1.00	1.00	1.00	2.20
2	−1.00	1.00	1.00	1.00	−1.00	1.00	1.00	−1.00	1.00	−1.00	−1.00	2.79
3	1.00	1.00	−1.00	1.00	1.00	−1.00	1.00	−1.00	−1.00	−1.00	1.00	3.79
4	−1.00	1.00	−1.00	−1.00	−1.00	1.00	1.00	1.00	−1.00	1.00	1.00	4.17
5	1.00	1.00	−1.00	1.00	−1.00	−1.00	−1.00	1.00	1.00	1.00	−1.00	4.07
6	−1.00	−1.00	−1.00	−1.00	−1.00	−1.00	−1.00	−1.00	−1.00	−1.00	−1.00	2.49
7	1.00	−1.00	−1.00	−1.00	1.00	1.00	1.00	−1.00	1.00	1.00	−1.00	5.34
8	1.00	−1.00	1.00	−1.00	−1.00	−1.00	1.00	1.00	1.00	−1.00	1.00	4.83
9	−1.00	−1.00	1.00	1.00	1.00	−1.00	1.00	1.00	−1.00	1.00	−1.00	4.07
10	1.00	−1.00	1.00	1.00	−1.00	1.00	−1.00	−1.00	−1.00	1.00	1.00	3.94
11	−1.00	−1.00	−1.00	1.00	1.00	1.00	−1.00	1.00	1.00	−1.00	1.00	3.78
12	1.00	1.00	1.00	−1.00	1.00	1.00	−1.00	1.00	−1.00	−1.00	−1.00	3.82

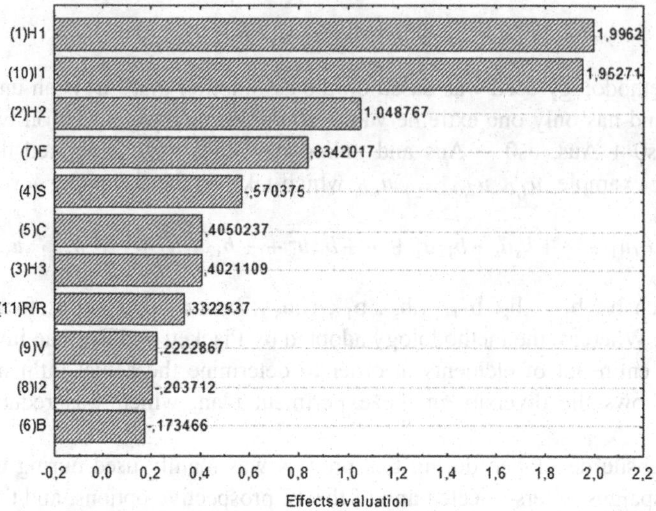

Fig. 2 Pareto charts for factors

output and influence were presented as R (Fig. 3). After further experimentation, we will estimate the probable influence on the final effect (Fig. 4), and the scope of interaction will be measured using interactive charts (Fig. 5). Figure 5 highlights the interactive charts of the elements as H1 and H2 and their influence on the system of productivity.

Fig. 3 Approximate values for main factors and output

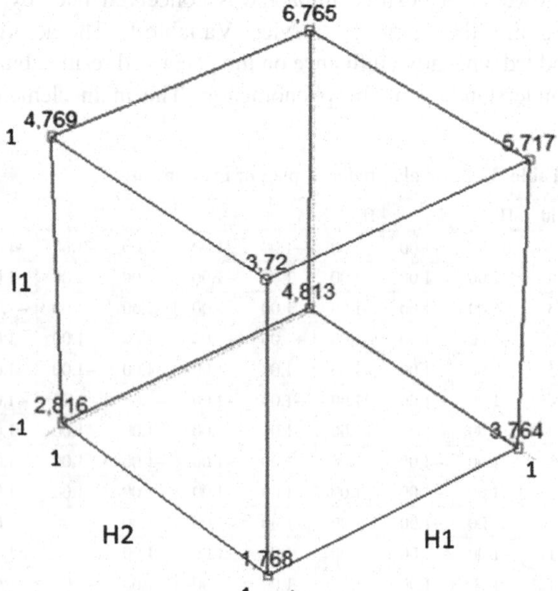

Fig. 4 Probability chart for effects

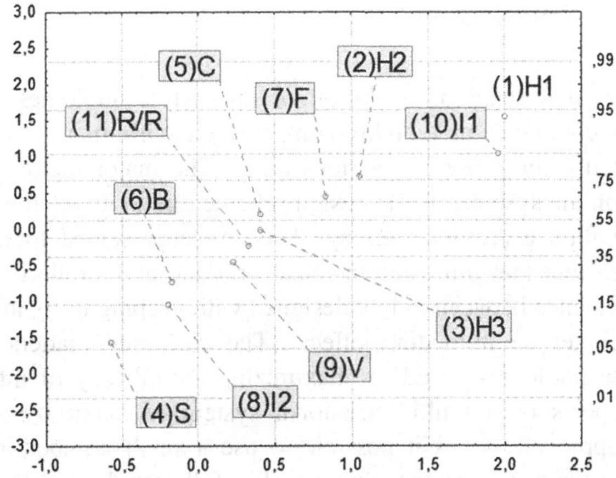

Fig. 5 Marginal average values

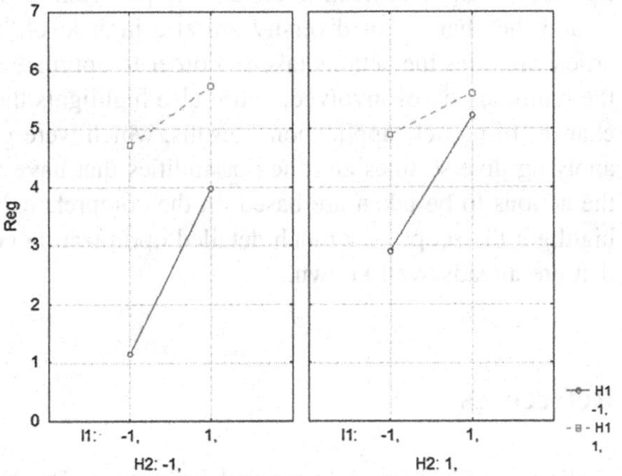

The mutual incident of H2 and I1 is turned on. This maybe a result of the user' perception after the addition of the graphical figures which minimize the effectiveness of H1-H3 type. All these elements will proceed further in the next phase of the system. Depending on the plan's execution and the addition of more testing and text diversity, testing can move to central plans. However, such processes will add more direct and indirect expenses, but in the long run, they will be not an issue. This fact should also be kept in mind that after a specific timeline, the effectiveness will also be affected as the standard of high effectiveness can lose its likeability by users.

5 Summary

In this stage of evolution of internet technologies and the immense focus on e-commerce as a highly competitive environment has led to decision support systems being used in various areas. The major factor, which affects the impact of online systems, is the design scheme due to its relationship to the results generated within a provided setting. The article presented focused primarily on the major factors that influence the final product, and similarly on the efforts to generate a balance between policy designed with keeping users in mind and those generated to center on immediate effects. The discussed matters also revolve around major technologies based on maximizing the efficacy of internet websites and the prospects of a multidimensional system to maximize performance. The proposed approach makes it possible to use a small number of variants selected on some grounds to test the efficacy of a full test. For instance, taking a complete plan reveals the substantiation of all the possible arrangements of the input figures, but the results can be difficult to understand particularly if there are huge values to add in and the chances of diversity are at a high level. The system presented in this article signifies the actions taken in order to optimize a website to full capacity and the multiple phases involved, and it also highlights the methods which do have the chances of particle application. Results, which were generated, reveal the utility of applying diverse lines and the possibilities that have implications for a website. If the actions to be taken are based on the comprehensive study of facts, then it will highlight the scope of a much detailed optimization course instead of explanations that are already well known.

References

1. Bayne, C.K., Rubin, I.B.: Practical Experimental Designs and Optimization Methods for Chemists. VCH Publisher, New York (1986)
2. Birgelen, M., Martin, G.M., Dolen, W.M.: Effectiveness of corporate employment web sites: How content and form influence intentions to apply. Int. J. Manpower **298**(8), 731–751 (2008)
3. Chairman-Hewett, T.T.: ACM SIGCHI Curricula for Human-Computer Interaction (1992)
4. Galitz, W.O.: The Essential Guide to User Interface Design: an Introduction to GUI Design Principles and Techniques. Wiley, Indianapolis (2007)
5. Goyal, P., Goyal, N., Gupta, A., Rahul, T.S.: Designing self-adaptive websites using online hotlink assignment algorithm. MoMM 579–583 (2009)
6. Hauser, J.C.R., Urban, G.L., Liberali, G., Braun, M.: Website morphing. Mark. Sci. **28**(2), 202–223 (2009)
7. Jankowski, J., Ziemba, P., Wątróbski, J., Kazienko, P.: Towards the tradeoff between online marketing resources exploitation and the user experience with the use of eye tracking. LNCS **9621**, 330–343 (2016)
8. Jansen, B.J.: The comparative effectiveness of sponsored and nonsponsored links for Web e-commerce queries. ACM Trans. Web **1**(1) (2007)

9. Jianming, H.E.: A social network-based recommender system. Ph.D. Dissertation. University of California at Los Angeles, Los Angeles, CA, USA. Advisor(s) Wesley W. Chu. AAI3437557 (2010)
10. Kaplanidou, K., Vogt, C.H.: Website Evaluation—Terminology and Measurement. http://www.michigan.org (2015). Accessed Aug 2015
11. Kelly, S.: Determining Effectiveness of Websites by Refining Transaction-Oriented Views through Supervised Competitive Clustering, Web Effectiveness Research Group, The University of Georgia (2008)
12. Krug, S.: Don't make me think: a common sense approach to web usability. Pearson Educ. India (2005)
13. Morgan, S.N.: Experimental Design: A Chemometric Approach. Elsevier Science Publishers B.V, Amsterdam (1993)
14. Morville, P., Rosenfeld, L.: Information Architecture for the World Wide Web. O'Reilly, Sebastopol (2006)
15. Nanou, T., Lekakos, G., Fouskas, K.: The effects of recommendations' presentation on persuasion and satisfaction in a movie recommender system. Multimedia Syst. **16**(4–5), 219–230 (2010)
16. Nielsen, J., Loranger, H.: Prioritizing Web Usability. New Riders, Berkeley (2006)
17. Scharl, A., Wober, W., Bauer, C.: An integrated approach to measure web site effectiveness in the european hotel industry. Inf. Technol. Tourism **6**, 257–271 (2008)
18. Sicilia, M., Ruiz, S.: The role of flow in web site effectiveness. J. Interact. Adv. **8**(1) (2007)
19. Tam, K.Y., Ho, S.Y.: Web personalization as a persuasion strategy: an elaboration likelihood model perspective. Inf. Syst. Res. **16**(3), 271–291 (2005)
20. Welch, E.W., Pandey, S.: Multiple measures of website effectiveness and their association with service quality in health and human service agencies. In: Proceedings of the 40th Hawaii International Conference on System Sciences. Hawaii (2007)
21. Ziemba, P., Jankowski, J., Wątróbski, J., Becker, J.: Knowledge management in website quality evaluation domain. LNCS **9330**, 75–85 (2015)
22. Ziemba, P., Jankowski, J., Wątróbski, J., Piwowarski, M.: Web projects evaluation using the method of significant website assessment criteria detection transactions on computational collective intelligence XXII. LNCS **9655**, 167–188 (2016)
23. Ziemba, P., Jankowski, J., Wątróbski, J., Wolski, W., Becker J.: Integration of domain ontologies in the repository of website evaluation methods. In: Proceedings of FedCSIS'2015, pp. 1585–1595 (2015). doi:http://dx.doi.org/10.15439/2015F297

How to Manage Keys and Reconfiguration in WSNs Exploiting SRAM Based PUFs

Domenico Amelino, Mario Barbareschi, Ermanno Battista and Antonino Mazzeo

Abstract A wide spectrum of security challenges were arose by Wireless Sensor Network (WSN) architectures and common security techniques used in traditional networks are impractical. In particular, being the sensor nodes often deployed in unattended areas, physical attacks are possible and have to be taken into account during the architecture design. Whenever an attacker enters in possession of a node, he/she can jeopardize the network by extracting cryptographic keys used for secure communication. Moreover, an attacker can also try to brute force the keys, hence they should be fully random and hard to guess. In this paper, we propose a novel solution based on generating keys from unique physical characteristics of a node integrated circuit without requiring additional hardware compared to common WSN node architectures. To this aim, we exploit the Static Random Access Memory based Physically Unclonable Functions and we show their applicability to the WSN by implementing a working prototype based on the STM32F4 microcontroller.

Keywords Wireless Sensor Network · Physically Unclonable Function · Static Random Access Memory · Reconfiguration

D. Amelino · M. Barbareschi (✉) · E. Battista · A. Mazzeo
DIETI - Department of Electrical Engineering and Information Technologies, University of Naples Federico II, Via Claudio 21, 80125 Naples, Italy
e-mail: mario.barbareschi@unina.it

D. Amelino
e-mail: domenico.amelino@unina.it

E. Battista
e-mail: ermanno.battista@unina.it

A. Mazzeo
e-mail: antonino.mazzeo@unina.it

M. Barbareschi · E. Battista · A. Mazzeo
CeRICT scrl - Centro Regionale Information Communication Technology, Naples, Italy

G. De Pietro et al. (eds.), *Intelligent Interactive Multimedia Systems and Services 2016*, Smart Innovation, Systems and Technologies 55,
DOI 10.1007/978-3-319-39345-2_10

1 Introduction

Wireless Sensor Networks (WSNs) are commonly devoted to collect environmental data which need to be transferred to a collection/central node for further processing and analysis. Security has a central role for WSN applications, data have to be transferred to central nodes is a secure way so as not be altered or crafted by malicious attackers. Due to the inherent wireless nature of WSN nodes, the computational resources available are scarce and security mechanisms must introduce a negligible overhead [2, 3, 7]. Integrity, confidentiality and authentication are commonly offered by means of symmetric cryptography where the data payload is encrypted/decrypted only by pairs of sensing node—collector. In such context, multiple keys are usually enforced so as to create a dedicated channel between a sensing node and the collector: each node has a key for encrypting/decrypting messages to/from the central node. The central node has to know the symmetric key for each sensing node.

This solution is actually not so secure if the node is subject to physical capture by malicious actors. Whenever the node is captured, the key can be extracted from its memory, and this allows the attacker to impersonate a legitimate node because he/she can gather all the information and security mechanisms stored in the node. To deal with this issue, the technology used for the key storage must be hardened to make worthless any access attempt. Moreover, involved keys have to characterized by a great randomness and hard to guess.

A novel approach is based on the generation of security keys from device intrinsic physical characteristics (like humans fingerprint) rather than storing this information in a non-volatile memory (NVM). The great advantage of intrinsic physical characteristics is their strict coupling with the device. Basing on this concept the literature defined the silicon Physically Unclonable Functions (PUFs) primitives, which compute a string of bits based on measurements of a physical parameter of the integrated circuit (IC). This string is know as *response* and it is intrinsically random and unique so as to be used to identify a device.

For these reasons, in this paper, we propose a key *generation* mechanism based on exploiting the PUF security properties. In particular, we describe the kind of PUF that are suitable for the WSN domain, namely the Static Random Access Memory (SRAM) PUF, we detail the process to generate a secret key from PUF responses and then we provide a working prototype deployed on a commercial microcontroller widely used in WSNs. Moreover, we discuss two case study scenarios related to the secure remote reconfiguration of WSN nodes and the symmetric key renewal process. Both of them are based on our PUF architecture capable of offering a secure, anti-tamper and trustworthy perimeter for the user application execution without requiring additional hardware resources.

2 Related Work

PUFs have been recently adopted for the physical protection of embedded devices, since they are able to guarantee attractive security properties in many application domains. As for WSNs, research efforts have been focusing on the development of novel authentication mechanisms. A mechanism to prevent nodes cloning is presented in [23]. Authors devise a mutual authentication scheme which relies on the PUF challenge/response mechanism and they demonstrate the resiliency of the approach against clone attack, replay attack, eavesdropping and tampering attempts. Similarly, in [15] a mutual authentication protocol is tailored for Wireless Body Sensor Networks, by adopting PUFs and one-way hashing functions. Those solutions, although presenting actual schemes based on PUFs and intended for WSNs, are not actualized by a concrete realization of a real PUF architecture.

Conversely, Liu et al. in [16] exploit PUFs for a trustworthy key generation. The technology which the PUF circuit is based on is the SDRAM type 3 (DDR3): indeed, they use the decay signature of the memory cells with one transistor/one capacitor structure. Commonly, wireless sensor nodes are resource constrained devices, and for the energy efficiency they do not employ power-consuming memory technologies, such as the DDR one.

On the contrary, this paper introduces the adoption of an SRAM-based PUF for WSNs taking into account that most of micro-controllers used for embedded nodes are inherently equipped with an SRAM. This allows for a security improvement requiring neither hardware substitution nor additional hardware resources, thus making it widely applicable. Our approach offers the same security guarantees of previously presented works and further improves their applicability by showing real use scenarios for node remote reconfiguration and cryptographic key renewal.

The remote reconfiguration is part of the Moving Target Defense (MTD) security approach [1] and we advance it by ensuring trustworthiness on the execution environment of the node. Modern reconfiguration schemes for WSNs, such as SIREN [8], tend to pre-load several application images into an external NVM. Due to the unattended nature of WSNs, nodes are subject to physical capture and, hence, cloning. Therefore, solely employing reconfiguration mechanisms is not completely effective.

It is worth to notice that some attempts to ensure a trustworthy software execution in WSNs have already been presented in [7, 14]. Though, those solutions require special hardware, such as the TPM-enabled hardware, in order to provide a root-of-trust and boot the node through a secure procedure. This is obviously costly due to the technological requirements (for instance, the chip Fritz or tamper resistant hardware) [11], and thus this limits their employment. With out approach we offer an analogous secure boot procedure without additional hardware cost.

3 Background

The opportunity of extracting physical characteristics from fabric induced variability for integrated circuits (ICs) has been representing a momentous breakthrough for the security of electronic devices. Silicon PUFs are circuits able to extract such physically imprinted characteristics, producing binary strings, namely *responses* that can be envisioned to be identifiers of ICs pretty much like the human fingerprints or the DNA. Being inherently random, manufacturing variations imprint random and unique physical effects. Moreover, any physical detail or parameter is hidden, i.e. cannot be predicted, and generated by uncontrollable process. Thus, PUFs responses are unique, unclonable and unpredictable. Moreover, PUF circuits are tamper-evident because any tamper attempt will alter the physical characteristics exploited by the PUF dramatically changing the PUF response.

PUFs can be categorized by considering their operational mechanism. Actually, some PUF architectures exploit delay measurements, such as the Arbiter PUF, Ring Oscillator PUF [21] and the Anderson PUF [4]. Other architectures exploit the pattern generated on the start-up of memory cells, such as the SRAM PUF [13] or the STT-MRAM PUF [22].

Interestingly, the SRAM PUF can be potentially exploited as secure primitive for a large amount of devices, as it is a very spread non-volatile memory technology. The mechanism behind the SRAM PUF is the value assumed by each SRAM cell when it is being powered-up (cold start). As shown in Fig. 1, the cell is realized through two symmetric halves, which are two cross-coupled inverters. They generate a feedback which makes the initial voltage point unstable and, hence, forces each cell to move towards one of the two stable voltage values (logic-0 or logic-1). Indeed, the feedback amplifies the differences of voltages between the two symmetrical halves caused by manufacturing variability.

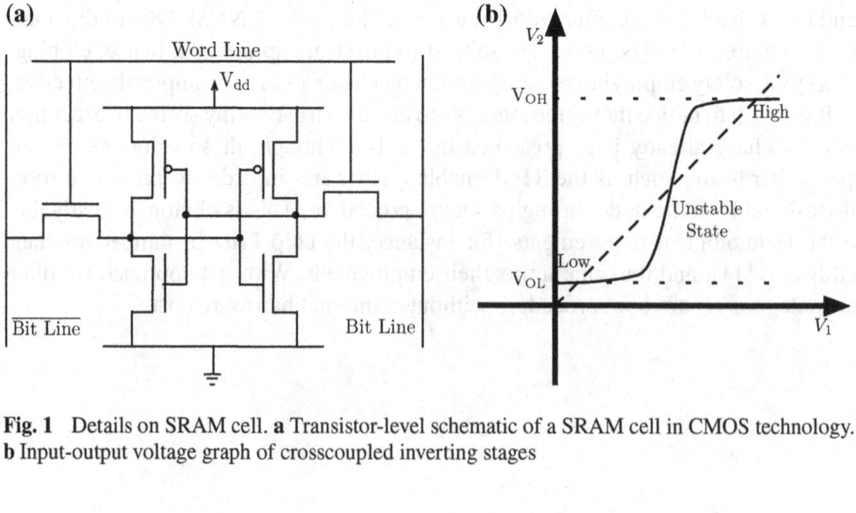

Fig. 1 Details on SRAM cell. **a** Transistor-level schematic of a SRAM cell in CMOS technology. **b** Input-output voltage graph of crosscoupled inverting stages

Ideally, a PUF should always be able to exactly reproduce the same output when queried. However, variations of electrical characteristics caused by environmental conditions, such as temperature variations or power supply instability, might change some response bits [5, 19]. Indeed, as shown in [6], some SRAM cells are less stable than others and more prone to be affected by external conditions.

In order to use the PUF response for cryptographic operations, it is required to have stable responses. The literature introduced post-processing techniques able to produce binary strings, eligible for cryptographic operations, from unstable PUF responses. Majority voting is a post-processing technique based on multiple measurements over a PUF circuit [17]. The technique can be accomplished by averaging *N* sequential measurements on a PUF (Temporal Majority Voting) or N simultaneous measurements on different PUFs (Spatial Majority Voting). Majority voting requires a threshold to establish the outcome of voting. Whenever the measurements to vote accumulates around the threshold, the outcome is not reliable. Conversely, the fuzzy extractor technique involves only one measurement and exploits an error correction code (ECC) in order to retrieve stable binary strings from noisy responses. In this way, PUF responses, as shown in the next Section, can be used as key provider for cryptographic-schemes.

4 Extracting Keys from Embedded SRAMs

The literature provides a great number of scientific paper, reporting experimental measurements on different SRAM circuits [6, 10]. PUFs guarantee random, unclonable and unpredictable responses these properties fulfill what is required for good cryptographic keys. Though, in order to exploit these benefits, the PUF mechanism has to be enriched with a post-processing technique.

Among recovering techniques previously illustrated, we adopt the fuzzy extractor approach. The fuzzy extractor is a recovery scheme composed of two primitives: (i) information reconciliation, which removes the error generated by the noise, and (ii) privacy amplification, which enhances the information bits entropy (randomness). Both the primitives rely on Helper Data, which is generated during the **enrollment phase**, described in Fig. 2a. A reference PUF response R is extracted and a secret symbol C_s is randomly obtained. Thus, Helper Data W is generated by a *xor* function, while the privacy amplification outputs the key K. To ensure the secrecy of K, this phase must be carried out within a trusted environment, before issuing the device, while Helper Data can be publicly available because it does not disclose secrets exploitable by attackers. Helper Data is used during the reconstruction phase, detailed in Fig. 2b, where K is reconstructed from the response R', which is the PUF response R of the device in the operating environment, thus may be noisy, and hence different. The **reconstruction phase** can be accomplished by the device, which embeds the PUF, at any time by using the Helper Data W.

Moreover, the fuzzy extractor scheme contains an ECC and a cryptographic hash function. The ECC is an algorithm developed to obtain a binary string such that a

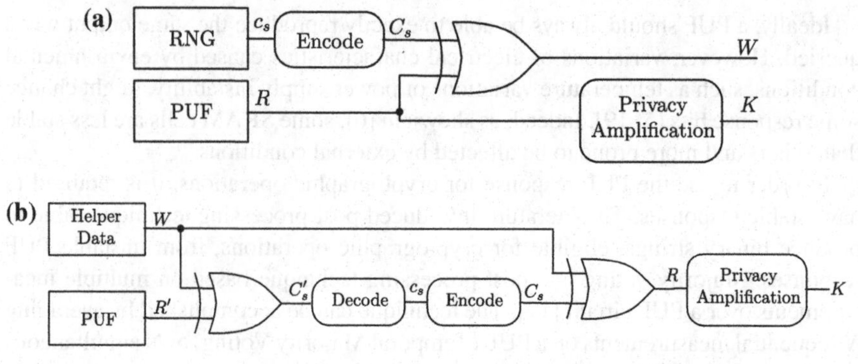

Fig. 2 The fuzzy extractor scheme. **a** Steps of the enrollment phase which generate the key and Helper Data. **b** Steps of the key reconstruction phase which regenerate the same key exploiting Helper Data

certain amount of errors caused by the external environment can be detected and corrected. Figure 2 shows the ECC primitives, namely the encode and decode functions, which are used in both the enrollment and the key reconstruction phases.

The fuzzy extraction design involves the crucial configuration of the ECC scheme. Indeed, an ECC is able to recover a maximum number of errors in the binary string that has to be processed. This amount is obviously correlated with the PUF stability. The maximum number of error t is determined by considering the bit error probability p_b, which specifies the probability that a transmitted information bit is received altered. The noise of PUF responses can be modeled as a binary symmetric channel and the error probability that a block of N bits has more than t error is:

$$P_{block} = 1 - \sum_{i=0}^{t} \binom{N}{i} p_b^i (1 - p_b)^{N-i} \tag{1}$$

Thus, P_{block} represents the failure rate of the decoder correction procedure of an N bits block.

The keys, extracted from PUF responses, have to be hard to guess, hence characterized by a high value of entropy. The entropy is related to the number of independent bits of an information source that can be generated. The entropy can be estimated by calculating the min-entropy, a lower bound of entropy that represents the worst-case measure of uncertainty for a random variable [12].

Let p_{max} be the maximum value between the occurrence probabilities of 0 and 1 bits. The min-entropy can be calculated as $H_{min} = -\log_2(p_{max})$. In particular, as Guajardo et al. assumed in [13], each bit of SRAM is independent, hence the min-entropy for N bits of SRAM can be calculated as sum of min-entropy for each bit:

$$(H_{min})_{total} = \sum_{i=1}^{N} -\log_2(p_{max_i}) \tag{2}$$

The value p_{max_i} can be derived by sampling SRAM bits from different devices. It is possible to calculate the average min-entropy as $\overline{H}_{min} = (H_{min})_{total}/B$, where B is the size of the memory block. The average min-entropy allows to determine the amount of source bits required to derive a significantly random key with a given length. Indeed, for a key of L bits, at least $\lceil L/((\overline{H}_{min})_{total} \rceil$ bits have to be extracted from the PUF.

For instance, to obtain **a key of 128 bits** from a SRAM PUF with an entropy of 0.8, the PUF has to provide **a response of 160 bits**.

5 Case Studies

In order to prove the feasibility for SRAM PUF to be employed in WSNs, this Section details a real implementation of a node based on a commercial microcontroller, namely the STM32F4 device, equipped with the previously illustrated PUF scheme. In particular, we choose the STM32F4 family based on ARM Cortex-M 32-bit microcontrollers, for its widespread availability and suitability for WSN applications. We use the STM32F407VGT6 which includes a ARM Cortex-M4F, 1 MB of NVM flash and 192 KB of SRAM.

First of all, the approach is based on the deployment of a secure bootloader, which represents the root-of-trust for applications that will run on the device. The bootloader has to be stored in the NVM, residing in a secure perimeter: the microcontroller has to be configured to deny any external access to memories. Furthermore, the bootloader is the only handler of the SRAM PUF and the only key manager for the node. This is required to ensure the correct extraction of the PUF response from the pristine state of the SRAM start-up pattern.

The STM32F4 microcontroller offers a memory protection mechanism to secure the perimeter with different memories read/write restrictions [20]. In particular, we adopt the level 2 read protection, which forbids all external accesses to flash sectors and also disables every debug interface, while level 2 write protection prevents flash memory overwrites. A joint use of these two approaches protects the bootloader ensuring its integrity, i.e. trustability.

The role of the bootloader is to load user application images which reside outside the secure perimeter and the PUF mechanism is used to extend trustability to them. Indeed, the bootloader firstly computes the PUF-based key using the fuzzy extraction procedure described in the previous Section. Then, the key is used to decrypt or verify the user application image. In case of success, the bootloader prepares to run the user application: the bootloader cleans the SRAM memory and loads the user software.

In our implementation, we use 128-bits key and a Reed Muller ECC scheme,[1] which requires a PUF response of 2816 bits. The primitives for decrypting and the

[1]The Reed-Muller ECC has a (128,8,63) configuration, which has probability error of 4.321086e-09.

privacy amplification are both implemented with Speck [9], a lightweight block cipher. In particular, in order to provide a hash digest from the Speck encryption primitive, we employed the Davis-Meyer scheme [18].

The bootloader so far described is perfectly suitable for WSNs nodes thanks to its significantly small footprint: the complete bootloader requires just 16 KB, equivalent to the 1.5 % of the total available flash memory of the STM32F407VGT6 device. Moreover, the time overhead introduced by the PUF extraction is about 104 ms.

5.1 Secure Reconfiguration

Being able to reconfigure a node in a WSN is a desirable feature so as to modify over time the specific tasks carried out by a node. Previous works, like in [8], have shown this advantage and the mechanisms to efficiently achieve remote reconfiguration A secure remote reconfiguration can be easily supported by our PUF based mechanism. Traditionally, to reconfigure a WSN node, a set of application images are preloaded once into an external memory at the deployment phase. During the WSN life cycle, as shown by SIREN, it is possible to implement a set of policies to reconfigure a node. The reconfiguration is based on the following phases: (i) choosing the next application to run from the image set; (ii) saving all the information required in the next application in a NVM; (iii) prepare the bootloader with a reference to the chosen next application; (iv) reboot the device. Once the bootloader starts, it will retrieve the pointer to the application to run from NVM and proceed to its loading.

By using the PUF-secured boot process described above, we can ensure that the application images are protected by means of cryptography. In the deployment phase, all the application images have to be encrypted with the device's PUF key extracted in the enrollment phase (Sect. 4) and the loaded in the external NVM that can be outside the secure perimeter. When the device boots, it generates the key to decrypt the application image by applying the reconstruction phase (Sect. 4), decrypts the application image and loads it in RAM, actually starting its execution.

On STM32F407VGT6 microcontroller, the execution time required by the bootloader to extract the key and build the root-of-trust is about *900* ms for an image file of 10 KB.

5.2 Key Renewal

A good security practice is to change the symmetric keys involved in communication over time. To avoid the transmission of keys on non-secure or potentially non-secure channel, a set of keys is pre-loaded on each node of the WSN. By applying policies to trigger the enforcement of another key in the set (e.g. max number of messages per key, time expiration, explicit key change message), they communication key is renewed.

Our PUF based solution, offers a key renewal procedure that generates high-entropy keys from the SRAM to be used by a trusted application running on the system. During device power-up, the boot loader can generate one or more keys by applying reconstruction phase (Sect. 4) on different SRAM areas and storing them into a specific memory location. At each start-up, the key given to the application can be renewed by executing the reconstruction phase on a different SRAM area, that can be configured/selected by using persistent information stored in an external NVM. It follows that the enrollment phase (Sect. 4) had to be accomplished on same SRAM areas in order to provide symmetric keys to the central node (offline procedure executed once at WSN deployments time). It is guaranteed that the software application running on the system is trusted (loaded with a secure boot procedure) and is able to retrieve the generated keys from a memory location conveniently indicated by the bootloader.

The key renewal procedure requires a device restart because it can only be executed on a pristine state of the SRAM start-up pattern. When the application requires to renew the communication key, the device can cold reboot itself.[2] The only penalty added is the actual time spent for reboot that can be considered negligible in most of the WSN sensing applications.

6 Conclusion

The tight constraints and hostile environmental working conditions of WSNs make the conventional computer security techniques inadequate for embedded sensor nodes. The main challenge is related to the unattended nature of the network, which exposes what is installed in the node memories to attacks, including secure protocols and their cryptographic keys. The key management has to guarantee strong keys and a secure perimeter where use them. In this paper, we moved in that direction, providing a key management based on the adoption of an SRAM PUF as key storage and provider. By exploiting the start-up pattern of an SRAM, the fuzzy extractor technique is able to generate bit strings which are eligible to be employed as cryptographic keys. We showed main steps involved in the design of the PUF and two meaningful case studies, focusing on the node reconfiguration and on the key renewal mechanism. The main advantages of our approach are listed below:

- Generated keys are tight coupled with physical parameters of SRAM cells, cannot be cloned and are hard to guess;
- PUF generated keys inherit the randomness from the fabric manufacturing variability and do not require an installation phase as they are retrieved on demand from physical parameters.
- The SRAM Based PUF does not require additional hardware resources and involved software procedures have to be executed once at the start-up.
- The approach requires only an SRAM accessible in a pristine state.

[2]The device goes into standby mode to power down the SRAM, before rebooting.

- The keys are generated by the SRAM PUF on demand and are not stored in an NVM; moreover, they are managed in a secure perimeter.
- The adoption of a PUF opens the possibility to make the execution environment trustworthy.

Acknowledgments The list of Authors is in alphabetical order. The corresponding authors are Mario Barbareschi and Ermanno Battista. This research work was partially supported by CeRICT for the project NEMBO—PONPE_00159.

References

1. Albanese, M., Battista, E., Jajodia, S., Casola, V.: Manipulating the attacker's view of a system's attack surface. In: 2014 IEEE Conference on Communications and Network Security (CNS), pp. 472–480. IEEE (2014)
2. Amato, F., Chianese, A., Moscato, V., Picariello, A., Sperli, G.: Snops: A Smart Environment for Cultural Heritage Applications, pp. 49–56 (2012)
3. Amato, F., Mazzeo, A., Moscato, V., Picariello, A.: Exploiting cloud technologies and context information for recommending touristic paths. Stud. Comput. Intell. **511**, 281–287 (2014)
4. Anderson, J.H.: A puf design for secure fpga-based embedded systems. In: Proceedings of Asia and South Pacific Design Automation Conference, pp. 1–6. IEEE Press (2010)
5. Barbareschi, M., Bagnasco, P., Mazzeo, A.: Supply voltage variation impact on anderson puf quality. In: 2015 10th International Conference on Design & Technology of Integrated Systems in Nanoscale Era (DTIS), pp. 1–6. IEEE (2015)
6. Barbareschi, M., Battista, E., Mazzeo, A., Mazzocca, N.: Testing 90 nm microcontroller sram puf quality. In: 2015 10th International Conference on Design & Technology of Integrated Systems in Nanoscale Era (DTIS), pp. 1–6. IEEE (2015)
7. Barbareschi, M., Battista, E., Mazzeo, A., Venkatesan, S.: Advancing wsn physical security adopting tpm-based architectures. In: 2014 IEEE 15th International Conference on Information Reuse and Integration (IRI), pp. 394–399. IEEE (2014)
8. Battista, E., Casola, V., Mazzeo, A., Mazzocca, N.: Siren: a feasible moving target defence framework for securing resource-constrained embedded nodes. Int. J. Crit. Comput.-Based Syst. **4**(4), 374–392 (2013)
9. Beaulieu, R., Shors, D., Smith, J., Treatman-Clark, S., Weeks, B., Wingers, L.: The simon and speck lightweight block ciphers. In: Proceedings of the 52nd Annual Design Automation Conference, p. 175. ACM (2015)
10. Böhm, C., Hofer, M., Pribyl, W.: A microcontroller sram-puf. In: 2011 5th International Conference on Network and System Security (NSS), pp. 269–273. IEEE (2011)
11. Cilardo, A., Barbareschi, M., Mazzeo, A.: Secure distribution infrastructure for hardware digital contents. CDT, IET **8**(6), 300–310 (2014)
12. Claes, M., van der Leest, V., Braeken, A.: Comparison of sram and ff puf in 65nm technology. In: Information Security Technology for Applications, pp. 47–64. Springer (2011)
13. Guajardo, J., Kumar, S.S., Schrijen, G.J., Tuyls, P.: FPGA Intrinsic PUFs and Their Use for IP Protection. Springer (2007)
14. Hu, W., Tan, H., Corke, P., Shih, W.C., Jha, S.: Toward trusted wireless sensor networks. ACM Trans. Sens. Netw. (TOSN) **7**(1), 5 (2010)
15. Lee, Y.S., Lee, H.J., Alasaarela, E.: Mutual authentication in wireless body sensor networks (wbsn) based on physical unclonable function (puf). In: 2013 9th International Wireless Communications and Mobile Computing, pp. 1314–1318. IEEE (2013)
16. Liu, W., Zhang, Z., Li, M., Liu, Z.: A trustworthy key generation prototype based on ddr3 puf for wireless sensor networks. Sensors **14**(7), 11542–11556 (2014)

17. Maes, R., Tuyls, P., Verbauwhede, I.: Intrinsic pufs from ip-ops on reconfigurable devices. In: Proceedings of Benelux Information and System Security, Eindhoven (2008)
18. Menezes, A.J., van Oorschot, P.C., Vanstone, S.A.: Handbook of Applied Cryptography. CRC press (1996)
19. Rampon, J., Perillat, R., Torres, L., Benoit, P., Di Natale, G., Barbareschi, M.: Digital right management for ip protection. In: IEEE Computer Society Annual Symposium on VLSI (ISVLSI), 2015, pp. 200–203. IEEE (2015)
20. STMicroelectronics: RM0090 Reference Manual, 10 (2015)
21. Suh, G.E., Devadas, S.: Physical unclonable functions for device authentication and secret key generation. In: Proceedings of the 44th annual Design Automation Conference, pp. 9–14. ACM (2007)
22. Vatajelu, I., Di Natale, G., Barbareschi, M., Torres, L., Indaco, M., Prinetto, P.: Stt-mram-based puf architecture exploiting magnetic tunnel junction fabrication-induced variability. ACM J. Emerg. Technol. Comput. Syst. *12*(4) (2015)
23. Yang, K., Zheng, K., Guo, Y., Wei, D.: Puf-based node mutual authentication scheme for delay tolerant mobile sensor network. In: 2011 7th International Conference on Wireless Communications, Networking and Mobile Computing, pp. 1–4. IEEE (2011)

Fast Salient Object Detection in Non-stationary Video Sequences Based on Spatial Saliency Maps

Margarita Favorskaya and Vladimir Buryachenko

Abstract In recent years, a number of methods of salient object detection in images have been proposed in the field of computer vision. However, sometimes the shooting conditions are far from the ideal, and the unpredicted camera jitters significantly impair the quality of video sequences. In this paper, the salient objects are roughly detected from the keyframes of non-stationary video sequences with two main purposes. First, the removal of salient objects helps to estimate a motion in background more accurately. Second, a visibility of salient objects can be improved after stabilization of video sequence. In this sense, the fast generation of multi-feature approximate saliency map is required. Various fast techniques suitable to extract intensity, color, contrast, edge, angle, and symmetry features from the keyframes are discussed. Some of them are based on Gaussian pyramid decomposition. The Law's 2D convolution kernels are applied for fast estimation of texture energy contrast and texture gradient contrast in particular. The experiments show the acceptable spatial saliency maps in order to obtain good background motion model of non-stationary video sequence.

Keywords Visual saliency · Salient object · Spatial saliency map · Background detection · Motion estimation · Video stabilization

1 Introduction

The recent achievements in cognitive science, explaining a human possibility to detect visually distinctive regions in images, attract a lot of interest for segmentation and recognition tasks in computer vision. The salient object detection in video

M. Favorskaya (✉) · V. Buryachenko
Institute of Informatics and Telecommunications, Siberian State Aerospace University,
31 Krasnoyarsky Rabochy av., Krasnoyarsk 660037, Russian Federation
e-mail: favorskaya@sibsau.ru

V. Buryachenko
e-mail: buryachenko@sibsau.ru

G. De Pietro et al. (eds.), *Intelligent Interactive Multimedia Systems and Services 2016*, Smart Innovation, Systems and Technologies 55,
DOI 10.1007/978-3-319-39345-2_11

sequences is widely used in many tasks including detection of structure from motion, object surveillance, active action recognition, abnormal active actions, unmanned aerial vehicles, among others. The mentioned tasks are conventional and many researches have been implemented in these scopes. In current research, the special task—detection of foreground salient objects is discussed under assumption that a video sequence is the non-stationary sequence. Poor predicted camera shakes and jitters of held-holding video captured without using a tripod or video obtained from the moving mobile platforms in rural territories are possible. As it is followed from the cognitive theory, a human attention is concentrated on the foreground moving objects, which ought to be processed more accurate, including the compensation of unwanted shakes and jitters, filtering, and segmentation.

According to the accepted classification, there are two main approaches for salient object detection called as the top-down and bottom-up approaches. Full historical survey of their development, covering 256 publications, was represented by Borji et al. in [1]. The top-down approach requires the supervised learning and the high-level processing of semantic data in order to obtain good salient map. The bottom-up approach use the low-level processing of color distributions, relative positions, contours, color/illumination contrast in local regions of image. Also some heuristic methods as additional methods can be mentioned.

The paper is structured as follows. The reviewing of prior work in this area is situated in Sect. 2. Fast approach for generation of saliency maps is described in Sect. 3. Section 4 includes a procedure of video stabilization based on the obtained saliency map. Experimental results are presented in Sect. 5. Finally, Sect. 6 provides conclusions and gives suggestions for future work.

2 Related Work

In the spatial domain, a salient region is understood as a part of an image that stands out from its surrounding and thus captures the attention of the end-user. In the temporal domain, a salient region is generally moving part of video sequence, which attacks the attention of the end-user by color, contrast, edge information, and its temporal behavior. However, there is no universally accepted definition of visual saliency, and many researchers bring their own definitions under developed algorithms.

Most methods produce a real valued saliency map, while often an actual segmentation of a salient object is needed in a view of gray-scale or binary masks. From a viewpoint of spatial domain, all computational methods of salient object detection can be roughly categorized into three groups. The pixel-based methods analyze the low level features, which are extracted from a surrounding of a central pixel, at single or multiple scales. Hereinafter, simple clustering techniques create the salient objects. The second group supports the frequency processing by mostly applying Gabor filters. These methods are often applied, when data from multispectral bands are fused. The region-based methods perform a clustering using the region growing or the conventional region segmentation with following feature extraction from these clusters.

Itti et al. [2] were the first, who introduced a purely bottom-up model, which used an iterative spatial competition for saliency with early termination. Colors, intensity, orientations, and other features, such as motion, stereo disparity, shape from shading, etc., are the initial data for feature maps, building on centre-surround differences. The fruitful development of pixel-based approach deals with so called superpixels analysis based on the Simple Linear Iterative Clustering (SLIC), which was proposed by Achanta et al. [3]. The superpixels' growing derives by pixels' clustering in the combined 5D color and image plane space that provides the efficient compactness of elements. The original approach for salient object detection was presented by Li et al. [4], when a hypergraph of an image utilized a set of hyperedges to capture the contextual properties of pixels or regions. The hyperedge in a hypergraph means a high-order edge associated with a vertex, linking with more than two vertices.

Hou et al. [5] formulated the principles of the spectral residual approach of the amplitude spectrum based on Fourier Transform. Guo et al. [6] extended this approach by the novel multi-resolution spatio-temporal saliency detection model called phase spectrum of quaternion Fourier transform. They created the spatio-temporal saliency map of an image by such quaternion representation. Le Meur et al. [7] proposed a coherent computational approach to the modelling of the bottom-up visual attention based on the current understanding of the human visual system behavior.

A fast salient object detection algorithm based on segment analysis was represented by Zhuang et al. [8]. First, an input image is segmented. Second, a multi-scale contrast, center-surround color histogram, and color spatial distribution maps based on segments are constructed in order to describe a salient object locally, regionally, and globally. These three feature maps are combined linearly into a saliency map with empirical weights as [0.22, 0.24, 0.54] and [0.22, 0.54, 0.24] for color and gray level images, respectively. Liu et al. [9] proposed a saliency detection framework in a view of saliency tree, using global contrast, spatial sparsity, and object prior information. The analysis of a saliency tree helps to obtain the final regional saliency measures.

3 Generation of Saliency Maps

The flow-chart scheme of saliency analysis in non-stationary video sequence is depicted in Fig. 1. The main goal is to find and exclude the saliency regions from non-stationary video sequence in order to receive the reliable motion estimations of background. When the unwanted motion estimations are obtained, the salient objects will be included in the stabilized video sequence with accurate post-processing. Suppose that the keyframes of scenes are extracted by any possible way, for example, as it was proposed in [10, 11].

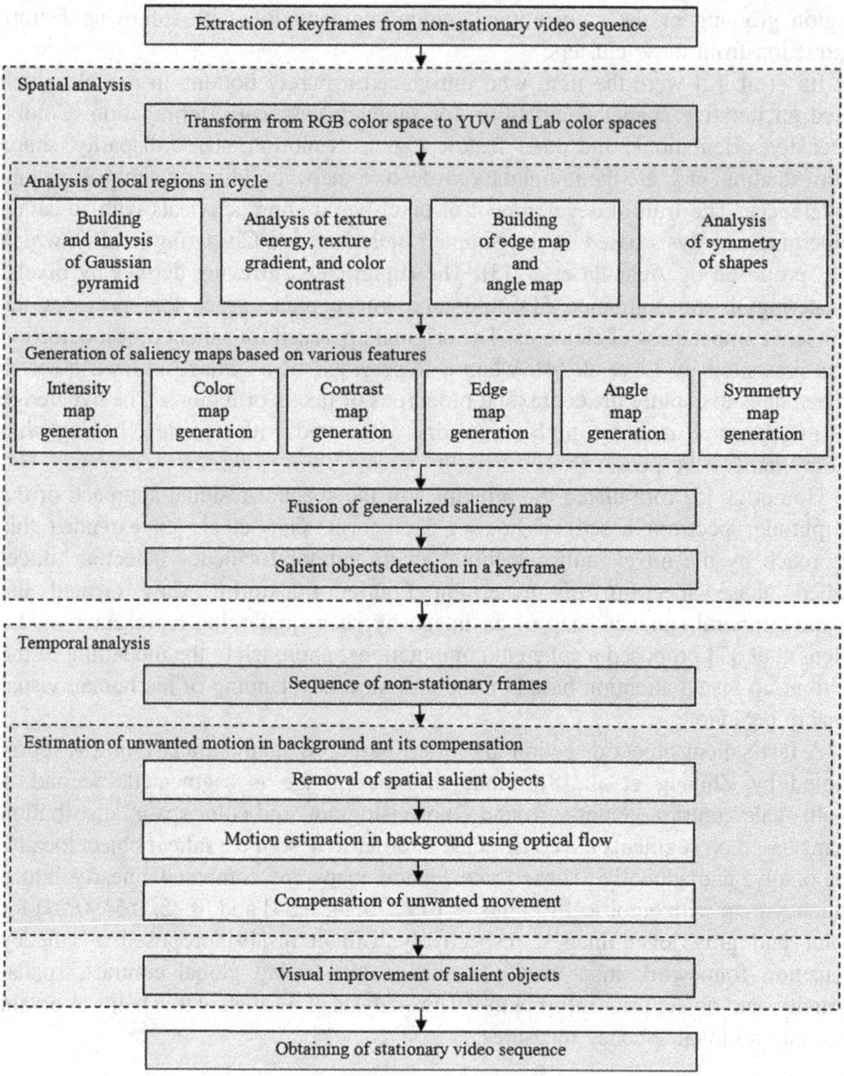

Fig. 1 Flow-chart scheme of saliency analysis on non-stationary video sequence

The fast generation of intensity and color saliency maps is discussed in Sect. 3.1, contrast saliency map is presented in Sect. 3.2, while the remained maps are described in Sect. 3.3. Section 3.4 provides a linear fusion of the obtained maps.

3.1 Intensity and Color Saliency Maps

Our approach in generation of intensity and color maps is close to propositions of Katramados and Breckon [12], whom suggested a simpler mathematical model called as DIVision of Gaussians (DIVoG). The DIVoG approach involves three following steps (consider a generation procedure of intensity map):

- Step 1. The Gaussian pyramid U is constructed from a set of bottom-up layers $U_1, U_2, \ldots, U_n$. The first layer U_1 has a resolution $w \times h$ pixels. Then a layer resolution is decreased by $(w/2^{n-1}) \times (h/2^{n-1})$ and derived via down-sampling using 5×5 Gaussian filter. Thus, U_n is a top layer of pyramid.
- Step 2. The reversed Gaussian pyramid D is generated, starting with a layer U_n, in order to derive its base D_1. Thus, the pyramid layers are derived via up-sampling using 5×5 Gaussian filter.
- Step 3. The minimum ratio matrix m_{nij}, performing the pixel by pixel division of U_n and D_n in each layer, is computed by Eq. 1.

$$m_{n_{ij}} = \min\left(D_{n_{ij}}/U_{n_{ij}}, U_{n_{ij}}/D_{n_{ij}}\right) m_{n-1_{ij}} \quad n \geq 1 \tag{1}$$

Then an element sm_{1ij} of saliency map SM_{In}, normalized into a range [0…1], is formed by Eq. 2.

$$sm_{1_{ij}} = \left(1 - m_{1_{ij}}\right) \in SM_{In} \tag{2}$$

To avoid a division by zero, all values in intensity map exist in range [0…1] with minimal pixel value k^n, where k is the size of the Gaussian kernel.

The color map is produced in the same manner. Each RGB-channel is processed separately. The Gaussian pyramid with $n = 5$ is constructed. The separated saliency maps are normalized to fit the range [0…255] with following conjunction into a single salience color map SM_{Cl}.

3.2 Contrast Saliency Map

Generally, three types of feature contrasts such as color contrast, texture energy contrast, and texture gradient contrast are available. In order to obtain a fast algorithm, the texture energy measures can be used, which were developed by Laws [13]. These measures are computed by applying convolution kernels to a digital image and then performing a non-linear windowing operation. The 2D convolution kernels are generated from the set of 1D convolution kernels of length three and five provided by Eq. 3, where L is a mask of average gray level, E is a mask of edge features, S is a mask of spot features, W is a mask of wave features, R is a mask of ripples features for kernels with dimensions 3 and 5.

$$
\begin{array}{lll}
L_3 = [1 \quad 2 \quad 1] & E_3 = [1 \quad 0 \quad -1] & S_3 = [1 \quad -2 \quad 1] \\
L_5 = [1 \quad 4 \quad 6 \quad 4 \quad 1] & E_5 = [-1 \quad -2 \quad 0 \quad 2 \quad 1] & S_5 = [-1 \quad 0 \quad 2 \quad 0 \quad -1] \\
W_5 = [-1 \quad 2 \quad 0 \quad -2 \quad 1] & R_5 = [1 \quad -4 \quad 6 \quad -4 \quad 1] &
\end{array}
\tag{3}
$$

Define the texture energy contrast in area Ω_i CE_i by Eq. 4, where $D(e_i, e_j)$ is Euclidian distance between texture energies e_i and e_j in the central region Ω_i and the surrounding regions Ω_j, respectively, w_{ij}^{en} is a spatial constraint of texture energy.

$$
CE_i = \sum_{j \in \Omega_j} D(e_i, e_j)^2 w_{ij}^{en} \tag{4}
$$

L^2-norm function $D(e_i, e_j)$ includes the convolutions between original image $I(x, y)$, where (x, y) are coordinates, and one of masks M_k determined by Eq. 3. This function is computed by Eq. 5.

$$
D(e_i, e_j) = \left| \sum_{(x,y) \in \Omega_i} |I(x,y) * M_k(x,y)| - \sum_{(x,y) \in \Omega_j} |I(x,y) * M_k(x,y)| \right|_2 \tag{5}
$$

A spatial constraint w_{ij}^{en} enhances the effect of nearer neighbor, it is an exponential function provided by Eq. 6, where α is a weighted coefficient controlling a spatial distance, en_i and en_j are the normalized texture energies (respect to area of regions) in the central region Ω_i and the surrounding regions Ω_j, respectively.

$$
w_{ij}^{en} = e^{-\alpha \| en_i - en_j \|^2} \tag{6}
$$

When $\alpha \to 0$, that $w_{ij}^{en} \to 1$ means, Eq. 6 degrades into global contrast calculation. Suppose that Eqs. 4–6 were used at each Gaussian pyramid level. Then the low-size levels ought to resize using 5×5 Gaussian filter, beginning from the top level. The texture gradient contrast can be estimated similar using Eqs. 4–6, but instead masks denoted by Eq. 3 the gradient masks are applied.

The color is the most informative one and conveys tremendous saliency information of a given image. Fu et al. [14] proposed the combined saliency estimator taking the advantages of color contrast and color distribution and obtained very promising results. The main idea is to compute in area Ω_i color contrast $\hat{C}_i$ saliency by Eq. 7, where $D(c_i, c_j)$ is Euclidian distance between colors c_i and c_j in the central region Ω_i and the surrounding regions Ω_j, respectively, w_{ij}^{c} is a spatial constraint of color contrast. (The given image in RGB-color space ought to be transformed in a color-opponent space with dimension L for lightness and a and b for the color-opponent dimensions—Lab-color space.)

$$\hat{C}_i = \sum_{j \in \Omega_j} D(c_i, c_j)^2 w_{ij}^c \quad (7)$$

The spatial constraint w_{ij}^c is computed similar to Eq. 6. According to the fact that a human attention has higher probability to fall onto the center area of the image, Eq. 7 is specified by prior distribution and then smoothed. Additionally, a color contrast CC_i is estimated by Eq. 8, where CD_i is a color distribution.

$$CC_i = \hat{C}_i \cdot CD_i \quad (8)$$

The experiments show that the color contrast estimated by Eq. 8 plays the significant role in contrast estimator of saliency. All three normalized contrast estimators join in a single map as a saliency contrast map SM_{Cn}.

3.3 Edge, Angle, and Symmetry Saliency Maps

The edge "stability" measures can be detected using the Canny operator at multiple scales and then are tracked from a line through to coarse scales with a fixed window centered at the finest scale. Another approach computes a Gaussian pyramid, for example, using Difference of Gaussians (DoG) was claimed by Rosin [15]. A multi-scale approach is necessary because the DoG level strongly depends on scale and an edge map is strongly effective for a single scale. The absolute values of the DoG responses are normalized in the range [0…255], and then the results are summed to form an edge feature map. At the same time, the edge density ought to be determined in any appropriate way.

The concept of angle saliency map generation is similar to edge detection, but instead of Canny or Sobel operator one can use the famous Harris detector.

The symmetry map can be considered as an additional ones based on the edge map, where the external boundaries of an object are well extracted. The main idea is to find a local mapping of the external boundaries on any symmetric shape, a circle in an ideal (also a rectangle is available). The area of salient object ought to be known previously. The external boundaries are represented in a polar system, which contains the discrete twelve or sixteen lines, passing through the center of gravity. The distance values of the opposite points of a boundary in each line are computed, and then their differences are summed for all lines and compared with a threshold. If a salient object has a symmetrical shape, then the positions of boundary points can be recalculated in order to reduce a noise and normalized in a range [0…1]. These data will be a basis to generate a binary symmetry map with salient objects.

3.4 Fusion of Spatial Saliency Maps

The normalized saliency maps based on intensity, color, and contrast are fused to the main saliency map SM_m linearly with empirical coefficients k_{In}, k_{Cl}, and k_{Cn}, respectively:

$$SM_m = k_{In} \cdot SM_{In} + k_{Cl} \cdot SM_{Cl} + k_{Cn} \cdot SM_{Cn}. \tag{9}$$

The edge, angle, and symmetry saliency maps are imposed on the main saliency map, if it is necessary, in order to receive the generalized saliency map SM_G. A visibility of the last map can be improved by multiply on coefficient k_G calculated by Eq. 10, where $d(p_i, p_j)$ is Euclidean distance from ith pixel p_i to a definite foreground or background jth pixel p_j, k is parameter, $k = 0.5$, Ω_F and Ω_B are the areas of foreground and background, respectively.

$$k_G = \exp\left(-k \frac{\min_{j \in \Omega_F}\left(d\left(p_i, p_j\right)\right)}{\min_{j \in \Omega_B}\left(d\left(p_i, p_j\right)\right)}\right). \tag{10}$$

The main idea of Eq. 10 is to give more weight to pixels, which are closer to the definite foreground region and vice versa for pixels that are closer to the definite background region. After obtaining of the generalized spatial saliency map, the salient objects can be marked by blobs, hiding any movement information about the salient objects from frame to frame.

4 Video Stabilization Based on Generalized Saliency Map

The salient objects are roughly detected from the keyframes of a non-stationary video sequences with two main purposes. First, the removal of salient objects helps to estimate a motion in background more accurately. Second, a visibility of salient objects can be improved after stabilization of video sequence. In general, a motion model of a static scene includes several types of movement:

- Random sensor noise with a motion vector $\overline{V}_n$, caused by video cameras. Usually this type of motion does not affect essentially on a scene representation.
- Periodical motion of dynamic textures in a background with a motion vector $\overline{V}_t$, which forms moving background regions (tree branches or rippling water). The effect of background motion is strongly determined by a scale of surveillance.
- Motion of foreground objects, describing by a vector $\overline{V}_o$. This type of motion introduces a real motion of objects with their possible shadows.
- Motion of unwanted camera displacements and/or shakes and jitters $\overline{V}_d$. This is an unpredicted high frequency motion with small magnitude.

Therefore, a motion vector in a static scene $\overline{V}_{ss}$ is forms from components, representing in Eq. 11 under illumination changes, which can enforce a visibility of a movement or instead of this hide some types of movement due to the changes of the Sun position in the outdoor scene or the light on/off sources in the indoor scene.

$$\overline{V}_{ss} = \overline{V}_n + \overline{V}_t + \overline{V}_o + \overline{V}_d \tag{11}$$

Random sensor noise can be neglected. If the areas of foreground objects are excluded by algorithm of salient object detection and the areas of dynamic textures are removed by algorithm of periodical motion detection, then a motion model of background will include only the unwanted shakes and jitters of camera. The detailed methodology of the unwanted jitters compensation is explained in [10, 16, 17].

5 Experimental Results

In order to obtain the objective comparison with known algorithms, two datasets containing images with different luminance and cluttered background were used. Also non-stationary video sequences with cluttered background and complex object motion were employed as test material. In both cases, more that 1,000 images and frames were analyzed. The short description of some tested images is situated in Table 1, while the comparative visual results with three algorithms are depicted in Fig. 2.

The objective estimation of algorithms based on precision–recall and F-measure values is depicted in Fig. 3. F-measure determines a successfulness of salient object detection respect to the chosen binary threshold provided by Eq. 12, where β is an empirical coefficient, $\beta^2 = 0.3$ as it is recommended in [22].

$$F(T) = \frac{(1 + \beta^2)\text{Precision}(T) \times \text{Recall}(T)}{\beta^2 \times \text{Precision}(T) + \text{Recall}(T)} \tag{12}$$

Table 1 Short description of tested images and video sequences

Name	Dataset	Resolution, pixels
3333.jpg [18]	MSRA10k	400 × 300
140927.jpg [18]	MSRA10k	400 × 300
174395.jpg [18]	MSRA10k	400 × 300
2_07-12-31.jpg [19]	FR-Dataset	400 × 400
Gleicher4.avi [20]	L1 robust optimal camera path	720 × 480
Sam_1.avi [20]	L1 robust optimal camera path	720 × 480

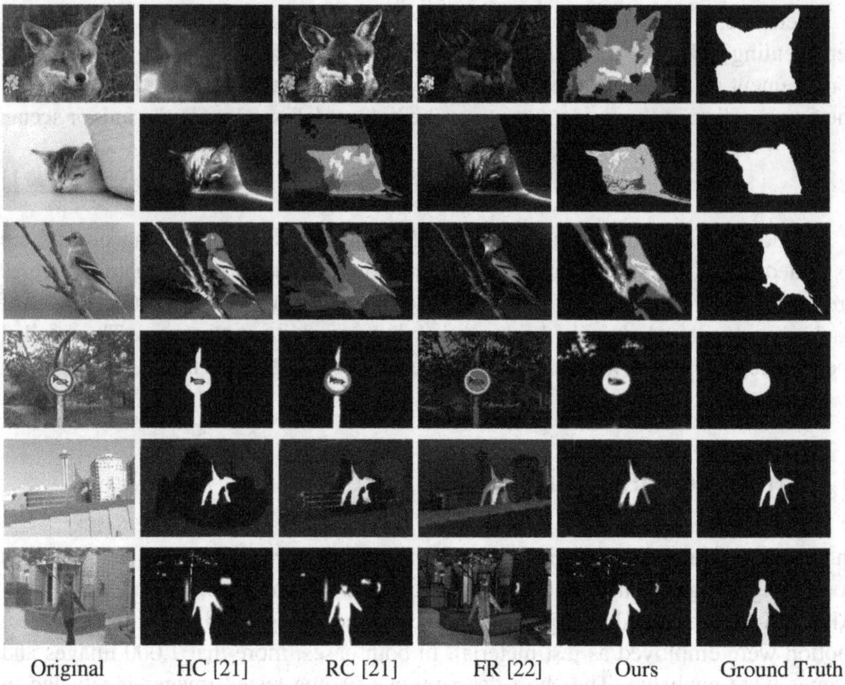

Fig. 2 Visual comparative results of algorithms

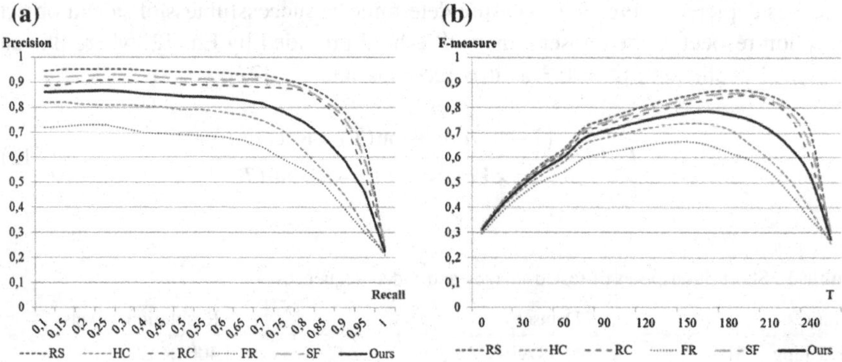

Fig. 3 Comparison of algorithms: **a** precision-recall estimators, **b** F-measures estimators

Good segmentation results are obtained with $T = 165$ that increases a computation time. The proposed algorithm has the close precision values to other algorithms, but can detect multiple salient objects in complex scenes better.

Table 2 Comparison of execution time

Method	RS [14]	HC [21]	RC [21]	FR [22]	SF [23]	Ours
Time (s)	2.7	0.111	0.144	1.33	0.153	0.088
Code	Matlab	C++	C++	C++	C++	C++

The results of computation time estimation for images with sizes 400 × 300 pixels are located in Table 2. For experiments, a computer Intel Core I5, Geforce GTX 560, 4GbRam DDR3 was used.

6 Conclusions

The proposed methodology excludes the saliency regions from a non-stationary video sequence in order to receive the reliable motion estimation of background. The multi-feature saliency map is based on the intensity, color, and contrast as the main features and edge, angle, and symmetry maps as the additional features. When the unwanted motion estimations have been obtained, the salient objects will be included in the stabilized video sequence with accurate post-processing. The experiments demonstrated fast processing of keyframes with a suitable accuracy for stabilization.

Acknowledgments This work was supported by the Russian Fund for Basic Researches, grant no. 16-07-00121 A, Russian Federation.

References

1. Borji, A., Cheng, M.-M., Jiang, H., Li, J.: Salient Object Detection: A Survey. http://arxiv.org/pdf/1411.5878.pdf (2014)
2. Itti, L., Koch, C., Niebur, E.: A model of saliency-based visual attention for rapid scene analysis. IEEE Trans. Pattern Anal. Mach. Intell. **20**(11), 1254–1259 (1998)
3. Achanta, R., Shaji, A., Smith, K., Lucchi, A., Fua, P., Susstrunk, S.: SLIC superpixels compared to state-of-the-art superpixel methods. J. Latex Class Files **60**(1), 1–8 (2011)
4. Li, X., Li, Y., Shen, C., Dick, A., van den Hengel, A.: Contextual hypergraph modelling for salient object detection. In: IEEE International Conference on Computer Vision (ICCV'2013), pp. 3328–3335. IEEE (2013)
5. Hou, X., Zhang, L.: Saliency detection: a spectral residual approach. In: IEEE Conference on Computer Vision and Pattern Recognition (CVPR'2007), pp. 1–8. IEEE (2007)
6. Guo, C., Zhang, L.: A novel multiresolution spatiotemporal saliency detection model and its applications in image and video compression. IEEE Trans. Image Process. **19**(1), 185–198 (2010)
7. Le Meur, O., Le Callet, P., Barba, D., Thoreau, D.: A coherent computational approach to model bottom-up visual attention. IEEE Trans. Pattern Anal. Mach. Intell. **28**(5), 802–817 (2006)

8. Zhuang, L., Tang, K., Yu, N., Qian, Y.: Fast salient object detection based on segments. In: International Conference on Measuring Technology and Mechatronics Automation (ICMTMA'2009), vol. 1, pp. 469–472. IEEE Computer Society (2009)
9. Liu, Z., Zou, W., Le Meur, O.L.: Saliency tree: a novel saliency detection framework. IEEE Trans. Image Process. **23**(5), 1937–1952 (2014)
10. Favorskaya, M., Jain, L.C., Buryachenko, V.: Digital video stabilization in static and dynamic scenes. In: Favorskaya, M.N., Jain, L.C. (eds.) Computer Vision in Control Systems-1, ISRL, vol. 73, pp. 261–309. Springer International Publishing Switzerland (2015)
11. Ejaz, N., Mehmood, I., Baik, S.W.: Efficient visual attention based framework for extracting key frames from videos. Sig. Process. Image Commun. **28**(1), 34–44 (2013)
12. Katramados, I., Breckon T.P.: Real-time visual saliency by division of Gaussians. In: IEEE International Conference on Image Processing, pp. 1741–1744. IEEE Press, New York (2011)
13. Laws, K.I.: Rapid texture identification. SPIE Image Process. Missile Guidance **238**, 376–380 (1980)
14. Fu, K., Gong, C., Yang, J., Zhou, Y., Gu, I.Y.-H.: Superpixel based color contrast and color distribution driven salient object detection. Sig. Process. Image Commun. **28**(10), 1448–1463 (2013)
15. Rosin, P.L.: A simple method for detecting salient regions. Pattern Recogn. **42**(11), 2363–2371 (2009)
16. Favorskaya, M., Buryachenko, V.: Fuzzy-based digital video stabilization in static scenes. In: Tsihrintzis, G.A., Virvou, M., Jain, L.C., Howlett, R.J., Watanabe, T. (eds.) Intelligent Interactive Multimedia Systems and Services in Practice, SIST, vol. 36, pp. 63–83. Springer International Publishing Switzerland (2015)
17. Favorskaya, M., Buryachenko, V.: Video stabilization of static scenes based on robust detectors and fuzzy logic. Frontiers Artific Intellig Appl **254**, 11–20 (2013)
18. MSRA10K Salient Object Database. http://mmcheng.net/msra10k/. Accessed 22 Dec 2015
19. IVRG—Images and Visual Representation Grow. http://ivrlwww.epfl.ch/supplementary_material/RK_CVPR09/. Accessed 22 Dec 2015
20. Auto-Detected Video Stabilization with Robust L1 Optimal Camera Paths. http://cpl.cc.gatech.edu/projects/videostabilization/. Accessed 22 Dec 2015
21. Cheng, M.-M., Mitra, N.J., Huang, X., Torr, P.H.S., Hu, S.-M.: Global contrast based salient region detection. In: IEEE Conference on Computer Vision and Pattern Recognition (CVPR'2011), pp. 409–416. IEEE Press, New York (2011)
22. Achanta, R., Hemami, S., Estrada, F., Süsstrunk, S.: Frequency-tuned salient region detection. In: IEEE International Conference on Computer Vision and Pattern Recognition (CVPR'2009), pp. 1597–1604. IEEE Press, New York (2009)
23. Perazzi, F., Krähenbühl, P., Pritch, Y., Hornung, A.: Saliency filters: contrast based filtering for salient region detection. In: IEEE Conference on Computer Vision and Pattern Recognition (CVPR'2012), pp. 733–740. IEEE Press, New York (2012)

Global Motion Estimation Using Saliency Maps in Non-stationary Videos with Static Scenes

Margarita Favorskaya, Vladimir Buryachenko and Anastasia Tomilina

Abstract The global motion estimation is a cornerstone of successful video stabilization. In current research, the stabilization task of non-stationary video sequence with static scenes is solved using saliency maps and the trajectories of feature descriptors. First, the feature descriptors are built in keyframes with the removed regions of moving foreground objects, which are considered the salient objects. The tracking results of feature descriptors form the feature trajectories. Second, a distinctiveness of a short-term trajectory is evaluated by histogram approach. Third, a temporal coherence of a long-term trajectory is exploited to verify the global motion through a whole video sequence. In this research, the constant flow and the affine flow are considered. The proposed algorithm permits to increase the peak signal to noise ratio up 4–7 dB on the average comparing with conventional stabilization methods in video sequences with static scenes.

Keywords Global motion estimation · Background estimation · Video stabilization · Static scene · Trajectory descriptor · Temporal coherence

1 Introduction

In video sequences, two types of motion, called as global and local motions, can be distinguished. The global motion is a motion of a camera, including panning, tilting, zooming and/or a more complex combination of these three components. The global motion impacts on the entire frame. The local motion is a motion of

M. Favorskaya (✉) · V. Buryachenko · A. Tomilina
Institute of Informatics and Telecommunications, Siberian State Aerospace University, 31 Krasnoyarsky Rabochy av., Krasnoyarsk 660037, Russian Federation
e-mail: favorskaya@sibsau.ru

V. Buryachenko
e-mail: buryachenko@sibsau.ru

A. Tomilina
e-mail: tomilina_ai@sibsau.ru

G. De Pietro et al. (eds.), *Intelligent Interactive Multimedia Systems and Services 2016*, Smart Innovation, Systems and Technologies 55,
DOI 10.1007/978-3-319-39345-2_12

visual objects, involving their scaling, rotations, and deformation often in a framework of affine or perspective models. In general, such 3D motions are mapped on 2D video sequences with information loss. The motion analysis becomes more difficult under unwanted shakes and jitters caused by the objective factors of non-successful camera shooting. All mentioned above motion types impact each other on 2D mapping and form a complex motion vector field, depending on scale, velocity, and visual artifacts of moving objects. Also a global motion can comprise a motion of large visual objects.

Methods for Global Motion Estimation (GME) can be categorized roughly as direct and indirect methods. The direct GME methods minimize the prediction errors in the pixel domain. These methods are computationally expensive due to the iterative processes in the non-linear (in general) estimation and the number of pixels involved. The indirect GME methods include two stages with obtaining a motion vector field in the first stage and performing the GME in the second stage. The key challenges regarding to the widely-used non-stationary video sequences can be declare as follows:

- The long-term robust motion tracking in a non-stationary pan-tilt-zoom video sequence fails under the observation modelling of foreground salient motions and the non-aligned backgrounds.
- In surveillance system, the assumption is given that the salient motions are sparse with respect to the cluttered background. However, this assumption may be violated due to the dynamically changing embedded background objects, illumination variation, occlusion, and so on.
- Usually the existing trackers ignore the contextual interactions between the intermittent moving targets and the background due to the high computational cost. This causes a necessity to integrate efficiently the complementary color, structure, and sparse residuals into a background.

Since the direct correspondences among the consecutive frames are impossible generally, various techniques were proposed to solve this problem. Our contribution deals with the short-term trajectories building based on background feature descriptors without regions of foreground salient objects and following temporal coherence of long-term trajectories. As a result, 3D trajectory of camera movement is built to recalculate each frame with new camera parameters in order to compensate unwanted shakes and jitters.

The paper is organized as follows. The objectivities of the GME are recalled in Sect. 2. Section 3 presents and analyzes the related work. The proposed GME of non-stationary background is discussed in Sect. 4. The experimental results obtained are presented and discussed in details in Sect. 5. Finally, Sect. 6 concludes the paper and opens perspectives of future work.

2 Objectivities

The GME, involving two frames $\mathbf{I}_n$ and $\mathbf{I}_{n+1}$, implies the minimization of differences $E_{n,\ n+1}$ between frame $\mathbf{I}_{n+1}$ and its predicted version $\hat{\mathbf{I}}_{n+1}$, providing by Eq. 1, where $e(i,\ j)$ is the difference of intensity values between a pixel with coordinates $(i,\ j)$ in the observable frame $\mathbf{I}_{n+1}$ and the corresponding pixel in the predicted frame $\hat{\mathbf{I}}_{n+1}$, $f_x(i,\ j)$ and $f_y(i,\ j)$ are the transform mapping functions.

$$\begin{aligned} E^{sq}_{n,n+1} &= \sum_{i,j} e^2(i,j) = \sum_{i,j} \left(I_{n+1}(i,j) - \hat{I}_{n+1}(i,j)\right)^2 \\ &= \sum_{i,j} \left(I_{n+1}(i,j) - I_n\left(f_x(i,j), f_y(i,j)\right)\right)^2 \rightarrow \min \end{aligned} \quad (1)$$

In Eq. 1, other functions can be applied, e.g. the absolute value of the intensity difference as it is show in Eq. 2.

$$E^{abs}_{n,n+1} = \sum_{i,j} |e(i,j)| \rightarrow \min \quad (2)$$

The transform mapping functions $f_x(i,\ j)$ and $f_y(i,\ j)$ ought to be chosen such that $E^{sq}_{n,n+1}$ or $E^{abs}_{n,n+1}$ values were minimized. Usually these mapping functions are described by affine or perspective transformations based on Lie operators [1]. However, the unwanted shakes and jitters add to the well-known models some random translation, which can be relatively constant in a short interval. While in many computer vision tasks affine and perspective models are applied for the projections of a visual objects, in stabilization task with static scenes these models are used for estimation of the background without foreground moving objects and regions with dynamic textures, if it is possible. The affine and perspective models have a view of equation systems in Eqs. 3 and 4, respectively, where $\{m_i\}$ are coefficients of a transformation.

$$\begin{aligned} f_x(i,j) &= m_1 i + m_2 j + m_3 \\ f_y(i,j) &= m_4 i + m_5 j + m_6 \end{aligned} \quad (3)$$

$$\begin{aligned} f_x(i,j) &= \frac{m_1 i + m_2 j + m_3}{m_7 i + m_8 j + 1} \\ f_y(i,j) &= \frac{m_4 i + m_5 j + m_6}{m_7 i + m_8 j + 1} \end{aligned} \quad (4)$$

Vector $\mathbf{m} = \{m_i\}$ can be computed by direct methods, among which Levenberg–Marquardt algorithm estimating iteratively the residuals [2, 3] or pixel subsampling patterns [4] are used, or indirect methods using, for example, the feature trajectories [5] or the interpolated camera path in 3D space [6].

Assume that a non-stationary video sequence with static scenes was preprocessed, and a non-stationary background was extracted. In this case, the direct methods to estimate vector $\mathbf{m} = \{m_i\}$ are worthless because a camera movement in static scenes is a superposition of a smooth movement with non-significant magnitude values in low frequency domain and a chaotic movement with non-significant magnitude values in high frequency domain. Often the indirect methods are applied. In general, a camera motion in static scenes involve rotations, translations, and scaling, but with restricted transforming parameters.

3 Related Work

The conventional way for video stabilization is to use three sequential steps—motion estimation, motion filtering, and motion correction. The most computationally expensive step is a motion estimation, which computes the local and global motion vectors between two consecutive frames. A motion filtering identifies the unwanted involuntary movements and separates them from movements of foreground objects and intended camera handling. This is the most complicated step of processing. Finally, a motion correction substitutes the current frame in order to suppress the involuntary movement. The main attention in literature is given to the first and second steps.

The well-known Levenberg–Marquardt algorithm can be used to minimize an objective Eqs. 3 and 4 iteratively. Since the region of the GME is the entire frame, the minimization process has very expensive computational cost. The complexity of the GME can be reduced significantly, if only a small subset of pixels is used in estimating. Alzoubi and Pan [4] proposed to estimate the global motion parameters based on several subsampling patterns and their combinations that permitted to reduce the computational time with insignificant accuracy loss (less than 0.1 dB).

The feature-based methods are widely applied for motion estimation, including Shift Invariant Feature Transform (SIFT) descriptors [7], Speeded Up Robust Features (SURF) descriptors [8], particle filter [9], among others. The region-based methods use a block matching approach or its modifications. For example, Xu and Lin [10] proposed to estimate a motion with custom rotation and translation invariant features in circular blocks with more precise results than a rectangular block matching and more complex logic of processing.

The motion filtering can be implemented as a simple accumulator or as the complex banks of adaptive filters. In general, a low frequency camera motion is smooth, while a high frequency unwanted motion indicates a random behavior. Commonly used techniques [7] apply a smoothing low-pass filter to the motion vector, which is then subtracted from the estimated motion to obtain the unwanted motion vector used for correction. The predicting filter such as Kalman filter [11] can be applied to separate types of motion. However, the predicting filter is a filter of high computational complexity. Fuzzy logic helps to separate the motion vectors, as it was shown in [12]. Hilbert–Huang transforms [13] decomposes adaptively the non-stationary signals through the process of empirical mode decomposition into the basic functions called intrinsic mode functions. The estimated local motion vector is decomposed into a finite number of intrinsic mode functions by applying the empirical mode decomposition. Then the energy content of decomposed signal is defined using Hilbert transformation. The summation of all intrinsic mode functions with the lower indices up to the specified intrinsic mode function approximates the unwanted jitter motion.

During the motion correction step, a subset of the original image is necessarily corrected, losing information along the frame boundaries in order to perform the image translation. As a result, a video with a high degree of stabilization, but with a lower resolution is obtained. The resolution depends directly on the maximum translations between two consecutive frames. The inpainting or interpolation algorithms can be used to restore a frame, but when the frame rate of the video is increased, the amount of lost information is naturally reduced; thus the negative impact of motion correction is limited.

4 Global Motion Estimation of Non-stationary Background

The motion estimation of a stationary video sequence with removed moving foreground objects is based on the observations that the trajectories containing background feature points are coherent with the dominant camera motion and have a consistent motion magnitude and orientation over a time. In the case of a non-stationary video sequence, these trajectories are distorted by high-frequency jitters but save coherent properties. First, the trajectory descriptors ought to be built (Sect. 4.1). Second, short-term trajectory distinctiveness is evaluated (Sect. 4.2). Third, a temporal coherence of long-term trajectory is exploited to verify the global motion through a whole video sequence (Sect. 4.3).

4.1 Trajectory Descriptors

Suppose that a video sequence was processed previously, and all moving foreground visual objects are removed from each frame using, for example, salient object detection technique. Suppose that each keyframe is divided by grid with 5 × 4 cells approximately. In first keyframe, a set of SURF descriptors is generated in such manner that 10–20 SURF descriptors are located into each cell. Then a tracking of SURF descriptors is starting. During the tracking, several SURFs can disappear, other SURFs can emerge. The main purpose is to support a uniform distribution of SURFs in background surveillance. The corresponding SURFs in the consecutive frames are detected using standard RANdom SAmple Consensus (RANSAC) algorithm and concatenated to construct a feature-level trajectory.

Suppose that N trajectories $\mathbf{TR} = \{\mathbf{tr}_1, \ldots, \mathbf{tr}_N\}$, $i = 1, \ldots, N$, are extracted from a whole video sequence, where each trajectory $\mathbf{tr}_i$ is represented by a set of short-term collections of corresponding SUFRs in consecutive frames from one keyframe to the following keyframe. A set of trajectories is provided by Eq. 5, where $KF_j\{\cdot\}$ is jth keyframe, $j = 1, \ldots, M$, M_j is a number of frames respect to the jth keyframe, FR_k is kth current frame, $fd_{i,k}$ is a feature detector concerning to ith trajectory in kth frame. A population of feature detectors has a constant value from frame to frame. Notice that frames FR_l are grouped in Eq. 5 respect to keyframes.

$$\begin{aligned}\mathbf{TR} = \{&KF_1\{FR_1\{fd_{11}, \ldots, fd_{N1}\}, \ldots, FR_{M_1}\{fd_{1M_1}, \ldots, fd_{NM_1}\}\}pt, \ldots, \\ &KF_j\{FR_1\{fd_{11}, \ldots, fd_{N1}\}, \ldots, FR_{M_j}\{fd_{1M_j}, \ldots, fd_{NM_j}\}\}, \\ &KF_{j+1}\{FR_1\{fd_{11}, \ldots, fd_{N1}\}, \ldots, FR_{M_{j+1}}\{fd_{1M_{j+1}}, \ldots, fd_{NM_{j+1}}\}\}, \ldots\}\end{aligned} \tag{5}$$

Due to different luminance, deformations, and occlusions, old feature detectors can disappear and new feature detectors can emerge. Therefore, the extracted trajectories are asynchronous with different lifespan. To keep a temporal coherence of the GME through the entire video sequence, the accumulated motion histogram was proposed to describe each short-term trajectory.

4.2 Short-Term Trajectories

Introduce the accumulative window as a set of a number of frames between two consecutive keyframes. Then each short-term trajectory $\mathbf{tr}_{ij}$ of ith feature detector respect to keyframe j has a view of Eq. 6.

$$\mathbf{tr}_{ij} = KF_j\{FR_1\{fd_{i1}\}, FR_2\{fd_{i2}\}, \ldots, FR_{M_j}\{fd_{iM_j}\}\} \tag{6}$$

Based on Eq. 6 and RANSAC technique, all corresponding SURFs $\{fd_{ik}\}$ can be connected each other and interpreted as a set of vectors with magnitudes $\{d_{ij}(b)\}$ and orientations $\{o_{ij}(b)\}$, where b is a bin of histogram. The accumulated global motion vector in each frame is characterized by the accumulated magnitude histogram AMH_{jk} and the accumulated orientation histogram AOH_{jk} of all SURFs in frame k provided by Eq. 7.

$$AMH_{jk}(b) = \sum_{i=1}^{K} d_{ijk}(b)/K \quad AOH_{jk}(b) = \sum_{i=1}^{K} o_{ijk}(b)/K \tag{7}$$

$AMH_{jk}(b)$ and $AOH_{jk}(b)$ are used to represent the normalized motion magnitude and the quantized orientation, respectively, of the bth bin in the accumulated motion histogram AMH_{jk}. Equation 7 provides a fast choice of the representative magnitude and orientation values for all SURF trajectories in a current frame. In a similar way, the accumulated magnitude histogram AMH_j and the accumulated orientation histogram AOH_j of all frames, concerning to jth keyframe, can be obtained. These histograms show the distribution of background motion between two consecutive keyframes. As a result, a set of the short-term accumulated magnitude histograms $\{AMH_j\}$ and the short-term accumulated orientation histograms $\{AOH_j\}$, $j = 1, \ldots, M$, concerning to all keyframes, will be built.

4.3 Temporal Coherence of Long-Term Trajectory

Generally, the background motion of a non-stationary video sequence includes three types of motion, such as camera motion, motion of dynamic textures, and unwanted shakes and jitters. The concept of temporal coherence forms the basis of many tracking algorithms. However, most of them focus either on incorporating data constancy assumptions to improve robustness under noise and varying illumination, or design the discontinuity-preserving smoothness constraints to ensure spatial coherence. These techniques can be classified into three groups, mentioned below:

- Temporal regularization was introduced by Murray and Buxton [14]. This technique is based on a spatio-temporal smoothness of the resulting motion along the temporal axis. Further improvements proposed the directional smoothness assumptions in a spatio-temporal volume, the spatio-temporal level sets, Kalman filtering with a temporally constant motion model. However, in the presence of larger displacements, the temporal derivatives in the smoothness term do not allow to capture the actual trajectories.
- Trajectorial stabilization is a completely different approach in contrast of temporal regularization, which was proposed by Black and Anandan [15]. This approach supposes to compute additionally motion vector fields between previous frames and register them in the current time instant that is suitable for estimation of larger displacements. However, there is no feedback between the different motion vector fields during the computation.
- Trajectorial regularization assumes that the motion vector fields are smooth along the trajectories of moving objects [16]. Few strategies were proposed that penalize first-order variations and/or second-order variations along motion trajectories with the soft constraints. This is very promising approach to smooth the trajectories of simultaneously moving visual objects in many frames. The approach suffers from two objective drawbacks dealing with difficult optimization of motion vector fields and the worse results in the presence of complex motion patterns.

For our task, a trajectorial stabilization is the best decision, more here a simplified case is possible based on the short-term accumulated magnitude histograms $\{AMH_{jk}\}$ and the short-term accumulated orientation histograms $\{AOH_{jk}\}$ between two consecutive keyframes. Black and Anandan [15] mentioned the correlation methods using the simple sum-of-square differences computing (similar to block-matching algorithm), the gradient methods taking the Taylor Series approximation of the first order, and the regression methods based on affine model.

Indicate these three cases, proposed by Black and Anandan, in the terms feature detectors. Then the following formulae can be obtained, represented by Eqs. 8–10, where $E_{cr}(u, v)$, $E_{gr}(u, v)$, and $E_{rg}(u, v)$ are the errors of correlation, gradient, and regression methods, respectively, gfd is a generalized feature detector in a frame, $p = AMH_{jk}(b_{\max}) \cdot \cos(AOH_{jk}(b_{\max}))$ and $q = AMH_{jk}(b_{\max}) \cdot \sin(AOH_{jk}(b_{\max}))$ are the projections of accumulative motion vector on OX and OY axis, respectively, $AMH_{jk}(b_{\max})$ and $AOH_{jk}(b_{\max})$ are maximum number of values of the motion magnitude and the quantized orientation in frame k respect to jth keyframe, (u, v) is the horizontal and vertical velocity of features' motion, δt is a time instant (equivalent to consecutive frames), ∇gfd is a local motion vector of generalized feature detector, $\mathbf{mv} = [u, v]^{T}$ is a motion vector, $\mathbf{m} = \{m_i\}$ is affine parameter vector from Eq. 3, $\mathfrak{R}$ is a spatio-temporal space.

$$E_{cr}(u, v) = \sum_{(p,q) \in \mathfrak{R}} (gfd(p, q, t) - gfd(p + u\delta t, q + v\delta t, t + \delta t))^2 \tag{8}$$

$$\begin{aligned} E_{gr}(u, v) &= \sum_{(p,q) \in \mathfrak{R}} (gfd_x(p, q, t)\, u + gfd_y(p, q, t)\, v + gfd_t(p, q, t))^2 \\ &= \sum_{\mathfrak{R}} \left((\nabla gfd)^{T} \mathbf{mv} + gfd_t \right)^2 \end{aligned} \tag{9}$$

$$E_{rg}(\mathbf{m}) = \sum \left((\nabla gfd)^{T} \mathbf{mv}(\mathbf{m}) + gfd_t \right)^2 \tag{10}$$

In this research, the constant flow (Eq. 9) and the affine flow (Eq. 10) are considered. Equations 9 and 10 reflect a camera trajectory in 3D space that permits to recalculate each frame with new camera parameters in order to compensate the unwanted shakes and jitters. The study of dynamic texture motion requires the following investigations.

5 Experimental Results

The experiments were executed by the designed software tool "DVS Analyzer", v. 3.04. The short description of tested non-stationary video sequences with cluttered background is represented in Table 1.

Table 1 Short description of tested video sequences

Title, URL	Snapshot	Resolution, pixels	Frames, number	Motion type
EllenPage_Juggling.avi http://www.youtube.com/watch?v=8YNUSCX_akk		1280 × 720	793	Non-contrast background, fast moving foreground objects
lf_juggle.avi http://cpl.cc.gatech.edu/projects/videostabilization/		480 × 360	460	Static camera, fast moving foreground objects
SANY0025_xvid.avi http://cpl.cc.gatech.edu/projects/videostabilization/		640 × 360	445	Slow camera motion, large foreground object
road_cars_krasnoyarsk.avi http://www.youtube.com/watch?v=pJ84Pwpbl_Y		852 × 480	430	Static camera, 3D-depth scene, fast moving objects

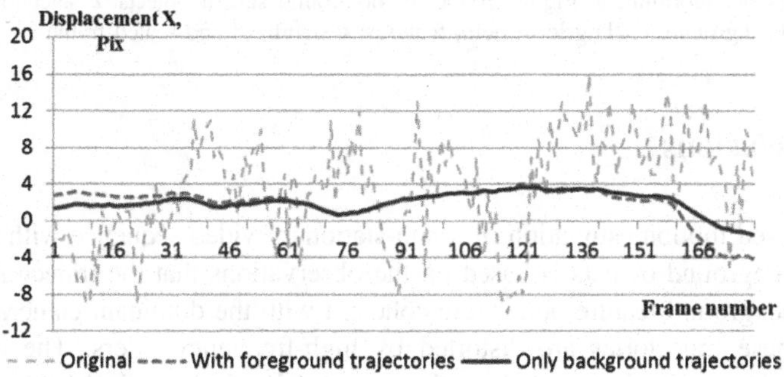

Fig. 1 Trajectories with and without foreground moving objects in videos EllenPage_Jugg.avi

The plots with three types of trajectories without stabilization, also with and without foreground moving objects are depicted in Fig. 1.

The detailed steps of proposed video stabilization algorithm using saliency maps and feature points trajectories for video sequence EllenPage_Jugg.avi are depicted in Fig. 2. Our approach increased the peak signal to noise ratio up 4–7 dB on the average, comparing with conventional stabilization methods in static scenes [17]. It is reasonable to expect the higher results in non-stationary video sequences with dynamic scenes.

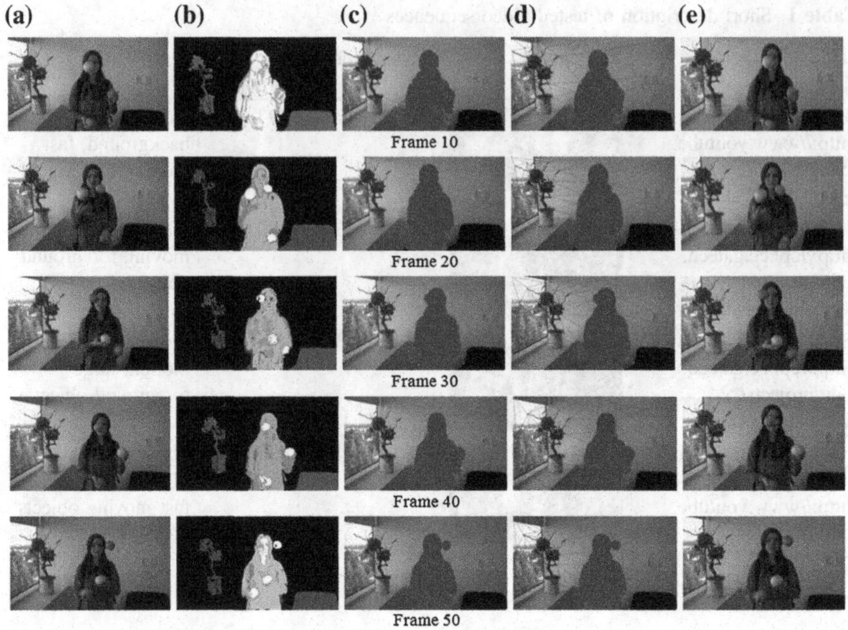

Fig. 2 Steps of algorithm: **a** original frame; **b** foreground salient objects; **c** selection of point features in background; **d** 2D grid of point features; **e** stabilized and scaled frame

6 Conclusions

The proposed motion estimation of a non-stationary video sequence with removed moving foreground objects is based on the observations that the trajectories, containing background feature points, are coherent with the dominant camera motion, even if these trajectories are distorted by high-frequency jitters. The model of trajectory descriptors is founded on the terms of keyframes. To keep a temporal coherence of the GME through the entire video sequence, the accumulated motion histograms were introduced to describe each short-term trajectory. For long-term trajectories, the constant flow and the affine flow were considered. Our approach permits to increase the peak signal to noise ratio up 4–7 dB comparing with conventional stabilization methods in video sequences with static scenes The study of dynamic texture motion, as well as the dynamic scenes, requires the following investigations.

Acknowledgments This work was supported by the Russian Fund for Basic Researches, grant no. 16-07-00121 A, Russian Federation.

References

1. Pan, W.D., Yoo, S.-M., Nalasani, M., Cox, P.G.: Efficient local transformation estimation using Lie operators. Inf. Sci. **177**(3), 815–831 (2007)
2. Levenberg, K.: A method for the solution of certain non-linear problems in least squares. Q. Appl. Math. **2**, 164–168 (1944)
3. Marquardt, D.W.: An algorithm for least-squares estimation of nonlinear parameters. J. Soc. Ind. Appl. Math. **11**(2), 431–441 (1963)
4. Alzoubi, H., Pan, W.D.: Fast and accurate global motion estimation algorithm using pixel subsampling. Inf. Sci. **178**(17), 3415–3425 (2008)
5. Lee, K.-Y., Chuang, Y.-Y., Chen, B.-Y., Ouhyoung, M.: Video stabilization using robust feature trajectories. In: IEEE 12th International Conference on Computer Vision, pp. 1397–1404. IEEE Press, New York (2009)
6. Torii, A., Havlena, M., Pajdla, T.: Omnidirectional image stabilization by computing camera trajectory. In: Wada, T., Huang, F., Lin, S. (eds.) Advances in Image and Video. LNCS, vol. 5414, pp. 71–82. Springer, Berlin, Heidelberg (2009)
7. Battiato, S., Gallo, G., Puglisi, G., Scellato, S.: SIFT features tracking for video stabilization. In: 14th International Conference on Image Analysis and Processing (ICIAP'2007), pp. 825–830. IEEE Press, New York (2007)
8. Huang, K.-Y., Tsai, Y.-M., Tsai, C.-C., Chen, L.-G.: Video stabilization for vehicular applications using SURF-like descriptor and KD-tree. In: 17th IEEE International Conference on Image Processing (ICIP'2010), pp. 3517–3520. IEEE Press, New York (2010)
9. Yang, J., Schonfeld, D., Mohamed, M.: Robust video stabilization based on particle filter tracking of projected camera motion. IEEE Trans. Circuits Syst. Video Technol. **19**(7), 945–954 (2009)
10. Xu, L., Lin, X.: Digital image stabilization based on circular block matching. IEEE Trans. Consum. Electron. **52**(2), 566–574 (2006)
11. Wang, C., Kim, J.-H., Byun, K.-Y., Ni, J., Ko, S.-J.: Robust digital image stabilization using the Kalman filter. IEEE Trans. Consum. Electron. **55**(1), 6–14 (2009)
12. Favorskaya, M., Buryachenko, V.: Fuzzy-based digital video stabilization in static scenes. In: Tsihrintzis, G.A., Virvou, M., Jain, L.C., Howlett, R.J., Watanabe, T. (eds.) Intelligent Interactive Multimedia Systems and Services in Practice, SIST, vol. 36, pp. 63–83. Springer International Publishing Switzerland (2015)
13. Ioannidis, K., Andreadis, I.: A digital image stabilization method based on the Hilbert-Huang transform. IEEE Trans. Instrum. Meas. **61**(9), 2446–2457 (2012)
14. Murray, D.W., Buxton, B.F.: Scene segmentation from visual motion using global optimization. IEEE Trans. Pattern Anal. Mach. Intell. **9**(2), 220–228 (1987)
15. Black, M.J., Anandan P.: Robust dynamic motion estimation over time. In: IEEE Computer Society Conference on Computer Vision and Pattern Recognition, pp. 292–302. IEEE Computer Society Press, New York (1991)
16. Valgaerts, L., Bruhn, A., Zimmer, H., Weickert, J., Stoll, C., Theobalt C.: Joint estimation of motion, structure and geometry from stereo sequences. In: Daniilidis, K., Maragos, P., Paragios, N. (eds.) Computer Vision—ECCV'2010, LNCS, vol. 6314, pp. 568–581. Springer, Berlin (2010)
17. Favorskaya, M., Jain, L.C., Buryachenko, V.: Digital video stabilization in static and dynamic scenes. In: Favorskaya, M.N., Jain, L.C. (eds.) Computer Vision in Control Systems-1, ISRL, vol. 73, pp. 261–309. Springer International Publishing Switzerland (2015)

SVM-Based Cancer Grading from Histopathological Images Using Morphological and Topological Features of Glands and Nuclei

Catalin Stoean, Ruxandra Stoean, Adrian Sandita, Daniela Ciobanu, Cristian Mesina and Corina Lavinia Gruia

Abstract The paper puts forward a new data set comprising 357 histopathological image samples obtained from colon tissues and distinguished into four cancer grades. At the same time, it proposes an automatic methodology for extracting knowledge from these images and discriminating between the disease stages on its base. The approach identifies the glands and nuclei and uses morphological and topological features related to these components to generate 76 attributes that are further used for classification via support vector machines. The values of one parameter used for the identification of the nuclei are tuned and surprisingly good results are reached when overlapping nuclei are identified as singular objects.

Keywords Image processing · Histopathological image · Feature extraction · Classification

C. Stoean (✉) · R. Stoean · A. Sandita
Faculty of Sciences, Department of Computer Science,
University of Craiova, Craiova, Romania
e-mail: catalin.stoean@inf.ucv.ro; cstoean@inf.ucv.ro

R. Stoean
e-mail: ruxandra.stoean@inf.ucv.ro

A. Sandita
e-mail: asandita@inf.ucv.ro

D. Ciobanu · C. Mesina
Faculty of Medicine, University of Medicine and Pharmacy of Craiova,
Craiova, Romania
e-mail: elada192@yahoo.com

C. Mesina
e-mail: mesina.cristian@doctor.com

C.L. Gruia
Emergency County Hospital of Craiova, Craiova, Romania
e-mail: paraschivdan65@yahoo.com

G. De Pietro et al. (eds.), *Intelligent Interactive Multimedia Systems and Services 2016*, Smart Innovation, Systems and Technologies 55,
DOI 10.1007/978-3-319-39345-2_13

1 Introduction

Histopathological images are obtained by staining body tissues that are subsequently examined under a microscope [1]. The current protocol for cancer diagnosis in general, and for the colorectal type, in particular, involves the human expert (pathologist) analysis of the histopathological image while the best treatment for cancer consists in its early diagnosis [2, 3]. Although the judgement of the pathologist is educated and based on a vast experience, it is subjective and may lead to serious variability [1, 4, 5]. Furthermore, the physicians are confronted with a vast amount of histopathological images, due to the important investments in developing advanced microscopy hardware and the persuasion of individuals to have medical examinations more often. A potential aid could come from quantitative image-based evaluation of digital pathology slides, as these could at least eliminate the most obvious cases and leave the pathologists to concentrate on the most difficult records.

There is an acknowledged lack of benchmark data sets in histopathology imaging for validating techniques and for their comparison. In this respect, a data set of 357 histopathological images is made available [6] and a diagnosis support methodology is proposed, which achieves an overall prediction accuracy of almost 80 % on distinguishing between cancer stages from information automatically extracted from the collection. Next section briefly describes the image data set and the proposed methodology to extract the features and set a diagnosis, Sect. 3 outputs the results and Sect. 4 contains concluding remarks.

2 Feature Extraction and Classification

The data set [6] contains histopathological images of normal tissue and for cancer grades G1, G2, and G3. For ease of reference, we will refer to them in the current paper as G0, G1, G2 and G3. The next subsection contains a detailed description of the collection. Subsequently, the feature extraction stage is concentrated on the identification of glands and nuclei and several measures derived from this information are further used to construct a numerical data set that is fed to a support vector machine (SVM) in order to accurately grade the tissues.

2.1 Histopathological Image Data Set

The images are obtained from colon tissue slides stained using hematoxylin and eosin (H&E) and obtained from an electron microscope at the ×10 magnification level at the Emergency County Hospital of Craiova, Romania. We depart from 30 histopathological images that represent cancer grades G1, G2 and G3 from 30 different patients. Most of them contain border regions between normal tissues and

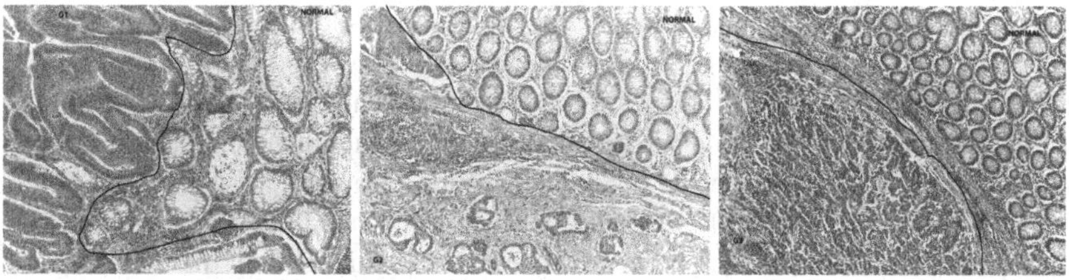

Fig. 1 Samples of the initial histopathological images. The borders are delineated between normal condition and grades 1, 2 and 3, respectively (from *left* to *right*)

malignant ones (see Fig. 1), so as in [7] representative parts could be extracted for G0, G1, G2 and G3. The obtained samples have a similar resolution of 800×600 pixels and they are 357 in total, distributed as follows: 62 cases as normal, 96 as G1, 99 as G2 and 100 as G3, respectively.

The images have various intensities, but they are all included in normal cases, as these are obtained from images contained in the representatives from the grades G1–G3.

2.2 Feature Extraction and Proposed Diagnosis Support Methodology

Usually, after the histopathological images are produced, computer-based grading follows several stages, i.e. image preprocessing, feature extraction, reduction of the number of feature and finally classification [1, 3–5, 8].

Although the images possess very different particularities, especially when the comparison is between distinct grades, the glands and the nuclei appear in most of them (there are exceptions with glands that are missing in some images denoting cancer grade 3). There are not many works that take into account both the glands and nuclei at the same time. Most of them concentrate on the nuclei and additionally tackle other textural features [3, 4, 7]. As opposed to the other studies, the focus of the current work thus becomes to detect and use these two common presences alike. For each found component, besides its counting, several characteristics were measured: the area and perimeter were computed, the enclosing circle was found and its radius was also taken into consideration. For every image and for the three measures the average, median, standard deviation and the ratio between the minimum and the maximum value were subsequently calculated.

Efforts for accurately identifying the gland interiors have been previously made ([9] and [10]) and the procedures that proved successful were employed in the current study. Gaussian smoothing is used as a preprocessing step as it showed to be more efficient as opposed to box, normalized box, median or bilateral filtering. It is then followed by a watershed algorithm [11] for boundary-based segmentation of the glands interiors. The watershed algorithm uses a set of marked pixels on the grayscale version of the image for avoiding the creation of too many contours. For that, a thresholding operator is used to transform the image into a binary one. Then, noise elimination has to take place on the black and white image, i.e. the discarding of very small white spots on the black background (erosion), as well as the opposite (dilation). In [9] and [10] an evolutionary algorithm is used to search for good input parameter values for the described methodology and among them the threshold, number of erosions and dilations were included. In the current work, values for these parameters were chosen as the ones that previously proved to be well-suited overall.

Besides the statistical measures referring to the area, perimeter and radius of the glands, the layout of the components is also considered via two graph-based techniques. Consequently, the interior points of the enclosing circles for the detected glands (seeds) are used to draw the Delaunay triangles and Voronoi diagrams. The Delaunay triangles have the property that no point in the initial set lies inside the circles that enclose the formed triangles. The Voronoi diagrams conduct a separation of the plane into regions where all included points are closer to the specific seed of their regions as opposed to the other ones. Euclidean distance is used in the current implementation. An example of the application of the two techniques for four images containing grades G0–G4 is illustrated in Fig. 2. For the obtained triangles and polygons once again the area, perimeter and radius of the circumcircle are calculated and the same average, median, standard deviation, and minimum to maximum ratio are considered for each one of them.

In the current work, various methods had been tested for identifying the nuclei [3, 4, 7], like color quantization, k-means, fitting ellipses to the found contours via Hough Transform, the watershed algorithm and different preprocessing techniques had been tried (various filtering options, the Laplacian to sharpen the image). However, experiments showed that all these in multiple combinations conducted to sub-

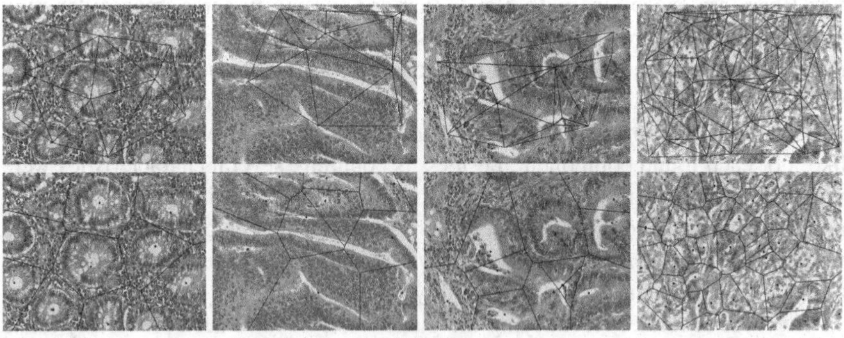

Fig. 2 Delaunay triangles (*first line*) and Voronoi diagrams (*second line*) found for the detected glands in images corresponding to a normal sample, grades 1, 2 and 3 (from *left* to *right*). The points that serve as inputs for the Delaunay triangles and Voronoi diagrams are found as central points for the *circles* that enclose the contours of the detected glands

optimal results as far as the visual appreciation of the physicians is concerned. Contours were obtained with higher precision when the distance transform algorithm [12] was applied for this purpose, as the method deals relatively well with overlapping nuclei [13–15].

The procedure flows as follows:

- The image is transformed into the grayscale variant.
- In a preprocessing step, the image is blurred using a normalized box filter.
- A threshold is applied to the obtained picture in order to obtain a binary one. As noticed in the experimental phase, the accuracy of the methodology depends at a high degree on the choice of this threshold parameter.
- The distance transform is applied on the resulting binary picture. The procedure calculates for every pixel the distance to the closest black one and thus produces a new image with the same size as the initial input. In our experiments an Euclidean distance is employed.
- The new image is then normalized.
- The image is later transformed into a binary one. On the resulting image, the contours of the detected objects represented the found nuclei.

Analogously to the case of gland detection, the area, perimeter, radius of the circumcircle of each detected object are computed and the associated statistical information associated to these are also calculated. The distribution of the found contours is also measured via the Delaunay triangulation and Voronoi diagrams with all the associated measurements as in the case of the glands. An example of the application of the two graph-based techniques for the detected nuclei can be seen for all the different grades in Fig. 3.

Table 1 recapitulates the features that are extracted from each image. As the number of attributes is relatively high, a SVM was employed to deal with the problem, as the methodology is acknowledged to be independent of the data dimensionality in

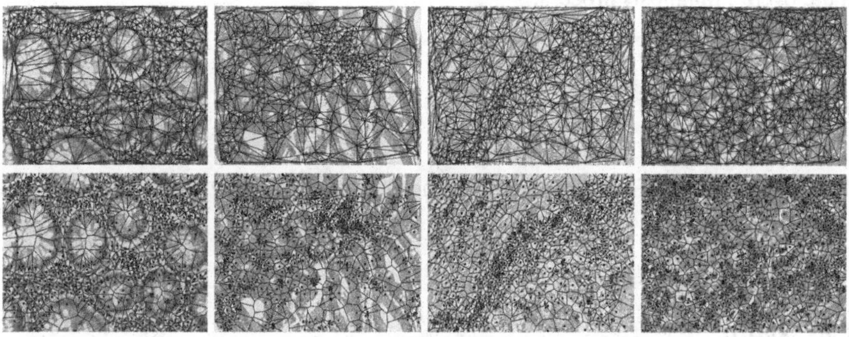

Fig. 3 Delaunay triangles (*first line*) and Voronoi diagrams (*second line*) found for the detected nuclei in another set of images corresponding to a normal sample, grades 1, 2 and 3 (from *left* to *right*). The points that serve as inputs for the Delaunay triangles and Voronoi diagrams are found as central points for the *circles* that enclose the contours of the detected nuclei

Table 1 The 76 features that are extracted from each histopathological image that enter into the subsequent classification process

Feature	Measures	Statistics	Total
Morphological	Area, perimeter, radius for glands and the same three for nuclei	Average, median, standard deviation, min/max	24
	Number of glands and of nuclei	–	2
Topological	Area, perimeter, radius for the Delaunay triangles and the same three for the Voronoi polygons for glands and repeated for nuclei	Average, median, standard deviation, min/max	48
	Number of Delaunay triangles for glands and for nuclei	–	2

The statistics are calculated for each measure on the line, for both glands and nuclei

a decision problem [16]. However, Principal Component Analysis (PCA) for reducing the number of features was also applied and SVM is afterwards utilized on the resulting data.

3 Experimental Results

The considered classification problem contains four classes to distinguish between, is characterized by 76 numerical attributes and contains 357 samples. Each sample is obtained from one histopathological image and the numerical values depend very much on the parameter values set for the procedures that derived them. In the current experiments, the interest lies in achieving a good overall accuracy, finding the accuracy between each pair of grades taken in turn, but also in observing how much does one parameter count (i.e. the threshold applied prior to the distance transform) in the overall diagnosis process.

3.1 Task

Observe the effect of the threshold parameter used to identify the nuclei over the overall accuracy of the classifier.

3.2 Setup

In the pre-experimental tests, it was observed that the SVM with a linear kernel conducted to significantly better results than when using a radial one. The threshold was set to 160, as the physicians visually appreciated the results on several images as reasonable. The PCA reduced the number of attributes from 76 to only 13, as this was the number of components required to explain at least 95 % of the variance. For the PCA-transformed data, the SVM with a radial kernel reached an accuracy that was around 4 percent better than when using a linear kernel.

The threshold that transforms the image into a binary one prior to the application of the distance transform value is tuned from 130 to 200, considering only multiples of 5. Therefore, there are 15 different numerical data sets for the histopathological images and each one is subject to classification by SVM.

The numerical data set is randomly split into 2/3 training samples that are used to instruct the SVM classifier and 1/3 test data that is used for computing the accuracy. The process is repeated 30 times in order to verify the significance of the results.

The overall accuracy is computed as the percent of samples in the test set that are correctly labeled. For a deeper investigation of the results, each grade is considered in turn as opposed to the other ones and several insights are made available through accuracy, sensitivity, specificity, precision, false negative rate and false discovery rate.

3.3 Results

Figure 4 contains an input image (*left*) and the number of nuclei discovered when the threshold parameter is varied from 130 to 200 (*center*). For the same input image as in Fig. 4 (*left*), Fig. 5 illustrates the output of the distance transform method for different threshold values. The plot in the right from Fig. 4 illustrates a comparison

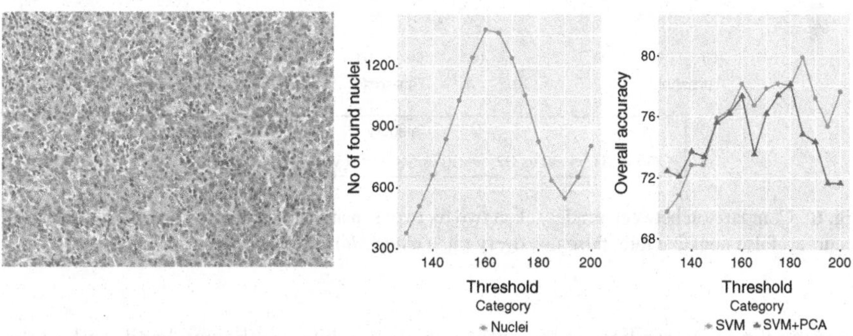

Fig. 4 The detected nuclei depend on the input threshold parameter before the distance transform method. The input (*left*) is a section of a histopathological image of G3 cancer, while the plot in the *middle* contains the number of nuclei there are detected when the input threshold parameter prior to the distance transform step is varied from 130 to 200. The plot on the *right* illustrates a comparison between the SVM applied to the complete data and the same classifier on the PCA-reduced data

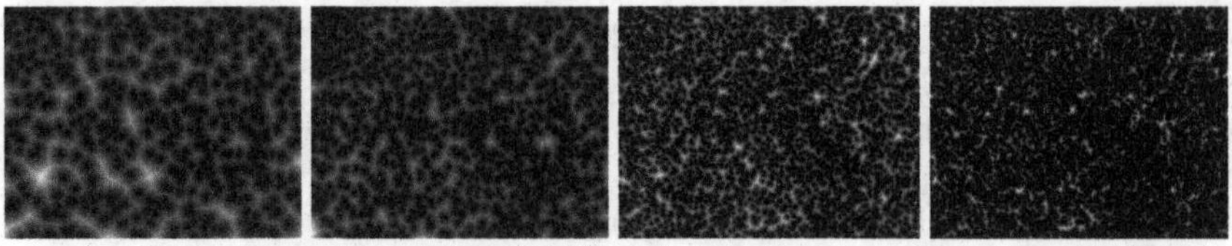

Fig. 5 The distance transform images when the threshold value is 130, 150, 170 and 190, respectively. The input image is the same from Fig. 4 (*left*)

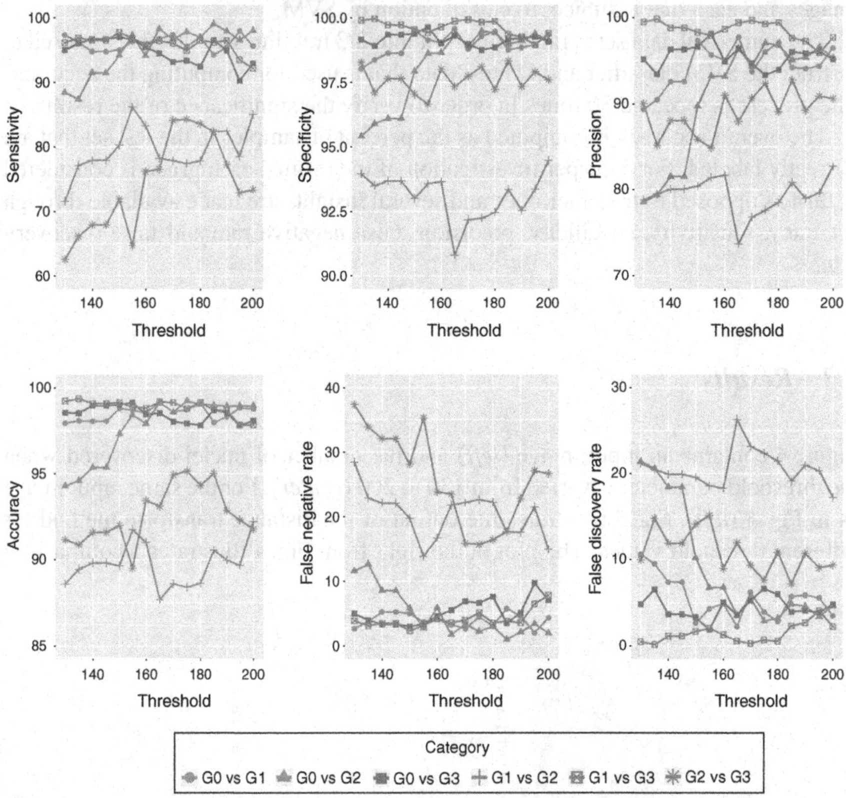

Fig. 6 Comparison between grades taken two by two: sensitivity, specificity, precision (*first line*), accuracy, false negative rate, false discovery rate (*second line*) after SVM classification

between the SVM applied on the entire numerical data, on the one hand, and on the principal components, on the other hand. As the results are generally superior in the case when the data is kept unaltered, the PCA is next left aside and Fig. 6 contains the detailed comparisons for each grade in turn in the direct SVM application only.

3.4 Observations and Discussion

Through its nature, the distance transform achieves a relatively good separation of nuclei that intersect or even marginally overlap. However, if two or more nuclei overlap at a great extent, they will be labeled as one object. As Fig. 4 (*right*) and Fig. 5 suggests, the higher the threshold value is (over 170), the less objects are detected. Although Fig. 5 indicates that more nuclei are detected as the threshold value increases, they overlap in the last image, conducting thus to a smaller number of found objects. Despite the general and natural opinion that the overlapping nuclei should be identified as separate entities in order to provide accurate data to the classifier, the best overall accuracy result (79.89 %) is obtained when the threshold is 185. In contrast to other studies, herein the focus does not lie into clinically assessing the automated identification capabilities, but instead a more objective measure is followed, that of choosing this proper threshold parameter for the detection of nuclei in correspondence to the prediction accuracy only.

The plots in Fig. 6 place the normal tissues (G0) very well as compared to the rest of the grades, especially when the threshold value is higher than 155. The distinction between G0 and the other grades represented a major concern for the current data set because the histopathological images for the normal tissue were all obtained by cutting pieces from larger images that belonged to G1 and mostly G2 and G3. This means that the same light conditions, the same amount of H&E and the exact same settings of the microscope were used when producing them. These results are very encouraging as regards the practical use of the proposed technique because, by removing even solely computer diagnosed G0 images, the pathologist can focus on the most difficult cases.

The weakest results are when comparing G1 versus G2 and G2 versus G3, as all the plots indicate. The specificity, precision and false discovery rate show that the hardest cases to be distinguished are between G1 and G2, as it actually occurs for the pathologists, as well [17].

A similar methodology where both nuclei and glands are taken into account for prostate cancer diagnosis, also with four grades to discriminate between, is presented in [17]. Although the measures used in the study are clearly described, the authors do not provide details about the manner of identifying the two types of components. The same SVM classifier is employed. The accuracies achieved between grades in [17] are between 76.9 and 92.8 %, while in the current study they are between 90.55 % (G1 vs. G2) and 98.98 % (G0 vs. G2) for threshold 185, in Fig. 6 (*first image, second line*). Naturally, the problem is not the same, hence the current image data set is made available [6] for future studies and direct comparison.

4 Conclusions and Future Work

A data set of 357 histopathological images is put forward, each having a resolution of 800×600 pixels, separated in 4, well balanced grades. Also, a diagnosis support methodology that uses morphological and topological features is proposed, which conducts to an overall accuracy of 79.89 %. The glands are discovered by applying a Gaussian filtering followed by a watershed algorithm, while the nuclei are found after a normalized box filter and distance transform are used.

The influence of a threshold parameter used for identifying the nuclei over the results indicates a surprising information: although in some pictures the discovered overlapping nuclei are merged in larger components, it is not only that the accuracy is not decreased, but some improvement is reached.

As the normal tissues are extracted from larger images that contained marginal separations between different grades of cancer and normal tissue, it is intended to add intensity-based features in a future work to see if this can boost the automatic diagnosis or, on the contrary, misleads the identification of the normal tissue from the other ones. Also, other topological information, like co-adjacency matrices could conduct to better results. Probably a gain in prediction could be obtained by fine tuning the parameter values of the methods involved in the identification of the main components.

Acknowledgments Present work was supported by the research grant no. 26/2014, code PN-II-PT-PCCA-2013-4-1153, entitled IMEDIATREAT—Intelligent Medical Information System for the Diagnosis and Monitoring of the Treatment of Patients with Colorectal Neoplasm—financed by the Romanian Ministry of National Education (MEN)—Research and the Executive Agency for Higher Education Research Development and Innovation Funding (UEFISCDI).

References

1. Mills, S.: Histology for Pathologists, 3rd edn. Lippincott Williams & Wilkins (2006)
2. Gorunescu, F., Belciug, S.: Evolutionary strategy to develop learning-based decision systems. Application to breast cancer and liver fibrosis stadialization. J. Biomed. Inf. **49**, 112–118 (2014)
3. Lee, H., Chen, Y.P.P.: Image based computer aided diagnosis system for cancer detection. Expert Syst. Appl. **42**(12), 5356–5365 (2015)
4. Gurcan, M., Boucheron, L., Can, A., Madabhushi, A., Rajpoot, N., Yener, B.: Histopathological image analysis: a review. IEEE Rev. Biomed. Eng. **2**, 147–171 (2009)
5. Demir, C.G., Kandemir, M., Tosun, A.B., Sokmensuer, C.: Automatic segmentation of colon glands using object-graphs. Med. Image. Anal. **14**(1), 1–12 (2010)
6. IMEDIATREAT Project. https://sites.google.com/site/imediatreat/
7. Sertel, O., Kong, J., Catalyurek, U.V., Lozanski, G., Saltz, J.H., Gurcan, M.N.: Histopathological image analysis using model-based intermediate representations and color texture: follicular Lymphoma grading. J. Sign. Process. Syst. **55**, 169–183 (2009)
8. He, L., Long, L.R., Antani, S., Thoma, G.R.: Histology image analysis for carcinoma detection and grading. Comput. Meth. Prog. Bio. **107**(3), 538–556 (2012)
9. Stoean, C., Stoean, R., Sandita, A., Mesina, C., Gruia, C.L., Ciobanu, D.: Evolutionary search for an accurate contour segmentation in histopathological images. In: ACM Genetic and Evolutionary Computation Conference, GECCO Companion, pp. 1491–1492. Spain, Madrid (2015)

10. Stoean, C., Stoean, R., Sandita, A., Mesina, C., Ciobanu, D., Gruia, C.L.: Investigation on parameter effect for semi-automatic contour detection in histopathological image processing. In: IEEE Post-Proceedings of the 17th International Symposium on Symbolic and Numeric Algorithms for Scientific Computing (in press). Timisoara, Romania (2015)
11. Beucher, S., Lantuej, C.: Use of watersheds in contour detection. In: International Workshop on Image Processing, Real-Time Edge and Motion Detection/Estimation. Rennes, France (1979)
12. Kimmel, R., Kiryati, N., Bruckstein, A.M.: Sub-pixel distance maps and weighted distance transforms. J. Math. Imaging Vis. **6**, 223–233 (1996)
13. Irshad, H., Veillard, A., Roux, L., Racoceanu, D.: Methods for nuclei detection, segmentation, and classification in digital histopathology: a review-current status and future potential. IEEE Rev. Biomed. Eng. **7**, 97–114 (2014)
14. Kong, J., Cooper, L., Kurc, T., Brat, D., Saltz, J.: Towards building computerized image analysis framework for nucleus discrimination in microscopy images of diffuse glioma. In: IEEE 33rd Annual International Conference of the IEEE Engineering in Medicine and Biology Society, pp. 6605–6608. Boston, MA, USA (2011)
15. Pang, Q., Yang, C., Fan, Y., Chen, Y.: Overlapped cell image segmentation based on distance transform. In: 6th World Congress on Intelligent Control and Automation, vol. 2, pp. 9858–9861. IEEE Press, Dalian (2006)
16. Joachims, T.: Text categorization with support vector machines: learning with many relevant features. In: 10th European Conference on Machine Learning, pp. 137–142. Springer, London (1998)
17. Doyle, S., Hwang, M, Shah, K., Madabhushi, A., Feldman, M.D., Tomaszeweski, J.E.: Automated grading of prostate cancer using architectural and textural image features. In: IEEE International Symposium on Biomedical Imaging: From Nano to Macro, pp. 1284–1287. Washington, DC (2007)

On Preservation of Video Data When Transmitting in Systems that Use Open Networks

Anton Vostrikov, Mikhail Sergeev and Nikolaj Solovjov

Abstract This paper considers the task of selection of an image processing method to be applied to frames of a video sequence to ensure the best quality at the receiving side of systems that use open networks. The basic definitions of masking and unmasking are provided. The advantages of the frame body pre-scaling method and the masking method based on the standard compression procedure using Hadamard-Mersenne quasi-orthogonal matrices are assessed.

Keywords Image quality • Image compression • Image masking • Image unmasking • Quasi-orthogonal matrices • Hadamard-Mersenne matrices

1 Introduction

The results of research studies show the steady growth of the share of video information in the network traffic of leading countries [1]. According to Cisco, traffic of the transferred "Video on Demand" (IP TV) has been doubling each year starting from 2008, and this trend will continue in the coming years [2]. It is assumed that the video streams will exceed 90 % of the user telecommunications traffic, including video conferences, mobile communications and distributed video

A. Vostrikov (✉) · N. Solovjov
Department of Computer Systems and Networks, Saint-Petersburg
State University of Aerospace Instrumentation, 67 Bolshaya Morskaya street,
Saint Petersburg 190000, Russian Federation
e-mail: vostricov@mail.ru

N. Solovjov
e-mail: famsol@yandex.ru

M. Sergeev
ITMO University, 49 Kronverksky Ave., Saint Petersburg 197101,
Russian Federation
e-mail: mbse@mail.com

G. De Pietro et al. (eds.), *Intelligent Interactive Multimedia Systems and Services 2016*, Smart Innovation, Systems and Technologies 55,
DOI 10.1007/978-3-319-39345-2_14

surveillance systems. This trend makes the task of preservation of video information when transferring over open communication channels an urgent one.

Definition 1 Preservation of video information means ensuring the best possible preservation of image quality in each frame, as well as protection from intentional or accidental distortion, and ensuring its protection from unauthorized access.

The main challenge of preservation is finding a reasonable compromise between user demands and the bandwidths of communication channels, taking into account the ever-increasing size of frames of video sequences.

The complexity of the problem lies in the fact that the video has to be compressed before transfer over today's communication channels. The compression rate is determined, on the one hand, by the channel bandwidth, and on the other hand by the size of frames and frequency of their receipt. The existing methods of lossless compression cannot provide the necessary compression ratio [3], forcing the use of lossy compression methods of two types.

The first type includes methods based on the analysis of adjacent frames to compensate the motion (for example, MPEG-4) while the other type includes methods that compress each frame individually (for example, MPEG-2, JPEG, JPEG2000). Methods of the first type allow for strong compression [4], but their symmetry—the ratio of time spent on compression and decompression, is significantly different from a value of 1.

The most widely used method of the second type (compressing each frame separately) in video data transfer, is the JPEG algorithm [5]. Its advantages are low hardware requirements and symmetry close to a value of 1. This is particularly important when transferring data in on-line mode from surveillance cameras or in interactive services, such as video conferences, where the tolerated end-to-end latency is severely restricted, and, as a rule, should not exceed 150 ms [6]. Moreover, each frame of video is compressed separately in JPEG, which allows for storage of important information in a frame, and extraction if necessary [7].

The main drawback of JPEG is the manifestation of the block frame structure when significant compression is used.

Below, we consider two problems: maintenance of the quality of the transferred video frames, and their protection from intentional distortion and unauthorized access when using the JPEG algorithm.

2 Preservation of Quality of Transferred Images

When the high JPEG compression rate is used on images, the borders of blocks on the decompressed image, to which JPEG breaks the compressed image, become visible. These borders are particularly visible in areas with almost constant or smoothly changing brightness (color), for example, water surfaces, clouds, etc. [8].

The deblocking filter that is used, for example, in the H.264 codec, which is based on the JPEG algorithm, helps to partly avoid the appearance of the block structure, when applied after decompression of a frame [9]. Deblocking operates on 4 × 4 or 8 × 8 blocks of pixels and is able to analyze the level of blockiness of both the whole image, and its part, which allows for processing of the single frame with different levels of filtering. The degree of deblocking and the processing time of each frame can be set by the user.

The main advantage of the deblocking filter is the lack of video stream delays, while its main drawback is the low effectiveness when used on a video stream with the smooth movement of objects [10].

Another approach to reducing the blockiness is scaling "down" (reducing the frame size using the interpolation algorithm) before applying the JPEG-compression [11]. A scaled "down" image is compressed and, after its transfer through the channel and subsequent decompression, it is restored to its original size—scaled "up" using the interpolation algorithm.

As an example, Fig. 1 shows a decompressed image, pre-compressed using the JPEG algorithm with the compression ratio = 1. With such a high compression ratio, the block structure is clearly visible on the image, especially in areas with smooth brightness change. Image 2 shows the result of decompression of an image that was scaled up after scaling down with scaling ratio = 2 and JPEG compression 14. To the human eye, the image on Fig. 2 clearly wins in comparison to the image on Fig. 1 because the block structure is less visible in Fig. 2, and the slight blurring after applying an interpolation algorithm for scaling up is barely visible.

The experiments for the quantitative comparison of the results of image processing by different methods of compression/decompression were performed using the

Fig. 1 Decompressed image after JPEG compression without scaling

Fig. 2 Decompressed image after JPEG compression with scaling

measurement of the structural similarity (SSIM) index proposed by Zhou Wang, an American scientist from New York University, for comparing grayscale images. This measurement is based on the calculation of three similarity components (brightness, contrast, and structure) and the merger of their values in the final result [12]:

$$SSIM = \left(\frac{\sigma_{XY}}{\sigma_X \sigma_Y}\right)\left(\frac{2\overline{X}\overline{Y}}{(\overline{X})^2 + (\overline{Y})^2}\right)\left(\frac{2\sigma_X 2\sigma_Y}{\sigma_X^2 + \sigma_Y^2}\right); \quad (1)$$

where:

$\overline{X} = \frac{1}{mn}\sum_{i=1,j=1}^{m,n} x_{ij}$, $\overline{Y} = \frac{1}{mn}\sum_{i=1,j=1}^{m,n} y_{ij}$—average brightness of pixels of the compared frames X and Y,

$\sigma_X^2 = \frac{1}{(m-1)(n-1)}\sum_{i=1,j=1}^{m,n}(x_{ij} - \overline{X})^2$, $\sigma_Y^2 = \frac{1}{(m-1)(n-1)}\sum_{i=1,j=1}^{m,n}(y_{ij} - \overline{Y})^2$—dispersion of pixel brightness,

$\sigma_{XY}^2 = \frac{1}{(m-1)(n-1)}\sum_{i=1,j=1}^{m,n}(x_{ij} - \overline{X})(y_{ij} - \overline{Y})$—frames covariance.

The first component of the expression is the correlation coefficient between images X and Y. The second component describes the similarity of the average brightness values of the two compared images. The third component characterizes the similarity of contrasts of the two compared images. The higher the SSIM index is, the better the compared images match. Lately, this index has been widely used in the comparative evaluation of image quality because it takes into account the peculiarities of the human eyes perception in the best possible way.

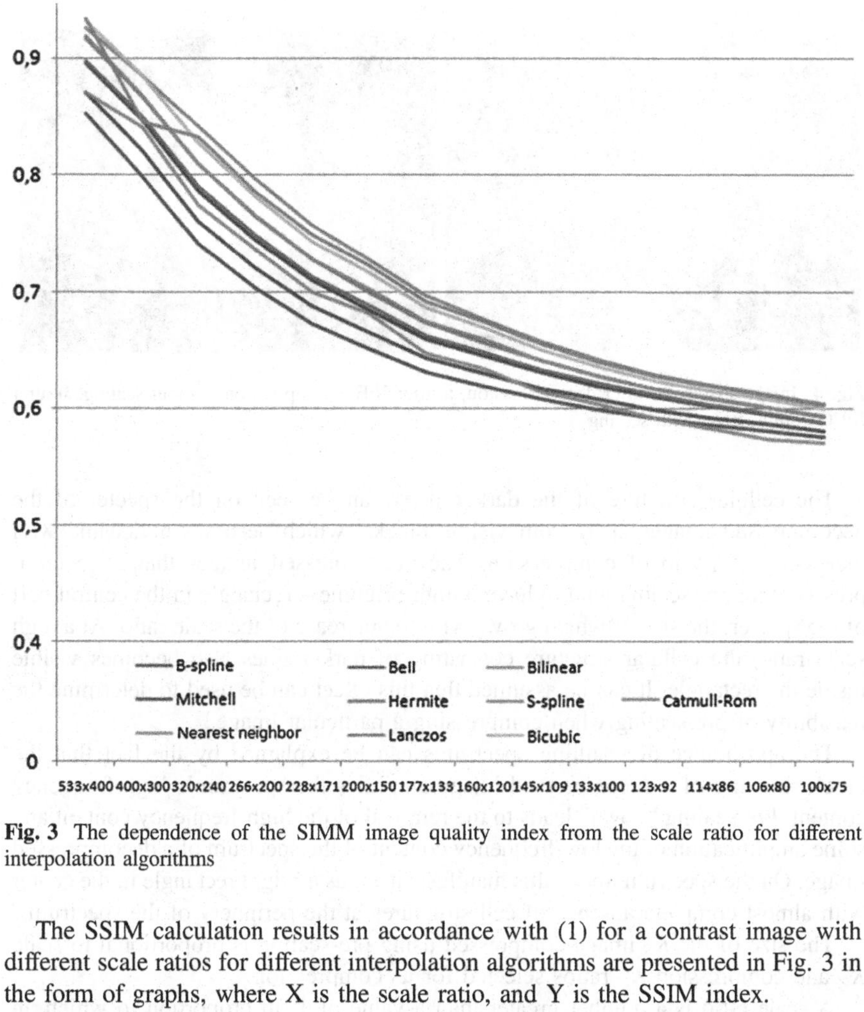

Fig. 3 The dependence of the SIMM image quality index from the scale ratio for different interpolation algorithms

The SSIM calculation results in accordance with (1) for a contrast image with different scale ratios for different interpolation algorithms are presented in Fig. 3 in the form of graphs, where X is the scale ratio, and Y is the SSIM index.

The graphs show the dependence of the SSIM index on the interpolation algorithm given the same scale ratio. With a decrease in scale ratio, the SSIM index is almost identical for interpolation algorithms after an image is restored to its original size. From this, it can be concluded that with an increase in scaling factor, the difference between them disappears. The experiments [11] showed that the bilinear interpolation algorithm is most suitable for scaling up in terms of quality of reconstruction and processing time.

Spectral analysis using the Fourier transformation of images decompressed with pre-scaling and without it demonstrates the dependence of the visual representation of the spectrum as a monochrome image, shown in Fig. 4, on the method and ratio of compression [13]. Figure 4 shows the specter of a decompressed image after JPEG compression without scaling, and Fig. 4b shows the specter of an image after JPEG compression with scaling.

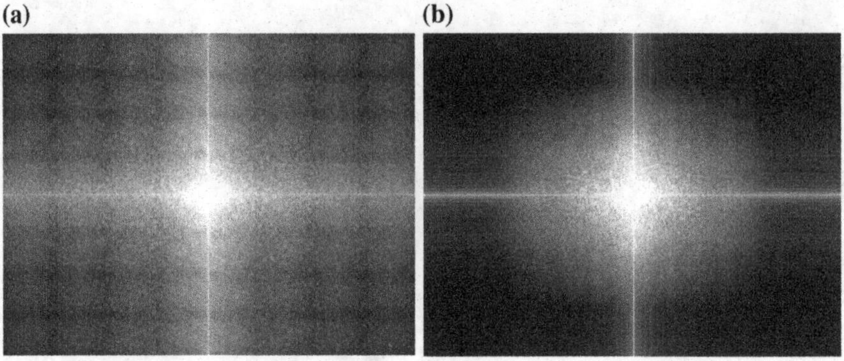

Fig. 4 Image spectrums after decompression, **a** after JPEG compression without scaling, **b** after JPEG compression with scaling

The cellular structure of the darker lines can be seen on the specter of the decompressed images apart from visible blocks, which becomes noticeable with increasing the ratio of compression. The decompressed images that were compressed with pre-scaling tend to have a high-brightness rectangle in the central part of the specter, the size of which grows with the increase of the scale ratio. At a high scale ratio, the cellular structure consisting of darker lines also becomes visible inside the rectangle. It can be assumed that this effect can be used to determine the feasibility of pre-scaling when compressing a particular image.

The appearance of resulting spectrums can be explained by the fact that the block structure of a decompressed image amplifies the spectrum's high-frequency content. Pre-scaling "down" leads to the removal of the high-frequency content and some amplification of the low-frequency content of the spectrum of a decompressed image. On the spectrum image this manifests itself as a bright rectangle in the center with almost complete absence of cell structures at the periphery of the spectrum.

The size of the K_S image compressed using pre-scaling is proportional to scale K_Z and compression K_C ratios selected for its compression.

A scale ratio is a number greater than a value of 1, in proportion to which an image size varies; a compression ratio is an image compression ratio ranging from 100 to 1. The higher the scale ratio is, the smaller is the image size, and the lower the compression ratio is, the higher is the image quality, i.e.,

$$K_S \rightarrow K_Z / K_C \tag{2}$$

Experiments [12] for each K_S value in accordance with (2) allowed for creating the tables showing the dependence of the quality of a restored image from the proportion of scale and compression ratios. Such dependencies were calculated for different types of images: high-contrast, blurred, monochrome and binary (text). They allowed determining the best K_Z and K_C values in terms of quality of the SSIM index given a particular K_S. The thresholds for JPEG compression ratios,

Table 1 Thresholds for JPEG compression ratios for comparison of quality of restored images with and without prescaling

Generation	High-contrast	Monochrome	Blurred	Text
JPEG quality ratio	13	16	11	46

with which the quality of a restored image after pre-scaling compression exceeds the quality of a restored image after compression without scaling, are shown in Table 1 for different types of images.

As seen from Table 1, the threshold value of the JPEG quality ratio, below which the use of pre-scaling is reasonable, depends on the image type. If these ratios for a high-contrast, monochrome and blurred image are similar, the threshold for a text image is much higher. This can be explained by the fact that text is a specific high-contrast image with a large quantity of fine and sharp details (black letters on white background). When images of this type are compressed, scaling followed by stretching leads to a significant blurring of a very large number of significant brightness jumps, which significantly affects the assessment of the quality of a reconstructed image. If the JPEG compression ratio is 46 or less, the quality deterioration due to the appearance of the block structure on a restored image that was compressed without scaling starts to exceed the quality deterioration caused by the blurring of a restored image compressed with pre-scaling.

Table 1 shows that the thresholds for JPEG compression ratios, with which the quality of a restored image after pre-scaling compression exceeds the quality of a restored image after compression without scaling, range from 11 to 16 for different types of images. Accordingly, the greatest effect can be achieved by pre-scaling ratio of 1.5–2.

Studies have shown [13], that the use of this method, in conjunction with the bilinear interpolation method, does not cause a delay in the video stream. Comparisons measuring the structural similarity of the results of decompression of the same frame with the deblocking filter or pre-scaling count in favor of the latter.

Another problem with the task of preserving the quality of the transferred video stream at a comfortable level for the viewer is the periodic display of black blocks on the screen. The reason for their occurrence is the absence of data on the decoder at the time of reproduction of a frame because of errors or delays in data transfer. It is often observed in the transmission of video over wireless networks [14], and even becomes critical in 4G networks because of the rapid growth of video traffic, outrunning the Cisco forecasts about it.

To reduce the probability of such events, the packet transfer of compressed data is used with different priorities and algorithms for data retransmission and delivery confirmation [15]. The most promising and robust technology for data transmission via wireless networks is the Ultra Wide Band (UWB) standard [16], which uses error control (noiseless) coding and data retransmission schemes to fight data transmission errors. The best video-data compression method, in this case, is the TinyProgressive algorithm—a simplified progressive version of JPEG, allowing for a splitting of data into high and low priority categories [17], and transferring it accordingly.

When such a data transmission scheme is used, low priority data is not always delivered, because if there were errors, which were not corrected by error control coding, in received high priority data, the packets are retransferred, which results in the lack of time for transmission of low-priority data.

In the UWB transfer protocol, the main parameter that affects the quality of playback, provided that the transfer speed is fixed, is the size of high-priority data packets (data units). This value can be set in the codec and can change during transmission, adjusting itself to the current number of transmission errors. Increasing the size of the high-priority data units helps increase the quality of playback, but also increases the probability of data unit errors. Consequently, the possibility of loss of low priority data is increased, which, in general, reduces the playback quality.

Theoretical studies, confirmed experimentally, have shown [15] that the dependence of image quality on the size of high-priority data units (packets) at a fixed probability of failure is of a stepped character. This allows for the selection of the maximum possible packet size for high-priority data at a given playback quality and known probability of failure in the transmission channel.

3 Protection of Video Data from Unauthorized Access

Today, there are a great number of methods to protect video-data from unauthorized access that use special channels, protocols, cryptographic methods or primitives [18–20]. However, some of them cannot be used in IP TV, while others cannot be used in real-time because of the great computing power they require. One of the prospective methods is the masking of frames [21] on the transmitting side, and their restoration at the receiving side.

Definition 2 *Masking* is the process of transformation of digital visual information with a short relevance period to a noise-like form in order to protect it from unauthorized access.

Definition 3 After being masked, the resulting data array is called *masked visual information* or just a *masked image.*

Definition 4 *Unmasking* is the process of reverse transformation of masked visual information by way of application of reversed masking operations in order to restore the original content.

Masking can be realized, for example, by way of replacement of well-known bases in the video data compression algorithms with bases originating from the new class of calculated quasi-orthogonal Hadamard-Mersenne, Hadamard-Euler, Hadamard-Fermat, and Mersenne-Walsh matrices [22–24].

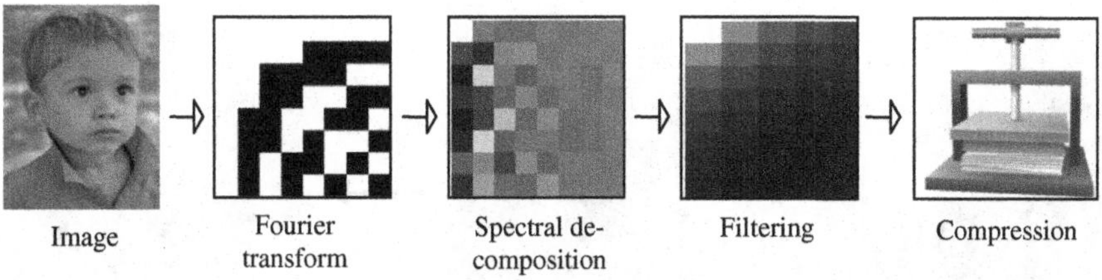

Fig. 5 An example of the sequence of operations in image processing

Definition 5 *Matrix masking and unmasking*—execution of an appropriate procedure of direct or reversed transformation based on the use of real quasi-orthogonal matrices.

The variety of orders of such matrices and the significant number of possible permutations of rows and columns allows for the elimination of redundancy in the video data, and its protection by way of masking.

The typical sequence of processing in matrix-based transformation procedures is shown in Fig. 5.

The use of the discrete Fourier transformation allows the image spectrum with the low-frequency part to be concentrated in the upper left corner of the transformed matrix. Applying a filter then removes the high-frequency part of the spectrum, and statistical Huffman processing eliminates the redundancy.

All phases, shown in Fig. 5, are the same in the masking algorithm, but the Fourier discrete transformation matrix is replaced by an original matrix of an orthogonal basis. This allows for, firstly, saving the possibility in principle of compression of masked information, for example, by adaptation of the filtering procedure to the structural peculiarities of the basis. This helps to make masked and unmasked video streams transmitted over a communication channel indistinguishable.

Secondly, the use of an unusual matrix and a masking key in the form of a vector of permutation of rows and columns, which are unknown to third parties, contribute, as shown by the studies, to the reliable protection of video from interception and spoofing. Examples of portraits of original orthogonal symmetric Hadamard-Mersenne matrices are given in [22]. These matrices are different from the Hadamard matrices because they have real values of elements, which depend on matrix dimension. Such matrices also exist in odd orders and generalize the Hadamard matrices.

Together, these differences lead to a significant complication of the task of unmasking the video by a third party. Figure 6 shows an example of the use of image masking and unmasking procedures.

An additional argument for the use of bases built on sequences of numbers [22–24], including Hadamard-Mersenne and Mersenne-Walsh matrices, is their similarity to Hadamard matrices. The algorithm for construction of Hadamard-Mersenne matrices is fractal, and with certain configurations, the matrices have increased sensitivity to the changes in the processor word length and the initial data.

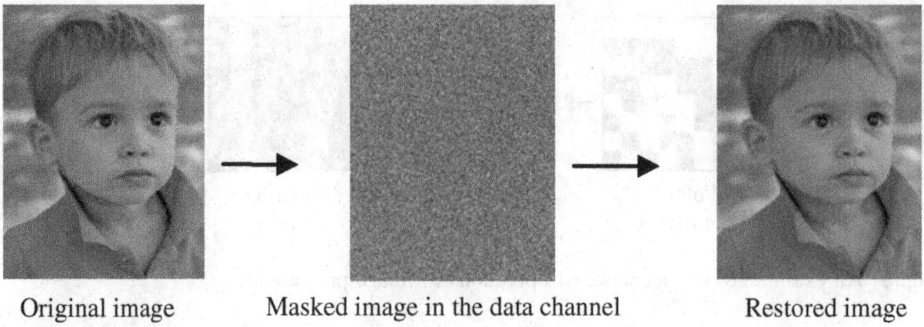

Fig. 6 An example of execution of image masking and unmasking procedures

The task of unmasking by the third party, trying to obtain unauthorized access, is further complicated by the fact that the orthogonal transformation matrix is not calculated in advance and is not transferred through a communication channel, but is a result of the work of the algorithm. Only the configuration of parameters for its calculation is sent as a key via an open communication channel. No less important in this case are the recursive procedures used to increase the matrix order.

4 Conclusion

Considering the above analysis, the following conclusions can be made about the current state of development of technology to preserve video information while transmitting it through public telecommunication networks.

The urgency of the problem persists. The reason is that the constant increase of channel capacity and speed of data processing by hardware and software is countered, in the first place, by the growth of the amount of both open and confidential video information, transferred in real time. Secondly, it's countered by the increasing requirements for the size and the frequency of change of frames being transferred. This leads to a persistent necessity of lossy compression of the video stream transferred in real time.

Improvement of the quality of decompressed frames in terms of reduction of the block effect can be achieved using the methods considered above, in particular scaling of the transmitted images before compression.

A considerable practical importance is to be found for the application of a new class of quasi-orthogonal matrices to increase the degree of noise immunity and protection of video transmission. It should be noted that matrix methods of data transformation are very practical, since they suggest effective implementation in modern microprocessor structures oriented to digital signal processing.

References

1. Cai, L., Shen, X., Mark, J.W.: Multimedia Services in Wireless Internet, Modeling and Analysis, pp. 52–57 (2009)
2. Cisco predicts a nearly 11-fold increase in global mobile data traffic from 2013 to 2018. http://www.infocity.az/?p=17567
3. Santa Cruz, D., et al.: An Analytical Study of JPEG 2000 Functionalities. ISO/IEC JTC1/SC29/WG1 N1816 Coding of Still Pictures, July (2000)
4. Krasilnikov, N.N.: Methods of increase of image compression rate by entropy encoders. In: Informatsionno-upravliaiushchie sistemy [Information and Control Systems], 2004, no. 1, pp. 10–13 (2004)
5. Vatolin, D., Ratushpjak, A., Smirnov, M., Jukin, V.: Methods of Data Compression. Archive Programs, Image and Video Compression. Dialog, Moscow (2003)
6. Apostolopoulos, J.G., Tan, W.T., Wee, S.: Video streaming: concepts, algorithms and systems. Technical Report HPL-2002-260. HP Laboratories, Sept 2002, pp. 1–34 (2002)
7. Sankin, P.S., Litvinov, M.Y.: Special characteristics of the compressed video stream content estimate. In: Informatsionno-upravliaiushchie sistemy [Information and Control Systems], 2009, no. 3, pp. 45–48 (2009)
8. Grigoryan, A.K., Avetisova, N.G.: Some methods of digital watermarks video stream integration. A review. In: Informatsionno-upravliaiushchie sistemy [Information and Control Systems], 2010, no. 2, pp. 38–45 (2010)
9. Lou, J., Jagmohan, A., He, D., Lu, L., Sun, M.: Statistical analysis based H.264 high profile deblocking speed up. In: IEEE International Symposium on Circuits and Systems, 2007 (2007)
10. Chang, S.C., Peng, W.H., Wang, S.H., Chiang, T.: A platform based bus-interleaved architecture for de-blocking filter in H.256/MPEG-4 AVC. IEEE Trans. Consum. Electron. **51**, 2005 (2005)
11. Solovjov, N.V., Shifris, G.V.: Improving the quality of compressed images using pre-scaling. In: Informatsionno-upravliaiushchie sistemy [Information and Control Systems], 2011, no. 3, pp. 15–23 (2011)
12. Wang, Z., Bovik, A.C.: Modern Image Quality Assessment. Morgan & Claypool, New York (2006)
13. Solovjov, N.V., Shifris, G.V.: Using Pre-scaling for video compression in real time. In: Informatsionno-upravliaiushchie sistemy [Information and Control Systems], 2011, no. 4, pp. 2–8 (2011)
14. Zhai, F., et al.: Rate-distortion optimized hybrid error control for real-time packetized video transmission. IEEE Trans. Image Process. **15**(1), 40–53 (2006)
15. Bashun, V.V., Sergeev, A.V.: A model and protocol of real time video data transmission over a wireless channel. In: Informatsionno-upravliaiushchie sistemy [Information and Control Systems], 2007, no. 6, pp. 20–27 (2007)
16. Standard ECMA_368 High Rate Ultra Wideband PHY and MAC Standard, Dec (2005)
17. Pennebaker, W.B., Mitchel, J.L.: JPEG Still Image Data Compression Standard. New York (1993)
18. Grigoryan, A.K., Litvinov, M.Y.: Using the wavelet transformation method for real-time DWM video stream embedding. In: Informatsionno-upravliaiushchie sistemy [Information and Control Systems], 2010, no. 4, pp. 53–56 (2010)
19. Gribunin, V.G., Okov, I.N., Turincev, I.V.: Digital Steganography. Solon-Press, Moscow (2002)
20. Cox, I.J., Kilian, J., Leighton, T., Shamoon, T.G.: Secure spread spectrum watermarking for images, audio and video. In: Proceedings of the IEEE International Conference on Image Processing, vol. 3. pp. 243–246 (1996)
21. Sergeev, A.: Generalized Mersenne matrices and Balonin's conjecture. Autom. Control Comput. Sci. **48**(4), 214–220 (2014). doi:10.3103/S0146411614040063

22. Balonin, Yu.N., Vostrikov, A.A., Sergeev, M.B.: Applied aspects of M-Matrix use. In: Informatsionno-upravliaiushchie sistemy [Information and Control Systems], 2012, no. 1, pp. 92–93 (2012)
23. Vostrikov, A., Balonin, N., Sergeev, M.: Mersenne-Walsh matrices for image processing. In: Intelligent Interactive Multimedia Systems and Services. Smart Innovations, Systems and Technologies, 2015, vol. 40, pp. 141–147. Springer (2015). doi:10.1007/978-3-319-19830-9_13
24. Sergeev, M., Vostrikov, A.: Expansion of the quasi-orthogonal basis to mask images. In: Intelligent Interactive Multimedia Systems and Services. Smart Innovations, Systems and Technologies, 2015, vol. 40, pp. 161–168. Springer (2015). doi:10.1007/978-3-319-19830-9_15

An Invariant Subcode of Linear Code

Sergei V. Fedorenko and Eugenii Krouk

Abstract An invariant subcode of a linear block code under the permutation is introduced. The concept of invariant subcode has two types of applications. The first type is decoding of linear block codes given the group of symmetry. The second type is the attack the McEliece cryptosystem based on codes correcting errors. Several examples illustrating the concept are presented.

Keywords Linear code · Permutation matrix · Quadratic residue code · Golay code

1 Introduction

The concept of invariant subcode was proposed by Krouk [1]. The methods for constructing the representation of linear block codes under permutation were reported in [1, Lemma 8.4] (via sequential constructing a basis for an invariant subcode of a linear block code) and in [2–5] (via block circulant representation of a linear block code). The different representations of linear block codes such as double circulant and quasi-cyclic codes are described in [6, Chapter 16.7].

The concept of invariant subcode has two types of applications. The first type is decoding of linear block codes given the group of symmetry [3, 5]. The second type is the attack the McEliece cryptosystem based on codes correcting errors [1].

The remainder of this paper is organized as follows. In Sect. 2, we propose the invariant subcode concept. In Sect. 3, we presented several examples illustrating the concept.

S.V. Fedorenko (✉) · E. Krouk
St. Petersburg State University of Aerospace Instrumentation, Saint Petersburg, Russia
e-mail: sergei.fedorenko@gmail.com

E. Krouk
e-mail: ekrouk@vu.spb.ru

G. De Pietro et al. (eds.), *Intelligent Interactive Multimedia Systems and Services 2016*, Smart Innovation, Systems and Technologies 55,
DOI 10.1007/978-3-319-39345-2_15

2 Invariant Subcode Concept

Let (n, k) code $\mathcal{G}$ be a binary linear block code with a codelength n and k information symbols. The code $\mathcal{G}$ has a generator matrix G and a parity-check matrix H. Let us introduce a permutation matrix for the code. The permutation matrix P for the code $\mathcal{G}$ has a property $GP = MG$ for some nonsingular matrix M. Let codeword $a \in \mathcal{G}$ be an invariant vector under the permutation matrix P such that $aP = a$. Then $aP = aI$ and $a(P - I) = 0$ where I is an identity matrix. All invariant codewords under the permutation matrix P form a subcode $\mathcal{S}$ of the code $\mathcal{G}$. Therefore

$$\begin{cases} aH^T = 0 \\ a(P - I) = 0, \end{cases}$$

$$a\left(\frac{H}{(P-I)^T}\right)^T = 0.$$

The matrix

$$S = \left(\frac{H}{(P-I)^T}\right)^T = 0.$$

is a parity-check matrix of the subcode $\mathcal{S}$. Finally, we obtain an invariant subcode $\mathcal{S} \subset \mathcal{G}$ under the permutation matrix P.

Proposition 1 *If the permutation matrix $P = (p_{i,j})$, $i,j = 1, \ldots, n$, has properties*
1. $p_{i,i} = 0$ *for* $i = 1, \ldots, n$,
2. $\operatorname{ord} P = l$, *l is a prime,*

then the invariant subcode $\mathcal{S} \subset \mathcal{G}$ under the permutation matrix P consists of l repeating submatrices. Thus a generator matrix G_P of code $\mathcal{S}$ has the form

$$G_P = (\underbrace{C \mid C \mid \cdots \mid C}_{l \text{ times}}).$$

Proof The proof is trivial. □

3 Examples

3.1 The Golay Code Under Order 2 Permutation

The generator matrix of the Golay code is given by

$$G = \begin{pmatrix} G_1 & G_2 \\ C & C \end{pmatrix}$$

$$
= \left(\begin{array}{c|c}
1\,0\,0\,0\,0\,0\,0\,1\,0\,1\,0\,1 & 0\,0\,0\,0\,0\,0\,0\,1\,1\,0\,1\,1 \\
0\,1\,0\,0\,0\,0\,0\,0\,0\,0\,1\,0 & 0\,0\,0\,0\,0\,1\,0\,1\,1\,1\,1\,1 \\
0\,0\,1\,0\,0\,1\,0\,1\,0\,1\,1\,0 & 0\,0\,0\,0\,0\,0\,0\,0\,1\,1\,0\,1 \\
0\,0\,0\,1\,0\,1\,0\,1\,1\,0\,0\,0 & 0\,0\,0\,0\,0\,0\,0\,1\,1\,1\,1\,0 \\
0\,0\,0\,0\,1\,1\,0\,0\,1\,1\,0\,0 & 0\,0\,0\,0\,0\,1\,0\,0\,1\,0\,1\,1 \\
0\,0\,0\,0\,0\,1\,1\,1\,0\,0\,0\,1 & 0\,0\,0\,0\,0\,1\,0\,0\,0\,1\,1\,1 \\
\hline
1\,0\,0\,0\,0\,0\,0\,0\,1\,1\,1\,0 & 1\,0\,0\,0\,0\,0\,0\,0\,1\,1\,1\,0 \\
0\,1\,0\,0\,0\,1\,0\,1\,1\,1\,0\,1 & 0\,1\,0\,0\,0\,1\,0\,1\,1\,1\,0\,1 \\
0\,0\,1\,0\,0\,1\,0\,1\,1\,0\,1\,1 & 0\,0\,1\,0\,0\,1\,0\,1\,1\,0\,1\,1 \\
0\,0\,0\,1\,0\,1\,0\,0\,0\,1\,1\,0 & 0\,0\,0\,1\,0\,1\,0\,0\,0\,1\,1\,0 \\
0\,0\,0\,0\,1\,0\,0\,0\,0\,1\,1\,1 & 0\,0\,0\,0\,1\,0\,0\,0\,0\,1\,1\,1 \\
0\,0\,0\,0\,0\,0\,1\,1\,0\,1\,1\,0 & 0\,0\,0\,0\,0\,0\,1\,1\,0\,1\,1\,0
\end{array}\right).
$$

Let permutation matrix be

$$
P = \begin{pmatrix} 0 & I_{12} \\ I_{12} & 0 \end{pmatrix},
$$

where I_{12} is the 12×12 identity matrix. The generator matrix of the invariant subcode under the permutation matrix P is

$$
G_P = (C|C).
$$

3.2 The Golay Code Under Order 3 Permutation

Let us consider the Turyn-construction of the Golay code [6, Chapter 18.7.4]. The generator matrix of the Golay code is given by

$$
G = \begin{pmatrix} G_1 & 0 & G_1 \\ 0 & G_1 & G_1 \\ C & C & C \end{pmatrix}
$$

$$
= \left(\begin{array}{c|c|c}
1\,0\,1\,1\,0\,0\,0\,1 & 0\,0\,0\,0\,0\,0\,0\,0 & 1\,0\,1\,1\,0\,0\,0\,1 \\
0\,1\,0\,1\,1\,0\,0\,1 & 0\,0\,0\,0\,0\,0\,0\,0 & 0\,1\,0\,1\,1\,0\,0\,1 \\
0\,0\,1\,0\,1\,1\,0\,1 & 0\,0\,0\,0\,0\,0\,0\,0 & 0\,0\,1\,0\,1\,1\,0\,1 \\
0\,0\,0\,1\,0\,1\,1\,1 & 0\,0\,0\,0\,0\,0\,0\,0 & 0\,0\,0\,1\,0\,1\,1\,1 \\
\hline
0\,0\,0\,0\,0\,0\,0\,0 & 1\,0\,1\,1\,0\,0\,0\,1 & 1\,0\,1\,1\,0\,0\,0\,1 \\
0\,0\,0\,0\,0\,0\,0\,0 & 0\,1\,0\,1\,1\,0\,0\,1 & 0\,1\,0\,1\,1\,0\,0\,1 \\
0\,0\,0\,0\,0\,0\,0\,0 & 0\,0\,1\,0\,1\,1\,0\,1 & 0\,0\,1\,0\,1\,1\,0\,1 \\
0\,0\,0\,0\,0\,0\,0\,0 & 0\,0\,0\,1\,0\,1\,1\,1 & 0\,0\,0\,1\,0\,1\,1\,1 \\
\hline
1\,1\,0\,1\,0\,0\,0\,1 & 1\,1\,0\,1\,0\,0\,0\,1 & 1\,1\,0\,1\,0\,0\,0\,1 \\
0\,1\,1\,0\,1\,0\,0\,1 & 0\,1\,1\,0\,1\,0\,0\,1 & 0\,1\,1\,0\,1\,0\,0\,1 \\
0\,0\,1\,1\,0\,1\,0\,1 & 0\,0\,1\,1\,0\,1\,0\,1 & 0\,0\,1\,1\,0\,1\,0\,1 \\
0\,0\,0\,1\,1\,0\,1\,1 & 0\,0\,0\,1\,1\,0\,1\,1 & 0\,0\,0\,1\,1\,0\,1\,1
\end{array}\right).
$$

Let permutation matrix be

$$P = \begin{pmatrix} 0 & I_8 & 0 \\ 0 & 0 & I_8 \\ I_8 & 0 & 0 \end{pmatrix},$$

where I_8 is the 8×8 identity matrix. The generator matrix of the invariant subcode under the permutation matrix P is

$$(G_P = C|C|C).$$

3.3 The Golay Code Under Order 4 Permutation

The generator matrix of the Golay code is given by

$$G = \begin{pmatrix} G_1 & G_2 & 0 & G_3 \\ G_3 & G_1 & G_2 & 0 \\ 0 & G_3 & G_1 & G_2 \\ C & C & C & C \end{pmatrix}$$

$$= \left(\begin{array}{c|c|c|c}
0\,1\,1\,0\,0\,1 & 1\,1\,1\,1\,0\,0 & 0\,0\,0\,0\,0\,0 & 0\,0\,0\,1\,0\,0 \\
0\,1\,0\,1\,1\,1 & 1\,0\,1\,0\,1\,0 & 0\,0\,0\,0\,0\,0 & 0\,0\,0\,0\,1\,0 \\
1\,1\,1\,0\,1\,0 & 1\,0\,1\,0\,0\,1 & 0\,0\,0\,0\,0\,0 & 0\,0\,0\,0\,0\,1 \\
\hline
0\,0\,0\,1\,0\,0 & 0\,1\,1\,0\,0\,1 & 1\,1\,1\,1\,0\,0 & 0\,0\,0\,0\,0\,0 \\
0\,0\,0\,0\,1\,0 & 0\,1\,0\,1\,1\,1 & 1\,0\,1\,0\,1\,0 & 0\,0\,0\,0\,0\,0 \\
0\,0\,0\,0\,0\,1 & 1\,1\,1\,0\,1\,0 & 1\,0\,1\,0\,0\,1 & 0\,0\,0\,0\,0\,0 \\
\hline
0\,0\,0\,0\,0\,0 & 0\,0\,0\,1\,0\,0 & 0\,1\,1\,0\,0\,1 & 1\,1\,1\,1\,0\,0 \\
0\,0\,0\,0\,0\,0 & 0\,0\,0\,0\,1\,0 & 0\,1\,0\,1\,1\,1 & 1\,0\,1\,0\,1\,0 \\
0\,0\,0\,0\,0\,0 & 0\,0\,0\,0\,0\,1 & 1\,1\,1\,0\,1\,0 & 1\,0\,1\,0\,0\,1 \\
\hline
1\,0\,0\,0\,0\,1 & 1\,0\,0\,0\,0\,1 & 1\,0\,0\,0\,0\,1 & 1\,0\,0\,0\,0\,1 \\
1\,1\,1\,1\,1\,1 & 1\,1\,1\,1\,1\,1 & 1\,1\,1\,1\,1\,1 & 1\,1\,1\,1\,1\,1 \\
0\,1\,0\,0\,1\,0 & 0\,1\,0\,0\,1\,0 & 0\,1\,0\,0\,1\,0 & 0\,1\,0\,0\,1\,0
\end{array}\right).$$

Let permutation matrix be

$$P = \begin{pmatrix} 0 & I_6 & 0 & 0 \\ 0 & 0 & I_6 & 0 \\ 0 & 0 & 0 & I_6 \\ I_6 & 0 & 0 & 0 \end{pmatrix},$$

where I_6 is the 6×6 identity matrix. The generator matrix of the invariant subcode under the permutation matrix P is

$$G_P = (C|C|C|C).$$

3.4 The Golay Code Under Order 6 Permutation

The generator matrix of the Golay code is given by

$$G = \begin{pmatrix} G_1 & 0 & 0 & G_2 & G_3 & G_4 \\ G_4 & G_1 & 0 & 0 & G_2 & G_3 \\ G_3 & G_4 & G_1 & 0 & 0 & G_2 \\ G_2 & G_3 & G_4 & G_1 & 0 & 0 \\ 0 & G_2 & G_3 & G_4 & G_1 & 0 \\ C & C & C & C & C & C \end{pmatrix}$$

$$= \left(\begin{array}{c|c|c|c|c|c} 0\,1\,1\,0 & 0\,0\,0\,0 & 0\,0\,0\,0 & 0\,0\,1\,0 & 1\,0\,1\,0 & 0\,1\,1\,1 \\ 0\,1\,0\,1 & 0\,0\,0\,0 & 0\,0\,0\,0 & 0\,0\,0\,0 & 0\,1\,1\,1 & 1\,1\,0\,1 \\ \hline 0\,1\,1\,1 & 0\,1\,1\,0 & 0\,0\,0\,0 & 0\,0\,0\,0 & 0\,0\,1\,0 & 1\,0\,1\,0 \\ 1\,1\,0\,1 & 0\,1\,0\,1 & 0\,0\,0\,0 & 0\,0\,0\,0 & 0\,0\,0\,0 & 0\,1\,1\,1 \\ \hline 1\,0\,1\,0 & 0\,1\,1\,1 & 0\,1\,1\,0 & 0\,0\,0\,0 & 0\,0\,0\,0 & 0\,0\,1\,0 \\ 0\,1\,1\,1 & 1\,1\,0\,1 & 0\,1\,0\,1 & 0\,0\,0\,0 & 0\,0\,0\,0 & 0\,0\,0\,0 \\ \hline 0\,0\,1\,0 & 1\,0\,1\,0 & 0\,1\,1\,1 & 0\,1\,1\,0 & 0\,0\,0\,0 & 0\,0\,0\,0 \\ 0\,0\,0\,0 & 0\,1\,1\,1 & 1\,1\,0\,1 & 0\,1\,0\,1 & 0\,0\,0\,0 & 0\,0\,0\,0 \\ \hline 0\,0\,0\,0 & 0\,0\,1\,0 & 1\,0\,1\,0 & 0\,1\,1\,1 & 0\,1\,1\,0 & 0\,0\,0\,0 \\ 0\,0\,0\,0 & 0\,0\,0\,0 & 0\,1\,1\,1 & 1\,1\,0\,1 & 0\,1\,0\,1 & 0\,0\,0\,0 \\ \hline 1\,0\,0\,1 & 1\,0\,0\,1 & 1\,0\,0\,1 & 1\,0\,0\,1 & 1\,0\,0\,1 & 1\,0\,0\,1 \\ 1\,1\,1\,1 & 1\,1\,1\,1 & 1\,1\,1\,1 & 1\,1\,1\,1 & 1\,1\,1\,1 & 1\,1\,1\,1 \end{array}\right).$$

Let permutation matrix be

$$P = \begin{pmatrix} 0 & I_4 & 0 & 0 & 0 & 0 \\ 0 & 0 & I_4 & 0 & 0 & 0 \\ 0 & 0 & 0 & I_4 & 0 & 0 \\ 0 & 0 & 0 & 0 & I_4 & 0 \\ 0 & 0 & 0 & 0 & 0 & I_4 \\ I_4 & 0 & 0 & 0 & 0 & 0 \end{pmatrix},$$

where I_4 is the 4×4 identity matrix. The generator matrix of the invariant subcode under the permutation matrix P is

$$G_P = (C|C|C|C|C|C).$$

3.5 The (48,24) Quadratic Residue Code Under Order 2 Permutation

The generator matrix of the (48,24) quadratic residue code is given by

$$G = \begin{pmatrix} G_1 & G_2 \\ C & C \end{pmatrix}$$

$$= \left(\begin{array}{c|c}
100000000000100111111011 & 000000000000111011101111 \\
010000000000110101001110 & 000000000000011100100000 \\
001000000000110101001001 & 000000000000110100111111 \\
000100000000111001111110 & 000000000000010100011110 \\
000010000000001111011111 & 000000000000110000011011 \\
000001000000000100001011 & 000000000000010111101100 \\
000000100000101010111001 & 000000000000110101010111 \\
000000010000011111000010 & 000000000000110011111101 \\
000000001000101000100110 & 000000000000111100000101 \\
000000000100001110101101 & 000000000000001000010101 \\
000000000010100010000001 & 000000000000111110001110 \\
000000000001100101110110 & 000000000000001001000011 \\
\hline
100000000000011100010100 & 100000000000011100010100 \\
010000000000101001101110 & 010000000000101001101110 \\
001000000000000001110110 & 001000000000000001110110 \\
000100000000101101100000 & 000100000000101101100000 \\
000010000000111111000100 & 000010000000111111000100 \\
000001000000010011100111 & 000001000000010011100111 \\
000000100000011111101110 & 000000100000011111101110 \\
000000010000101100111111 & 000000010000101100111111 \\
000000001000010100100011 & 000000001000010100100011 \\
000000000100000110111000 & 000000000100000110111000 \\
000000000010011100001111 & 000000000010011100001111 \\
000000000001110111110101 & 000000000001110111110101
\end{array}\right).$$

Let permutation matrix be

$$P = \begin{pmatrix} 0 & I_{24} \\ I_{24} & 0 \end{pmatrix},$$

where I_{24} is the 24×24 identity matrix. The generator matrix of the invariant subcode under the permutation matrix P is

$$G_P = (C|C).$$

3.6 The (48,24) Quadratic Residue Code Under Order 3 Permutation

The generator matrix of the (48,24) quadratic residue code is given by

$$G = \begin{pmatrix} G_1 & G_2 & 0 \\ 0 & G_1 & G_2 \\ C & C & C \end{pmatrix}$$

$$
= \left(\begin{array}{c|c|c}
1000001001010101 & 0011110110011011 & 0000000000000000 \\
0100001000101010 & 1000101101000110 & 0000000000000000 \\
0010001001100001 & 1110001101000001 & 0000000000000000 \\
0001001001100110 & 1000011110100000 & 0000000000000000 \\
0000101000110111 & 1100110010101011 & 0000000000000000 \\
0000010001110011 & 0110000010011001 & 0000000000000000 \\
0000000100101011 & 0110010101010010 & 0000000000000000 \\
0000000011111101 & 1010101111100100 & 0000000000000000 \\
\hline
0000000000000000 & 1000001001010101 & 0011110110011011 \\
0000000000000000 & 0100001000101010 & 1000101101000110 \\
0000000000000000 & 0010001001100001 & 1110001101000001 \\
0000000000000000 & 0001001001100110 & 1000011110100000 \\
0000000000000000 & 0000101000110111 & 1100110010101011 \\
0000000000000000 & 0000010001110011 & 0110000010011001 \\
0000000000000000 & 0000000100101011 & 0110010101010010 \\
0000000000000000 & 0000000011111101 & 1010101111100100 \\
\hline
1011111111001110 & 1011111111001110 & 1011111111001110 \\
1100100101101100 & 1100100101101100 & 1100100101101100 \\
1100000100100000 & 1100000100100000 & 1100000100100000 \\
1001010111000110 & 1001010111000110 & 1001010111000110 \\
1100011010011100 & 1100011010011100 & 1100011010011100 \\
0110010011101010 & 0110010011101010 & 0110010011101010 \\
0110010001111001 & 0110010001111001 & 0110010001111001 \\
1010101100011001 & 1010101100011001 & 1010101100011001
\end{array}\right).
$$

Let permutation matrix be

$$
P = \begin{pmatrix} 0 & I_{16} & 0 \\ 0 & 0 & I_{16} \\ I_{16} & 0 & 0 \end{pmatrix},
$$

where I_{16} is the 16×16 identity matrix. The generator matrix of the invariant subcode under the permutation matrix P is

$$
G_P = (C|C|C).
$$

3.7 *The (48,24) Quadratic Residue Code Under Order 4 Permutation*

The generator matrix of the (48,24) quadratic residue code is given by

$$
G = \begin{pmatrix} G_1 & 0 & G_2 & G_3 \\ G_3 & G_1 & 0 & G_2 \\ G_2 & G_3 & G_1 & 0 \\ C & C & C & C \end{pmatrix}
$$

$$
= \left(\begin{array}{c|c|c|c}
100000110010 & 000000000000 & 000000100000 & 001101010111 \\
000110101011 & 000000000000 & 000000010000 & 000000110111 \\
001100001010 & 000000000000 & 000000001101 & 011000110001 \\
000000111101 & 000000000000 & 000000000010 & 001001101101 \\
101111001100 & 000000000000 & 000000000000 & 100101011000 \\
111000001001 & 000000000000 & 000000000000 & 000011011111 \\
\hline
001101010111 & 100000110010 & 000000000000 & 000000100000 \\
000000110111 & 000110101011 & 000000000000 & 000000010000 \\
011000110001 & 001100001010 & 000000000000 & 000000001101 \\
001001101101 & 000000111101 & 000000000000 & 000000000010 \\
100101011000 & 101111001100 & 000000000000 & 000000000000 \\
000011011111 & 111000001001 & 000000000000 & 000000000000 \\
\hline
000000100000 & 001101010111 & 100000110010 & 000000000000 \\
000000010000 & 000000110111 & 000110101011 & 000000000000 \\
000000001101 & 011000110001 & 001100001010 & 000000000000 \\
000000000010 & 001001101101 & 000000111101 & 000000000000 \\
000000000000 & 100101011000 & 101111001100 & 000000000000 \\
000000000000 & 000011011111 & 111000001001 & 000000000000 \\
\hline
101101000101 & 101101000101 & 101101000101 & 101101000101 \\
000110001100 & 000110001100 & 000110001100 & 000110001100 \\
010100110110 & 010100110110 & 010100110110 & 010100110110 \\
001001010010 & 001001010010 & 001001010010 & 001001010010 \\
001010010100 & 001010010100 & 001010010100 & 001010010100 \\
111011010110 & 111011010110 & 111011010110 & 111011010110
\end{array}\right).
$$

Let permutation matrix be

$$
P = \begin{pmatrix} 0 & I_{12} & 0 & 0 \\ 0 & 0 & I_{12} & 0 \\ 0 & 0 & 0 & I_{12} \\ I_{12} & 0 & 0 & 0 \end{pmatrix},
$$

where I_{12} is the 12×12 identity matrix. The generator matrix of the invariant subcode under the permutation matrix P is

$$
G_P = (C|C|C|C)
$$

3.8 The (48,24) Quadratic Residue Code Under Order 6 Permutation

The generator matrix of the (48,24) quadratic residue code is given by

$$
G = \begin{pmatrix} G_1 & 0 & 0 & G_2 & G_3 & G_4 \\ G_4 & G_1 & 0 & 0 & G_2 & G_3 \\ G_3 & G_4 & G_1 & 0 & 0 & G_2 \\ G_2 & G_3 & G_4 & G_1 & 0 & 0 \\ 0 & G_2 & G_3 & G_4 & G_1 & 0 \\ C & C & C & C & C & C \end{pmatrix}
$$

$$
= \left(\begin{array}{c|c|c|c|c|c}
11110100 & 00000000 & 00000000 & 00001000 & 00001011 & 11000001 \\
11001001 & 00000000 & 00000000 & 00000100 & 10100011 & 10010001 \\
11000001 & 00000000 & 00000000 & 00000010 & 10011010 & 00111010 \\
11000111 & 00000000 & 00000000 & 00000001 & 10001001 & 10100100 \\
\hline
11000001 & 11110100 & 00000000 & 00000000 & 00001000 & 00001011 \\
10010001 & 11001001 & 00000000 & 00000000 & 00000100 & 10100011 \\
00111010 & 11000001 & 00000000 & 00000000 & 00000010 & 10011010 \\
10100100 & 11000111 & 00000000 & 00000000 & 00000001 & 10001001 \\
\hline
00001011 & 11000001 & 11110100 & 00000000 & 00000000 & 00001000 \\
10100011 & 10010001 & 11001001 & 00000000 & 00000000 & 00000100 \\
10011010 & 00111010 & 11000001 & 00000000 & 00000000 & 00000010 \\
10001001 & 10100100 & 11000111 & 00000000 & 00000000 & 00000001 \\
\hline
00001000 & 00001011 & 11000001 & 11110100 & 00000000 & 00000000 \\
00000100 & 10100011 & 10010001 & 11001001 & 00000000 & 00000000 \\
00000010 & 10011010 & 00111010 & 11000001 & 00000000 & 00000000 \\
00000001 & 10001001 & 10100100 & 11000111 & 00000000 & 00000000 \\
\hline
00000000 & 00001000 & 00001011 & 11000001 & 11110100 & 00000000 \\
00000000 & 00000100 & 10100011 & 10010001 & 11001001 & 00000000 \\
00000000 & 00000010 & 10011010 & 00111010 & 11000001 & 00000000 \\
00000000 & 00000001 & 10001001 & 10100100 & 11000111 & 00000000 \\
\hline
00110110 & 00110110 & 00110110 & 00110110 & 00110110 & 00110110 \\
11111111 & 11111111 & 11111111 & 11111111 & 11111111 & 11111111 \\
01100011 & 01100011 & 01100011 & 01100011 & 01100011 & 01100011 \\
11101011 & 11101011 & 11101011 & 11101011 & 11101011 & 11101011
\end{array}\right).
$$

Let permutation matrix be

$$
P = \begin{pmatrix}
0 & I_8 & 0 & 0 & 0 & 0 \\
0 & 0 & I_8 & 0 & 0 & 0 \\
0 & 0 & 0 & I_8 & 0 & 0 \\
0 & 0 & 0 & 0 & I_8 & 0 \\
0 & 0 & 0 & 0 & 0 & I_8 \\
I_8 & 0 & 0 & 0 & 0 & 0
\end{pmatrix},
$$

where I_8 is the 8×8 identity matrix. The generator matrix of the invariant subcode under the permutation matrix P is

$$G_P = (C|C|C|C|C|C).$$

4 Conclusion

The invariant subcode concept is introduced. The two types of applications (decoding of linear block codes and the attack the McEliece cryptosystem) are pointed out.

Acknowledgments The reported study was funded by RFBR according to the research project No. 16-01-00716 a.

References

1. Kabatiansky, G., Krouk, E., Semenov, S.: Error Correcting Coding and Security for Data Networks: Analysis of the Superchannel Concept. Wiley, West Sussex (2005)
2. Krouk, E.A., Fedorenko, S.V.: Decoding by generalized information sets. Problemy Peredachi Informatsii **31**(2), 54–61 (1995) (in Russian); English translation in Problems of Information Transmission **31**(2), 143–149 (1995)
3. Fedorenko, S., Krouk, A.: About block circulant representation of linear codes. In: Proceedings of Sixth International Workshop on Algebraic and Combinatorial Coding Theory at Pskov, Russia, pp. 116–118 (1998)
4. Fedorenko, S.: On the structure of linear block codes given the group of symmetry. In: Proceedings of IEEE International Workshop on Concatenated Codes, Schloss Reisensburg by Ulm, Germany (1999)
5. Fedorenko, S., Krouk, A.: The table decoders of quadratic-residue codes. In: Proceedings of Seventh International Workshop on Algebraic and Combinatorial Coding Theory at Bansko, Bulgaria, pp. 137–140 (2000)
6. MacWilliams, F.J., Sloane, N.J.A.: The Theory of Error-Correcting Codes. North-Holland Publishing Company, Amsterdam-New York-Oxford (1977)

Development Prospects of the Visual Data Compression Technologies and Advantages of New Approaches

Anton Vostrikov and Mikhail Sergeev

Abstract The static image and video information compression algorithms development over the last 15–20 years, as well as standardized and non-standardized formats for data storage and transmission have been analyzed; the main factors affecting the further development of approaches that eliminate the redundancy of transmitted and stored visual information have been studied. The conclusion on the current prospects for the development of image compression technologies has been made. New approaches that use new low-level quasi-orthogonal matrices as transform operators have been defined. The advantages of such approaches opening new fundamentally different opportunities in the field of applied processing of digital visual information have been identified and presented.

Keywords Image compression · Video compression · Protection against unauthorized access · M-matrices · Audiovisual information coding · Metadata

1 Introduction

The compression of images and video information (successive images of the same scene) has been under research and constant application for the last 40 years. The following factors were the main stimulus to that:

- Great importance of visual information as such for people;
- Continuous development of microelectronics and computers, as well as information recording, storage, and playback technology;

A. Vostrikov (✉) · M. Sergeev
Department of Computer Systems and Networks, Saint-Petersburg State University of Aerospace Instrumentation, 67 Bolshaya Morskaya street, Saint Petersburg 190000, Russian Federation
e-mail: vostricov@mail.ru

M. Sergeev
e-mail: mbse@mail.ru

G. De Pietro et al. (eds.), *Intelligent Interactive Multimedia Systems and Services 2016*, Smart Innovation, Systems and Technologies 55,
DOI 10.1007/978-3-319-39345-2_16

- Emergence and development of network technology providing high-speed information exchange not only for personal or local use, but internationally.

The last stimulus led to the emergence of planetary-scale services that allow people to instantly share images and videos or provide their own visual materials for public display for long periods of time. Last but not least, of course, are the so-called "social media" services, which also seek to provide users with all the necessary features to use visual information along with the text. For example, the number of new images posted on Facebook is doubling every 2 years [1].

Not only the number of new images, but their resolution also increases, which ultimately results in the almost exponential growth of the volume of stored and transmitted data, which is a digital representation of visual images. As a result, the unprecedented increase in storage volumes and channel capacities is continuously competing with the unprecedented increase in the volume of new digital information.

This paper attempts to assess the current situation, and study the trends and prospective areas of research and development in the field of image and video compression. The paper is based on the analysis of existing domestic and foreign publications and the authors' experience obtained in cooperation with a team of scientists and engineers of one of the R&D companies during the research and development of video image transmission and recording systems, as well as with employees of Department of Computing systems and networks of Saint-Petersburg State University of Aerospace Instrumentation.

2 Standards and Non-standardized Solutions for the Digital Visual Information Compression

2.1 Images Compression

The discrete cosine transformation (DCT) was the base for images compression algorithms for decades. The algorithm, consolidated by the JPEG group standard and which received the name JPEG, became the most popular among them. The main disadvantages of the similar algorithms are known both to experts, and ordinary users. In particular, reticulated character artifacts at significant compression coefficients and contrast boundaries blur.

With the computing systems performance development and, thanks to certain achievements in the "wavelet bases" theory, to replace DCT (or in addition to DCT) the discrete wavelet transform (DWT) came. The JPEG group was noted by the DWT basis standard creation with considerable functionality. The standard appeared at the end of the 2000th year and was called JPEG2000. The increase in compressed and then the restored images quality in comparison with JPEG is noted as the various tests result at identical compression coefficients approximately for 20 %. Especially obvious distinction is noted at big coefficients, because JPEG2000 doesn't make distortion in the restored images leading to the "blocks" emergence.

It is also necessary to note, that DWT was applied in different variations and also in non-standardized form, including in the hardware codecs form, which perform the real-time processing (for example, integrated circuits ADV601/ADV611 of Analog Devices corporation). They also had rather good results, however, over time JPEG2000 were ousted. In a very short time, after the JPEG2000 standard publication, Analog Devices company begins to produce popular to date hardware single-chip JPEG2000-codec ADV202/ADV212.

It should be noted that there is so called VTC mode as a part of the video compression applied standard—MPEG4, (Visual Texture Coding—coding of the visual textures), which also allows to compress "still images", using DWT. However, in comparison with JPEG2000, it has a very little functionality.

Something unusual among the applied bases for the images transformation to be compressed, is Walsh functions system and, respectively, Walsh-Hadamard transformation [2]. In particular, WebP [3] format, developed by the Google Inc. company, together with DCT, uses Walsh functions basis. WebP is considered by community as progressive replacement of the JPEG format, however at the time of the article writing, WebP distribution and popularity didn't become a little more significant. Moreover, some authors prove their skepticism concerning the new algorithm advantages [4]. Presumably, the reasons of failure should be looked for in unprecedented JPEG format prevalence, which have being used for a long time for the large images number storage.

Nevertheless, it should be noted the existence of the additional mechanisms, entered into WebP to increase the compression coefficient. There are several prediction modes, allowing in certain cases to transfer not the separate compressed images blocks (areas), but only the difference (a prediction error) between the next blocks among them. Together all the applied approaches, according to the developers approval at "lossless compression" increase the compression efficiency in WebP relatively to PNG by 26 %, and for the mode "lossy"—relatively to JPEG—at 25–34 %. The equivalence quality assessment is based on SSIM metrics [5].

Actually the algorithms based on DCT and DWT, together with the contents prediction mechanisms of the adjoining spatial areas or without them, are urged to eliminate redundancy in the transferred/stored data. It concerns not purely information redundancy (in this case, more than 2–3 multiple compression can't be reached), but redundancy in the sense of the human visual system perception. For example, it is known that small details of the visual images (spatially high-frequency) are much less important for the quality picture perception, than large ones (spatially low-frequency). That's why it is possible to eliminate high-frequency components with the necessary "roughness" degree from the image, making the information losses, but not too disturbing the perception. In chromaticity components such distortions are noticeable significantly less, than in the brightness one, and this fact is also used.

Thus, the images compression algorithms, using different tools, operate the "painless" small details removal idea with some additional receptions, giving a dozen or two percent gain. Meanwhile, the revolutionary ideas, which give a quantum leap of images compression characteristics algorithms aren't present at the moment and, perhaps, they won't appear without approach change to the analysis level of this information type in the near future.

2.2 Video Compression

The playing video is presented in the consistently replacing each other images form with a particular frequency. Therefore images compression algorithms are quite applicable for the individual video frames, as evidenced by the existence of the storage video files formats "Motion-JPEG" (compressed video frames as separate image algorithm JPEG) and "Motion-JPEG2000" (video frames are compressed with the algorithm JPEG2000). However all of them aren't so effective for the video, as the algorithms, operating inter-frame (temporary) redundancy idea, when the difference between the adjacent frames, rather than the whole video frame, is exposed to compression. It is obvious, that in this case it is possible to reach much higher coefficients, since the difference between the video frames is generally very small, because of the small time interval.

There is a limited set of the applications, demanding separate compression of each video frame, however this set is limited. For example, in the cases, when each frame has to represent the complete information unit, but not to be synthesized from some sequence of initial arrays. Or when changes in the recorded picture are essential, but thus it is necessary to keep both quality of the video image, and intensity of the kept information transferred flow is continuous at the set level.

Various mechanisms of the changes prediction in the video sequence and additional reduction need to compress and keep all taken information are actively used in addition to the reduction of inter-frame redundancy. To date the most famous video compression algorithms are that ones, which utilize the mentioned above ideas—H.261, H.263, H.264 and algorithms of MPEG group: MPEG1, MPEG2, MPEG4. It is important to note, that expediency and sufficiency of DCT application (but not DWT) to eliminate the spatial redundancy is repeatedly shown for all these algorithms.

In 2013 the new video images compression standard H.265 [6] was released. Its developers predict approximately double capacity reduction, which are necessary for the transfer, in comparison with the best existing codecs. When developing the standard, the following requirements were formulated:

- Possibility of compression support without information losses and visible distortions.
- Frame formats support from QVGA (320 × 240) to 4K and 8K (UHDTV).
- Color sampling to 4:4:4, broad color coverage and alpha channel, increased color depth support.
- Constant and variable frame frequency support (up to 60 frames per second and more).

The founders paid special attention for the codec insistence restriction question to the computing resources. Moreover, it is claimed that with the image dimension growth, compression coefficient has to increase a little.

All the statements and estimates certainly demand a practical check on the various platforms to confirm them, and also to except any essential negative sides. Now it is possible to say, that this standard incorporated all the possible algorithmic opportunities, aimed to provide the consumer with qualitative video with the minimum expenses.

2.3 Opportunities to Improve the Compression Algorithms Characteristics

During the last 15–20 years, all the attempts to improve the existing images compression algorithms were expressed in considerable number of the published scientific works and the defended dissertations. The results generalization shows, that the quantitative assessment of "improvements" at objective approach makes 10–20 %, and more rare 30 %, it varies depending on taken image type (gray or color photos, black-and-white text, cartographical images, etc.). For example, mesh artifacts elimination on the image, which passed direct and return JPEG transformation with rather big compression coefficients [7–9] was a classical task during some time. Besides, interesting results were achieved with using nonlinear decomposition by means of wavelet-functions [10], at color transformations application for color images [11] and with use of images perception model by the human vision [12].

To replace DCT by DWT in video information compression technologies was done considerable work in two various directions. The first option is the prediction error coding by means of DWT [13], the second is so-called 3D wavelet decomposition [14, 15]. And though these technologies showed the quality increase in comparison with the existing approaches at the similar compression coefficients, nevertheless, the majority of them are intended to ensure functionality development; for example, scalability and progressive transfer entering (resolution reduction or blur part of the overall image).

When concentrating only on the search of compression coefficients increase opportunities, a number of modern scientific and technical achievements is ignored. In connection with prompt mobile technologies development and the increasing need of Internet users in the video information, besides actually breakthrough compression technology search, development and deployment of the transferred images effective scaling technologies in real time, depending on the information consumer (receiver) characteristics are seemed to be more actual to date. For an illustration it is possible to give an example, when the portable device with a relatively low resolution display transmits Web page images of the Full HD format. This is obviously impractical and, therefore, the transmitted content becomes dependent on the terminal properties. Thus, when defining the receiver content characteristics, at the level of the company, providing the Web server, or the Internet service provider, or even at the routers level it is possible to reduce the demanded information traffic significantly.

3 An Opportunities Assessment in the Compression Technologies Development and the New Approach Advantages

3.1 Possibilities for Compression Algorithms Improvement

In the work [16] the authors give approaches classification to the coding problem (compression) solution of images and video, and also the corresponding consecutive technologies development (see Table 1). In the current work context, the greatest interest will be represented by the 4th and 5th generations, therefore it is necessary to consider in details.

Recognition and restoration involves the content type determining (house, car, landscape, person, etc.) to implement the coding method, focused on certain content. Rather big breakthrough is made in this direction in the MPEG4 algorithm, which applies specific technology of recognition, coding and further "animation" of the person face image.

One more step on the way to the 5th generation is taken in the MPEG7 standard. The certain standard way of the audiovisual information various types description is specified in it. (The elements, which are the description of audiovisual content, are known also as "metadata".) As soon as the audiovisual content is described in the metadata terms, the image is prepared for the coding process. It should be noted, that in this case, metadata, describing the image, will be coded instead of the image. For example, in a case with the person face—it is enough to set his attributes massif for synthesis on the playback side. The background, on which the person is shown after restoration, can be coded rather roughly or not to be coded at all, and on playback side to create any neutral option.

Table 1 Approaches classification to images and video coding

Generation	Approach	Technology
0	Direct coding of an analog signal	Pulse Code Modulation (PCM)
1	Redundancy elimination	DCT, DWT, …
2	Structural coding	Image segmentation
3	The analysis and synthesis	Coding on the models basis
4	Recognition and restoration	Coding on the basis of the knowledge base
5	Intellectual coding	Semantic analysis and coding

Since the descriptive content characteristics have to make sense in the application context, they will be various for different appendices. It means that the same content can be described variously depending on the concrete application. For the visual content part, for example, the description of form, size, texture, color, movement (trajectory), position (for example, "where on a scene the object can be placed"), etc. will be the lowest abstraction level. And for the audio-content: key, tonality, speed, speed variations, situation in a sound space. The highest representation level is semantic information like such description: "It is a scene with the barking brown dog at the left and the blue ball falling on the right with a background sound of the passing cars". Existence of intermediate abstraction levels is allowed. It is possible to examine MPEG7, for example, in [17] in more detail.

It is interesting, that in the simpler option one-dimensional—sound—signals coding algorithms passed the similar way: from the simple "digitization" to the elimination redundancy, then—to the processing on the basis of the hearing person model aid, and at last, to the separate phonemes (for the speech) recognition to transfer only their codes for the further synthesis on the reception side.

3.2 New Quasi-orthogonal Bases in the Visual Information Compression Problems

The research area, related to the recent discovery of matrix bases [18–24] is actively developing nowadays. Emergence of the minimax structured low-level quasi-orthogonal matrices (Hadamard-Mersenne, Hadamard-Euler, Hadamard-Fermat, Hadamard-Walsh matrices)—so-called "M-matrices", which algorithmic creating is possible for any orders, open essentially new opportunities for reversible images transformation with various purposes. In this case, using a single device, some actual problems in the images processing field, which, as a rule, separated from each other, can be solved at once. The researches show, that at the same time it is possible to carry out: compression, protection against unauthorized access and protection against deliberate distortion of images [25].

DCT and DWT matrices and appropriate algorithms, implemented for images compression, were intended for digitized pictures of television standards and thus do not take into account the progress made in increasing the size of images up to Full-HD, 4K, 8K, obtained from contemporary video sources. Moreover, some sources can form an image of quite arbitrary size using so called "quality-box" technology [26]. In contrast to the matrix of DCT and DWT, which orders are usually multiples of 8, the quasi-orthogonal matrices exist for different orders. This fact opens the possibility of selecting the optimal matrix for each specific application. In the Table 2 there are values or formulas for calculating such matrices elements, given for orders corresponding to different prime numbers, covering the set of natural numbers. As it is seen from the table, the number of different elements modulo in the pointed matrices is not more than three, which shows its simplicity.

Table 2 M-matrices types and the values of its elements [26]

Matrices order n	Matrices types	Elements values
$4t$	Hadamard	1, −1
$2t$, $4t$	Belevitch	1, −1, 0
t, $2t$, $3t$, $4t$	Weighed (Toski-Seberry)	1,−1, 0
$4t-1$	Mersenne	1, $-b$, where $b=\frac{t}{t+\sqrt{t}}$
$4t-2$	Euler	1, $-b$, where $b=\frac{t}{t+\sqrt{2t}}$
$4t-3$	Seidel	1, $-b$, d, where $b=1-2d$, $d=\frac{t}{t+\sqrt{n}}$
$4t-3$	Fermat	1, $-b$, s, where $q=n-1=4u^2$, $p=q+\sqrt{q}$, $b=\frac{2n-p}{p}=1-\frac{2u-1}{2u-1}\times\frac{1}{u}$, $s=\frac{\sqrt{nq-2\sqrt{q}}}{p}=\frac{\sqrt{nu-1}}{2u+1}\times\frac{1}{\sqrt{u}}$

The structure of these matrices can be determined iteratively [21] and involves different configurations of the elements in them: cyclic, bicyclic, negacyclic, symmetrical and others. It determines both the effectiveness of computing software implementation and the possibility of building efficient structural solutions. Thus the various matrices number of the certain class, various orders and possibility to implement procedure of rows and columns permutation allow speaking about emergence of rich applied tools in the visual information processing field that can be embedded into well-known compression algorithms.

3.3 The Combined Solution of Compression and Protection Tasks of Visual Information at the New Approaches Application

So, one of the new approach advantages, connected with quasi-orthogonal low-level matrices application for the visual information compression is possibility of simultaneous protection against unauthorized access. Application of the mentioned quasi-orthogonal matrices bases to protect images isn't necessary to be identified with the cryptographic data security methods in general. Though from the work scheme point of view this approach is identical to information security methods "with the closed key". The matter is that for images, by definition to excess information type, application of traditional protection ways against unauthorized access often doesn't bring desirable results. Actually the cryptography can hide the valid numerical pixels value, but thus to make insufficient change to the reproduced picture. The high resource intensity of cryptographic algorithms is not a smaller restriction. Taking into account the need of transfer, for example, of high resolution video images (Full-HD, 4K, 8K) in real time to implement such approach an extremely performed computer will needed. It is necessary to note that for this restriction overcoming, the researches with the positive result, allowing to be limited by cryptographic primitives use—the computing procedures, which are the part of any cryptographic algorithm were conducted [27].

The terms "masking" and "unmasking" for the procedures, which are carried out by means of these bases, are used in the works on opportunities and restrictions research in the new quasi-orthogonal bases application. Thus, digital static images and video images with the small relevance time are the target information for them. Further there are the main definitions, connected with it.

Definition 1 *Masking* is the process of transformation of digital visual information with a short relevance period to a noise-like form in order to protect it from unauthorized access.

Definition 2 After being masked, the resulting data array is called *masked visual information* or just a *masked image*.

Definition 3 *Unmasking* is the process of reverse transformation of masked visual information by way of application of reversed masking operations in order to restore the original content.

There are matrix masking, with low-level quasi-orthogonal matrices use, and cryptographic, with cryptographic primitives use. The cryptographic masking advantage is higher resistance to attacks in comparison with the matrix one, the disadvantage is higher computing performance demands [28].

It is important to note that masking with the low-level quasi-orthogonal matrices application isn't reversible from the point of view of numerical pixels values information integrity. Even without truncation of frequency components bits of the image to compress. It is connected with existence of irrational values as a part of matrix elements, which in the conditions of a limited number of bits for coding inevitably bring inaccuracies at the backward transformation. However for images this fact isn't a restriction. For an illustration in Fig. 1 there are shown initial ("a"), masked ("b") and restored ("from") images, and also the difference reproduced among the initial and restored picture form ("d").

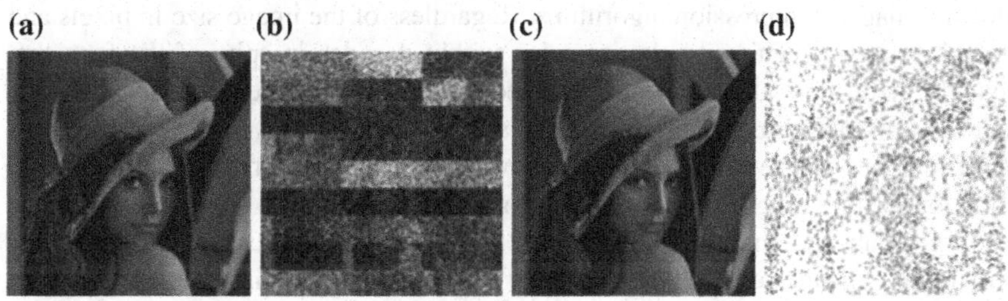

Fig. 1 Masking, unmasking procedures illustration

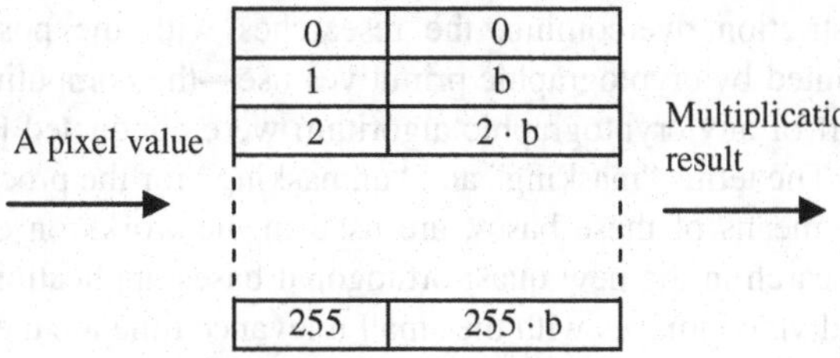

Fig. 2 An illustration to the multiplication operations replacement way with the tabular selection for the two-level matrix (with values 1, b) and at image pixel depth, equal to 8

3.4 *The Advantage of New Approaches Concerning Calculating Units Implementation Efficiency*

The small levels number (various values of elements of a matrix) allows the matrices authors to implement the computations process very effectively, because the multiplication operation, as the most resource-intensive, can be replaced with the table selection operation (see Fig. 2). As a result, the multiplication is reduced to the fast memory selection operation. The necessary tables number in this case will be equal (k − 1), where k—is the levels number (since one of the elements—is always ‘1’), and each table size—to amount of possible various image pixels values (or color components).

Such computation way is equally simply implemented, both by means of program-controlled systems (on the basis of microprocessors), and programmable logical devices (PLD). That’s why it is difficult to overestimate this advantage in continuously growing dimension of the stored and transferred images the last time.

3.5 *The Advantage of the Possibility to Choose the Transformative Matrix Order*

Application of the matrices operators intended for the image transfer into the spatial frequency area, with the exclusively fixed size, is a traditional approach for the modern images compression algorithms. Regardless of the image size in pixels and from its contents. This situation remains within decades in spite of the fact, that actually there have already been some technological revolutions during this time that led to multiple increasing of images size.

M-matrices tools existence provides advantage in any reformative matrices order variation (with a single step), when performing compression, based on the image resolution data and on the spatial-frequency characteristics of its separate areas. It is possible to consider critically expediency of the operator 8 dimension application and for lower resolution images, and also for HD resolutions for a primitive illustration of this consideration. And also for the image areas, rich in the small details and large plain areas.

4 Conclusion

The further development of algorithms to reduce channel capacity required for data transmission or amount of free space required for data storage is likely to be governed by the area of application of these algorithms. The current trend of "mobilization" of personal computers that expands the markets of laptops, ultra-books, communicators and tablet computers, as well as the tendency of the dominance of visual information over the other types of information on the Internet will require the change in the set of approaches to video and image compression. Besides, the traditional methods using the specific aspects of the human visual perception are almost at the end of their development options.

It should be kept in mind that finding more and more efficient ways to reduce the size of data transmitted over the open communication channels is only one of the challenges. It is also necessary to solve the problem of confidentiality with a reasonable increase in the required computing power. In this sense, the emergence of bases of quasi-orthogonal low-level matrices (M-matrices) is very timely. These bases have opened yet inaccessible application possibilities in the field of digital visual information processing, such as:

- Simultaneous combination of data transformation procedures in order to compress and protect the information from unauthorized access.
- Reduced performance requirements for computers that perform direct and reverse transformation.
- Selection of transform matrix dimensions depending on the dimension of pixel matrix of an original image and the data on frequency spatial distribution in its areas.

References

1. Internet of the Future: We Need a "smart" Visual Search. http://www.cnews.ru/reviews/index.shtml?2012/11/14/509824
2. Walsh Function. https://en.wikipedia.org/wiki/Walsh_function
3. WebP, A New Image Format for the WEB. http://blog.chromium.org/2010/09/webp-new-image-format-for-web.html
4. On H.264. http://x264dev.multimedia.cx/archives/541
5. A New Image Format for the Web. https://developers.google.com/speed/webp/
6. High Efficiency Video Coding. https://en.wikipedia.org/wiki/High_Efficiency_Video_Coding
7. Pong, K.K., Kan, T.K.: Optimum loop filter in hybrid coders. IEEE Trans. Circuits Syst. Video Technol. **4**(2), 158–167 (1997)
8. O'Rourke, T., Stevenson, R.L.: Improved image decompression for reduced transform coding artifacts. IEEE Trans. Circuits Syst. Video Technol. **5**(6), 490–499 (1995)
9. Llados-Bernaus, R., Robertson, M.A., Stevenson, R.L.: A stochastic technique for the removal of artifacts in compressed images and video. In: Recovery Techniques for Image and Video Compression and Transmission. Kluwer (1998)

10. Wajcer, D., Stanhill, D., Zeevi, Y.: Representation and coding of images with nonseparable two-dimensional wavelet. In: Proceedings of the IEEE International Conference on Image Processing. Chicago, USA (1998)
11. Saenz, M., Salama, P., Shen, K., Delp, E.J.: An evaluation of color embedded wavelet image compression techniques. In: Proceedings of the SPIE/IS&T Conference on Visual Communications and Image Processing (VCIP), 23–29 Jan 1999, pp. 282–293. San Jose, California (1999)
12. Jayant, N.S., Johnston, J.D., Safranek, R.J.: Signal compression based on models of human perception. Proc. IEEE **81**(10), 1385–1422 (1993)
13. Shen, K., Delp, E.J.: Wavelet based rate scalable video compression. IEEE Trans. Circuits Syst. Video Technol. **9**(1), 109–122 (1999)
14. Podilchuk, C.I., Jayant, N.S., Farvardin, N.: Three-dimensional subband coding of video. IEEE Trans. Image Process. **4**(2), 125–139 (1995)
15. Taubman, D., Zakhor, A.: Multirate 3-D subband coding of video. IEEE Trans. Image Process. **3**(5), 572–588 (1994)
16. Harashima, H., Aizawa, K., Saito, T.: Modelbased analysis synthesis coding of videotelephone images—conception and basic study of intelligent image coding. Trans. IEICE **E72**(5), 452–458 (1989)
17. http://book.itep.ru/2/25/mpeg_7.htm
18. Balonin, N.A., Mironovskii, L.A.: Hadamard matrices of odd order. In: Informatsionno-upravliaiushchie sistemy [Information and Control Systems], 2006, no. 3, pp. 46–50 (2006)
19. Balonin, Yu.N., Sergeev, M.B.: M-Matrix of the 22nd order. In: Informatsionno-upravliaiushchie sistemy [Information and Control Systems], 2011, no. 5(54), pp. 87–90 (2011)
20. Balonin, N.A., Sergeev, M.B., Mironovsky, L.A.: Calculation of Hadamard-Mersenne Matrices. In: Informatsionno-upravliaiushchie sistemy [Information and Control Systems], 2012, no. 5, pp. 92–94 (2012)
21. Balonin, N.A., Sergeev, M.B., Mironovsky, L.A.: Calculation of Hadamard-Fermat matrices. In: Informatsionno-upravliaiushchie sistemy [Information and Control Systems], 2012, no. 6, pp. 90–93 (2012)
22. Balonin, N.A., Sergeev, M.B.: Two ways to construct Hadamard-Euler matrices. In: Informatsionno-upravliaiushchie sistemy [Information and Control Systems], 2013, № 1, pp. 7–10 (2013)
23. Balonin, N.A.: Existence of Mersenne matrices of 11th and 19th orders. In: Informatsionno-upravliaiushchie sistemy [Information and Control Systems], 2013, no. 2, pp. 90–91 (2013)
24. Balonin, N.A., Balonin, YuN, Vostrikov, A.A., Sergeev, M.B.: Computation of Mersenne-Walsh matrices. Vestnik komp'iuternykh i informatsionnykh tekhnologii (VKIT) **2014**(11), 51–55 (2014)
25. Vostrikov, A.A., Chernyshev, S.A.: Implementation of novel quasi-orthogonal matrices for simultaneous images compression and protection. In: Smart Digital Futures 2014. Intelligent Interactive Multimedia Systems and Services (IIMSC-2014), pp. 451–461. IOS Press (2014). doi:10.3233/978-1-61499-405-3-451
26. Balonin, N.A., Sergeev, M.B.: Initial approximation matrices in search for generalized weighted matrices of global or local maximum determinant. In: Informatsionno-upravliaiushchie sistemy [Information and Control Systems], 2015, № 6, pp. 2–9 (2015)
27. Litvinov, M.Y., Sergeev, A.M.: Problems on formation protected digital images. In: XI International Symposium on Problems of Redundancy in Information and Control Systems: Proceeding, pp. 202–203. Saint-Petersburg (2007)
28. Vostrikov, A.A., Sergeev, M.B., Litvinov, M.Yu.: Masking of digital visual data: the term and basic definitions. In: Informatsionno-upravliaiushchie sistemy [Information and Control Systems], 2015, no. 5(78), pp. 116–123 (2015)

A Near-Far Resistant Preambleless Blind Receiver with Eigenbeams Applicable to Sensor Networks

Kuniaki Yano and Yukihiro Kamiya

Abstract BRAKE has been proposed as a preambleless blind receiver (PBR) applicable to spread spectrum (SS) signals. However, the performance is degraded under the near-far problem. In this paper, we propose an eigenbeam BRAKE, i.e., the combination of BRAKE with the pre-beamforming using the eigenvectors derived from the correlation matrix. This scheme is to avoid the performance degradation under the near-far problem. Although this combination is expected to be effective, a new algorithm for controlling BRAKE is required to make it work with eigenbeams. So this paper proposes the BRAKE control algorithm as well. The performance is verified through computer simulations.

Keywords Blind signal processing · BRAKE · Eigenbeams

1 Introduction

Preambleless blind receivers (PBRs) which do not require preambles for the channel estimation and the timing detection is interesting for the implementation of the base station in wireless sensor networks. It is expected that PBRs enable sensor nodes to reduce the power consumption since the sensor nodes do not have to send the preambles. At the same time, PBRs also contribute to simplify the sensor node structures.

BRAKE was proposed in [1] as a PBR for spread spectrum signals. It blindly and concurrently achieves beamforming by using multiple antennas, RAKE combining and the timing detection based on the constant modulus algorithm (CMA) [2].

K. Yano · Y. Kamiya (✉)
Graduate School of Computer Science and Technology,
Aichi Prefectural University, Nagakute, Japan
e-mail: kamiya@ist.aichi-pu.ac.jp

G. De Pietro et al. (eds.), *Intelligent Interactive Multimedia Systems and Services 2016*, Smart Innovation, Systems and Technologies 55,
DOI 10.1007/978-3-319-39345-2_17

However, the performance of BRAKE is degraded under the near-far problem. To overcome this problem, in this paper, we propose to combine BRAKE with eigenbeams (EBs) [3] which are generated by using eigenvectors of the correlation matrix.

In addition, we propose a control scheme of BRAKE which is necessary to combine with EBs as well. This scheme is required since weight matrices of BRAKE has to be adjusted depending on the situation in which the received signals are dispersed over several EBs or concentrates in an EB. None of [4–8] deals with the application of EBs over the space-domain signal processing. The performance will be verified through computer simulations.

2 Preliminaries

2.1 System Overview

Figure 1 shows the system overview in which there are U transmitters. Suppose that the u $(1 \leq u \leq U)$th transmitter sends L_{B} bits of data expressed as a vector as

$$\mathbf{b}_u = [b_{1,u} \quad b_{2,u} \quad \cdots \quad b_{L_{\mathrm{B}},u}]^{\mathrm{T}} \tag{1}$$

This is modulated and we obtain L_{Q} symbols as

$$\mathbf{q}_u = \left[q_{1,u} \quad q_{2,u} \quad \cdots \quad q_{L_{\mathrm{Q}},u}\right]^{\mathrm{T}} \tag{2}$$

Fig. 1 System overview

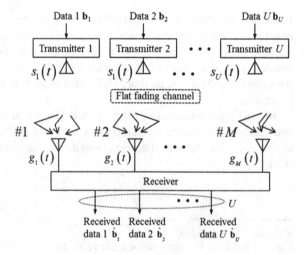

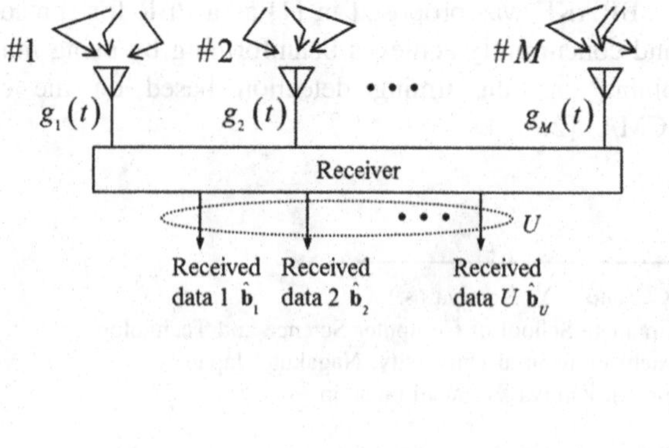

The unit-power signal sent by the transmitter is formulated as follows:

$$s_u(t) = \sum_{l_Q - 1}^{L_Q} q_{l_Q,u} \sum_{l_C - 1}^{L_C} c_{l_C,u}\delta(t - (l_Q L_C + l_C)T_{\mathrm{CHIP}}) \tag{3}$$

where $q_{l_Q,u}$ and $c_{l_C,u}$ denote the l_Qth symbol and the l_Cth chip of the spreading code, respectively. Let T_{CHIP} denote the chip duration while $\delta(t)$ is the impulse response of the band-limit filter. Finally, γ is defined as $\gamma = l_Q L_C + l_C$.

The signals go through flat fading channels and received by M antennas equipped with the receiver. The received signals are formulated as follows:

$$\mathbf{g}(t) = \mathbf{H}\mathbf{s}(t) + \sqrt{\frac{P_{\mathrm{N}}}{2\mathbf{n}(t)}} \tag{4}$$

$$\mathbf{g}(t) = [\, g_1(t) \quad g_2(t) \quad \cdots \quad g_M(t) \,]^{\mathrm{T}} \tag{5}$$

$$\mathbf{s}(t) = [\, s_1(t) \quad s_2(t) \quad \cdots \quad s_U(t) \,]^{\mathrm{T}} \tag{6}$$

$$\mathbf{n}(t) = [\, n_1(t) \quad n_2(t) \quad \cdots \quad n_M(t) \,]^{\mathrm{T}} \tag{7}$$

where $\mathbf{g}_m(t)$ is the received signal at the mth $(=1, \ldots, M)$ antenna, obtained through the RF-front ends (RF-E/Fs) while $\mathbf{n}(t)$ of size $(M \times 1)$ contains the unit-power complex AWGN. Also, P_{N} and $\mathbf{H}$ are the noise power and the matrix $(M \times U)$ containing the channel coefficients, respectively.

Next, the received signals are sampled as follows:

$$\mathbf{g}[k] = \mathbf{g}(kT_{\mathrm{SMP}} + T_{\mathrm{OFF}}) \tag{8}$$

where T_{SMP} and T_{OFF} denote the sampling duration and the timing offset, respectively, while k $(=0, 1, \ldots, K-1)$ is an integer as a timing index. The largest number of k is given by K as follows:

$$K = L_C L_Q \frac{T_{\mathrm{CHIP}}}{T_{\mathrm{SMP}}} \tag{9}$$

2.2 Receiver Configuration

Receiver Configuration Overview

Figure 2 shows the receiver configuration. The samples are fed into the eigenvector selector and beam generator (ESBG).

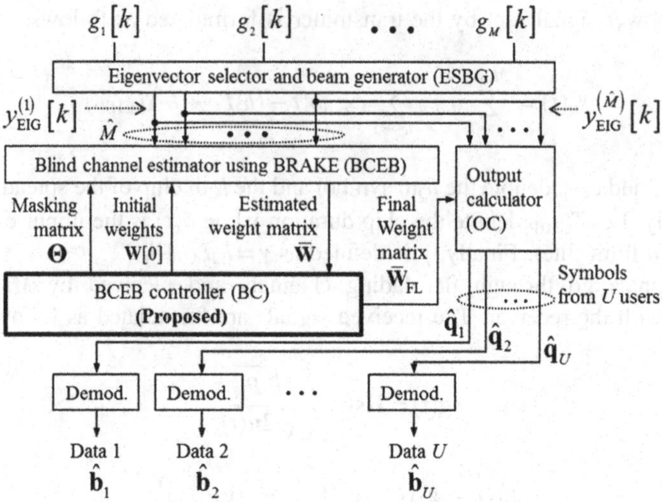

Fig. 2 Receiver configuration overview

In this part, $\widehat{M}$ beam outputs are obtained using the selected eigenvector of the correlation matrix, as explained in the next section, followed by the explanation of the blind channel estimator using BRAKE (BCEB), estimating blindly the channel coefficients and the timing using BRAKE. The control algorithm is located in the BCEB controller (BC). This is the algorithm necessary to combine the BCEB with the ESBG. It sets the initial weight matrix for BCEB, or gives constraints by the masking matrix, as explained in Sect. 3. In addition, BC decides the final weight matrix and send it to the output calculator (OC) to obtain the received symbol vectors $\hat{\mathbf{q}}_1, \hat{\mathbf{q}}_2, \ldots, \hat{\mathbf{q}}_U$ which corresponds to $\mathbf{q}_1, \mathbf{q}_2, \ldots, \mathbf{q}_U$. Finally the OC is briefly explained.

Eigenvector Selector and Beam Generator (ESBG)

By using $\mathbf{g}[k]$, the correlation matrix $\mathbf{R}$ of size $(M \times M)$ is obtained as follows:

$$\mathbf{R} = \frac{\mathbf{G}\mathbf{G}^{\mathrm{H}}}{K} \tag{10}$$

where $\mathbf{G} = [\mathbf{g}[0] \;\; \mathbf{g}[1] \;\; \cdots \;\; \mathbf{g}[K-1]]^{\mathrm{T}}$.

We obtain the eigenvectors of $\mathbf{R}$ through the eigenvalue decomposition (EVD) as follows:

$$\mathbf{R} = \mathbf{V}\mathbf{D}\mathbf{V}^{\mathrm{H}} \tag{11}$$

$$\mathbf{D} = \mathrm{diag}[\rho_1, \quad \rho_2, \quad \ldots, \quad \rho_M] \tag{12}$$

$$\rho_1 > \rho_2 > \cdots > \rho_{U+1} = \rho_{U+2} = \cdots = \rho_M = P_{\mathrm{N}} \tag{13}$$

$$\mathbf{V} = \begin{bmatrix} \mathbf{v}_1 & \mathbf{v}_2 & \cdots & \mathbf{v}_M \end{bmatrix} \tag{14}$$

$$\mathbf{v}_m = \begin{bmatrix} v_{1,m} & v_{2,m} & \cdots & v_{M,m} \end{bmatrix}^{\mathrm{T}} \tag{15}$$

The eigenvectors $\mathbf{v}_1, \mathbf{v}_2, \ldots, \mathbf{v}_{\widehat{M}}$ are corresponding to the $\widehat{M}$ eigenvalues selected according to the following criterion

$$\rho_1 > \rho_2 > \cdots > \rho_{\widehat{M}} \geq \beta \rho_M \tag{16}$$

where β is a constant. These are employed to generate the following signals.

$$\mathbf{y}_{\mathrm{EIG}}[k] = \widehat{\mathbf{V}}^{\mathrm{H}} \mathbf{g}[k] \tag{17}$$

$$\mathbf{y}_{\mathrm{EIG}}[k] = \begin{bmatrix} y_{\mathrm{EIG}}^{(1)}[k] & y_{\mathrm{EIG}}^{(2)}[k] & \cdots & y_{\mathrm{EIG}}^{(\widehat{M})}[k] \end{bmatrix}^{\mathrm{T}} \tag{18}$$

$$\widehat{\mathbf{V}} = \begin{bmatrix} \mathbf{v}_1 & \mathbf{v}_2 & \cdots & \mathbf{v}_{\widehat{M}} \end{bmatrix} \tag{19}$$

Blind Channel Estimator Using BRAKE (BCEB)

BCEB blindly estimates the channel coefficients and the timing based on BRAKE [1]. Figure 3 shows its configuration.

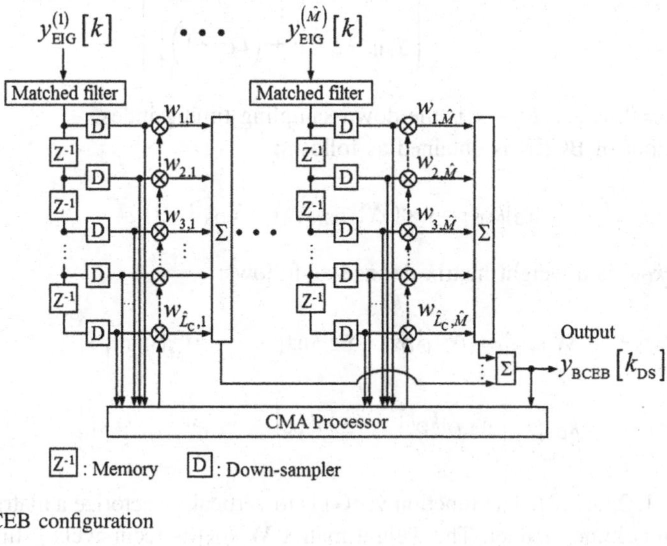

Fig. 3 BCEB configuration

The signals contained in $\mathbf{y}_{\mathrm{EIG}}[k]$ are fed into the matched filters as follows:

$$\mathbf{y}_{\mathrm{MF}}\left[\tilde{k}\right] = \sum_{\zeta=1}^{\widehat{L_{\mathrm{C}}}} \hat{c}_{\zeta}\mathbf{y}_{\mathrm{EIG}}[\tilde{k}-\zeta], \quad \tilde{k}-\zeta>0 \tag{20}$$

$$\mathbf{y}_{\mathrm{MF}}\left[\tilde{k}\right] = \left[y_{\mathrm{MF}}^{(1)}\left[\tilde{k}\right] \quad y_{\mathrm{MF}}^{(2)}\left[\tilde{k}\right] \quad \cdots \quad y_{\mathrm{MF}}^{(\widehat{M})}\left[\tilde{k}\right] \right]^{\mathrm{T}} \tag{21}$$

where $\hat{c}_{\zeta}$ is the ζth weight coefficient of the matched filter.

The variable $\hat{c}_{\zeta}$ is equal to the samples of the band-limited waveform of the spreading code $\mathbf{c}$. Let us define such weight coefficients of the matched filter as a vector $\hat{\mathbf{c}}$ of size $\left(\widehat{L_{\mathrm{C}}} \times 1\right)$ as follows:

$$\hat{\mathbf{c}} = \left[\hat{c}_1 \quad \hat{c}_2 \quad \cdots \quad \hat{c}_{\widehat{L_{\mathrm{C}}}} \right]^{\mathrm{T}}, \quad \widehat{L_{\mathrm{C}}} = L_{\mathrm{C}} \frac{T_{\mathrm{CHIP}}}{T_{\mathrm{SMP}}} \tag{22}$$

In addition, $\tilde{k}=0, 1, \ldots, \tilde{K}-1$ is the timing index where $\tilde{K}$ is given as follows:

$$\tilde{K} = \widehat{L_{\mathrm{C}}} L_{\mathrm{Q}} + \left(\widehat{L_{\mathrm{C}}} - 1\right) \tag{23}$$

The matched filters are followed by the tapped delay lines (TDLs) as depicted in Fig. 3. The stored samples in the TDLs are down-sampled as follows:

$$\mathbf{Y}_{\mathrm{DS}}[k_{\mathrm{DS}}] = \begin{bmatrix} \mathbf{y}_{\mathrm{MF}}\left[k_{\mathrm{DS}}\widehat{L_{\mathrm{C}}}\right] \\ \mathbf{y}_{\mathrm{MF}}\left[k_{\mathrm{DS}}\widehat{L_{\mathrm{C}}}+1\right] \\ \vdots \\ \mathbf{y}_{\mathrm{MF}}\left[k_{\mathrm{DS}}\widehat{L_{\mathrm{C}}} + \left(\widehat{L_{\mathrm{C}}}-1\right)\right] \end{bmatrix} \tag{24}$$

where $k_{\mathrm{DS}}=0, 1, \ldots, L_{\mathrm{Q}}-1$ is the down-sampling timing index.

The output of BCEB is obtained as follows:

$$y_{\mathrm{B}}[k_{\mathrm{DS}}] = \mathrm{vec}(\mathbf{W}[k_{\mathrm{DS}}])^{\mathrm{H}}\mathrm{vec}(\mathbf{Y}_{\mathrm{DS}}[k_{\mathrm{DS}}]) \tag{25}$$

where $\mathbf{W}[k_{\mathrm{DS}}]$ is a weight matrix defined as follows:

$$\mathbf{W}[k_{\mathrm{DS}}] = \left[\mathbf{w}_1[k_{\mathrm{DS}}] \quad \mathbf{w}_2[k_{\mathrm{DS}}] \quad \cdots \quad \mathbf{w}_{\widehat{M}}[k_{\mathrm{DS}}] \right] \tag{26}$$

$$w_{\hat{m}}[k_{\mathrm{DS}}] = \left[w_{1,\hat{m}}[k_{\mathrm{DS}}] \quad w_{2,\hat{m}}[k_{\mathrm{DS}}] \quad \cdots \quad w_{\widehat{L_{\mathrm{C}}},\hat{m}}[k_{\mathrm{DS}}] \right]^{\mathrm{T}} \tag{27}$$

where $\hat{m}=1, 2, \ldots, \widehat{M}$. The function vec(•) is to vertically vectorize a matrix, piling column-on-column manner. The weight matrix $\mathbf{W}[k_{\mathrm{DS}}]$ is recursively estimated by the following equation based on CMA criterion [1].

$$\mathbf{W}[k_{DS}+1]=\mathbf{W}[k_{DS}]\circ\mathbf{\Theta}-\nabla_{\mathbf{W}} \tag{28}$$

$$\nabla_{\mathbf{W}}=4\mu\mathbf{Y}_{DS}[k_{DS}]y_{B}^{*}[k_{DS}]\Delta[k_{DS}] \tag{29}$$

$$\Delta[k_{DS}]=\left(|y_{B}[k_{DS}]|^{2}-\sigma^{2}\right) \tag{30}$$

where μ and σ are constant. The matrix $\mathbf{\Theta}$ whose size is identical to that of $\mathbf{W}[k_{DS}]$ is called the masking matrix which will be explained in Sect. 3. The operator $\circ$ indicates the Hadamard product, i.e., the element-wize product.

Output Calculator (OC)

The OC calculates the output using the final weight matrix based on the Eq. (25). The details will be given at [Step 6] in Sect. 3.

3 Proposed Algorithm—BCEB Controller (BC)

In this chapter, we propose an algorithm necessary to combine the blind channel estimator using BRAKE (BCEB) with EBs. The algorithm will be explained as follows:

[Step 1] Setting of initial values:
Set $\psi=1$. Set the initial weight matrix and the masking matrix as follows:

$$[\mathbf{W}[0]]_{\hat{l}_C,\hat{m}}=\begin{cases}1 & \text{if } \hat{l}_C=1\\ 0 & \text{else}\end{cases} \tag{31}$$

In addition,

$$\mathbf{\Theta}|_{\mathrm{STEP1}}=\mathbf{I}\left(\widehat{L}_C\times\widehat{M}\right) \tag{32}$$

where $\mathbf{I}(L\times M)$ denotes a matrix of size $(L\times M)$ in which all entities are 1.
[Step 2] Send $\mathbf{W}[0]$ and $\mathbf{\Theta}|_{\mathrm{STEP1}}$ to BCEB. So we can obtain the weight matrix after the convergence of (28). Let us call the obtained weight matrix as $\overline{\mathbf{W}}_{\psi}$.
[Step 3] Signal timing detection: Identify the entity whose amplitude is maximum as follows:

$$\left|\bar{w}_{\max}^{\psi}\right|=\max_{1\le\hat{l}_C\le\widehat{L_C},\,1\le\hat{m}\le\widehat{M}}\left(\left|\overline{\mathbf{W}}_{\psi}\right|_{\hat{l}_C,\hat{m}}\right) \tag{33}$$

Suppose that the maximum amplitude entity is located on $(\hat{l}_{\max},\hat{m}_{\max})$ in $\overline{\mathbf{W}}_{\psi}$.

[Step 4] Decision: dispersion or concentration of the signal power over the EBs. Extract the $\hat{l}_{\max}$th row of $\overline{\mathbf{W}}_{\psi}$ as follows:

$$\left[\bar{w}^{(\psi)}_{\hat{l}_{\max},1} \quad \cdots \quad \bar{w}^{(\psi)}_{\hat{l}_{\max},\hat{m}_{\max}} \quad \cdots \quad \bar{w}^{(\psi)}_{\hat{l}_{\max},\widehat{M}}\right] \tag{34}$$

Note $\left|\bar{w}^{(\psi)}_{\max}\right| = \left|\bar{w}^{(\psi)}_{\hat{l}_{\max},\hat{m}_{\max}}\right|$. Check if there is any entity whose amplitude exceeds more than $\alpha\left|\bar{w}^{(\psi)}_{\max}\right|$ where $\alpha \le 1$, except the entity $\bar{w}^{(\psi)}_{\hat{l}_{\max},\hat{m}_{\max}}$. So we decide that the signal power is dispersed over the EBs. Then, go to Step 4-1. If not, it is decided that the signal power concentrates in the $\hat{m}_{\max}$th beam, and go to Step 4-2.

[Step 4-1] Set $\mathbf{W}[0]$ as

$$\mathbf{W}[0] = \mathbf{I}\left(\widehat{L}_{\mathrm{C}} \times \widehat{M}\right) \tag{35}$$

and the masking matrix as follows:

$$\left[\boldsymbol{\Theta}|_{\mathrm{STEP3}}\right]_{\hat{l}_{\mathrm{C}},\hat{m}} = \begin{cases} 1 & \text{if } \hat{l}_{\mathrm{C}} = \hat{l}_{\mathrm{C}}^{(\max)}[\psi] \pm 3 \text{ and } 1 \le \hat{m} \le \widehat{M} \\ 0 & \text{else} \end{cases} \tag{36}$$

[Step 4-2] Set $\mathbf{W}[0]$ as (35), and the masking matrix as follows:

$$\left[\boldsymbol{\Theta}|_{\mathrm{STEP3}}\right]_{\hat{l}_{\mathrm{C}},\hat{m}} = \begin{cases} 1 & \text{if } \hat{m} = \hat{m}_{\max}, \text{and } 1 \le \hat{l}_{\mathrm{C}} \le \widehat{L}_{\mathrm{C}} \\ 0 & \text{else} \end{cases} \tag{37}$$

[Step 5] Send $\mathbf{W}[0]$ and $\boldsymbol{\Theta}|_{\mathrm{STEP3}}$ to BCEB. BCEB returns the weight matrix. Store it as $\overline{\mathbf{W}}^{(\psi)}_{\mathrm{FL}}$.

[Step 6] Output calculation: Send $\overline{\mathbf{W}}^{(\psi)}_{\mathrm{FL}}$ to OC so that we obtain the output as follows:

$$\hat{q}_{\psi}[k_{\mathrm{DS}}] = \mathrm{vec}\left(\overline{\mathbf{W}}_{\mathrm{FL}}[\psi] \circ \boldsymbol{\Theta}|_{\mathrm{STEP3}}\right)^{\mathrm{H}} \mathrm{vec}(\mathbf{Y}_{\mathrm{DS}}[k_{\mathrm{DS}}]) \tag{38}$$

[Step 7] If $\psi = \widehat{M}$, Finish. Otherwise, renew ψ as $\psi \leftarrow \psi + 1$ and perform (35). In addition, renew the masking matrix $\boldsymbol{\Theta}|_{\mathrm{STEP1}}$ as follows:

$$\boldsymbol{\Theta}|_{\mathrm{STEP1}} \leftarrow \boldsymbol{\Theta}|_{\mathrm{STEP1}} \circ \mathrm{inv}\left(\boldsymbol{\Theta}|_{\mathrm{STEP3}}\right) \tag{39}$$

where inv(A) denotes the function which performs the inversion of the entity of A as $1 \to 0$ or $0 \to 1$. This renewal is done to prohibit BCEB to seek the signal timing of which the signal is already detected. Back to Step 2.

4 Computer Simulations

Table 1 summarizes the simulation conditions. The parameter of CMA, namely μ defined in (30) is heuristically determined as follows: First, we set $=\mu_1 \times 10^{-\mu_2}$ ($\mu_1 \in [1,3,5,7,9]\, \mu_2 \in [11,12,13,14]$). Second, run BCEB by (28)–(30) with μ. By trying all of the μ values, we finally select the value which minimizes the following.

$$\Delta_{\mathrm{ave}} = \frac{1}{100} \sum_{k_{\mathrm{DS}} = L_{\mathrm{Q}} - 100}^{L_{\mathrm{Q}} - 1} \Delta[k_{\mathrm{DS}}] \tag{40}$$

The simulations are performed 100 times and the output SNR is statistically evaluated in Fig. 4 which shows the output SNR versus the probability $P(\text{Output SNR[dB]} \leq \text{Abscis})$, for the signal of the lowest SNR, 0 [dB/antenna/bit], comparing the proposed method with the ideal and the compared method. In this figure, the ideal method uses the correct signal timing and the true channel coefficients. The compared method is the BRAKE without EB, i.e., the version removing BCEB and BC from the configuration shown in Fig. 2 while the final weight matrix $\overline{\mathbf{W}}_{\mathrm{FL}}$ is directly handed to OC from BCEB. In this case, $\overline{\mathbf{W}}_{\mathrm{FL}} = \mathbf{W}[L_{\mathrm{Q}} - 1]$ obtained by (28).

It is clearly observed that the proposed method achieves SNR around 1 dB less than that of the ideal method at $P(\text{Output SNR[dB]} \leq \text{Abscis}) = 0.5$, even though the compared method stays at very low output SNR. It should be emphasized that the proposed method successfully detects the weakest signal even under the near-far problem.

Table 1 Simulation conditions

The number of transmitters (U)	5
Modulation	BPSK
Symbol length (L_{Q})	5000
Spreading code	127-chip M-seq.
Average SNR of each signal [dB/antenna/symbol]	0, 10, 20, 30, 40
The number of antennas (M)	20
Sampling [sample/chip]	4
Band limit filter	Cosine roll-off (roll-off factor: 0.5)
The number of weight coefficients at an antenna ($\widehat{L}_{\mathrm{C}}$)	508 (= 127 × 4)
Constant σ in (30)	1000
Constant α in [Step 4]	0.3
Constant β in (16)	1.03

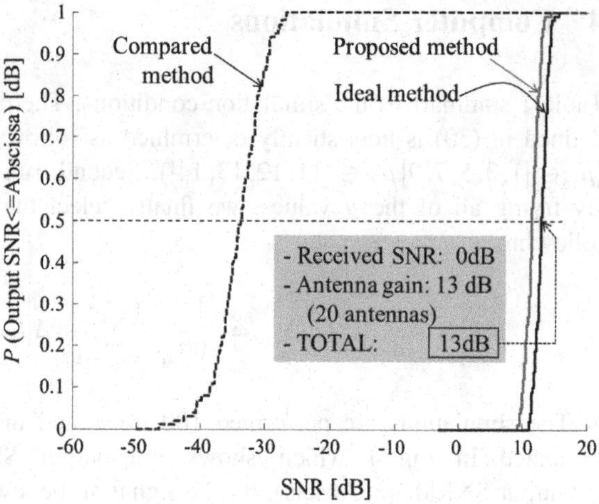

Fig. 4 Simulation result

5 Conclusions

In this paper, we proposed to combine BRAKE with EB associated with the control algorithm. Through computer simulations, we verified that the proposed method detected the weak signal even under the near-far problem.

Further considerations will apply this method to the base station of sensor networks, and will evaluate the reduction of the power consumption at the sensor nodes.

References

1. Takayama, K., Kamiya, Y., Fujii, T., Suzuki, Y.: A new blind 2D-RAKE receiver based on CMA criteria for spread spectrum systems suitable for software defined radio architecture. IEICE Trans. Commun. **E91-B**(6), 1906–1913 (2008)
2. Agee, B.G.: The least-squares CMA: a new technique for rapid correction of constant modulus signals. In: Proceedings of International Conference on Acoustics, Speech, and Signal Processing, pp. 953–956 (1986)
3. Al-Neyadi, H.M.: Successive blind recursive constant modulus detectors for DS/CDMA signals with BPSK modulation. In: 2006 IEEE GCC Conference (GCC) (2006)
4. Elnashar, S.E., Elmikati, H.: A robust linearly constrained CMA for adaptive blind multiuser detection. In: 2005 IEEE Wireless Communications and Networking Conference (2005)
5. Xue, Q., Jiang, X., Wu, W.: A new CMA-based blind adaptive multiuser detection. In: IEEE VTS 53rd Vehicular Technology Conference (2001)
6. Gelli, G., Paura, L., Verde, F.: A two-stage CMA-based receiver for blind joint equalization and multiuser detection in high data-rate DS-CDMA systems. IEEE Trans. Wirel. Commun. (2004)
7. Bahng, S., Host-Madsen, A.: Block CMA-based blind and group-blind multiuser detectors. In: Proceedings of ICASSP'04 (2004)
8. Cheung, P.K.P., Rapajic, P.B.: CMA-based code acquisition scheme for DS-CDMA systems. IEEE Trans. Commun. **48**(5) (2000)

A New Approach for Subsurface Wireless Sensor Networks

Hikaru Koike and Yukihiro Kamiya

Abstract Subsurface wireless sensor network is a sensor network in which sensors are installed under water or soil. This is to monitor earthquakes or Tsunami, to secure a society against natural disasters. The subsurface wireless communication is not easy since the power mitigation of the signal is severe due to the water and soil. In addition, the battery capacity in the sensor node is very limited. In this paper, based on the spread spectrum (SS) which is robust against low signal-to-noise power ratio (SNR), we propose a new transmitter and receiver configurations achieving code division multiplexing (CDM) for the subsurface wireless sensor networks. The proposed method is advantageous in terms of the following points. First, it realizes CDM by using a single spreading code. Second, the receiver does not need the training sequence to adjust the multiple antenna for the SNR improvement. This is effective to reduce the battery consumption at the sensor node since it does not need to send the training sequence. The performance is verified through computer simulations.

Keywords Wireless ad hoc network · Digital signal processing · Spread spectrum · Medium access

1 Introduction

To realize a secure society against natural disasters, sensor network can play an important role by collecting data for monitoring earthquakes, Tsunami and so on. For such applications, the sensors are installed under water or soil. So we need to think about subsurface wireless communications i.e., the wireless data transfer through the water or the soil, not through the air.

H. Koike · Y. Kamiya (✉)
Aichi Prefectural University, Nagakute, Japan
e-mail: kamiya@ist.aichi-pu.ac.jp

G. De Pietro et al. (eds.), *Intelligent Interactive Multimedia Systems and Services 2016*, Smart Innovation, Systems and Technologies 55,
DOI 10.1007/978-3-319-39345-2_18

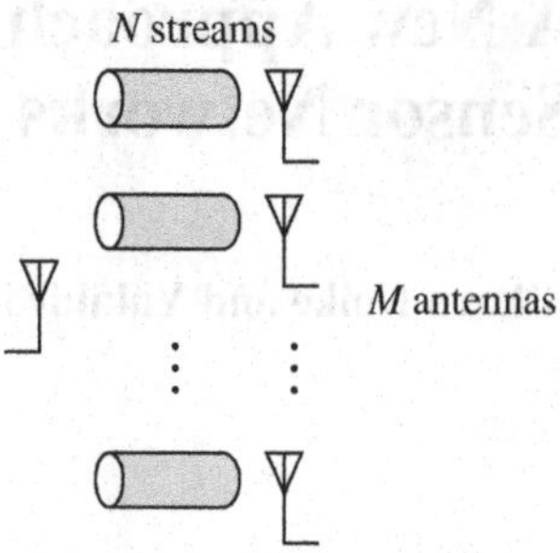

Fig. 1 An image of multiplexing for multiple data streams

The primary concern of the subsurface communications is the serious degradation of the signal-to-noise power ratio (SNR) at the receiver [1, 2]. This is due to the absorption of the signal power by water and soil.

Spread spectrum (SS) is known by its robustness against low-SNR [3]. Based on SS, the code division multiplexing (CDM) can be a solution for the data rate acceleration. In addition, the antenna diversity allows the receiver to collect the signal power to improve SNR. This image is depicted in Fig. 1. This system enables us to send N streams simultaneously so that we can accelerate the data rate even though the data rate of the single stream is not fast.

Although CDM and the antenna diversity are very classic and mature, these solutions cause the following problems in the context of the subsurface sensor networks.

1. CDM consumes a lot of spreading codes. Since we need to extend the system to deal with a number of sensors, a single transmitter-receiver pair should not need many spreading codes.
2. Since the receiver is equipped with multiple antennas, it obtains multiple signals. In order to combine the signals so as to improve SNR, the receiver needs a training sequence [4]. Because the channel suffers from the low-SNR, the training sequence must be sufficiently long. As a result, not only it degrades the throughput, but also it consumes a lot of battery power even though it is very limited in the sensor.

In this paper, focusing on the subsurface wireless sensor networks, we propose a new scheme for CDM solving the above-mentioned two problems. The proposed method employs only one spreading code. In addition, the receiver can obtain the space diversity gain using the multiple antennas, without the training sequence. Therefore the transmitter is released from the transmission of the training sequence, and it saves the battery life.

However, the data rate in the proposed method is slightly lower than that of the conventional CDM using multiple spreading codes. This is because of the following principle. The proposed transmitter sends the multiple streams into the same channel, i.e., the same frequency, the same spreading code and the same timing but with slightly different symbol rate. This difference is very slight. Suppose that the data rate of the stream #0 which is the fastest is set at $R_0 = 1/T_S^{(0)}$ where $T_S^{(0)}$ is the symbol duration. So the symbol rate of the next stream # $n(= 0, 1, \ldots, N)$ denoted as R_n is defined as follows:

$$R_n = \frac{1}{T_S^{(n)}}, T_S^{(n)} = T_S^{(0)} + \frac{n}{T_{SMP}} \tag{1}$$

where R_n and $T_S^{(n)}$ is the symbol rate and the symbol duration of the stream #n, respectively. Furthermore T_{SMP} is the sampling duration. For example, assume that the processing gain is set at 100 and the sampling is performed so as to take 4 samples per chip. So the ratio between R_1 and R_2 is as follows:

$$\frac{R_1}{R_0} = \frac{T_S^{(0)}}{T_S^{(1)}} = \frac{T_S^{(0)}}{T_S^{(0)} + \frac{1}{T_{SMP}}} = \frac{T_S^{(0)}}{T_S^{(0)} + \frac{1}{400}}. \tag{2}$$

This data rate loss is not serious compared with the advantage of the proposed method as mentioned above. The performance improvement in terms of the output SNR is evaluated through computer simulations.

2 Proposed Method

This section provides in-depth explanations of the transmitter and receiver configurations in the proposed method. Signals are also formulated through this section.

2.1 Transmitter Configuration

Figure 2 shows the system overview in which there is a transmitter-receiver pair. Suppose that the transmitter is sending a data vector $\mathbf{b} \in (0, 1)$ of size $\left(L_D \times 1\right)$ where L_D denotes the data length. The data vector is modulated and divided by a serial-parallel converter (SPC) into N symbol streams denoted by $\mathbf{q}_n$ $(n = 1, \ldots, N-1)$ of size $\left(L_S/N \times 1\right)$ where L_S is the symbol length at the modulation output.

The nth symbol stream $\mathbf{q}_n$ is multiplied with a spreading code vector $\mathbf{c}_n$ $(n = 0, \ldots, N-1)$ of size $\left(L_C \times 1\right)$, where L_C is the spreading code length, as follows:

$$\mathbf{s}_n = \mathrm{vec}\left(\mathbf{c}_n \mathbf{q}_n\right) \tag{3}$$

where vec $(\mathbf{A})$ denotes the vectorization of a matrix $\mathbf{A}$, i.e.,

$$\mathrm{vec}\left(\begin{bmatrix} a_1 & a_3 \\ a_2 & a_4 \end{bmatrix}\right) = \left[a_1\ a_2\ a_3\ a_4\right]^{\mathrm{T}} \tag{4}$$

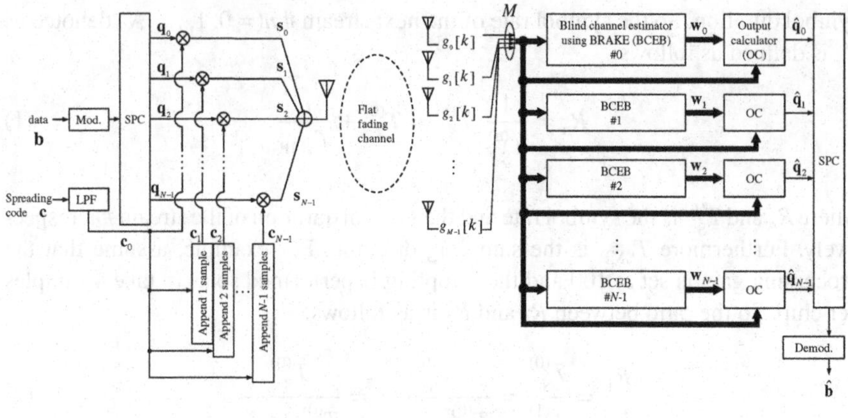

Fig. 2 System image

The generation of the nth spreading code is a key of the proposed method. This is generated through the following steps.

Step 1 Generate a spreading code, for example, an M-sequence $\mathbf{c}$.

Step 2 Let $\mathbf{c}$ go through a root roll-off filter for the band-limitation. So we obtain

$$\mathbf{c}_0 = \mathbf{h}_{\text{RROF}} * \mathbf{c} \tag{5}$$

where $\mathbf{h}_{\text{RROF}}$ denotes the impulse response of the root roll-off filter while the symbol $*$ denotes the convolution. Let $\hat{L}_{\text{C}}$ denote the length of the filter output. So the size of $\mathbf{c}_0$ is $\left(\hat{L}_{\text{C}} \times 1\right)$

Step 3 Append 0 at the tail of $\mathbf{c}_0$ so that we obtain $\mathbf{c}_1$. Likewise we can obtain $\mathbf{c}_n$ by appending n bits of 0 at the tail of $\mathbf{c}_0$. Therefore the size of $\mathbf{c}_n$ is $\left(L_n \times 1\right)$ where $L_n = \hat{L}_{\text{C}} + n - 1$. Figure 3 compares $\mathbf{c}_0$ with $\mathbf{c}_1$ and $\mathbf{c}_2$ as an example. It is visually shown that $\mathbf{c}_1$ and $\mathbf{c}_2$ are generated by appending 0 s at the tail of $\mathbf{c}_0$.

Finally all of the streams are put together as follows:

$$\mathbf{s} = \sum_{n=0}^{N-1} \mathbf{s}_n \tag{6}$$

It should be noted that the lengths of the vectors $\mathbf{s}_0, \dots, \mathbf{s}_{N-1}$ are not identical. The longest is $\mathbf{s}_{N-1}$ of size $\left(\left(\hat{L}_{\text{C}} + N - 2\right) \times 1\right)$ but the others are shorter than this. To realize the addition in (6), we append 0 s at the tail of the vectors that are shorter than $\mathbf{s}_{N-1}$ to adjust the length.

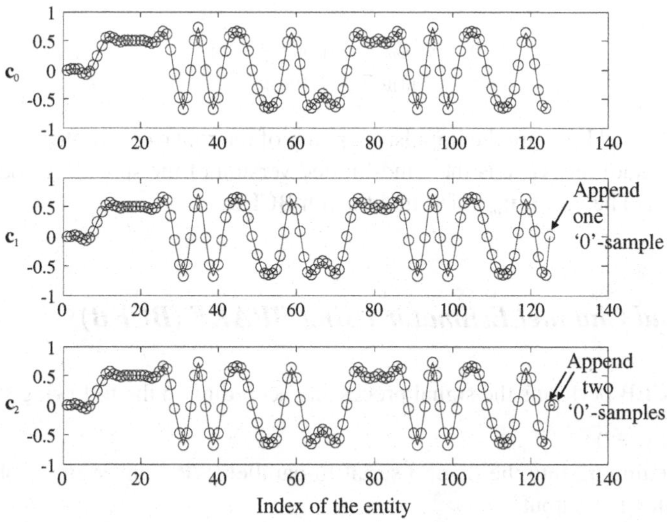

Fig. 3 Spreading codes

2.2 *Receiver Configuration*

The overview of the receiver configuration is shown in Fig. 2. It consists of N pairs of a blind channel estimator using BRAKE (BCEB) and an output calculator (OC) in order to recover the N streams $\mathbf{q}_n$ $(n = 1, \ldots, N-1)$. The following sections provide in-depth explanations about each part of the receiver.

2.3 *Received Signal Modelling*

Suppose that the receiver is equipped with M antennas receiving the signals arriving through flat fading channels. The received signal $\mathbf{g}_m$ of size $\left(L_S\left(\hat{L}_C + N - 2\right) \times 1\right)$ is expressed as

$$\mathbf{g}_m = \alpha_m \mathbf{s} + \sqrt{\frac{P_\eta}{2}} \eta_m, m = 0, 1, \ldots M1 \tag{7}$$

where α_m is a channel coefficient while η_m is a noise vector containing unit-power complex additive white Gaussian noise (AWGN). The size is identical to that of $\mathbf{s}$. The noise power is defined by P_η. These are divided and fed into the nth BCEB. The received signal vector $\mathbf{g}_m$ is successively fed into the root roll-off filter and the matched filter so that we obtain $\mathbf{u}_m$ $(m = 0, \ldots, M-1)$ of size $\left(L_S\left(\hat{L}_C + N - 2\right) \times 1\right)$ as follows:

$$\mathbf{u}_m = \left(\mathbf{g}_m * \mathbf{h}_{\mathrm{RROF}}\right) * \mathbf{c}_{\mathrm{ROF}} \\ \mathbf{c}_{\mathrm{ROF}} = \mathbf{c} * \mathbf{h}_{\mathrm{ROF}} \tag{8}$$

where $\mathbf{h}_{\mathrm{RROF}}$ and $\mathbf{h}_{\mathrm{ROF}}$ are the impulse response of the root roll-off filter and the roll-off filter. In addition, $\mathbf{c}_{\mathrm{ROF}}$ is the band-limited version of the spreading code by the roll-off filter. The vector $\mathbf{u}_m$ is fed into the nth BCEB.

2.4 Blind Channel Estimator Using BRAKE (BCEB)

The nth BCEB performs the signal processing according to the following steps.

Step 1 Set $v = 0$.

Step 2 Feeding $\mathbf{u}_m$ into the L_n-port serial-to-parallel (S/P) converter, we obtain the following output

$$\mathbf{h}_m[v] = \left[u_{v,m}\ u_{v+1,m}\ \cdots\ u_{v+L_n-1,m}\right]^{\mathrm{T}} \tag{9}$$

of size $\left(L_n \times 1\right)$ where L_n is the length of $\mathbf{c}_n$.

Step 3 The weight matrix

$$\mathbf{w}_m^{(n)}[0] = \left[1\ \mathbf{0}^{1\times(L_n-1)}\right]^{\mathrm{T}}$$

where $\mathbf{0}^{n\times m}$ is a matrix of size $(n \times m)$ filled with 0. Therefore the size of $\mathbf{w}_m^{(n)}[0]$ is $\left(L_n \times 1\right)$.

Step 3 Put $\mathbf{u'}_m[v]$, $m = 0, \ldots, M-1$ together into a matrix $\mathbf{U}[v]$ of size $\left(L_n \times M\right)$, as shown below.

$$\mathbf{U}[v] = \left[\mathbf{u'}_0[v]\ \mathbf{u'}_1[v]\ \cdots\ \mathbf{u'}_{M-1}[v]\right] \tag{10}$$

Likewise, put $\mathbf{w}_m^{(n)}[0]$, $m = 0, \ldots, M-1$ together into a matrix $\mathbf{W}^{(n)}[0]$ of size $\left(L_n \times M\right)$, as shown below.

$$\mathbf{W}^{(n)}[0] = \left[\mathbf{w}_0^{(n)}[0]\ \mathbf{w}_1^{(n)}[0]\ \cdots\ \mathbf{w}_{M-1}^{(n)}[0]\right] \tag{11}$$

Step 4 Renew the weight matrix $\mathbf{w}_v$ as follows [5]:

$$\mathbf{W}^{(n)}[v+1] = \mathbf{W}^{(n)}[v] - 4\mu\left(y^{(n)}[v]\right)^{*}\mathbf{U}[v]\,\nabla^{(n)}[v] \tag{12}$$

$$y^{(n)}[v] = \mathrm{vec}\left(\mathbf{W}^{(n)}[v]\right)^{\mathrm{H}}\mathrm{vec}\left(\mathbf{U}[v]\right) \tag{13}$$

$$\nabla^{(n)}[v] = \left|y^{(n)}[v]\right|^2 - \sigma^2 \tag{14}$$

where the superscripts $\cdot^*$ and $\cdot^{\mathrm{H}}$ denote the complex conjugate and the complex conjugate transpose, respectively. In addition, and are constants controlling the convergence of (12).

Step 5 If $v = V - 1$, Finish. Otherwise, set $v = v + 1$ and Go to Step 3.

2.5 Selection Strategy of μ and σ of BCEB

In (12), μ and σ are keys for the successful estimation of the weight matrix $\mathbf{W}^{(n)}$. Since (14) is non-linear, it is difficult to find the optimal values of μ and σ [6]. Therefore we employ the following algorithm to realize better selections of μ and σ as follows:

1. Define μ and σ as:

$$\mu = \mu_1 \times 10^{\mu_2}, \text{and } \sigma = 2^{\sigma_1} \tag{15}$$

2. Set the candidate of μ_1, and by the following vector as:

$$\vec{\mu}_1 = \left[\mu_1^{(\min)} \cdots \mu_1^{(\max)}\right] \tag{16}$$
$$\vec{\mu}_2 = \left[\mu_2^{(\min)} \cdots \mu_2^{(\max)}\right] \tag{17}$$
$$\vec{\sigma}_1 = \left[\sigma_1^{(\min)} \cdots \sigma_2^{(\max)}\right] \tag{18}$$

3. Set μ and σ by applying one of the entities of $\vec{\mu}_1$, $\vec{\mu}_2$ and $\vec{\sigma}_1$ and perform Step 4 in section. Then calculate the following quantity:

$$\Psi = \frac{1}{100} \sum_{\varepsilon=V-100}^{V} \left| \nabla^{(n)} [\varepsilon] \right| \tag{19}$$

   This means the average of the last 100 values in the convergence of (14).
4. Take $\mathbf{W}^{(n)}$ obtained with μ and σ which minimizes (19).

2.6 Output Calculator (OC)

The output calculator calculates the output of the proposed method by using the weight vector estimate by the BCEB through the following steps.

$$\hat{q}_n [v] = \mathrm{vec}\left(\mathbf{W}^{(n)} [V - 1]\right)^{\mathrm{H}} \mathrm{vec}\left(\mathbf{U} [v]\right), v = 0, \ldots, V - 1 \tag{20}$$

where $\mathbf{W}^{(n)}[V-1]$ is the final weight vector obtained in the nth BECB. We can obtain the received symbol vector of the nth stream

$$\hat{\mathbf{q}}_n = \left[\hat{q}_n[0]\ \hat{q}_n[1] \cdots \hat{q}_n[V-1]\right]^{\mathrm{T}}.$$

After the parallel-to-serial conversion (PSC), we can recover the received data vector $\hat{\mathbf{b}}$.

3 Computer Simulations

We verify the performance of the proposed method through computer simulations. Figure 4 depicts the compared method (CM), i.e., the transmitter-receiver configuration which will be compared with the proposed method. Although it looks complicated, the CM is based on the following simple and trivial idea.

1. The multiplexing depicted in Fig. 1 is realized by multiplying each data stream with a cyclic-shifted version of the spreading code.
2. The CM receiver detects the multiplexed streams by using the matched filters corresponding with the cyclic-shifted version of the spreading code.
3. We assume that the CM receiver perfectly knows the information regarding the received signal, including the bit- and the chip-timings and channel coefficients, even though the proposed method does not have this information.

Table 1 lists simulation conditions. The computer simulations are performed 100 times by statistically generating the channel coefficients α_m, data sequences and

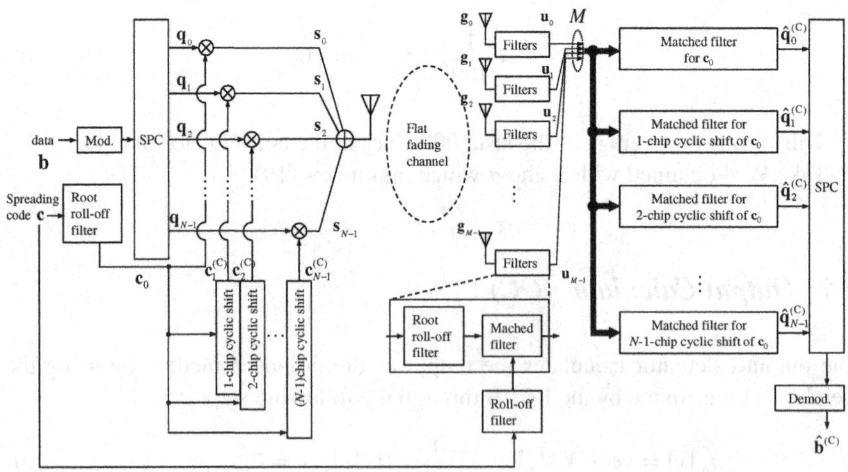

Fig. 4 Configuration of the compared method

Table 1 Simulation conditions

Transmitter	
The length of a stream [bit]	1000
Spreading code length (L_C) [chip]	31-chip (M-sequence)
The number of the streams (N)	10, 30
Channel	
Channel	Rayleigh fading channel
Receiver	
The number of antennas (M)	20
Sampling	4 [sample/chip]
Average SNR	−15 [dB] before despreading
The number of iteration of (12) (V)	1000
Parameters for the search of μ and σ	$\vec{\mu}_1 = [1 \cdots 9]$
	$\vec{\mu}_2 = [-24\ -23\ -22]$
	$\vec{\sigma}_1 = [22\ 23\ 24]$

noise. The performance of the proposed method is compared with that of CM in terms of the output SNR obtained in each snapshot, calculated as follows:

$$\Gamma_{n,\lambda} = \frac{\left|r_{n,\lambda}\right|^2}{1 - \left|r_{n,\lambda}\right|^2} \tag{21}$$

where $\Gamma_{n,\lambda}$ is the output SNR of the nth stream at the λth snapshot in the simulation, while $r_{n,\lambda}$ is the correlation coefficient which is defined for the proposed method as:

$$r_{n,\lambda} = \frac{\hat{\mathbf{q}}_{n,\lambda}^{\mathrm{H}} \mathbf{q}_{n,\lambda}}{\sqrt{\hat{\mathbf{q}}_{n,\lambda}^{\mathrm{H}} \hat{\mathbf{q}}_{n,\lambda}} \sqrt{\mathbf{q}_{n,\lambda}^{\mathrm{H}} \mathbf{q}_{n,\lambda}}} \tag{22}$$

and for CM,

$$r_{n,\lambda} = \frac{\left(\hat{\mathbf{q}}_{n,\lambda}^{(\mathrm{C})}\right)^{\mathrm{H}} \mathbf{q}_{n,\lambda}}{\sqrt{\left(\hat{\mathbf{q}}_{n,\lambda}^{(\mathrm{C})}\right)^{\mathrm{H}} \hat{\mathbf{q}}_{n,\lambda}^{(\mathrm{C})}} \sqrt{\mathbf{q}_{n,\lambda}^{\mathrm{H}} \mathbf{q}_{n,\lambda}}}. \tag{23}$$

Figure 5 shows SNR versus the probability $\Pr\left(\Gamma_{n,\lambda} \leq \text{Abscissa}\right)$ comparing the proposed method with CM. According to Table 1, the maximum output SNR is 13 dB (the SNR before despreading: −15 dB + the spreading gain: 15 dB + the antenna gain: 13 dB). It is observed that the proposed method improves the output SNR for 1.1 dB, $\Pr\left(\Gamma_{n,\lambda} \leq \text{Abscissa}\right) = 0.5$, compared with CM realizing 10 streams. In addition, the proposed method improves the output SNR for 3.0 dB, compared with

Fig. 5 System image

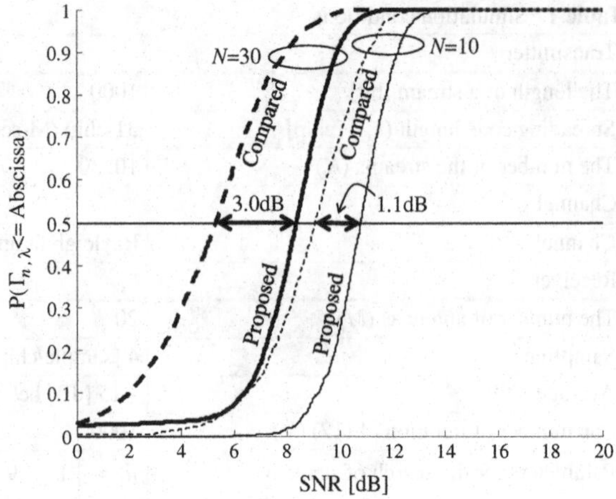

CM realizing 30 streams. It should be emphasized that the CM exploits the perfect information about the chip- and the symbol-timings as well as the channel coefficients, while such information is not given to the proposed method.

4 Conclusions

In this paper, we proposed a new transmitter and receiver configurations for subsurface wireless sensor networks. The advantages are summarized as follows:

1. The proposed method achieves CDM using a single spreading code. This is advantageous for the future extension to the multiple access of the sensors which sends multiple streams by CDM.
2. The proposed method does not need the training sequence to obtain the antenna diversity gain. Since the transmitter does not have to send the training sequence, it can save the battery power and can improve the throughput, simultaneously.

Further considerations include the extension of the proposed system in order to manage multiple sensors based on code division multiple access (CDMA).

References

1. Yoon, S., Cheng, L., Ghazanfari, E., Wang, Z., Zhang, X., Pamukcu, S., Suleiman, M.T.: Subsurface monitoring using low frequency wireless signal networks. In: 2012 IEEE International Conference on Pervasive Computing and Communications Workshops (PERCOM Workshops), pp. 443–446 (2012). doi:10.1109/PerComW.2012.6197530

2. Silva, A., Moghaddam, M.: Adaptive sub-MHz magnetic induction-based system for mid-range wireless communication in soil. In: 2015 IEEE-APS Topical Conference on Antennas and Propagation in Wireless Communications (APWC), pp. 1627–1630 (2015). doi:10.1109/APWC.2015.7300222
3. Simon, M.K., Omura, J.K., Scholtz, R.A., Levitt, B.K.: Spread Spectrum Communications. Computer Science Press (1985)
4. Compton, R.T.: Adaptive Antennas: Concepts and Performance. Prentice Hall (1988)
5. Treichler, J.R., Agee, B.G.: A new approach to multipath correction of constant modulus signals. IEEE Trans. Acoust. Speech Signal Process. **ASSP-32**(2), 459–472 (1983)
6. Haykin, S.: Adaptive Filter Theory. Prentice Hall (2013)

Implementation of Mobile Sensing Platform with a Tree Based Sensor Network

Katsuhiro Naito, Shunsuke Tani and Daichi Takai

Abstract This paper develops a new mobile sensing platform employing a tree based sensor network. The mobile sensing platform consists of mobile sensor devices, relay devices, and a sink device. We assume that robots, UAVs, etc. carry a mobile sensor device to measure environment. Therefore, the mobile sensor device can be easily relocated and can perform sensing at any locations. The relay devices can construct a tree based route to the sink device. Functions of the relay devices are data collection from the mobile sensor devices and data forwarding to the sink device. They also implement our special routing protocol and a media access control mechanism to avoid interference of radio signals in a sensor network and to reduce power consumption. We have developed special software for wireless module System on Chip (SoC) for IEEE 802.15.4 because our research target is to design a feasible and reasonable sensor network system. The consumed power of the SoC is 15 mA in a transmission, 17 mA in a reception, and 6 μA in a sleep mode. Therefore, our mobile sensing platform can work with a solar cell and a Li-Po battery. The evaluation results show that our protocol can synchronize timing among relay devices, and can create a tree based route to a sink device. Additionally, they can find that mobile sensor devices can inform measured data to a sink device through relay devices.

Keywords Mobile sensor networks · Tree-based routing · Media access control · Wireless module SoC

K. Naito (✉) · S. Tani · D. Takai
Department of Information Science, Aichi Institute of Technology,
1247 Yachigusa, Yakusa, Toyota, Aichi 470-0392, Japan
e-mail: naito@pluslab.org

S. Tani
e-mail: shunsuke@pluslab.org

D. Takai
e-mail: takai@pluslab.org

G. De Pietro et al. (eds.), *Intelligent Interactive Multimedia Systems and Services 2016*, Smart Innovation, Systems and Technologies 55,
DOI 10.1007/978-3-319-39345-2_19

1 Introduction

Mobile sensing systems are a new type of flexible sensor networks that employ mobile devices such as an unmanned aerial vehicle (UAV), a robot, etc. [1–3]. As a direct transmission from mobile devices is limited, mobile devices require a platform network to extend the measurement area. Traditionally, sensor network systems use multi-hop communication to extend a network area [4–6]. Platform networks also employ multi-hop communication to cover a network area, and should operate on low power devices [7–9]. Therefore, requirements for platform networks for mobile sensing systems are flexible data collection and low power operation [10–12].

Since an important function in sensor networks is a long life operation with small capacity battery, almost all systems require an efficient data collection scheme with low power consumption. The schemes are generally classified into a media access layer approach [13, 14] and a routing layer approach [15, 16]. A media access layer approach typically employs a special frame format to realize a sleep operation for turning off all circuits on a device. A routing layer approach for sensor networks sometimes employs a tree-based route because a sink device is an only device for data collection. Traditional sensor networks assume a fixed and static devices for data collection. Therefore, a system requires various numbers of sensor devices to cover a whole sensing area.

Mobile sensor networks are extended sensor networks where some devices can move to a measurement location. According to a flexible mobility of sensor devices, a fewer number of devices can cover a whole measurement area comparing to a traditional sensor network [17, 18]. Some researchers have considered efficient data gathering scheme to realize efficient resource usages. On the contrary, an implementation of these schemes on real hardwares is difficult due to the limitation of a computational resource and overhead of an optimization for data collection.

This paper proposes a simple mobile sensor network system for practical wireless system on chips (SoC). The proposed system consists of an original media access control scheme and an original tree-based routing to reduce required hardware resource and to realize a low power operation. We have developed our original communication protocol on a SoC to show the feasibility of the proposed protocol on real devices. The evaluation results show that the proposed protocol can work on the real SoC hardware to realize a mobile sensor network system.

2 Mobile Sensing Platform

Figure 1 shows the overview of the mobile sensing platform. The platform consists of a sink device, relay devices, mobile sensor devices. The sink device is a fixed device in a sensor network, and collects measured information from mobile sensor devices. The relay devices are a fixed device, and construct a route to the sink device with a routing protocol. They also transfer measured information from mobile sensor

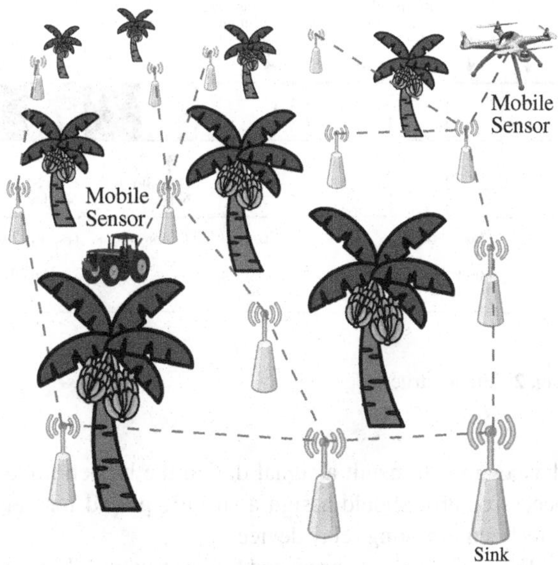

Fig. 1 Mobile sensing platform

devices to the sink device. The mobile sensor devices are a mobile device such as a robot, UAV, etc. They have a few sensors to measure environment or specific objects, and transfer measured information to a neighbor relay device.

Typical sensor network systems construct a route from all sensor devices to a sink device. The constructed route is usually fixed because all devices do not move in a short period. On the contrary, mobile sensor devices frequently move in a sensing area in the mobile sensing platform. Therefore, the mobile sensing platform requires a new route management scheme for mobile sensor devices. In the proposed platform, relay devices construct a tree based route to a sink device because they are fixed in a sensing area. Mobile sensor devices have no route to a sink device and only ask a neighbor relay device to transfer its own measured information. The benefit of the proposed scheme is to reduce control messages for the routing management and to reduce power consumption for updating routing information.

2.1 Media Access Control

Sensor networks usually require low power operation because each sensor devices work by a battery power. Relay devices in the mobile sensing platform also require low power consumption due to the same reason. It is common knowledge that a deep sleep operation is the best method to reduce consumed power in a micro-processor chip. Therefore, media access control also considers a deep sleep period to turn off circuits on a micro-processor chip. Additionally, multi-hop networks usually suffer from interference from simultaneous wireless transmissions because neighbor relay

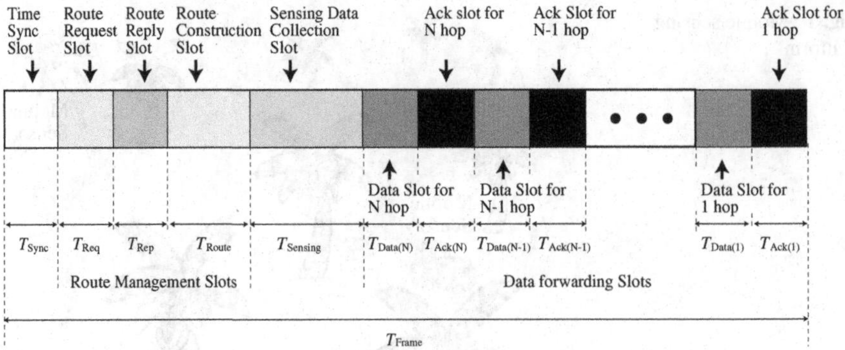

Fig. 2 Frame structure

devices may transmit a signal due to the hidden node problem. As a result, media access control should assign a specific period for neighbor relay devices to avoid interference among relay devices.

Figure 2 shows the proposed frame structure. The frame structure consists of some slots for specific purposes: time synchronization, route management, sensing data collection, and data forwarding. The proposed frame structure indicates a specific communication period depending on the purposes or hop count from the sink device. Therefore, relay devices can easily sleep according to the frame structure. Additionally, neighbor devices with a different hop count transmit a signal in a different period to avoid interference. The following is the detail of each slot.

- Time synchronization slot
  Time synchronization scheme is a key function to maintain timing among a sensor network to realize a sleep operation for reducing consumed power. Typical quartz on a microcomputer board is not sufficiently accurate to maintain high-precision time for a frame synchronization. The proposed scheme employs a time synchronization scheme between nearby devices. Each device has two time information: absolute time and relative time in a frame. The absolute time starts when a device boots up. The relative time repeats in a frame period. An upstream device calculates relative time in a frame by checking a current time in the time synchronization slot. It transmits a time synchronization message to downstream devices. The time synchronization message includes relative time. Therefore, the downstream devices can regulate the relative time according to the relative time in the time synchronization message.
- Route management slots
  Relay devices are almost fixed device in the proposed system. Therefore, routes from each relay device to a sink device are almost static. The proposed scheme assigns a special period for a route management to decrease power consumption. As a result, each relay device can sleep during the route management slots when it has an available route. The route management slots consist of three slots: a route request slot, a route reply slot, and a route construction slot. The route request slot

is a special period to request a new route to neighbor devices. The route reply slot is a special period to inform an available route information to the requested device. The route construction slot is a period to construct a route between an upstream device and the requested device in a three-way handshake procedure.

- Sensing data collection slot
  Mobile sensor devices move in a sensing area. Therefore, reconstructing of a route to the sink device is a considerable overhead. The proposed frame assigns a unique time slot for collecting sensing data. All relay devices listen the channel during the sensing data collection slot to receive a data message from neighbor mobile sensor devices. The received data messages are forward to the sink device.
- Data forwarding slots
  Sensor network systems typically employ a large number of relay devices. Therefore, transmission range of each relay device is typically overlapped. As consequence, signal interference among neighbor relay devices may deteriorate communication performance. The proposed frame assigns data transmission slot and acknowledgement slot in the data forwarding slots, and divides each slot to subslots for each hop count from the sink device. As a result, relay devices, where their hop count is different, transmit a data message in different timing.

2.2 Routing Control

A routing control is an important function to realize a multi-hop network. Practical sensor networks require low-power operation for hardware devices to realize a long lifetime. Therefore, almost all hardware devices support a deep sleep mode that realizes a quite low-power operation. Additionally, they have a limited size of memory space to decrease power consumption. As a result, typical hardware devices require a simple routing control mechanism to implement a routing protocol.

The proposed routing protocol realizes a tree based route from relay devices to a sink device because all relay devices forward a data message to the sink device. Each relay device can construct a route to its own upstream device by exchanging routing control messages. Each routing control message is presented as follows.

2.3 Routing Control Messages

The proposed routing protocol uses five types of routing control messages: route request message, route reply message, route construction request message, route construction reply message, and route construction acknowledgement message.

- Route request
  Route request(RREQ) control messages are to find a device with a route to a sink

device. Downstream devices without a route broadcast a RREQ message to neighbor upstream devices. The message includes a source address of a device.

- Route reply
  Route reply(RREP) control messages are to inform an available route to a sink device to a downstream device that requests a route. Upstream devices with an available route can transmit a RREP message to a requested downstream device. The message includes a destination address, a source address, and a hop count to a sink device. The requested downstream device can select a shortest route by evaluating the hop count in a RREP message.
- Route construction request
  Route construction request(RCREQ) messages are to request a route construction to an upper device. A downstream device selects an adequate upstream device based on a hop count, and transmits a RCREQ message to the selected upstream device. The upstream device that receives the RCREQ message starts a route construction process. The message includes a destination address and a source address.
- Route construction reply
  Route construction reply(RCREP) control messages are to inform the start of the route construction process to the downstream device. The upstream device transmits a RCREP message to the requested downstream device. The message includes a destination address and a source address.
- Route construction acknowledgement
  Route construction acknowledgement(RCACK) control messages are to complete the route construction process to the upstream device. The downstream device transmits a RCACK message to the own upstream device. The message includes a destination address and a source address.

2.4 Routing Procedures

Figure 3 shows the route construction procedures for one-hop devices in the proposed protocol. The figure assumes that a sensor network consists of a sink device and five relay devices.

In a route request slot, relay devices without an available route try to find a route to the sink device. In Fig. 3i, relay device R2 and R3 transmit a RREQ message to neighbor devices. In this situation, the sink device is an only device with an available route.

In a route reply slot, a device with an available route to the sink device reply a route reply message to the requested devices. In Fig. 3ii, the sink device replies a RREP message to relay device R2 and R3. The RREP message contains the address of R2 or R3 as a destination address, the address of the sink device as a source address, and 0 as a hop count to a sink device.

In a route construction slot, a device that finds an available route constructs a route to the sink device. In Fig. 3iii, relay device R2 and R3 transmit a RCREQ message to the sink device. The sink device replies a RCREP message to relay device R2 and

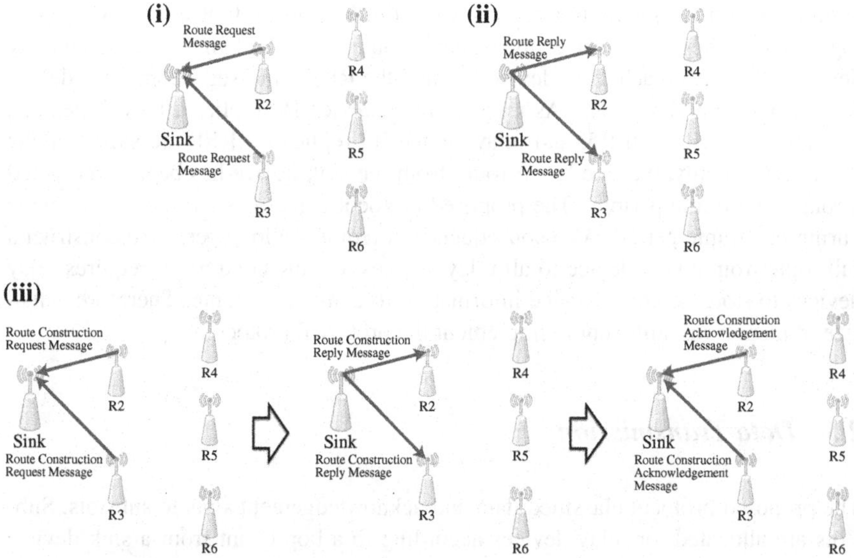

Fig. 3 Route construction procedures. **i** Route request slot **ii** Route reply slot **iii** Route construction slot

Fig. 4 Routing procedures for each hop count. **i** Ist frame **ii** 2nd frame

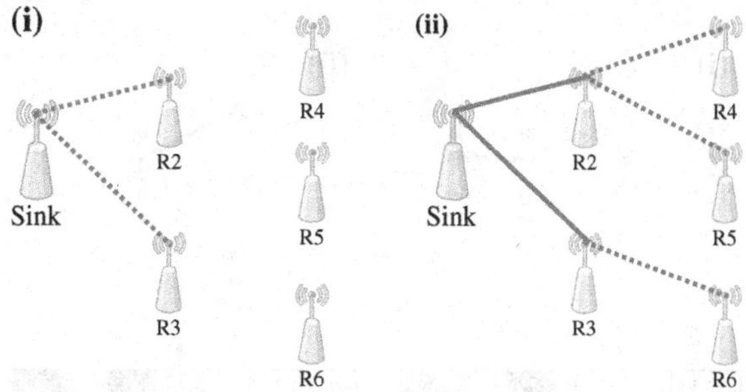

R3. Finally, relay device R2 and R3 reply a RCACK message to the sink device. The proposed protocol can build the routes from relay device R2 and R3 to the sink device through the three-way handshake procedure.

Figure 4 shows the routing procedures for each hop count. The proposed protocol constructs a route hop by hop to reduce a required memory space for routing information. Therefore, the protocol can construct a route for one hop during one frame period. In Fig. 4i, all relay devices transmit a RREQ message to neighbor devices. Therefore, the RREQ message from relay device R2 reaches the sink device, relay device R4 and R5, and the RREQ message from relay device R3 reaches the sink device and relay device R6. Since relay devices R4, R5 and R6 do not have an available route, the sink device replies a RREP message to device R2 and R3. Finally, relay devices R2 and R3 can construct a route during the 1st frame period.

In the second frame period, relay devices R4, R5 and R6 transmit a RREQ message because they do not have any available route. The RREQ messages from relay device R4 and R5 reach relay device R2, and the RREQ message from relay device R6 reaches relay device R3. As a result, relay device R2 replies a RREP message to relay device R4 and R5, and relay device R3 replies a RREP message to relay device R6. Finally, the tree based route from the sink device has been constructed through two frame periods. The proposed protocol constructs a route for one hope during one frame period. As a consequence, it requires a long period to construct a full route from a sink device to all relay devices. On the contrary, it requires relay devices to store neighbor device information to construct a route. Therefore, small size of memory is sufficient to implement the proposed protocol.

2.5 Data Transmission

The proposed protocol classified data and acknowledgement slots to subslots. Subslots are allocated for relay devices according to a hop count from a sink device. Therefore, neighbor relay devices with a different hop count use another subslots to avoid collisions.

Figure 5 shows the example operation of data relaying. The mobile sensor transmits a data message including measured information to neighbor relay devices in

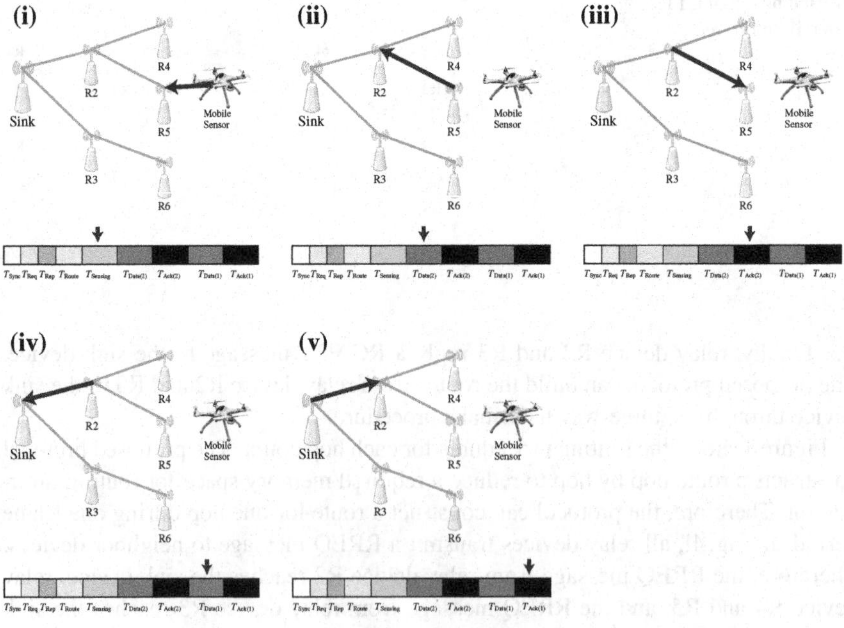

Fig. 5 Data transmission procedures. **i** Mobile sensors **ii** Relay at 2 Hop device **iii** Acknowledgment for 2 Hop devices **iv** Relay at 2 hop devices **v** Acknowledgment for 1 hop devices

the sensing data collection slot. Therefore, all relay devices should listen the wireless channel to receive a data message. In the example, relay device R5 receives a data message from the mobile sensor. Since relay device R5 is a two hop count device, it relays the data message in the subslot for two hop count. Relay device R2 receives a data message from relay device R5 because it knows relay device R5 is the downstream device. It also replies the acknowledgement message to relay device R5 in the acknowledgement subslot for two hop counts. Then, it relays the data message in the data subslot for one hop count. The sink device receives a data message from relay device R2 and replies the acknowledgement message to relay device R2.

3 Experimental Results

We have implemented the proposed protocol on a system-on-chip (SoC) for radio communication. Table 1 shows the specification of the SoC and Xbee which is a well known wireless module. Xbee typically requires a master micro-computer board

Table 1 Specifications of TWE-Lite and Xbee

Device	TWE-Lite Dip	Xbee S1
Development environment	TWE-Lite Developing Software	None
Microcontroller	32 bit RISC (4, 8, 16, 32 [MHz])	None
Memory	32 [KB] RAM, 4 [KB] EEPROM, and 160 KB Flash Memory	None
Input DC	2.0–3.6 V [V]	2.8–3.4 [V]
UART	2	None
I2C	1	None
SPI	3	None
AD	10 bit, 4 Channels	None
PWM	4	None
GPIO	20	None
Wireless module	IEEE 802.15.4	IEEE 802.15.4
Frequency	2.4 [GHz]	2.4 [GHz]
Number of channels	16	16
Modulation scheme	O-QPSK, DSSS	QPSK DSSS
Bit rate	250 [kbps]	250 [kbps]
Transmission power	+2.5 [dBm]	+0 [dBm]
Receiver sensitivity	−95 [dBm]	−92 [dBm]
Consumed power (TX)	15 [mA]	45 [mA]
Consumed power (RX)	17 [mA]	50 [mA]
Consumed power (Sleep)	6 [μA]	10 [μA]

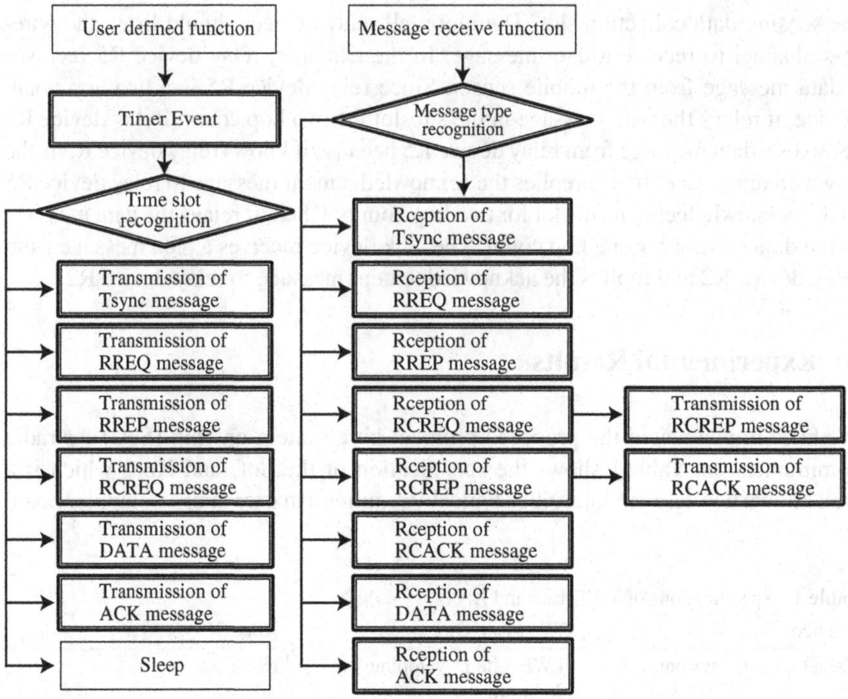

Fig. 6 Flowchart for message processing

such as Arduino. Therefore, additional power is required to work with Xbee in practical implementations. The SoC has a RISC based microcomputer and IEEE 802.15.4 based wireless module. Figure 6 shows the flowchart for the message processing. Table 2 shows the detail parameters for the proposed frame structure. The developer software provides a timer event mechanism and a trigger function for receiving a message. Therefore, the developed software implements the slot based operation with the timer event mechanism to realize a sleep operation. Additionally, it also classified the original messages by using the trigger function for receiving a message.

We assumed the device location in Fig. 7. We have set up a small sensor network system due to a limitation of an experimental space. The dot lines mean the constructed routes in the experiment. The mobile sensor device moves on the solid line, and transmits data messages to neighbor relay devices. It transmits 122 data messages during the moving of the mobile sensor device. The sink device receives 223 data messages from the network. In the proposed system, a mobile sensor device does not construct a route to a network. Therefore, some relay device receives a same data message from a mobile sensor device. As a result, duplicate data messages may reach a sink device. In the experiment, the sink device can receive 104 unique data messages from the mobile sensor device. Therefore, the arrival rate of each data message is 85 [%] because the developed software does not perform a retransmission for a lost

Table 2 Frame format parameters

Device	TWE-Lite Dip
Number of devices	Sink device:1, Relay devices:6, Mobile device:1
Measurement interval	10 [s]
Frame period	10 [s]
TSYNC Slot	200 [ms]
RREQ Slot	200 [ms]
RREP Slot	200 [ms]
RCON Slot	600 [ms]
Sensing data collection Slot	200 [ms]
Data Subslot	300 [ms]
ACK Subslot	100 [ms]
Maximum hop count	10 [hop]

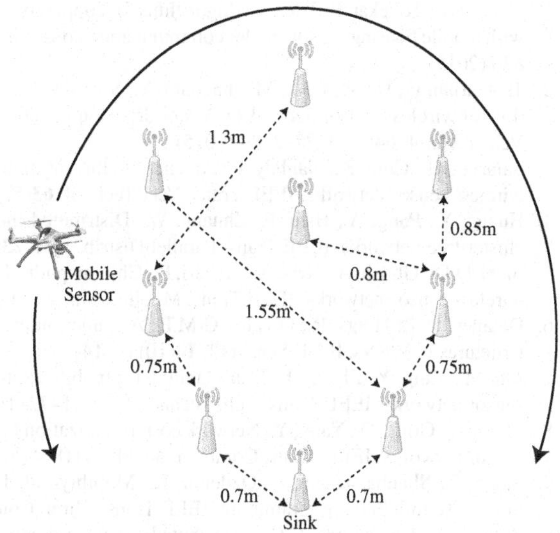

Fig. 7 Device location in experiment

message to evaluate the communication performance. Through the experiment, we can find that the proposed mechanism can implement on typical SoC modules for IEEE 802.15.4 communication. Therefore, the proposed mechanism is a one of a feasible protocol for resource-constrained devices.

4 Conclusion

This paper has developed a new mobile sensing platform employing a tree based sensor network. The mobile sensing platform consists of mobile sensor devices, relay devices, and a sink device. Relay devices can construct a tree based route to the sink device, and can forward data message from mobile sensor devices. We have developed a special software for wireless module System on Chip (SoC) for IEEE 802.15.4. The evaluation results show that our protocol can synchronize timing among relay devices, and can create a tree based route to a sink device. Additionally, they can find that mobile sensor devices inform measured data to a sink device through relay devices.

Acknowledgments This work is supported in part by Grant-in-Aid for Scientific Research (B)(15H02697) and Grant-in-Aid for Scientific Research (C)(26330103), Japan Society for the Promotion of Science (JSPS) and the Integration research for agriculture and interdisciplinary fields, Ministry of Agriculture, Forestry and Fisheries, Japan.

References

1. Hook, J.V., Tokekar, P., Isler, V.: Algorithms for cooperative active localization of static targets with mobile bearing sensors under communication constraints. IEEE Trans. Robot. **31**(4), 864–876 (2015)
2. Tashtarian, F., Hossein, M., Moghaddam, Y., Sohraby, K., Effati, S.: On maximizing the lifetime of wireless sensor networks in event-driven applications with mobile sinks. IEEE Trans. Veh. Technol. **64**(7), 3177–3189 (2015)
3. Salarian, H., Chin, K., Naghdy, F.: An energy-efficient mobile-sink path selection strategy for wireless sensor networks. IEEE Trans. Veh. Technol. **63**(5), 2407–2419 (2014)
4. Huang, Y., Pang, A., Hsiu, P., Zhuang, W.: Distributed throughput optimization for ZigBee cluster-tree networks. IEEE Trans. Parallel Distrib. Syst. **23**(3), 513–520 (2012)
5. Incel, O.D., Ghosh, A., Krishnamachari, B., Chintalapudi, K.: Fast data collection in tree-based wireless sensor networks. IEEE Trans. Mobile Comput. **11**(1), 86–99 (2012)
6. Delaney, D.T., Higgs, R., O'Hare, G.M.P.: A stable routing framework for tree-based routing structures in WSNs. IEEE Sensors J. **14**(10) (2014)
7. Ma, M., Yang, Y., Zhao, M.: Tour planning for mobile data-gathering mechanisms in wireless sensor networks. IEEE Trans. Veh. Technol. **62**(4), 1472–1483 (2013)
8. Zhao, M., Gong, D., Yang, Y.: Network cost minimization for mobile data gathering in wireless sensor networks. IEEE Trans. Commun. **63**(11), 4418–4432 (2015)
9. Salari, S., Shahbazpanahi, S., Ozdemir, K.: Mobility-aided wireless sensor network localization via semidefinite programming. IEEE Trans. Wirel. Commun. **12**(12), 5966–5978 (2013)
10. Zhong, M., Cassandras, C.G.: Distributed coverage control and data collection with mobile sensor networks. IEEE Trans. Autom. Control **56**(10), 2445–2455 (2011)
11. Shih, Y., Chung, W., Hsiu, P., Pang, A.: A mobility-aware device deployment and tree construction framework for ZigBee wireless networks. IEEE Trans. Veh. Technol. **62**(6), 2763–2779 (2013)
12. Velmani, R., Kaarthick, B.: An efficient cluster-tree based data collection scheme for large mobile wireless sensor networks. IEEE Sensors J. **15**(4), 2377–2390 (2015)
13. Huang, P., Xiao, L., Soltani, S., Mutka, M.W., Ning, X.: The evolution of MAC protocols in wireless sensor networks: a survey. IEEE Commun. Surv. Tutorials **15**(1), 101–120 (2013)
14. Khanafer, M., Guennoun, M., Mouftah, H.T.: A survey of beacon-enabled IEEE 802.15.4 MAC protocols in wireless sensor networks. IEEE Commun. Surv. Tutorials **16**(2), 856–876 (2014)
15. Chiwewe, T.M., Hancke, G.P.: A distributed topology control technique for low interference and energy efficiency in wireless sensor networks. IEEE Trans. Ind. Inf. **8**(1), 11–19 (2012)
16. Naito, K., Ehara, M., Mori, K., Kobayashi, H.: Implementation of field sensor networks with SunSPOT devices. In: IPSJ The Fifth International Conference on Mobile Computing and Ubiquitous Networking (ICMU 2010), Apr 2010

17. Du, R., Chen, C., Yang, B., Lu, N., Guan, X., Shen, X.: Effective urban traffic monitoring by vehicular sensor networks. IEEE Trans. Veh. Technol. **64**(1), 273–286 (2015)
18. Hodge, V.J., O'Keefe, S., Weeks, M., Moulds, A.: Wireless sensor networks for condition monitoring in the railway industry: a survey. IEEE Trans. Intell. Transp. Syst. **16**(3), 1088–1106 (2015)

Prototype Implementation of Actuator Sensor Network for Agricultural Usages

Takuya Wada and Katsuhiro Naito

Abstract Information technology (IT) agricultural systems for field observation and environmental control have attracted considerable attention. Recently, various kinds of dedicated devices for IT agricultural systems have been released for large-scale farmers. It is well known that introducing IT agricultural systems can enhance agricultural production. On the contrary, an initial cost of the system is quite expensive even if typical dedicated devices are introduced. Therefore, reducing an initial cost of the system is a big challenge to disseminate the IT agricultural systems among small and medium-sized farmers. This paper proposes a prototype of IT agricultural systems that can measure an environment and can control the environment according to the measured information. To mitigate the initial cost of the system, our prototype uses Arduino compatible boards that are one of well-known micro-computer boards for general purposes. Therefore, it has a flexibility for designing a hardware and a capability of software development on the Arduino integrated development environment. The proposed system consists of sensing and control devices, a gateway device, and a web service. The sensing and control devices have some sensors such as a temperature, humidity, and soil moisture sensors, and a control function of a water sprinkling. The web service provides a user interface to manage the information in the database system. Experimental results show that the developed prototype system can realize a periodic environmental monitoring and environmental control according to the measured information. Additionally, users can observe the environment visually through the web service.

Keywords IT Agriculture systems · Environmental measurement · Controlling environment · Arduino · Sensor networks

T. Wada (✉) · K. Naito
Department of Information Science, Aichi Institute of Technology,
1247 Yachigusa, Yakusa, Toyota, Aichi 470-0392, Japan
e-mail: wada@pluslab.org

K. Naito
e-mail: naito@pluslab.org

G. De Pietro et al. (eds.), *Intelligent Interactive Multimedia Systems and Services 2016*, Smart Innovation, Systems and Technologies 55,
DOI 10.1007/978-3-319-39345-2_20

1 Introduction

Recent agriculture has serious issues due to various kinds of problems such as a shortage of water, diseases of plants etc. [1, 2]. Information technology (IT) has focused to solve these issues because detail information can support farmers to enhance agricultural productivity [3–7]. Recent IT agriculture systems can measure a field environment, growing conditions of plants by using some environmental sensors. Wireless sensor networks have been also focused to collect measured information in a field because a multi-hop communication technology is suitable to extend communication area [8–10].

Typical wireless sensor networks make up of a lot of sensor devices, a gateway device, and a data management system. The sensor devices have some sensors and a wireless communication module. Therefore, they can measure an environment by sensors and transmit measured information to the gateway device. The gateway device has a wireless communication module and a network interface module because its function is a relaying of measured information to the data management system. The data management system has a data base function and a data receiving function from the gateway device. Since sensor networks require a routing protocol for multi-hop communication, various routing protocols have been proposed [9, 11]. Additionally, due to a difficulty of using a commercial power supply in a field, a low-power operation is also an important function in sensor networks. As a result, a low-power micro-computer is required to develop sensor networks.

Commercial IT agriculture systems have been released recently in practical agricultural fields. With these systems, agricultural workers can monitor an environment of fields remotely. Therefore, they can estimate diseases of plants and to enhance agricultural productivity when they can obtain dense information at many positions. On the contrary, an installation fee of these systems is quite expensive. For example, a smart plastic greenhouse employing sensors is two or three times as expensive as a conventional one. Additionally, the approximate amount of the conventional system is more than $10,000 even if the system includes only some sensors, a communication function, and a data management function [12, 13]. Therefore, it is difficult for almost all farmers to install a useful IT agriculture systems.

This paper develops a prototype IT agriculture system that employs a general microcomputer board. The developed system consists of sensing and control devices, a gateway device and a web service. The sensing and control device has some sensors such as temperature, humidity, brightness and soil moisture sensors and IEEE 802.15.4 based wireless module. They employ an Arduino compatible board as a hardware. The gateway device also employs the same microcomputer board as a hardware and implements an ethernet interface and IEEE 802.15.4 based wireless module. Therefore, it can forward a data message from IEEE 802.15.4 network to an IP network. The web service has two main function: a database system and a web application The database system can store measured information and management information of the sensor network. The web application has two interfaces: a data upload interface and a user interface. The data upload interface is used to receive a data message from the gateway device. The user interface provides management views of the sensor network and measurement data for users.

In the experimental evaluation, we have developed the prototype system. The prototype system employs an ATmega2560 low-power Atmel 8-bit AVR RISC-based micro-controller to realize a low-power operation. Additionally, we have employed Xbee Series 1 as IEEE 802.15.4 based wireless module. The experimental results demonstrate that the prototype can measure an environment of a field periodically and can upload the measurement data to the web service through the gateway device. In addition, we have confirmed that the sensing and control device can control the environment of the field depending on the measurement data. Finally, the prototype can be developed with a general Arduino board that is a quite reasonable price comparing to a special hardware for agricultural use. Therefore, the cost of each sensor device and gateway device is lower than $100. Then, our prototype can be a candidate system to promote a reasonable IT agricultural system.

2 Actuator Sensor Network for Agricultural Usages

2.1 System Model

Figure 1 shows the system model of the development system. The development system is composed of sensing and control devices, a gateway device, and a Web service. The sensing and control devices are installed in agricultural fields to measure an environment. They measure temperature, humidity, brightness and soil moisture by external sensors, and control the growing condition by water sprinkling. Additionally, they also transmit the measurement data to the gateway device, that is placed in an office, by using the wireless communication module based on IEEE 802.15.4. The gateway device receives the measurement data and uploads them to the web service by the HTTP communication. The Web service can store the uploaded data into the database. In addition, the Web service also provides a user interface to handle the system and confirm the stored data for users.

2.2 Hardware

The development system uses Arduino based micro-controller for general purposes. The development system employs an original Arduino compatible board called Comduino. Since Comduino is designed as Arduino-compatible board using ATmega2560, an integrated development environment of the Arduino (IDE) is available. Figure 2 shows the overview of the board. The board supports a solar battery, a Li-Po battery, two Xbee radio interfaces, and connectors for general Arduino shields.

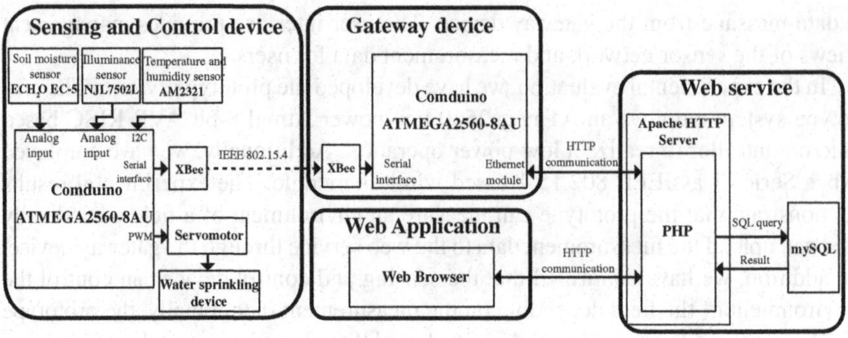

Fig. 1 System model

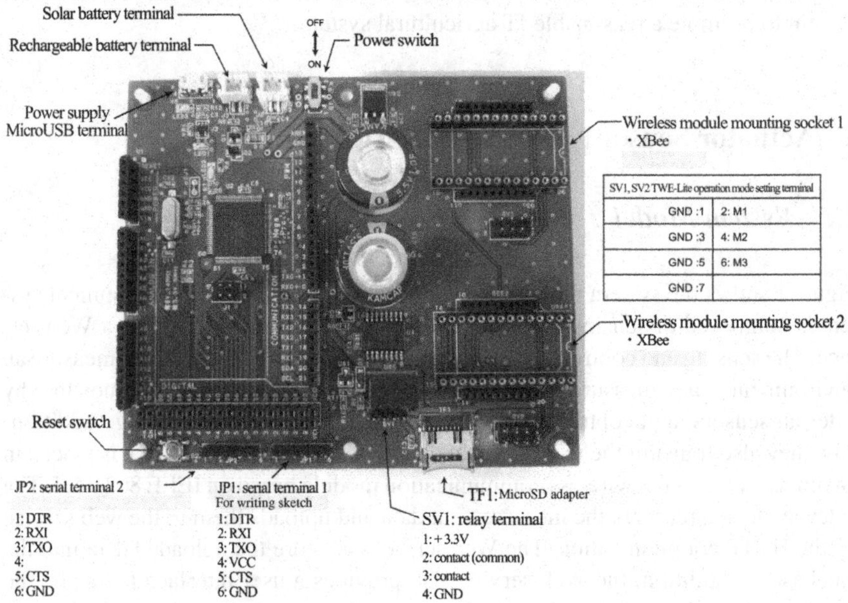

Fig. 2 Overview of Comduino

Table 1 shows the comparison of the well-known microcomputer board and Comduino. We can find that the microcomputer board of Arduino-based including Comduino have a limited hardware resource: small size of memories, a low clock CPU comparing to Raspberry Pi and BeagleBone. On the other hand, their consumed power is quite low because they use a resource constrained hardware and have a sleep mode. Low-power operation is quite important function in practical sensor networks due to a limitation of a battery. Therefore, Arduino-based microcomputer is suitable hardware for sensor networks.

Table 1 Specification of microcomputer boards

Name	Arduino Uno R3	Comduino	Raspberry Pi (Model B+)	BeagleBone Black
Processor	Atmega328	Atmega2560	ARM11	AM3358
Clock speed	16 MHz	8 MHz	700 MHz	1 GHz
RAM	2 KB	8 KB	512 MB	512 MB
Flash	32 KB	256 KB	MicroSDCard	4 GB (8 bit eMMC)
EEPROM	1 KB	4 KB		
Power	51 mA	21 mA	250 mA	210–460 mA
Sleep	40.9 mA	8.15 mA	n/a	n/a
Digital GPIO	14	54	40	65
Analog input	6	16	n/a	7
PWM	6	14		8

2.3 Sensing and Control Device

The circuit diagram of a sensing and control device is shown in Fig. 3. The modules for sensing and control devices are shown as Table 2.

The prototype device employs various kinds of interfaces for sensors. The soil moisture and illuminance sensors use an analog input interface. The temperature and humidity sensor uses an I2C interface. The servo motor is controlled by a PWM interface. Since the servo motor requires 5 V DC source, we have implemented a 5 V regulator. Therefore we have confirmed that various sensors can be attached to our prototype board. Xbee is a well known wireless module for a low-power operation. The prototype has two connectors for Xbee modules. Therefore, we can design a wireless network flexibly. In the prototype implementation, we utilize Xbee Series 1 that is an IEEE 802.15.4 based module.

Figure 5 shows the flowchart of the prototype device. The prototype implements a sleep operation by a watch-dog timer. Owing to the limitation of the hardware specification, the prototype wakes up every 8 s. It also measures an environment by using the sensors every predefined interval period. In addition, it also evaluates the measurement data of the soil moisture sensor to control the water sprinkling system. The water sprinkling system is activated when the measurement data are lower than a predefined threshold value.

2.4 Gateway Device

Figure 4 shows the circuit diagram of the gateway device, and Table 3 shows the additional module for the gateway device. The function of the gateway device is relaying a measurement data from sensing and control devices to the Web service. There-

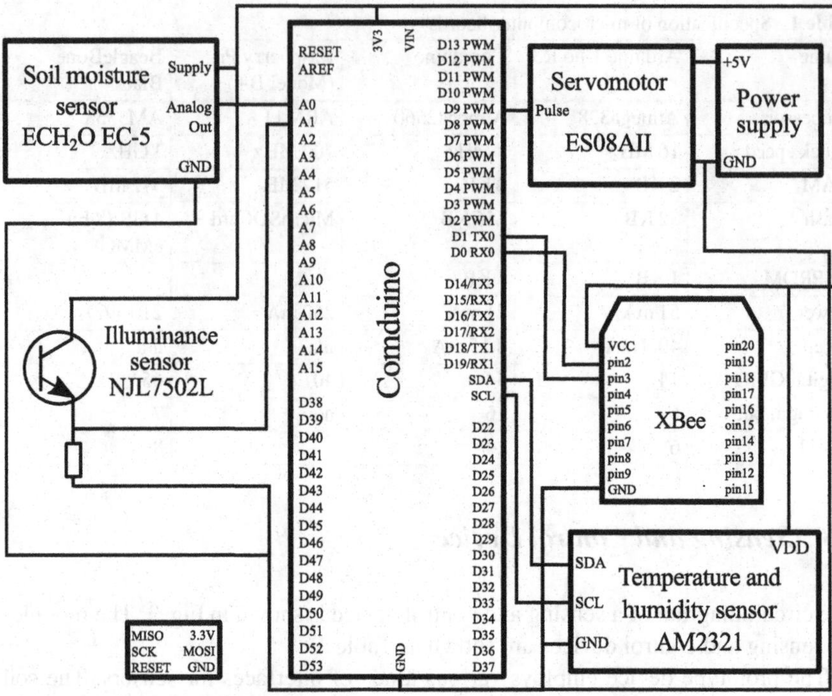

Fig. 3 Circuit diagram of sensing and control device

Table 2 Modules for the sensing and control device

Name	Type
Soil moisture sensor	ECH_2O EC-5
Illuminance sensor	NJL7502L
Temperature and humidity sensor	AM2321
Control servo for water sprinkling system	ES08AII
Wireless communication module based on IEEE 802.15.4.	XBee Series 1

fore, the gateway device receives measurement data through Xbee module that is an IEEE 802.15.4 based wireless module, and transmits the measurement data through the Ethernet module. It also employs HTTP communication to the web service.

Figure 6 shows the flowchart of the software in the gateway device. Since the Xbee module is connected through a serial interface, the serial interface is initialized for the Xbee module in the initialization process. Additionally, a Serial Peripheral Interface (SPI) is also initialized for the Ethernet module. As accurate time is important in sensor network systems, it also obtains accurate time by Network Time Protocol

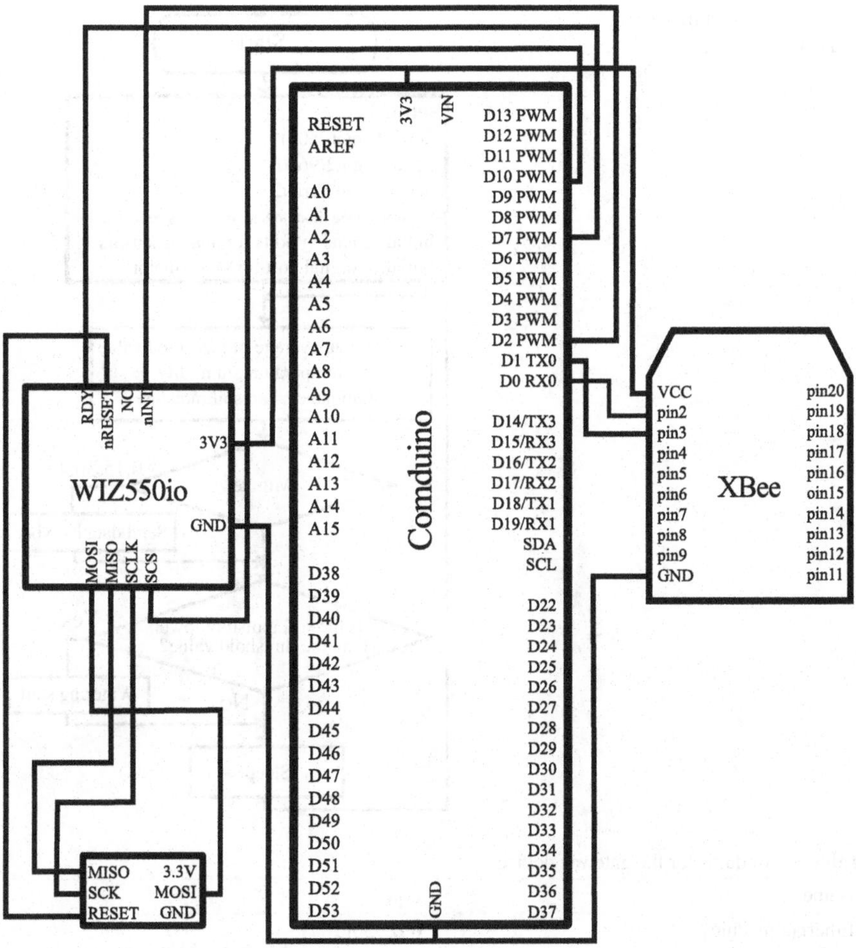

Fig. 4 Circuit diagram of gateway device

(NTP). In the normal operation, it receives a message from the Xbee module when the receive queue of the Xbee module has a message. Then, it also relays the message to the web service every 15 min. The message is transmitted by HTTP with POST format. Further, it sends a time request message to a NTP server every hour to adjust the local time.

2.5 *Web Service*

The web service has a database function and a web application function. The web application has two web interfaces for a gateway device and users. The gateway

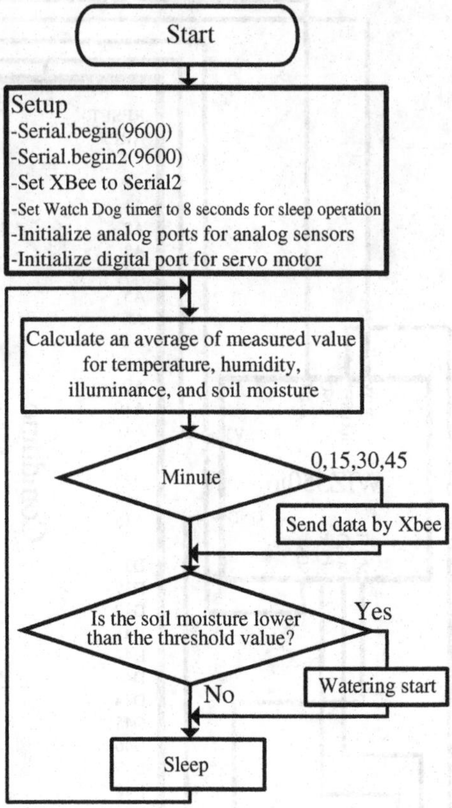

Fig. 5 Flowchart of sensing and control device

Table 3 Modules for the gateway device

Name	Type
Ethernet module	WIZ550io
Wireless communication module based on IEEE 802.15.4.	XBee Series 1

device uses a data upload interface to post a measurement data by HTTP. Users use a user interface to manage the system and to check the measurement data.

Figure 7 shows the communication sequence of the data upload interface. The sensing and control device transmits a measurement data to the gateway device. The gateway device transmits the measurement data to the data upload interface on the Web service by HTTP. The Web service stores the received data in the database. The prototype system employs PHP and MySQL to implement the software. Table 4 shows the database table in the prototype system.

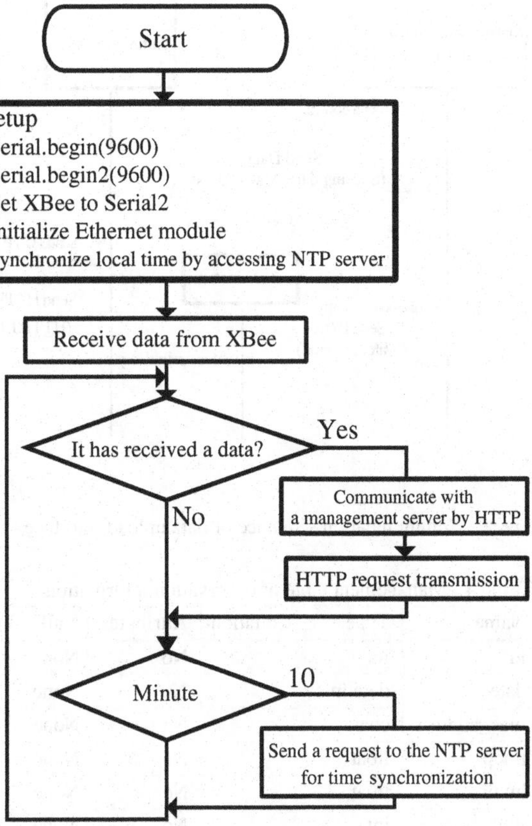

Fig. 6 Flowchart of gateway device

Figure 8 shows the communication sequence of the user interface. The web application can create a visual image of the measurement data by JQuery. In the HTTP communication, the web browser sends an HTTP request to the software on the web service by asynchronous communication. We use a JSON format to transfer the measurement data from the Web service. Finally, the Web browser can draw some graph according to the received JSON data.

3 Experimental Results

We have implemented the proposed system on the Arduino compatible board and cloud server. The sensing and control device is installed at a pot for tomato in an indoor environment. Since the cloud server is installed on our local server, the gateway device is connected to the Web service through a local network.

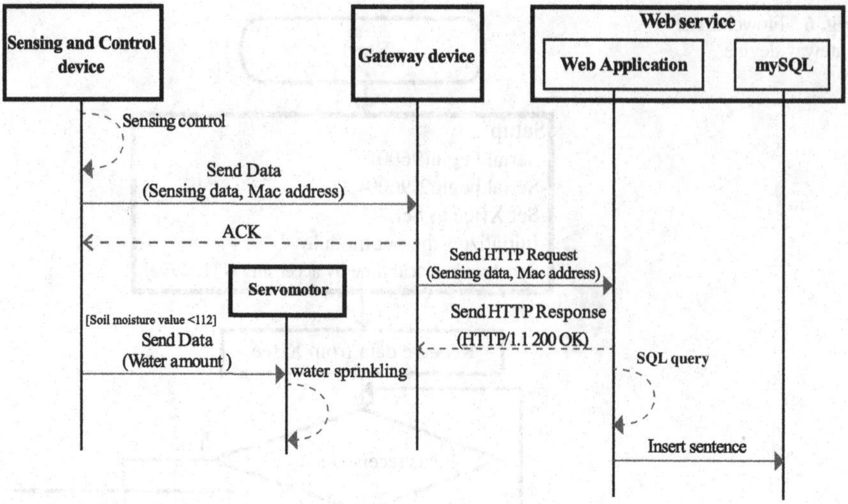

Fig. 7 Communication sequence of data upload interface

Table 4 Management table of observation information

Name	Type	Collation	Attributes	Null	Default	Extra action
id	int		No	None	AUTO_INCREMENT	Primary
date	datetime		No	None		
mac_address	char		No	None		
tmep	float		No	None		
humi	float		No	None		
illu	int		No	None		
water	int		No	None		

The experimental results show that the developed system can measure the environment by temperature, humidity, illuminance and soil moisture sensors, and can inform the measurement data to the Web service every 15 min. The sensing and control device can also control the water sprinkling function according to the measurement value by the soil moisture sensor. Additionally, the developed Web service can provide a visualization of the measurement data. Figure 9 shows the example of the graph of the measurement data.

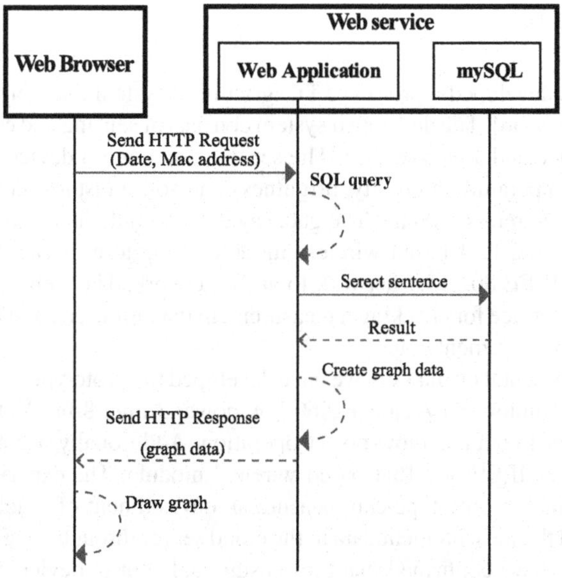

Fig. 8 Communication sequence of user interface

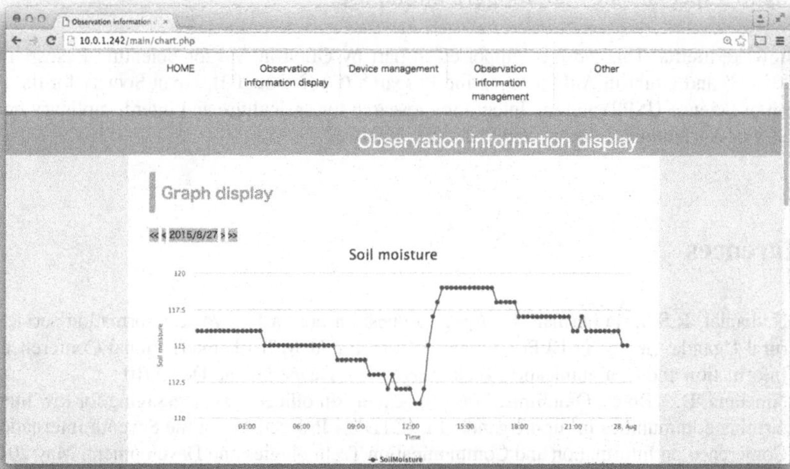

Fig. 9 Example view of Web service

4 Conclusion

This paper has developed a prototype IT agriculture system that employs a general microcomputer board. The developed system consists of sensing and control devices, a gateway device and a web service. The sensing and control device has some sensors such as temperature, humidity, brightness and soil moisture sensors and IEEE 802.15.4 based wireless module. The gateway device implements an ethernet interface and IEEE 802.15.4 based wireless module. Therefore, it can forward a data message from IEEE 802.15.4 network to an IP network. The web service can provide a visual interface for checking a measurement data for users, and a management function of a measurement data.

In the experimental evaluation, we have developed the prototype system. The prototype system employs an ATmega2560 low-power Atmel 8-bit AVR RISC-based micro-controller to realize a low-power operation. Additionally, we have employed Xbee Series 1 as IEEE 802.15.4 based wireless module. The experimental results demonstrate that the prototype can measure an environment of a field periodically and can upload the measurement data to the cloud server through the gateway device. In addition, we have confirmed that the sensing and control device can control the environment of the field depending on the measurement data. Finally, the prototype can be developed with a general Arduino board that is a quite reasonable price comparing to a special hardware for agricultural use.

Acknowledgments This work is supported in part by Grant-in-Aid for Scientific Research (B) (15H02697) and Grant-in-Aid for Scientific Research (C) (26330103), Japan Society for the Promotion of Science (JSPS) and the Integration research for agriculture and interdisciplinary fields, Ministry of Agriculture, Forestry and Fisheries, Japan.

References

1. Nathaniel, R.S.S., Futterman, F.: AppLab question box: a live voice information service in rural Uganda. In: ICTD'10 Proceedings of the 4th ACM/IEEE International Conference on Information and Communication Technologies and Development, Dec 2010
2. Ninsiima, D.: "Buuza Omulimisa" (ask the extension officer): text messaging for low literate farming communities in rural Uganda. In: ICTD'15 Proceedings of the Seventh International Conference on Information and Communication Technologies and Development, May 2015
3. Kumar, A., Kamal, K., Arshad, M., Mathavan, S., Vadamala, T.: Smart irrigation using low-cost moisture sensors and XBee-based communication. In: Global Humanitarian Technology Conference (GHTC) 2014, Oct 2014, pp. 333–337
4. Cai, T., Abbott, E., Bwambale, N.: The ability of video training to reduce agricultural knowledge gaps between men and women in rural Uganda. In: ICTD'13: Proceedings of the Sixth International Conference on Information and Communications Technologies and Development, Dec 2013, pp. 13–16
5. Francis Dittoh, V.d.B., van Aart, C.: Voice-based marketing for agricultural products: a case study in rural Northern Ghana. In: ICTD'13 Proceedings of the Sixth International Conference on Information and Communications Technologies and Development, Dec 2013

6. Matharu, G.S.. Mishra, A., Chhikara, P.: A framework to leverage cloud for modernization of indian agricultural produce marketing system. In: ICTCS'14 Proceedings of the 2014 International Conference on Information and Communication Technology for Competitive Strategies, Nov 2014
7. Honda, K., Ines, A.V.M., Yui, A., Witayangkurn, A., Chinnachodteeranun, R., Teeravech, K.: Agriculture information service built on geospatial data infrastructure and crop modeling. In: IWWISS'14: Proceedings of the 2014 International Workshop on Web Intelligence and Smart Sensing, Sept 2014, pp. 1–9
8. Zeni, M., Ondula, E., Mbitiru, R., Nyambura, A., Samuel, L., Fleming, K., Weldemariam, K.: Low-power low-cost wireless sensors for real-time plant stress detection. In: *DEV'15 Proceedings of the 2015 Annual Symposium on Computing for Development*, Nov 2015, pp. 51–59
9. Luo, J., Hu, J., Wu, D., Li, R.: Opportunistic routing algorithm for relay node selection in wireless sensor networks. IEEE Trans. Ind. Inf. (2015)
10. Fukatsu, T., Hirafuji, M.: Web-based sensor network system "Field Servers" for practical agricultural applications. In: IWWISS'14: Proceedings of the 2014 International Workshop on Web Intelligence and Smart Sensing, Sept 2014, pp. 1–8
11. Liu, F., Tsui, C.Y., Zhang, Y.: Joint routing and sleep scheduling for lifetime maximization of wireless sensor networks. IEEE Trans. Wirel. Commun. 2258–2267 (2010)
12. eLAB Experience: Field Server. http://www.elab-experience.com/fieldserver. Accessed Dec 2015
13. Corp, P.S.: e-kakashi. https://www.e-kakashi.com/. Accessed Dec 2015

Development of Multi-hop Field Sensor Networks with Arduino Board

Tomoya Ogawa and Katsuhiro Naito

Abstract Wireless sensor networks have been focused to enhance agricultural production because environmental information is useful to estimate growing conditions of plants, diseases of plants etc. A density of sensors is also an important factor to obtain an accurate estimation. Therefore, a lot of sensors are required in a practical field sensing. In conventional systems, the number of sensors is limited due to a cost of an installation and management. Additionally, their devices typically use a single-hop communication. Hence, it is difficult to cover a larger area due to the limit of a communication distance. This paper proposes a wireless sensor network for agricultural usages. The proposed protocols for the wireless sensor network can work with a limited memory resource and CPU performance. Therefore, a developed system can employ Arduino micro-computer boards that have a limited hardware resource. Since a price of Arduino boards is usually inexpensive, it is easy to deploy a lot of sensors in a field to improve an accuracy of estimation. Additionally, our network supports multi-hop communication to extend a sensing area. It also supports a sleep operation at all nodes to mitigate consumed power even if typical multi-hop routing protocols do not support a sleep operation at relay nodes. Experimental results show that the developed sensor network can construct a multi-hop network by the proposed protocols, and work well on a resource constrained hardware.

Keywords Sensor networks · Multi-hop communication · Field sensing · IT agriculture · Arduino · IEEE 802.15.4

T. Ogawa (✉) · K. Naito
Department of Information Science, Aichi Institute of Technology,
1247 Yachigusa, Yakusa, Toyota, Aichi 470-0392, Japan
e-mail: ogawa@pluslab.org

K. Naito
e-mail: naito@pluslab.org

G. De Pietro et al. (eds.), *Intelligent Interactive Multimedia Systems and Services 2016*, Smart Innovation, Systems and Technologies 55,
DOI 10.1007/978-3-319-39345-2_21

1 Introduction

Sensor networks have been used to collect environmental information from a wide area effectively. Typical sensor network systems consist of a lot of sensor devices with a wireless communication module and some sensors, and a sink device that collects the data from the sensor devices. Since replacement cost of a battery is a huge issue in the system, the system requires the sensor devices to reduce required electric power for long-time operation.

Conventional researches have been classified into two research areas: media access control mechanisms and routing protocols. Conventional MAC mechanisms typically control transmission timing by using a special frame structure [1–3]. They also assign an active and a sleep period to realize a sleep operation that is a quite low-power operation mode of micro-computer [4–6]. Since synchronization of frame timing is important, time synchronization mechanisms have been also proposed [7–9]. As routing protocols, various routing protocols have been proposed [10–12]. Since long operation is an important function in sensor networks, a lot of complex algorithms have been proposed to extend the lifetime. The complex algorithms usually require various information to optimize parameters for a routing. Therefore, traffic for control messages also increases according to an increase in complexity. Additionally, they also require a large size of memories to store the collected information and to calculate the optimum parameters. As a result, almost all conventional researches consider performance by simulations because almost all simulators do not consider a limited resource of memory space on hardware.

On the contrary, practical hardware for sensor networks is resource constrained devices that have a small size of memories and a limited computational resource because resource constrained devices are suitable to reduce power consumption of hardware. These sensor devices require a simple multi-hop communication mechanism to execute software on the limited resource hardware. Additionally, a typical quartz on a micro-computer board does not have enough accuracy to maintain synchronization timing even if time synchronization is an important issue to realize a sleep operation. Therefore, multi-hop communication mechanisms also are tolerant of rough time synchronization.

This paper focuses on Arduino boards that are well-known resource constrained devices. Typical Arduino boards implement only 2 KB SRAM and 32 KB flash Memory. Therefore, complex algorithms for multi-hop communication are difficult to execute on the limited hardware resource. This paper studies a simplified algorithm for multi-hop communication to implement on resource constrained devices. The proposed routing protocol can create a tree-based route between a device and a sink device by exchanging routing information between neighbor devices. The benefit of the proposed routing protocol is to reduce the number of control packets and required memory resource for route construction processes. In addition, the proposed media access mechanism provides an original frame structure to realize a sleep operation and to mitigate packet collisions. Since synchronization of a sleep period and an active period is required to reduce power consumption, the proposed protocol also deploys a time sharing mechanism to share the frame timing among neighbor devices. The proposed multi-hop communication mechanism is tolerant of rough time synchronization because it assumes enough long slot period comparing to the time synchronization accuracy. Experimental results show that the authors can implement the proposed mechanisms on Arduino compatible boards, and the proposed mechanisms can construct a sensor network with low-power consumption.

2 Proposed System

The proposed wireless sensor network comprises of a lot of sensor devices and a sink device. As sensor devices work by a battery, they use a sleep operation to mitigate consumed power. A sleep operation should be carefully considered in multi-hop networks because message relaying process is also performed in active periods. Hence, this paper proposes a special frame structure that assigns a unique period for a time synchronization, route construction, data transmission and an acknowledgement transmission. The proposed routing protocol can construct a tree-based route from sink device to sensor devices. Therefore, each sensor device can manage its route by storing an address of its upstream device.

2.1 Frame Structure

Figure 1 shows the proposed frame structure. The period in the frame structure is classified into a time synchronization slot, a routing slot, data transmission slot, and acknowledgement transmission slot. The details are described as follows.

- Time synchronization slot
  The proposed mechanism requires devices to share the frame timing in the whole network. Therefore, the start timing of the frame should be synchronized among devices. Each device adjusts local time according to its upstream device during the time synchronization slot.

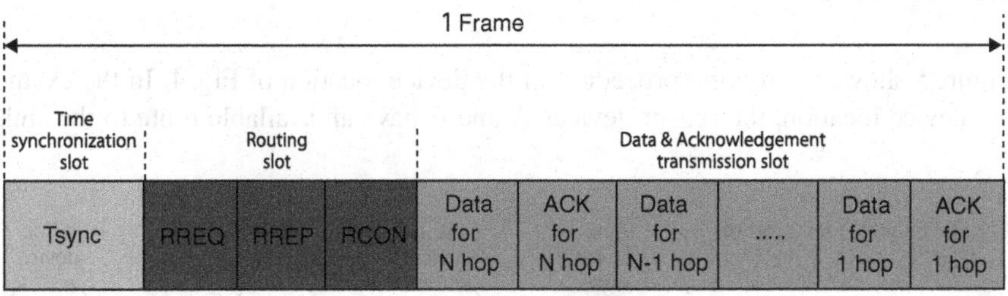

Fig. 1 Frame structure

- Routing slot
  Since a topology of typical sensor networks is almost static, dynamic changes of routes rarely happen. Hence, each device constructs a route during the routing slot. As a result, devices with a route can reduce power consumption during the routing slot by using a sleep mode.
- Data transmission slot
  Signal arrival area of each sensor device is usually overlapped in practical situation because it's difficult to optimize a location of each sensor device. Therefore, mitigating interference among sensor devices is important to improve communication performance. The proposed frame structure assigns a sub-slot classified by a hop-count from a sink device. Therefore, neighbor sensor devices usually use a different sub-slot to transmit a message. Additionally, our structure is easy to maintain the operation of a sleep mode.
- Acknowledgement transmission slot
  The acknowledgement transmission slot is also classified to sub-slots according to the same reason for the data transmission slot. The frame structure can avoid interference among a data message and an acknowledgement message by dissociating the slot.

2.2 *Time Synchronization*

Figure 3 shows the communication sequence for time synchronization in the network shown in Fig. 2. At the start time of the time synchronization slot, the sink device broadcasts a TSYNC message including time information T_1. The one-hop device receiving the message calculates T_2 that is a different time to the sink device by reducing a propagation delay and a processing delay. Then, it adjusts the local time according to the difference between T_2 and T_1. After the adjustment of the local time, the one-hop device also broadcast a TSYNC message including time information T_3. The second-hop device also performs the same processing for the time synchronization.

2.3 *Routing*

Figure 5 shows the routing procedure in the device location of Fig. 4. In the example device location, the sensor devices A and C have an available route to the sink

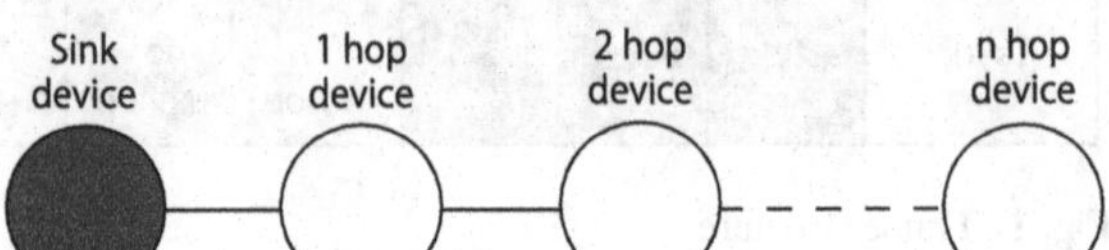

Fig. 2 Example location of devices

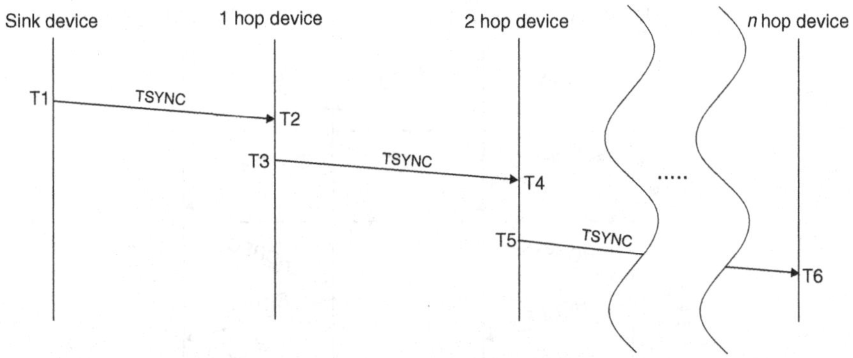

Fig. 3 Communication sequence for time synchronization

Fig. 4 Example device location for routing construction process

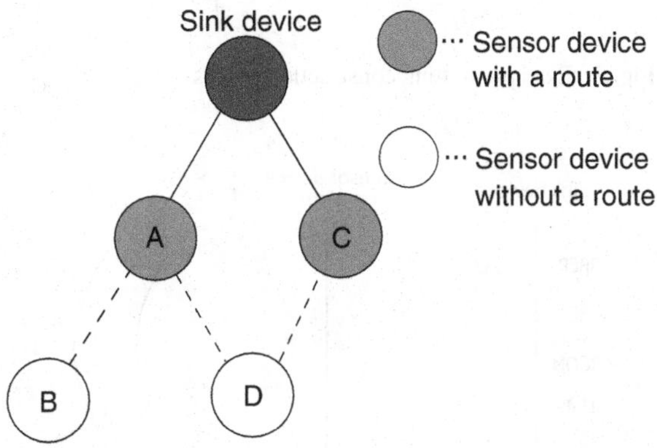

device. The routing slot has some sub-slots: a route request sub-slot, a route reply sub-slot, and a route construction sub-slot. Sensor devices B and D without any available route transmit a RREQ (Route REQuest) message to the neighbor device in the route request sub-slot. As a reply message to the RREQ messages, sensor devices A and C reply the RREP (Route REPly) message in the route reply sub-slot. Then, sensor device B and D start a route construction process by transmitting the RCREQ (Route Construction REQuest) message to sensor device A and C respectively. Sensor device A and C reply the RCREP (Route Construction REPly) messages to sensor device B and D respectively. Finally, sensor device B and D confirm the route construction by transmitting the RCACK (Route Construction ACKnowledgement) message. As a result, our routing protocol can construct a tree-based route from a sink device.

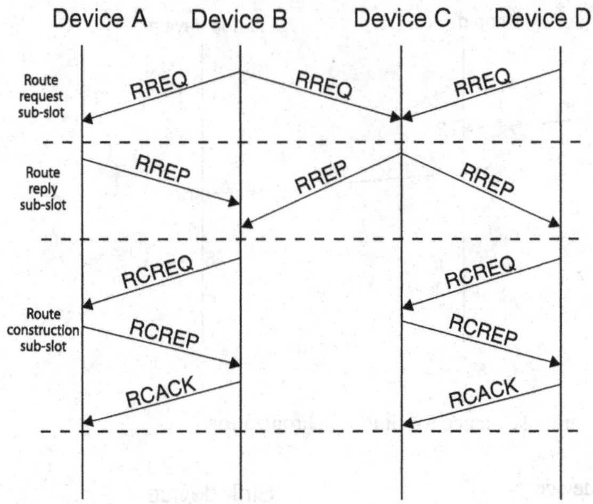

Fig. 5 Example routing construction process

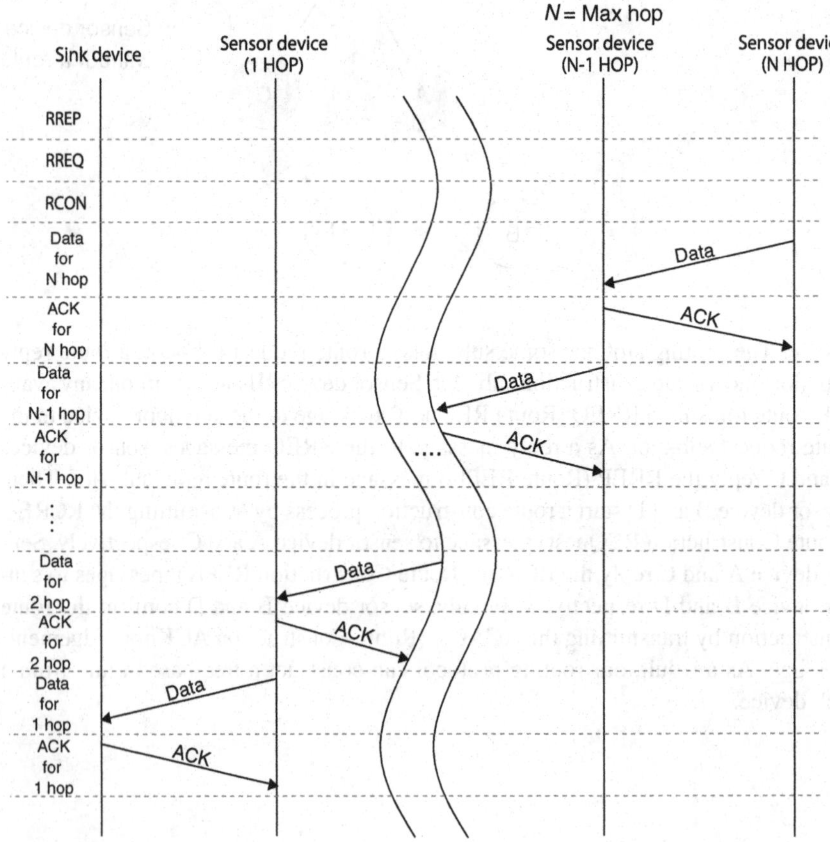

Fig. 6 Data collection sequence

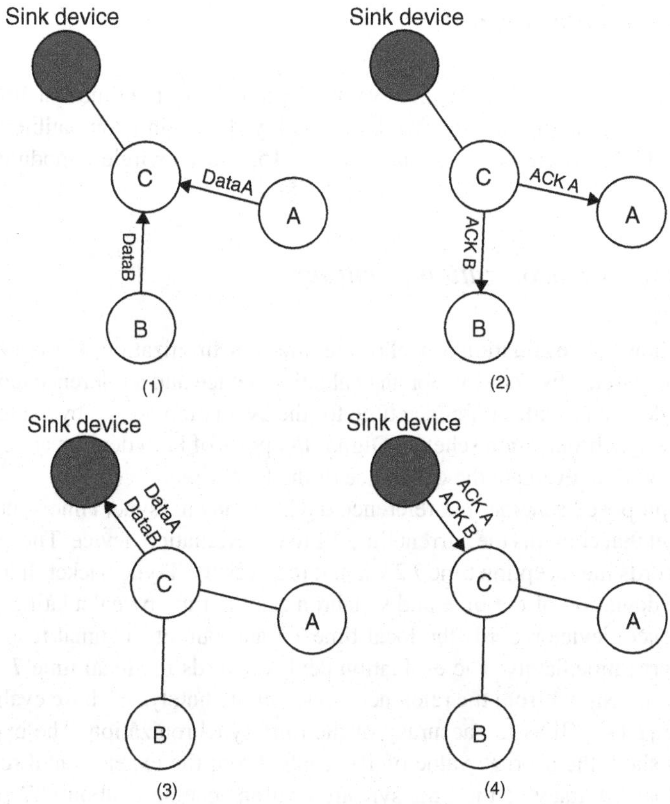

Fig. 7 Operation example of data collection

2.4 *Data Collection*

Figure 6 shows the data collection sequence in the location of Fig. 7. Since the unique sub-slot is assigned according to a hop-count from a sink device, each sensor device transmits a data message and an acknowledgement message during its sub-slot period. In Fig. 7, sensor devices A and B transmit a data message during the data sub-slot for two-hop devices. Then, sensor device C replies the acknowledgement messages to sensor device A and B during the acknowledgement sub-slot for two-hop devices. It also transmits these data messages to the sink device during the data sub-slot for one-hop devices. Finally, the sink device also replies the acknowledgement messages to the sensor device C during the acknowledgement sub-slot for one-hop devices.

3 Experimental Results

In the experimental results, we have evaluated accuracy of the time synchronization scheme and packet arrival ratio. We have employed Arduino compatible board for devices and Xbee Series 1 that is an IEEE 802.15.4 based wireless module.

3.1 Time Synchronization Accuracy

Figure 8 shows the evaluation model of the time synchronization. In the evaluation, we have prepared a two devices for an evaluation device and a reference device. The reference device provides reference time for the evaluation device by using the proposed time synchronization scheme. Digital I/O ports of both devices are connected by jumper wire to evaluate the difference of the local time.

In the proposed scheme, the reference device sends a Tsync(Time synchronization) packet that contains the current time $T1$ to the evaluation device. The evaluation device records the reception time $T2$ when it receives the Tsync packet. It also determines the deviation of the time and synchronizes the time by calculating $T2 - T1$. The reference device records the local time $T3$ and outputs a signal from the digital I/O port immediately. The evaluation device records the local time $T4$ when it recognizes the signal from the reference device immediately. We have evaluated the difference $|T4 - T3|$ as the accuracy of the time synchronization. The experimental results show the average value of 100 trials. From the experimental results, we can find that accuracy of the time synchronization scheme is about 4.7 [ms]. The proposed frame structure requires synchronization among neighbor devices, not in whole network devices. Therefore, neighbor devices can share the frame structure within about 5 [ms] even if the number of multi-hop increases.

3.2 Packet Arrival Ratio

We have evaluated the packet arrival ratio of the proposed system in the two network topologies in Fig. 9. We have calculated the packet arrival ratio by the following equation.

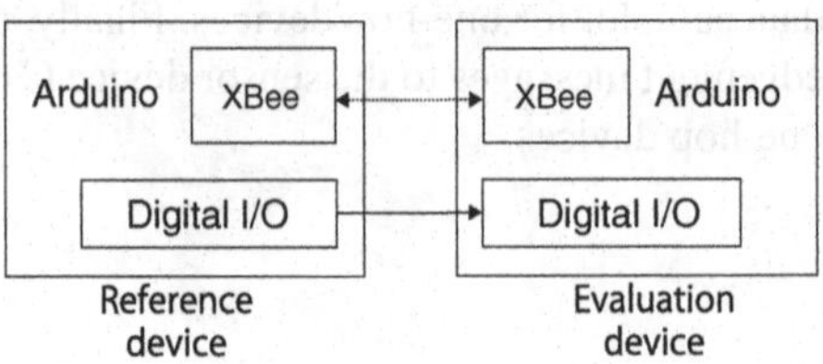

Fig. 8 System model for time synchronization accuracy evaluation

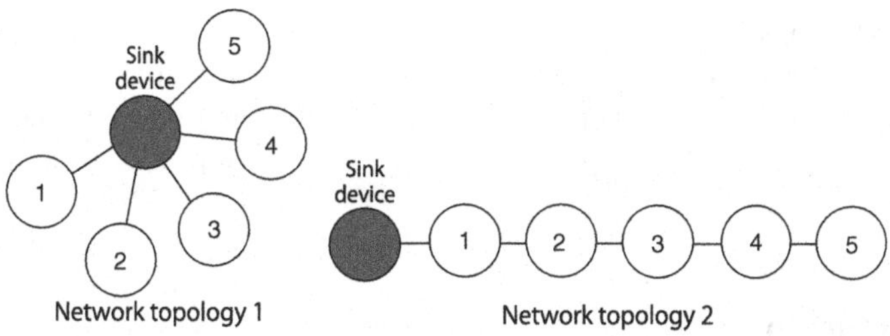

Fig. 9 Topologies for packet arrival ratio evaluation

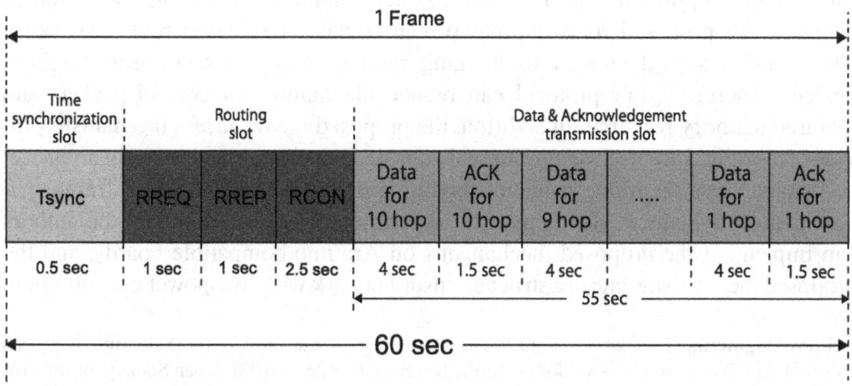

Fig. 10 Parameters of frame structure

$$Packet\,arrival\,rate = \frac{The\,number\,of\,packets\,that\,the\,sink\,node\,receives}{The\,number\,of\,packets\,that\,each\,node\,has\,generated} \quad (1)$$

Figure 9 shows the network topologies in the experiments. In the network topology 1, the five sensor devices transmit data to the sink device in one-hop communication. In the network topology 2, each sensor device transmits data to the sink device by multi-hop communication. In the experiment, each device is placed at 2 m intervals, and we have set the parameters for frame structure in Fig. 10. The maximum number of hops is set to 10 hops, and the transmission power of XBee is 10 mW. In order to evaluate the static condition of the route, we have built the route before the experiment.

Table 1 shows the packet arrival ratio. In the network 1, each device had a high packet arrival ratio because the performance of the one-hop communication is enough reliable for a packet delivery. In the network 2, the packet arrival ratio decreases according to the increase of the hop counts because continuous successful deliveries are required to deliver a packet in multi-hop communication. Since wireless communication usually suffers from random errors due to noise, more reliable communication can be performed by implementing retransmission mechanisms for packets.

Table 1 Packet arrival ratio

• •	Node1 (%)	Node2 (%)	Node3 (%)	Node4 (%)	Node5 (%)
Network 1	100	99.1	100	98.2	100
Network 2	100	100	100	99.1	95.0

4 Conclusion

This paper proposes a sensor network system suitable for resource constrained devices. The proposed routing protocol can create a tree-based route between a device and a sink device by exchanging routing information between neighbor devices. Therefore, our protocol can reduce the number of control packets and required memory resource. In addition, the proposed media access mechanism provides an original frame structure to realize a sleep operation and to mitigate packet collisions. Since the frame structure should be shared among devices, a frame synchronization scheme is also deployed. Experimental results show that the authors can implement the proposed mechanisms on Arduino compatible boards, and the proposed mechanisms can construct a sensor network with low-power consumption.

Acknowledgments This work is supported in part by Grant-in-Aid for Scientific Research (B)(15H02697) and Grant-in-Aid for Scientific Research (C)(26330103), Japan Society for the Promotion of Science (JSPS) and the Integration research for agriculture and interdisciplinary fields, Ministry of Agriculture, Forestry and Fisheries, Japan.

References

1. Yang, C., Chin, K.-W.: Novel algorithms for complete targets coverage in energy harvesting wireless sensor networks. IEEE Commun. Lett. **18**(1), 118–121 (2014)
2. Hu, J., Ma, Z., Sun, C.: Energy-efficient MAC protocol designed for wireless sensor network for IoT. In: 2011 Seventh International Conference on Computational Intelligence and Security (CIS), pp. 721–725, Dec 2011
3. Zhao, Z., Zhang, X., Liu, P.S.P.: A traffic queue-aware MAC protocol for wireless sensor networks. In: 2007. CHINACOM'07. Second International Conference on Communications and Networking in China, pp. 841–844, Aug 2007
4. Jeon, J.-H., Byun, H.-J., Lim, J.-T.: Joint contention and sleep control for lifetime maximization in wireless sensor networks. IEEE Commun. Lett. **17**(2), 269–272 (2013)
5. Fuemmeler, J., Veeravalli, V.: Smart sleeping policies for energy efficient tracking in sensor networks. IEEE Trans. Signal Process. **56**(5), 2091–2101 (2008)
6. Wu, F.-J., Tseng, Y.-C.: Distributed wake-up scheduling for data collection in tree-based wireless sensor networks. IEEE Commun. Lett. **13**(11), 850–852 (2009)
7. Du, X., Guizani, M., Xiao, Y., Chen, H.-H.: Secure and efficient time synchronization in heterogeneous sensor networks. IEEE Trans. Veh. Technol. **57**(4), 2387–2394 (2008)
8. He, J., Cheng, P., Chen, J., Shi, L., Lu, R.: Time synchronization for random mobile sensor networks. IEEE Trans. Veh. Technol. **63**(8), 3935–3946 (2014)

9. Cheng, K.-Y., Lui, K.-S., Wu, Y.-C., Tam, V.: A distributed multihop time synchronization protocol for wireless sensor networks using Pairwise Broadcast Synchronization. IEEE Trans. Wirel. Commun. **8**(4), 1764–1772 (2009)
10. Liu, F., Ying Tsui, C., Zhang, Y.: Joint routing and sleep scheduling for lifetime maximization of wireless sensor networks. IEEE Trans. Wirel. Commun. **9**(7), 2258–2267 (2010)
11. Du, X., Guizani, M., Xiao, Y., Chen, H.-H.: Two tier secure routing protocol for heterogeneous sensor networks. IEEE Trans. Wirel. Commun. **6**(9), 3395–3401 (2007)
12. Lou, W., Kwon, Y.: H-SPREAD: a hybrid multipath scheme for secure and reliable data collection in wireless sensor networks. IEEE Trans. Veh. Technol. **55**(4), 1320–1330 (2006)

Communication Simulator with Network Behavior Logging Function for Supporting Network Construction Exercise for Beginners

Yuichiro Tateiwa and Naohisa Takahashi

Abstract Interconnecting virtual machines realizes computer networks on ordinary personal computers. Such a technique enables each student instead of a group to construct networks in network exercises for beginners. In the exercises, students may ask teachers to judge the correctness/incorrectness of their networks and to support the debugging for their networks. The waiting time of students can be long because the number of teachers is less than the number of students. An effective solution to this problem is to develop a system that can judge whether students' networks are correct and visualize the behavior of students' networks as hints. Detail logs of network behavior are necessary for realizing such a system. Here, we propose a communication simulator to record network behavior in detail during request/response communications, which are the transmissions of request data (e.g., icmp echo request) and the corresponding response data (e.g., icmp echo reply).

Keywords Communication simulator · Network construction exercise · Network behavior log

1 Introduction

It is important to increase the number of network engineers who administer computer networks as an infrastructure to a ubiquitous society and provide new services to the society. The experience of basic network construction is useful for not only network administrators but also network application programmers and network system designers.

Y. Tateiwa (✉) · N. Takahashi
Nagoya Institute of Technology, Gokiso-cho, Showa-ku, Nagoya, Aichi, Japan
e-mail: tateiwa@nitech.ac.jp
URL: http://tk-www.elcom.nitech.ac.jp/

N. Takahashi
e-mail: naohisa@nitech.ac.jp

G. De Pietro et al. (eds.), *Intelligent Interactive Multimedia Systems and Services 2016*, Smart Innovation, Systems and Technologies 55,
DOI 10.1007/978-3-319-39345-2_22

Network engineers define network specifications based on the network usage specified by their clients, and then, these engineers construct networks that satisfy these specifications. Therefore, there are network exercises where problems are based on frequent usage examples and students must construct networks for the available examples. We call these examples **communication examples**; they include "Client A can browse web sites on server B" and "Communications from client C to server A are intercepted by firewall D."

In traditional exercises for beginners, students are first divided into several groups; then, they construct networks with physical network devices. In recent years, however, it has become possible to realize networks by an interconnection of virtual machines running on an ordinary personal computer because of the performance improvement of personal computers and the evolution of virtual machine technology. We call such networks of virtual machines **VMN**. VMNs realize an e-learning environment where each student can construct his/her own network without considering time and placement (e.g., [1–5]).

In traditional exercises, each group attempts to construct networks based on exercise problems. When they want to confirm the correctness of their networks or cannot solve certain problems related to these networks, they approach their teachers for help. In e-learning, however, their waiting time is increased because students construct networks individually and the number of teachers is less than the number of students. An effective solution to this problem is to develop a system that can judge whether students' networks (we call them **answers**) are correct and visualize the behavior of these networks as hints.

Detail logs of network behavior are necessary for realizing such a system. For example, the logs needed for the judgment system are passing points of communication data (e.g., network interface Y on node X) and a terminated operation (e.g., discarding communication data because of a mismatch of the destination MAC address in the communication data to the MAC address assigned to the network interface that received the communication data). By comparing such data between the answers and the correct answers, the system judges whether the answers are correct. The logs needed for the hint generation system are operations for communication data in each node, one of which is packet forwarding caused by a mismatch of the destination IP address in the communication data to the IP addresses in its received node. It is not possible to gather such data by using packet capturing of tcpdump [6] or by using the trace data of network simulator ns-3 [7].

Therefore, here, we propose a communication simulator to record network behavior in detail during **request/response communications**, which are transmissions of request data (e.g., icmp echo request) and the corresponding response data (e.g., icmp echo reply). Fourteen proposed operations realize request/response communication in the simulator. The simulator receives network configurations and communication attributes (e.g., source node and destination IP address) as its input and then, computes the transmission of request data and response data. During such computation, the simulator writes the executed operations into a file (called **log file**) one by one. The simulator has the following features:

- The target of simulation is limited to the transmissions of requests and responses because it is used for judging whether the communication examples are successful in the networks.
- Each operation uses a few configurations as its input in order to obtain detailed relations between configurations and the success/failure of the communication examples.

2 Preparation

This section describes the operators to operate data structures.

- operator []: It returns a sequence whose elements are its arguments. The arguments are of variable length and are separated by commas. For example, "[1, 2, 3]" returns a sequence whose first element is "1," whose second element is "2," and whose last element is "3."
- operator <>: It returns a structure whose elements are its arguments. The arguments are of variable length and are separated by commas.
- operator { }: It returns a set whose elements are its arguments. The arguments are of variable length and are separated by commas.
- operator +: It returns a merged sequence between its arguments. For example, "[1] + [2] + [3]" returns a sequence whose first element is 1, whose second element is 2, and whose last element is 3.
- operator .: It accesses the value of a member valuable in a structure. For example, when a structure s has a member variable e and there is a variable y and whose data type is s, "y.e" accesses the value of valuable e.

3 Network Construction Exercise

3.1 Exercise Purpose

The target students have studied TCP/IP but have never constructed networks. The exercise target involves the students getting accustomed to designing networks that satisfy the following requirements and construct networks based on these designs.

1. IP address and subnet mask
2. IP network in a segment
3. Default route
4. Static routing
5. Server services (web page publishing on WWW)

Table 1 Data structure used by network devices

Name	Meaning	Data structure
nd	Node identifier	String
work	State of working	Boolean (if it is working, the value is true; otherwise, the value is false)
proc	A set of running processes	A set consisting of structures (process name **name**, listening protocol name **prot**, listening port number **port**, listening IP address **ip**)
ni	A set of network interface configurations	A set consisting of structures (network interface name **name**, assigned IP address **ip**, assigned subnet mask **mask**, assigned MAC address **mac**)
rtable	Routing table	A set consisting of structures (destination IP address *ip*, destination network address *nwaddr*, next hop IP address *nh*, subnet mask of *nwaddr* **mask**, sender network interface name *ni_name*)
atable	ARP table	A set consisting of structure (IP address **ip**, MAC address **mac**)
ni_num	Number of network interfaces	Integer

Table 2 Data structure of network

Name	Data structure
Network	A structure (a host set **hs**, a router set **rt**, a switching hub set **sw**, a repeater hub set **rp**, a cable set **cb**)
Host	A structure (nd, work, proc, ni, rtable, atable)
Router	A structure (nd, work, ni, rtable, atable)
Switching hub	A structure (nd, work, ni_num)
Repeater hub	A structure (nd, work, ni_num)
Cable	A structure (node identifier **nd1**, network interface name of nd1 **ni1_name**, node identifier **nd2**, network interface name of nd2 **ni2_name**)

Table 3 Actions of network devices

Name	Actions
Host	It receives a destination IP address and a destination port number as its input, and gets web pages with HTTP. It receives a destination IP address as its input, and checks a network continuity by using icmp echo messages
Router	It relays IP packets. It replies to ICMP echo requests
Switching hub	It relays Ethernet frames
Repeater hub	It relays Ethernet frames
Cable	It transmits data only between a network interface (host or router) and another network interface (switching hub or repeater hub)

3.2 Network

Tables 1, 2, and 3 list the specifications of networks in the exercise.

3.3 Structure of Exercise Problem

Exercise problems consist of **communication examples** and **configuration requirements**. The former is an example of communication available in students' networks. The latter consists of nodes that must be installed into networks, and setting values that must be set in nodes, and cables that must connect the assigned nodes.

Communication examples consist of items in Table 4. Configuration requirements have the same data structure as the network, as given in Table 2. Node identifiers (e.g., the node identifier of host hs_0 is $hs_0.nd$) and network interface names (e.g., the network interface name of router rt_0 is $ni_0.name(ni_0 \in rt_0.ni)$) must be concrete values, and the other items can be either '$*$' (denoting arbitrary values) or concrete values.

3.4 Example of Exercise Problem

Communication examples are expressed using natural language in actual exercise problems. For example, when a communication example consists of the following values, it is expressed by sentences given at the top of Fig. 1.

Table 4 Data of communication example

Name	Data structure
kind	A string where "web-get" represents a communication getting web pages from WWW servers and "icmp-echo" denotes the communication checking network continuity
source	A string that denotes a node identifier
destination	A structure (destination node identifier **nd**, process name running on node nd **proc**, listening IP address of proc **ip**, listening port number of proc **port**, listening protocol name of process proc **prot**)
communication route	A labeled rooted tree whose vertices are nodes that send or receive data, labels are identifiers of the nodes, edges are data transmissions between nodes, and root is the source node. Note that multiple children exist when the data are broadcast by repeater hubs and switching hubs
process reception information	A set consisting of structures (**proc**, **t**), where proc denotes a process reading requests or responses, t represents a communication timepoint, i.e., a time point when the communication data remain at a node of communication). Values of communication timepoints are vertices in communication routes

```
Construct the following network: After you
send icmp-echo requests with their
destination IP addresses 192.168.0.2 at
svr1, svr2 receives them, svr3 and svr4
drop them. And then svr2 sends icmp-echo
replies, and svr1 receives them, svr3 and
svr4 drop them.
```

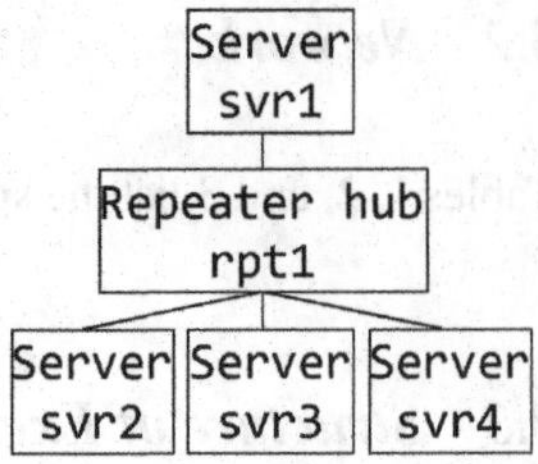

Fig. 1 Exercise problem

- kind $k =$"*icmp − echo*"
- source $s =$"*svr1*"
- destination $d = <$"", "", "192.168.0.2", "*icmp*", "">
- communication route $R =< v_0, V, E, L, F >, V = [v_0, v_1, v_2, v_3, v_4, v_5, v_6, v_7, v_8, v_9, v_{10}], E = \{\{v_0, v_5\}, \{v_5, v_2\}, \{v_5, v_3\}, \{v_5, v_4\}, \{v_6, v_{10}\}, \{v_{10}, v_7\}, \{v_{10}, v_8\}, \{v_{10}, v_9\}, \{v_1, v_7\}\}, L =$ ["*svr1*", "*svr2*", "*svr3*", "*svr4*", "*rpt1*", "*svr1*", "*svr2*", "*svr3*", "*svr4*", "*rpt1*"], $F(V_i) returns L_i$
- process reception information $P = \{<v_1,$"*kernel*">, $<v_6,$ "*client*">}

Configuration requirements are expressed using natural language and figures in actual exercise problems. For example, when configuration requirements $cr =< hs, rt, sw, rp, cb >$ consist of the following values, they are expressed by a figure given at the bottom of Fig. 1, where letters at the top of the squares denote node type and letters at the bottom represent the node identifier.

- $hs = \{n1, n2, n3, n4\}$, $rt = \{\}$, $sw = \{\}$, $rp = \{n5\}$, $cb = \{l1, l2, l3, l4\}$
- $n1 =$ <"*svr1*", ∗, ∗, ∗, ∗, ∗>, $n2 =$ <"*svr2*", ∗, ∗, ∗, ∗, ∗>, $n3 =$ <"*svr3*", ∗, ∗, ∗, ∗, ∗>, $n4 =$ <"*svr4*", ∗, ∗, ∗, ∗, ∗>, $n5 =$ <"*rpt1*", ∗, 5>
- $l1 =$ <"*svr1*", ∗,"*rpt1*", ∗ >, $l2 =$ <"*rpt1*", ∗,"*svr2*", ∗ >, $l3 =$ <"*rpt1*", ∗,"*svr3*", ∗ >, $l4 =$ <"*rpt1*", ∗,"*svr4*", ∗ >

3.5 Exercise Process

Each student constructs networks with physical devices and virtual devices on the basis of exercise problems. Then, they write reports on the executed steps and configurations in the constructions. After solving all problems, they submit their reports to their teachers. In the case they cannot comprehend certain sections of the exercises, they approach their teachers or teaching assistants for help.

4 Communication Simulator

4.1 Function Definition

The deta structure of communication data **pkt** used in the following functions is a structure (protocol name **prot**, destination IP address **dip**, source IP address **sip**, destination port number **dp**, source port number **sp**, payload **pl**, destination MAC address **dmac**, source MAC address **smac**).

4.1.1 Functions to Get Data from Configurations

- RT(nd): It returns a routing table from the node whose identifier is nd.
- AT(nd): It returns an ARP table from the node whose identifier is nd.
- MAC(nd): It returns a set consisting of structures (**ni**, **mac**), where ni is a network interface name in the node whose identifier is nd, and mac is its MAC address.
- IP(nd): It returns a set consisting of structures (**ni**, **ip**), where ni is a network interface name in the node whose identifier is nd, and ip is its IP address.
- NI(nd): It returns a set consisting of all network interfaces in the node whose identifier is nd.
- PROC(nd): It returns a set consisting of listening processes in the node whose identifier is nd.
- WORK(nd): It returns the working state of the node whose identifier is nd.
- ARP(nd, ni, ip): If the node whose identifier is nd gets a MAC address corresponding to an IP address ip by an ARP communication at the network interface whose name is ni, it returns the MAC address; otherwise, it returns "".
- KIND(nd): If the node of nd is a host, it returns "hs"; else, if the node of nd is a router, it returns "rt". Further, if the node of nd is a switching hub, it returns "sw"; else, if the node of nd is a repeater hub, it returns "rp".
- PairNI(nd, ni_name): Here, nd denotes a node identifier, and ni_name representes a network interface name. It returns a tupple $(nd_{pair}, ni\_name_{pair})$ whose ni_name_{pair} is the name of the network interface connected to a network interface whose name is ni_name in the node whose identifier is nd, and whose nd_{pair} is an identifier of the node that has the network interface ni_name_{pair}.

4.1.2 Functions Consisting Operations in Node

Every function writes its name, its inputs, and its outputs at the end of a **log file** during its execution.

- ReqPkt(prot, pl, dip, dp): Here, prot denotes the protocol name; pl, the payload; dip, the destination IP address; and dp, the destination port number. If $prot =$"*tcp*" is true, it inputs at random an ephemeral port number into source port *sp*; otherwise, it sets $sp =$"". Then, it returns <*prot*, *dip*, "", *dp*, *sp*, *pl*, "", ""> as the communication data.

- RepPkt(proc, pkt): Here, proc denotes a process name and pkt represents the communication data. If *proc* =“*apache*” $\wedge$*pkt.pl* = “*HTTP GET REQUEST*” is true, it sets *pl* = “*HTTP GET RESPONSE*”; else, if *proc* =“*kernel*”$\wedge$*pkt.pl* = “ICMP *ECHO REQUEST*” is true, it sets *pl* = “*ICMP ECHO REPLAY*”. Then it returns <*pkt.prot, pkt.sip, pkt.dip, pkt.sp, pkt.dp, pl*,“”, “”> as the communication data.
- ChkDIP(dip, nd): Here, dip denotes the destination IP address and nd represents the node identifier. If there is i whose $i.ip = dip(i \in IP(nd))$ is true, it returns true; otherwise, it returns false.
- ChkDMac(dmac, nd, ni): Here, dmac denotes the destination MAC address; nd, the node identifier; and ni, the network interface name. If there is m whose $m.ni = ni \wedge m.mac = dmac(m \in MAC(nd))$ is true, it returns true; otherwise, it returns false.
- Proc(prot, dp, dip, nd): Here, proc denotes the process name; dp, the destination port number; dip, the destination IP address; and nd, the node identifier. If there is proc whose $proc.prot = prot \wedge proc.port = dp \wedge proc.ip = dip$ in $proc \in PROC(nd)$ is true, it returns *proc.name*; otherwise, it returns “”.
- L1Rtng(nd, ni): Here, nd denotes the node identifier and ni represents the network interface name. It returns a queue whose elements are *ni* and whose $ni \neq ni_s \wedge WORK(nd_{adj}) = true \wedge nd_{adj} \neq$ “” is true in $(nd_{adj}, ni_{adj}) = PairNI(nd, ni_s)(ni_s \in NI(nd))$. The elements are sorted in an ascending order by nd_{adj} as the first key and ni_{adj} as the second key.
- L2Rtng(dmac, ni, nd): Here, dmac denotes the destination MAC address; ni, the network interface name; and nd, the node identifier. If there is ni_s whose $m.ni = ni_{adj} \wedge m.mac = dmac \wedge ni \neq ni_s \wedge WORK(nd_{adj}) = true \wedge nd_{adj} \neq$ “” in $m = MAC(nd_{adj}), (nd_{adj}, ni_{adj}) = PairNI(nd, ni_s), ni_s \in NI(nd)$ is true, it returns a queue whose elements are ni_s sorted in the ascending order by nd_{adj} as the first key and ni_{adj} as the second key; otherwise, it returns $L1Rtng(nd, ni)$.
- L3Rtng(dip, nd): Here, dip denotes the destination IP address, and nd represents the node identifier. If it finds $rtable \in RT(nd)$ whose *rtable.ip* matches *dip* by the longest prefix match or whose *rtable.nw* and *rtable.mask* match *dip* by the longest prefix match, it returns a tupple (*rtable.nh, rtable.ni*); otherwise, it returns a tupple (“”, “”).
- Rcv(proc, pkt): Here, proc denotes a process name and pkt represents communication data. If (*proc* =“*apache*” $\wedge$*pkt.pl* =“*HTTP GET REQUEST*”) $\vee$ (*proc* = “*kernel*”$\wedge$*pkt.pl* =“*ICMP ECHO REQUEST*”) is true, it returns *false*; otherwise, it returns *true*.
- SIP(nd, ni): Here, nd denotes the node identifier and ni represents the network interface name. If there is i whose $i.ni = ni(i \in IP(nd))$ is true, it returns i.ip; otherwise, it returns “”.
- DMac(ni, dip, nd): Here, ni denotes the network interface name; dip, the destination IP address, and nd, the node identifier. If *record.mac* $\neq$ “” ($record \in AT(nd)$) is true, it returns *record.mac*; otherwise, it returns $ARP(nd, ni, dip)$.
- SMac(nd, ni): Here, nd denotes the node identifier and ni represents the network interface name. If there is m whose $m.ni = ni(m \in MAC(nd))$ is true, it returns m.mac; otherwise, it returns “”.

- Transmit(nd, ni): Here, nd denotes the node identifier and ni represents the network interface name. It returns *PairNI(nd, ni)*.
- Dsc(pkt): Here, pkt denotes the communication data. It discards pkt. It returns nothing.

4.2 Pesudo Code

Procedures 1–4 show the pesudo code of the simulator. When you execute simulation by this simulator on the basis of communication example *E*, you execute function Communicate(d.prot, pl, dip, d.port, s), where d denotes the destination of *E*; pl is set as "HTTP GET REQUEST" (type of *E* is "web-get"); pl is set as "ICMP ECHO REQUEST" (type of *E* is "icmp-echo"); dip is set as "d.ip" (d.nd=""); dip is set as $ni.ip(ni \in IP(d.nd))$ ($d.nd \neq$ ""); and s is source of *E*.

Procedure 1 Communicate

Input: protocol name **prot**, payload **pl**, destination IP address **dip**, destination port number **dp**, source node identifier **nd**

```
if WORK(nd) then
    pkt = ReqPkt(prot, pl, dip, dp)
    HsRtOut(pkt, nd)
else
    Dsc(pkt)
end if
```

5 Prototype System

5.1 Communication Simulator

Let us consider an incorrect network for the exercise problem described in Sect. 3.4. Its incorrect configurations are the IP addresses of svr4 (192.168.0.2) and svr2 (192.168.0.254). Figure 2 shows a part of the log file generated by our simulator by using the network. Each line consists of a function name, input values, and output values. The symbol "\\" on the right side denotes hyphenation.

```
ReqPkt,icmp,icmp-echo,192.168.0.2,,<icmp,192.168.0.2,,,,icmp-echo,,>
L3Rtng,192.168.0.2,svr1,true,{<kernel,icmp,,>},{<dummy0,,,AE:10:94:\\
E3:51:E7>,<eth0,192.168.0.1,255.255.255.0,FE:FD:98:82:21:3A>,<lo,12\\
7.0.0.1,255.0.0.0,>,<tunl0,,,>},{<192.168.0.0,0.0.0.0,255.255.255.0\\
,U,0,0,eth0>},{},<0.0.0.0,eth0>
```

Fig. 2 Log file

Procedure 2 HsRtOut

Input: sent communication data **pkt**, identifier of sender node *nd*

```
(ip, ni) = L3Rtng(pkt.dip, nd)
if ni ≠ "" then
    if pkt.sip = "" then
        pkt.sip = SIP(nd, ni)
    end if
    pkt.smac = SMac(nd, ni)
    pkt.dmac = DMac(ni, ip, nd)
    if pkt.dmac ≠ "" then
        (nd_r, ni_r) = Transmit(nd, ni)
        if KIND(nd) = "rp" ∨ KIND(nd) = "sw" then
            SwRp(pkt, nd_r, ni_r)
        else
            Dsc(pkt)
        end if
    else
        Dsc(pkt)
    end if
else
    Dsc(pkt)
end if
```

Procedure 3 SwRp

Input: received data **pkt**, identifier of receiver node *nd*, name of receiver network interface *ni*

```
if WORK(nd) then
    if KIND(nd) = "rp" then
        NI_s = L1Rtng(nd, ni)
    else if KIND(nd) = "sw" then
        NI_s = L2Rtng(pkt.dmac, ni, nd)
    end if
    if NI_s ≠ {} then
        repeat
            pickupanelementfromtheheadofNI_sandsetittoni_s
            (nd_r, ni_r) = Transmit(nd, ni_s)
            if KIND(nd_r) = "hs" ∨ KIND(nd_r) = "rt" then
                HsRtIn(pkt, nd_r, ni_r)
            else
                Dsc(pkt)
            end if
        until NI_s ≠ {}
    else
        Dsc(pkt)
    end if
else
    Dsc(pkt)
end if
```

Procedure 4 HsRtIn

Input: received data *pkt*, identifier of receiver node *nd*, name of receiver network interface *ni*

```
if WORK(nd) then
    if ChkDMac(pkt.dmac, nd, ni) = true then
        if ChkDIP(pkt.dip, nd) = true then
            proc = Proc(pkt.prot, pkt.dp, pkt.dip, nd)
            if proc ≠ "" then
                if Rcv(proc, pkt) = false then
                    pkt_s = RepPkt(proc, pkt)
                    HsRtOut(pkt_s, nd)
                end if
            else
                Dsc(pkt)
            end if
        else
            HsRtOut(pkt, nd)
        end if
    else
        Dsc(pkt)
    end if
else
    Dsc(pkt)
end if
```

The first log implies that the execution of function ReqPkt with prot = “icmp,” pl = “icmp-echo,” dip = “192.168.0.2,” and dp = “” as its input arguments returned communication data (prot = “icmp,” dip = “192.168.0.2,” sip = “”, dp = “”, sp = “”, pl = “icmp-echo,” smac = “”, and dmac = “”). The second log implies the execution of function L3Rtng, and in particular, the log stores the node information of its argument nd.

5.2 Application Example of Log

The visualization of communication is helpful for students to debug their networks. We visualized the executed operations during communication by getting several data from the log file. Figure 3 expresses the executed operations, their orders, and their nodes by a directed graph. The vertices consist of “an operation name,” “@,” and “identifier of its execution node.” The direction of the edges denotes the execution order of the operations on the vertices. This network is incorrect because svr4 receives communication data while the exercise problem requires that svr2 receives the data according to the graph.

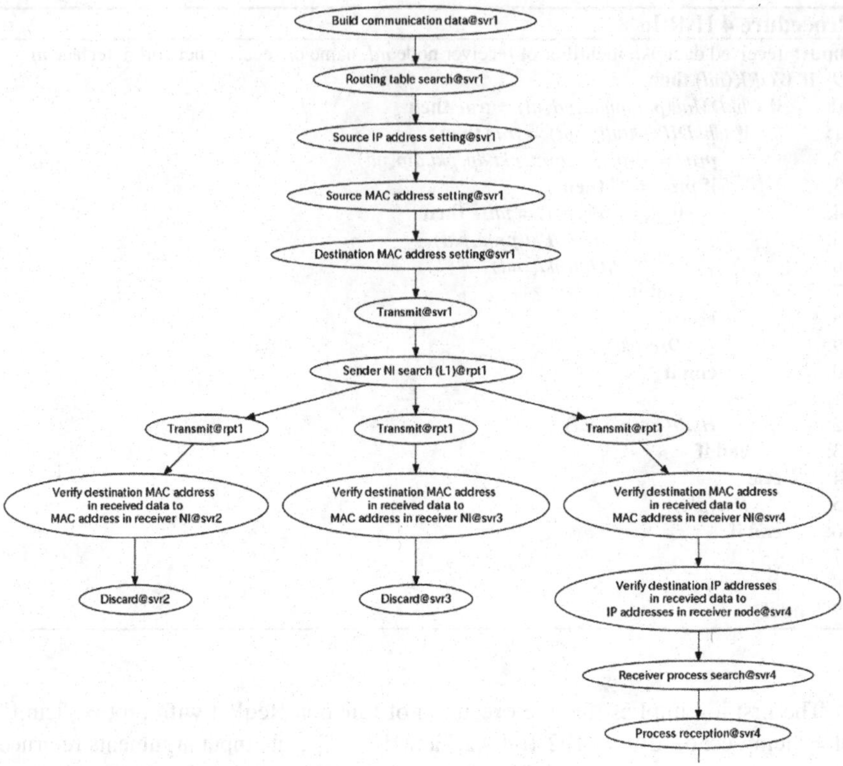

Fig. 3 Graph of execution operations

6 Evaluation Experience

This experience aims at clarifying the basic performance of our simulator. We implemented our simulator in C++, and executed it with ping executions of several networks on a computer with an Intel core i7-3770K 3.50-GHz CPU and 4-GB main memory. Table 5 shows the result. "Transmissions" denotes all counts at which the nodes transmitted communication data. "Routing entries" representes the counts of the routing table entries in all nodes. According to the result, we conclude that our simulator is available for beginners' exercises because the number of routers in the exercises is less than 10, and the consumption of the computer resources in the case is enough small to current computers.

Table 5 Measurement result

Network	Communication	Average CPU usage (%)	Exec. time (s)	Max. RSS (KB)	Transmissions	Routing entries	Log file size (Bytes)	Log counts
A correct network (Fig. 4)	Execute ping to svrR at svrL	64	0.006	2,980	8	4	5,594	36
Fig. 4 (N = 10)	Execute ping to svrR at svrL	81	0.02	3,488	44	112	91,061	184
Fig. 4 (N = 100)	Execute ping to svrR	96	1.228	37,076	404	10,102	6,718,846	1,624
Fig. 4 (N = 254)	Execute ping to svrR	96	7.22	213,204	1,020	64,772	43,354,830	4.088

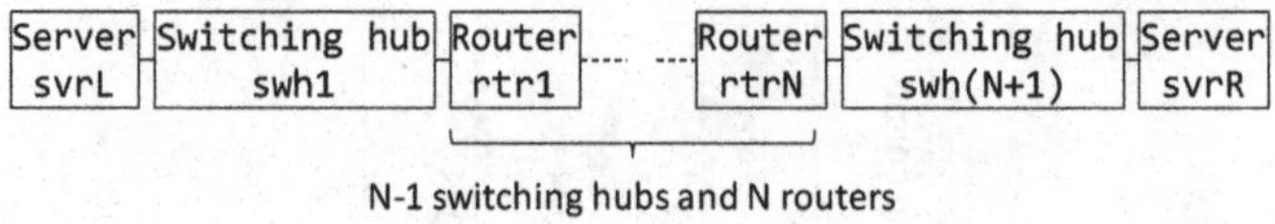

Fig. 4 Network for the experiment. It is possible to transmit data between svrL and svrR

7 Conclusion

In this paper, we proposed a simulator that stores network behavior for request/response communications. We consider that the stored data are useful for the correct/incorrect judgment and hint generation. We showed a graph to express network behavior as an example of hints. We carried out an evaluation experiment where we measured the basic features of the simulator. According to its result, we concluded that the simulator can sufficiently work on computers with ordinary performance in computing small networks that beginners construct in the exercises. Our future works include an expansion of the simulation target, e.g., firewalls and NAT.

Acknowledgments This study was partially funded by the Grants-in-Aid for Scientific Research Foundation (25750082) and public interest TATEMATSU foundation.

References

1. Tateiwa, Y. et al.: LiNeS: virtual network environment for network administrator education. In: Proceedings of Information and Control (ICICIC-2008) (2008)
2. Iguchi, N. et al.: Development of hands-on IP network practice system with automatic scoring function. In: Proceedings of 2013 Seventh International Conference on Complex, Intelligent, and Software Intensive Systems, pp.704–709 (2013)
3. Le, Xu, Huang, Dijiang, Tsai, Wei-Tek: Cloud-based virtual laboratory for network security education. IEEE Trans. Educ. **57**(3), 145–150 (2014)
4. Wannous, Muhammad, Nakano, Hiroshi: NVLab, a networking virtual web-based laboratory that implements virtualization and virtual network computing technologies. IEEE Trans. Learn. Technol. **3**(2), 129–138 (2010)
5. Ruiz-Martínez, Antonio, Pereñíguez-García, Fernando, Marín-López, Rafael, Ruiz-Martínez, Pedro M., Skarmeta-Gómez, Antonio F.: Teaching advanced concepts in computer networks: VNUML-UM virtualization tool. IEEE Trans. Learn. Technol. **6**(1), 85–96 (2013)
6. TCPDUMP_LIBPCAP public repository. http://www.tcpdump.org/ (2016). Accessed 28 Jan 2016
7. ns-3. https://www.nsnam.org/ (2016). Accessed 28 Jan 2016

A New Method to Apply BRAKE to Sensor Networks Aiming at Power Saving

Minori Kinose and Yukihiro Kamiya

Abstract Wireless sensor network is a powerful solution for improving the efficiency of agricultural industries. A main concern is the battery life of sensors. To reduce the power consumption of the sensors, we want to shorten the length of the signal as much as possible. BRAKE [7], a blind receiver for spread spectrum systems, is interesting to cope with this matter since it does not need preambles which are usually put at the head of packets for enabling the receiver to detect the packets and its timing. However, BRAKE is not suitable for sensor networks as it is, because BRAKE cannot work with short signals. In this paper, we propose a new simple method to apply BRAKE to the sensor networks. Computer simulations verify the drastic effect of the proposed method.

Keywords Blind signal processing · BRAKE · Sensor networks

1 Introduction

In agricultural industries, it is expected that the sensor networks must be a powerful solution to improve the productivity. From the practical point of view, the battery life of sensors are concerned [1–6]. It is particularly concerned in sensor networks deployed over a wide area because it is not easy to replace the batteries of many sensors dispersed over the wide area.

Simply, we can think about the following two possibilities for the battery saving of the sensors. One of the simplest way to reduce the power consumption at the sensors is to reduce the transmission power of the signal. This causes to decrease the signal-to-noise power ratio (SNR) at the receiver. Another way is to shorten the signal length transmitted by the sensors. However, this is not easy.

M. Kinose · Y. Kamiya (✉)
School of Information Science and Technology, Aichi Prefectural University, Nagakute, Aichi, Japan
e-mail: kamiya@ist.aichi-pu.ac.jp

G. De Pietro et al. (eds.), *Intelligent Interactive Multimedia Systems and Services 2016*, Smart Innovation, Systems and Technologies 55,
DOI 10.1007/978-3-319-39345-2_23

BRAKE [7], a digital signal processing scheme for the receiver, can be a solution to let the sensors shorten and weaken the signal. This is because BRAKE realizes the blind signal detection. It does not need preambles which is usually put at the head of packets for enabling the receiver to detect the signal and its timing. Therefore, using BRAKE, we can shorten the length of the signal by removing the preamble without causing any problems on the signal detection, and finally we can accomplish the battery life extension. Moreover, BRAKE can combine the signals received by multiple antennas without the preambles [7]. It means that BRAKE can recover the weakened signals since it puts multiple signals together so as to improve SNR. It seems that BRAKE is a key to achieve the battery life extension. As far as our best knowledge, there is no conventional work which tries to remove the preambles in sensor networks.

However, there is a problem for this application of BRAKE. BRAKE can blindly detect spread spectrum signals, i.e., it does not need preambles for the signal detection, even though it requires the knowledge about the length of the spreading code [7]. However, it is not applicable to the sensor networks as it is, since BRAKE cannot cope with short signals even if it does not need preambles. The reason is as follows: BRAKE consists of transversal filters and estimates their weights adaptively, based on CMA [7]. As a result, it takes long time taking a number of signal samples to converge.

This paper proposes a new simple method which enables BRAKE to work with short signals. This method must be a key to extend the battery life of sensors and facilitate the realization of large-scale networks. It is promising that the proposed method improves the productivity of agricultural industries.

2 System Image and Conventional Methold

This section briefly explains the conventional BRAKE. This algorithm is precisely defined in [7]. Figure 1 shows a system image. Sensors installed on trees send a short data such like the temperature while a data collection node receives the data. BRAKE is implemented in the data collection node.

Recall that this algorithm is for spread spectrum signals. So the transmitter in the sensor nodes is configured as shown in Fig. 2. The modulated symbols are spread by a spreading code whose length is set at L_{SP} [chip].

Figure 3 illustrates the block diagram. The signal received by the multiple antennas at the data collection node is sampled after the down-conversion and the band limitation. The samples are expressed as follows:

$$\mathbf{x}[\ell] = \mathbf{x}(\ell T_{\text{sample}}) \tag{1}$$

where $\mathbf{x} \in C^{1 \times N_A}$ contains the signals while N_A denotes the number of the receiving antennas. In addition, ℓ is an integer while T_{sample} is the sampling duration. It is assumed that T_{sample} is equal to the chip duration T_{chip}.

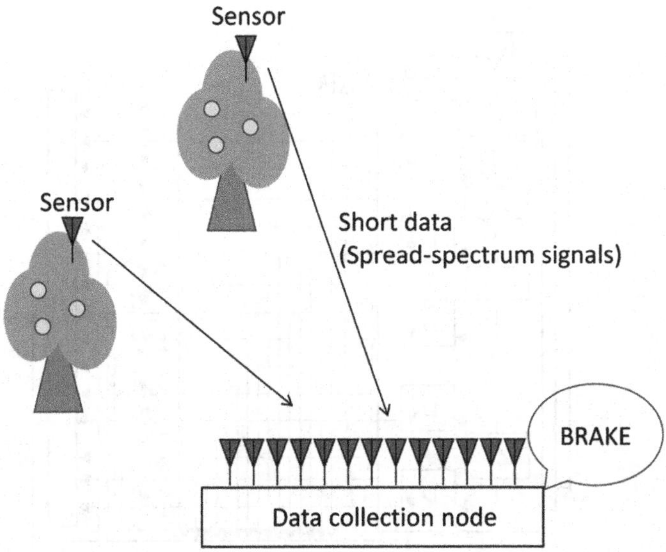

Fig. 1 A system image

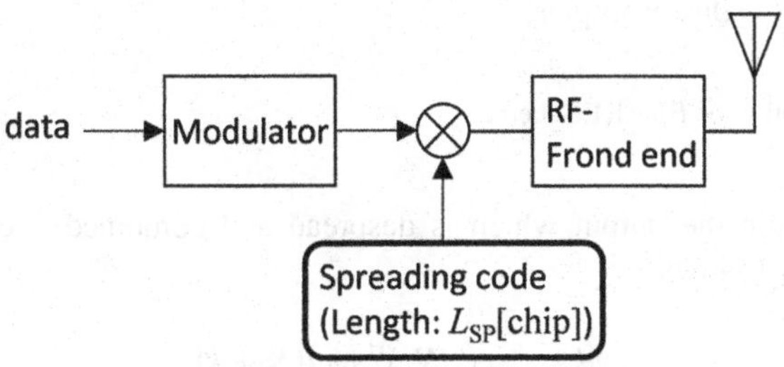

Fig. 2 Configuration of a sensor node

Next, the samples are fed into tapped-delay lines consisting of L_{SP} taps. The stored samples are expressed by $\mathbf{X}_{TDL}[\ell] \in C^{(2L_{SP} \times N_A)}$ defined as:

$$\mathbf{X}_{TDL}[\ell] = \begin{bmatrix} \mathbf{x}[\ell] \\ \mathbf{x}[\ell - 1] \\ \vdots \\ \mathbf{x}[\ell - 2L_{SP} + 1] \end{bmatrix} \tag{2}$$

This is down-sampled as:

$$\mathbf{X}_{DS}[k] = \mathbf{X}_{TDL}[kL_{SP}] \tag{3}$$

where $k = 0, 1, 2, \ldots, \infty$.

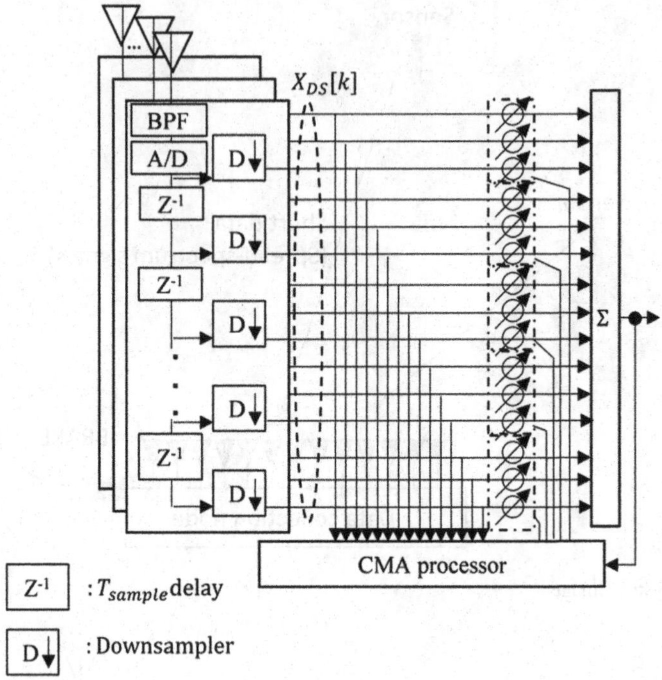

Fig. 3 Configuration of BRAKE scheme

We can obtain the output which is despread and combined over the multiple antennas [7] as follows:

$$y[k] = \mathrm{vect}(\mathbf{W})^{\mathrm{H}}\mathrm{vect}(\mathbf{X}_{\mathrm{DS}}[k]) \tag{4}$$

where $\mathbf{W} \in C^{(2L_{\mathrm{SP}} \times N_{\mathrm{A}})}$ is a weight matrix while the superscript ^{H} denotes the complex conjugate. The operator vect($\cdot$) vectorizes matrices as follows:

$$\mathrm{vect}\left(\begin{bmatrix} a_1 & a_3 \\ a_2 & a_4 \end{bmatrix}\right) = \begin{bmatrix} a_1 & a_2 & a_3 & a_4 \end{bmatrix}^{\mathrm{T}} \tag{5}$$

where the superscript ^{T} denotes the transpose.

The weight matrix **W** is controlled by the steepest-descent method based on the CMA criteria as follows:

$$\mathrm{vect}(\mathbf{W}[k+1]) = \mathrm{vect}(\mathbf{W}[k]) - 4\mu\mathrm{vect}(\mathbf{X}_{\mathrm{DS}}[k])\left(|y[k]|^2 - \sigma^2\right) \tag{6}$$

where μ and σ are parameters which affect the convergence of the steepest-decent method.

Through after conversion of Eq. (6), we can obtain the output of BRAKE by Eq. (4). This weight matrix **W** achieves the despreading, beamforming and the RAKE-combining at the same time [7].

3 Proposed Method

Our goal is to let BRAKE work with the short signals, as mentioned above. In the following, we introduce a basic idea of the proposed method. It is just simple. Suppose that the weights of the transversal filter in BRAKE and signal samples are expressed by vectors and, respectively, as illustrated in Fig. 4.

The basic idea is to conduct the procedures namely Stage 1 to Stage MS as shown in the following.

Stage 1: Set an initial value of **w**, such as $\mathbf{w}^{(0)} = [1 \;\; 0 \;\; \cdots \;\; 0]$

Perform BRAKE, then $\mathbf{w}^{(1)}$ is obtained. But inaccurate.

Stage 2: Set an initial value of **w**, such as $\mathbf{w}^{(0)} = \mathbf{w}^{(1)}$

Perform BRAKE, then $\mathbf{w}^{(2)}$ is obtained.

•
•
•

Stage M_S: Set an initial value of **w**, such as $\mathbf{w}^{(0)} = \mathbf{w}^{(M_S-1)}$

Perform BRAKE, then $\mathbf{w}^{(M_S)}$ is finally obtained.

After conducting these stages, finally we can obtain $\mathbf{w}^{(M_S)}$ which is much more accurate than $\mathbf{w}^{(0)}$ even though the number of samples is limited.

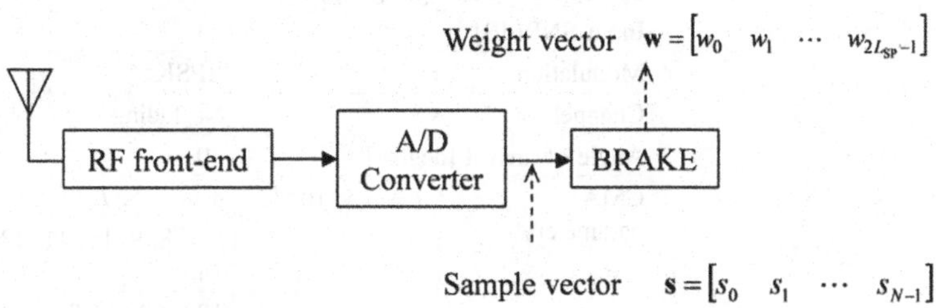

Fig. 4 Overview of the proposed method

4 Computer Simulations

We verify the performance of the proposed method through computer simulations. Table 1 summarizes simulation conditions. We assume that a sensor sends a BPSK-modulated signal to the receiver which impinge on the receiver antenna with 30 degrees of the angle of arrival. Although the input SNR of the signal is set at −3 dB, very low value, the receiver can recover the weak signal using an antenna consisting of 50 antennas with the half-wavelength spacing. Hence, theoretically, we can finally improve the SNR of the signal up to 14[dB] if we can optimally combine the signals received by the 50 antennas. It should be emphasized that such a drastic SNR improvement is very hard to achieve if there is no preamble signals, i.e., there is no channel estimations.

It is well-known that the performance of CMA [8] is affected by the parameters μ and σ [2], even though it is impossible to theoretically determine the optimal values. This is because of its non-linearity [9]. So we tried to set the parameters by trying several settings shown in Table 1, in order to take the parameters which achieves the highest SNR.

In addition, there is no fading in the channel, so a line-of-sight signal is received. The number of stages MS is set at 100. Under these conditions, we calculate the average output SNR after 50 trials. This is repeated varying the number of input samples.

Figure 5 compares the proposed method with the conventional BRAKE in terms of the output SNR as a function of the length of the sample sequence. It is obvious that the proposed method achieves a high SNR such as around 12 dB using 400 samples only, which is achieved by the conventional method using more than 1500 samples. It is remarkable that the proposed method reduces the length of the input sample sequence 3 times less than that of the conventional method in order to achieve the same output SNR. It should be noted that the 400 samples represent 50

Table 1 Simulation conditions

The number of antenna		50
Antenna spacing (Normalized by the wavelength)		0.5
Spreading code length [chip] L_{SP}		15
Input SNR[dB]		−3
Modulation		BPSK
Channel		No fading
Angle of arrival [degree]		30
CMA parameters	$\mu = \alpha \times 10^{-\beta}$	$\alpha = 3, 5, 7, 9$
		$\beta = 8, 9, 10, 11, 12, 13, 14$
	σ	100, 500, 1000
The number of input samples N		100~4300
The number of stages M_S		100

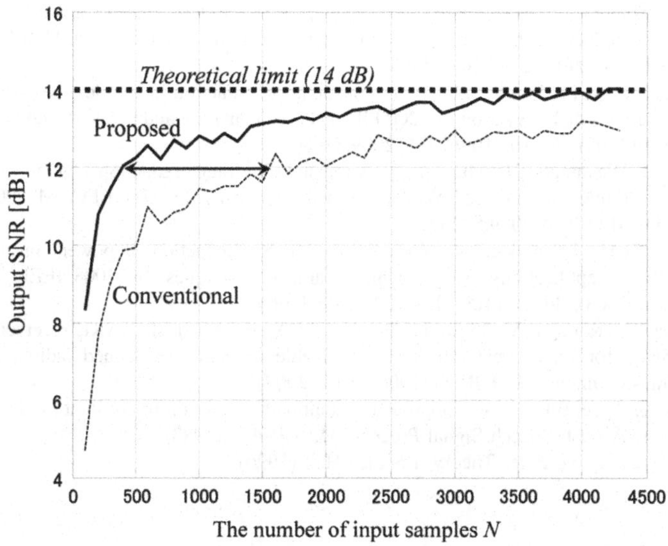

Fig. 5 The number of input samples versus output SNR

bits if we assume the sampling of 8 [sample/bit] at the receiver. It means that the receiver achieves sufficient performance for only less than the packet of which the length is less than 8 bytes.

5 Conclusions

In this paper, we proposed a simple method to apply BRAKE to the receiver for sensor networks aiming at the reduction of the power consumption. Through the computer simulations, we verified that the proposed method reduces the number of samples 3 times less than the conventional method to achieve the same performance in terms of the output SNR.

References

1. Ramachandran, V.R.K., Sanchez Ramirez, A., van der Zwaag, B.J., Meratnia, N., Havinga, P.: Energy-efficient on-node signal processing for vibration monitoring. In: 2014 IEEE Ninth International Conference on Intelligent Sensors, Sensor Networks and Information Processing (ISSNIP) (2014)
2. BenSaleh, M.S.: Loop transformations for power consumption reduction in wireless sensor networks memory: a review. In: 2013 European Modelling Symposium (EMS) (2013). doi:10.1109/EMS.2013.108

3. Itoh, T., Zhang, Y., Matsumoto, M., Maeda, R.: Wireless sensor network for power consumption reduction in information and communication systems. In: 2009 IEEE Sensors (2009). doi:10.1109/ICSENS.2009.5398308
4. Macii, D., Ageev, A., Somov, A.: Power consumption reduction in Wireless Sensor Networks through optimal synchronization. In: 2009 IEEE Instrumentation and Measurement Technology Conference (2009). doi:10.1109/IMTC.2009.5168665
5. Sakhaee, E., Wakamiya, N., Murata, M.: A transmission range reduction scheme for reducing power consumption in clustered Wireless Sensor Networks. In: GLOBECOM 2009 (2009). doi:10.1109/GLOCOM.2009.5425772
6. Shafiullah, G.M., Thompson, A., Wolfs, P.J., Ali, S.: Reduction of power consumption in sensor network applications using machine learning techniques. In: 2008 IEEE Region 10 Conference (2008). doi:10.1109/TENCON.2008.4766574
7. Takayama, K., Kamiya, Y., Fujii, T., Suzuki, Y.: A new blind 2D-RAKE receiver based on CMA criteria for spread spectrum systems suitable for software defined radio architecture. IEICE Trans. Commun. **E91-B** (6), 1906–1913 (2008)
8. Treichler, J., Agee, B.G.: A new approach to multipath correction of constant modulus signals. IEEE Trans. Acoust. Speech Signal Process. **31**, 459–472 (1983)
9. Haykin, S.: Adaptive Filter Theory, Prentice Hall (1996)

A Tool for Visualization of Meteorological Data Studied for Integration in a Multi Risk Management System

Emanuele Cipolla, Riccardo Rizzo, Dario Stabile and Filippo Vella

Abstract This paper presents a tool for visualization of meteorological data integrated in an application for management of a multi risk situation. As part of a much more complex and demanding visualization system, the tool was developed taking into account some constraints as the scarcity of screen space and processing power in the considered system. Due to these limitations the tool was developed as web application optimized for mobile devices and is capable to present time series data in a compact and effective visualization. The tool also can process the available data in order to detect exceptional events and to put them in the contest allowing a very fast visual analysis.

1 Introduction

The development of multimedia computer simulations, communications and scientific visualizations have deeply changed the techniques for risk and the emergency management. Simulation of emergency situations and advanced visualizations provide help to practitioners and decision makers in Emergency Management (E.M.), while multimedia communications, social networks and web applications are useful tools for emergency operators [1, 2].

E. Cipolla · R. Rizzo · D. Stabile · F. Vella (✉)
Institute for High Performance Computing and Networking - ICAR,
National Research Council of Italy, Palermo, Italy
e-mail: vella@pa.icar.cnr.it; F.vella@pa.icar.cnr.it

E. Cipolla
e-mail: cipolla@pa.icar.cnr.it

R. Rizzo
e-mail: ricrizzo@pa.icar.cnr.it

D. Stabile
e-mail: stabile@pa.icar.cnr.it

G. De Pietro et al. (eds.), *Intelligent Interactive Multimedia Systems and Services 2016*, Smart Innovation, Systems and Technologies 55,
DOI 10.1007/978-3-319-39345-2_24

Mobile devices, through multimedia communication, are the perfect receivers for multimedia broadcast information, for example, in critical situations they can show evacuations routes to people in risk area [3]. Conversely, the mobile devices allow the operators to obtain more information from the ground about the emergency situations [4] and transform witness in diffuse information sources [5]. This huge information flow coming from multiple sources is organized, processed and visualized in an emergency control room, where the information from the field is integrated with the one obtained from time series processing.

The project SIGMA[1] is an Italian research project aimed to the development of a system that collects data from many sensor networks (rain, phreatic, seismic, volcanic) in order to monitor the environment and to provide useful data for prevention and management of multi-risk situation.

The system presented in this paper is targeted to the visualization of weather data and outliers of time series precipitation data [8]. The system is designed as part of the visualization system of the whole project, that comprises also many other parts related to seismic risk, volcano monitoring, and industrial risk, and was designed to be simple and easy to navigate. The system was built as a web application and the information is presented in a very compact form for an important reason: both the monitor space in the control room and the display occupation in a mobile device are a limited resource. Some systems have been proposed for the visualization of the emergency situation occurring in a given place. We describe here some examples of systems and applications of visualization in emergency situations employing techniques and solutions for specific critical situations.

In [6] a system for emergency management in flood emergency is presented. The work is related to a project involving the emergency management in Portuguese dam system. The system is provided with a three dimensional terrain visualization that represents the involved valleys. The specific particularity of the work is that the authors cast this application as a real time strategy game where a user can analyze the single involved entity clicking on the item and reading the corresponding information. The geographic information is managed with a Geographic Information System. A geographic data stack is used to manage terrain features such as roads, water streams and buildings. The proposed interface is bound to geographic data and provides information for the values of the rivers and the dams. The system does not provide an analysis of the previous values and local point by point information is presented. The action to be taken and the visualization of the relevant information is left to the user who directs his attention to the items assessed as the most relevant. The visualization on a map is very useful, while the evaluation relies on the user experience and his heuristics due to the lack of data analysis.

In [7] an approach for the management of emergencies in public transportation is described, using the Valencia metro as a study case. Their approach considers the emergency as a process and combines the information from different sources to describe the applied plans. The proposed approach presents different emergency resolution paths to the users who can choose the most adapt ones according to the

[1] http://www.sigma.pa.icar.cnr.it/.

collected information. Information is retrieved and managed according to its validity and the emergency status. Communication and collaboration are described as fundamental aspects of the visualization system. A key aspect for the authors is the system capability to generate valuable information and to spot what is important out the data collected by all the sensors. As in this case, our system is aimed at visualizing information that come from a net of heterogeneous sensors. We don't consider the emergency plans as we are focused on data collection rather than on the execution of an emergency plan. Notwithstanding we consider of great importance the capability, for the visualization system, to provide information coming from multiple sources and visualize important aspect of information highlighting the most relevant and significative information.

2 System Description

A prior analysis of the operating environment and the recent strong presence of mobile devices has greatly influenced the choice of architecture, shown in the Fig. 1, the adopted technologies and the development of user interfaces. We chose to develop a web application because of the following reasons. *Data and computing capacity*: as we manage heterogeneous data, coming from several remote sources, in our application we used a relational database as common storage, warehouse for both row and processed data (analysis, filtering, merging the data, etc.). Server side computing capacity is greater than the one of the mobile device and this allow for an easier and more accessible front-end to the user. *Availability and update propagation*: web applications are immediately available to users through a browser on a wide range of devices (smartphone, tablet, laptop, PC, etc.), without downloads and

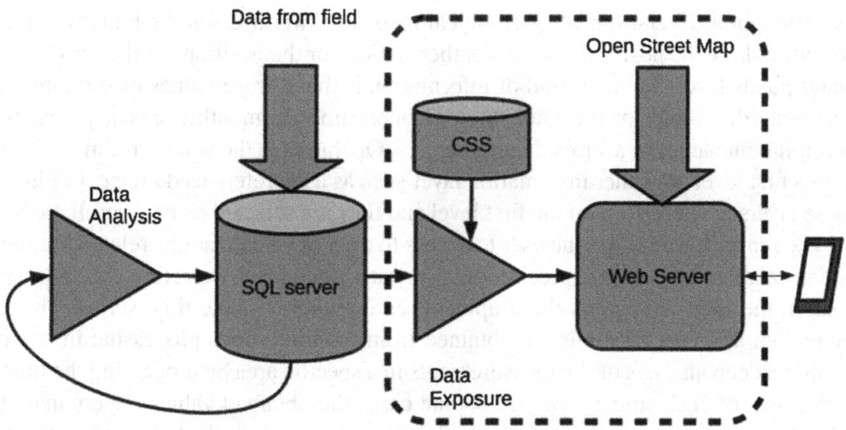

Fig. 1 Block diagram of web application system architecture

installations and can update their content in real time. *User-experience*: the progress and evolution of web standards and the strong improvement of interfaces and connectivity to mobile devices, have reduced the distance between web and mobile applications feature-wise. Our application needs to convey a wealth of information in very limited time and space. We have considered a set of principles to be adopted to create an effective interface to show relevant information and create an interface able to vehiculate these contents in a critical environment and situations. These principles are: *Clarity*: the interface should be as easy as possible. We strove to encapsulate all the relevant information in the visual medium. Some visualizations embed statistical information, such as the size of the circles representing the number of exceptional events for each station. *Modularity:* a user can switch from a visualization tool to another maintaining the attention focus on the relevant information. Multiple information can be obtained clicking on the menus obtained clicking on the stations. *Information Density:* we concentrated all the useful informations in few screenshots allowing a rich visualization in a glance. For example, we used the visualization of some averaged values (daily precipitation in Fig. 4) as a background for the visualization of the location of exceptional events, in order to put these points in a context. *Consistency:* the interface maintains the same underlying structure in all the visualization layers. The dots representing the stations can be clicked to open further menus that show multiple graphs. The outliers can be analysed in detail clicking on the outlier symbol and opening a detailed map. Every visualization can be accessed with a touch on a mobile device and all informations can be reached with few clicks. The proposed visualization system has a modular structure, organized in levels of visualization, that can be composed using many stacked layers. On the device screen (a laptop or a mobile phone) each level overlaps the previous ones and the final visualization is composed by all the information coming from multiple layers. Furthermore, a reduced number of visualization levels is also important for mobile devices. Figure 2 shows the logical structure of the visualizations implemented in the system. The first visualization level is depicted on the left of Fig. 2 and it is composed by overlapping layers over a geographical map. The two layers at the bottom of the structure show the positions of the weather stations or the positions of the sensor for water ponds level. Another kind of information is the average values of some measurements distributed on the whole region; for example the monthly or yearly average precipitation values or averaged temperatures. Operators in the control room can add to this first level any other information layer such as a georeferenced image. In Fig. 3 these layers are at the top on the first level and they are selected using a scroll menu.

The sensor layer allows the user to access to a set of visualizations related to time series data for temperature, precipitation or water pond level. Selecting one sensory station, the user can access the graphical representation of the time series. These more complex visualizations are obtained from the time series processing in order to offer a deep analysis of the measurements in a specific area by processing the time series data of each single station, in some cases the obtained values are compared with the surrounding measurements. In the following sections the two levels and the visualizations are described in details.

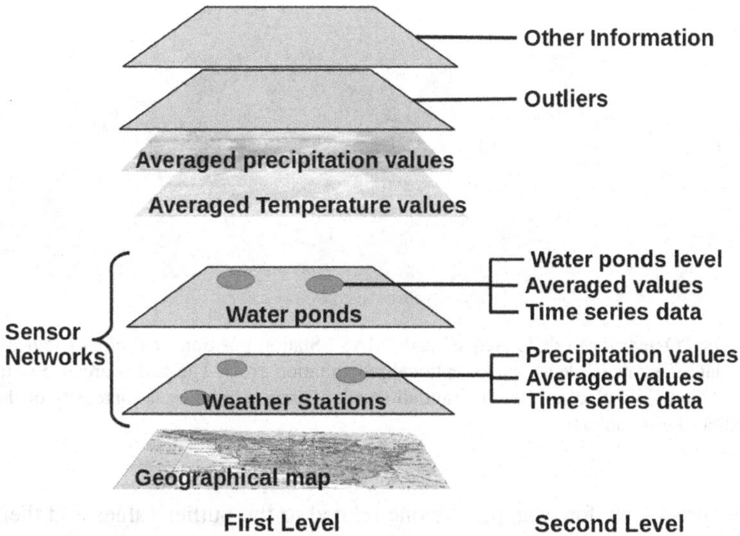

Fig. 2 The two visualization level obtained from the proposed system

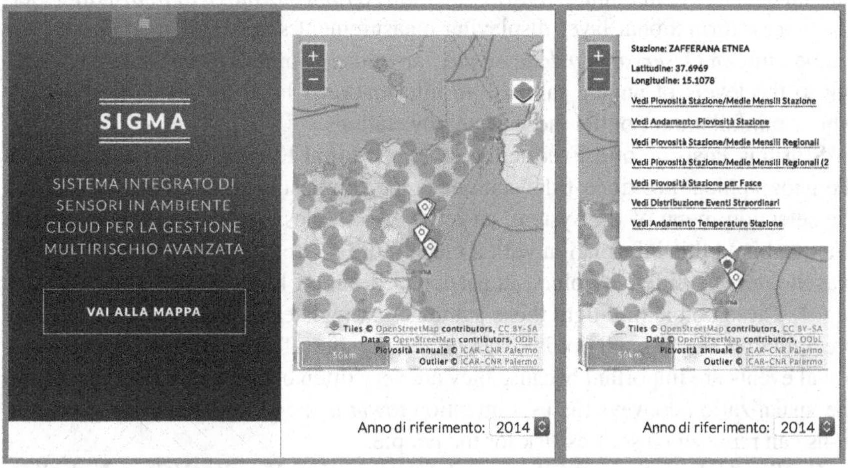

Fig. 3 On the *left* and *center* part of the image there is the homepage of the mobile version, that provides the first level visualization; on the *right* part there is the second level visualization, with the set of visualizations related to time series data, available for each station

2.1 First Visualization Level

This level shows information obtained processing meteorological data for the whole region. For example it shows the averaged precipitation values that allow the user to spot the areas with heavy rainfalls. Other visualizations are extracted through

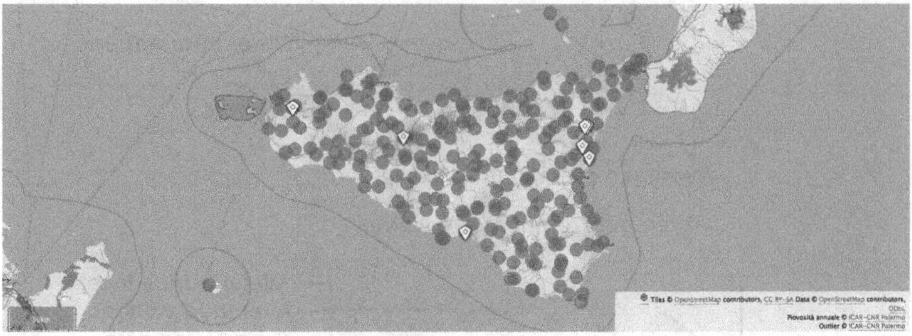

Fig. 4 The "Osservatorio delle Acque" and "SIAS" Station position. *Red circles* indicates the stations that registrated at least one exceptional precipitation event. The outlier are also visualized using the "diamond" placeholder. The rainfall density is represented by the intensity of the *blue color* (colour figure online)

complex processes, for example the one related to the outlier values and their frequency. In detail we find: a base layer displaying a geographical map, from OpenStreetMap or MapQuest, an informational layer to display measurement stations of rainfall data and temperature of *SIAS - Sicilian Agrometeorological Information Service*,[2] an informational layer displaying measurement stations of rainfall data and temperature of *Osservatorio delle Acque*,[3] an informational layer showing data relating to the levels of annual rainfall, an informational layer for displaying stations whose rainfall values outlie those of nearby stations.

Stations Position and Frequency of Exceptional Events. The visualization of the geographical position of the stations is the simplest information obtained from the selection menu of the system, and it is also the less informative visualization. We combined this information with a much more interesting visualization by using placeholders of different colors, in particular different levels of red, as reported in Fig. 4. These colors highlight those stations that registered at least one exceptional precipitation event. Each dot is scaledon the number of these events. These exceptional events are important because they are very often connected to flood events, so the visualization conveys the user attention towards the places where the precipitations can represent a serious risk for the people.

Outlier Calculation and Averaged Precipitation Density Values Visualization. Informally, an outlier is a point in a complex system where the behaviour of a given phenomenon is inconsistent with other points in its neighbourhood. Assuming that the presence of an outlier could help the detection of critical situations, we previously proposed an algorithm to spot the outlier from time series data [8] in complex geographical scenarios. As sensor networks can be modelled using graphs, the connections among the nodes can be used to convey the influence that each node has on its neighbours, provided that it can be expressed using a distance function.

[2]http://www.sias.regione.sicilia.it/.

[3]http://www.osservatorioacque.it/.

In the proposed algorithm the distance function used is Mutual Information, and is computed for each couple of nodes after a uniform discretization of the relevant time series values that allows the construction of probability densities, according to the frequentist interpretation of probability. Outliers are marked on the geographical map, as shown in Fig. 4. In this figure the outliers overlap to the averaged precipitation density values already calculated using the whole time series for the region, the density is represented by the intensity of the blue color.

2.2 *Second Visualization Level*

The visualizations on this level deal with the time series of the measurements for a single station.

Station Daily Mean Precipitation Values: "Tiles" visualization. This visualization shows, in a compact form, the daily mean precipitation value for a time span of several years. The overall visualization resembles a tile mosaic and reports a whole year on each horizontal strip and the weeks over the columns of the visualization. The color of each tile is related to the average precipitation value, referred to the global mean of the precipitation value for the station. Figure 5 shows an example of this visualization. On the left there is the visualization format for the mobile devices. This can be enlarged and a single day average value can be obtained hovering on the single "tile". This visualization was used since it is more readable than overlapped graphs. It allows the user to compare many years in a glance, and to spot patterns

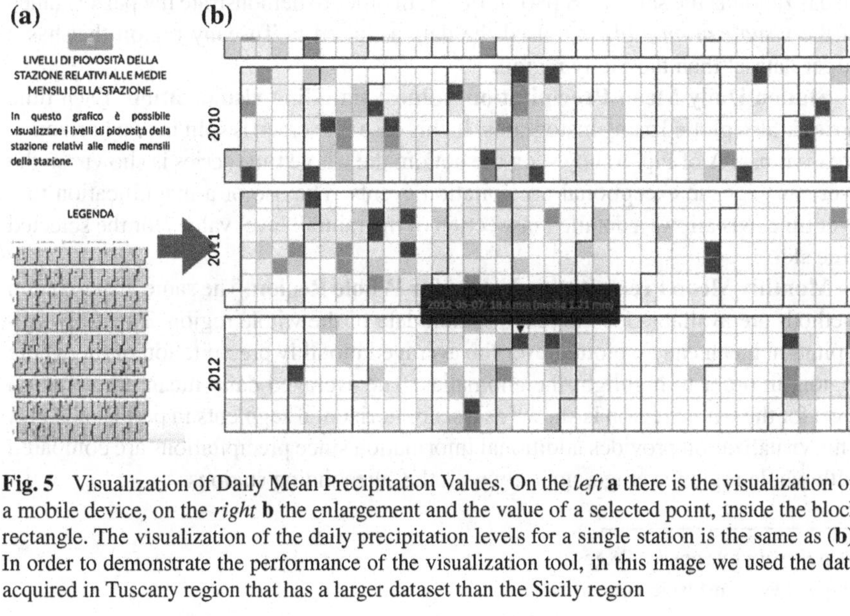

Fig. 5 Visualization of Daily Mean Precipitation Values. On the *left* **a** there is the visualization on a mobile device, on the *right* **b** the enlargement and the value of a selected point, inside the block rectangle. The visualization of the daily precipitation levels for a single station is the same as (**b**). In order to demonstrate the performance of the visualization tool, in this image we used the data acquired in Tuscany region that has a larger dataset than the Sicily region

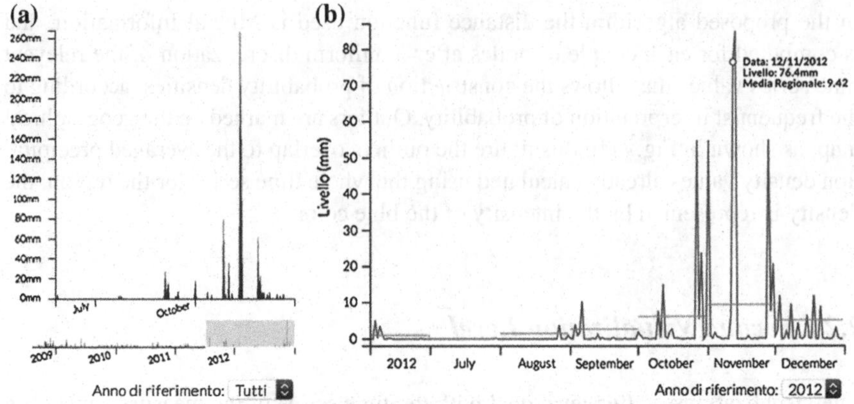

Fig. 6 **a** The whole time series of the averaged precipitation level for one station and zoom window. The listbox allows the user to select the year of interest, in this case all the measurements available are visualized. In order to demonstrate the performance of the visualization tool, in this image we used the data acquired in Tuscany region that has a larger dataset than the Sicily region. **b** The visualization of the daily precipitation mean against the monthly precipitation mean of the whole region

of events, if present. The graph visualization is also available and will be presented in the next section. The compact look of this visualization is effective on mobile devices and allows the user to have a global view on the time series data; it is also possible to zoom using the pinch-in gesture. The tiles are colored according the seven standardized levels by the Civil Protection[4]: C1: <1 mm, C2: 1 – 10 mm, C3: 10.1 – 20 mm, C4: 20.1 – 30 mm, C5: 30.1 – 40 mm, C6: 40.1 – 50 mm, C7: >50 mm. This visualization is the same as B part of Fig. 5. In order to demonstrate the performance of the *visualization tool*,[5] we used the data acquired in Tuscany region that has a larger dataset than the Sicily region.

Station Daily Mean Precipitation Values: time-line visualization. Each time series is also plotted using histograms. In this case the screen is split in two horizontal areas, as part A of Fig. 6 shows: on the bottom the whole time series is shown, so the user can spot the exceptional precipitation events. The use of a magnification tool over time axis allows a detailed inspection of the rainfall level values for the selected time slot.

Monthly Mean Precipitation Values for Whole Region. The same visualization methods are available for the precipitation data of the whole region. The daily precipitation mean can be plotted over the averaged monthly precipitation of the whole region, in order to highlight the anomalies. The averaged daily mean precipitation level for the whole region is shown as a set of horizontal segments in part B of Fig. 6. This visualization provides additional information since precipitations are compared with whole region average, not only with the surrounding stations.

---

[4] http://www.protezionecivile.gov.it/.

[5] http://www.sigma.pa.icar.cnr.it/toscana/.

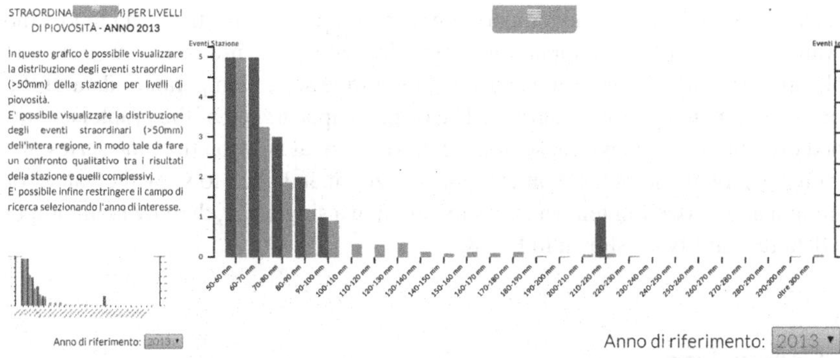

Fig. 7 Chart of the extraordinary events, *left* viewing on mobile, *right* on the PC/tablet

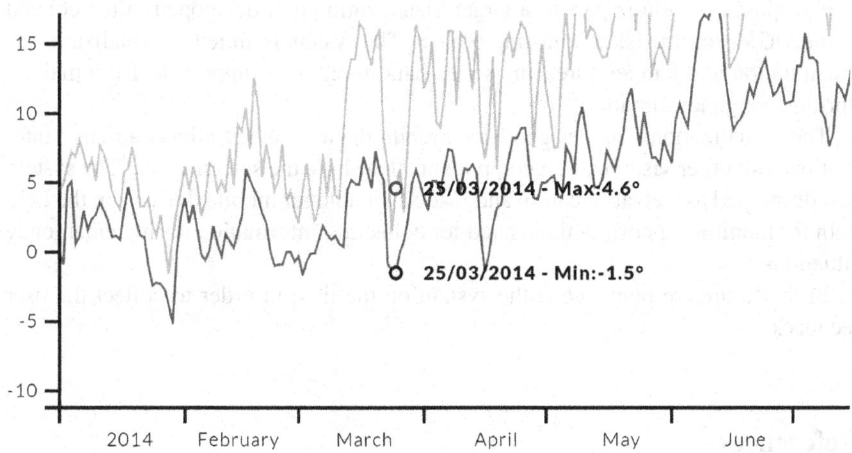

Fig. 8 Detail of the temperature value at a specific date

Extraordinary Events Distribution. A further view, shown in Fig. 7, allows the user to see a chart showing the distribution of the number of extraordinary events of the monitoring station (precipitation level greater than 50 mm) grouped by rainfall intervals. The chart, with a left scale of reference for the single station and a right scale for all stations in the region, shows the number of detected extraordinary events grouped by range of rainfall. Clicking on each bar the user can read the number of extraordinary events recorded in the individual station (blue bars) and in the whole region (orange bars). This graph displays the bands of greater concentration of the extraordinary events and allows a qualitative comparison of the overall behaviour across the region.

Temperature Trend. The temperature trend graph displays the variation of the maximum and minimum temperatures recorded by a station during the year. The application performs a real-time analysis of the time series temperature data of the selected station and plot it in a graph. The original appearance of the graph is maintained even on mobile devices, so you can have a global view on the time series just like on a personal computer. You can pinch to zoom and click to view the values of minimum and maximum temperatures recorded in a single day, thereby maintaining full data readability, as shown in Fig. 8.

3 Conclusions

The proposed system is part of a larger visualization tool developed in the context of the *SIGMA* multi-risk monitoring project. The system is aimed to visualization of precipitation and temperature time series data in order to support decision making during emergency situations.

The visualization was designed for mobile devices and it allows an easy integration with other visualization components that share the screen space. The system was developed as web application and made available for mobile devices on the field both for monitoring normal trends and for collecting information during emergency situations.

In the future we plan to test the system on the field in order to collect the user feedbacks.

References

1. Sanders, R., Tabuchi, S., et al.: Decision support system for ood risk analysis for the river thames, united kingdom. Photogramm. Eng. Remote Sens. **66**(10), 1185–1193 (2000)
2. Leskens, J.G., Kehl, C., Tutenel, T., Kol, T., de Haan, G., Stelling, G., Eise-mann, E.: An interactive simulation and visualization tool for ood analysis usable for practitioners. In: Mitigation and Adaptation Strategies for Global Change, pp. 1–18 (2015)
3. Onorati, T., Aedo, I., Romano, M., Daz, P.: Emergensys: Mobile technologies as support for emergency management. In: Caporarello, L., Di Martino, B., Martinez, M. (eds.) Smart Organizations and Smart Arti-facts, LNISO, vol. 7, pp. 37–45. Springer International Publishing (2014)
4. Luyten, K., Winters, F., Coninx, K., Naudts, D., Moerman, I.: A situation-aware mobile system to support_re brigades in emergency situations. In: Meersman, R., Tari, Z., Herrero, P. (eds.) On the Move to Meaningful Internet Systems 2006: OTM 2006 Workshops. LNCS, vol. 4278, pp. 1966–1975. Springer, Berlin (2006)
5. Gmez, D., Bernardos, A., Portillo, J., Tarro, P., Casar, J.: A review on mobile applications for citizen emergency management. Highlights on Practical Applications of Agents and Multi-Agent Systems. Communications in Computer and Information Science, vol. 365, pp. 190–201. Springer, Berlin (2013)

6. Nobrega, R., Sabino, A., Rodrigues, A., Correia, N.: Flood emergency interaction and visualization system. In: Visual 2008—Proceedings of the 10th International Conference on Visual Information Systems. LNCS, p. 09. Springer, Berlin (2008)
7. Can_os, J.H., Alonso, G., Ja_en, J.: A multimedia approach to the efficient implementation and use of emergency plans. MultiMed. IEEE **11**(3), 106–110 (2004)
8. Cipolla, E., Vella, F.: Identification of spatio-temporal outliers through minimum spanning tree. In: Tenth International Conference on Signal-Image Technology and Internet-Based Systems, SITIS 2014, pp. 248–255. Marrakech, Morocco, 23–27 Nov 2014

An Algorithm for Calculating Simple Evacuation Routes in Evacuation Guidance Systems

Koichi Asakura and Toyohide Watanabe

Abstract This paper proposes an algorithm for calculating simple evacuation routes in evacuation guidance systems. In order to ensure that evacuees move to the shelters safely in disaster situations, our algorithm produces the simple routes in which the minimum number of turns is included. Evacuees can move to the shelters by following simple routes with low risk of making a mistake. For calculating simple routes, a road network is transformed so that an edge in the road network is converted into a vertex and a vertex is converted into several edges. Experimental results show that the length of the produced simple routes is not so different from that of the shortest routes.

Keywords Evacuation guidance systems · Simple routes · Disaster simulation · Evacuee agent

1 Introduction

In disaster situations such as major earthquakes, it is extremely important for people in the affected areas to safely and correctly follow evacuation routes to the nearest shelters. Recent developments with smartphones and their applications mean we can now use many kinds of map applications, anywhere and anytime. With online map applications such as Google Maps [1], we can identify the best routes to a destination while considering the current situations in the areas. Also, offline map applications such as MapsWithMe [2] enables us to reference map data without a network

K. Asakura (✉)
Department of Information Systems, School of Informatics, Daido University,
10-3 Takiharu-cho, Minami-ku, Nagoya 457-8530, Japan
e-mail: asakura@daido-it.ac.jp

T. Watanabe
Nagoya Industrial Science Research Institute, 1-13 Yotsuya-dori, Chikusa-ku,
Nagoya 464-0819, Japan
e-mail: watanabe@nagoya-u.jp

G. De Pietro et al. (eds.), *Intelligent Interactive Multimedia Systems and Services 2016*, Smart Innovation, Systems and Technologies 55,
DOI 10.1007/978-3-319-39345-2_25

connection. These map applications are enormously useful in terms of helping people in the disaster areas to move to the nearest shelters for evacuation.

However, it is possible that the usual evacuation routes retrieved by the above map applications may not be feasible for evacuation since roads may be impassable due to rubble from collapsed buildings, the spread of fire, and so on in the disaster areas. In response to this issue, we have already proposed a construction system for navigational maps [3]. In this system, we adopt a mechanism based on ant colony systems in order to take recentness of information into account [4, 5].

In addition, in such affected areas, special considerations for selecting evacuation routes may be required. For example, in confusing areas, evacuation routes must be simple enough to lower the risk of evacuees making a navigational mistake, rather than just being the shortest. People tends to clump together and move to the shelters as a group in evacuation activities, and in order to avoid being separated from the group, they should select simple routes rather than short but difficult ones.

In this paper, we propose an algorithm for calculating simple evacuation routes for evacuation guidance systems. Here, we define a simple route as the route in which the number of turns at intersections is the fewest. In our algorithm, a road network in the map is transformed into another network and then Dijkstra's algorithm is applied to the transformed network.

The rest of this paper is organized as follows. Section 2 describes the main issue that we tackle in this paper, namely, providing the simplest routes. Section 3 gives an algorithm for calculating simple routes and Sect. 4 evaluates the proposed algorithm with simulation experiments. We conclude in Sect. 5 concludes with a brief summary and details on our future work.

2 Problem Description

Our proposed evacuation guidance systems helps evacuees in disaster areas to take safe and secure routes to the shelters. For this purpose, we define a simple route as the route to the shelter along which evacuees have to turn at intersections at the minimum number of times. Figure 1 shows an example of the shortest route and the

Fig. 1 The shortest route and the simple route

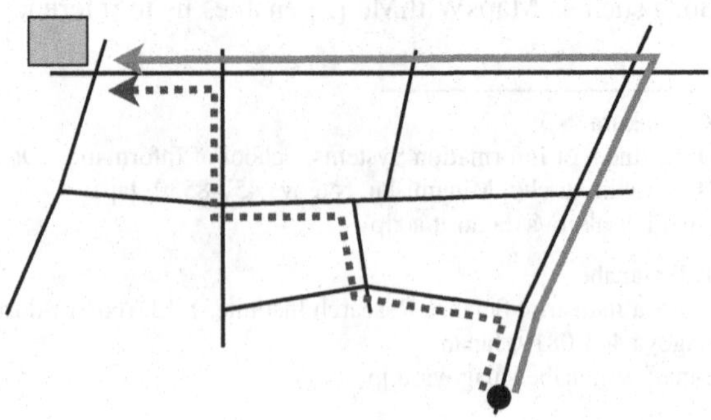

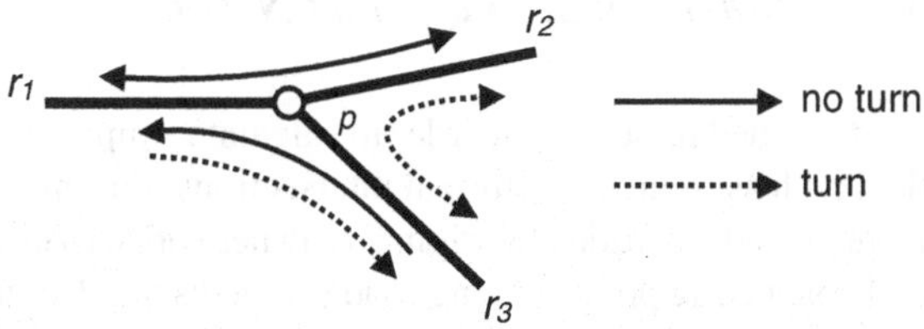

Fig. 2 Asymmetry of turn at an intersection

simple route to the shelter. In this figure, a shelter is denoted as a square, and an evacuee is denoted as a bullet. Line segments show the road network. The dashed blue arrow shows the shortest route from the position of the evacuee to the shelter. Conventional route calculating algorithms such as Dijkstra's algorithm, the Bellman-Ford algorithm, and the A* algorithm [6], can produce the shortest route easily, but as shown in the figure, the evacuee has to turn at the intersections many times along the shortest route. In contrast, the red arrow shows the simple route. We can observe that this route has only one turn. Although the length of the simple route is slightly longer than that of the shortest one, the simple routes enables evacuees to move to the shelter easily.

In order to calculate the simple route in a road network, we have to consider asymmetry of turn at intersections. In other words, it depends on the movement direction whether an evacuee has to turn at an intersection or not even if the evacuee passes through the same two road segments at the same intersection. Figure 2 shows an example of this situation. In this figure, the road segments r_1, r_2 and r_3 connect at the intersection p. When an evacuee passes through at p from r_1 to r_2 and from r_2 to r_1, the evacuee moves along the road without turning. Furthermore, when the evacuee takes routes from r_2 to r_3 and from r_3 to r_2, one turn is counted at these movements. However, there is an asymmetrical relationship between the route from r_1 to r_3 and the route from r_3 and r_1. When evacuees move from r_1 to r_3, they have to turn right at the intersection p. However, no turns are required for the movement from r_3 to r_1 because evacuees just follow the road and thus the navigation has no difficulties.

In order to calculate simple routes, this issue has to be resolved. In the next section, we propose a transformed road network for handling asymmetry of turn.

3 Algorithm

In this section, we propose an algorithm for calculating simple routes for easy evacuation in disaster areas. For the sake of resolving the issue of asymmetry discussed above, road networks are transformed into another type of network. Section 3.1 discusses a transformed road networks and Sect. 3.2 presents the algorithm for calculating simple evacuation routes.

3.1 Transformation of Road Networks

As described in Sect. 2, in order to calculate simple routes for evacuation, we have to deal with asymmetry of turn at intersections. For this issue, we propose using transformed road networks. In original road networks, road segments are denoted as edges and connecting points among road segments are denoted as vertices. In transformed road networks, the edges and the vertices in the original road network are switched into vertices and edges, respectively. Thus, one connecting point in the original road network is transformed into several edges between vertices. Additionally, these edges have a cost 0 or 1, where cost 0 means that people can walk along two road segments without turning, and cost 1 means that they have to turn at the intersection between two road segments. The transformed road networks are directed networks, namely, the edges in the transformed road network are directed edges.

Figure 3 shows an example of transformed road networks. An original road network is shown in Fig. 3a. In this road network, four road segments are connected at one intersection. Since this intersection is a general four-way intersection, there are no turns between r_1 and r_3, and between r_2 and r_4. Also, there are turns between r_1 and r_2, between r_1 and r_4, and so on. The transformed road network is shown in Fig. 3b. In this network, one intersection is transformed into 12 directed edges each with a cost 0 or 1. From this transformed network, we can assume that it costs 1 from r_3 to r_4, namely, we have to turn at the intersection from r_3 to r_4.

An example of asymmetry of turn at the intersection is given in Fig. 4. In Fig. 4a, people turns at the intersection from r_1 to r_3, but no turns are required from r_3 to r_1. As shown in Fig. 4b, our transformed road network can handle this situation since the transformed road network is a directed network.

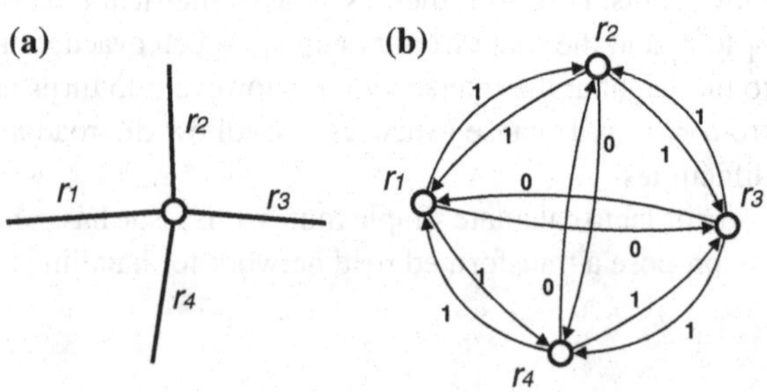

Fig. 3 Example of a transformed road network. **a** original road network **b** transformed network

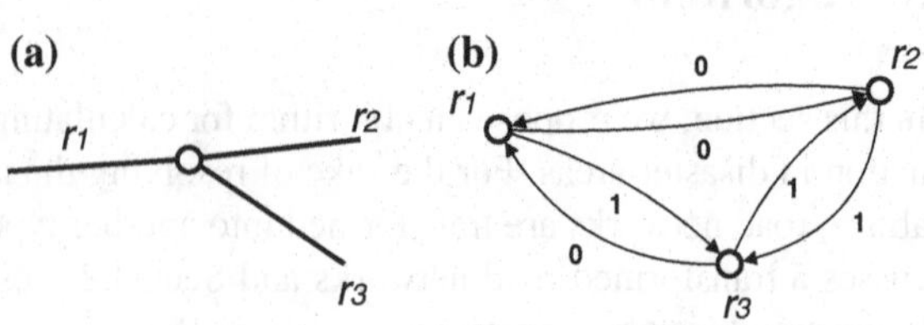

Fig. 4 A transformed road network for asymmetry of turn. **a** original road network **b** transformed network

3.2 Calculation of Simple Routes

After constructing the transformed road network, we can easily calculate a simple route to the shelter by using an algorithm for calculating the shortest path, such as Dijkstra's algorithm, the Bellman-Ford algorithm, or the A* algorithm [6]. The calculated cost represents the minimum number of turns from each road segment to the shelter.

4 Evaluation

In this section, we evaluate the proposed algorithm. Section 4.1 compares the simple route with the shortest route with a practical road network. Section 4.2 analyzes the time complexity of the proposed algorithm.

4.1 Experimental Results

For experimental evaluation, we used a map of the area surrounding our university (Fig. 5). This map covers an area 2.0 km wide and 1.5 km high. There is a residential

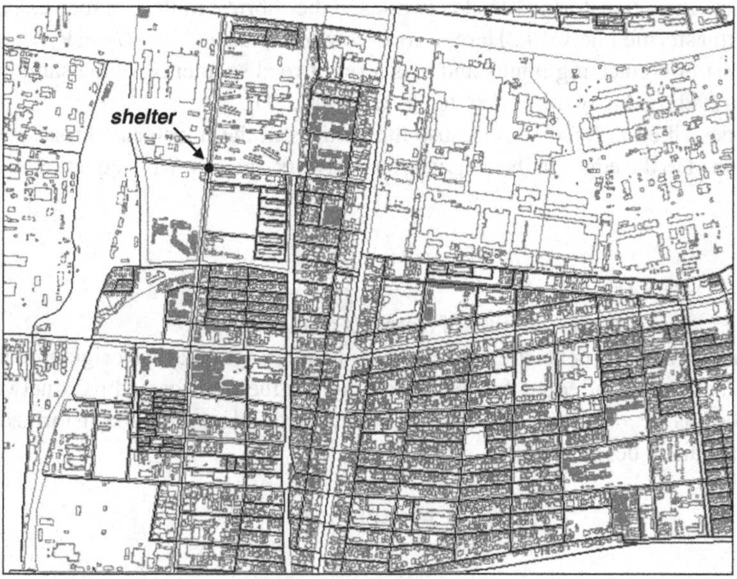

Fig. 5 A practical road network for experimental evaluation

area at the bottom right of the map. Thus, we can observe a road network of a fine-grained checkerboard pattern. We provide one shelter on the map and calculate shortest routes and simple routes from each connecting point on the map. We then compare the number of turns in the routes and the length of the routes. We have 591 connecting points on the map.

Figures 6 and 7 show typical comparative results. In these figures, blue lines denote shortest routes and red lines denote simple routes. In Fig. 6a, the shortest route forms a zigzag pattern, since it passes through the residential area at the bottom right of the map. In contrast, with the calculated simple route in Fig. 6b, evacuees can pass through the residential area without turning. Also, we can observe from Fig. 7 that only one turn is required for evacuation with the calculated simple route and that the length of the two routes is almost the same.

Table 1 summarizes the experimental results statistically. The average values of the number of turns in the routes and the length of the routes are shown. From these results, it is clear that our proposed algorithm can ably calculate the simple routes. Also, we can observe that the average length of the simple routes is not so different from that of the shortest routes. This demonstrates that the proposed algorithm can produce better evacuation routes.

4.2 Computational Complexity

In the proposed algorithm, in order to calculate simple routes, a road network is reformed into a transformed road network and the shortest routes are then calculated in the transformed network. Here, a road network is denoted as $G = (V, E)$, where V is a set of connecting points and E is a set of road segments. In the same way, a transformed road network is denoted as $G' = (V', E')$.

As described in Sect. 3.1, because the vertices in the transformed road network correspond to the edges in the original road network, the following equation is established:

$$|V'| = |E|, \tag{1}$$

where $|\cdot|$ expresses the number of elements of the set.

When n road segments are connected at one intersection in the original road network, the number of edges for the intersection in the transformed road network is described as ${}_nP_2 = n \cdot (n-1)$. Thus, the number of edges in the transformed road network can be described as

$$|E'| = \sum_{v \in V} (deg(v) \cdot (deg(v) - 1), \tag{2}$$

where $deg(v)$ denotes the degree of the vertex v [7]. Here, as an approximate value for $deg(v)$, we adopt the average degree:

Fig. 6 Experimental result 1. **a** The shortest route. Turns: 14. Length: 2142 m **b** The shortest route. Turns: 5. Length: 2271 m

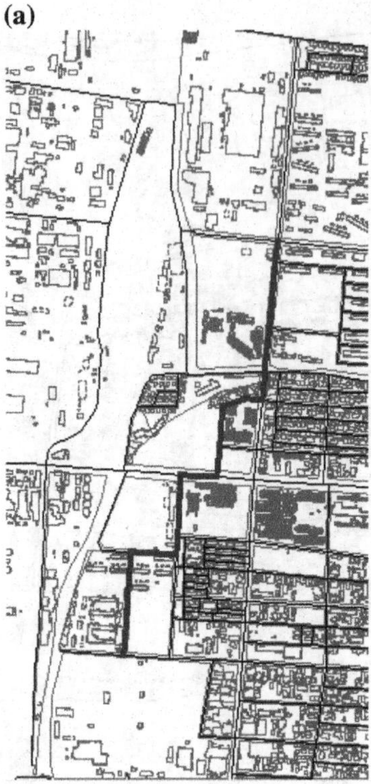

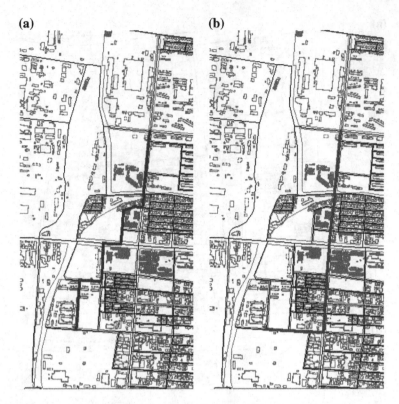

Fig. 7 Experimental result 2. **a** The shortest route. Turns: 5. Length: 1077 m **b** The shortest route. Turns: 1. Length: 1086 m

Table 1 Statistical results for experiments

	Turns	Length [m]
Shortest routes	2.53	1064.40
Simple routes	1.38 (54.5 %)	1152.90 (108.3 %)

$$deg(v) = \frac{2|E|}{|V|}. \quad (3)$$

Thus, the number of edges in the transformed road network is approximated as

$$|E'| = \sum_{v \in V} \frac{2|E|}{|V|} \cdot \left(\frac{2|E|}{|V|} - 1 \right) = 2|E| \cdot \left(\frac{2|E|}{|V|} - 1 \right). \quad (4)$$

For $G = (V, E)$, the computational complexity of Dijkstra's algorithm is denoted as

$$O(|E| + |V| \log |V|). \quad (5)$$

Thus, the computational complexity for calculating simple routes is denoted as

$$O(|E'| + |V'| \log |V'|) = O\left(\left(2|E| \cdot \frac{2|E|}{|V|} - 1\right) + |E| \log |E|\right)$$
$$= O\left(\frac{|E|^2}{|V|} + |E| \log |E|\right). \tag{6}$$

5 Conclusion

In this paper, we have proposed an algorithm for calculating simple evacuation routes for evacuation guidance systems in disaster areas. With simple routes, evacuees can move to shelters easily without any mistakes because they only have to turn at intersections the minimum number of times. Experimental results show that our proposed algorithm can produce simple routes and that the length of the simple routes is not so different from that of the shortest routes.

In our future work, we will develop a route calculation algorithm that takes the trade-off between length of the route and simplicity of the route into account. This issue is a kind of optimization problem with multi-variable functions.

References

1. Google Inc.: Google Maps. https://maps.google.com/
2. My.com: MAPS.ME. http://maps.me/
3. Asakura, K., Watanabe, T.: Construction of navigational maps for evacuees in disaster areas based on ant colony systems. Int. J. Knowl. Web Intell. **4**(4), 300–313 (2013)
4. Dorigo, M., Stützle, T.: Ant Colony Optimization. Bradford Company (2004)
5. Blum, C.: Ant colony optimization: introduction and recent trends. Phys. Life Rev. **2**(4), 353–373 (2005)
6. Cormen, T.H., Leiserson, C.E., Rivest, R.L., Stein, C.: Introduction to Algorithms. MIT Press (2009)
7. Diestel, R.: Graph Theory. Springer (2010)

SkyCube-Tree Based Group-by Query Processing in OLAP Skyline Cubes

Hideki Sato and Takayuki Usami

Abstract *SkyCube*-tree has been developed to realize efficient range query processing for Skyline Cube (SC). Apart from range queries, this paper demonstrates that *SkyCube*-tree can be made use of efficient group-by query processing, though it is originally designed to realize efficient range query processing. Since a group-by query for SC includes the entire dataset as its processing range, the query processing time is potentially large. From the experimental evaluation, the followings are clarified:

- The size of *SkyCube*-trees is sufficiently allowable, since it is at most 2.5 times as large as that of materialized view.
- The time of *SkyCube*-tree based sequential processing is nearly equal to that of materialized view based one, regardless of its dedication to range query processing.
- The time of *SkyCube*-tree based parallel processing is comparatively small and stable. Even though *cell-granularity* is over 80 %, its processing time is around 10 % of that of materialized view based one.

Keywords Aggregate function · *Skyline* operator · Skyline cube · Group-by query · *SkyCube*-tree · GPGPU

1 Introduction

In Data WareHouse (DWH) environments, On-Line Analytical Processing (OLAP) [1] tools have been extensively used for a wide range of decision-support applications. These tools are built upon Data Cube [2], a collection of data cuboids which

H. Sato (✉)
School of Informatics, Daido University, 10-3 Takiharu-cho, Minami-ku, Nagoya
457-8530, Japan
e-mail: hsato@daido-it.ac.jp

T. Usami
System Development Department, Soft Valley Corporation, Hachioji, Japan

G. De Pietro et al. (eds.), *Intelligent Interactive Multimedia Systems and Services 2016*, Smart Innovation, Systems and Technologies 55,
DOI 10.1007/978-3-319-39345-2_26

(a)

name	year	team	position	hits	HR	SB
a	2000	X	C	96	11	12
b	2000	X	1B	89	5	13
c	2000	Y	C	70	19	10
d	2000	Y	2B	63	20	5
e	2001	X	1B	101	3	8
f	2001	Y	C	77	6	4
g	2001	Z	C	101	1	10
h	2001	X	2B	122	14	15
i	2001	Y	1B	91	8	2
k	2001	Z	2B	80	10	6
l	2002	X	C	58	2	11
m	2002	X	1B	70	4	9
n	2002	Z	2B	85	6	13

(b)

<table>
<tr><th>name</th><th>year</th><th>team</th><th>hits</th><th>HR</th></tr>
<tr><td>a</td><td>2000</td><td>X</td><td>96</td><td>11</td></tr>
<tr><td>c</td><td rowspan="2">2000</td><td rowspan="2">Y</td><td>70</td><td>19</td></tr>
<tr><td>d</td><td>63</td><td>20</td></tr>
<tr><td>h</td><td>2001</td><td>X</td><td>122</td><td>14</td></tr>
<tr><td>i</td><td>2001</td><td>Y</td><td>91</td><td>8</td></tr>
<tr><td>g</td><td rowspan="2">2001</td><td rowspan="2">Z</td><td>101</td><td>1</td></tr>
<tr><td>k</td><td>80</td><td>10</td></tr>
<tr><td>m</td><td>2002</td><td>X</td><td>70</td><td>4</td></tr>
<tr><td>n</td><td>2002</td><td>Z</td><td>85</td><td>6</td></tr>
</table>

Fig. 1 Skyline Cube (SC) and group-by query result. **a** Skyline Cube. **b** Outstanding players

are implemented with Multi-Dimensional DataBase (MDDB) [3]. In MDDB to represent a data cuboid, tuples are partitioned into different cells based on the values of their dimension attributes, where an aggregate function (e.g., SUM) is applied to a measure attribute (e.g., sales) for the tuples partitioned in each cell and the resulting value is assigned to the cell.

Skyline Cube (SC) [4, 5] has been proposed as an extension of Data Cube by using the *skyline* operator [6] to aggregate tuples in a cell instead of the conventional aggregate functions. The *skyline* operator has been received considerable attention in the database and data mining fields. Given a set S of skyline attributes, a tuple t is said to *dominate* another tuple t', denoted by $t \succ_S t'$, if Eq. (1) is satisfied. It is assumed that smaller values are preferable over larger ones. Here, $t[A_i]$ is used to represent the value of the attribute A_i of the tuple t. Given a set D of tuples, Eq. (2) defines the *skyline* operator Ψ on D.

$$(\exists A_i \in S, t[A_i] < t'[A_i]) \wedge (\forall A_j \in S, t[A_j] \leq t'[A_j]) \quad (1)$$

$$\Psi(D, S) = \{t \in D | \nexists t' \in D, t' \succ_S t\} \quad (2)$$

Figure 1a shows an example to explain the SC concept. Each tuple has attributes of *Baseball batters' statistics database*. Regarding *hits*, *HR*(*HomeRun*), and *SB* (*StolenBase*), higher scores are supposed to be preferable over lower ones. If we want to find the outstanding players for each combination of *team* and *year*, a group-by query with dimension attributes $G = \{year, team\}$ and skyline attributes $S = \{hits, HR\}$ can be issued toward the table of Fig. 1a. The table of Fig. 1b is the query result.

In addition to the query type mentioned in the above (See Fig. 2b), range query as another type (See Fig. 2c) can be available for SC, where the *skyline* operator is applied to aggregate tuples satisfying a range condition regarding dimension

(a)

Skyline Cube
F={G_1, G_2, … , S_1, S_2, … , O_1, O_2, …}
dimension attribute $G_{pi} \in (G=\{G_1, G_2, \ldots\})$
skyline attribute $S_{qj} \in (S=\{S_1, S_2, \ldots\})$
other attribute $O_{rk} \in (O=\{O_1, O_2, \ldots\})$
skyline preference $m_{qj} \in \{MAX, MIN\}$

(b)

select G_{p1}, G_{p2}, … ,
  S_{q1}, S_{q2}, … , O_{r1}, O_{r2}, …
from F
group by G_{p1}, G_{p2}, …
skyline of S_{q1} m_1, S_{q2} m_2, …

(c)

select S_{q1}, S_{q2}, … , O_{r1}, O_{r2}, …
from F
where $l_1 \leqq G_{p1} \leqq h_1$
  and $l_2 \leqq G_{p2} \leqq h_2$
  ....
skyline of S_{q1} m_1, S_{q2} m_2, …

Fig. 2 Skyline Cube (SC) and SQL statements toward SC. **a** Schema of Skyline Cube. **b** Group-by query. **c** Range query

attributes.[1] Although a range query for SC provides users with flexible querying function, it brings much burdens upon a query processing system. To overcome the difficulty, the work [7] has proposed *SkyCube*-tree, *R*-tree [8] like hierarchical index structure, to implement an efficient range query processing system. Apart from range queries, this paper demonstrates that *SkyCube*-tree can be made use of efficient group-by query processing, though it is originally designed to realize efficient range query processing.

The rest of this paper is organized as follows. Section 2 presents the *SkyCube*-tree. Section 3 discusses group-by query processing methods based on *SkyCube*-tree. Section 4 experimentally evaluates group-by query processing methods. Section 5 mentions the related work. Finally, Sect. 6 concludes the paper.

2 *SkyCube*-Tree

SkyCube-tree is organized in a *R*-tree like hierarchical structure to realize efficient range query processing for SC. In *SkyCube*-tree, a cell of all dimension attributes is treated as a point in a multi-dimensional space. *Skyline* operation result over tuples belonging to a cell is associated to it. In order to process a range query, points located inside the rectangle specified by a range condition are searched first, then *skyline* operation results associated to the found points are used to compute the query result. For a dimension attribute unreferenced in a query, a range is set between $-\infty$ and $+\infty$.

In order to design *SkyCube*-tree, it is required to consider that the *skyline* operator is *holistic* in nature. That is to say, Eq. (3) holds on a set D of tuples regarding a set S of skyline attributes. In other words, the skyline set regarding a subset $S'(\subseteq S)$ cannot be necessarily derived from the skyline set regarding S. *SkyCube*-tree makes use of the *extended skyline* operator [9] to cope with the problem. The *extended skyline* operator is based on *strongly dominance* relation defined in Eq. (4), which strengthens *dominance* relation defined in Eq. (1). According to Eq. (4), a tuple t cannot *dominate* another tuple t' unless all the skyline attribute values of the former

[1]It is possible to impose a range condition on a categorical dimension attribute by converting its domain into a numerical one.

are preferable over those of the latter. t is said to *strongly dominate* t' (denoted by $t \succ_S^+ t'$), if t and t' satisfy Eq. (4). Based on *strongly dominance* relation, Eq. (5) defines the *extended skyline* operator Ψ^+ on D regarding S. Equation (6) holds on D regarding S. By using Ψ^+, the skyline set regarding a subset $S'(\subseteq S)$ is derivable from the skyline set regarding S.

$$\Psi(D, S') \supseteq \Psi(\Psi(D, S), S') \; for \; S' \subseteq S \tag{3}$$

$$(\forall A_i \in S, \; t[A_i] < t'[A_i]) \tag{4}$$

$$\Psi^+(D, S) = \{t \in D \mid \nexists \, t' \in D, \; t' \succ_S^+ t\} \tag{5}$$

$$\Psi(D, S') = \Psi(\Psi^+(D, S), S') \; for \; S' \subseteq S \tag{6}$$

Figure 3 is the storage structure of *SkyCube*-tree. An *extended skyline* set is variable-length and would occupy a large space region. Therefore, an *extended skyline* set is stored in *extended skyline set space*, which is separated from the body of a *SkyCube*-tree and organized as a sequential file. An *extended skyline* set is pointed from *SkyCube*-tree with its address.[2] A leaf node of a *SkyCube*-tree stores a list of point records, each of which consists of *point coordinates* and *extended skyline set* address. It is noted that a leaf node links to its next leaf node via *nextLink*. An internal node of *SkyCube*-tree stores a list of Minimum Bounding Rectangle (MBR) records, each of which consists of *child node address*, *MBR coordinates*, and *extended skyline set* address. MBR is a minimum rectangular region containing a set of points (or MBRs) which its child node stores and is specified by a pair of its left-bottom point coordinates and its right-upper point coordinates.

R-tree, whose structure is similar to *SkyCube*-tree, obtains an efficient retrieval performance, by balancing its height on insertion/deletion of data. As a result, *R*-tree is not necessarily superior in space efficiency because of an increasing number of nodes. Since a query covering a huge area of SC can be easily expressed, it is inevitably required to reduce the number of I/O operations for accessing to *SkyCube*-tree. Therefore, *SkyCube*-tree should be made compact to reduce the number of I/O operations. While a sizable quantity of data are added to DWH at a certain interval, dynamic insertion/deletion of data is not generally performed until a next interval. From this point of view, *SkyCube*-tree is constructed in a bottom-up way as follows.

(1) A set of tuples is arranged in the order following a space-filling curve[3] for a multi-dimensional space based on all the dimension attributes.
(2) Following the curve, tuples with each combination of dimension attribute values stand in line and can be put together. For these tuples, *extended skyline* set is computed[4] and stored in *extended skyline set space*. Then, a point record is

[2] The size of address data is 8 bytes.

[3] *Z*-curve [10] is employed for the work.

[4] Block Nested Loop (BNL) algorithm [11] is employed for the work.

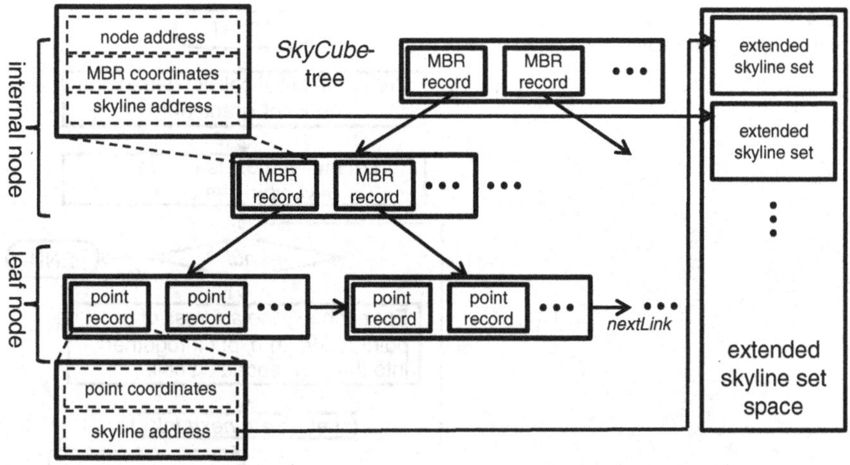

Fig. 3 Storage structure of *SkyCube*-tree

created and inserted into the current leaf node. If an enough space is not left in the node for storing the point record, a new leaf one is created and made current. The new one is referenced from the previous one via *nextLink*.

(3) Following the list of nodes created, MBR is calculated for all the points (or MBRs) stored in each node. Also, *extended skyline* set is computed based on all the *extended skyline* sets stored in the node and stored in *extended skyline set space*. Then, a MBR record is created and inserted into the current internal node. If an enough space is not left in the node for storing the MBR record, a new internal one is created and made current.

(4) If the number of nodes created is one, this unique one is made the root of the *SkyCube*-tree and construction process is finished. Otherwise, **(3)** is repeated.

3 Group-By Query Processing

Regarding a non-empty subset of dimension attributes, a non-empty subset of skyline attributes, and an arbitrary granule of cells, the group-by query result can be computed by accessing to *SkyCube*-tree. Figure 4 shows the flow chart of group-by query processing. All the leaf nodes of *SkyCube*-tree can be traversed rightward via *nextLink* from the leftmost one, which can be accessed from the root node. Each *skyline address* of a point record stored in a leaf node is put together into a corresponding cell. As soon as all the *skyline addresses* of a cell are collected, the *skyline* set of the cell is computed based on the *skyline addresses*, each of which points an *extended skyline* set in the *extended skyline set space*.

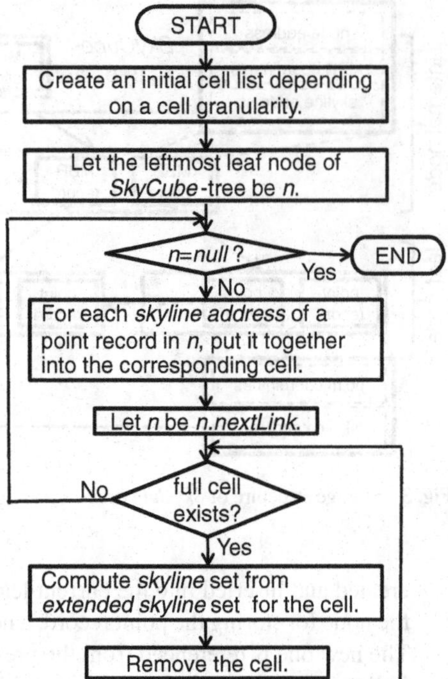

Fig. 4 Flow chart of group-by query processing

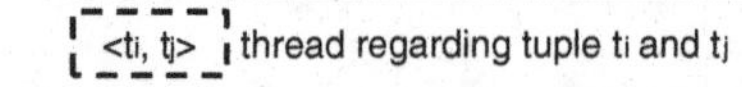

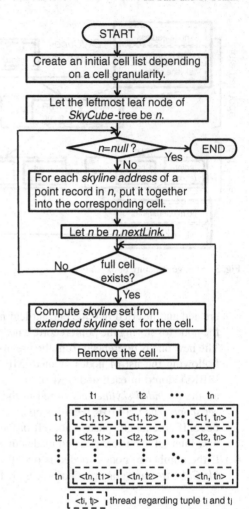

Fig. 5 Threads for parallel skyline computation

Like range query processing for SC [7], parallel processing scheme can be applied to the process "Compute *skyline* set from *extended skyline* set for the cell" shown in Fig. 4. To this end, GPGPU (General-Purpose computing on Graphics Processing Units) [12] has been used to implement parallel processing, where massive light-weight threads can be dealt with comparatively easily.

Let U be *extended skyline* set. For tuple $t(\in U)$, whether t belongs to a *skyline* set or not can be judged by checking if $t'(\in U - \{t\})$ *dominates* t (See Eq. (2)). Furthermore, *dominance* relation over t and t' can be checked independently from other combinations. Based on the property, parallel processing can be executed for the process "Compute *skyline* set from *extended skyline* set for the cell". Figure 5 shows a set of threads to be executed in parallel with GPU. For each combination of $t_i, t_j(\in U)$, a thread is dedicated to checking *dominance* relation over t_i and t_j. If $t_j \succ_S t_i$ holds, t_i is removed from the final *skyline* set. Although *dominance* relation over t_k and $t_k(k = 1, \ldots, n)$ is self-evident, a thread is also generated for simplicity.

4 Experiments

In this section, *SkyCube*-tree and group-by query processing methods are experimentally evaluated. First, the settings of experiments are described. Then, the size of *SkyCube*-trees is presented. Finally, the processing time for group-by queries is presented.

4.1 Experimental Settings

The experiments are conducted on an Intel Core i5-3550 3.3 GHz PC with 8 GB memory and a NVIDIA Quadro K4000 with 768cores and 3 GB memory. C language is used to implement programs. Also, CUDA [12], designed by NVIDIA, is employed as GPGPU for working with C.

A synthetic dataset is generated to carry out the experiments. It contains 100,000 tuples, with a total of 14 attributes $(a_1, a_2, \ldots, a_{14})$: 5 attributes $a_1, \ldots, a_5$ are for dimension, and 9 attributes $a_6, \ldots, a_{14}$ are for skyline. The domain size of each dimension attribute is 10, where attribute values are generated from the uniform distribution. On the other hand, the domain size of each skyline attribute is 1,000. Specifically, a_6, a_7, a_8 are independently generated from Gaussian distribution, a_9, a_{10}, a_{11} are correlated with a_6, and a_{12}, a_{13}, a_{14} are anti-correlated with a_6.

In the experiments, *SkyCube*-tree and the proposed query processing methods are compared with the materialized view based one. A materialized view is a cuboid which is precomputed and stored for answering queries. When a query is issued, it can be processed by using a materialized view whose attribute set covers dimension attributes and skyline attributes referenced in the query. For the experimental comparison, a materialized view organized in a sequential file is constructed. However, it is noted that the materialized view is different from that of the work [4, 5], because the former maintains *extended skyline* sets.

4.2 Storage Size of Skyline Cube

Figure 6 shows the storage size of the *SkyCube*-tree and the materialized view, each of which is constructed for the synthetic dataset. In the experiments, the page size is set to 4KB and the number of nodes in *SkyCube*-tree is 253. The storage structure of *SkyCube*-tree consists of *tree body* and *extended skyline set space*. The size of the

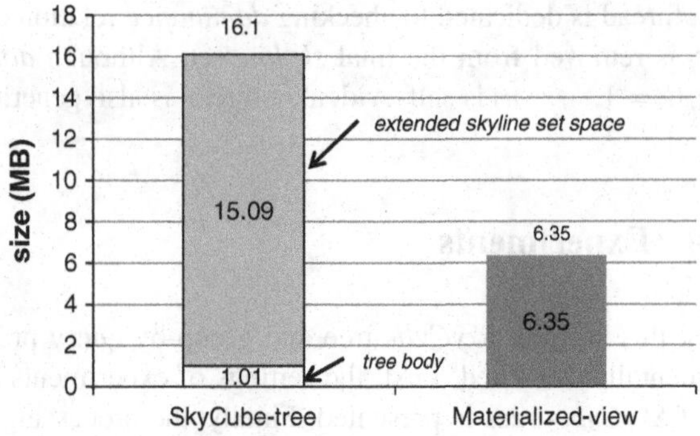

Fig. 6 Storage size of *SkyCube*-tree and materialized view

SkyCube-tree is 16.1 MB, whose *tree body* occupies 1.01MB and whose *extended skyline set space* occupies 15.09MB. Meanwhile, the size of the materialized view is 6.35MB. Given a dimension attribute set G and a skyline attribute set S, the number of possible cuboids regarding a non-empty subset of G and a non-empty subset of S is $(2^{|G|} - 1)\ (2^{|S|} - 1)$. The materialized view is constructed regarding all the dimension attributes and all the skyline attributes, which corresponds to the largest cuboid among all the possible ones.

The total size of the *SkyCube*-tree is 2.5 times as large as that of the materialized view. Although a unique *SkyCube*-tree is enough to make answers of all the queries, the number of materialized views to be potentially needed is $(2^{|G|} - 1)\ (2^{|S|} - 1)$, as is mentioned in the above. Therefore, the storage size of *SkyCube*-trees is sufficiently allowable. The main difference of storage size between the *SkyCube*-tree and the materialized view constructed resides in the back portion of *extended skyline set space*, where a family of *extended skyline set* referenced from internal nodes of the *SkyCube*-tree are stored. These *extended skyline sets* are prepared for efficiently processing a range query whose size is that of MBR corresponding to an internal node or more. It is noted that the portion is not used for group-by query processing and extra I/O times are not needed consequently.

4.3 Group-By Query Processing Time

Regarding group-by query processing time, *SkyCube* based sequential processing (SG-) method, *SkyCube* based parallel processing (PG-) method, and materialized view based sequential processing (MG-) method are compared. Given the number of dimension attributes (*dimension#*), the number of skyline attributes (*skyline#*), and *cell-granularity*, 100 queries are generated and query processing times are averaged. For each query, a subset of dimension attributes and a subset of skyline attributes are randomly chosen. Each dimensional size of a cell is set to the nth root of *cell-granularity*, if *dimension#* is n.

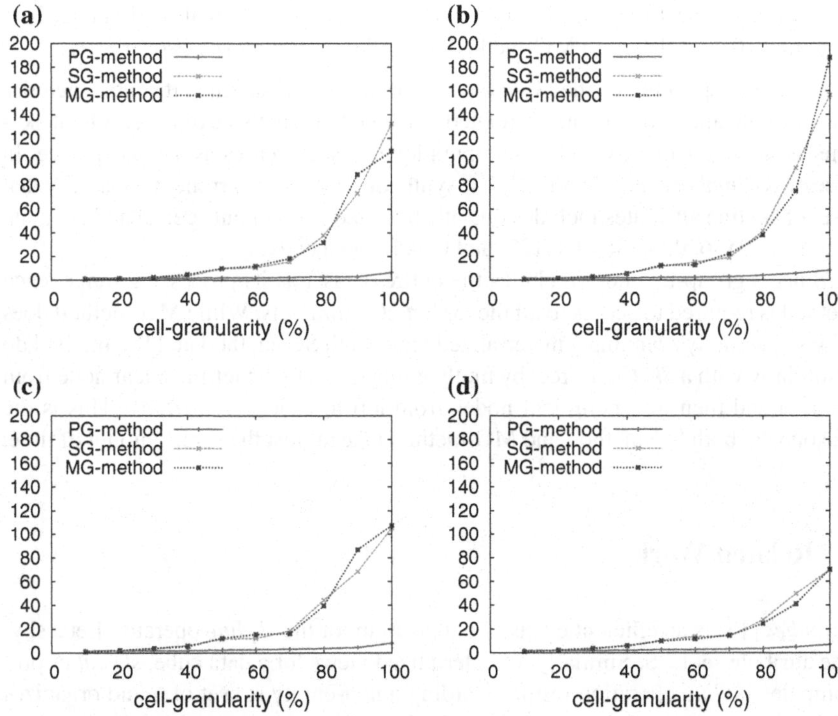

Fig. 7 Group-by query processing time for synthetic dataset (seconds). **a** *dimension#* = 3, *skyline#* = 2. **b** *dimension#* = 3, *skyline#* = 3. **c** *dimension#* = 3, *skyline#* = 4. **d** *dimension#* = 3, *skyline#* = 5

For MG-method, the materialized view is firstly constructed regarding all the dimension attributes and all the skyline attributes. To make the experiments fair, additional materialized views are constructed regarding 3 or more dimension attributes and 2 or more skyline attributes, until the total size of materialized views becomes not less than that of the corresponding *SkyCube*-tree. The number and each dimension (skyline) attribute are randomly chosen. MG-method processes a query by using the materialized view whose attributes set minimally covers dimension attributes and skyline attributes referenced in the query.

Figure 7 shows the experimental results where *dimension#* is set at 3, *skyline#* is varied from 2 to 5, and *cell-granularity* is varied from 10 to 100 % with increments of 10 %. The tendency of each graph is roughly the same as follows:

- The processing time with SG-method is nearly equal to that with MG-method, though *SkyCube*-tree is originally designed for efficient range query processing.
- The graphs rise to the right. The larger *cell-granularity* becomes, the more processing time is needed.

- The processing time with PG-method[5] is comparatively small and stable. It is around 10 % of that with MG-method, even if *cell-granularity* is over 80 %.

Generally speaking of the *skyline* operator, if a set of skyline attributes includes one attribute and another anti-correlated attribute both, the size of the resultant skyline set increases rapidly and its computational cost increases as a consequence. In case of *skyline#* between 2 and 5 for the synthetic dataset, occurrence probability that a set of skyline attributes includes one attribute and another anti-correlated attribute both is 0.30556, 0.64286, 0.84127, and 0.94444 respectively.

Since a group-by query includes the entire dataset as its processing range, each method is required to access to all the *extended skyline* sets. While MG-method does it by sequentially scanning a materialized view, both SG-method and PG-method do it similarly with a *SkyCube*-tree, by firstly going down to the leftmost leaf node from the root and then traversing leaf nodes from left to right via *nextLink*. This is the reason why both SG-method and MG-method take almost the same amount of time.

5 Related Work

Skycubes [13] is an efficient evaluation algorithm for the *skyline* operator. Let a skyline attribute set be S. Similarly to materialized views for a data cube, *skycubes* precomputes *skyline* operation result regarding a non-empty subset of *S* and organizes all the operation results of distinct skyline attribute subsets in a lattice structure. When a skyline operation is requested, the corresponding query result in the lattice is returned as an answer. However, *skycubes* is simply dedicated to evaluating the *skyline* operator. Meanwhile, queries for SC must take an additional set of dimension attributes into consideration.

Range-Sum (*Range-Max*) [14] is an algorithm for processing range queries which use function SUM (MAX) to aggregate tuples belonging to each cell of a data cube. To perform query processing efficiently, the algorithms rely on respective auxiliary index structures *prefix-Sum* and *b-ary* tree. Both index structures partition each dimension attribute domain by an even interval, which leads to efficient searching of index regions relating to query processing. However, they might be inferior in space efficiency, if value distribution of a dimension domain is to some extent uneven and/or sparse.

Ag+-tree [15, 16] is a hierarchical index structure dedicated to processing range queries regarding a data cube. Similarly to *SkyCube*-tree, a unique Ag+-tree is constructed for a data cube. Let the number of dimension attributes be *n*. A node of Ag+-tree consists of *n* pages, each of which is dedicated to a corresponding attribute, and a single page possessing several kinds of aggregation values, where page is a unit of I/O operations. However, it lacks high space efficiency, while *SkyCube*-tree is compactly constructed in a bottom-up way.

[5]Note that it includes the time to transfer data between main memory of CPU and RAM memory of GPU.

As mentioned before, the work [4, 5] proposed SC as an extension of Data Cube. While materialized views are used to process group-by queries in the work, *SkyCube*-trees are used to do that in our work. Additionally, another difference resides in that range queries can be processed by using the same *SkyCube*-trees in our work [7].

6 Conclusion

Since a group-by query for SC includes the entire dataset as its processing range, the query processing time is potentially large. Regardless of the difficulty, this paper demonstrates that group-by queries for SC are processed efficiently with *SkyCube*-tree, though it is originally designed to realize efficient range query processing. While its storage size is sufficiently allowable, the processing time with SG-method is nearly equal to that with MG-method and the processing time with PG-method is comparatively small and stable. Even though *cell-granularity* is over 80 %, the processing time with PG-method is around 10 % of that with MG-method.

A hierarchy consisting of several levels of a dimension attribute is supposed to implement traditional OLAP tools. For example, {"the first half of the year", "the second half of the year"} might be one level, {"the first quarter of the year", "the second quarter of the year", …} might be another level, and the others so on for a fiscal period attribute. *Drill-up* (*Drill-down*) operator ascends (descends) a hierarchy, which makes group-by queries executed under *cell-granularity* corresponding to some level of a hierarchy. However, it is noted that *SkyCube*-tree based query processing methods are more flexible, since they can make group-by queries executed under any of *cell-granularity*.

Our future work is related to reduction of a *skyline* set which is potentially huge for a large number of skyline attributes. To this end, two kinds of operators might be available. One is the *k-dominant skyline* operator [17] and the other is the *top-k* query [18].

References

1. Codd, E.F.: Providing OLAP (On-Line Analytical Processing) to User-Analysts: An IT Mandate, Technical report. E.F.Codd and Associates (1993)
2. Gray, J., Bosworth, A., Layman, A., Pirahesh, H.: A relational aggregation operator generalizing group-by, cross-tabs and subtotals. In: Proceedings of the 12th Int'l Conference on Data Engineering, pp. 152–159 (1996)
3. Agrawal, R., Gupta, A., Sarawagi, S.: Modeling multidimensional databases, In: Proceedings of the 13th Int'l Conference on Data Engineering (1997)
4. Yiu, M.L., Lo, E., Yung, D.: Measuring the sky: on computing data cubes via skylining the measure. IEEE Trans. Knowl. Data Eng. **24**(3), 492–505 (2012)
5. Luk, M.N., Yiu, M.L., Megiddo, N., Lo, E.: Group-by skyline query processing in relational engines. In: Proceedingsc of the 18th ACM Conference on Information and Knowledge Management, pp. 1433–1436 (2009)

6. Borzsonyi, S., Kossmann, D., Stocker, K.: The skyline operator. In: Proceedings of the 17th Int'l Conference on Data Engineering, pp. 421–430 (2001)
7. Sato, H., Usami, T.: Range query processing in OLAP skyline cubes. IEEJ Trans. Electron. Inf. Syst. **136**(4) (2016). (in Japanese, to be published)
8. Guttman, A.: R-Trees: a dynamic index structure for spatial searching. In: Proceedings of the 1984 ACM SIGMOD Int'l Conference on Management of Data, pp. 45–57 (1984)
9. Vlachou, A., Doulkeridis, C., Kotidis, Y., Vazirgiannis, M.: SKYPEER: efficient subspace skyline computation over distributed data. In: Proceedings of the 23rd Int'l Conference on Data Engineering, pp. 416–425 (2007)
10. Orenstein, J., Merrett, T.H.: A class of data structures for associative searching. In: Proceedings of the 3rd ACM SIGACT-SIGMOD Symposium on Principles of Data Base Systems, pp. 181–190 (1984)
11. Papadias, D., Tao, Y., Fu, G., Seeger, B.: Progressive skyline computation in database systems. ACM Trans. Database Syst. **30**(1), 41–82 (2005)
12. Sanders, J., Kandrot, E.: CUDA by example: an introduction to general-purpose GPU programming, Addison-Wesley Professional (2010)
13. Yuan, Y., Lin, X., Liu, Q., Wang, W., Yu, J.X., Zhang, Q.: Efficient computation of the skyline cube. In: Proceedings the 2005 Int'l Conference on Very Large Data Bases, pp. 241–252 (2005)
14. Ho, C.T., Agrawal, R., Megiddo, N., Srikant, R.: Range queries in OLAP data cubes. In: Proceedings of the 1997 ACM SIGMOD Int'l Conference on Management of Data, pp. 73–88 (1997)
15. Feng, Y., Makinouchi, A.: Indexing for range-aggregation queries on large relational datasets. Int'l J. Database Theory Appl. **3**(4), 1–14 (2010)
16. Feng, Y., Makinouchi, A.: Ag+-tree: an index structure for range-aggregation queries in data warehouse environments. Int'l J. Database Theory Appl. **4**(2), 51–64 (2011)
17. Chan, C-Y., Jagadish, H.V., Tan, K-L., Tung, A.K.H., Zhang, Z.: Finding k-dominant skylines in high dimensional space. In: Proceedings of the 2006 ACM SIGMOD Int'l Conference on Management of Data, pp. 503–514 (2006)
18. Ilyas, H.F., Beskales, G., Soliman, M.A.: A survey of top-k query processing techniques in relational database systems. ACM Comput. Surv. **40**(4) (2008). Article 11

Trends in Teaching/Learning Research Through Analysis of Conference Presentation Articles

Toyohide Watanabe

Abstract Information Technology (IT) has been changing the procedural means/methods used in social activities, and this effect has influenced directly or indirectly the behaviors of human beings. Of course, in teaching/learning activities, this observation is nothing out of the ordinary. Currently, IT plays an important role in supporting the teaching and learning processes both effectively and effectually. In this paper, we investigate current research trends, examining various articles published in conferences so as to extract the features correlated with specific research interests and objectives. We discuss these research features on the basis of our "knowledge transfer scheme" learning principle.

Keywords Knowledge transfer scheme • Knowledge composition • Knowledge acquisition • Knowledge understanding • Information technology

1 Introduction

Information Technology (IT) has been changing the procedural means/methods used in social activities and is also influencing our daily lives, working styles, communication means, etc. Although in terms of historical development we can only consider the last 70 years since the appearance of the computer, the effects can be observed in a wide range of human activities, environments, social organizations, etc. In the realm of teaching/learning activities, IT has had a profound influence on the methods, means, processes, interactions, and environments used. IT has encouraged teachers/learners to free themselves from spatio-temporal constraints. The resultant ubiquitous environment has made it possible for everyone who wants to learn to do so anytime, anywhere.

In the IT-based evolution of teaching/learning activity, the starting point was to make the teaching procedure more powerful/effective by means of data processing

T. Watanabe (✉)
Nagoya Industrial Science Research Institute, Nagoya, Japan
e-mail: watanabe@nagoya-u.jp

G. De Pietro et al. (eds.), *Intelligent Interactive Multimedia Systems and Services 2016*, Smart Innovation, Systems and Technologies 55,
DOI 10.1007/978-3-319-39345-2_27

(such as CAI), and the current viewpoint is now to intelligently support processes and environments for learning as well as for the conventional teaching procedure. This trend may be dependent on the concepts of personality and virtualization in a cyber society. The powerful expansion of the Internet is one of the major factors here, since the features of the Internet can provide IT-based functionality for individual learners ubiquitously.

In this paper, we focus on articles that have been presented at various international/domestic conferences and analyze the research points from the viewpoints of investigation style, research topic, practical/experimental effect, etc. In our view, the basic platform is a knowledge transfer scheme, proposed by us, as a framework for learning [1]. With this framework, our discussion clarifies the learner's role in an IT-based environment and unifies the architectural views for teaching and learning actions.

2 Knowledge Transfer Scheme

We have already defined learning as the transfer of knowledge from the outside of the learner, such as textbooks, teaching/learning contents, and authoring resources, to in the side of the learner, such as his/her brain and notes. Under this definition, we can consider our learning mechanism, namely, knowledge composition, knowledge acquisition, and knowledge understanding (or utilization), and illustrate the knowledge transfer scheme with simple/basic actions. Figure 1 gives a basic overview of our knowledge transfer scheme.

(1) Composition transforms knowledge from a scattered structure to an aggregated one.
(2) Acquisition moves the composed knowledge to the learner's brain or notes.
(3) Understanding refines the moved knowledge with already acquired knowledge.

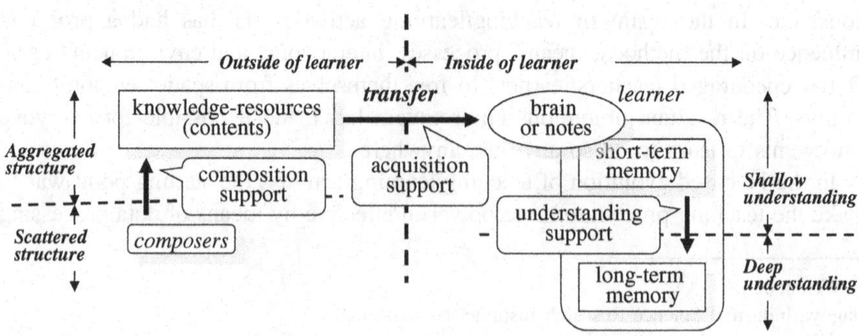

Fig. 1 Knowledge transfer scheme

These phases play important roles in making the learning procedures for each learner effectual and effective. If they can work successively, the learner will be successful in managing the knowledge in his/her learning process. Of course, the composition process is not the responsibility of the learner but rather of the teacher, the author of the textbook, etc. Functions and mechanisms that support smart learning need to be designed. Thus, the research topics we discuss in this paper concern these phases, and we also focus on the interaction/interdependency among these phases. Of course, the basic functionality will remain the domain of traditional teacher behaviors. Our concept is that we can look upon traditional teaching actions as a support means of knowledge transfer scheme. We regard these phases as keys in research topics during our discussion of the current trends of research issues.

3 Conference Articles

In this paper, we examine three conference articles:

(1) Ed-Media (World Conference on Educational Media and Technology), managed by AACE (Association for the Advancement of Computing in Education) [2]
(2) E-Learn (World Conference on E-Learning), managed by AACE [3]
(3) Domestic Annual Conference of JSiSE (Japanese Society of Information Systems in Education) [4]

3.1 Ed-Media and E-Learn

First, let us consider Ed-Media. Table 1 lists the number of articles presented at this conference in various research categories. As shown, there were 853 articles in 2008, 497 in 2011, and 408 in 2014. These numbers have been decreasing year by year: for 6 years from 2008, the decreasing ratio is 43.3 % (47.8 % when poster articles etc. are included), and in 2015 it is 35.3 % (40.9 %). The cause of this decrease is uncertain, and we are unable to properly investigate this phenomenon even when IT-based promotions/managements play important roles. In Ed-Media, the significant points are that non-traditional presentation styles (best practices sessions, virtual sessions, etc.) are introduced in addition to conventional oral sessions with two categories (full (long) and brief (short) articles). Best practices sessions focus on the effects/results/plans of case studies and large-scale experiments. Virtual sessions make it possible for even presenters who cannot physically participate in a conference to join virtually through the Internet.

Category no. 3 in Table 1, “New Roles of the Instructor & Learner”, occupies the highest ratio for all articles through 2011, 2014, and 2015, and every year the ratio is over 50 % (except 36.8 % in 2008). Each category includes various

Table 1 Trends of articles in Ed-Media

Category	2015				2014				2011			2008		
	Full	Brief	Other	Total (%)	Full	Brief	Other	Total (%)	Full	Brief	Total (%)	Full	Brief	Total (%)
1: Infrastructure	18	10	18	46 (15.3)	15	23	9	47 (12.7)	35	51	86 (17.3)	41	35	76 (8.9)
2: Tools & Content-oriented Applications	16	5	7	28 (9.3)	15	20	10	45 (12.2)	22	22	44 (8.9)	80	117	197 (23.1)
3: New Roles of the Instructor & Learner	59	38	68	165 (54.8)	85	75	58	218 (59.1)	136	176	312 (62.8)	110	204	314 (36.8)
4: Human-Interaction (HCL/CHI)	6	2	2	10 (3.3)	5	4	5	14 (3.8)	9	4	13 (2.6)	39	50	89 (10.4)
5: Cases & Projects	18	10	7	35 (11.6)	17	15	5	37 (10.0)	17	20	37 (7.4)	54	76	130 (15.3)
6: Universal Web Accessibility (Special Strand)	4	1	3	8 (2.7)	2	1	5	8 (2.2)	0	2	2 (0.4)	21	18	39 (4.6)
7: Indigenous Peoples & Technology	1	2	2	5 (1.7)	0	0	0	0 (0.0)	1	2	3 (0.6)	3	5	8 (0.9)
Subtotal of above columns				296				369			497			853
0-1: ET	1	1		2 (0.7)										
0-2: Design	1			1 (0.3)										
0-3: LD		1	1	2 (0.7)										
Subtotal of above columns	124	70	107	301	139	138	92	369	220	277	497	348	505	853
Poster				34				38						
Others				14				1						
Total of all				349				408			497			853

Table 2 Categorical keywords in Ed-Media

1: Infrastructure	4: Human-computer Interaction (CHI/HCI)
1.1: Architectures for Educational Technology Systems	4.1: Computer-Mediated Communication
1.2: Design of Distance Learning Systems	4.2: Design Principles
1.3:Distributed Learning Environments	4.3: Usability/User Studies
1.4: Methodologies for System Design	4.4: User Interface Design
1.5: Multimedia/Hypermedia Systems	5: Cases & Projects
1.6: WWW-based Course-Support Systems	5.1: Corporate
2: Tools & Content-oriented Application	5.2: Country-Specific Developments
2.1: Agents	5.3: Exemplary Projects
2.2: Authoring Tools	5.4: Institution-Specific Cases
2.3: Evaluation of Impact	5.5: Virtual Universities
2.4: Groupware & WWW-based Tools	6: Universal Web Accessibility (Special Strand)
2.5: Interactive Learning Environments	6.1: Emerging technologies & Accessibility
2.6: Multimedia/Hypermedia Applications	6.2: Infrastructure, Technology & Techniques
2.7: Research Perspectives	6.3: International Challenges
2.8: Virtual Reality	6.4: New Roles for Teachers/Learners
2.9: WWW-based Course Sites & Learning Resources	6.5: Other: Research, Library Issues, etc.
3: New Roles of the Instructor & Learner	6.6: Policy and Law
3.1: Constructivist Perspectives	6.7: Site Management Considerations
3.2: Cooperative/Collaborative Learning	7: Indigenous Peoples & Technology
3.3: Implementation Experiences	7.1: Indigenous Peoples & Technology Issues/Applications
3.4: Improving Classroom Teaching	
3.5: Instructor Networking	
3.6: Instructor Training and Support	
3.7: Pedagogical Issues	
3.8: Teaching/Learning Strategies	

individual keywords, as shown in Table 2. This conference (Ed-Media) focuses mainly on topics related to education, as reflected by the "educational media and technology" in the conference name, and the media, tools/systems, methods, and practical studies/performances for teaching and teachers are investigated. The idea is to show how the functionality and mechanisms in IT-based tools/systems help support teaching activities and teachers' pedagogical means/methods.

E-Learn, which was also organized by AACE, focuses on the learning and learner. Table 3 shows the categorical features of these topics. When we compare the categories in Table 3 with those in Table 2, the difference is clearly recognizable. E-Learn focuses on practical/functional problems in the learning process, learning efficiency, and learning support. Table 4 shows the number of articles in

Table 3 Categories in E-Learn

No	Code	Topics	No	Code	SIG Topics
1	CD	Content Development	1	ELEARN-DEV	E-Learning in Developing Countries (Opportunities, Trends, Challenges, and
2	EVAL	Evaluation			
3	IMPL	Implementation Examples & Issues			Success Stories
4	INDE	Instructional Design	2	ELEARN-TREND	Innovations (Social Learning, Mobile, Augmented etc.)
5	PI	Policy Issues			
6	RES	Research	3	ELEARN-DESIGN	Designing, Developing, and Assessing E-Learning
7	SOC	Social & Cultural Issues			
8	STD	Standards & Interoperability			
9	TOOL	Tools & Systems			
10	OTH	Other			

Table 4 Number of articles in E-Learn 2015

	Virtual	Full	Brief	Roundtable	Other	Total	Ratio
CD	5	20	59	4	1	89	22.3
EVAL	4	13	18	1		36	9.0
IMPL	9	23	69	14	4	119	29.8
INDE	9	6	24	7		46	11.5
PI		2	2			4	1.0
RES	4	8	18	1	2	33	8.3
SOC	1	1	5	2		9	2.3
STD		1				1	0.3
TOOL	1	3	7	2		13	3.3
OTH	8	2	10	2		22	5.5
Subtotal	41	79	212	33	7	372	
ELEARN-DEV	1	2	1			4	1.0
ELEARN-TREND		1	6	1		8	2.0
ELEARN-DESIGN		6	7	2		15	3.8
Total	42	88	226	36	7	399	

(Note) Number of poster presentations: 19

each category, with the highest ratio occupied by Implementation Examples and Issues (IMPL). There are more articles in E-Learn2015 than in Ed-Media2015. The focus of E-Learn is learning while that of Ed-Media is education (or teaching), and as such, the categories in E-Learn concentrate on industrial criteria while those in Ed-Media relate to supporting-based functionality or system-oriented mechanisms. It is safe to say that the topics in E-Learn are dependent on the system-oriented feature and those in Ed-Media relate to human activities.

3.2 JSiSE

Next, we refer to the domestic annual conference JSiSE in Japan. Here, we discuss the features of each article in addition to the trends for interest points. Table 5 shows the categories of articles in the JSiSE2015 conference. These categories completely separate the development of IT-based functionality/mechanisms from the teaching/learning performance. The annual JSiSE conference combines the features of Ed-Media, which focuses mainly on teaching/learning performance, with those of E-Learn, which are based globally on IT-based functionality/mechanisms, with respect to the characteristics of “support” and “technology”.

(a) ***Presentation category***

Table 6 shows the affiliation of the first author for 212 articles in 2013 and 235 in 2014. The ratio by which authors belong to universities is 89.2 % in 2013 and 89.8 % in 2014. The highest ratio is one of features attended inherently with this academic association in the teaching/learning domains. Table 7 shows to which field the articles are related. Here, we define three domains:

(1) Information engineering, which is the viewpoint focused on the advanced development or effective management of IT-based tools/functions with respect to teaching/learning.
(2) Cognition, which is the viewpoint concentrated on thinking/recognition processes or the ability related to such processes.
(3) Education, which is the viewpoint dependent on the behaviors of teachers/learners with respect to performance, observation, evaluation, investigation, etc.

There are more articles in the education domain than in the information engineering domain. Specifically, the ratio of the information engineering domain is 44.1 % and that of the education domain is 52.1 %. Of course, the same speaker may present two or more articles.

(b) ***Article research view***

Table 8 shows the total number of technical key terms used in each article. Of course, one article may contain several terms at once, while another may not include even one term. These terms are extracted from the title, abstract, keywords,

Table 5 Categories in JSiSE

	Category	Clue terms
Support	Design	Teaching design, Instructional design
		Learning environment design
	Teaching/Learning methods	Distance of learning
		Blended learning
		Cooperative teaching
		Collaborative learning
		Active learning
	Evaluation	Analysis of learner's features and behavior
		Learning evaluation, Assessment
	Special support teaching	HRD, Life learning
		University
		Elementary/Middle/High school
		FD
	Course-oriented teaching	Programming
		Language
		Information
		Special support
		Skill
		Medical, Nurse, and Hospital
		Information literacy
Technology	ICT utilization	Multi-media utilization
		Social-media utilization
		Device utilization
	Development/Management	Platform
		Infrastructure
		Authoring-support
	Advanced learning support technology	Intelligent learning support system
		Analysis
		Modelling
		Interface

and article body. Terminals such as mobile devices (tablets, smartphones, etc.) are currently popular learning media, and the Japanese government is promoting their use in primary/middle school classrooms as part of a pedagogical policy to establish an IT-based environment. The composition issue of learning contents is also considered an important project and is strongly supported by the Japanese government. Other key terms include Learning Management System (LMS) and portfolio; these terms indicate that the teaching/learning management plays a basic and effective role in managing/maintaining the learning progress of each learner successively,

Table 6 Affiliation of first author

	2013	2014	Total
University	181	209	390
College of technology in Japan	11	6	17
Junior college	8	11	19
Industry	2	2	4
High school		2	2
Middle school		1	1
Special school		1	1
Other	10	3	13
Total	212	235	447

Table 7 Research domain

		2013	2014	Total
1	Information engineering	96	101	197
2	Cognition	4	13	17
3	Education	112	121	233
	Total	212	235	447

Table 8 Key terms in articles

		2013	2014	Total
1	Terminal (Smartphone, Tablet, etc.)	27	26	53
2	Contents	24	16	40
3	LMS	16	15	31
4	Collaborative learning	14	12	26
5	Skill	9	14	23
5	Portfolio	13	10	23
7	Problem composition	4	13	17
7	Evaluation (Peer evaluation, Self evaluation)	8	9	17
7	Game	9	8	17
7	VR/AR	11	6	17
11	Presentation	6	10	16
12	Cooperation	4	9	13
13	Instructional design	3	8	11
13	PBL	8	3	11
15	Carrier	6	3	9
16	Distance	3	5	8
17	Mining	3	3	6
18	Flip teaching	2	3	5
18	Rubric	2	3	5

ultimately providing a successful teaching/learning strategy suitable for individual learners.

(c) ***Article description point***

Table 9 shows the description position of each article with nine types of classification. As shown, the categorical position of an article is not always consistent with the externalized representation of the authors. For example, even if the authors describe an article as addressing the experimental results, we may look upon the article as an "enforcement" if it describes only the experimental process without any detail analyses or considerations. The classification "enforcement" accounts for 20.6 % of all articles in Table 9 and refers to articles that report the results and situation of projects, experiences, and trials. The classification "implementation" occupies 19.3 %. The articles in the education domain are mainly composed on the basis of views for "enforcement", "consideration", and "survey" while those in the information engineering domain discuss "implementation", "proposal", and "management" with a view to making the functionality.

Table 9 was chiefly arranged on the basis of description phrases from the title, abstract, and introduction in the article body. Additionally, we analyzed the features of these articles by ourselves (i.e., without input from the authors), with the results shown in Table 10. Table 11 explains classification features in Table 10. In Table 10, the classifications are divided into more detailed fields than in Table 9. Table 10 is organized with domain-specific terms while Table 9 is arranged basically with description phrases of the authors from the title, abstract, and introduction. Table 10 is arranged in accordance with our own judgment, target objects, and description objectives gleaned from phrases in the article body.

Table 9 Description points of authors

		2013	2014	Total	Note
1	Enforcement	47	45	92	Perform plans in practice
2	Implementation	40	47	87	Develop IT-based functions/tools/systems
3	Consideration	44	26	70	Estimate next direction after evaluation or with new fact
4	Proposal	11	49	60	Propose new trial/plan
5	Survey	21	32	53	Investigate results, collected by questionnaire, interview, etc.
6	Experiment	18	15	33	Perform plan for verification, discovery, refinement, etc.
7	Management	14	8	22	Manage systems/works for teaching/learning
8	Composition	10	8	18	Compose teaching/learning contents resources
9	Analysis	7	5	12	Analyze existing/newly acquired information
		212	235	447	

Table 10 Classification features of articles

		2013				2014				Total
		inf.	cog.	edu.		inf.	cog.	edu.		
1	Function development	38		4	42	25		9	34	76
2	Enforcement report	10		34	44	7		24	31	75
3	Method proposal	12	1	6	19	6	10	5	21	40
4	Contents composition	6		15	21	4	1	14	19	40
5	Analysis report	6	1	18	25	5		8	13	38
6	Survey report		1	12	13			25	25	38
7	Teaching method	2		10	12	2	1	15	18	30
8	Skill training	5			5	15			15	20
9	Processing technology	2		1	3	13		2	15	18
10	Teaching support	7		3	10	6		1	7	17
11	Learning management	1		2	3	6		3	9	12
12	Terminal utilization			1	1			9	9	10
13	Learning environment	3			3	4		3	7	10
14	Experiment result	3	1	2	6				0	6
15	Learning functionality	1			1	4		1	5	6
16	Interaction support			2	2		3		3	5
17	Learning support	1			1	1			1	2
18	Learning method				0	1			1	1
19	Industry-university cooperation				0			1	1	1
				1	1			1	1	2
	Total	97	4	111	212	99	15	121	235	447

Table 11 Explanation of classification features in Table 10

	Classification feature	Note
1	Function development	Develop functions, tools and systems
2	Enforcement report	Describe performed plans
3	Method proposal	Propose new plans/ideas
4	Contents composition	Compose teaching/learning-contents resources
5	Analysis report	Analyze phenomenon from experiments and investigation
6	Survey report	Describe results, collected by questionnaire, interview, etc.
7	Teaching method	Propose teaching means, methods, and procedures
8	Skill training	Acquire special skill
9	Processing technology	Design and develop IT-based functions and mechanisms

(continued)

Table 11 (continued)

	Classification feature	Note
10	Teaching support	Support teaching process
11	Learning management	Manage and maintain learning process
12	Terminal utilization	Utilize newly developed devices and tools
13	Learning environment	Develop effective learning environment
14	Experiment result	Describe results, performed and arranged in experiment
15	Learning functionality	Research learning functions and mechanisms
16	Interaction support	Develop interface between human beings and systems
17	Learning support	Support learning process
18	Learning method	Propose learning means, methods and procedures
19	Industry-university cooperation	Cooperate and interact between industries and universities

The projects that have been financially supported by official organizations are arranged owing to the classification in Table 10. "Function development", "method proposal", "skill training", "teaching support", and "processing technology" are accepted as the most appreciated projects in the information engineering domain, while "survey report", "teaching method", and "contents composition" received official funds in the education domain. Table 12 shows the ratio of fund-supported

Table 12 Financially supported articles

		2013				2014				Total
		inf.	cog.	edu.		inf.	cog.	edu.		
1	Function development	14		3	17	11		4	15	32
2	Method proposal	8	1	2	11	5	2	2	9	20
3	Enforcement report	2		4	6	3		3	6	12
4	Survey report			4	4			7	7	11
5	Analysis report	1		7	8	2			2	10
6	Teaching method			6	6	1		3	4	10
7	Contents composition			4	4			5	5	9
8	Skill training	2			2	5			5	7
9	Teaching support	4			4	2			2	6
10	Learning environment	2			2	1		2	3	5
11	Processing technology	1			1	3			3	4
12	Experiment result	2	1	1	3				0	3
13	Terminal utilization				0			3	3	3
14	Learning management	1			1				0	1
15	Interaction support			1	1				0	1
16	Learning functionality	1			1				0	1
		38	2	31	71	33	2	29	64	135

Table 13 Ratio of funded articles

		Ratio (%)	Note
1	Method proposal	50	= 20/40
1	Learning environment	50	= 5/10
1	Experiment result	50	= 3/6
4	Function development	42	= 32/76
5	Skill training	35	= 7/20
5	Teaching support	35	= 6/17
7	Teaching methods	33	= 10/30
8	Terminal utilization	30	= 3/10
9	Survey report	29	= 11/38
10	Analysis report	26	= 10/38
11	Contents composition	23	= 9/40
12	Processing technology	22	= 4/18
13	Interaction support	20	= 1/5
14	Learning functionality	17	= 1/6
15	Enforcement report	16	= 12/75
16	Learning management	8	= 1/12
17	Learning support	0	= 0/2
17	Learning method	0	= 0/1
17	Industry-university collaboration	0	= 0/1
	Other	0	= 0/2

projects for all articles in descending order. The highest supported research areas are “method proposal”, “learning environment”, and “experiment result”, while “enforcement report”, “learning management”, and “learning support” tend not to be supported. Table 13 shows the ratio of funded articles in ascending order.

4 Consideration

Our knowledge transfer scheme not only represents the movement of knowledge from the outside of the learner to the inside but also indicates that learners should to make use of acquired knowledge in applicable cases. However, this requirement is not always satisfied in many trials. Key terms such as “problem composition” and “evaluation” may indicate topics that can lead to knowledge understanding. The teaching/learning activity feature is not enough in itself for one trial; “cooperation”, “continuousness”, “sustainability”, and others must also be established. For this viewpoint, the management and maintenance approach, which has links to several situations, becomes increasingly important. We arrange the current research issues from such viewpoint:

(1) unified management of various learning situations (e.g., LMS, portfolio);
(2) integration/systemization of contents (e.g., CMS);

(3) cooperation among universities (e.g., communization of system management, development and sharing of open resources, etc.); and
(4) cooperation/coordination among industries and universities (e.g., practical usability, refinement of functions and contents, etc.).

We feel it is necessary to support and use the environment in addition to developing systems (functions, tools, contents, etc.) and using tools.

5 Conclusion

We extracted the features from articles presented at two international conferences (Ed-Media and E-Learn) and one domestic conference (JSiSE) and investigated the current trends. Of course, this paper is limited in some sense because the behavior of human beings is a vast factor that would take several articles to discuss. Moreover, it is not easy to estimate and evaluate phenomena from behavior.

As future work, we intend to discuss how to extend the successive scenario for design, implementation, and utilization, or how they can be extended. To this end, it is crucial to present a paradigm for understanding and using acquired knowledge. The year 2045 is being predicted as the year of technological singularity. Really, will we observe the evolution of teaching/learning performance in just this time? What insights can we gain from our learning strategy? How can we best facilitate this process? We will address these questions in our future work.

References

1. Watanabe, T.: Learning support specification based on viewpoint of knowledge management. In:Proceedings of E-Learn2012 (AACE), pp. 1596–1605 (2012)
2. AACE: Proceedings of Ed-Media2008, Proceedings of Ed-Media2011, Proceedings of Ed-Media2014, and Proceedings of Ed-Media2015
3. AACE: Proceedings of E-Learn2015
4. JSiSE: Proceedings of Annual Conference 2013 and Proceedings of Annual Conference 2014

Motion Prediction for Ship-Based Autonomous Air Vehicle Operations

Ameer A. Khan, Kaye E. Marion, Cees Bil and Milan Simic

Abstract A ship operating in an open sea environment undergoes stochastic motions which make deployment and landing of UAVs and other vehicles on a ship difficult and potentially dangerous. There is always a delay between the decision to commit and the moment of actual launch or recovery. This paper presents an artificial neural network trained using singular value decomposition, genetic algorithm and conjugate gradient method for the real time prediction of ship motions. These predictions assist in determining the best moment of commitment to launch or to recover. Predictions generated using these algorithms allow improvements in safety as well reducing the number of missed or aborted attempts. It is shown that the artificial neural network produces excellent predictions and is able to predict the ship motion satisfactorily for up to 7 s ahead.

Keywords Ship motion · Prediction · ANN · Sea state · Artificial neural network

1 Introduction

This paper presents the results of an investigation into the development of an algorithm capable of predicting ship motion in various sea states to support deployment and recovery of UAVs and other vehicles operated off their decks. After the command is given, the algorithm provides important future ship motion information automatically to the UAV to determine the best moment for launch or recovery. This information will also facilitate the computation of the correct UAV flight path and help identify dangerous or problematic situations that could potentially cause damage to the ship or UAV.

A.A. Khan · C. Bil (✉) · M. Simic
School of Engineering, RMIT University, Melbourne, Australia
e-mail: bil@rmit.edu.au

K.E. Marion
School of Mathematics and Geospatial Sciences, RMIT University, Melbourne, Australia

G. De Pietro et al. (eds.), *Intelligent Interactive Multimedia Systems and Services 2016*, Smart Innovation, Systems and Technologies 55,
DOI 10.1007/978-3-319-39345-2_28

Ship motion in an open water environment is the result of complex hydrodynamic forces acting between the ship, the water and other random processes. This leads to the use of statistical prediction methods rather than a deterministic analysis, which would result in a ship-specific model that involves highly complex calculations and many assumptions and approximations [1].

Past research in ship motion prediction [2–5] has shown that traditional statistical prediction techniques such as autoregressive moving average models and Kalman filters are unable to maintain a high degree of accuracy when the prediction interval exceeds 3–4 s in sea states of 5 and above. Traditional statistical techniques used for time series prediction have difficulty dealing with noisy data, do not have much parallelism and fail to adapt to changing circumstances.

This paper explores the use of Artificial Neural Networks (ANN) which is a form of artificial intelligence used to develop algorithms capable of predicting ship motion. Artificial neural networks, in contrast to traditional statistical techniques, promise to produce predictions with high accuracy, as well as high efficiency, due to their ability to learn and adapt to prevailing conditions.

The ability to predict ship motion reliably, in any sea state, will enable more intelligent automated air vehicles operations off ship platforms. Landing and take-off of helicopters and aircraft, manned or unmanned, from ship decks, in rough sea conditions, can be difficult and dangerous. If ship motion can be predicted ahead with reasonable accuracy and communicated to the vehicle, touchdown dispersion can be improved on landing, and a smoother aircraft trajectory can be achieved on take-off. The benefits of reducing the touchdown dispersions is that, the time to failure of the aircraft frames and the number of aborted landing attempts can be increased.

Prediction of ship motion is important for the safe deployment for any vehicle, such as missiles, UAVs and decoys from ship platforms, such as shown in Fig. 1, for correct timing and initializing trajectory calculations [5].

In some cases, there is a launch "lock-out" condition where the missile, or remote piloted vehicle, cannot be launched safely if, for example, the ship's roll

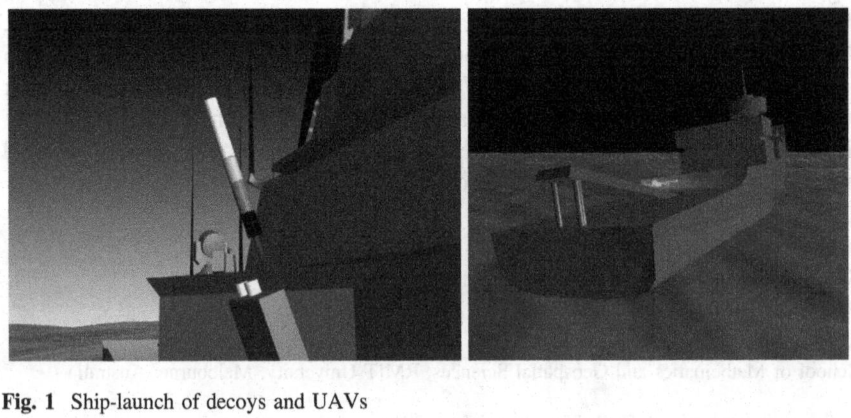

Fig. 1 Ship-launch of decoys and UAVs

angle exceeds a predefined operational limit, as there is a delay between launch commit and actual launch. The algorithm developed in this investigation can determine if the air vehicle is in the launch "lock-out" condition by predicting ship motion by at least 7 s ahead. In this investigation the operational limit is when the ship's roll angle exceeds a certain magnitude. It is important that the predicted angles are of a high accuracy as the batteries for the system are "one shot" batteries, which means that the process of deployment, once activated, cannot be reversed.

2 Artificial Neural Networks

Artificial Neural Networks (ANN) form a class of systems that are inspired by biological neural networks [6]. An ANN is simply a series of neurons that are interconnected to create a network. They are a class of non-linear systems and there are a wide variety of different approaches that can be used. The use of ANN is particularly appealing due to its ability to learn and adapt. This is important for our investigation as one of the underlying goals is to create an algorithm that is able to work in all conditions and environments. The ANN architecture that was used to create network for time series prediction was a multi-layer feed-forward ANN.

To validate the ANN, a data set of measured data was divided into two. One part of the data set was designated for training purposes and the second part was designated as the validation data set. Once trained, the validation data set was put in the ANN and the resulting predictions based on the validation set were used to measure the effectiveness of the ANN. The basic model for time series prediction is shown in Fig. 2.

In Fig. 2 it can be seen that there are seven points. Lag 0 represents the current sample while the past six values are represented by lags 1–6. In this investigation lags of up to 30 were used as inputs into the ANN. The output of the ANN is the prediction. It can be noticed that there is only one output shown. For every lead prediction interval, it is advisable to use a single ANN. If multiple predictions are required, then for every prediction interval a separate ANN should be used as the weights for an optimal prediction will vary according to the prediction interval desired. By having the ANN create multiple predictions, the overall optimal

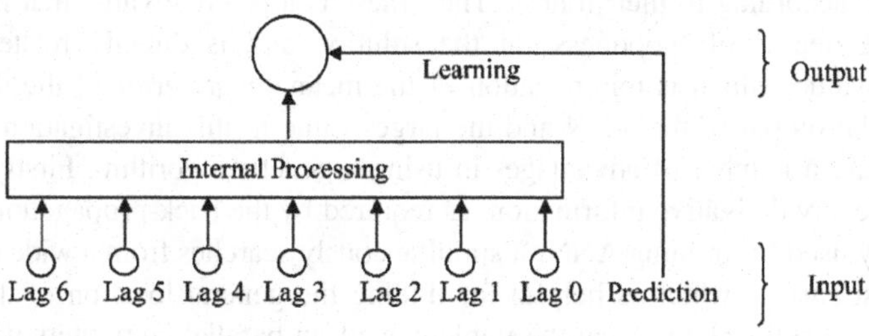

Fig. 2 Basic model for time series prediction using ANN

prediction cannot be made. When separate ANNs create separate predictions, the optimal weight configuration can be obtained for each prediction and therefore, higher accuracy can be expected.

2.1 Training the Artificial Neural Network

Training of the ANN can be viewed as a minimization process where the weights in the ANN are systematically adjusted in a manner that reduces the error between the output of the ANN and the desired output. Therefore, the process of ANN training becomes an optimization problem. Performance of the ANN will be dependent upon the quality of the solution found after the training process has been completed. The advantage of using ANN for ship motion prediction is that it is neither ship specific nor condition specific. In addition, training the ANN can be done on a conditions and autonomous basis as new ship motion data is measured and input to the ANN for training. This way the ANN can adapt to changing conditions. In the following sections three techniques are discussed that were used to determine ANN weights.

2.2 Genetic Algorithm (GA)

The genetic algorithm (GA) is a part of a rapidly growing area of artificial intelligence called evolutionary computing. The term 'evolutionary computing' is based on Darwin's theory of evolution, which states that problems are solved by an evolutionary process resulting in a best solution. It is basically survival of the fittest where the 'fittest' (best) 'survivor' (solution) evolves to create the next population. Solution to a problem solved by the GA uses an evolutionary process based on the principles of genetics and natural selection [7].

The algorithm begins with a population of solutions. Solutions from one population are taken and used to form a new generation of solutions or the next population of solutions. The expectation is that the new population will be better than the old one. Solutions or individuals are then selected to form new solutions or 'offspring' according to their fitness. The fitness is a positive value that is used to reflect the degree of 'goodness' of the solution and is directly related to the objective value, which is minimization of the mean square error of the difference between the output of the ANN and the target value in this investigation [8].

There are a number of advantages in using a genetic algorithm. Firstly, it does not require any derivative information, as required by the back-propagation method commonly used for training ANN. It simultaneously searches from a wide sampling of the cost surface which is helpful for finding the general location of the global minimum quickly. The GA can be implemented on parallel computers which will allow the solution to be found more rapidly than if it were implemented on a single processor which is an important consideration for real time prediction as required for this project. Also, as it provides a list of optimum variables and not just a single solution, if the global minimum is difficult to locate a good alternative solution can be returned.

2.3 Singular Value Decomposition (SVD)

The aim of this investigation is to develop a methodology to predict ship motion in real time. Singular Value Decomposition (SVD) is a linear regression technique that can quickly obtain an approximate set of optimum weights which is far superior to randomly generated weights [9]. The values returned from SVD can be used as the initial starting points for a selection of the population for the GA algorithm, discussed previously. They can, sometimes, be of such a high standard that this method can be used alone. A detailed description of the SVD technique is beyond the scope of this paper but essentially it can be expressed by matrix X which satisfies the function:

$$A.X = B \tag{1}$$

When A and B are known and can be calculated efficiently using SVD. When applying it to the ANN process the weights between the input layer and the hidden layer are initially randomly generated. The training samples are then inserted into the ANN and the hidden layer activation functions are calculated creating a matrix equivalent to A. Also, the values for the inverse transfer function of the output are also calculated creating a matrix equivalent to B. Applying SVD and solving Eq. 1, the approximate optimal weights X are found.

2.4 Conjugate Gradient Method (CG)

The conjugate gradient (CG) algorithm created for ANN used in this investigation was based on the Polak-Ribiere algorithm. The mathematical justifications for the algorithm and a detailed description can be found in [10]. In a general sense, the algorithm generates a sequence of vectors and search directions. The exact minimum will be attained if the multi-dimensional function can be expressed as a quadratic. The ANN error function is quadratic close to the minimum so convergence to the local minimum will be very rapidly [11].

In this investigation the CG method was used as a means of improving the solutions returned by the singular value decomposition methods. The CG method can be used independently of the other techniques however there are distinct advantages in using the CG method jointly with the other training techniques. The CG method was used in conjunction with the SVD method because the SVD returns only linear optimal weights. The results of the SVD method are inserted into the CG algorithm as the initial starting point for its search which can potentially yield values for the weights that are far superior to the solutions developed by the CG method when a randomly generated initial starting point is used.

3 Experimental Design

In this section an investigation is presented into the application of ANN for the prediction of actual ship motion and to assess the levels of accuracy that can be attained when using the ANN to predict a ship's roll motion up to 7 s. The ANN algorithms developed were applied to measure ship roll angle data taken from a cruiser size vessel operating in sea states 5–6 based upon the Pierson-Moskowitz Sea Spectrum. The term sea state is a description of the properties of sea surface waves at a given time and place [12]. Greater the sea state indicate rougher conditions. There were four databases of roll angle data, each 562 s in length, sampled at 2 Hz. The training data was set to two thirds of the data sets and the validation set was designated as the final third of the data sets. All results shown are the predictions made using the validation set only.

In order to increase the likelihood of obtaining predictions of higher accuracy the data was preprocessed using two separate preprocessing schemes. The first preprocessing scheme, which will be referred to as PP1, was chosen to remove the trends and seasonality by taking the difference between consecutive points and subtracting the mean. Due to the nature of the ANN being sensitive to subtle variations, it may overemphasize the trends and seasonality at the expense of focusing on more important characteristics in the data. The ANN may focus too much on the obvious trends at the expense of extracting the more subtle variations. Therefore, all trends that can be eliminated and easily brought back after the predictions are made should be purged before being entered into the ANN.

The difference performed by PP1 eradicates constant trends and any seasonal trends with a large period if present. The drawback when taking a difference of the data is that low frequency information that may be important to the network may be lost as the differencing is a potent high-pass filter. The ANN is actually capable of dealing with the trends and seasonality but it may be better for the ANN to focus on its main objective which is to predict the future values and not be hampered by being made to find trends and seasonality. The second preprocessing scheme is to simply subtract the mean from the data before it is entered into the ANN and will be referred as PP2. The results of the investigation are shown in Table 1. All results were generated on a Pentium 4, 2.8 GHz processor. To assess the performance of the ANN three performance criteria were defined:

- Criterion 1 is the percentage of predictions accurate within the 95 % confidence interval. The success criterion for this investigation was to develop an algorithm that can predict when a predefined angle is exceeded.

Table 1 Results of ANN simulation

Preprocessing technique	Database	Training method	Criterion (%)			CPU time (s)
			1	2	3	
PP1	**1**	SVD	40.64	70.59	28.48	0.03
		SVD/GA	44.75	55.00	26.64	194.92
		SVD/CG	49.40	66.67	22.92	60.13
	2	SVD	41.67	67.86	22.32	0.02
		SVD/GA	45.03	50.00	19.68	191.08
		SVD/CG	45.70	71.43	17.86	62.19
	3	SVD	46.15	71.15	30.56	0.02
		SVD/GA	57.28	53.49	18.55	186.58
		SVD/CG	68.05	90.20	17.58	59.09
	4	SVD	29.17	69.77	41.42	0.02
		SVD/GA	39.80	59.52	26.67	186.98
		SVD/CG	38.05	67.44	25.88	60.72
	Average for all databases	**SVD**	**39.41**	**69.84**	**30.70**	**0.02**
		SVD/GA	**46.72**	**54.50**	**22.89**	**189.89**
		SVD/CG	**50.30**	**73.94**	**21.06**	**60.53**
PP2	1	SVD	53.80	52.94	11.64	0.02
		SVD/GA	52.60	17.65	2.12	184.20
		SVD/CG	59.36	17.65	2.90	59.05
	2	SVD	63.14	14.29	4.97	0.02
		SVD/GA	56.55	21.43	1.43	192.38
		SVD/CG	65.50	53.57	12.66	61.11
	3	SVD	68.93	50.00	10.54	0.02
		SVD/GA	70.00	41.46	9.38	187.73
		SVD/CG	75.00	78.72	7.40	61.08
	4	SVD	42.51	19.05	3.92	0.02
		SVD/GA	55.26	35.14	8.44	188.81
		SVD/CG	54.97	40.54	9.81	59.39
	Average for all databases	**SVD**	**57.10**	**34.07**	**7.77**	**0.02**
		SVD/GA	**58.60**	**28.92**	**5.34**	**188.28**
		SVD/CG	**63.71**	**47.62**	**8.19**	**60.16**

- Criterion 2 is the percentage forecasts that correctly predicted that the roll angle would exceed 7°.
- Criterion 3 is the percentage of forecasts that incorrectly predicted the lock-out conditions.

The number of neurons in the input layer (NNIL) was varied from 5 to 30 and the number of neurons in the hidden layer (NNHL) was varied from 2 to 5. The output layer consisted of only 1 neuron in all cases. Results shown in Table 1 are the best predictions of 7 s in advanced generated from these ANN architectures for each of the four databases labeled 1 to 4 in Table 1. First, the singular value decomposition method was used to generate 30 solutions for the weights. The remainder of the population was randomly generated. The genetic algorithm was then used to find a solution. Also, the best weights returned from the SVD training algorithm were also inserted into the CG algorithm as the initial starting point for the CG search.

4 Discussion of Results

Based on our findings, ANN is capable of predicting ship motion in real time. Against performance criteria 1, 2 and 3 the ANN was able to predict ship motion to an approximate accuracy level of 64 %, 74 % and 5.34 % respectively.

Figures 3 and 4 show that the predictions basically mirror the actual recorded motion of the validation set which is separate and distinct from the data set of data used to train the ANN. The highest accuracy was obtained using the SVD/CG trained ANN which took on average 60 s to train. Considering that the validation set is approximately 170 s in length, at least 110 s of the validation set was genuinely predicted in real time. Therefore, there is reasonable evidence that the ANN is capable of producing predictions in the real time. There are two noticeable ANN characteristics that can be observed in Table 1. The first important characteristic is that for all four databases the ANN performed better based upon performance

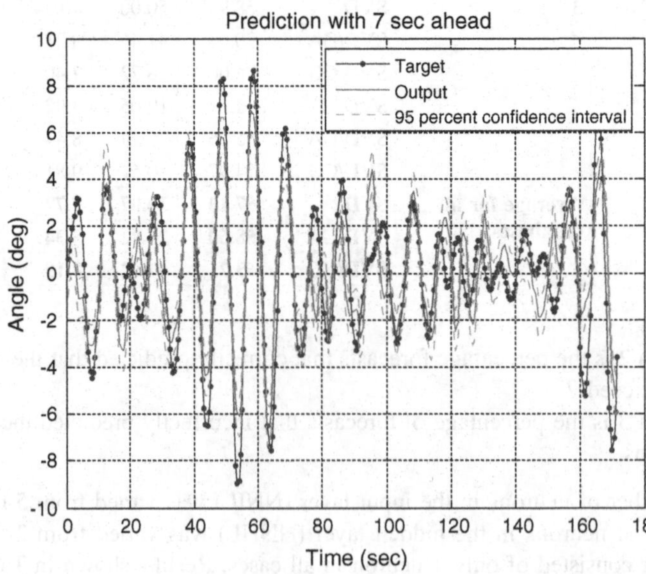

Fig. 3 NNIL = 10 and NHIL = 3 (ANN used PP2 and was trained using SVD/CG)

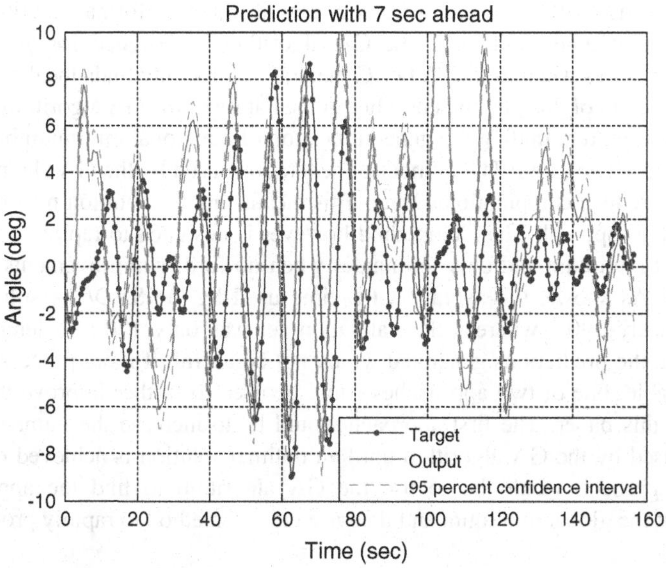

Fig. 4 NNIL = 25 and HHIL = 5 (ANN used PP1 and was trained using SVD/CG)

criteria 1 and 3 when PP2 was used rather than PP1. This implies that a higher percentage of predictions were within the 95 % confidence interval and there was a lower percentage inaccurate lock-out predictions. The other evident characteristic is that the ANN which accepted the data preprocessed using PP1 outperformed the ANN using PP2 based on performance criterion 2.

Figures 3 and 4 show the 7 s predictions made by an ANN using PP2 and PP1 respectively and give an understanding for these two characteristics highlighted in the above discussion. Figure 3 shows that the ANN trained using PP2 is able to produce very high quality predictions. Motion is represented well despite rapid influxes in the amplitudes of the roll motion. The only problem with the predictions, however, is that they are conservative in nature and tend to understate the true amplitude of the motion. This explains why the resulting performance levels were high with regards to criteria 1 and 3 but not against criteria 2.

Figure 4 shows the prediction of the ANN using PP1 and shows that the ANN tends to overstate the roll motion amplitudes. This explains why the ANN using PP1 is able to outperform the ANN using PP2 based on performance criteria 2. The problem associated with the predictions generated by the ANN using PP1 is that, because the predictions are overstated, they will be predisposed to yield more false lock-outs. This is confirmed in Table 1 which shows that the predictions generated by the ANN using PP1 had a higher percentage of lock-out conditions accurately predicted and also a higher percentage of lock-out conditions incorrectly predicted.

Table 1 also shows that the combination of the SVD and CG (SVD/CG) training methods is superior to that of the SVD method alone and the combined SVD and

GA algorithms (SVD/GA) based upon nearly all three performance criteria. This seems to be counterintuitive as the GA algorithm is designed for global minimization searches whereas the SVD/CG method usually only finds local minimums. The explanation of this phenomenon lies in the nature of the GA algorithm. The GA algorithm is good at finding the general location of the global minimum but is slow to find an exact solution. The number of generations used by the GA algorithm was cropped to reduce computational time. This means that the solution may only have been close to optimal. The CG method however is designed to rapidly progress to the nearest minimum and only terminates when no further improvements could be generated. As shows, the average CPU time used by the SVD/GA solution was approximately 190 s whereas the validation set was only 170 s in length which means that the predictions generated would not be useful. If more processor power was available, one of two approaches could be used to further improve the results shown in this paper. The first approach would be to increase the number of generations used by the GA algorithm until an optimal solution is achieved or a more efficient approach would be to use the GA algorithm to find the approximate location of the global minimum and then use the CG method to rapidly progress this minimum.

Due to the small size of the data samples used to test the ANN algorithms it is yet to be determined how long the ANN will remain valid without requiring a new set of weights to be generated. As stated in Sect. 2.2, the GA algorithm can be processed on parallel computers which would reduce CPU time. It may be feasible to train the ANN with the SVD/GA method but as the SVD/GA method can take up to three times longer to arrive at a set of weights that result in inferior predictions to those generated by the ANN trained using the SVD/CG method it would be prudent to train the ANN using the SVD/CG method.

The most efficient method for generating the necessary weights was the SVD method. The CPU time required was only a fraction of a second on average and it is this efficiency that meant that the SVD method could be used with the other two training methods. Indeed, if one chooses to predict future ship motion with the ANN that uses PP2, it is well worth considering using the SVD method alone due to the computational efficiency of the SVD algorithm.

Table 1 shows that the ANN trained using SVD method produced predictions that, on average, performed only 6 % worse than the ANN that was trained using the SVD/CG methods based on performance criteria 1 when PP2 was used and based on criteria 3, the ANN trained solely with the SVD algorithm performed better producing less false lock-out predictions. Furthermore, the SVD/CG method takes on average more than 60 s longer to arrive at a solution. If the ANN is trained using the SVD algorithm alone then the computer processor would not need to be as powerful and the weights could be updated more regularly. This is especially useful when the ship performs maneuvers or when there are sudden dramatic changes in the behavior of the roll motion which would require a new set of weights to be generated.

5 Conclusion

An Artificial Neural Network (ANN) was developed and trained using the singular value decomposition method, the conjugate gradient method and the genetic algorithm was presented and applied to the prediction of ship roll motion. It was shown that the artificial neural network was capable of predicting ship motion up to 7 s in advance. The artificial neural network trained using the combination of the singular value decomposition and conjugate gradient method produced the most accurate prediction when the data was pre-processed by subtracting the mean. The artificial neural network trained using the singular value decomposition method proved to be able to predict ship motion reliably, accurately and quickly which makes it a suitable candidate for predicting ship motion when conditions are expected to change rapidly requiring the retraining of the network. Overall the use of the artificial neural network is an effective technique for the prediction of ship roll motion.

Acknowledgments The authors would like to thank BAE Systems for providing funding and support for the project.

References

1. Evelyn, R., Bishop, R.E.D., Price, W.G.: Probabilistic Theory of Ship Dynamics. Springer Science + Business Media, New York (1974). ISBN: 978-1-5041-1879-8
2. Yumori, I.: Real time prediction of ship response to ocean waves using time series analysis. OCEANS **13**, 1082–1089 (1981)
3. Cortes, N.B.: Predicting ahead on ship motions using Kalman filter implementation. Thesis, RMIT University, Melbourne (1999)
4. Crump, M.R.: The dynamic and control of catapult launching unmanned air vehicles from moving platforms. Thesis, RMIT University, Melbourne (2002)
5. Sidar, M.M., Doolin, B.F.: On the feasibility of real-time prediction of aircraft carrier motion at sea. IEEE Trans. Autom. Control **28**(3), 350–356 (1983)
6. Suykens, J.A.K., Vandewalle, J.P.L., Moor, B.D.: Artificial neural networks for modeling and control of non-linear systems. Kluwer Academic Publishers, Dordrecht (1996)
7. Haupt, R.L., Haupt, S.E.: Practical Genetic Algorithms. John Wiley and Sons (1998)
8. Man, K.F., Tang, K.S., Kwong, S.: Genetic Algorithms. Springer Verlang, London (1999)
9. Yanai, H., Takeuchi, K., Takane, Y.: Projection Matrices, Generalized Inverse Matrices, and Singular Value Decomposition. New York, Springer Science and Business Media (2011). ISBN: 978-1-4419-9886-6
10. Polak, E.: Computational Methods in Optimisation. Academic Press, New York (1971)
11. Press, W.H., Flannery, B.P., Vetterling, W.T., Teukolsky, S.A.: Numerical recipes in Fortran: The Art of Scientific Computing. Cambridge University Press, New York (1992)
12. Keefer, T.: Glossary of Meteorology. Allen Press, Boston (2005). http://amsglossary.allenpress.com/glossary. Accessed 2005

The Effect of Receding Horizon Pure Pursuit Control on Passenger Comfort in Autonomous Vehicles

Mohamed Elbanhawi, Milan Simic and Reza Jazar

Abstract Passengers in autonomous vehicles are prone to motion sickness. Receding horizon control of pure pursuit tracking algorithms has been shown to improve path tracking performance. In this paper we present a numerical study on the effect of the receding horizon pure pursuit controller on passenger comfort. Three standard cases at the different speeds are utilized to compare the effect of traditional and receding horizon pure pursuit control on passenger comfort. The results show improvements in passenger comfort at higher speeds using receding horizon control and that path continuity is more influential that optimal tracking control.

Keywords Self-driving vehicle · Autonomous cars · Path tracking · Passenger comfort

1 Autonomous Cars and the Loss of Controllability

1.1 Introduction

Cars are the most popular mode of transportation for personal use in terms of time spent and distance travelled [1–4]. Autonomous cars, or self-driving cars, are an attractive solution to the existing limitations of transport systems. The use of autonomous fleets is expected to improve transportation by improving fuel efficiency, reducing pollution, reducing traffic congestion and managing accident risk [5, 6]. In 2015, it is estimated that there were around 1.2 million fatalities due to car accidents [7, 8]. Several researches have concluded that road accidents are linked to unintended lane departure [9], driver distraction [10], and other human caused errors [11].

M. Elbanhawi (✉) · M. Simic · R. Jazar
School of Engineering, Bundoora East Campus, RMIT University,
Plenty Road, Bundoora, Melbourne, VIC 3083, Australia
e-mail: mohamed.elbenhawi@rmit.edu.au

G. De Pietro et al. (eds.), *Intelligent Interactive Multimedia Systems and Services 2016*, Smart Innovation, Systems and Technologies 55,
DOI 10.1007/978-3-319-39345-2_29

Operation of autonomous agents is generally described using the Sense Plan Act (SPA) stages. This article is concerned with analyzing the role of tracking control, i.e. act portion of SPA, on passenger comfort as defined in [12].

1.2 Related Work

1.2.1 Path Tracking Control

Several researchers have studied tracking of the planned paths for mobile robots. The motion of car-like robots is constrained due to mechanical limitations of steering system (see Sect. 2). Hence, path planning and tracking have been shown to be intractable problems in literature [13, 14]. Several tracking controllers were developed for autonomous cars e.g. pure pursuit (PP) tracking [15], Stanley controller [16] a critically damped controller [17, 18].

Pure pursuit controller is the most commonly used controller for car like tracking [19]. Snider [20] has empirically proven that pure pursuit outperformed Stanley, error based kinematic tracker and linear quadratic dynamic controllers, under actuation noise and discontinuous paths. The critically damped controller [17] is limited by its variable velocity which is not suitable for urban on-road scenarios. It was extended to bounding the velocity however, when the velocity is fixed (for lateral control purposes) the controller oscillates and is rendered unstable [18].

In our earlier work, we proposed a Receding Horizon Controller (RHC) for lateral pure pursuit tracking [21]. Receding Horizon Control (RHC) is an iterative method of implementing the first step of the Model Predictive Control (MPC) control horizon, revaluating the state and the desired control. MPC manipulates the upcoming steering variable to optimize the predicted future vehicle performance. The optimized control and prediction timeframes are set to particular windows, referred to as control and prediction horizons.

Our RHC-PP controller was motivated by the effectiveness of MPC and the limitations of kinematic controllers at higher speeds and for discontinuous paths. In fact, MPC has been mostly successful as a strategy to improve the performance of multivariate dynamic nonlinear control systems [22, 23]. MPC has been utilized in mobile robots path tracking. Gu and Hu [24] used MPC for kinematic tracking of wheeled robots. Attia et al. [25] proposed MPC for lateral control. Vehicle stability over slippery driving surfaces was improved using MPC [26, 27].

1.2.2 Passenger Comfort in Autonomous Cars

Motion sickness is caused by the conflict of inertial and visual systems in humans [28]. In essence, passive drivers (or passengers) in autonomous vehicles, will be more prone to motion sickness, as they will not be focused on the road [29].

More than 2 in every 9 autonomous vehicle passengers is expected to experience motion sickness [30]. The loss of controllability theory has been proposed as the principal cause of motion sickness in autonomous vehicles [31]. Le Vine et al. [32] predicated that reducing vehicle yaw rate and traction speeds, to manage passenger comfort, would increase traffic congestion and inefficiency. Therefore, the need to understand, quantify and improve passenger comfort is critical for the wide development of self-driving technology.

1.3 Contribution

The authors' earlier research hypothesized, the effect of parametrically continuous passenger vehicles and optimal path tracking control in improving passenger comfort in autonomous vehicles [31]. We expect these factors to reduce the resulting lateral acceleration, which is directly linked to passenger motion sickness [33].

Therefore, path planning algorithm [14, 34] that generate parametrically continuous paths, and RHC-PP [21] controller, were developed. Parametrically continuous paths were shown to improve passenger comfort and path tracking performance [13, 35]. RHC-PP improves path tracking performance [21]. We investigated effect of RHC-PP controller application on passenger comfort.

2 Vehicle Model

2.1 Kinematic Model

Vehicle lateral motion is investigated, due to its direct causal relation to passenger comfort. Hence, roll and pitch motions are ignored by assuming that the road is flat and accordingly by ignoring suspension reaction effects. Independent longitudinal control is a valid assumption on structured streets as constant velocity is utilized. Therefore, the Bicycle model is adopted when representing a front wheel steering vehicle's planar motion [36]. This model is widely adopted for autonomous vehicle path tracking control [17, 18, 37].

Bicycle model is summarized in this section. Vehicle's kinematics is described by Cartesian position, (x, y) and heading, θ, i.e. vehicle pose. The pose is measured between the global frame and a body frame in the center of rear axle. Front wheels control the steering and the rear wheels are fixed. The steering angle, φ, is used to represent the average left and right steering wheels' angles. Velocity, v, and steering angle, φ, are two control commands for the Bicycle model. Model is given by the set of Eq. (1a, 1b). Curvature, k, is limited to k_{max}, as steering is mechanically limited to a maximum angle φ_{max}, as expressed in Eq. (2). L is wheel base, ρ is the radius of curvature, β is the side slip angle and Δt is the time step.

$$\begin{cases} x_{i+1} = x_i + v_i \cdot \cos(\theta_i) \cdot \Delta t \\ y_{i+1} = y_i + v_i \cdot \sin(\theta_i) \cdot \Delta t \\ \theta_{i+1} = \theta_i + v_i \cdot \frac{\tan(\phi_i)}{L} \cdot \Delta t \end{cases} \tag{1a}$$

$$\begin{cases} x_{i+1} = x_i + v_i \cdot \cos(\theta_i + \beta_i) \cdot \Delta t \\ y_{i+1} = y_i + v_i \cdot \sin(\theta_i + \beta_i) \cdot \Delta t \\ \theta_{i+1} = \theta_i + v_i \cdot \frac{\tan(\phi_i)}{L} \cdot \Delta t \end{cases} \tag{1b}$$

$$k_{max} = \frac{1}{\rho_{min}} = \frac{\tan(\phi_{max})}{L} \tag{2}$$

$$v_x \cdot \sin(\theta) - v_y \cos(\theta) = 0 \tag{3}$$

2.2 Lateral Dynamic Model

Bicycle model can be used to accurately model vehicle lateral motion, at lower speeds, as shown earlier [37–39]. Path modeling errors are caused by ignoring vehicles dynamic effects on side slip.

For the planar motion study, vehicle is modeled as a rigid three degree of freedom system, operating in a two dimensional environment. Three motion modes are considered: longitudinal, lateral and yaw, which are expressed by vx, vy and ω, three time dependent variables. The corresponding dynamic equations of motion are formulated by Eqs. (4), (5) and (6), as in [36], where m is the vehicle mass, a is the axle distance from the center of gravity, C is the wheel coefficient and $\boldsymbol{F}$ is the longitudinal force vector.

$$\dot{v}_x = \frac{F_x}{m} + \omega v_y \tag{4}$$

$$\dot{v}_y = \frac{1}{m v_x}(-a_1 C_{af} + a_2 C_{ar})\omega - \frac{1}{m v_x}(C_{af} + C_{ar}) v_y + \frac{1}{m} C_{af}\phi - \omega v_x \tag{5}$$

$$\dot{\omega} = \frac{1}{I_z v_x}(-a_1^2 C_{af} - a_2^2 C_{ar})\omega - \frac{1}{I_z v_x}(\alpha_1 C_{af} - a_2 C_{ar}) v_y + \frac{1}{I_z} a_1 C_{af}\phi \tag{6}$$

As previously mentioned, for on road urban driving a fixed forward velocity is assumed. The majority fast highway maneuvers considered in this study, such as lane changes [40–42], are conducted at a fixed speed. Hence, the lateral vehicle control is analytically solvable under this assumption. It is formulated as a two-degree of freedom, linear time-invariant (LTI) model.

2.3 Steady State Responses

Steady state conditions could be assumed for a turn with fixed speed and steering angle. The input output relations are referred to as steady state responses [36]. Side slip and yaw steady state responses, R_β, R_ω, to the steering angle are given by Eqs. (9) and (10). *C*-coefficients proportionally relate the total lateral forces to yaw rate, and steering angle and *D*-coefficients relate the yaw moment. For a variable steering angle, the transient equations of motion must be solved.

Recent research has shown that steady state steering and transient steering were sufficiently close at higher speeds for [37, 43]. Therefore, in this study we rely steady state responses to estimate the vehicle side slip and yaw rate based on the desired optimal control.

$$R_\beta = \frac{\beta}{\phi} = \frac{D_\phi(C_\omega - mv_x) - D_\omega C_\phi}{D_\omega C_\beta - C_\omega D_\beta + mv_x D_\beta} \tag{9}$$

$$R_\omega = \frac{\omega}{\phi} = \frac{C_\phi D_\beta - C_\beta D_\phi}{D_\omega C_\beta - C_\omega D_\beta + mv_x D_\beta} \tag{10}$$

3 Controllers

Tracking controllers attempt to generate control signals, speed v and steering angle ϕ, for a vehicle to track a predesigned trajectory. It is defined as a piecewise linear B-spline path, $c(u)$, of J number of consecutive coordinates. Accordingly, for a lateral controller, inputs are reference (x_j y_j θ_i) positions along the path. The vehicle's state, at step j, is calculated using Eq. (1b) (x'_j y'_j θ'_j), where $j \in [1, J]$. Vehicle location is assumed to be fully observable and deterministic.

3.1 Pure Pursuit Controller

PP controller generates control signal, i.e. steering command, proportionally to the heading error, given in Eq. (11). Heading at any distance is derived using curvilinear path parameters. The desired heading is defined from a fixed forward distance i.e. look-ahead distance. This distance is a tuning parameter, empirically derived based on the fixed velocity, v [20]. Steering output is calculated based on the current and desired heading error, as given by Eq. (12), where heading gain f_θ is positive $f_\theta > 0$.

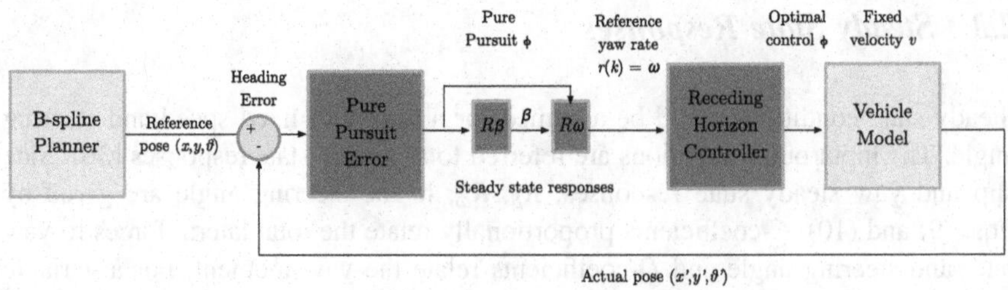

Fig. 1 Receding Horizon Control (RHC) Pure Pursuit (PP) control scheme [21]

$$\theta_j = a\tan\frac{y_j - y_{j-1}}{x_j - x_{j-1}} \tag{11}$$

$$\phi_j = f_\theta(\theta_{j+1} - \theta_j') \tag{12}$$

3.2 Receding Horizon Controller

Proposed RHC-PP [21] relied on three main changes to the PP controller. Initially, the PP heading error, Eq. (11), is used to calculate a desired yaw rate. It is estimated using steady state responses given by Eqs. (9) and (10). This compensates for the predicted slip and lateral motion of the vehicle. Finally, the optimal steering angle can be calculated for the prediction, N_p, and control, N_c, horizons (time steps) to reach the desired yaw rate. RHC-PP control is illustrated in Fig. 1.

Equation (13) gives the optimal change in steering angle, in order to achieve a desired yaw rate, from the current state. It is shown in Elbanhawi et al. [21], based on the work by Wang [23]. The current vehicle state is $x(k)$, while $r(k)$ is the reference yaw rate based on pure pursuit heading error and steady state responses, at a fixed velocity

$$\phi_k^* = \Delta\phi^* + \phi_{k-1} \tag{13}$$

4 Methodology

4.1 Benchmarks

We used three case studies: Step change (SC), Lane Change (LC) and Double Lane Change (DLC), see [21]. Parametric continuous paths are generated for the case studies using cubic B-splines [13], as shown in Fig. 2.

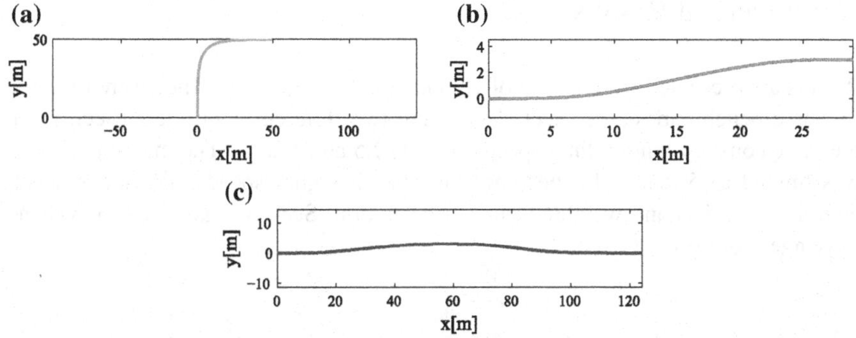

Fig. 2 Reference B-spline paths **a** Step Change (SC), **b** Lane Change (LC), and **c** Double Lane Change (DLC)

4.2 Parameters

Parameters used in the numerical study are listed in Table 1.

4.3 Passenger Comfort Metrics

Planar (lateral and longitudinal) accelerations are used to estimate load disturbances as shown in Eqs. (14) and (15) [12],. Since fixed velocity is adopted in this study longitudinal acceleration is zero. We utilize standard root mean square (r.m.s) weighting of acceleration to estimate passenger comfort based on the resulting acceleration [44].

$$a_{lateral} = \frac{\omega^2}{k} \tag{14}$$

$$a_{longitudinal} = \dot{v} = 0 \tag{15}$$

Table 1 Vehicle parameters' selection and values

Parameter	m [kg]	I_z [kg/m^2]	L [m]	a_1 [m]	a_2 [m]	C_{ar} [N/rad]	C_{af} [N/rad]	ϕ_{max} [°]
	Mass	Moment of inertia	Wheel base	Front to centre of gravity	Rear to centre of gravity	Rear wheel coefficient	Front wheel coefficient	Maximum steering angle
Value	1000	1650	2.6	1.0	1.6	3000	3000	25

5 Numerical Results

Results are presented from a total of 18 numerical experiments. They were repeated for three benchmark cases (Sect. 4.1), using two different controllers (Sect. 3), at the three constant longitudinal speeds (v = 1, 2.5 and 5 m/s). The maximum speed was limited to 5 m/s as PP becomes unstable at higher speeds. The mean acceleration (Sect. 5.1) and weighted r.m.s. acceleration (Sect. 5.2) are used to evaluate passenger comfort.

5.1 Mean Acceleration

See Fig. 3.

5.2 Weighted r.m.s. Acceleration

See Fig. 4.

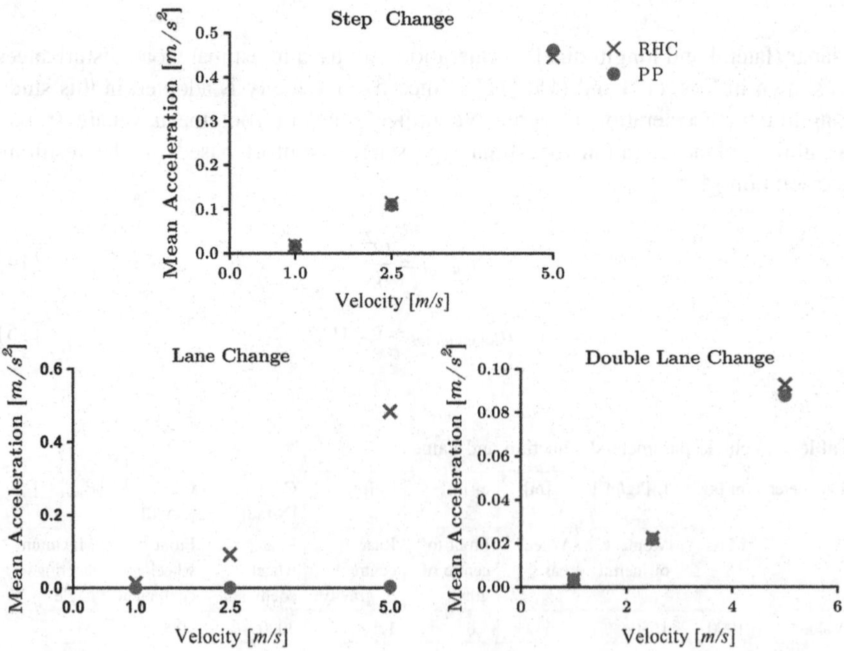

Fig. 3 Mean acceleration results for SC (*top*), LC (*centre*) and DLC (*bottom*) using PP (*red cross*) and RHC (*blue dot*) controllers (color figure online)

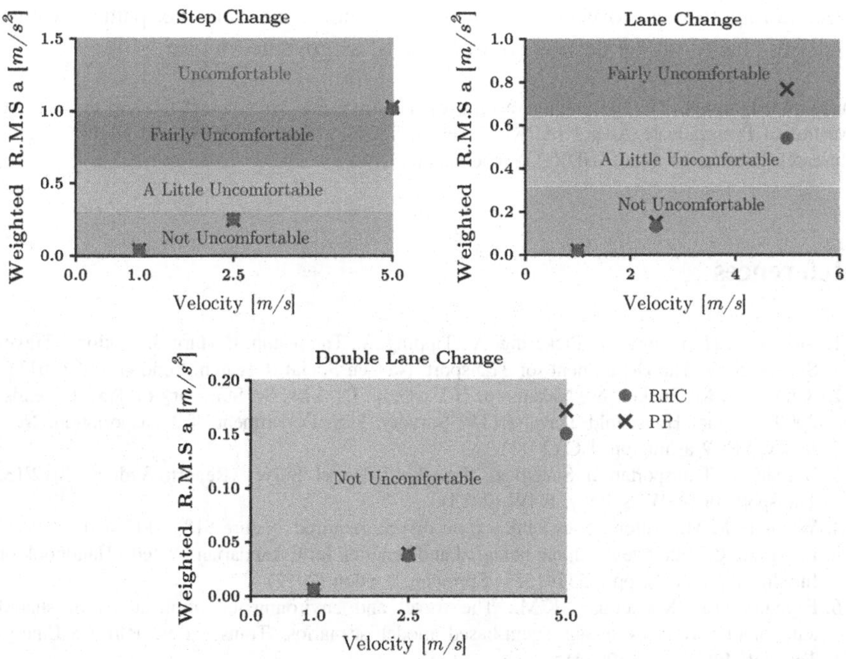

Fig. 4 Passenger comfort levels for weighted r.m.s acceleration results for SC (*top left*), LC (*top right*) and DLC (*bottom*) using PP (*red cross*) and RHC (*blue dot*) controllers (color figure online)

6 Discussion

A numerical study was conducted, on the bicycle model of the passenger vehicle, using two lateral controllers PP and RHC-PP. Planar acceleration was measured to evaluate passenger comfort using the aforementioned controllers. RHC-PP was shown to improve path tracking and vehicle stability at higher speeds. Parametrically continuous paths were used, as reference paths, since they improve both path tracking and passenger comfort.

Mean acceleration results given in Fig. 3, show that there are no significant differences between both controllers at lower speeds. This is because of the negligible effect of side slip at lower speeds. At v = 5 m/s, RHC-PP results in lower acceleration values. However, both controllers could not be compared at higher speeds due to the instability of PP and its consequential failure to follow reference paths.

Standard r.m.s. weighted acceleration results are given in Fig. 4. These results show that RHC maintains acceleration values below the Uncomfortable region of 1 m/s^2. Similar to the mean acceleration results, no significant changes are noted at lower speeds. Comparing results presented in this paper, with our earlier work, we conclude that parametric continuity in reference paths is more influential to passenger comfort. A further study is needed to combine and then isolate the effects of parametric continuity and tracking control. This study also highlights the need to select an appropriate forward velocity for different maneuvers (i.e. path curvature and yaw) based on the desired comfort levels, as governed by Eq. (14).

Acknowledgments The first author recognizes the Australian government funding through the Australian Postgraduate Award (APA), Research Training Scheme (RTS) and Higher Degree Research Publication Grant (HDRPG) scholarships.

References

1. Morris, S., Humphrey, A., Pickering, A., Tipping, S., Templeton, I., Hurn, J.: National Travel Survey 2013. The Department for Transport, NatCen Social Research, London, UK (2013)
2. Santos, A., McGuckin, N., Nakamoto, H.Y., Gray, D., Liss, S.: Summary of Travel Trends: 2009 National Household Travel (NTS) Survery. U.S. Department of Transportation, New Jersey, SE, Washington, DC (2011)
3. Bureau of Transportation Statistics, Household Travel Survey Report: Sydney 2012/13. Transport for NSW, Sydney, NSW (2013)
4. Waldrop, M.M.: Autonomous vehicles: no drivers required. Nature **518**, 20 (2015)
5. Laurgeau, C.: Intelligent vehicle potential and benefits. In: Eskandarian, A. (ed.) Handbook of Intelligent Vehicles, pp. 1537–1551. Springer, London (2012)
6. Fagnant, D.J., Kockelman, K.M.: The travel and environmental implications of shared autonomous vehicles, using agent-based model scenarios. Transp. Res. Part C: Emerg. Technol. **40**(3), 1–13 (2014)
7. Bureau of Infrastructure Transport and Regional Economics (BITRE), Road Deaths Australia, Canberra ACT2014
8. Bureau of Transportation Statistics, National Transportation Statistics, U.S. Department of Transportation 2014
9. Saleh, L., Chevrel, P., Claveau, F., Lafay, J.F., Mars, F.: Shared steering control between a driver and an automation: stability in the presence of driver behavior uncertainty. IEEE Trans. Intell. Transp. Syst. **14**, 974–983 (2013)
10. He, J., McCarley, J.S., Kramer, A.F.: Lane keeping under cognitive load: performance changes and mechanisms. Human Factors: J. Human Factors Ergon. Soc. **56**, 414–426 (2014)
11. Lee, J.D.: Fifty years of driving safety research. Human Factors: J. Human Factors Ergon. Soc. **50**, 521–528 (2008)
12. Gonzalez, D., Perez, J., Lattarulo, R., Milanes, V., Nashashibi, F.: Continuous curvature planning with obstacle avoidance capabilities in urban scenarios. In: 2014 IEEE 17th International Conference on Intelligent Transportation Systems (ITSC), pp. 1430–1435 (2014)
13. Elbanhawi, M., Simic, M., Jazar, R.: Improved manoeuvring of autonomous passenger vehicles: simulations and field results. J. Vib. Control (2015)
14. Elbanhawi, M., Simic, M., Jazar, R.: Randomized bidirectional B-Spline parameterization motion planning. IEEE Trans. Intell. Transp. Syst. 1–1 (2015)
15. Craig, C.R.: Implementation of the pure pursuit path tracking algorithm. Carnegie Mellon University, Pittsburgh, Pennsylvania, USA CMU-R1-TR-92-01 (1992)
16. Thrun, S., Montemerlo, M., Dahlkamp, H., Stavens, D., Aron, A., Diebel, J. et al.: Stanley: the robot that won the DARPA grand challenge. In: Buehler, M., Iagnemma, K. Singh, S. (eds.) The 2005 DARPA Grand Challenge, vol. 36, pp. 1–43. Springer, Berlin, Heidelberg (2007)
17. Cheein, F.A., Scaglia, G.: Trajectory tracking controller design for unmanned vehicles: a new methodology. J. Field Robot. **31**, 861–887 (2014)
18. Serrano, M.E., Scaglia, G.J.E., Cheein, F.A., Mut, V., Ortiz, O.A.: Trajectory-tracking controller design with constraints in the control signals: a case study in mobile robots. Robotica FirstView 1–18 (2014)

19. Corke, P.: Robotics, Vision and Control: Fundamental Algorithms in MATLAB. Springer (2011)
20. Snider, J.M.: Automatic steering methods for autonomous automobile path tracking. Technical Report CMU-RITR-09-08, Robotics Institute, Pittsburgh, PA (2009)
21. Elbanhawi, M., Simic, M., Jazar, R.: Receding horizon lateral vehicle control for pure pursuit path tracking. J. Vib. Control (2016) (In press)
22. Mayne, D.Q.: Model predictive control: recent developments and future promise. Automatica **50**(12), 2967–2986 (2014)
23. Wang, L.L.: Model Predictive Control System Design and Implementation Using MATLAB. Springer Publishing Company, Incorporated (2009)
24. Gu, D., Hu, H.: Receding horizon tracking control of wheeled mobile robots. IEEE Trans. Control Syst. Technol. **14**, 743–749 (2006)
25. Attia, R., Orjuela, R., Basset, M.: Combined longitudinal and lateral control for automated vehicle guidance. Veh. Syst. Dyn. **52**, 261–279 (2014)
26. Falcone, P., Borrelli, F., Asgari, J., Tseng, H.E., Hrovat, D.: Predictive active steering control for autonomous vehicle systems. IEEE Trans. Control Syst. Technol. **15**, 566–580 (2007)
27. Beal, C.E., Gerdes, J.C.: Model predictive control for vehicle stabilization at the limits of handling. IEEE Trans. Control Syst. Technol. **21**, 1258–1269 (2013)
28. Murdin, L., Golding, J., Bronstein, A.: Managing motion sickness. BMJ **343** (2011)
29. Diels, C., Bos, J.E.: Self-driving carsickness. Appl. Ergonom. (2015)
30. Schoettle, B., Sivak, M.: Motion sickness in self-driving vehicles. The University of Michigan Sustainable Worldwide Transportation, The University of Michigan Transportation Research Institute, 2901 Baxter Road, Ann Arbor, Michigan 48109-2150 U.S.A. UMTRI-2015-12 (2015)
31. Elbanhawi, M., Simic, M., Jazar, R.: In the passenger seat: investigating ride comfort measures in autonomous cars. Intell. Transp. Syst. Mag. IEEE **7**, 4–17 (2015)
32. Le Vine, S., Zolfaghari, A., Polak, J.: Autonomous cars: the tension between occupant experience and intersection capacity. Transp. Res. Part C: Emerg. Technol. **52**(3), 1–14 (2015)
33. Turner, M., Griffin, M.J.: Motion sickness in public road transport: passenger behaviour and susceptibility. Ergonomics **42**, 444–461 (1999)
34. Elbanhawi, M., Simic, M.: Randomised kinodynamic motion planning for an autonomous vehicle in semi-structured agricultural areas. Biosyst. Eng. **126**(10), 30–44 (2014)
35. Elbanhawi, M., Simic, M., Jazar, R.: The role of path continuity in lateral vehicle control. Proc. Comput. Sci. **60**, 1289–1298 (2015)
36. Jazar, R.N.: Vehicle Dynamics: Theory and Application. Springer (2008)
37. Marzbani, H., Jazar, R., Fard, M.: Steady-state vehicle dynamics. In: Dai, L., Jazar, R.N. (eds.) Nonlinear Approaches in Engineering Applications, pp. 3–30. Springer International Publishing (2015)
38. Rajamani, R.: Lateral vehicle dynamics. In: Vehicle Dynamics and Control, pp. 15–46. Springer, US (2012)
39. Soudbakhsh, D., Eskandarian, A.: Vehicle lateral and steering control. In: Eskandarian, A. (ed.) Handbook of Intelligent Vehicles, pp. 209–232. Springer, London (2012)
40. Petrov, P., Nashashibi, F.: Adaptive steering control for autonomous lane change maneuver. In: Intelligent Vehicles Symposium (IV), 2013, pp. 835–840. IEEE (2013)
41. van Winsum, W.: Speed choice and steering behavior in curve driving. Human Factors **38**, 434+ (1996)
42. van Winsum, W., de Waard, D., Brookhuis, K.A.: Lane change manoeuvres and safety margins. Transp. Res. Part F: Traffic Psychol. Behav. **2**(9), 139–149 (1999)
43. Marzbani, H., Ahmad Salahuddin, M.H., Simic, M., Fard, M., Jazar, R.N.: Steady-state dynamic steering. Front. Artif. Intell. Appl. **262**, 493–504 (2014)
44. ISO 2631-1 (International Organisation for Standardisation): Mechanical Vibration and Shock—Evaluation of Human Exposure to Whole-Body Vibration—Part 1: General Requirements. ISO 2631-1 International Organisation for Standardisation, Geneva (1997)

From Automotive to Autonomous: Time-Triggered Operating Systems

Maria Spichkova, Milan Simic and Heinz Schmidt

Abstract This paper presents an approach for application of time-triggered paradigm to the domain of autonomous systems. Autonomous systems are intensively used in areas, or situations, which could be dangerous to humans or which are remote and hardly accessible. In the case when an autonomous system is safety critical and should react to the environmental changes running within a very limited time frame, we deal with the same kind of problems as automotive and avionic systems: timing properties and their analysis become a crucial part of the system development. To analyse timing properties and to show the fault-tolerance of the communication, a predictable timing of the system is needed.

1 Introduction

The trend in the automotive and avionics industry is to shift functionality from mechanics and electronics to software. Subsystems in automotive and avionics engineering express increasing level of autonomy as described in [12]. The goal of this changes is to make vehicles and aircrafts more reliable and fault-tolerant, focusing on the tasks that could be complicated for the human to solve quickly, precisely and safely. In avionics, fly-by-wire are used since many decades. These systems were developed to replace the manual (human) flight controls of an aircraft with an electronic interface. An example of a fly-by-wire system is Unmanned Aerial Vehicle (UAV) landing [22]. Many automotive manufacturers have presented drive-by-wire prototypes for premium cars. An example of these prototypes are Parking Assists Systems [9].

M. Spichkova (✉) · M. Simic · H. Schmidt
RMIT University, Melbourne, Australia
e-mail: maria.spichkova@rmit.edu.au

M. Simic
e-mail: milan.simic@rmit.edu.au

H. Schmidt
e-mail: heinz.schmidt@rmit.edu.au

G. De Pietro et al. (eds.), *Intelligent Interactive Multimedia Systems and Services 2016*, Smart Innovation, Systems and Technologies 55,
DOI 10.1007/978-3-319-39345-2_30

To analyse timing properties of a distributed system and to show the fault-tolerance of its behaviour, we require a predictable timing of the system. This can be solved using the time-triggered paradigm (TTP). In avionics TTP was applied successfully to distributed systems. This idea has been later propagated to the automotive domain.

In a time-triggered system (TTS), all actions are executed at predefined points in time. This ensures a deterministic timing behaviour of the system: task execution times and their order, as well as message transmission times are deterministic. Having a deterministic timing behaviour, we can predict and to prove formally the timing properties of the system with a reasonable effort.

In [36], we introduced a formal framework for modelling and analysis of autonomous systems (AS) and their compositions, especially focusing on the adaptivity modelling aspects and reasoning about adaptive behaviour. In our current work we extend this framework with the core features of a framework for the verification of properties for automotive TTS, presented in [20, 21]. In this paper, we suggest to apply the well-developed ideas of time-triggered systems within the domain of autonomous robotic systems, as timing properties and their verification are crucial for safety-critical systems.

Outline: The paper is organised as follows. Section 2 provides a brief overview of the related work. In Sect. 3, we discuss the application of TTP within automotive domain, time-triggered operating system OSEKtime as well as our previous research in this area. Section 4 introduces our TTP Model for AS, which is the core contribution of the paper. Section 5 introduces our model for OSEKtime. Section 6 present the core directions of our future work. In Sect. 7 we summarise the paper and propose directions for future research.

2 Related Work

The adaptation and context-awareness can have many forms: navigation applications to guide users to a given destination, robot motion [10], keyless entry systems [15], driver assistance applications [27], adaptive cruise control systems [14], etc. Next step from driver assistance and similar applications is complete takeover of the vehicle control. Reliable, real time robot (vehicle) operating system is extremely important for the safe and comfortable ride. A concise survey of concepts, architectural frameworks, and design methodologies that are relevant in the context of self-adapting and self-optimizing systems is presented in [2].

Rushby [26] introduced an approach to derive a time-triggered implementation from a fault-tolerant algorithm specified as a functional program. Nolte et al. [23] presented an overview of wireless automotive communication technologies. The goal of Nolte et al. was to identify the strong candidates for future in-vehicle and inter-vehicle automotive applications. There are also a number approaches on analysis of worst case execution time (WCET) and the corresponding properties of the software components. For example, Fredriksson et al. [16] presented a contract-based technique to achieve reuse of WCETs in conjunction with reuse of software components. A number of frameworks and methodologies for the formal analysis of time-triggered automotive systems were presented in [5–7, 17, 34]. The core features of these frameworks and methodologies can be applied within the field of AS, to increase reliability and to allow formal verification of the functional properties.

Both in the case of automotive and in the case of autonomous systems, we to take into account *cyber-physical* nature of these systems. Many approaches on *cyber-physical systems* do not include an abstract logical level of the system modelling, missing the advantages of the abstract representation. In our previous work [33], we introduced a platform-independent architectural design in the early stages of CPS development. The results of our ongoing work on simulation, validation and visualization of CPSs in industrial automation [3, 4] provide basis for the analysis and simulation of autonomous systems, as a special kind of CPSs.

3 TTP: OSEKtime Operating System

An operating system OSEKtime was developed by the European Automotive Consortium OSEK/VDX in accordance to the time-triggered paradigm. OSEK[1] is a standards body, founded by German automotive company consortium, in 1993. The consortium included many industrial partners (such as BMW, Robert Bosch GmbH, DaimlerChrysler, Opel, Siemens, and Volkswagen Group) as well as the University of Karlsruhe. The French automotive manufacturers Renault and PSA Peugeot Citroen had a similar consortium, VDX.[2] In 1994, a new consortium OSEK/VDX[3] was created, based on OSEK and VDX.

Thus, OSEKtime OS is time-triggered and supports static cyclic scheduling based on the computation of the WCETs of its tasks. The verification of an OSEKtime OS is being performed in the Verisoft-XT project [38]. WCET of the tasks can be estimated from a compiled program and the processor the program runs on, using the corresponding tool, e.g. an *aiT* analyser [1]. aiT statically compute tight bounds for WCETs of tasks in real-time systems, directly analyses binary executables, and is independent of the compiler and source code language.

As per OSEKtime specification [25], the core properties of this operating system are

- predictability,
- modular concept as a basis for certification,
- dependability,
- compatibility to the OSEK/VDX.

[1]German: Offene Systeme und deren Schnittstellen fr die Elektronik in Kraftfahrzeugen; English: Open Systems and their Interfaces for the Electronics in Motor Vehicles.

[2]Vehicle Distributed eXecutive.

[3]http://www.osek-vdx.org.

Thus, the system has a priori known behaviour, even under defined peak load and fault conditions. The dependability property is insured by a fault-tolerant communication layer FTCom (Fault-Tolerant Communication, cf. [24]) is This layer was introduced to insure interprocess communication within the OS and make task distribution transparent. It allows reliable operation through fault detection and fault tolerance.

In our previous research [20, 21], we presented a formal framework for the verification of application properties for time triggered systems, based on FlexRay and FTCom [19, 28]. FlexRay is a time-triggered communication protocol, developed by the FlexRay Consortium.[4] Primary application domain of the FlexRay protocol is distributed real-time systems in vehicles. A comparison of an established protocol TTP/C and FlexRay was presented by Kopetz in [18]. Kopetz came to the conclusion that FlexRay and TTP/C were designed against the same set of automotive requirements, but that there is a difference in goals: "*The inherent conflict between flexibility and safety is tilted towards flexibility in FlexRay and safety in TTP/C.*".

A distributed automotive system is built from a number of nodes. The difference between the nodes are the configuration data of each node and the applications running on them. On each node, there are three layers:

(1) Micro controller and FlexRay controller. The network cable connects the FlexRay controllers of all nodes.
(2) OSEKtime OS and OSEK FTCom.
(3) A number of applications, implementing the desired behaviour of the automotive system.

When we adopt these ideas to the domain of autonomous systems, a node becomes a single autonomous robot, and the connection between the robots has to be wireless. This implies that the FlexRay communication protocol might be not applicable for autonomous systems even after a number of modifications: FlexRay provides high-speed, deterministic and fault-tolerant communication, but it was designed with the focus on in-vehicle networking, to support x-by-wire applications. Thus, only the architecture of the layers (2) and (3) can be adapted for AS.

4 TTP Model for Autonomous Systems

The modelling language that we use in our approach is FOCUSST [32], which is an extension of the FOCUS language used in [20, 21]. It allows us to create concise but easily understandable specifications and is appropriate for application of the specification and proof methodology presented in [29, 37]. The FOCUSST language was inspired by FOCUS [8], a framework for formal specification and development of interactive systems. In both languages, specifications are based on the notion of

[4]Core members of the consortium are Freescale Semiconductor, Robert Bosch GmbH, NXP Semiconductors, BMW, Volkswagen, Daimler, and General Motors.

streams and channels (a channel is in effect a name for a stream). The FOCUSST specification layout also differs from the original one: it is based on human factor analysis within formal methods [30, 31, 35].

A system in our model consists of N autonomous robots, communicating with each other. On the software level, a robot $Robot^i$ ($1 \leq i \leq N$) has Mi applications $AppRobot^i_1, \ldots, AppRobot^i_{Mi}$ that are running under OSEKtime OS, cf. Fig. 1. The *FTComCNI* component of each robot consists of two subcomponents: the *FTCom* component and a *CNIbuffer* (buffer of the Communication Network Interface), cf. Fig. 2. *CNIbuffer* is used to store messages that have to be sent to other robots via the communication protocol. The local communication among the applications $AppRobot^i_1, \ldots, AppRobot^i_{Mi}$ is conducted directly via *FTCom*.

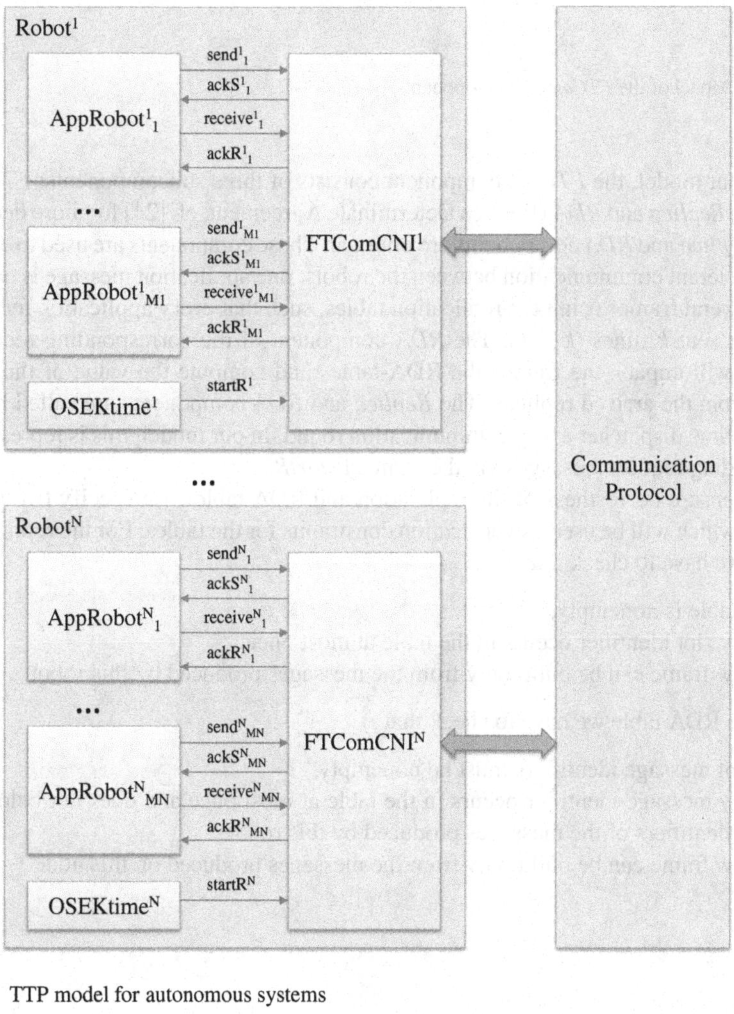

Fig. 1 TTP model for autonomous systems

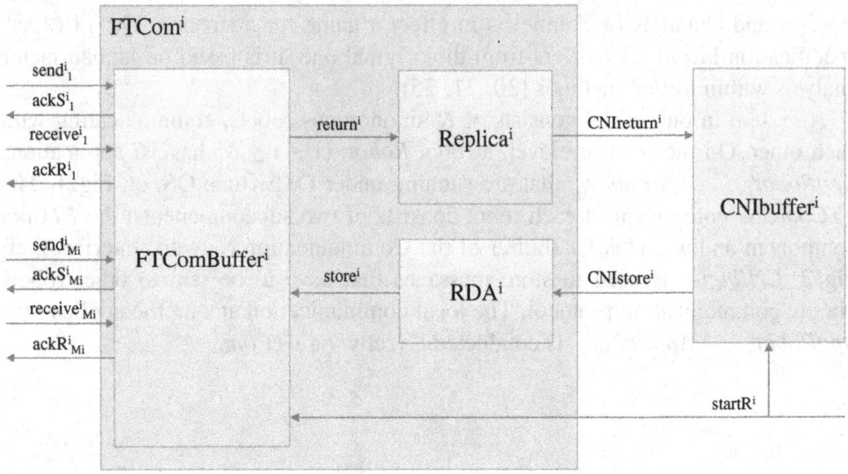

Fig. 2 Model of the *FTComCNI* component

In our model, the *FTCom* component consists of three subcomponents: *FTComBuffer*, *Replica* and *RDA* (Replica Determinate Agreement, cf. [24] for more details). The *Replica* and *RDA* components are optional. These components are used to ensure fault-tolerant communication between the robots: one application message is packed into several frames using the replication tables, such that every application message will be sent k times ($k > 1$). The *RDA* component of the corresponding receiver-robot, will unpack the frames the RDA-tables and compute the value of the message from the arrived replicas. The *Replica* and *RDA* components are called by the OSEKtime dispatcher every communication round. In our model, this is represented by sending request messages via the channel *startR*.

To ensure correctness of the replication and RDA tables, we specify two predicates, which will be used as verification constraints for the tables: For the replication table we have to check that

- the table is nonempty,
- every slot identifier occurs in the table at most once,
- a new frame can be build only from the messages produced by this robot.

For the RDA table we have to check that

- list of message identifiers must be nonempty,
- every message identifier occurs in the table at most once and does not belong to the identifiers of the messages produced by this robot.
- a new frame can be build only from the messages produced on this node.

5 Model of OSEKtime OS

We model OSEKtime as a tuple

$$OSEKtime = \langle Tasks,\ DT,\ State,\ SynchrP \rangle$$

where *Tasks* denotes the set of application tasks with the corresponding sets *State* of their internal states, *DT* is a dispatcher table, and *SynchrP* is a set of synchronisation parameters. An OSEKtime task can be in one of three states (*running*, *preempted*, *suspensed*), cf. also Fig. 3:

$$State = \{running,\ preempted,\ suspensed\}$$
$$SynchrP = \{synchr,\ asynchrH,\ asynchrS\}$$

OSEKtime allows two start-up techniques, which depend on synchronisation methods and availability of global system time [25]:

- Synchronous start-up (we model it as the parameter *synchr*). In this case, the tasks are not executed until a global time is available. For automotive systems based on the FlexRay protocol, the synchronous start-up has to be preferred, as FlexRay properties rely on a precise synchronisation. In the case of autonomous systems, we have to apply one of the options of asynchronous start-up.
- Asynchronous start-up: Tasks are executed according to the local time without waiting for the synchronisation to the global time.
  - In the case of a *hard synchronisation* (we model it as the parameter *asynchrH*), the synchronisation of the local time to the global time has to be performed at the end of a dispatcher round by delaying the start of the next dispatcher round.
  - In the case of a *smooth synchronisation* (we model it as the parameter *asynchrS*), the synchronisation of the local time to the global time is done during several dispatcher rounds by limiting the delay of the start of the next dispatcher round according to pre-defined configuration.

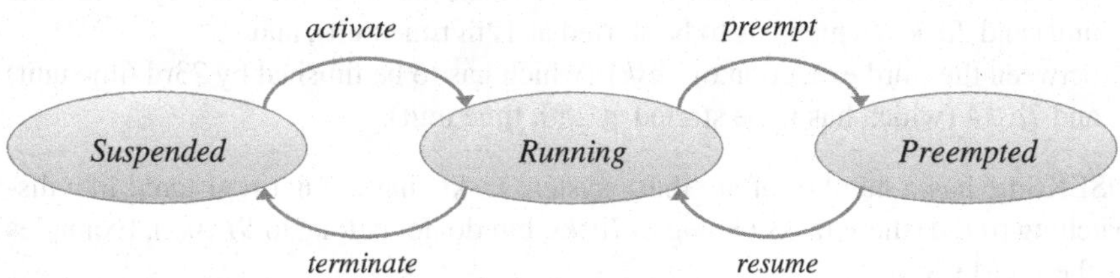

Fig. 3 Time-triggered model of OSEKtime tasks

Table 1 Example of an OSEKtime dispatcher table

DTentryID	Task	start	WCET
1	*Task*1	0	3
2	*Task*2	3	4
3	*Task*1	7	3
4	*Task*3	12	8
5	*Task*1	20	3
6	*Task*4	25	7
7	*Task*1	32	3

The specifications for set of tasks and the dispatcher table have to be introduced for each particular application. In general, they can be specified as follows:

$$DT = \langle N,\ STasks,\ start,\ wtime \rangle$$

where N is the length of the dispatcher table (number of items in it);

STasks is a subset of *Tasks*, which contains only the tasks to be scheduled, i.e., all the tasks from *Tasks* except auxiliary system tasks (such as *ttIdleTask*, *ErrorHook*, etc., cf. below);
$start = \{1 \dots N\}, STasks \rightarrow \mathbb{N}$ defines the mapping of tasks to their start times within the dispatcher round;
$wcet = \{1 \dots N\}, STasks \rightarrow \mathbb{N}$ defines the mapping of tasks to their worst-case execution times (WCETs).

Each task from *STasks* can also appear in *DT* as a number of subtasks. OSEKtime has a static scheduling, which allows specification based on WCETs of the subtasks.

Example An example of a dispatcher table $STasks = \{Task1, Task2, Task3, Task4\}$, $N = 7, T_{CYCLE} = 35$ is presented in Table 1. It is easy to see that while executing this dispatcher table, *ttIdleTask* will start at least twice in the middle of each round:

- between the second execution of *Task*1 (which has to be finished by 10th time unit) and *Task*3 (which has to be started at 12th time unit), and
- between the third execution of *Task*1 (which has to be finished by 23rd time unit) and *Task*4 (which has to be started at 25th time unit).

OSEKtime has a number of auxiliary system tasks that are not registered in a dispatching round (these tasks belong to *Tasks*, but do not belong to *STasks*). Examples of these tasks are

- *ttStartOS* routine starts the operating system,
- *ttIdleTask* acts as the idle task of the OSEKtime OS. It is the first task started by the OSEKtime dispatcher, and is always running if there is no other task ready. As this task is not registered in *DT*, it is not periodically restarted and does not

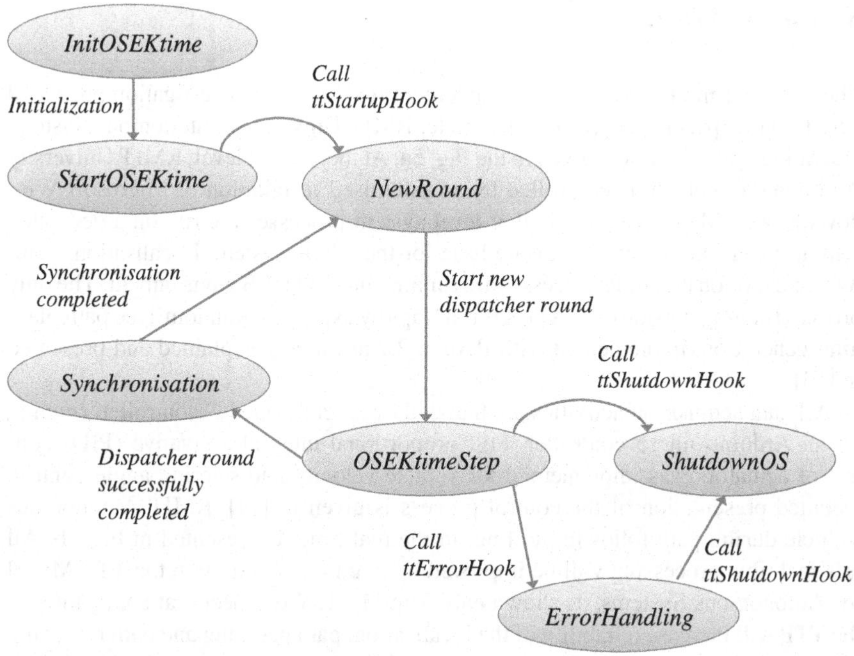

Fig. 4 OSEKtime as a state machine

have any deadline. *ttIdleTask* has the lowest priority and it can be interrupted by all interrupts handled by OSEKtime.

- *ErrorHook* task is responsible for error handling for deadline violations;
- *ttShutdown* task is responsible for shutdown of the system.

OSEKtime as an abstract state machine is presented in Fig. 4.

FTCom layer has the following core functions: *ttSyncTimes*, *ttSendMessage* and *ttReceiveMessage*. The *ttSyncTimes* function specifies the routine to synchronise the local time with the global time in the system. This function will be called by OSEKtime OS in the state *Synchronisation*. The functions *ttSendMessage* and *ttReceiveMessage* are called by tasks. *ttSendMessage* stores a message in the FT-CNI buffer, which is used to communication between nodes (in the case of a time-triggered automotive system, the content of the FT-CNI buffer has to be sent via FlexRay). *ttReceiveMessage* checks the FT-CNI buffer for the requested by an application message (i.e., the message with the requested id) and stores this message to the local buffer.

6 Future Work

The next step in our research is to implement results of the investigation presented here to our experimental electrical vehicle, RMIT University Autonomous System. Operating systems view is given in the Fig. 5a. At the highest level, RMIT University Autonomous Vehicle is controlled by a customised installation of Microsoft Windows Server. Majority of the higher level system processes are run on a dedicated Laptop PC as a computer hardware base for the whole system. Localisation, mapping and motion planning processes are running in MATLAB environment. The only process running outside of MATLAB is a Rapidly exploring Random Tree path planning generation, implemented with Python 2.7 as already explained and presented in [13].

All data acquisition activities are handled by an embedded C controller, running on an Arduino micro controller. Full proportional-integral-derivative (PID) control of actuators was implemented for vehicle velocity and steering angle control. Detailed presentation of the control process is given in [11]. RMIT Autonomous Vehicle during path following testing, in the real time, is presented in Fig. 5b. All this real time processing will be improved by aligning system with the TTP Model for Autonomous Systems, as shown earlier in Fig. 1. We expect that application of the TTP will increase reliability of the localisation, path planning and path following features, as it would allow us verify the corresponding timing properties in a formal way, also using semi-automated theorem provers. This would allow to use the RMIT University Autonomous System for safety-critical tasks, especially focussing on the interaction between humans and the AS.

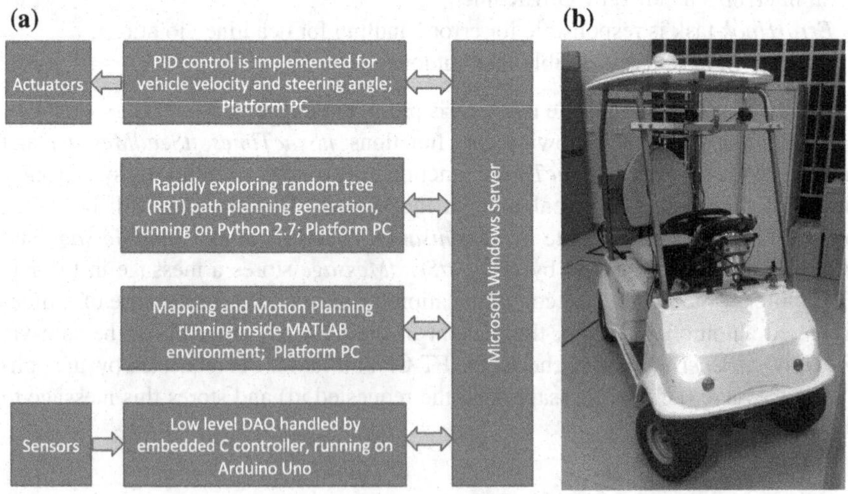

Fig. 5 RMIT autonomous vehicle. **a** Model to be aligned with TTP. **b** Path following testing

7 Conclusions

This paper presents an approach for application of time-triggered paradigm (TTP) to the domain of autonomous systems. TTP has been applied successfully to distributed systems in avionics and automotive domain. In this paper, we apply the well-developed ideas of time-triggered systems within the domain of autonomous robotic systems. The core feature of TTP is that all actions within the system have to be executed at predefined points in time. This ensures a deterministic timing behaviour of the system, which makes the system behaviour more predictable and provides many benefits for the analysis of the system properties.

We introduce a TTP model of an autonomous robotic system as well as the corresponding models of OSEKtime operating system and its Fault-Tolerant Communication layer. We also briefly discuss our future work on the application of the presented ideas to the experimental electrical vehicle, RMIT University Autonomous System.

References

1. aiT WCET Analyzer: Worst-Case Execution Time Analyzers. http://www.absint.com
2. Bauer, V., Broy, M., Irlbeck, M., Leuxner, C., Spichkova, M., Dahlweid, M., Santen, T.: Survey of modeling and engineering aspects of self-adapting and self-optimizing systems. Technical Report TUM-I130307, TU München (2013)
3. Blech, J.O., Spichkova, M., Peake, I., Schmidt, H.: Cyber-virtual systems: Simulation, validation and visualization. In: 9th International Conference on Evaluation of Novel Approaches to Software Engineering (ENASE 2014) (2014)
4. Blech, J.O., Spichkova, M., Peake, I., Schmidt, H.: Visualization, simulation and validation for cyber-virtual systems. In: Evaluation of Novel Approaches to Software Engineering, pp. 140–154. Springer International Publishing (2015)
5. Botaschanjan, J., Broy, M., Gruler, A., Harhurin, A., Knapp, S., Kof, L., Paul, W., Spichkova, M.: On the correctness of upper layers of automotive systems. Formal Aspects Comput. **20**(6), 637–662 (2008)
6. Botaschanjan, J., Gruler, A., Harhurin, A., Kof, L., Spichkova, M., Trachtenherz, D.: Towards modularized verification of distributed time-triggered systems. In: FM 2006: Formal Methods, pp. 163–178. Springer (2006)
7. Botaschanjan, J., Kof, L., Kühnel, C., Spichkova, M.: Towards verified automotive software. SIGSOFT Softw. Eng. Notes **30**(4), 1–6 (2005)
8. Broy, M., Stølen, K.: Specification and Development of Interactive Systems: Focus on Streams, Interfaces, and Refinement. Springer (2001)
9. Elbanhawi, M., Simic, M.: Examining the use of B-splines in parking assist systems. Appl. Mech. Mater. **490491** (2014)
10. Elbanhawi, M., Simic, M.: Sampling-based robot motion planning: a review. IEEE Access, **30**(99) (2014)
11. Elbanhawi, M., Simic, M., Jazar, R.: Improved manoeuvring of autonomous passenger vehicles: Simulations and field results. J. Vib. Control (2015)
12. Elbanhawi, M., Simic, M., Jazar, R.: In the passenger seat: investigating ride comfort measures in autonomous cars. IEEE Intell. Transp. Syst. Mag. **7**(3), 4–17 (2015)
13. Elbanhawi, M., Simic, M., Jazar, R.: Randomized bidirectional b-spline parameterization motion planning. IEEE Trans. Intell. Transp. Syst. **17**(2), 406–419 (2016)

14. Feilkas, M., Fleischmann, A., Hölzl, F., Pfaller, C., Scheidemann, K., Spichkova, M., Trachtenherz, D.: A top-down methodology for the development of automotive software. Technical Report TUM-I0902, TU München (2009)
15. Feilkas, M., Hölzl, F., Pfaller, C., Rittmann, S., Schätz, B., Schwitzer, W., Sitou, W., Spichkova, M., Trachtenherz, D.: A refined top-down methodology for the development of automotive software systems—the KeylessEntry system case study. Technical Report TUM-I1103, TU München (2011)
16. Fredriksson, J., Nolte, T., Nolin, M., Schmidt, H.: Contract-based reusableworst-case execution time estimate. In: Embedded and Real-Time Computing Systems and Applications, pp. 39–46. IEEE (2007)
17. Hölzl, F., Spichkova, M., Trachtenherz, D.: Autofocus tool chain. Technical Report TUM-I1021, TU München (2010)
18. Kopetz, H.: A comparison of TTP/C and FlexRay. Technical Report, TU Wien (2001)
19. Kühnel, C., Spichkova, M.: FlexRay und FTCom: Formale Spezifikation in FOCUS. Technical Report TUM-I0601, TU München (2006)
20. Kühnel, C., Spichkova, M.: Upcoming automotive standards for fault-tolerant communication: FlexRay and OSEKtime FTCom. In: EFTS 2006 International Workshop on Engineering of Fault Tolerant Systems (2006)
21. Kühnel, C., Spichkova, M.: Fault-tolerant communication for distributed embedded systems. In: Software Engineering of Fault Tolerance Systems, vol. 19, p. 175. World Scientific Publishing (2007)
22. Lu, K., Li, Q., Cheng, N.: An autonomous carrier landing system design and simulation for unmanned aerial vehicle. In: Guidance, Navigation and Control Conference (CGNCC), IEEE Chinese, pp. 1352–1356 (2014)
23. Nolte, T., Hansson, H., Bello, L.L.: Wireless automotive communications. In: Proceedings of the 4th International Workshop on Real-Time Networks (RTN?05), pp. 35–38 (2005)
24. OSEK/VDX: Fault-Tolerant Communication. Specification 1.0. http://portal.osek-vdx.org (2001)
25. OSEK/VDX: Time-Triggered Operating System. Specification 1.0. http://portal.osek-vdx.org (2001)
26. Rushby, J.: Systematic formal verification for fault-tolerant time-triggered algorithms. In: Dependable Computing for Critical Applications, vol. 11. IEEE (1997)
27. Simic, M.: Vehicle and public safety through driver assistance applications. In: Proceedings of the 2nd International Conference Sustainable Automotive Technologies (ICSAT 2010), vol. 490491, pp. 281–288 (2010)
28. Spichkova, M.: FlexRay: verification of the FOCUS specification in Isabelle/HOL. A case study. Technical Report TUM-I0602, TU München (2006)
29. Spichkova, M.: Specification and seamless verification of embedded real-time systems: FOCUS on Isabelle. Ph.D. thesis, TU München (2007)
30. Spichkova, M.: Human factors of formal methods. In: In IADIS Interfaces and Human Computer Interaction 2012. IHCI 2012 (2012)
31. Spichkova, M.: Design of formal languages and interfaces: "formal" does not mean "unreadable". In: Blashki, K., Isaias, P. (eds.) Emerging Research and Trends in Interactivity and the Human-Computer Interface. IGI Global (2014)
32. Spichkova, M., Blech, J.O., Herrmann, P., Schmidt, H.: Modeling spatial aspects of safety-critical systems with FocusST. In: 11th Workshop on Model Driven Engineering, Verification and Validation MoDeVVa 2014 (2014)
33. Spichkova, M., Campetelli, A.: Towards system development methodologies: from software to cyber-physical domain. In: First International Workshop on Formal Techniques for Safety-Critical Systems (FTSCS'12) (2012)
34. Spichkova, M., Hölzl, F., Trachtenherz, D.: Verified system development with the AutoFocus tool chain. In: 2nd Workshop on Formal Methods in the Development of Software, pp. 17–24. EPTCS (2012)

35. Spichkova, M., Liu, H., Laali, M., Schmidt, H.: Human factors in software reliability engineering. In: Workshop on Applications of Human Error Research to Improve Software Engineering. WAHESE'15 (2015)
36. Spichkova, M., Simic, M.: Towards formal modelling of autonomous systems. In: Intelligent Interactive Multimedia Systems and Services: 2015, KES-IIMSS, pp. 279–288. Springer (2015)
37. Spichkova, M., Zhu, X., Mou, D.: Do we really need to write documentation for a system? In: Model-Driven Engineering and Software Development (2013)
38. Verisoft XT Project. http://www.verisoftxt.de

Active Suspension Investigation Using Physical Networks

Milan Simic

Abstract Active suspension systems consist of mechanical, electrical, electronics and other subsystems. Different types of physical signals and forms of energy circulate between subsystems which make system design a challenging task. Various physical systems can be expressed using the same type of ordinary differential equations. This gives us a common platform for the complex systems' design. Using physical networks' approach modeling of an active suspension system is conducted and evaluated for various on road scenarios. Active suspension contributes to the ride comfort, safety, better and easier car control both by the human driver, or autonomous system.

Keywords Active suspension • Flat ride • Passenger comfort • Vehicle vibrations • Roll vibrations • Pitch vibrations • Physical network • Simulation

1 Introduction

While performing translatory motion, passenger car is subjected to road imperfections of different type, size and shapes, resulting in unwanted vibrations. They may cause serious damages to the vehicle. Main objective of this investigation is to ensure flat ride over road imperfections, by performing modeling of the moving vehicle and its suspension system. Suspension is interfacing between vehicle and the road. Systems could be relatively simple, i.e. built using passive spring type components. Car body movement, as reaction to the road conditions, is determined by the road and fixed suspension parameters: stiffness, mass and friction.

When adaptive, i.e. semi-active suspension is applied, suspension parameters are controlled by onboard computer. With active suspension implemented, movements of the wheels, along yaw axis, are detected by sensors and controlled by computer system via actuators mounted on the wheels. Actions performed improve ride

---

M. Simic (✉)
School of Engineering, RMIT University, Melbourne, Australia
e-mail: milan.simic@rmit.edu.au

G. De Pietro et al. (eds.), *Intelligent Interactive Multimedia Systems and Services 2016*, Smart Innovation, Systems and Technologies 55,
DOI 10.1007/978-3-319-39345-2_31

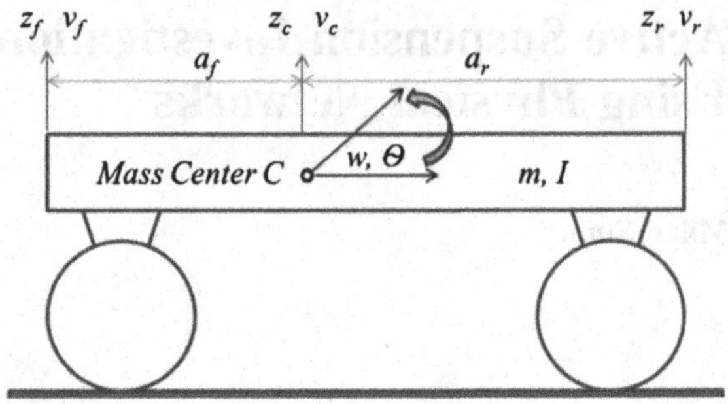

Fig. 1 Bicycle model of the car as a beam of mass *m* and moment of inertia *I*

comfort and enable better car control. Modern active suspension systems minimize roll and pitch vibrations. It is important when the vehicle performs a curvature trajectory, or when it increases, or decreases translatory motion speed. All vibrations, around pitch, roll, or yaw axis, disturb ride comfort in driver operated vehicles and even more in autonomous vehicles where passengers are subjected to loss of controllability experience, as presented in [1].

A conventional passenger vehicle, with four wheels, is a complex mechatronics system, but, it is often presented using simplified, two-wheel bicycle model as shown in Fig. 1. There is a large number of factors to consider when developing new suspension systems [2]. This model is used for vibration studies and flat ride investigations, as given in [3]. It is also used for autonomous vehicles' path planning in order to achieve maximum ride comfort [1]. Bicycle model is again used for autodriver algorithm development and implementation as explained in [4, 5]. Depending on the application, different views, descriptions and parameters of the bicycle model are considered. There are other models, like multi-body system where wheels are the end effectors on the mobile chassis [6]. On the other side we have quarter-car active suspension approach [7, 8]. One of the interesting approaches is the evaluation of active, energy regenerative suspension systems, using optimal control [9]. Bicycle model is used in our research to present the concept. It can easily be extended to full body model, or used for energy harvesting investigation. Model selection depends on the research objective targeted.

In our model, from Fig. 1, vehicle is represented as a beam of mass *m*, equally distributed along its length, $l = a_f + a_r$, and with a moment of inertia *I*. Length portion associated to the front is denoted as a_f while the length portion associated to the rear of the vehicle is denoted as a_r. This representation is a two degree-of-freedom (DOF) system. Mass center is located in *C* and it is center of rotation, i.e. bounce and pitch motion, with angular speed of *w* and angle *Θ*. We will investigate car body vertical velocities in three key points, front, mass center and rear of the vehicle. They are labeled as v_f, v_c, v_r, while displacements are z_f, z_c, z_r along vertical, *z* axes. Interfacing between the vehicle body and the road is conducted through suspension system and tires. Those components of the system will be presented in the next, revised, mechanical model of the vehicle.

In addition to that, further simplification, used in flat ride vibration studies, is to decouple bicycle model, i.e. separate systems equations that describe behavior of the front and the rear of the vehicle. Application of physical networks enables us to conduct more comprehensive vibration studies while taking into account mutual influences from forces acting on front, or rear vehicle's axles. Physical network modeling is possible thanks to the widespread analogies of ordinary differential equations used to represent various physical systems. Short introduction on physical network modeling of mechanical systems is given in the next section, while more on various engineering systems' modeling solutions could be found in [10–13].

2 Physical Networks

Physical systems are classified based on the type of equations used to express system behavior for the given input. Systems equations' coefficients are defined by physical system's parameters. Having in mind difficulties to solve some types of differential equations, we usually restrict our approach to investigations of systems which could be described by ordinary, linear differential equations with constant coefficients.

An ordinary differential equation (ODE) is a relation between two variables, an independent variable *t*, for time, dependent variable $y = y(t)$, together with the derivatives of *y*, being $\frac{dy}{dt}; \frac{d^2y}{dt}, \ldots, \frac{d^ny}{dt}$, as shown in Eq. (1):

$$A_n \frac{d^n y}{dt^n} + A_{n-1} \frac{d^{n-1} y}{dt^{n-1}} + \cdots + A_1 \frac{dy}{dt} + A_0 y = f(t) \tag{1}$$

Equation (1) is ordinary because only one independent variable exists, which is time. It is linear because only first exponent of dependent variable, or its derivatives, is present. Physical network is a class of linear graphs that represent physical system equations. Basic definitions, principles and mathematical rules for solving system of differential equations are independent of the physical system. There are two time dependent variables in each physical network: flow, *f*, and potential, *p*. Flow is a variable that streams through network elements and connection lines. Potential is a variable established and measured across network elements, or between any two network points. Potential of a point in the network depends on the chosen referent point. Each network has a basic reference i.e. ground point.

There are two types of network elements interconnected in a physical network: *active* and *passive*. Active elements are flow, or potential sources, whose operations are expressed as functions of time: $F = F(t)$ and $P = P(t)$. They supply energy to the system. Passive elements, with two connection points, define relationships between basic network variables: *flow* and *potential*. They cannot supply more energy to the system than what was already accumulated, if they can store energy.

Table 1 Physical network elements

Description	Generic	Electrical	Translation
Trough variable	*Flow—f*	*Current—i*	*Force—F*
Across variable	*Potential—p*	*Voltage—u*	*Velocity—v*
Element	A	Inductivity L	Stiffness k
Integration	$f = A\int pdt$	$i = \frac{1}{L}\int udt$	$F = k\int vdt = kx$
Stored energy	$\frac{f^2}{2A}$	$\frac{Li^2}{2}$	$\frac{F^2}{2k}$
Element	B	Conductivity G/Resistivity R	Damping constant B
Proportion	f = Bp	$i = Gu = \frac{1}{R}u$	F = Bv
Power dissipation	$fp = \frac{f^2}{B} = p^2B$	$iu = i^2R = \frac{u^2}{R}$	$Fv = \frac{F^2}{B} = v^2B$
Element	C	Capacity C	Mass m
Differentiation	$f = C\frac{dp}{dt}$	$i = C\frac{du}{dt}$	$F = m\frac{dv}{dt}$
Stored energy	$\frac{Cp^2}{2}$	$\frac{Cu^2}{2}$	$\frac{mv^2}{2}$

That is expressed through initial conditions. There are three types of passive elements which specify relationships between network variables. Those are *proportion, integration* and *differentiation*.

Examples of all three types of passive elements in a generic physical network are given in column 1 of the Table 1. Examples for an electrical network, i.e. electrical circuit, and a mechanical system with translation are given in the following columns of the same table. We could illustrate analogies in other systems, like mechanical with rotation, acoustical [12] and hydraulic.

Expressions for the accumulated energy and power dissipation in each of the systems are also given. In a generic system, elements that define relations between two network variables, as integration, proportion and differentiation are labeled as *A, B* and *C*, respectively, as shown in the table. Common names for all passive network elements are impedance, *Z*, or admittance *Y*. In electrical systems we have electrical current and voltage as network variables. Their relationships are expressed through inductivity, *L*, conductivity *G*, i.e. resistivity, $R = 1/G$, and capacity *C*. Inductivity is associated to integration, conductivity to proportion and capacity to differentiation.

For a mechanical system with translation we have *k* for stiffness, related to the integration, *B* for friction, as a proportional element, and *m* for the mass of the object expressing differentiation. Basic network variables in translatory systems are force, *F*, and velocity *v*. Equations for stored energy, by elements that perform integration, or differentiation of network variables, and power losses for the proportional elements, are also presented.

3 Modeling

3.1 Mechanical Model

Referring to the basic bicycle model from Fig. 1 we will analyze vehicle behavior when driving over a bump on the road. Road bump of amplitude A will generate vertical forces F_f *and* F_r defined by the Hooke's law, acting on the front and the rear of the vehicle. In our study we assume that they are the same, given by Eq. (2):

$$F_f = F_r = F = kA \tag{2}$$

Vehicle is shown as a beam of mass m and moment of inertia I, sitting on two sets of springs and friction elements representing two wheels. Vertical forces will cause rotation represented by an angle θ and angular speed w. Moment of inertia of a beam with mass m and length l is given by Eq. (3):

$$I = \frac{ml^2}{12} \tag{3}$$

Beam rotation radius, R, is given by Eq. (4):

$$R = \sqrt{\frac{I}{m}} \tag{4}$$

Vertical forces acting on the front axle, F_f, and on the rear axle, F_r, are together moving body mass center as defined by Eq. (5):

$$F_f + F_r = m\frac{dv_C}{dt} \tag{5}$$

For the further study we have to include moments of inertia, in reference to the mass center, as given by Eq. (6):

$$F_r a_r - F_f a_f = I\frac{dw}{dt} = mR^2\frac{dw}{dt} \tag{6}$$

Total mass of the vehicle can be split into masses that correspond to the front, m_f, and the rear part of the body, m_r, together with a mutual mass, m_m, between them, as shown in the Fig. 2. Forces acting on the front and rear end of the vehicle are shown as ideal force sources. Mutual mass is introduced to express influences of the forces acting on one part of the vehicle to the other and vice versa. It is presented in [14], that all three masses can be determined through calculations given by the set of following Eqs. (7)–(9):

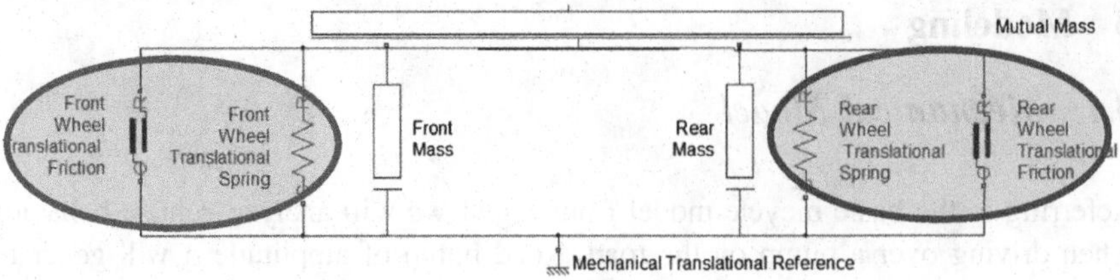

Fig. 2 Mechanical model showing coupled system through mutual mass

$$m_f = m\frac{a_r^2 + R^2}{l^2} \tag{7}$$

$$m_r = m\frac{a_f^2 + R^2}{l^2} \tag{8}$$

$$m_m = m\frac{a_f a_r - R^2}{l^2} \tag{9}$$

From the last equation, when $m_m = 0$, we have decoupling condition. In that case, specified by Eq. (10), we can treat parts of the vehicle's body separately.

$$R^2 = a_f a_r \tag{10}$$

The same condition could be derived by the evaluation of motion vibration equations, as shown in [3]. This study is more comprehensive and we will not decouple the system.

System description as given by Eq. (11). We concentrate on mass of the vehicle modeling, because of the mutual influences. Other elements, circled in Fig. 2 will be included later.

$$\begin{bmatrix} m_f & m_m \\ m_m & m_r \end{bmatrix} \begin{bmatrix} \frac{dv_f}{dt} \\ \frac{dv_r}{dt} \end{bmatrix} = \begin{bmatrix} F_f \\ F_r \end{bmatrix} \tag{11}$$

We will perform modeling of the vehicle, together with the road conditions and their interaction. In our scenario vehicle is riding, with a constant speed, over a road bump. That will generate two upward forces acting on front and on the rear axle. Actions are delayed in time, defined by the translatory speed of the vehicle. In this study we have assumed that a pulse function is good representation of the bump. It has amplitude of $A = 0.05$ m and the length of 1 m. Kinodynamic characteristics of the vehicle are given as following: vehicle mass is $m = 1500$ kg, length is $l = 4$ m, front and rear interfacing springs are the same type and equal to $K_f = K_r = K = 2000$ N/m, friction components are also the same and have value of $B_f = B_r = 10$. Translatory speed of the vehicle is $v_t = 10$ m/s.

We assume that the center of the mass is in the middle of the beam. Radius of rotation can be found using Eqs. (3) and (4). It is equal to $R = l/(2\sqrt{3}) = 2\sqrt{3}/3$. In physical networks *meter, kilogram* and *second* (MKS) system of units is used. That simplifies transitions between systems. Using Eqs. (7)–(9) values for the masses m_f, m_m, m_r are calculated and presented by a vector given in Eq. (12).

$$\begin{bmatrix} m_f \\ m_m \\ m_r \end{bmatrix} = \begin{bmatrix} 500 \\ 250 \\ 500 \end{bmatrix} \tag{12}$$

3.2 Electrical Model

As a next step, for the simplicity of modeling, we have transferred our system into an electrical circuit as given in Fig. 3a. Mechanical force corresponds to electrical current, while velocity corresponds to voltage. Stiffness k corresponds to inductivity, L, as $k \sim 1/L$. Damping constant B corresponds to conductivity G, i.e. resistivity R, as $B \sim G = 1/R$. Finally, mass m (kg), corresponds to capacity $C(F)$, as $m \sim C$. Numerical values for elements and variables are the same across various networks, all in MKS system. Forces, acting on axles, are represented as electrical current sources, while three mass elements are given as capacitors.

Equation (13) define electrical circuit from Fig. 3a:

$$\begin{bmatrix} C_f & C_m \\ C_m & C_r \end{bmatrix} \begin{bmatrix} \frac{dv_f}{dt} \\ \frac{dv_r}{dt} \end{bmatrix} = \begin{bmatrix} I_f \\ I_r \end{bmatrix} \tag{13}$$

Next model transformation is to simplify it as shown in the Fig. 3b. New system equations are (14) to (16). We have to find correspondence between two circuits.

$$C_f^{'}\left(\frac{dV_f}{dt} - \frac{dV_m}{dt}\right) = I_f \tag{14}$$

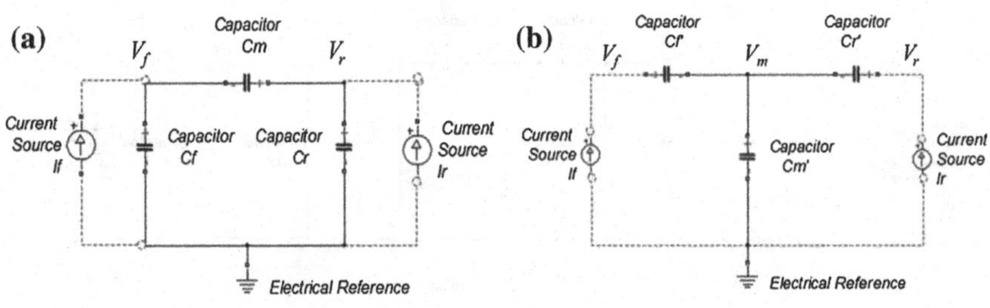

Fig. 3 **a** Vehicle as an equivalent electrical network. **b** Simplified electric circuit

$$C_r'\left(\frac{dV_r}{dt}-\frac{dV_m}{dt}\right)=I_r \tag{15}$$

$$C_m'\frac{dV_m}{dt}=I_f+I_r \tag{16}$$

By comparing Eq. (13) with Eqs. (14)–(16) we can derive analytical expressions for circuit elements from Fig. 3b, as given in [14]. Corresponding numerical values for our particular system are given by Eq. (17).

$$C_f'=C_r'=250;\quad C_m'=-750 \tag{17}$$

4 Final Model for Vibration Studies

Model, which now includes wheel parameters, is given in Fig. 4. We have friction and stiffness interfacing components to the road. Road is represented as electrical ground. Stiffness is given as its analog quantity, inductivity ($L = 1/k$). Friction is represented with its analog quantity, resistivity ($R = 1/B$). Mass is given as capacity ($m = C$).

Mass of the wheels is represented as C_{wf} and C_{wr}. Energy is supplied to the system by front and rear controlled current sources that correspond to forces acting on the axles. Force sources, as per Hooke's law, are given by Eq. (2). Models presented up to this point are not functional models. They are used to demonstrate steps and concepts in the system design. Following models are fully functional, running in Simulink environment, based on MATLAB R2012b.

Complete vehicle model, including scenario of the ride over the road bump, is given in Fig. 5. In addition to system components, already described, data acquisition (DAQ) and display devices are also presented. Integrators are added to calculate displacements based on vertical velocities.

Front vertical velocity is shown in Fig. 6, while displacements are presented in Fig. 7.

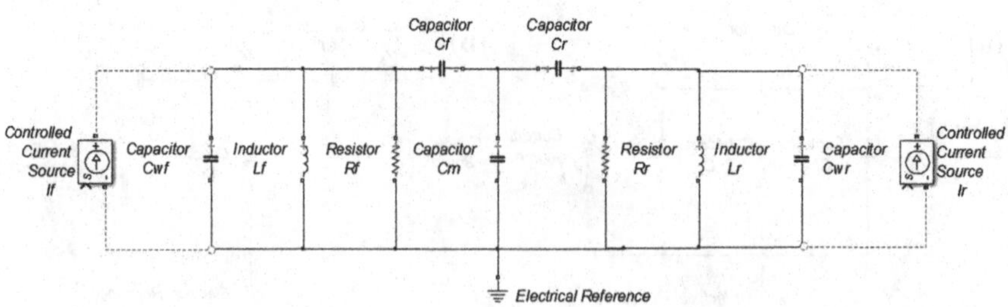

Fig. 4 Model of the vehicle on the road represented as an equivalent electrical circuit

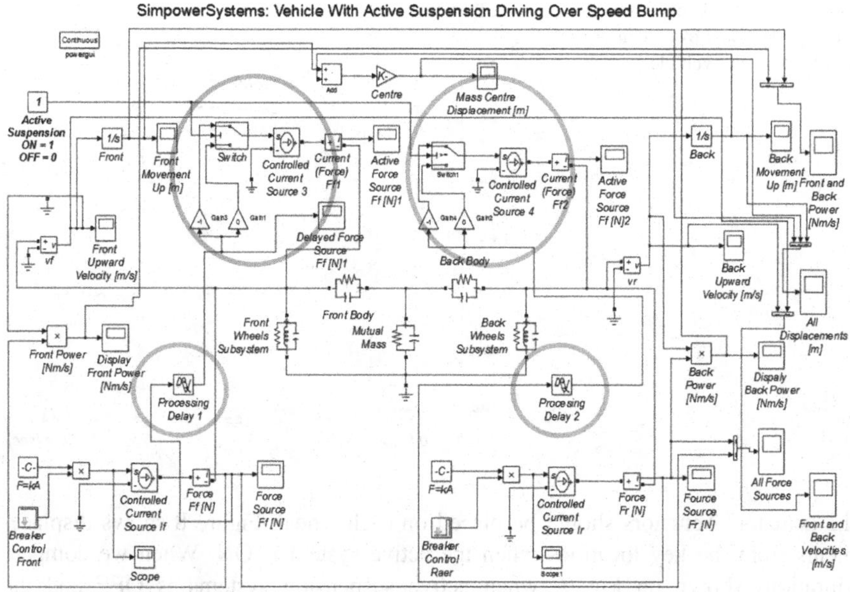

Fig. 5 Final model of the vehicle with active suspension components circled

Figures 6 and 7 show behavior of a vehicle with standard passive suspension system. Model need fine tuning and testing with sets of real suspension data. Our next experiment is to apply active suspension by detecting the bump and applying counter force, using actuators. Elements of the active suspension system are circled in Fig. 5. Signal is coming from the DAQ current sensor. It goes through delay block simulating DAQ electronics processing delay. Switch is used for the selection of the suspension mode of operation. After the switch, controlled force is applied to

Fig. 6 Vertical velocity at the front of the vehicle generated by a bump on the road

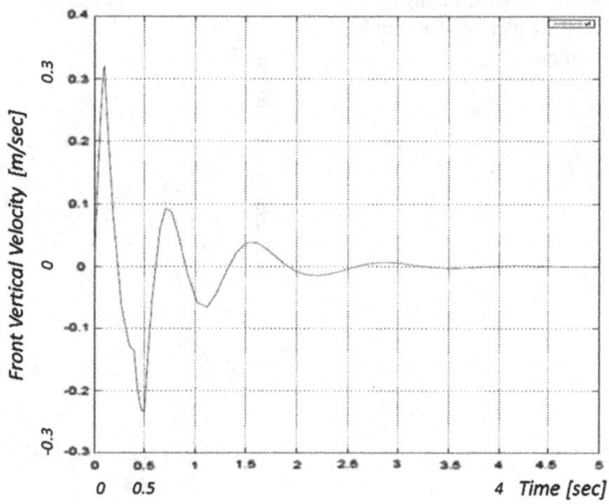

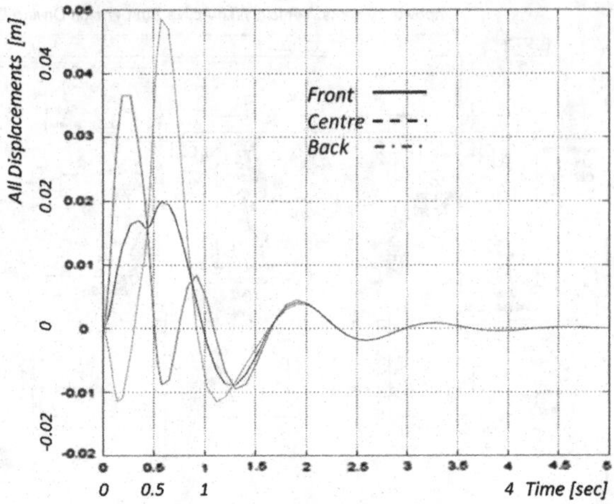

Fig. 7 Displacements from the key locations: front, center and back of the vehicle

the actuator. Actuators should be placed on each wheal. Figure 8 shows displacements from the key locations when the active system is ON. When we compare vibrations shown on Fig. 7, when active suspension systems is OFF and the vibrations from the Fig. 8, when active suspension systems is ON, we can see that the shapes of the signals are similar, but the main difference is in the magnitude. With the active suspension displacements are 10 times smaller than without it. We assumed that the DAQ processing delay and actuator delay are in the range of *ms*. For quicker responses, i.e. less vibrations, we need shorter delays. It is interesting to mention here that sliding mode control uses inertial delay for active suspension with full car model [15]. On the other side quarter-car active suspension model is used for adaptive tracking control of vehicle suspensions with actuator saturations [7].

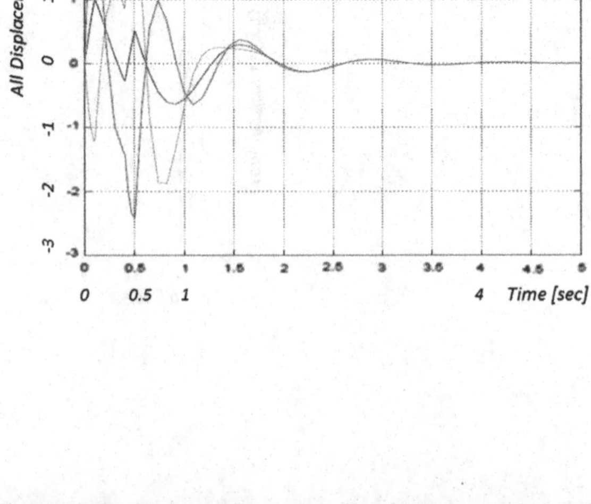

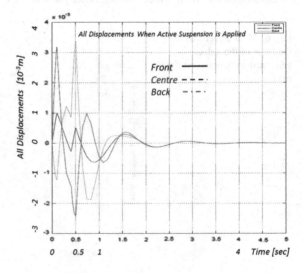

Fig. 8 Displacements with active suspension engaged are for the order of magnitude smaller

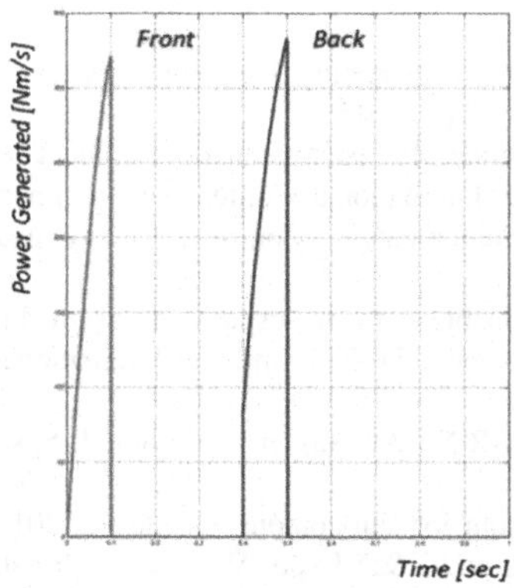

Fig. 9 **a** Power generated by riding over a road bump **b** Vehicle vibration lab test rig

Solar and thermal energy harvesting are already subjects of intensive research [16], but recovery of the energy dissipated while driving over the bumpy road, will be subject to another investigation. Modeling approach presented here gives us a powerful tool for that research as well. Instant power in mechanical systems is product of force and velocity. Power from any location on the vehicle body, can easily be calculated and presented for the investigation [10]. Power generated as result of the driving over the road bump, from front and from the back of the vehicle is shown in Fig. 9a.

Figure 9b shows School of engineering vehicle vibration lab test rig. It will be used for further research and our modeling and simulation methodology evaluation, ride comfort studies and power harvesting.

5 Conclusion

Engineering systems' modeling using physical networks' is a new approach with widespread potential. Vehicle suspension systems, which include various physical systems, can easily be represented by electrical circuits' models. This modeling approach covers both passive and active systems. Active systems design is based on the latest development in sensing and actuation technology. Model enables investigations with simulated dips, or bumps on the road. Various scenarios can easily be examined, with, or without active suspension engaged. Further investigations will be realized by conducted hardware in the loop testing and tuning in the RMIT University vibration lab. Flat ride is important for driver operated vehicle, but probably even more for autonomous vehicles where passengers develop loss of controllability experience.

References

1. Elbanhawi, M., Simic, M., Jazar, R.: In the passenger seat: investigating ride comfort measures in autonomous cars. Intell. Transp. Syst. Mag. IEEE **7**, 4–17 (2015)
2. Srinivasan, S., Shanmugam, P., Solomon, U., Sivakumar, P.: Factors in development of semi-active suspension for ground vehicles. In: 2015 International Conference on Smart Technologies and Management for Computing, Communication, Controls, Energy and Materials (ICSTM), pp. 581−589 (2015)
3. Marzbani, H., Simic, M., Fard, M., Jazar, R.N.: Sustainable flat ride suspension design. In: Intelligent Interactive Multimedia Systems and Services, pp. 251–264. Springer International Publishing, Switzerland (2015)
4. Bourmistrova, A., Simic, M., Hoseinnezhad, R., Jazar, R.N.: Autodriver algorithm. J. Syst. Cybern. Inf. **9**, 8 (2011)
5. Jazar, R.N., Simic, M., Khazaei, A.: Autodriver algorithm for autonomous vehicle. In: 2010 ASME International Mechanical Engineering Congress and R&D Expo, Vancouver, British Columbia, Canada (2010)
6. Max, G., Lantos, B.: Active suspension, speed and steering control of vehicles using robotic formalism. In: 2015 16th IEEE International Symposium on Computational Intelligence and Informatics (CINTI), pp. 53–58 (2015)
7. Zhang, J., Wang, J.: Adaptive tracking control of vehicle suspensions with actuator saturations. In: 2015 34th Chinese Control Conference (CCC), pp. 8051–8056 (2015)
8. Mittal, R., Bhandari, M.: Design of robust PI controller for active suspension system with uncertain parameters. In: 2015 International Conference on Signal Processing, Computing and Control (ISPCC), pp. 333–337 (2015)
9. Jun, Y., Xinbo, C., Jianqin, L., Shaoming, Q.: Performance evaluation of an active and energy regenerative suspension using optimal control. In: 2015 34th Chinese Control Conference (CCC), pp. 8057–8060 (2015)
10. Simic, M.: Vehicle modelling using physical networks. In: Small Systems Simulation Symposium 2016, Nis, Serbia (2016)
11. Simic, M.: Physical networks' approach in train and tram systems investigation. In: Nonlinear Approaches in Engineering Applications, 4th edn. Springer (2016)
12. Simic, M.: Exhaust system acoustic modeling. In: Dai, L., Jazar, R.N. (eds.) Nonlinear approaches in engineering applications, pp. 235–249. London Springer International Publishing, Cham, Heidelberg, New York, Dordrecht (2015)
13. Sanford, R.S.: Physical Networks. Prentice-Hall, Englewood Cliffs, N.J. (1965)
14. Simic, M.: Physical systems analysis using ECAP. Masters Research, Electronics Engineering, University of Nis, Nis, Serbia, Yugoslavia (1978)
15. Bhowmik, A., Marar, A., Ginoya, D., Singh, S., Phadke, S.B.: Inertial delay control based sliding mode control for active suspension with full car model. In: 2016 IEEE First International Conference on Control, Measurement and Instrumentation (CMI), pp. 376–380 (2016)
16. Royale, A., Simic, M.: Research in vehicles with thermal energy recovery systems. In: Procedia Computer Science, vol. 60, pp. 1443–1452 (2015)

Automatic Generation of Trajectory Data Warehouse Schemas

Nouha Arfaoui and Jalel Akaichi

Abstract A mobile object is a spatial object that changes the form and the location permanently over the time. Each displacement creates a trajectory that reflects the evolution of its position in space during a given time interval. It generates, then, a huge amount of trajectory data that are stored into trajectory data warehouse because it is the only tool that can analysis the historical trajectory data. In this work, we focus on the design of trajectory data warehouse schema and we propose automating this task to reduce human intervention since it is done manually and requires good knowledge of the domain. To achieve this goal, firstly, we automate the extraction of trajectory data mart schemas from a moving data base. Then, we merge them to get the trajectory data warehouse schema using a new schema integration methodology that is composed by schema matching and schema mapping.

Keywords Trajectory data warehouse schema · Trajectory data mart schema · Schema integration · Animal movement

1 Introduction

Thanks to the emergence of location-aware devices, mobile communication systems, GPS, etc., a huge amount of TrD is generated once following the moving object. The latter is defined as a spatial object that changes the form and the location permanently over the time. It is distinguished by spatial components that evolve over time. The movement of such object creates a trajectory that corresponds to "the evolution of the position (perceived as point) of an object that is moving in space during a given time interval in order to achieve a given goal" [1]. Trajectory Data (TrD) is stored

N. Arfaoui (✉) · J. Akaichi
Department of Information Systems, BESTMOD - Institut Suprieur de Gestion,
King Khalid University, Tunis, Tunisia
e-mail: arfaoui.nouha@yahoo.fr

J. Akaichi
e-mail: jalel.akaichi@kku.edu.sa

G. De Pietro et al. (eds.), *Intelligent Interactive Multimedia Systems and Services 2016*, Smart Innovation, Systems and Technologies 55,
DOI 10.1007/978-3-319-39345-2_32

into Trajectory Data Warehouse (TrDW). The traditional data warehouse [2] and the spatial data warehouse [3] are not created to deal with this kind of data because of its continuous evolution over the time and space.

To ensure a good TrDW, it is important to start by generating its corresponding schema. Many software tools exist, but they describe the schemas manually making this task tedious, error-prone and time-consuming [4].

In this work, we focus on the generation of TrDW schemas, and we opt then for the bottom up approach. It starts by generating automatically the Trajectory data mart schemas (TrDM) from the moving data bases. Then, it automates the generations of the TrDW by merging the TrDW schemas using the new schema integration methodology. The latter generates the mapping rules taking into account the semantic similarities of the elements of the schemas and the existing conflicts. Then, it transforms the generated rules into queries and executes them.

In order to well understand the proposed approach, we take a herd of animals as a moving object. It moves continuously to satisfy its nutritional needs. By this way, it can develop a grazing habit reflecting a form of adaptation and complex dynamic interactions with its environment especially the vegetation. The intensive exploitation of natural resources provokes the rarefaction of species of high value pastoral which influences badly the ecological balance and aggravates the fragility of the environment until a large part of those areas risk irreversible desertification. The different data collected from this movement should be stored into TrDW. This can help later to understand the habits and then to intervene, for example, by changing its future trajectory to protect the vegetation into specific place.

The outline of this work is as following. In Sect. 2, we focus on the state of the art where we present some work related to the TrDW. In Sect. 3, we define the model of the trajectory of a herd. Then, we move to the generation of TrDM from a moving data base. Next, we define the followed steps to build the final schema of TrDW. We finish this work with conclusion and future work.

2 Related Works

In the literature, many works have been proposed to deal with TD and/or TrDW. In the following, we present some of them.

Indeed, the authors in [5], introduce the notion of TrD and proposes the TrDW to storage such data to transform the trajectory raws into useful information.

Braz et al. in [6], introduce its model and its corresponding issues.

The authors, in [7], discuss the loading phase of the DW that has to deal with overwhelming streams of trajectory observations.

In [8], a framework is proposed to transform the traditional data cube model into a trajectory warehouse.

The authors, in [9], present a framework for improving the design and the implementation of TrDWs for analyzing traffic data.

In [10], the authors develop an application to receive the stream data set, to store and compute the pre-aggregation values and to present final results in order to reveal the knowledge about the trajectories.

Also, there are different works that propose a real application of TrDW. For example, we can mention, [11] who study the human movements' behavior, [12] who followed the displacement of the seal, [13] who model the trajectory of a mobile hospital, etc.

3 Trajectory Data Modeling

In this section, we focus on modeling TrD. We use, then, the Entity-Relationship (ER) model to present the different components that can influence any trajectory as well as their relationships (Fig. 1). Indeed, a herd is a set of animals belonging to

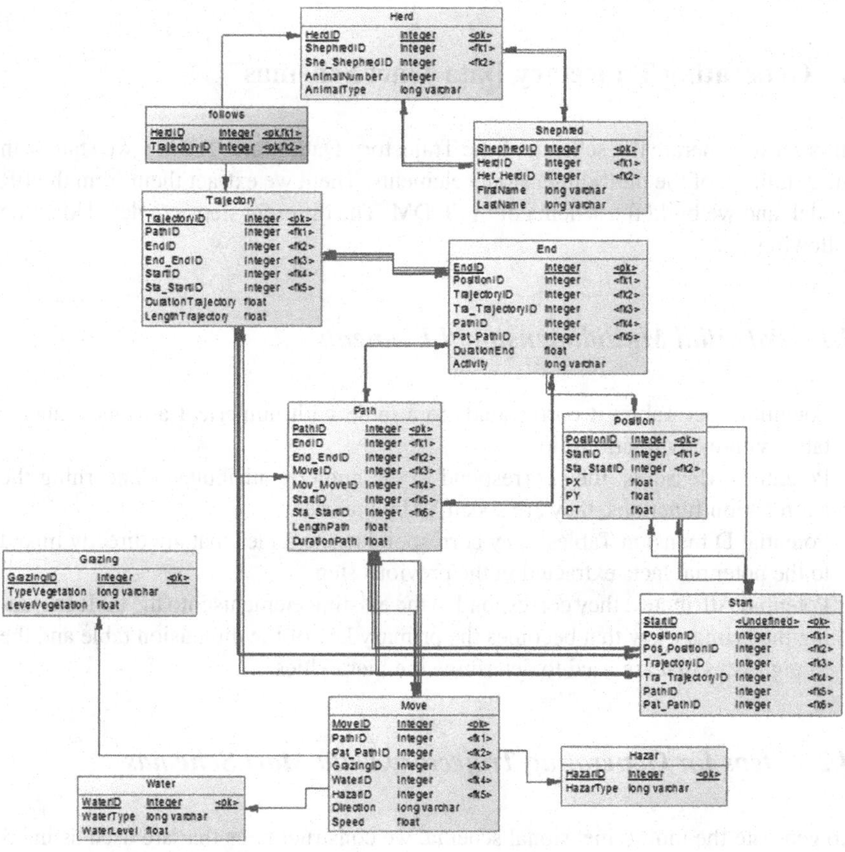

Fig. 1 A moving data base schema

the same type. It can be supervised by a shepherd in the case of domestic animals or not in the case of wild animals. The herd follows one specific trajectory. In this case, a trajectory is considered as a set of paths having different directions, lengths, and duration. So the duration of the trajectory is the sum of the duration of its composed paths. Each path is defined by "start" and "end" and the same thing concerning the trajectory which has a start that represents the start of the first path and an end that represents the end of the last path. Talking about the start, it gives information about the start position thanks to "PX" and "PY" and start time thanks to "PT". Concerning the end, it gives information about the final position thanks to "PX" and "PY" and the final time that is presented by an interval. The end corresponds to a stop where the animals can have different activities such as eating, drinking, or resting. So that, we have $TStart_{(N)} = TEnd_{(N-1)} +$ duration of $stop_{(N-1)}$.

Between a "start" and an "end", the herd moves. This movement is in function of the food so it happens generally close to the water especially during the dry seasons and in grazing to allow the animal eating. The animal can find during their displacement some hazards which may be natural (sea, river, lake, mountain or desert) or artificial (road, complex, or railway).

4 Generating Trajectory Data Mart Schemas

In order to generate the schema of the Trajectory Data Mart (TrDM), we start with the definition of the multidimensional elements. Then, we extract them from the ER model, and we build the schema of the TrDM. The different steps are detailed in the following.

4.1 *Potential Multidimensional Elements*

- Potential Fact Table: it corresponds to a table with numerical attributes and/or table with n-ary relationship.
- Potentials Measures: they correspond to the numeric attributes. Concerning the aggregation functions, they are specified by the user.
- Potential Dimension Tables: they correspond to the tables that are directly linked to the potential facts extracted in the previous step.
- Potential Attributes: they correspond to the existing elements into the table including the primary key that becomes the primary key of the dimension table and the foreign keys that are used to determine the hierarchies.

4.2 *Steps for Generating Trajectory Data Mart Schemas*

To generate the multidimensional schema, we construct trees that are used as intermediate models. They contain the extracted multidimensional elements. Next, each tree is transformed to multidimensional schema.

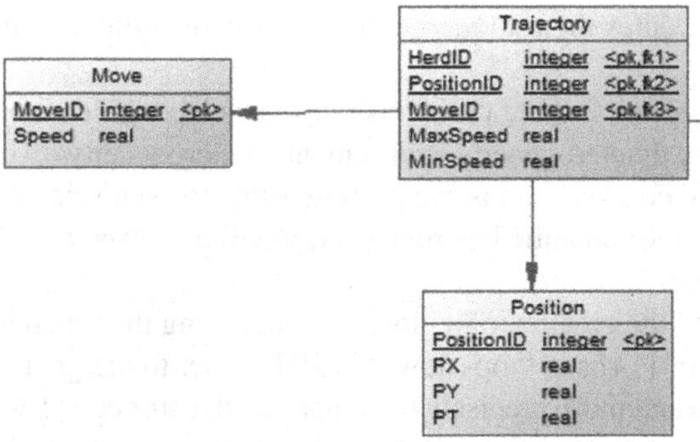

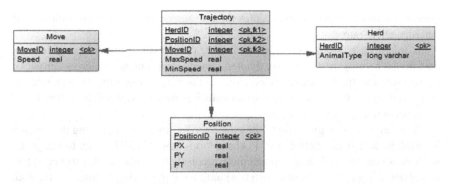

Fig. 2 First example of TrDM schema

Step 1: Build the trees from ER model

From the ER model, we extract the entities (Ef) having n-ary relationships with other entities and/or those having numerical attributes. They represent the potential facts. Every "Ef" becomes the root of the tree. The number of trees corresponds to the number of "Ef" entities.

From the ER, we extract the entities (E) that are directly linked to "Ef" corresponding to the potential dimensions. They form the first level of the tree.

For every "E", we extract the set of entities that are bonded via "-to-one" relationship.

Step 2: Transform the trees to multidimensional model

- The fact is created having as name "Trajectory".
- The existing numeric attributes become the potential measures.
- The root becomes a dimension keeping its attributes.
- The identifier of the root becomes a foreign key in the new fact table.
- The measures are defined by an aggregation functions that are specified by the user.
- The nodes that are directly linked to the roots are transformed to dimensions keeping their attributes and their identifiers.

Figure 2 presents an example of TrDM schema generated from Fig. 1.

5 Building Trajectory Data Warehouse Schema

The construction of TrDW schema is about merging the set of TrDM schemas. This task is achieved by using the new schema integration methodology that deals mainly with star schemas. It is composed by schema matching and schema mapping. The schema matching is used to extract the semantically closest elements as well as the conflicts, and presents them as mapping rules. The schema mapping transforms the generated rules to queries and executes them to merge the schemas regardless their heterogeneity source.

In order to use the schema integration three problems must be solved which are: data model heterogeneity, structural heterogeneity, and semantic heterogeneity.

To overcome the first and the second problems, we propose using the star schema as a common model. Concerning the semantic heterogeneity, it will be solved using the schema matching and mapping.

There are two strategies behind the creation of a global schema using the schema integration, which are "bottom-up" [14] and "top- down" [15]. In order to merge the schema, we opt for the bottom-up and more precisely the binary ladder strategy [16]. It is about merging two schemas each time until having one single schema at the end.

5.1 Schema Matching

The schema matching is used to find semantic correspondence between the elements of the two schemas and presents them as mapping rules. The latter will be used to facilitate the merging of schemas. This phase takes as input two or many schemas to get as output set of mapping rules facilitating the merge of the schemas.

The schema matching is composed by the following steps:

- Categorization: It specifies the category of each element. This reduces the risk of error which provides a gain of time.
  - Schema 1 (Fig. 2):
    Fact: {Trajectory}; FactKeys {HerdID (integer), PositionID (integer), MoveID (integer)}.
    Measure: {MaxSpeed (real), MinSpeed (real)}.
    Dimension, DimensionKey and Attribute: {Move {MoveID (integer), Speed (real)}, Position{PositionID (integer), PX (real), PY (real), PT (real)}, Herd {HerdID (integer), AnimalType (long varchar)}}.

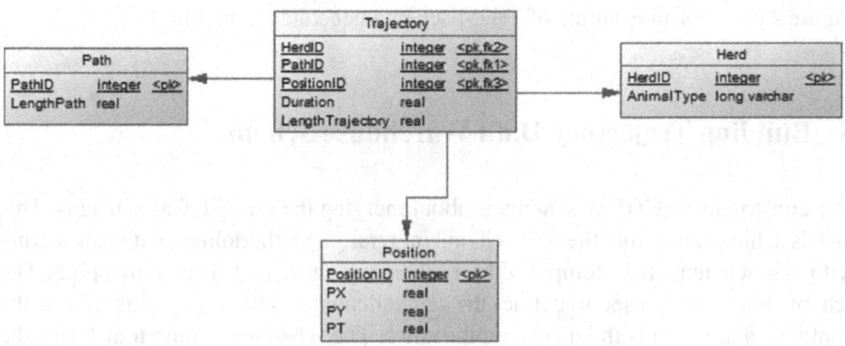

Fig. 3 Second example of TrDM schema

– Schema 2 (Fig. 3):
  Fact: {Trajectory}; FactKeys {HerdID (integer), PositionID (integer), PathID (integer)}.
  Measure: {Duration (real), LengthTrajectory (real)}.
  Dimension, DimensionKey and Attribute: {Path {PathID (integer), LengthPath (real)}, Position{PositionID (integer), PX (real), PY (real), PT (real)}, Herd {HerdID (integer), AnimalType (long varchar)}}.

- Comparison and construction of the matrix of similarity: It is to assign a coefficient to the elements belonging to the same category. To compare the names of schemas elements, we should take into consideration their linguistic matching. For this reason, we suggest the following formula (1) that calculates the degree of similarity of two elements belonging to the same category. It returns "1" if the two elements are similar, and "0" if they are not.

$$\begin{aligned}\mathrm{DeSim}(e1, e2) = \mathrm{DeId}(e1, e2) + \mathrm{DeSy}(e1, e2) + \mathrm{SeTy}(e1, e2) + \\ \mathrm{DePost}(e1, e2) + \mathrm{DePre}(e1, e2) + \mathrm{DeAbb}(e1, e2)\end{aligned} \tag{1}$$

  With:

  – DeId (e1, e2) = 1 if e1 and e2 are identical and 0 else.
  – DeSy (e1, e2) = 1 if e1 and e2 are synonymous, and 0 else.
  – DeTy (e1, e2) = 1 if e1 and e2 are the same with the existence of typing error, and 0 else.
  – DePost (e1, e2) = 1 if one of the two elements is the postfix of the other, and 0 else.
  – DePre (e1, e2) = 1 if one of the elements is the prefix of the other, and 0 else.
  – DeAbb (e1, e2) = 1 if one of the elements is the abbreviation of the other, 0 else.

  At this level, we take, also, into consideration the comparison of the types of the elements. If they are different we need human intervention to specify what should be kept.
  In the following, we apply this step to measures, dimensions and attributes because the two schemas have the same fact name "Trajectory" that will be the same in the final schema.
  Table 1 presents the similarity matrix used to compare the measures. According to the values of "Max", there is no similar measures. In such case, the final schema contains all of them.

Table 1 Measure comparison

Measure	MaxSpeed (real)	MinSpeed (real)	Max
Duration (real)	0	0	0
LengthTrajectory (real)	0	0	0

Table 2 Dimension tables comparison

Dimension	Position	Move	Herd	Max
Position	1	0	0	1
Path	0	0	0	0
Herd	0	0	1	1

Table 3 Attributes comparison

Attributes	HerdID (integer)	AnimalType (long varchar)	Max
HerdID (integer)	1	1	1
AnimalType (long varchar)	1	1	1

Table 2 presents the similarity matrix used to compare the dimension tables. "Move" and "herd" have two similar dimension tables. The rest are different. Table 3 presents the similarity matrix used to compare the attributes belonging to the dimension "Herd". All the attributes are similar and they have the same types. We keep then for each similar couple, only one. The same thing is done to "herd".

- Generation of the mapping rules: The rules visualize the conditional relationships between the instances of different categories. They are expressed as: "If Similar (X, Y) then Keep (X or Y) and Save (X, Y)", with:
  - X and Y: They are the two elements that belong to the same category.
  - Similar ( ): It is a function that specifies if the two inputs are similar or not. It uses the similarity matrix determined in the previous step.
  - Keep ( ): It keeps one of the two elements of the input that will be included in the generated schema.
  - Save ( ): It saves the rule.

  An example:

  - If Similar (AnimalType (long varchar), AnimalType (long varchar)) Then Keep (AnimalType (long varchar) or AnimalType (long varchar)) and Save (AnimalType (long varchar), AnimalType (long varchar))

  The different rules are stored into rules database. But, before making the comparison, it is crucial to start by verifying if there is any rule in the database that contains the two elements "X" and "Y". If it is not the case, the similarity measure is calculated to determine their similarity degree. If they are similar, the database is updated, else, the next elements is treated.

5.2 Schema Mapping

Once the mapping rules are extracted; we move to the next where they are transformed into queries. In the following, we present some of them.

Concerning the fact tables, the two schemas have the same name so the resulting schema contains also the same fact table name.

- Query = "Insert into Schema (idSchema) values ("+schemaId+")";
- Query1 = "Insert into Fact (FactName, idSchema) values ('Trajectory', "+ schemaId +")";

Since the fact tables are similar, we move to compare their measures. If the measures are similar we choose one of them, if not, we add both of them.

- Query2 = "Insert into Measure (MeasureName, MeasureType, idFact) values ('MaxSpeed', "+" 'real' "+ factId +")";

For the dimensions, we compare their names, if they are similar, we move to their attributes. We keep then the similar ones and we add the different. If the dimensions are different we keep them with their attributes without modification.

- Query3 = "Insert into Dimension (DimensionName, DimensionPKName, DimensionPK-Type, idSchema) values ('Move', 'MoveID', 'integer', "+schemaId +")";

Figure 4 presents the star schema related to the trajectory of the herd. It is the result of the application of the previous rules.

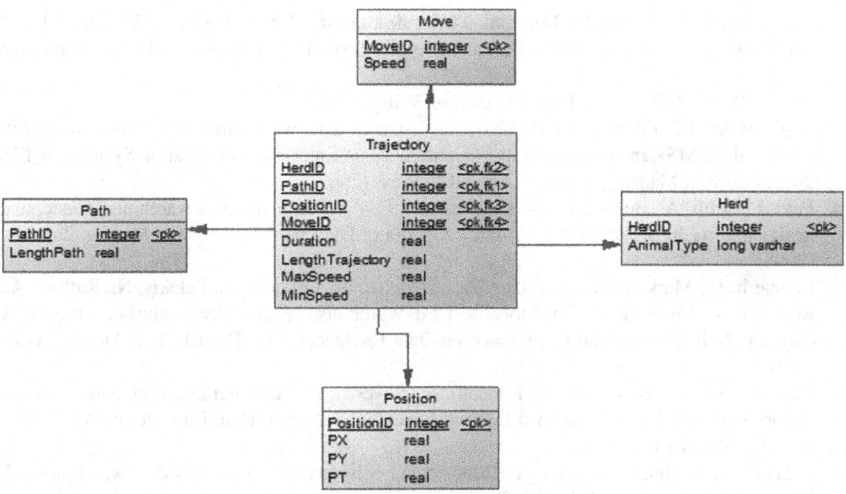

Fig. 4 The TrDW schema

6 Conclusion

In this work, we focus on the displacement of the herd that we took as example to study the trajectory of a moving object. The last one is defined as a spatial object that changes the form and the location permanently over the time. During its movement, the herd generates huge amount of TrD. We use the TrDW as a powerful tool to storage this kind of data to transform it later to useful knowledge. The problem at this level concerns the generation of its schema. For this reason, we proposed, in this work, an approach that serves to automatically generate the TrDW schema. Indeed, from the mobile data base, the multidimensional schemas are generated. They correspond to TrDM schemas. The latter are merged to build the final schema of the TrDW using the schema integration methodology that is composed by two steps. The first step extracts the similar elements from two schemas to generate the mapping rules. The second step transforms the rules into queries and executes them to merge the schemas.

As future work, we propose using some data mining techniques to predict the future movement of animals so we will be able to intervene for example by changing the direction of the movement in case where there is a threat to the existing vegetation.

Acknowledgments I would like to thank everyone.

References

1. Spaccapietra, S., Parent, C., Damiani, M.-L., de Macedo, J.A.F., Porto, F., Vangenot, C.: A conceptual view on trajectories. Technical report, Ecole Polytechnique Federal de Lausanne (2007)
2. Inmon, W.H.: Building the Data Warehouse. Wiley (1992)
3. Malinowski, E., Zimnyi, E.: Implementing spatial data warehouse hierarchies in object-relational DBMSs. In: 9th International Conference on Enterprise Information Systems, ICEIS 2007, Funchal, Madeira, Portugal. INSTICC Press (2007)
4. Feki, J., Nabli, A., Ben-Abdallah, H., Gargouri, F.: An Automatic data warehouse conceptual design approach. In: Wang, J. (ed.) Encyclopedia of Data Warehousing and Mining, 2nd edn. (2008)
5. Leonardi, L., Marketos, G., Frentzos, E., Giatrakos, N., Orlando, S., Pelekis, N., Raffaet, A., Roncato, A., Silvestri, C., Theodoridis, Y.: T-Warehouse: visual olap analysis on trajectory data. In: 26th International Conference on Data Engineering (ICDE'10), Los Angeles, USA (2010)
6. Braz, F., Orlando, S., Orsini, R., Roncato, A., Silvestri, C.: Approximate aggregations in trajectory data warehouses. In: 23rd International Conference on Data Engineering Workshop, pp. 536–545. IEEE (2007)
7. Orlando, S., Orsini, R., Raffaet, A., Roncato, A., Silvestri, C.: Trajectory data warehouses. J. Comput. Sci. Eng. **1**(2), 211–232 (2007)
8. Leonardi, L., Marketos, G., Frentzos, E., Giatrakos, N., Orlando, S., Pelekis, N., Raffaet, A., Roncato, A., Silvestri, C., Theodoridis, Y.: T-warehouse: visual olap analysis on trajectory data. In: 26th International Conference on Data Engineering (ICDE), pp. 1141–1144. IEEE (2010)

9. Campora, S., de Macedo, J.A.F., Spinsanti, L.: St-toolkit: a framework for trajectory data warehousing. In: 14th AGILE Conference on Geographic Information Science (2011)
10. Braz, F. J., Orlando, S.: Trajectory data warehouses: proposal of design and application to exploit data. In: GeoInfo, pp. 61–72 (2007)
11. Giannotti, F., Trasarti, R.: Mobility, data mining and privacy: the GeoPKDD paradigm. In: 14th SIAM Conference on Mathematics for Industry (MI09), pp. 10–18, San Francisco, California (2009)
12. Sakouhi, T., Akaichi, J., Malki, J., Bouju, A., Wannous, R.: Inference on semantic trajectory data warehouse using an ontological approach. In Foundations of Intelligent Systems, pp. 466–475. Springer International Publishing (2014)
13. Oueslati, W., Akaichi, J.: Trajectory data warehouse modeling based on a Trajectory UML profile: medical example. In: International Work-Conference on Bioinformatics and Biomedical Engineering, pp. 1527–1538 (2014)
14. Batini, C., Lenzerini, M., Navathe, S.B.: A comparative analysis of methodologies for database schema integration. ACM Comput. Surv. **18**(4), 323–364 (1986)
15. Lenzerini, M.: Data integration: a theoretical perspective. In: Symposium on Principles of Database Systems (PODS02), pp. 233–246. ACM (2002)
16. Batini, C., Lenzerini, M.: A methodology for data schema integration in the entity relationship model. IEEE Trans. Softw. Eng. **SE-lo**, 650–663 (1984)

9. Campora, S., de Macedo, J., Spinsanti, L.: St-toolkit: a framework for trajectory data warehousing. In: 14th AGILE Conference on Geographic Information Science (2011)
10. Braz, F.J., Orlando, S.: Trajectory data warehouses: proposal of design and application to exploit data. In: GeoInfo, pp. 61–72 (2007)
11. Giannotti, F., Trasarti, R.: Mobility data mining and privacy: the GeoPKDD paradigm. In: SIAM Conference on Mathematics for Industry (CM09), pp. 10–18, San Francisco, California (2009)
12. Sakouhi, T., Akaichi, J., Malki, J., Bouju, A., Wannous, R.: Inference on semantic trajectory data warehouse using an ontological approach. In: Foundations of Intelligent Systems, pp. 466–475. Springer International Publishing (2014)
13. Oueslati, W., Akaichi, J.: Trajectory data warehouse modeling based on a Trajectory UML profile: medical example. In: International Work-Conference on Bioinformatics and Biomedical Engineering, pp. 527–538 (2014)
14. Batini, C., Lenzerini, M., Navathe, S.B.: A comparative analysis of methodologies for database schema integration. ACM Comput. Surv. 18(4), 323–364 (1986)
15. Lenzerini, M.: Data integration: a theoretical perspective. In: Symposium on Principles of Database Systems (PODS), pp. 233–246. ACM (2002)
16. Batini, C., Lenzerini, M.: A methodology for data schema integration in the entity relationship model. IEEE Trans. Softw. Eng. SE-10(6), 650–664 (1984)

A Taxonomy of Support Vector Machine for Event Streams Classification

Hanen Bouali, Yasser Al Mashhour and Jalel Akaichi

Abstract Radio Frequency Identification technologies have been widely used in various domain. They allowed experts to automatically record on-line data of every-day life at a rapid rate. Consequently, in order to extract knowledge, these data should be treated rigorously. Among these treatments, we cite the classification that is paramount. Generally, for this purpose several classification' technologies exists. Particularly, Support Vector Machine has been applied to several domains to improve results efficiency. But, with the changes and the evolution of event streams and to support such changes, the SVM technique evolution is seen in many researches. The goal of this paper is to present an overview of different works related to SVM evolution, then we propose a comparative study between those works. As a result we obtain a taxonomy which shows in details support vector machine types and correlations between different types.

Keywords Machine learning · Support Vector Machine · Taxonomy · Event streams

1 Introduction

There are millions of events resulting from mobile devices and studying such a huge number without classification would make it almost impossible to make any decision. Therefore, scientists use classification systems to help them make sense of the world around them. They use it to recognize information and objects. Thus, sorting

H. Bouali (✉)
Bestmod, ISG Tunis, Université de Tunis, Tunis, Tunisia
e-mail: hanene.bouali@gmail.com

Y. Al Mashhour · J. Akaichi
King Khalid University, Guraiger, Abha, Saudi Arabia
e-mail: almshhour@kku.edu.sa

J. Akaichi
e-mail: Jalel.akaichi@kku.edu.sa

G. De Pietro et al. (eds.), *Intelligent Interactive Multimedia Systems and Services 2016*, Smart Innovation, Systems and Technologies 55,
DOI 10.1007/978-3-319-39345-2_33

events into groups, make them easier to understand and to see relationships between them. Also, the classification is gaining importance as a crucial tool for simplifying complex real life applications. Consequently, several methods have been proposed to solve classification's issues. Among them, we find hierarchical artificial neural network [1], motifs and their temporal relations [2], Support vector machine [3], Bayesian Network [4], fuzzy pattern trees [5], and decision trees [6].

The selection of the appropriate techniques depends on application' parameters and mainly based on advantages and limits of each technique. In the best of our knowledge and by digging the literature, we find that Support vector machine is the most powerful technique used for event stream classification [7]. But with the changes, dynamism and the evolution of data streams and to support such characteristics, the SVM technique evolution is seen in many researches. Since 1970's Vapnik et al. have applied themselves to the study of statistical learning theory. Until the early of the 1990, a new kind of learning machine, Support Vector Machine was presented based on statistical learning theory. Lots of researches are going on this technique for the improvement of results efficiency. The SVM usually deals with classification of different types of classes. Since SVMs demonstrated to be very powerful and strong algorithm, they have been applied to many domains: bio-informatics [8–10], credit card fraud detection [11, 12], document categorization [13–16], recognition system [17–19], intrusion detection, feature selection [20, 21]. Therefore, this paper intends to present an overview of different works related to support vector machine evolution to judge their accuracy, then, a comparative study between those works is proposed.

The rest of this paper is organized as follows: In Sect. 2, we present SVM techniques for binary classification problems. Afterwards, Sect. 3 introduces SVM for multi classification. As a result, in Sect. 4, we report a taxonomy encompassing different detailed techniques in order to help users in choice task.

2 SVM for Binary Classification

Given a set of training examples, each marked as belonging to one of two categories, an SVM training algorithm builds a model that assigns new examples into one category or the other, making it a binary classifier. Two major types of classes are seen: linear and non-linear. Linear classification problems are easily distinguishable or can be easily separated in low dimension whereas non-linear problems are not easily distinguishable or cannot be easily separated. When comparing linear and non-linear classification, we see that linear SVM has minimal complexity (O(n_{sv})) compared to non-linear SVM (O(n_{sv} $*$ d $*$ M_k)) where

- n_{sv}: the number of support vectors
- d: the sample dimension
- M_k: the cost of evaluating kernel function

In addition to the complexity, non-linear SVM has high classification and training cost but higher classification accuracy. Because of its low classification cost and complexity, linear SVM is better adapted to real time prediction while scarifying accuracy. But, most researchers take advantages from both linear and non-linear algorithm.

2.1 Binary Classification for Linearly Separable Data

In real world application, there are many irrelevant and redundant features among the dataset; it makes the accuracy of the model unsatisfactory. One important way to avoid that is linear SVM. In [22] authors propose the twin SVM (TWSVM). The TWSVM is two parallel proximal hyper planes such that each hyper plane is closer to one of the two classes and in at least one distance from the other. This method has as inconvenient the loss of sparseness by using a quadratic loss function making the proximal hyperplanes close enough to the class itself. To handle the sparseness problem, authors in [23] propose non-parallel support vector machine (NPSVM) which has the valuable sparseness similar with the standard SVM. It has several advantages over existing TWSVMs used for large scale data. It can reach the same sparseness with the standard SVM.

2.2 Non Linear Binary Classification

Due to non-linearities or noise, real world data is usually not linearly separable. In the case of imperfectly separable input space, where noise in the input data is considered, there is no enforcement that there be no data points between the planes H1 and H2 mentioned in the previous section. For small datasets, SVM can solve it reasonably fast and can be performed on a reasonably configured computer. For large datasets, solving SVM functions requires large computing power and large memory for storage. Hence, a number of methods of SVM training have been developed over the years to improve on the memory requirements issue, speed up the training time and finding the best training model using appropriate kernel function and hyper plane parameters. To handle the problem of speed up, authors in [24] introduce an extreme SVM which provide very good generalization performance in relatively short time. However, it is inappropriate to deal with large scale data set due to the highly intensive computation. Thus, authors propose a parallel extreme SVM (PESVM) based on parallel programming. Furthermore, for new coming training data, it is brutal for PESVM to always retain a new model on all training data including old and new coming data. PESVM can tackle, in addition to the large scale problems, the evaluation metrics of speed up, size up and scale up. Actually, many real life machine learning problems can be more naturally viewed as on-line rather than batch learning problems, the data is often collected continuously in time. To support such training data change, researchers develop an incremental learning algorithm which can meet the requirement of on-line problem at the same time. PESVM is proposed to give Extreme SVM a good scalability in order to run faster when the input data set is very large. To support such training data change, researchers develop an incremental learning algorithm named parallel incremental extreme SVM (PIESVM) which can meet the requirement of on-line problem at the same time. PIESVM [25] is capable of adding new data to generate an appropriately altered classifier. PIESVMs are very useful when the training data can be not obtained in one time or the number of input data points is very large. PIESVM improve the training speed of the classifier.

3 SVM for Multi Classification

The SVM is a powerful tool for binary classification, capable of generating very fast classifier functions following a training period. There are several approaches to adopting SVMs to classification problems with three or more classes:

3.1 One Against All

One Against All (OAA) consists of building one SVM per class, trained to distinguish the samples in a single class from the samples in all remaining classes. The initial formulation of the one-against-all method required unanimity among all SVMs: a data point would be classified under a certain class if and only if that class's SVM accepted it and all other classes' SVMs rejected it. This approach constructs k SVM models where k is the number of classes. The *i*th SVM is trained with all of the examples in the *i*th class with positive labels, and all other example with negative labels. To handle the multi-class problems, the scenario is to decompose an M-class problem into a series of two-class problems. The drawbacks of this method, however, is that when the results from the multiple classiers are combined for the nal decision, the outputs of the decision functions are directly compared without considering the competence of the classiers. To overcome this limitation, authors in [26] introduce reliability measures into the multi-class framework. Two measures are designed: Static reliability measures and dynamic reliability measures. One Against All approach was used in many real world applications. Among those, the credit rating model [27] propose the ordinal pairwise partitioning (OPP) approach as a tool for upgrading conventional multi-class SVM models for appropriately deals with ordinal classes. In this study, the OPP approach partitions the dataset into data subsets with reduced classes. Hence, the combining of the OPP with MSVM produce a new hybrid type called an ordinal multi-class SVM (OMSVM). On the other hand, OAA approach was used for pattern recognition application, it is necessary to provide the detection ability to identify hidden classes that do not appear in the collection. Due to the nature of OAA it can detect the hidden class directly. This work [28] extends the smooth SVM [29] from binary k-class classification based on a single machine approach to obtain a multi-class smooth SVM (MSSVM). This study implements MSSVM for a ternary SVM classification and labels it as TSSVM. OAA decomposes the problem into k sub-problems solved in a series of TSSVM. Another broad of application is medical fields especially in cancer classification. OAA approach has the ambiguity in cases of ties (multiple SVMs satisfier) and rejections (no SVM satisfies). To cope with these problems, Bayesian network is used by estimating the posterior probability for classes. This paper [30] proposes a novel multi-class classification approach integrating SVM and nave Bayes classifier for multi-class cancer classification called OVRSVM. When classifying a sample, the method first estimates the probability of each class by using Bayesian Networks, and then organizes OVRSVMs as the sub-suptions architecture according to the probability. Finally, a sample is evaluated sequentially until an OVRSVM is satisfied. One Against All approach has been improved in [31]. Authors consider an improved technique of OAA using a new classification method. Some improvement must be considered: the number of SVMs should not be increased. On the other hand, the OAA needs k classes. So, the number will not change. Once the number of classes is fixed, we should use the hyper planes to gain higher classification accuracy. Authors introduce [32] a simplified multi-class SVM (SIMSVM) that directly solves a multi-class classification problem. SIMSVM reduces the size of the dual variable from $l * k$ to l where l and k are the size of training data and number of classes, respectively. Real world datasets shows that SIMSVM can greatly speed up the training process and achieve competitive classification accuracy.

3.2 *One Against One*

One Against One (OAO) is also known under the name of pairwise coupling, it consists of building one SVM for each pairs of classes. This approach constructs $k(k-1)/2$ classifiers where each one is trained on data from two classes. The growth of the intern information delivery has made automatic text categorization essential [16]. This investigation explores the challenges of multi-class text categorization using OAO Fuzzy SVM. The proposed multi-class text categorization system composed of two modules namely the processing documents module and the classifying module. The OAO in this approach is adopted to create multi-maximum margins hyper planes by the voting strategy to classify documents.

3.3 *DAGSVM*

The directed acyclic graph SVM (DAGSVM) constructs $k * (k-1)/2$ hyper-planes. The training phase of this algorithm is similar to that of the One-Against-One method that employs multiple binary classifiers; however, DAGSVM uses a graph-visiting strategy for testing. The testing phase of DAGSVM requires the construction of a rooted binary decision directed acyclic graph (DDAG) that uses C classifiers. Each node of this graph represents a binary SVM for a pair of classes, e.g., (p, q). At the topologically lowest level, there are k leaves corresponding to k classification decisions. Every non-leaf node (p, q) has two branches: the left branch corresponds to the decision not p, while the right branch corresponds to not q. this method [30] is computationally less costly because a smaller number of binary classification problem have to be resolved. This technique is used in many real world applications. We cite classifying slate tile. In the slate production process, the final stage is the manual classification of tiles in commercial categories. This technique reduces the number of binary classifier from $(k * k - 1)/2$ to $k - 1$. Because of the fewer classification have to be computed, there are fewer interior nodes in the DAG. This technique also avoids the propagation of errors in each binary classifier unlike the OAO approach. And each element class is always part of the binary classifier. A key aspect of this DAG-based methodology is its generalizability: a perceptron for each node in the graph establishes more general results for the error rates.

3.4 OAO Versus OAA

The OAA approach represents the earliest and most common SVM multi-class approach and involves the division of an N-class dataset into N two classes' cases. The OAO approach involves constructing a machine for each pairs of classes which involves $N (N - 1)/2$ machines. The OAO its performance can be compromised due to unbalanced training datasets. The OAA schema determines the class label of x using the majority voting strategy for the rest-class. The OAO schema does not have the merit of this option. OAO suffers from the restriction of binary sub classifier and each sub classifier can only make the decision about one pair of binary choices.

4 SVM Taxonomy

Taxonomy is the science of classification according to a pre-determined system, with the resulting catalog used to provide a conceptual framework for discussion, analysis or information retrieval. Moreover, the concept of taxonomy combines terminology and the science of classification. Taxonomy is not only a tool for systematic storage efficient and effective technique and recall for usage of knowledge but it is also a net way of pointing to knowledge expansion and building. The key to taxonomy effectiveness rest on criteria of comprehensiveness and usefulness. Obviously, to be effective, a taxonomy must represent the full spectrum of data stream. Thus, comprehensiveness is a necessary condition for effectiveness. It is, however, not sufficient. To further be effective, a taxonomy must be parsimonious. It should not include unnecessary categories (classes). Finally, to be considered effective, tax-

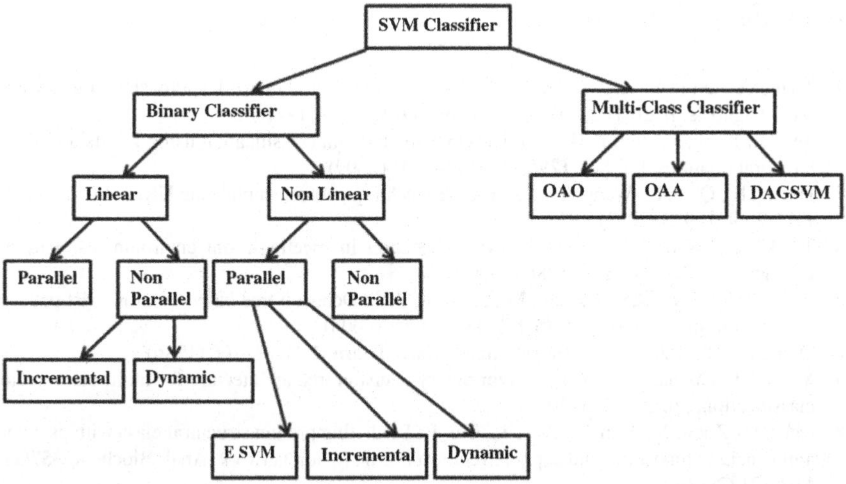

Fig. 1 SVM taxonomy

onomy should be robust and generally useful. Support vector machine is divided essentially into two major classifiers: Binary classifier and Multi class classifier. Each classifier is subdivided into sub-classifier. Hence, the user faces the task of selecting the appropriate technique from among several methods. Therefore, the taxonomy presented in this paper, tries to gives information to users to decide which is the suitable technique to solve his problem and the most appropriate one to the main criteria of the problem. The figure below (Fig. 1) shows in details Support vector machine types and correlations between different types. The taxonomy is described as a tree. Each tree node represents the SVM type and edges represent type's correlation. Level n in the tree represent subclasses of the level n – 1.

5 Conclusion

With the changes and the evolution of data streams, the SVM technique development is seen in many researches field such as text categorization, bio-informatics, and pattern recognition. Accordingly, we reviewed in this paper different types of SVM by investigating the benefits and drawbacks of main important algorithm in the literature. This investigation led to a taxonomy of different event streams SVM' classification techniques which are divided essentially into binary classification and multi-class classification. Each class is sub-divided into subclasses and each one support data types. The choice of the appropriate SVM technique is based on the problem criteria and properties. As a future work, we suggest to apply these introduced approaches on real application problems and to study their performance and applicability.

References

1. Zhang, Y., Ding, X., Liu, Y., Griffin, P.J.: An artificial neural network approach to transformer fault diagnosis. IEEE Trans. Power Delivery **11**(4), 1836–1841 (1996)
2. Seo, S., Kang, J., Ryu, K.H.: Multivariable stream data classification using motifs and their temporal relations. Inf. Sci. **179**(20), 3489–3504 (2009)
3. Huang, H., Qian, L., Wang, Y.: A svm-based technique to detect phishing URLs. Inf. Technol. J. **11**(7), 921–925 (2012)
4. Temko, A., Nadeu, C.: Acoustic event detection in meeting-room environments. Pattern Recogn. Lett. **30**(14), 1281–1288 (2009)
5. Wang, C.H., Guo, R.S., Chiang, M.H., Wong, J.Y.: Decision tree based control chart pattern recognition. Int. J. Prod. Res. **46**(17), 4889–4901 (2008)
6. Quinlan, J.R.: Induction of decision trees. Mach. Learn. **1**(1), 81–106 (1986)
7. Vavrek, J., Cizmar, A., Juhar, J.: Svm binary decision tree architecture for multi-class audio classification, pp. 202–206 (2012)
8. Chen, C., Zhou, X., Tian, Y., Zou, X., Cai, P.: Predicting protein structural class with pseudo-amino acid composition and support vector machine fusion network. Anal. Biochem. **357**(1), 116–121 (2006)
9. Yu, X., Cao, J., Cai, Y., Shi, T., Li, Y.: Predicting rrna-, rna-, and dna-binding proteins from primary structure with support vector machines. J. Theor. Biol. **240**(2), 175–184 (2006)
10. Zhang, G., Ge, H.: Support vector machine with a pearson vii function kernel for discriminating halophilic and non-halophilic proteins. Comput. Biol. Chem. **46**, 16–22 (2013)
11. Hens, A.B., Tiwari, M.K.: Computational time reduction for credit scoring: an integrated approach based on support vector machine and stratified sampling method. Expert Syst. Appl. **39**(8), 6774–6781 (2012)
12. Ngai, E., Hu, Y., Wong, Y., Chen, Y., Sun, X.: The application of data mining techniques in financial fraud detection: a classification framework and an academic review of literature. Decis. Support Syst. **50**(3), 559–569 (2011)
13. Bratko, A., Filipič, B.: Exploiting structural information for semi-structured document categorization. Inf. Process. Manage. **42**(3), 679–694 (2006)
14. Hao, P.Y., Chiang, J.H., Tu, Y.K.: Hierarchically svm classification based on support vector clustering method and its application to document categorization. Expert Syst. Appl. **33**(3), 627–635 (2007)
15. Lee, K.S., Kageura, K.: Virtual relevant documents in text categorization with support vector machines. Inf. Process. Manage. **43**(4), 902–913 (2007)
16. Wang, T.Y., Chiang, H.M.: One-against-one fuzzy support vector machine classifier: an approach to text categorization. Expert Syst. Appl. **36**(6), 10030–10034 (2009)
17. Wu, D., Shao, L.: Multi-max-margin support vector machine for multi-source human action recognition. Neurocomputing **127**, 98–103 (2014)
18. Xanthopoulos, P., Razzaghi, T.: A weighted support vector machine method for control chart pattern recognition. Comput. Ind. Eng. **70**, 134–149 (2014)
19. Zhai, S., Jiang, T.: A novel sense-through-foliage target recognition system based on sparse representation and improved particle swarm optimization-based support vector machine. Measurement **46**(10), 3994–4004 (2013)
20. Li, H., Li, C.J., Wu, X.J., Sun, J.: Statistics-based wrapper for feature selection: an implementation on financial distress identification with support vector machine. Appl. Soft Comput. **19**, 57–67 (2014)
21. Moustakidis, S.P., Theocharis, J.: Svm-fuzcoc: a novel svm-based feature selection method using a fuzzy complementary criterion. Pattern Recogn. **43**(11), 3712–3729 (2010)
22. Yang, Z.M., He, J.Y., Shao, Y.H.: Feature selection based on linear twin support vector machines. Procedia Comput. Sci. **17**, 1039–1046 (2013)
23. Mehrkanoon, S., Huang, X., Suykens, J.A.: Non-parallel support vector classifiers with different loss functions. Neurocomputing **143**, 294–301 (2014)

24. Liu, Q., He, Q., Shi, Z.: Extreme support vector machine classifier. In: Advances in Knowledge Discovery and Data Mining, pp. 222–233. Springer (2008)
25. He, Q., Du, C., Wang, Q., Zhuang, F., Shi, Z.: A parallel incremental extreme svm classifier. Neurocomputing **74**(16), 2532–2540 (2011)
26. Liu, Y., Zheng, Y.F.: One-against-all multi-class svm classification using reliability measures. In: Proceedings of 2005 IEEE International Joint Conference on Neural Networks, IJCNN'05, vol. 2, pp. 849–854. IEEE (2005)
27. Kim, K.J., Ahn, H.: A corporate credit rating model using multi-class support vector machines with an ordinal pairwise partitioning approach. Comput. Oper. Res. **39**(8), 1800–1811 (2012)
28. Chang, C.C., Chien, L.J., Lee, Y.J.: A novel framework for multi-class classification via ternary smooth support vector machine. Pattern Recogn. **44**(6), 1235–1244 (2011)
29. Purnami, S.W., Embong, A., Zain, J.M., Rahayu, S.: A new smooth support vector machine and its applications in diabetes disease diagnosis. J. Comput. Sci. **5**(12), 1003 (2009)
30. Hong, J.H., Cho, S.B.: A probabilistic multi-class strategy of one-vs.-rest support vector machines for cancer classification. Neurocomputing **71**(16), 3275–3281 (2008)
31. Liu, Y., Wang, R., Zeng, Y.S.: An improvement of one-against-one method for multi-class support vector machine. In: 2007 International Conference on Machine Learning and Cybernetics, vol. 5, pp. 2915–2920. IEEE (2007)
32. He, X., Wang, Z., Jin, C., Zheng, Y., Xue, X.: A simplified multi-class support vector machine with reduced dual optimization. Pattern Recogn. Lett. **33**(1), 71–82 (2012)

24. Tao, Q., He, Q., Shi, Z.: Extreme support vector machine classifier. In: Advances in Knowledge Discovery and Data Mining, pp. 222–233. Springer (2008)
25. He, Q., Du, C., Wang, Q., Zhuang, F., Shi, Z.: A parallel incremental extreme svm classifier. Neurocomputing 74(16), 2532–2540 (2011)
26. Tian, Y., Zhang, Y.-Q.: One-against-all multi-class svm classification using reliability measures. In: Proceedings of 2005 IEEE International Joint Conference on Neural Networks, IJCNN 2005, vol. 2, pp. 849–854. IEEE (2005)
27. Kim, K., Ahn, H.: A corporate credit rating model using multi-class support vector machines with an ordinal pairwise partitioning approach. Comput. Oper. Res. 39(8), 1800–1811 (2012)
28. Chang, C.-C., Chien, L.-J., Lee, Y.-J.: A novel framework for multi-class classification via ternary smooth support vector machine. Pattern Recogn. 44(6), 1235–1244 (2011)
29. Purnami, S.W., Embong, A., Zain, J.M., Rahayu, S.P.: A new smooth support vector machine and its applications in diabetes disease diagnosis. J. Comput. Sci. 5(12), 1003 (2009)
30. Liong, L.H., Chen, Y.-C.: A probabilistic multi-class strategy of one-versus-rest support vector machine for cancer classification. Neurocomputing 71(16), 3275–3281 (2008)
31. Liu, Y., Wang, R., Zeng, Y.-S.: An improvement of one-against-one method for multi-class support vector machine. In: 2007 International Conference on Machine Learning and Cybernetics, vol. 5, pp. 2915–2920. IEEE (2007)
32. He, X., Wang, Z., Jin, C., Zheng, Y., Xue, X.: A simplified multi-class support vector machine with reduced dual optimization. Pattern Recogn. Lett. 33(1), 71–82 (2012)

Social Networks Security Policies

Zeineb Dhouioui, Abdullah Ali Alqahtani and Jalel Akaichi

Abstract Social networks present useful tools for communication and information sharing. While these networks have a considerable impact on users daily life, security issues are various such as privacy defects, threats on publishing personal information, spammers and fraudsters. Consequently, motivated by privacy problems in particular the danger of sexual predators, we seek in this work to present a generic model for security policies that must be followed by social networks users based on sexual predators identification. In order to detect those distrustful users, we use text mining techniques to distinguish suspicious conversations using lexical and behavioral features classification. Experiments are conducted comparing between two machine learning algorithms: support vector machines (SVM) and Nave Bayes (NB).

Keywords Privacy protection · Security policies · Social networks · Predators · Text mining · Machine learning · Support Vector Machine

1 Introduction

A social network is a set of entities that may have relationships and is modeled generally as a graph where nodes refers to entities and edges to relations. While facilitating much needed users requirements, social networks enlarge privacy concerns. Among privacy risks, we can mention fraudsters, spammers malicious users especially sexual predators that target children and teenagers. In fact, according to the report of the National Center for Missing and Exploited Children (NCMEC) in 2008, there is 1 in

Z. Dhouioui (✉)
Bestmod, ISG Tunis Université de Tunis, Tunis, Tunisia
e-mail: dhouioui.zeineb@hotmail.fr

A.A. Alqahtani · J. Akaichi
King Khaled University, Guraiger, Abha, Saudi Arabia
e-mail: aalrabaa@kku.edu.sa

J. Akaichi
e-mail: Jalel.akaichi@kku.edu.sa

G. De Pietro et al. (eds.), *Intelligent Interactive Multimedia Systems and Services 2016*, Smart Innovation, Systems and Technologies 55,
DOI 10.1007/978-3-319-39345-2_34

7 teenagers are approached for sexual purposes. The widespread use of the internet and the lack of parental control have brought the cyber-crime and social networks are considered as a new opportunities for pedophiles [3] Social networks are a reach source for sexual predators since these sites are popular for teenagers. In fact, online sexual solicitation raise relatively to the degree of privacy of user-generated content such as profile, photos or statutes [10].

Additionally, Instant messaging is a part of everyday life habits. Thus, the huge flow of instant messages in social networks affect the accuracy of the existing methods proposed to detect sexual predators. Another challenging task is that conversations content is always inaccessible. Moreover, malicious users mask their real identity since it is easy to give false personal information in social networks.

The ability to detect suspicious conversations can help in the improvement of security policies and the detection of such offenders is a primordial security issue. In this work, we present a method for picking out malicious users and identifying suspicious conversations from a set of conversations. Lexical and behavioral features are used to flag sexual predators. Finally, we experiment with SVM and Naive Bayes classification to filter conversations. Ultimately, we find that SVM classifier outperformed the Naive Bayes. The rest of this paper is organized as follows: in Sect. 2 we review existing approaches in this field. Section 3 presents in detail our proposed method based on lexical and behavioral features. The experiments and discussions are outlined in Sect. 4. In Sect. 5, conclusion is depicted and future works are presented.

2 State of the Art

Social networks pedophiles target minor victims [8]. Thus, it is primordial to assure adolescents and promote security policies in social networks mainly by automatically recognizing suspicious conversations. In this section, we focus on briefly reviewing existing methods that handle the identification of sexual predators. Authors in [2] have proposed a two steps approach in order to detect sexual predators using social network dialogues. This approach starts by detecting suspicious conversations where predators participate; then the sexual predators are identified. In [4], sexual predator detection in chat conversation is dealt based on a sequence of classifiers. Indeed, documents are split into three parts according to different stages of predation. To deal with the security issue, Rahman Miah et al. introduce a method able to differentiate child-exploitation, adult-adult and general-chatting dialogues using text categorization approach and psychometric information [9]. In the same way, Bogdanova et al. are convinced that standard text-mining features are relevant to distinguish general-chatting from child-exploitation conversations, but are unable to distinguish between child exploitation and adult-adult conversations [3]. A probabilistic method has been introduced in [7] giving three classes of the chat interventions followed by a sexual predators which are gaining access referring to the intention of predators to approach the victim. Secondly, the deceptive relationship is the preliminary to a sexual attack.

Finally, the sexual affair refers to a sexual affair intention of the predator towards the victim. Since detecting sentiment in texts is helpful to identify online sexual predators, authors in paper [3] present a list of sentiment features. They also used a corpus containing predators chats obtained from http://www.perverted-justice.com. Additionally, authors handle this task via the natural language processing (NLP) techniques. McGhee and his colleagues classified possible sexual predators strategies and introduce the following sexual features [6]:

- Percentage of approach words using verbs like meet, and nouns such hotel
- Percentage of relationship words such as dating words
- Percentage of communicative desensitization words including family members names
- Percentage of words expressing sharing information related to the age, location or also sending photos

The Chat Coder can identify lines in the chat log which include predatory language. The term grooming was defined as a strategy followed by sexual offenders to force their victims to admit the sexual affair. In most cases, the predator tries to isolate the victim in order to easily and more exploit the victim. Actually, grooming is used to describe malicious behavior with the intention of sexual exploitation with a children and is classified as follows: the first class is the communicative desensitization in which vulgar sexual language is used, pornographic con-tents are sent and sexual slang terms are used. The second class refers to the reframing that aims to gain the victim trust by showing online sexual advances. Features extraction [5] is frequently used such as lexical and behavioral ones. Doubtful online text chat can be categorized into two types according to Pendar [8]: interaction between predator and victim and consensual interaction between two adults.

3 The Proposed Method

Defining privacy policies is a challenging concept which requires innovative techniques. Privacy policies are closely related to privacy preserving and users protection and include access control and confidentiality techniques. The new paradigm for security policies require new mechanisms based on trust [1]. Acquiring a high level of trust is primordial before sharing information and build new relationships. Indeed, malicious users appear usually as honest to gain the trust of other users. Users must verify their security parameters and social networks sites are charged with improving security policies to safeguard users from malicious one. Thus, identifying unfaithful users such as sexual predators is a crucial task. Given a social network G(E, V), a member U_i must have a high security level. U_i checks access control mechanisms. Then, he verifies relationships policies based on executing a trust management system based on the detection of sexual predators. Privacy concern includes security, reputation and credibility and finally profiling. In our context, we start by the profiling users to detect credible users to ensure security (Fig. 1).

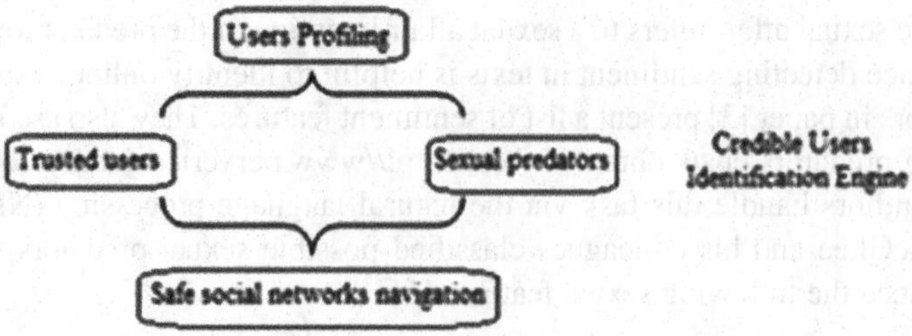

Fig. 1 The overall framework for security policies

The SPD algorithm (Sexual Predators Detection) can be applied to detect suspect conversation and predators.

Algorithm 1 SPD

Input: conversation sets C
Output: list of sexual predators
1: **while** $C \neq \emptyset$ **do**
2:     *featureclassification*(keywords)
3:     *Lexicondevelopment*()
4:     *Conversationclassification*()
5:     *Sum(feature, weights)*
6: **end while**

To detail the previous algorithm, we describe the following steps:

Step 1: raw data collection: this step consists of collecting conversations
Step 2: Features classification. We distinguish two types of features [3]:

- Behavioral features
  - The number of times a user initiates a conversation
  - The number of questions asked
  - The number of intimate conversations
  - The frequency of turn-taking
  - Intention of grooming or hooking
- Lexical features
  - Percentage of approach words
  - Percentage of relationship words
  - Percentage of communicative desensitization words: these words refer to family members names, for instance, (mm, dad.)
  - Percentage of words expressing sharing information
  - Percentage of isolation
  - Number of emoticons

Step 3: Lexicon development: this step focuses on the informal language of online social networks. For this reason, 3 types of lexicon were created:

- Exchange of personal information
- Grooming
- Approach

Step 4: Suspicious conversation detection: based on features we can classify conversation into two classes positive and negative
Step 5: Flagging users according to predator degree by summing the features weights.

There is no pre-processing stage of conversations texts due to the special characteristics of these conversation texts such as neglecting grammar rules, using abbreviations and emotions. However, pre-filtering is crucial to reduce the computational task by eliminating conversations that contains only one participant or very short ones (less than 10 interventions for both users), or those containing many and several unrecognized characters. Our method can face the challenging task of sexual predators identification as a classification using lexical and behavioral features using a supervised learning. Ideally, we can detect suspicious conversations and distinguish between the victim and the predator. We aimed to anticipate the detection of sexual offenders that can approach teenagers reducing the number of victims and improving social networks security policies. We hypothesize that features are weighted and hence the predaterhood score is the sum of these weights. We believe that weights are appropriate in reflecting the danger of sexual predators. The most interesting from this work is the use of features which flag predatory messages and misbehavior and consequently users can be notified to be aware of suspicious users.

4 Experiments

4.1 Metrics

In what follows, we will focus our experimentation on comparing the performance of the proposed classification method using the following metrics:

$$Precision = \frac{a}{a + b}$$

$$Recall = \frac{a}{a + c}$$

$$F\text{-}measure = \frac{2 \times Precision \times Recall}{Precision + Recall}$$

where:

- *a*: the number of conversations appropriately assigned
- *b*: the number of conversations inaccurately assigned
- *c*: the number of conversations inaccurately rejected

We compared the performances of different feature sets using the most famous and used learning algorithms: Nave Bayes and SVM classifiers. Actually, SVM is used for binary classification with vastly dimensional space and well perform with text classification. Naive Bayes is simple and relevant for nominal features.

4.2 Results and Discussions

For this study, we were able to exploit data from the web site http://www.perverted-justice.com, but we preferred to gather real conversations from Facebook. Real data gathering was the task requiring the greatest effort. Due to the intimacy of these conversations, we have only collected 30 conversations, where 23 conversations will be used for the training model and 7 for the cross validation. The training model contains 17 conversations of sexual predators and 6 of non-sexual predators classified manually. Instant conversations are characterized by a particular vocabulary. The features extraction refers to find the main characteristics of the text. Features are composed of the most frequent expressions in the collected data expressing misbehavior. We assign weights to behavioral and lexical features (Figs. 2 and 3) and we have also computed the frequency of these features in the following figures (Figs. 4 and 5).

Classification algorithms such as Naive Bayes and Support Vector Machine (SVM) are used. For the evaluation, we used: the precision (*P*), Recall (*R*) and the F-measure using the weka tool. Figures 6 and 7 show the results obtained for the previously mentioned classifier. The best performance was obtained using the SVM.

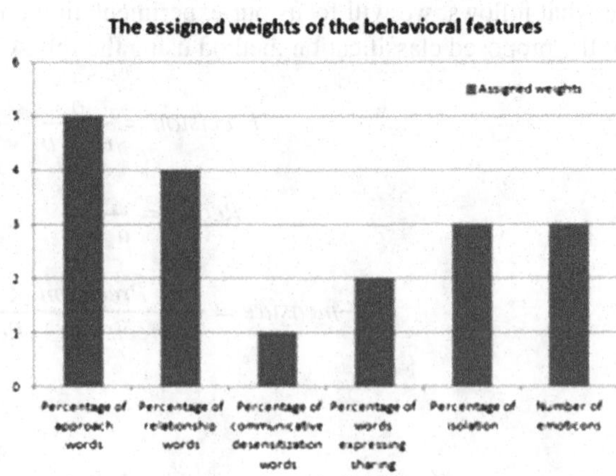

Fig. 2 The assigned weights of the behavioral features

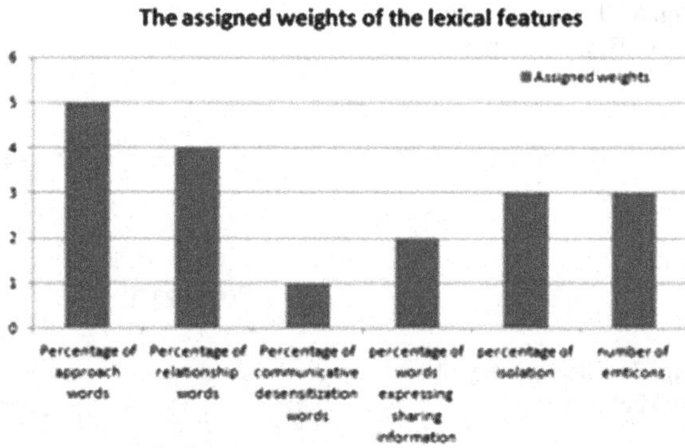

Fig. 3 The assigned weights of the lexical features

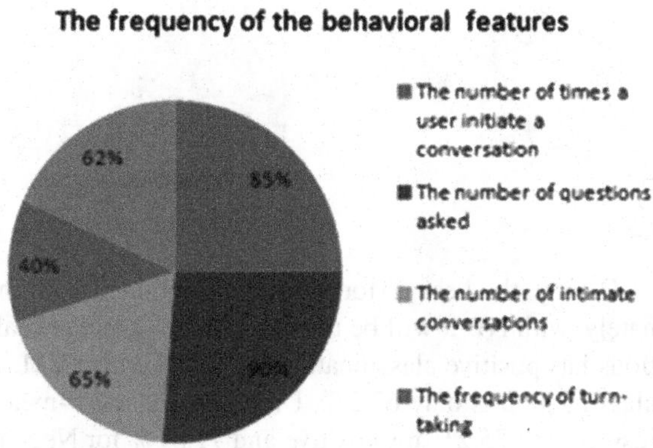

Fig. 4 The frequency of the behavioral features

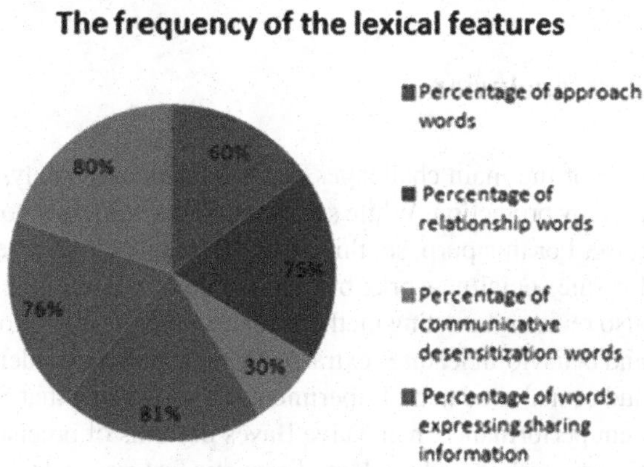

Fig. 5 The frequency of the lexical features

Results can be reported as follows: SVM outperform Naive Bayes for the different settings we considered. Collected data contains both types of conversations, those including sexual predators and conversations between normal users. Therefore, we eliminate conversations with only one participant or containing less than 10 interventions or those including insignificant characters.

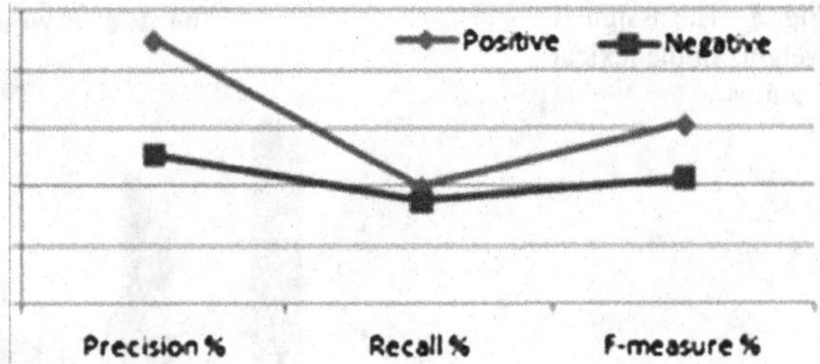

Fig. 6 The accuracy of Naive Bayes classifier

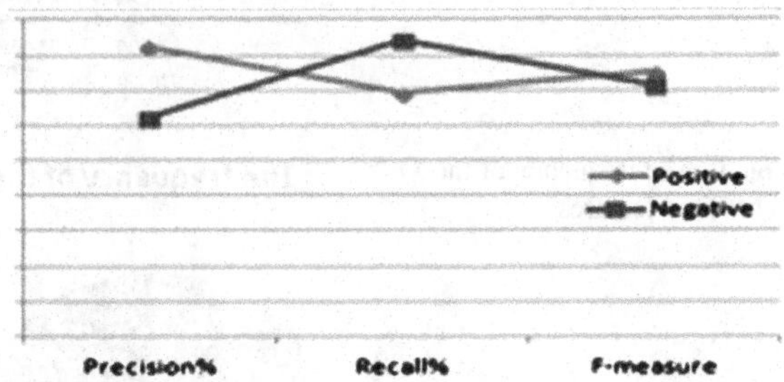

Fig. 7 The accuracy of SVM classifier

Clearly, the best performance was obtained with the SVMs algorithm. Unfortunately, with NB it will be unclear how to detect sexual predators. In fact, conversations has positive classification is 82.2 % with SVM. However, we obtain negative classification is only 62.2 %. Undoubtedly, the F-measure rates shows encouraging results with 75.2 % for Positive and 71.57 % for Negative.

5 Conclusion

One of the main challenges in social networks analysis is to provide high level of privacy protection. While social networks witness a non-stop popularity, real threats raise. For this purpose, this work presents a global overview of some good practice for safe social networks browsing. We discussed possible privacy threats, we have also reviewed existing methods to detect sexual predators and finally based on lexical and behavioral features extracted from texts we can identify suspicious conversations and sexual predators. Experimental results show that SVM classifier present prominent performance than Naive Bayes in terms of precision, recall and the F-measure. Future work includes adding linguistic features and to exploit largest number of conversations.

References

1. Ahn, G.J., Shehab, M., Squicciarini, A.: Security and privacy in social networks. Internet Comput. IEEE **15**(3), 10–12 (2011)
2. Alemán, Y., Vilarino, D., Pinto, D.: Searching sexual predators in social networks
3. Bogdanova, D., Rosso, P., Solorio, T.: Modelling fixated discourse in chats with cyberpedophiles. In: Proceedings of the Workshop on Computational Approaches to Deception Detection, pp. 86–90. Association for Computational Linguistics (2012)
4. Escalante, H.J., LabTL, I., No, L.E.E., ú Villatoro-Tello, E., Juárez, A., Montes-y-G6mez, M.: Sexual predator detection in chats with chained classifiers. In: WASSA 2013, p. 46 (2013)
5. Inches, G., Crestani, F.: Overview of the international sexual predator identification competition at pan-2012. In: CLEF (Online Working Notes/Labs/Workshop), vol. 30 (2012)
6. McGhee, I., Bayzick, J., Kontostathis, A., Edwards, L., McBride, A., Jakubowski, E.: Learning to identify internet sexual predation. Int. J. Electron. Commer. **15**(3), 103–122 (2011)
7. Michalopoulos, D., Mavridis, I.: Utilizing document classification for grooming attack recognition. In: 2011 IEEE Symposium on Computers and Communications (ISCC), pp. 864–869. IEEE (2011)
8. Pendar, N.: Toward spotting the pedophile telling victim from predator in text chats. In: null, pp. 235–241. IEEE (2007)
9. RahmanMiah, M.W., Yearwood, J., Kulkarni, S.: Detection of child exploiting chats from a mixed chat dataset as text classification task. In: Proceedings of the Australian Language Technology Association Workshop, pp. 157–165 (2011)
10. Savirimuthu, J.: Online sexual grooming of children, obscene content and peer victimisation: legal and evidentiary issues. In: Online Child Safety, pp. 61–158. Springer (2012)

Recommending Multidimensional Spatial OLAP Queries

Olfa Layouni, Fahad Alahmari and Jalel Akaichi

Abstract Huge volumes of data are stored in a spatial data warehouse. Stored data is explored and analyzed by different users. The users interrogate spatial data cube through Spatial OLAP queries which are generally misspoke, leading to non-pertinent results. Therefore, we propose in this paper a Spatial OLAP queries recommendation system based on the SOLAP server query log; that helps and guides users in their exploration. Compared to the users needs, the proposed queries return more relevant results.

Keywords Spatial OLAP · Spatial data cube · SOLAP · Query · Recommendation

1 Introduction

Nowadays, the popularity of spatial data, such as names of cities, postal codes, the position of individual object in space, the maps created from satellite images, can be stored in a spatial data warehouse. According to Franklin [13] spatial data representing about 80 % of all data stored in databases has a spatial or location component, therefore, data are stored in a spatial data warehouse. Stefanovic et al. [28] defined a spatial data warehouse as a collection of spatial and thematic, integrated, nonvolatile and historical data to support the spatial decision making process. In addition, the analysis of spatial data warehouse allows the historical data; the integration and the storage of large volumes of spatial and non-spatial data from multiple sources. A spatial data warehouse contains both a spatial and an alphanumeric data type. It

O. Layouni (✉)
BESTMOD Laboratory, Université de Tunis, ISG - Tunis, Tunis, Tunisia
e-mail: layouni.olfa89@gmail.com

F. Alahmari · J. Akaichi
College of Computer Science, King Khalid University, Abha, Saudi Arabia
e-mail: fahad@kku.edu.sa

J. Akaichi
e-mail: jalel.akaichi@kku.edu.sa

G. De Pietro et al. (eds.), *Intelligent Interactive Multimedia Systems and Services 2016*, Smart Innovation, Systems and Technologies 55,
DOI 10.1007/978-3-319-39345-2_35

is made of a multidimensional spatial model which defines the concepts of spatial measures and dimensions in order to take into account the spatial component. In the literature [4, 19, 21], the authors identified three types of spatial dimensions hierarchies based on the spatial references of the hierarchy members: non-geometric spatial dimensions, geometric-to-non-geometric spatial dimensions and fully geometric spatial dimensions. Also, a spatial data warehouse supports two types of spatial measures: the first type of spatial measures is a set of all the geometries representing the spatial objects corresponding to a particular combination of dimension members. The second type of spatial measures results from the computation of spatial metric or topological operators [2, 3]. In order to analyze and explore a spatial data warehouse, we need a SOLAP server to help the user to make the best decisions.The SOLAP has been identified as an effective means to explore the contents of a spatial data warehouse. The SOLAP is the result obtained after the combination of Geographic Information Systems (GIS), with OLAP tools. To navigate in the spatial data cube the user launches a sequence of SOLAP queries over a spatial data warehouse. A spatial data cube can be queried by using the MDX *(Multi-Dimensional eXpressions)* with spatial extensions query language [2].

SOLAP users interactively navigate a spatial data cube by launching sequences of SOLAP queries over a spatial data warehouse. The problem appeared when the user may have no idea of what the forthcoming SOLAP queries should be. As a solution and to help the user in his navigation, we need a recommendation system. This system gives the possibility to recommend SOLAP queries.

This paper is organized as follows: Sect. 2 introduces related works. Section 3 enlighten our approach of recommending SOLAP queries. Section 4 presents our experimentation. We discuss future work in Sect. 5.

2 Related Work

In this section we focus on the related works of the exploration of spatial data warehouse to provide recommendations to the user. For this reason, we begin by presenting a recommendation system concept, in order to help the user in his exploration of data. Then, we introduce the various methods that have been proposed to explore data.

A recommendation system is usually categorized into a content-based method, a collaborative method and a hybrid method [1, 20].

- *Content-based method*: The user is recommended elements similar to the ones the user preferred in the past.
- *Collaborative method*: The current user is recommended elements similar to the preferences of the previous users and the preferences of the current user.
- *Hybrid method*: This method combines both the content-based and the collaborative method.

In various studies [20, 25–27], we find that the authors described the characteristics of the general algorithm of a recommender system for the exploration of data. These characteristics are the inputs, outputs and the recommendation steps. The inputs of the algorithm can be a log of sessions of queries, a schema, an instance of the relational or multidimensional database, a current session and a profile. The outputs of the algorithm can be a query, a set of ordered queries and a set of tuples. An algorithm of recommendation is decomposed into three steps. The first step consists in choosing an approach for evaluating the used scores. In fact, in this step we can choose one of the categories of recommendation: a content-based or a collaborative or a hybrid method. The second step is the filter; this step consists in selecting the candidates' recommendations. The last step is the guide; this step consists in organizing the candidates' recommendations. Furthermore, we describe some methods that recommend queries for helping users to explore data. In fact, those methods can be classified into two categories, the first category exploits the profile in [5, 6, 25–27] and so does the second category with the log of queries [20, 22–24]. In fact, we find that the methods proposed by [5, 6, 25–27] take as inputs of the algorithm the profile of the current user. Also, the methods proposed by [22–24] take as inputs the log of OLAP sessions of queries. Indeed, we find that the inputs of the algorithm can be: a schema, an instance, the current OLAP query, the current OLAP session, the previous OLAP sessions [5, 6, 20, 22–27]. Besides, we find that the output of the methods proposed by [22–24] is a query, those proposed by [5, 6, 20] the output is a set of queries and only the method proposed by [25–27] the output is a set of tuples. So far, we have found that the methods proposed by [5, 6, 25–27] are content based methods and the methods proposed by [20, 22–24] are collaborative methods. Adding to that, we have discovered that, in the filter step of the proposed algorithm, the method proposed by [5, 6, 20, 25–27] gives the possibility to select the candidate's queries recommendation. Besides, for computing the candidate's recommendations the methods proposed by [25–27] apply the maximum entropy theory; also the methods proposed by [5, 6] use a graphic model; as well the methods proposed by [22–24] apply the Markov model. However, the guiding step was applied in the methods proposed by [20, 22–24]. And finally, we find that the methods proposed by [22–27] used the SQL language, the methods proposed by [5, 6, 20] used the MDX language.

3 Recommendation System

In this section we detail our approach for recommending SOLAP queries. To help the user to go forward in his exploration of the spatial data cube, we propose an approach for recommending SOLAP queries. Our approach consists of three steps. The first step consists in computing all the generalized sessions of SOLAP queries of the log. The second step is the filter which consists in predicting the candidates SOLAP queries. The last step is the guide that consists in ordering the candidates SOLAP queries. But in order to recommend queries, we need to compute the similarity between queries. So, in our context, we need to compute the similarity between SOLAP queries by taking into account spatial relations as topological, orientation and distance metric relations.

3.1 Spatial Similarity Measures

In SOLAP queries, we found the use of the three main categories of spatial relations, which are defined in the literature as follows: topological relation, direction relation and metric distance relation [9, 14, 16]. So, to compare between two SOLAP queries, we need to measure the similarity between the topological relation, direction relation and metric distance relation, invoked in each query.

3.1.1 Topological Distance

A spatial scene could be a topological relation between spatial objects [10, 17, 18]. The topological distance between two SOLAP queries takes into account not only the two spatial topological relations *TR* invoked by a SOLAP query q and TR' invoked by a SOLAP query q'; with *TR* and $TR' \in \{disjoint, meet, overlap, coveredBy, equal, covers, contains, inside\}$; but also spatial objects (a, b) and (a', b') invoked in each query q and q', respectively.

If $a \neq a'$ and $b \neq b'$ then we measure the similarity distance between *TR* and TR', and the similarity distance between objects a, a', b and b' invoked in each query q and q', respectively. The distance between objects represents the similarity between references according to the schema of the spatial data warehouse. In this case, to measure the topological distance, we use the distance proposed by [11], in order to measure the similarity between *TR* and TR'. So, the $Dist_{topR}(TR, TR')$ represents this similarity and takes a value by comparing between *TR* and TR' lake as indicated in table, which represents in Fig. 1.

	disjoint	meet	equal	inside	coveredBy	contains	covers	overlap
disjoint	0	1	6	4	5	4	5	4
meet	1	0	5	5	4	5	4	3
equal	6	5	0	4	3	4	3	6
inside	4	5	4	0	1	6	7	4
coveredBy	5	4	3	1	0	7	6	3
contains	4	5	4	6	7	0	1	4
covers	5	4	3	7	6	1	0	3
overlap	4	3	6	4	3	4	3	0

Fig. 1 Topological distance [11]

In this case, the topological distance between two SOLAP queries $Dist_{top}$ is calculated as follows:

$$Dist_{top}(TR(a,b), TR'(a',b')) = Dist_{topR}(TR, TR') + \sum_{i=1}^{n} Dist_{objects}(r_1, r_2) \quad (1)$$

If $a = a'$ and $b = b'$ then we measure the similarity distance between TR and TR' which represents the topological relationships invoked in each query q and q', respectively. In this case, the topological distance is computed as follows:

$$Dist_{top}(TR(a,b), TR'(a',b')) = Dist_{topR}(TR, TR') \quad (2)$$

3.1.2 Direction Distance

To compute the direction distance between two spatial direction scenes invoked in SOLAP queries q and q'. We propose to adopt the direction distance proposed by [16]; which based on the 9-direction system north, northwest, west, southwest, south, southeast, east, northeast, and equal to represent directions [11, 16]. This distance measure represents the transformation cost from any direction to any other, each transformation is equal to 2 as shown in Fig. 2.

The direction distance between two SOLAP queries $Dist_{Dir}(q,q')$ is calculated as follows:

$$Dist_{Dir}(q,q') = \sum_{i=1}^{2} Cost(TOR(a,b), TOR'(a',b')) \quad (3)$$

With:

- $TOR(a,b)$ and $TOR'(a',b')$ represent two spatial direction scenes invoked in q and q', respectively.
- *Cost* is the transformation cost.

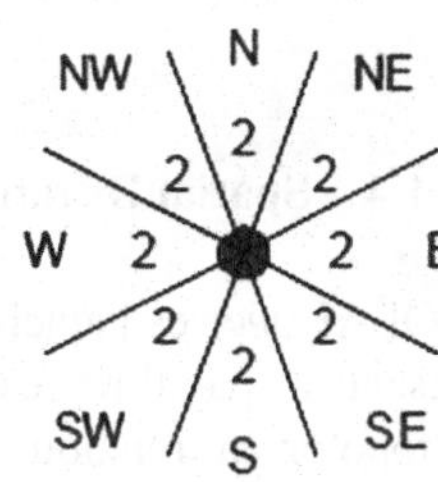

Fig. 2 Direction distance [16]

Fig. 3 Metric distance [16]

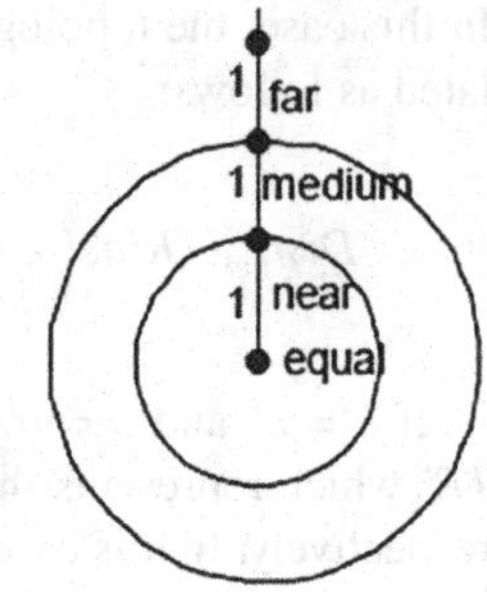

3.1.3 Distance in Term of the Metric Distance

To compute the metric distance between two spatial metric distance scenes invoked in SOLAP queries q and q'. We propose to adopt the distance proposed by [16]; which is based on the traditional 4-granularity metric distance equal, near, medium, far [11, 16]. This distance measure represents the cost of transitions for one granularity to another,each transformation is equal to 1 as shown in Fig. 3.

The metric distance between two SOLAP queries $Dist_{MetD}(q, q')$ with two spatial scenes based on the metric distance $TD(a, b)$ and $TD'(a', b')$ invoked in q and q', respectively, is calculated as follows:

$$Dist_{MetD}(q, q') = Dist_{MetD}(TD(a, b), TD'(a', b')) = \sum_{i=1}^{2}\sum_{j=1}^{2} a_{ij} \tag{4}$$

With:

- a, b, a' and b' are spatial objects invoked in each query q and q'.
- i and j are the number of objects invoked in q and q', respectively.
- $$a_{i,j} = \begin{cases} = 0 & if \quad i \quad is \quad equal \quad to \quad j \\ = 1 & if \quad i \quad is \quad near \quad to \quad j \\ = 2 & if \quad i \quad is \quad medium \quad to \quad j \\ = 3 & if \quad i \quad is \quad far \quad to \quad j \end{cases}$$

3.1.4 Spatial Distance

SOLAP queries launched by users in the cube have spatial scenes. Each scenes represents a spatial representation of one or more spatial objects. This scene could be a topological relation, direction relation and metric distance relation between them [10, 17, 18]. Given two SOLAP queries q and q'; the spatial similarity measure between them $Dist_{SpatialR}(q, q')$ is modeled as follows:

$$Dist_{SpatialR}(q, q') = Dist_{top}(q, q') + Dist_{Dir}(q, q') + Dist_{MetD}(q, q') \tag{5}$$

3.2 SOLAP Queries Recommendation System

The first step is to generalize sessions of SOLAP queries saved in the log, which contains all the previous sessions of SOLAP queries, in order to obtain a generalized log. In this step, we propose to use our spatial similarity measures proposed in the previous subsection, to compute the similarity between SOLAP queries. Also, we propose to use the method of TF-IDF (Term Frequency-Inverse Document Frequency) [8] to evaluate the importance of terms like spatial measures, spatial dimensions, etc. In order to know which dimensions or measures are spatial types. It is used to extract all the spatial dimensions and measures from the schema of the spatial data cube. Finally, for the classification of SOLAP queries, we choose the Hierarchical Ascendant Classification (HAC) [12]. It is used for classifying queries in two different classifications one contains the SOLAP queries without spatial data: OLAP queries and the other contains the SOLAP queries. The second step uses the result obtained in the first step to search the most similar sessions to the generalized current session. That's why, we need to search among the set of generalized sessions of SOLAP queries the ones that are the most similar to the generalized current session and also the query representing its session. This step gives the possibility to move from a candidate SOLAP session to a SOLAP candidate query. The results obtained in this step can be a set of candidates SOLAP queries or an empty set. If we obtain an empty set, the recommendation of queries is done by the default function, which implements the idea proposed by [15]. And if we obtain a set of candidates SOLAP queries, we sort this set in the order of the most similar to the query that represents the current session. So, we order this set first by calculating the distance between candidates SOLAP queries and the SOLAP query representing the current session and that is by using the quick sort in order to order them.

Algorithm 1 Spatial-OLAP-RS (*Sc*, *Log*)

Input: *Sc*: The current session,
Log: the log of sessions of SOLAP queries.
Output: An ordered set of SOLAP queries recommendations.

Loggeneralized ← *Generalize*(*Log*)
CandidatesSOLAP ← *Filter*(*Sc*, *Loggeneralized*)
if (*CandidatesSOLAP* $\neq \phi$) **then**
    Guid(*CandidatesSOLAP*, *QuickSort*)
else
    Default(*Loggeneralized*)
end if

---

4 Experimentation

In this section, we present the results of the experiment that we have conducted to assess the capabilities of our approach. This system gives the possibility to recommend an ordered set of SOLAP queries for the user, after launching the current session. First, to navigate in the spatial data cube the current user launched a sequence of SOLAP queries by using the SOLAP server over a spatial data warehouse. All the previous sessions of SOLAP queries are stored in the log. In fact, our *Spatial-OLAP-RS* system recommends an ordered set of SOLAP queries to the current user. Our experiment evaluates the efficiency of our approach proposed to recommend SOLAP queries. The performance is presented in Fig. 4 according to various log sizes. These log sizes are obtained by playing with two different parameters that change over the time. Those two parameters are as follows:

- X: represents the number of sessions store in the log, it ranges from 10 to 150.
- Y: represents the maximum number of queries that could be launched for per session, it ranges from 10 to 50.

The goal of our experimentation is to measure the execution time taken by applying our proposed approach. So, we conclude that the time taken to recommend SOLAP queries increases with the log size but remains highly acceptable as shown in Fig. 4.

To evaluate the performance of our approach, we choose to use the precision indicator. It's widely used to assess the quality and performance of recommendations systems [7]. In fact, the precision indicator reflects the fraction of recommendations which are the most relevant to the user preferences. In our case, the precision indicator is obtained as follows:

$$Precision = |\{RelevantSOLAPQueriesRecommendation$$

$$\cap ProposedSOLAPQueriesRecommendation\}|$$

$$/|\{ProposedSOLAPQueriesRecommendation\}|$$

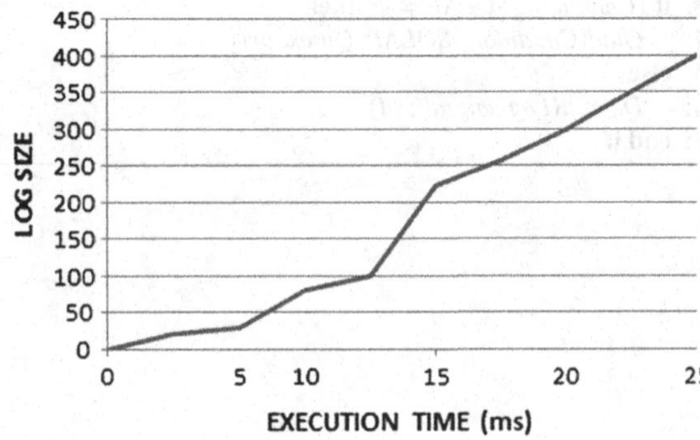

Fig. 4 Performance of our approach

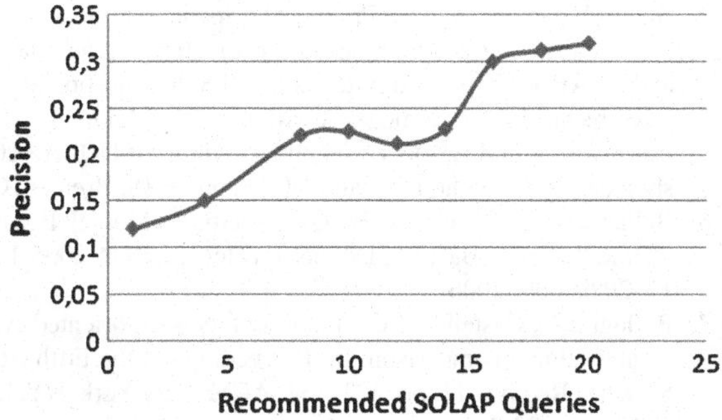

Fig. 5 Precision of *Spatial-OLAP-RS*

In order to calculate the precision indicator, we propose to use a human evaluation technique. We asked 10 persons with SOLAP skills; to launch a set of 5 SOLAP queries to achieve some objectives of data analysis, and retrieve the need information. So, we launch the same SOLAP queries proposed by the 10 persons in our *Spatial-OLAP-RS* system, we obtain a set of recommended SOLAP queries, for each person proposition. The result obtained of the computed precision with our *Spatial-OLAP-RS* system is presented in the Fig. 5.

We note that when the number of recommended queries is less then 5 queries, the precision is low, it's represent 15 %; but it's increase more and more with the augmentation of the number of the recommended queries.

5 Conclusions

In this paper, we proposed a recommendation system in order to help a current user in his exploration of a spatial data cube. For that purpose, we suggested an approach for generating recommendations of Spatial OLAP queries to the current user. Adding to that, to validate our approach, we evaluated the efficiency of our approach proposed to recommend SOLAP queries. Future work consists in going further in the recommendations. We would like to improve our proposed approach by recommending SOLAP queries based on users behaviors and profiles.

References

1. Adomavicius, G., Tuzhilin, A.: Toward the next generation of recommender systems: a survey of the state-of-the-art and possible extensions. IEEE Trans. Knowl. Data Eng. **17**(6), 734–749 (2005)
2. Badard, T.: Lopen source au service du gospatial et de lintelligence daffaires. Geomatics Sciences Department, avril 2011
3. Badard, T., Dubé, E.: Enabling geospatial business intelligence. Geomatics Sciences Department, Sept 2009

4. Bdard, Y., Rivest, S., jose Proulx, M.: Spatial on-line analytical processing (solap): Concepts, architectures, and solutions from a geomatics engineering perspective. In: Data Warehouses and OLAP: Concepts, Architecture, and Solutions, pp. 298–319. Press (2006)
5. Bellatreche, L., Giacometti, A., Marcel, P., Mouloudi, H., Laurent, D.: A personalization framework for OLAP queries. In: Proceedings of DOLAP 2005, ACM 8th International Workshop on Data Warehousing and OLAP, pp. 9–18, Bremen, Germany, 4–5 Nov 2005
6. Bellatreche, L., Mouloudi, H., Giacometti, A., Marcel, P.: Personalization of MDX queries. In: 22èmes Journées Bases de Données Avancées, BDA 2006, Lille, Actes (Informal Proceedings), 17–20 octobre 2006
7. Bellogin, A., Castells, P., Cantador, I.: Precision-oriented evaluation of recommender systems: an algorithmic comparison. In: Proceedings of the Fifth ACM Conference on Recommender Systems, RecSys '11, pp. 333–336. ACM, New York, NY, USA (2011). http://doi.acm.org/10.1145/2043932.2043996
8. Brouard, C.: Comparaison du modèle vectoriel et de la pondération tf*idf associée avec une méthode de propagation d'activation. In: CORIA, pp. 1–10. Neuchâtel, France, Apr 2013
9. Bruns, H.T., Egenhofer, M.J.: Similarity of spatial scenes. In: 7th Symposium on Spatial Data Handling, pp. 31–42 (1996)
10. Clementini, E., Felice, P.D.: A comparison of methods for representing topological relationships. Inf. Sci. Appl. **3**(3), 149–178 (1995)
11. Egenhofer, M.J., Al-Taha, K.K.: Reasoning about gradual changes of topological relationships. In: Frank, A.U., Campari, I., Formentini, U. (eds.) Theories and Methods of Spatio-Temporal Reasoning in Geographic Space: Proceedings of the International Conference GIS, pp. 196–219. Springer, Berlin, Heidelberg (1992)
12. Fablet, R., Bouthemy, P.: Motion-based feature extraction and ascendant hierarchical classification for video indexing and retrieval. In: Proceedings of the 3rd International Conference on Visual Information Systems, VISual99. Lecture Notes in Computer Science, vol. 1614, pp. 221–228. Springer Verlag, Amsterdam, The Netherland (1999)
13. Franklin, C.: An introduction to geographic information systems: linking maps to databases. Database **15**(2), 12–21 (1992)
14. Glorio, O., Mazón, J.N., Garrigós, I., Trujillo, J.: A personalization process for spatial data warehouse development. Decis. Support Syst. **52**(4), 884–898 (2012)
15. Kleinberg, J.M.: Authoritative sources in a hyperlinked environment. J. ACM **46**(5), 604–632 (1999)
16. Li, B., Fonseca, F.: Tdd: a comprehensive model for qualitative spatial similarity assessment. Spat. Cogn. Comput. **6**(1), 31–62 (2006)
17. Li, J.Z., Zsu, M.T., Tamer, M.: Point-set topological relations processing in image databases. In: In First International Forum on Multimedia and Image Processing, pp. 1–54 (1998)
18. Egenhofer, M.J., Franzosa, R.D..: Point-set topological spatial relations. Int. J. Geogr. Inf. Syst. **2**(5), 161–174 (1991)
19. Marketos, G.: Data warehousing & mining techniques for moving object databases. Ph.D. thesis, Department of Informatics, University of Piraeus (2009)
20. Negre, E.: Exploration collaborative de cubes de donnes. Ph.D. thesis, Universit Franois Rabelais of Tours, France (2009)
21. Rivest, S., Bdard, Y., Proulx, M., Nadeau, M., Hubert, F., Pastor, J.: Solap technology: merging business intelligence with geospatial technology for interactive spatio-temporal exploration and analysis of data. J. Photogram. Remote Sens. (ISPRS) **60**(1), 17–33 (2005)
22. Sapia, C.: On modeling and predicting query behavior in olap systems. In: Proceedings of INTL Workshop on Design and Management of Data Warehouses (DMDW 99), SWISS LIFE, pp. 1–10 (1999)
23. Sapia, C.: Promise: predicting query behavior to enable predictive caching strategies for olap systems. In: Kambayashi, Y., Mohania, M., Tjoa, A. (eds.) Data Warehousing and Knowledge Discovery. Lecture Notes in Computer Science, vol. 1874, pp. 224–233. Springer, Berlin Heidelberg (2000)

24. Sapia, C., Alexander, F., Erlangen-nrnberg, U.: Promise: modeling and predicting user behavior for online analytical processing applications. Ph.D. thesis, Technische Universitt Mnchen (2001)
25. Sarawagi, S.: Explaining differences in multidimensional aggregates. In: Proceedings of the 25th International Conference on Very Large Data Bases, VLDB '99, pp. 42–53. Morgan Kaufmann Publishers Inc., San Francisco, CA, USA (1999)
26. Sarawagi, S.: User-adaptive exploration of multidimensional data. In: VLDB, pp. 307–316. Morgan Kaufmann (2000)
27. Sathe, G., Sarawagi, S.: Intelligent rollups in multidimensional olap data. In: Proceedings of the 27th International Conference on Very Large Data Bases, VLDB '01, pp. 531–540. Morgan Kaufmann Publishers Inc., San Francisco, CA, USA (2001)
28. Stefanovic, N., Han, J., Koperski, K.: Object-based selective materialization for efficient implementation of spatial data cubes. IEEE Trans. Knowl. Data Eng. **12**(6), 938–958 (2000)

Modeling Moving Regions: Colorectal Cancer Case Study

Marwa Massaâbi and Jalel Akaichi

Abstract Objects in space have to be represented in order to be stored and analyzed. The three basic abstractions of spatial moving objects are moving point, line or region. The first two abstractions are highly handled. However, moving regions have always been a challenge due to their unstable shape and movement. Researchers are not giving enough attention for managing and querying this particular type of spatial data in order to solve real world problems. Motivated by this fact, we present in this paper an overview on moving regions. We survey region's modeling aspects. Then, we support our research by studying a biomedical case to highlight the importance of using moving regions. The case study illustrates the conceptual aspect of the movement of the colorectal cancer. We also use fuzzy logic thanks for its simplicity and its easy understanding. The combination offers an easier understanding for decision makers.

Keywords Moving regions · Data modeling · Fuzzy logic · Fuzzy UML

1 Introduction

A region is an object whose extent is changing over time (for example, air polluted zones, shrinking balloon, volcanic lava eruption, forests, lakes). This object has to meet certain criteria: it must have a changing extent (grow/shrink) and location (move continuously) over time. There are two types of moving regions: a simple region and a complex one (having holes inside). The difficulty in this case is when supporting and dealing with this particular type of moving objects. In other words, it is hard to capture every single movement state in every time unit and between them.

M. Massaâbi (✉)
BESTMOD Laboratory, Université de Tunis, ISG - Tunis, Tunis, Tunisia
e-mail: massaabi.marwa@gmail.com

J. Akaichi
College of Computer Science, King Khalid University, Abha, Saudi Arabia
e-mail: jalel.akaichi@kku.edu.sa

G. De Pietro et al. (eds.), *Intelligent Interactive Multimedia Systems and Services 2016*, Smart Innovation, Systems and Technologies 55,
DOI 10.1007/978-3-319-39345-2_36

Some of this region data can be imprecise or incomplete due to an inconstant sampling rate for example (every few seconds or every few days), or erroneous points (a point in the ocean). The uncertainty and the imprecision can also be caused by bad reception condition of GPS receivers. Despite the progress in this research area, the present works still require further improvements in order to make modeling moving regions an easier and more effective task. Therefore, in our work, we chose to use Fuzzy Logic to overcome this imperfection. In fact, Fuzzy Logic was introduced by Lotfi Zadeh in 1965 [18]. Fuzzy logic is an extension of the Boolean logic. It is a mathematical model distinguished by its ability to represent and process uncertain and imprecise information. In this paper, we survey the existing methods for modeling moving regions. Then, we take colorectal cancer as an example and we propose a conceptual model that represents its movement. This paper is organized as follows: Sect. 2 surveys and classifies moving regions modeling techniques. Section 3 presents a detailed case study. Finally, Sect. 4 concludes and proposes research directions.

2 Related Works

Several researchers took interest in modeling moving regions. Therefore, in this section, we recall the elaborated works in this field in order to overview its evolution over the last years. Hence, we classified the surveyed works into three categories: Crisp, Fuzzy and Conceptual modeling.

2.1 Crisp Modeling

In [16], the authors used the polygon representation to model every object moving both discretely and continuously. These objects are characterized by a dynamic shape and extent (i.e. region). They took special interest in the continuously moving regions having their data stored in a discrete way. Therefore, their work aims to estimate the in-between state of the object to restore its continuous feature. In [10], the authors presented their big data based system "Storm DB" used to answer rain fall amount queries in the continental United States. The moving regions in their case are the rain clouds. Storm DB is composed of five steps: They started by extracting the regions from radar images provided by the US national weather service. Then, they classified the regions according to their belonging to an existing storm or a new one. The next step is to calculate the duration of a point in a region in order to obtain the rainfall amount list. The fourth step aims to accelerate the algorithm used in the previous one. They concluded by the Big Data architecture. It consists of a map/reduce algorithm that supports parallelism and processes a massive amount of data. Their approach, indeed deals with moving regions. However, the modeling aspect was not their strength point since they considered regions as simple points. In [17], the authors presented a data model for route planning in the case of forest fire. They developed an application that suggests safe circuitous routes to avoid the moving flames. To do so, they used a fire simulation model. This model takes into consideration the wind properties since it influences the fire state and direction. It also simulates and gives information about the fire. This model is taken from [9]. It consists of a real time algorithm that simulates and predicts forest fire spreading. Some authors preferred well-known techniques, whether some other authors chose algebra for managing moving regions. The following works [1–3, 7] treated the same issue from almost the same point of view. Well, in [1], the authors presented their approach for modeling moving objects and discussed some issues. Among the issues, they discussed the level of abstraction suited for an algebra. For instance, a region could be represented as a polyhedral shape in the space or as a continuous function. As they consider moving objects as geometries changing over time, they defined corresponding algebraic operations. In order to specify the algebraic operations, they began by proposing an abstract algebraic data model, i.e., a model that describes data in a simple and sometimes unrealistic way. The abstract model was followed by the according discrete one. The same data structures and algorithms have already been used for spatial databases in the Rose algebra [4, 5]. In [3], the authors proposed a data model for moving objects in the two-dimensional space. They presented their system design and presented its associated operations. They specified the types' semantics and operations in the abstract model. While in [2], the authors completed the previous work [3] by presenting the discrete model. They transformed the abstract model into a discrete one for which they provided data structures and illustrating algorithms. The work elaborated in [7] combines between the three previously cited works [1–3] and enriches them. In fact, the authors developed powerful algorithms able to deal with a large set of operations and created a basis for the development of a reliable DBMS extension package for moving objects.

2.2 *Fuzzy Modeling*

The previously presented methods used crisp methods to deal with crisp spatial objects. However, some authors preferred to use the Fuzzy Set theory for a better representation of fuzzy spatial objects. In fact, this theory offers several properties and operations that correspond to some real life objects specificities (i.e. spatial objects with vague and imprecisely determined boundaries or interiors). In [6], the authors proposed an approach that models regions and their trajectories, and predicts their future movements as well. In this work, regions are delimited by a rectangle, also named enclosing box, to which was assigned a center point. The trajectory, in this case, is the link between the center points. After modeling regions, the authors predicted its future positions using the linear regression and the recursive motion function. They tested their application using two datasets concerning fires and storms. In his works, [11–14], Markus Schneider was interested in fuzzy spatial databases and fuzzy moving objects. We begin with [11], the researcher offered a fuzzy spatial data type for fuzzy points, fuzzy lines and fuzzy regions. He also proposed novel fuzzy spatial operations. In [12], he continued his work by proposing metric operations on fuzzy spatial objects. Then, in [13] he defined a conceptual model of fuzzy spatial objects with its implementation. The particularity of this work is the use of the discrete geometric domain (grid partition) instead of the Euclidean space. In [14], the author began by presenting a detailed overview on spatial data types and a survey concerning spatial fuzziness. Then, he proposed a fuzzy algebra to represent fuzzy data types. He focused on treating spatial vagueness.

2.3 Conceptual Modeling

Conceptual modeling is important. It offers a clearer vision and understanding of the treated issues. The conceptual diagrams and data models indeed depict and describe complex objects and their relations. In [8], the authors proposed a UML data model that combines fuzzy sets and possibility theory to overcome data fuzziness and manage complex objects. They presented their contributions at various levels: fuzzy class, fuzzy generalization, fuzzy aggregation, fuzzy association and fuzzy dependency. Then, they presented the corresponding fuzzy relational schema with the transformation steps. Another work, [15], in which the authors presented a conceptual model for fuzzy object-oriented databases using UML. Their model targets imprecise data, which they called fuzzy data. In fact, they defined fuzzy objects, classes and the relations between them, together with the corresponding notations.

2.4 Recapitulation

To recapitulate, we present the following table (Table 1).

Table 1 classifies and resumes the previously cited methods for modeling moving regions. The works are cited in the table in the chronological order.

According to the presented state of the art, several techniques were used to model moving regions. Among the years, researchers took advantage of fuzzy set theory, algebra, big data, etc. Each and every one of these works has its advantages and drawbacks. Therefore, in this paper, we introduce a UML conceptual model to represent the movement of cancer cells. In fact, constructing an accurate corresponding model allows capturing the basic concepts and relations between them. Building the model is in other way describing the problem in order to understand and solve it.

Table 1 Classification of the presented methods

Cited works	Crisp modeling	Algebraic modeling	Fuzzy modeling	Conceptual modeling
[1]		Algebra		
[11]			Fuzzy Sets	
[3]		Algebra		
[12]			Fuzzy Sets	
[16]	Polygon representation			
[7]		Algebra		
[13]			Fuzzy sets	
[14]			Fuzzy sets	
[6]	Enclosing box technique			
[8]				UML
[10]	Big data approach			
[15]				Fuzzy UML
[17]	Simulation algorithm			

3 Case Study

Analyzing moving objects trajectories is a very interesting and general field. One of its major advantages is that it can be applied in several domains. In this case, we are interested in the medical field: observing illnesses, examining their behavior from a patient to another, and why not, predicting their effects on different patients too. Well, diseases move in regions, like fires, they neither have an exact shape nor follow a definite path. For this reason, we chose one specific disease (cancer). In fact, there are several types of cancers, but we specifically chose colorectal cancer. It starts in the colon then travels to the whole body in evolving speed. It moves in a group of countless microscopic cells characterized by irregular shape, size and borders. In order to clarify the fuzzy aspect in the transformation phase, we propose a conceptual model that describes the moving cancer cells. The proposed conceptual model is characterized by the fuzzy class. A fuzzy class is a class that describes the fuzzy object with at least one fuzzy attribute. A fuzzy attribute has linguistic values instead of numerical ones. The moving object is represented by a fuzzy class named 'Cancer' (Fig. 1).

It is defined by the following attributes: Speed, Size and Age. 'Speed' describes the tumor's propagation speed. The tumor propagation speed is very slow in the first fifteen or twenty years which, unfortunately, makes the early detection very difficult. In the following ten years the speed increases and becomes noticeable. In the final stage, i.e., the last two years, the tumor spreads widely in a high speed. 'Size' provides the cancer cells size which tells about the gravity of the situation. And 'Age'

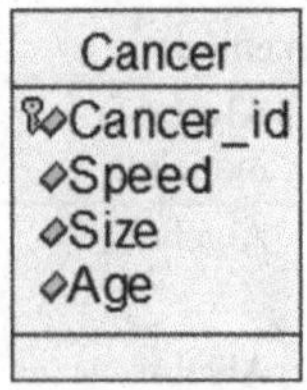

Fig. 1 The class 'Cancer'

represents the diagnosed age of the illness. Among these attributes, only two are represented fuzzily: 'Speed' and 'Size' are better identified using linguistic values. The set of linguistic values defining the attribute 'Speed' is {low, medium, high}. The set of linguistic values defining the attribute 'Size' is {tiny, small, big}. At this level, the membership function should be calculated for each linguistic value.

The membership function describing the fuzzy attribute 'Speed' is given below:

$$\mu_{low}(Speed) = \begin{cases} 1 & \text{if } Speed \leqslant 20 \\ 0 & \text{if } Speed \geqslant 25 \\ \frac{25-Speed}{25-20} & \text{if } 20 < Speed < 25 \end{cases}$$

$$\mu_{medium}(Speed) = \begin{cases} 0 & \text{if } Speed \leqslant 20 \\ 0 & \text{if } Speed \geqslant 40 \\ 1 & \text{if } 35 \leqslant Speed \leqslant 25 \\ \frac{Speed-20}{25-20} & \text{if } 20 < Speed \leqslant 25 \\ \frac{40-Speed}{40-35} & \text{if } 40 \leqslant Speed < 35 \end{cases}$$

$$\mu_{high}(Speed) = \begin{cases} 0 & \text{if } Speed \leqslant 35 \\ 1 & \text{if } Speed \geqslant 40 \\ \frac{Speed-35}{40-35} & \text{if } 35 < Speed < 40 \end{cases}$$

The membership function of the fuzzy attribute 'Speed' is graphically represented by Fig. 2. The curve defines to each point a membership value between 0 and 1.

The membership function describing the fuzzy attribute 'Size' is given:

$$\mu_{tiny}(Size) = \begin{cases} 1 & \text{if } Size \leqslant 0.5 \\ 0 & \text{if } Size \geqslant 1 \\ \frac{1-Size}{1-0.5} & \text{if } 0.5 < Size < 1 \end{cases}$$

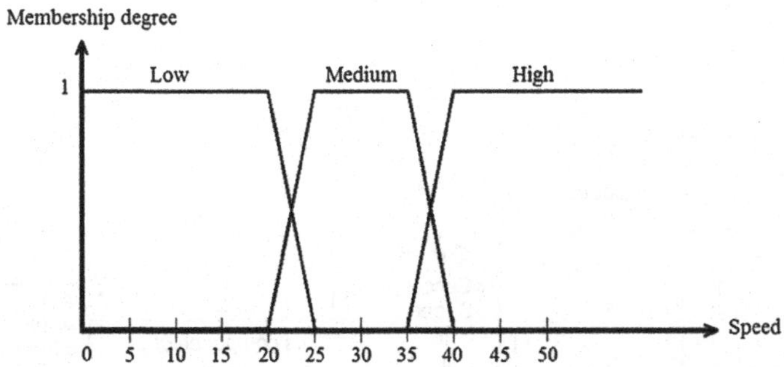

Fig. 2 The membership function of the attribute 'Speed'

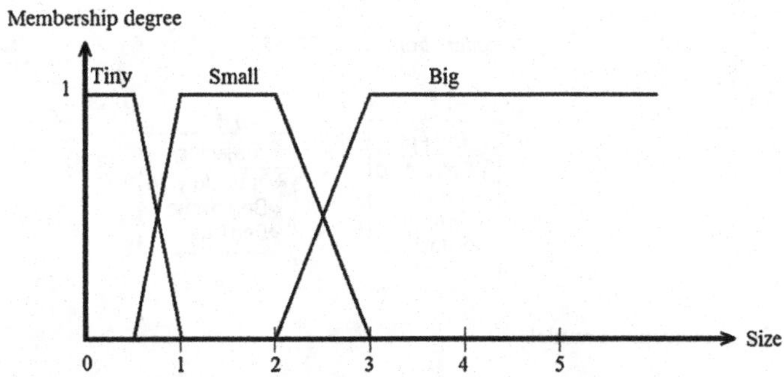

Fig. 3 The membership function of the attribute 'Size'

$$\mu_{small}(Size) = \begin{cases} 0 & \text{if } Size \leqslant 0.5 \\ 0 & \text{if } Size \geqslant 3 \\ 1 & \text{if } 1 \leqslant Size \leqslant 2 \\ \frac{Size-0.5}{1-0.5} & \text{if } 0.5 < Size \leqslant 1 \\ \frac{4-Size}{4-3} & \text{if } 4 \leqslant Size < 3 \end{cases}$$

$$\mu_{big}(Size) = \begin{cases} 0 & \text{if } Size \leqslant 2 \\ 1 & \text{if } Size \geqslant 3 \\ \frac{Size-2}{3-2} & \text{if } 2 < Size < 3 \end{cases}$$

The membership function of the fuzzy attribute 'Size' is graphically represented by Fig. 3. The curve defines to each point a membership value between 0 and 1.

After defining the attributes and their membership functions, we created the fuzzy general conceptual model of the cancer's movement using a Unified Modeling Language diagram (Fig. 4).

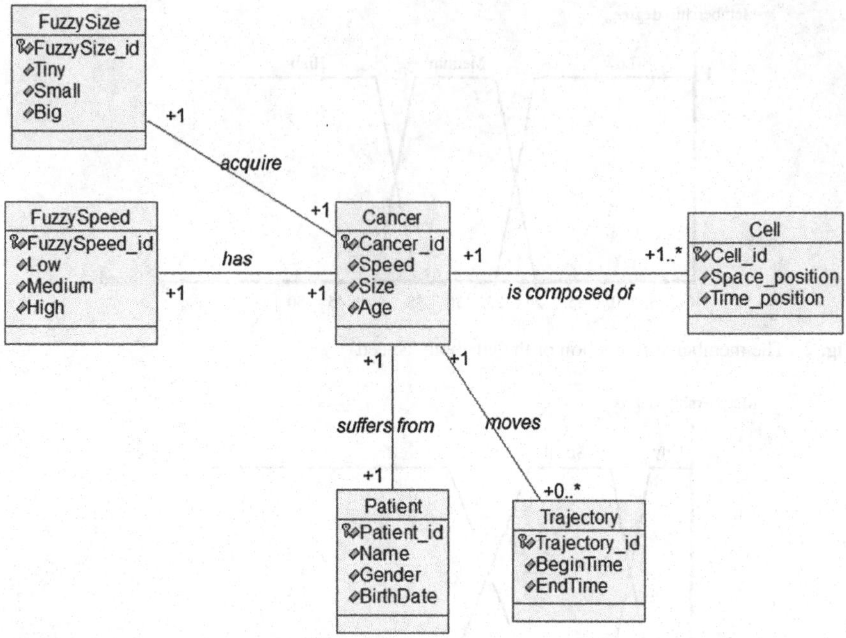

Fig. 4 The fuzzy conceptual model

The model describes the interaction between the tumor itself and the surrounding objects that could affect it. The class 'Cancer' is related to three classes: 'Patient' which describes the patient having the tumor, 'Trajectory' which describes the trajectory of the tumor in the human body, and more importantly 'Cell'. Actually, the 'Cell' class is the one that identifies the cancer as a region. It is in this case represented by a set of points not one single point. This representation reflects in a more accurate way the nature of this moving object. In addition to the diagram classes, we added two fuzzy classes: 'FuzzySize' and 'FuzzySpeed'. In fact, every linguistic attribute of the class 'Cancer' is represented in the conceptual model by a fuzzy class. Each fuzzy class is characterized by its primary key and its linguistic values as attributes.

Once the conceptual model is done, we moved to next one. The logical model of the cancer's movement using the snowflake schema is presented by Fig. 5. It describes the relations between the tables and it is inspired by the conceptual model. Both model were elaborated by Rational Rose which is a software published by the company Rational, and acquired by IBM, to create and edit UML diagrams.

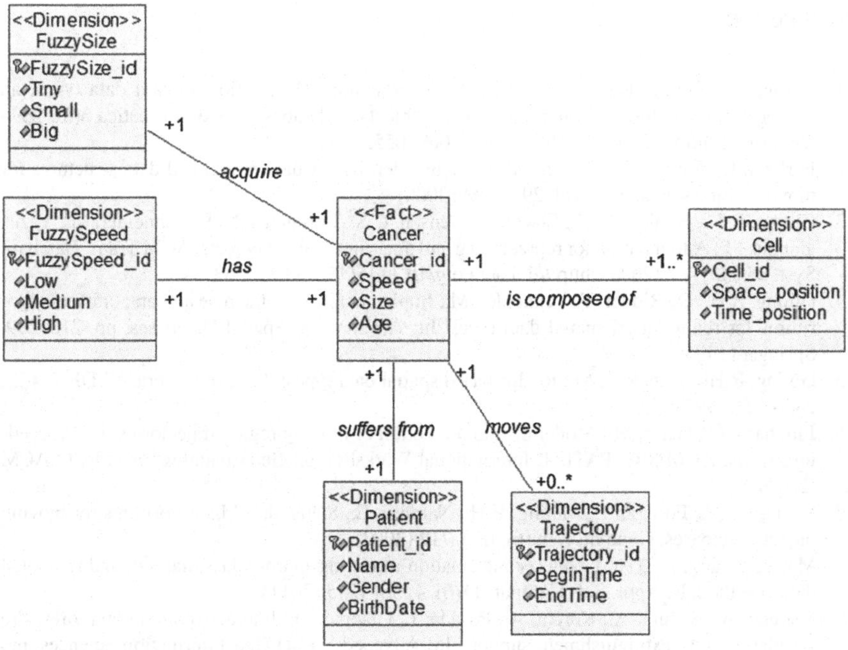

Fig. 5 The fuzzy logical model

4 Conclusions and Future Works

In the recent years, researchers around the world are more and more interested in studying moving object data (points, lines and regions). Moving regions, in particular, are characterized by their continuously changing shape and path together with some possible holes inside sometimes. The regions' specifications inspired us to use fuzzy logic to have a better representation, storage and analysis. In this paper, we surveyed the issue of modeling moving regions. We also reviewed limitations of the current systems. Unfortunately, the current generation of moving regions modeling systems still requires further improvements to obtain better results. Hence, we began by presenting a moving cancer conceptual model. This model is a first step to a complete model that deals with moving regions. We presented the integration of fuzzy logic into the regions' modeling using fuzzy membership functions, also, by providing both conceptual and logical fuzzy models. The next step will be to enrich our model by a complete framework dealing with moving regions' data. The framework will be able to analyze and predict the behavior of this particular kind of data in order to support the decision making process.

References

1. Erwig, M., Güting, R.H., Schneider, M., Vazirgiannis, M.: Spatio-temporal data types: an approach to modeling and querying moving objects in databases. Geoinformatica **3**(3), 269–296 (1999). http://dx.doi.org/10.1023/A:1009805532638
2. Forlizzi, L., Güting, R.H., Nardelli, E., Schneider, M.: A data model and data structures for moving objects databases, vol. 29. ACM (2000)
3. Güting, R.H., Böhlen, M.H., Erwig, M., Jensen, C.S., Lorentzos, N.A., Schneider, M., Vazirgiannis, M.: A foundation for representing and querying moving objects. ACM Trans. Database Syst. **25**(1), 1–42 (2000). http://doi.acm.org/10.1145/352958.352963
4. Güting, R.H., De Ridder, T., Schneider, M.: Implementation of the rose algebra: efficient algorithms for realm-based spatial data types. In: Advances in Spatial Databases, pp. 216–239. Springer (1995)
5. Güting, R.H., Schneider, M.: Realm-based spatial data types: the rose algebra. VLDB J. **4**(2), 243–286 (1995)
6. Junghans, C., Gertz, M.: Modeling and prediction of moving region trajectories. In: Proceedings of the ACM SIGSPATIAL International Workshop on GeoStreaming, pp. 23–30. ACM (2010)
7. Lema, J.A.C., Forlizzi, L., Güting, R.H., Nardelli, E., Schneider, M.: Algorithms for moving objects databases. Comput. J. **46**(6), 680–712 (2003)
8. Ma, Z., Zhang, F., Yan, L.: Fuzzy information modeling in uml class diagram and relational database models. Appl. Soft Comput. **11**(6), 4236–4245 (2011)
9. Moreno, A., Segura, Á., Korchi, A., Posada, J., Otaegui, O.: Interactive urban and forest fire simulation with extinguishment support. In: Advances in 3D Geo-Information Sciences, pp. 131–148. Springer (2011)
10. Olsen, B., McKenney, M.: Storm system database: a big data approach to moving object databases. In: 2013 Fourth International Conference on Computing for Geospatial Research and Application (COM. Geo), pp. 142–143. IEEE (2013)
11. Schneider, M.: Uncertainty management for spatial datain databases: fuzzy spatial data types. In: Advances in Spatial Databases, pp. 330–351. Springer (1999)
12. Schneider, M.: Metric operations on fuzzy spatial objects in databases. In: Proceedings of the 8th ACM international symposium on Advances in geographic information systems, pp. 21–26. ACM (2000)
13. Schneider, M.: Design and implementation of finite resolution crisp and fuzzy spatial objects. Data Knowl. Eng. **44**(1), 81–108 (2003)
14. Schneider, M.: Fuzzy spatial data types for spatial uncertainty management in databases. In: Handbook of Research on Fuzzy Information Processing in Databases, vol. 2, pp. 490–515 (2008)
15. Singh, S., Agarwal, K., Ahmad, J.: Conceptual modeling in fuzzy object-oriented databases using unified modeling language. Int. J. Latest Res. Sci. Technol. **3**, 174–178 (2014)
16. Tøssebro, E., Güting, R.H.: Creating representations for continuously moving regions from observations. In: Advances in Spatial and Temporal Databases, pp. 321–344. Springer (2001)
17. Wang, Z., Zlatanova, S., Moreno, A., van Oosterom, P., Toro, C.: A data model for route planning in the case of forest fires. Comput. Geosci. **68**, 1–10 (2014)
18. Zadeh, L.A.: Fuzzy sets. Inf. Control **8**(3), 338–353 (1965)

A UML/MARTE Extension for Designing Energy Harvesting in Wireless Sensor Networks

Raoudha Saida, Yessine Hadj Kacem, M.S. BenSaleh and Mohamed Abid

Abstract Power supply is the major concern in the wireless sensor networks (WSNs) applications. Currently, the node lifetime is limited by a battery supply which is a short lifetime, unmanageable and uneconomical. Energy Harvesting was proposed as a promising alternative to power sensor nodes in many application fields. Several energy harvesting concepts are considered in WSNs systems such as solar, vibration, thermal, kinetic, acoustic noise, radio frequency (RF), biochemical and hybrid energy sources. The existing modeling design for the power supply section of sensor nodes is limited to the design of solar energy harvesting method which is mostly employed in outdoor applications with sufficient sun light. However, other energy harvesting concepts are potential ambient sources of energy which offer an enough amount of power for sensor nodes. In this paper, we propose a high level methodology based on UML/MARTE standard to model specifications of outlined energy harvesting devices in the WSNs. We define new packages extending the "HW_Harvesting" package which is extending the "HW_PowerSupply" package. A case study of a WSNs system regarding leak detection in water pipeline monitoring is used to evaluate the practical use of our proposal.

Keywords WSN · Energy harvesting · MDE · MARTE

R. Saida (✉) · M. Abid
CES Laboratory, National Engineering School of Sfax, Sfax, Tunisia
e-mail: saidaraoudha@yahoo.fr

R. Saida · M. Abid
Digital Research Centre of Sfax (CRNS), Sfax, Tunisia

Y. Hadj Kacem
College of Computer Science, King Khalid University, Abha, Saudi Arabia

M.S. BenSaleh
National Electronics, Communications and Photonics Center, KACST,
Riyadh, Saudi Arabia

G. De Pietro et al. (eds.), *Intelligent Interactive Multimedia Systems and Services 2016*, Smart Innovation, Systems and Technologies 55,
DOI 10.1007/978-3-319-39345-2_37

1 Introduction

During the last decade, there has been a considerable research area in the development of WSNs. These WSNs systems are used in a variety of applications such as health care systems [1], environmental management [2] and many others. The common focus in WSNs applications is the energy efficiency. Typically, sensor nodes have a limited power resource for operating the network requirements. Currently, they are supplied by a local battery which is a short lifetime and its replacement becomes time consuming, uneconomical and unfeasible in some cases. Several techniques are investigated in the literature to maximize the lifetime of sensor nodes. Among these approaches, energy harvesting is an interesting technique. Energy harvesting is a promising alternative of meeting the network energy requirements, in which ambient energy is extracted from the environment and is converted into electricity to power the sensor nodes. A review of the possible sources of energy which could be extracted from the environment is given in [3]. There are numerous sources of energy to power the sensor network such as solar power, kinetic energy, mechanical vibrations, temperature differences, magnetic fields, etc. The identification of the energy harvesting strategy is according to the operating environment and the system's energy requirements. Since energy harvesting approach offers an adequate power supply for sensor nodes, designers need to integrate this powerful technique in the development cycle of WSNs systems. The Model Driven Engineering (MDE) [4] paradigm is commonly being adopted to design complex systems. The use of the Unified Modeling Language (UML) profiles, to deal with real time constraints and performance issues of embedded systems, reduces the design complexity. In this context, the Modeling and Analysis of Real Time and Embedded systems (MARTE) [5] standard offers a rich support for modeling and analyzing real time embedded systems. The MARTE profile provides the concepts needed to describe real time features and the semantics of embedded systems at different abstraction levels for enabling performance analysis. MARTE standard proposes in its Non Functional Properties (NFPs) package the HW_Power package for annotating power issues and heat dissipation of hardware (HW) components. MARTE also provides HW _PowerSupply and HW_Battery packages to supply energy. However, MARTE standard does not provide a support to specify the diversity of possible harvesting energy sources in WSNs applications. An earlier work proposed in [6], in which authors modeled the power supply section of sensor nodes and proposed a MARTE extension of the HW_PowerSupply package by adding the HW_Harvesting stereotype to define energy harvesting devices and specially focus on solar energy (the HW_PV stereotype). Solar energy is probably the most exploited to power WSNs, but it is not always available due to its strong dependency on weather, day and night. While solar energy is limited to outdoor applications with sufficient sun light, other potential ambient sources of energy has been considered for energy harvesting. Therefore, exploiting and introducing these energy harvesting kinds by extending MARTE profile is necessary. In this work, we aim to extend MARTE profile to support possible

WSNs energy harvesting techniques including vibration, thermal, kinetic, acoustic noise, RF, biochemical and hybrid energy sources to provide new unlimited energy sources enhancing sensor network capabilities to power itself and achieve a longer lifetime. We will add therefore other MARTE extensions to satisfy outlined energy harvesting types. An interesting domain application of WSNs is the water pipeline monitoring to detect leaks and control the water quality. A case study of a WSNs system regarding water pipeline monitoring using a vibration energy harvesting device is used to evaluate the practical use of our extension. In the present paper, we present a high level methodology based on UML/MARTE standard to model specifications of various energy harvesting devices in the wireless sensor network. In Sect. 2, we outline MARTE capabilities for power section modeling. Section 3 describes our proposed extensions. Section 4 describes the behavioral analysis of the proposed extensions. In Sect. 5, we give an overview about energy harvesting methods used in water pipeline monitoring and we give a case study illustrating our proposed extensions. We end our work with conclusions and perspectives in Sect. 6.

2 MARTE Capabilities for Designing Power Section

MARTE profile is a standard UML profile promoted by the Object Management Group (OMG). UML/MARTE is decomposed into several sub-profiles such as Global Resource Modeling (GRM), Hardware Resource Modeling (HRM) and Software Resource Modeling (SRM), etc., which each one has its own specifications. MARTE profile offers a support for annotating NFPs of embedded systems. This UML profile defines in its HRM and NFPs packages, the generic package HW_Power for modeling power consumption and heat dissipation of HW components according to its specifications [5]. In particular, the HW_PowerSupply package specifies the energy supplies. While our major constraint in WSNs design is the energy efficiency, this entity should be extended and refined to contain more properties necessary to perform energy sources. MARTE standard represents a model for energy harvesting which focuses only on solar energy scavenging device, described in [6], in which "HW_Harvesting" stereotype describes the harvester device. This stereotype is extended with "HW_PV" to describe the harvesting done by a solar panel. However, this proposed model lacks an explicit and rich support to exploit the other energy harvesting kinds in order to maintain the sensor network lifetime. Therefore, MARTE needs to be refined and extended to offer a powerful high level model to design the power section of a WSNs node. In the rest of paper, we propose to develop a generic model that focuses on different energy harvesting techniques including vibration, thermal, wind, acoustic noise, RF, biochemical and hybrid sources.

3 The Proposed Extensions

We propose to extend the "HW_Harvesting" package to add the other energy harvesting devices: vibration, thermal, kinetic energy, acoustic noise, RF, biochemical and hybrid energy harvesting types.

The structure of the extended profile is shown in red in Fig. 1. To include all necessary properties to design outlined energy scavenging types, seven new classes are introduced in MARTE profile. The content of each of these new stereotypes is described in the following subsections.

3.1 Vibration Energy Harvesting: HW_Vibration

One class of energy harvesting systems is the vibration energy source. Vibrations are mostly present in built environments such as bridges, roads and rail tracks. Vibration harvester captures the energy from the mechanical motion and convert it into electricity. Three basic mechanisms [7] are used for transforming vibration into electrical energy: piezoelectric [8], electrostatic [9] and electromagnetic [10]. This extracted energy from the vibration source depends on the amplitude of vibration and the frequency of the source vibration object. Figure 2 shows the "HW_Harvesting" extension with the "HW_Vibration" stereotype to describe a vibration device.

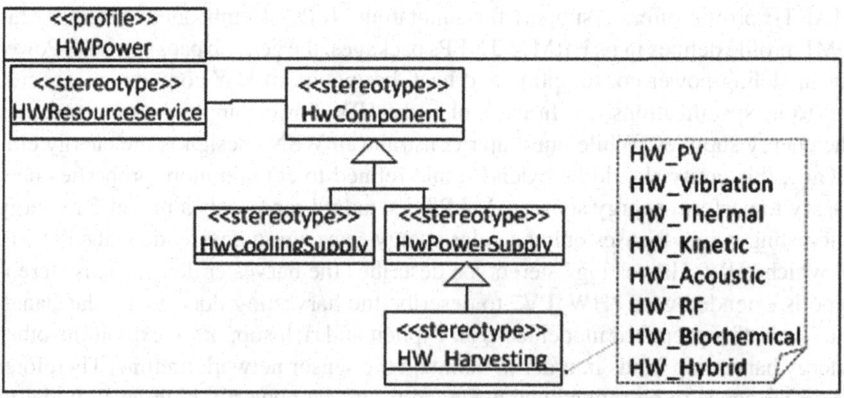

Fig. 1 Structure of the extended HW_Harvesting package

Fig. 2 Proposed MARTE extension for vibration energy harvesting modeling

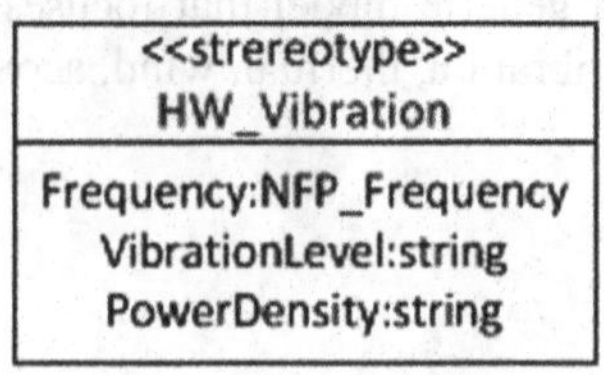

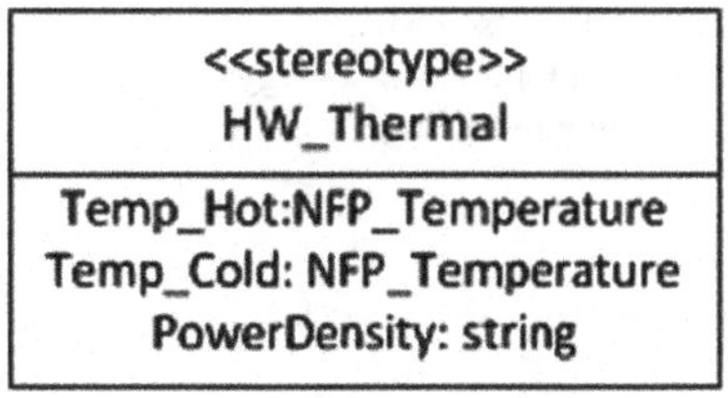

Fig. 3 Proposed MARTE extension for thermal energy harvesting modeling

3.2 Thermal Energy Harvesting: HW_Thermal

Another method is the thermal energy harvesting [11]. This technique produces electricity using the Seebeck effect when there are differences in temperature or gradients. Thermal energy sources can be machines, animals, human body and others. The thermoelectric conversion depends on the temperature's difference between the two sides of the transducer and the power density achieved by the harvester. Figure 3 shows the "HW_Harvesting" extension with the "HW_Thermal" stereotype to describe a thermal device.

3.3 Kinetic Energy Harvesting: HW_Kinetic

Among existing kinds of renewable energy resources, the kinetic energy is an important source for powering sensor network. Kinetic energy can be harvested in the environment from the wind flow and moving water in rivers or in pipes. The technological concepts associated with energy harvesting from flowing water or from the wind are similar. Kinetic energy is mostly harvested using turbines. The importance of harvesting energy from water and wind to power sensor network is highlighted in [12]. The extracted energy depends on the density of the source, its velocity and the cross area. Figure 4 shows the "HW_Harvesting" extension with the "HW_Kinetic" stereotype to describe a kinetic device.

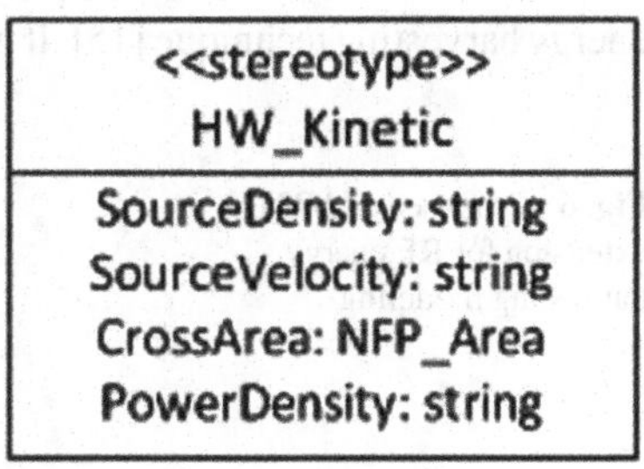

Fig. 4 Proposed MARTE extension for kinetic energy harvesting modeling

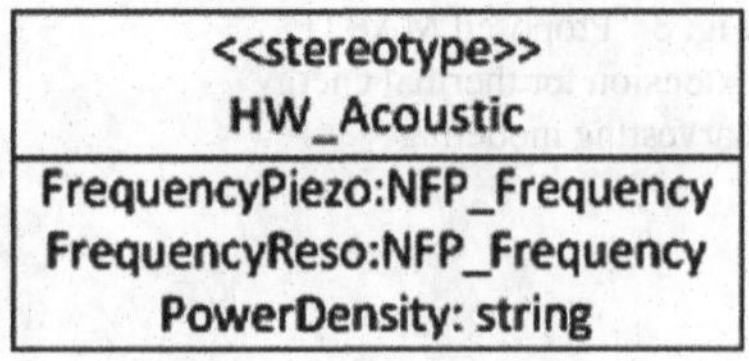

Fig. 5 Proposed MARTE extension for acoustic energy harvesting modeling

3.4 Acoustic Energy Harvesting: HW_Acoustic

Acoustic energy [13] is another viable form of ambient energy. It is the process of converting continuous and high acoustic waves from the environment into electrical energy by using an acoustic transducer or resonator. The extracted energy depends on the frequency of piezoelectric material and the frequency of resonator. Figure 5 shows the "HW_Harvesting" extension with the "HW_Acoustic" stereotype to describe an acoustic device.

3.5 Radio Frequency Energy Harvesting: HW_RF

Other technique of energy harvesting is the RF [14] energy harvesting. The RF energy is captured by a receiving antenna which is attached to each sensor node. The antenna is characterized by its frequency, the impedance,the size and the directivity. Then, this captured energy is converted through a power conversion circuit into DC energy which is stored in an energy storage device to power the sensor node. Figure 6 shows the "HW_Harvesting" extension with the "HW_RF" stereotype to describe a RF device.

3.6 Biochemical Energy Harvesting: HW_Biochemical

Another harvesting strategy employed to power sensor nodes is the biochemical energy harvesting technique [15]. It indicates the process of generating the electricity

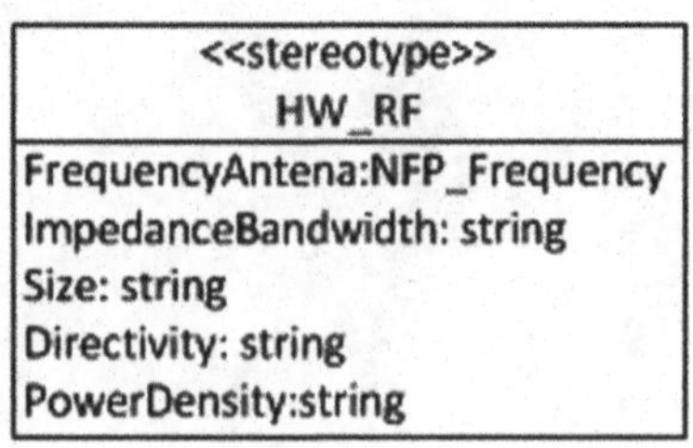

Fig. 6 Proposed MARTE extension for RF energy harvesting modeling

Fig. 7 Proposed MARTE extension for biochemical energy harvesting modeling

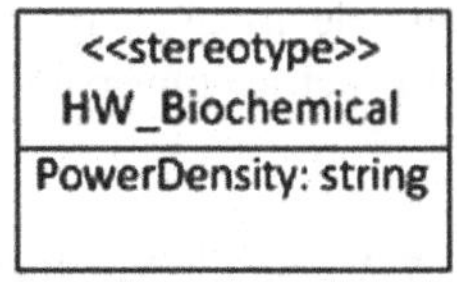

by converting the chemical energy of a fuel (e.g. hydrogen) through a chemical reaction with an oxidizing agent. The harvested energy is mainly used for small-scale powering applications. Figure 7 shows the "HW_Harvesting" extension with the "HW_Biochemical" stereotype to describe a biochemical device.

3.7 *Hybrid Energy Harvesting: HW_Hybrid*

Several existing approaches for energy harvesting are focusing on one type of energy while other types of energy are wasted. Developing hybrid models with the combination of several energy harvesting strategies in the same device can increase the capability to power the sensor node from more than one source which leads to store more energy and have persistent batteries. We present four hybrid approaches: Hybrid solar and thermal cell, hybrid solar and mechanical cell, hybrid solar and biochemical cell and hybrid biochemical and biomechanical cell. We have extended the "HW_Harvesting" stereotype with the "HW_Hybrid" stereotype to describe an hybrid device which is characterized by the output voltage which represents the sum of multi energy harvesters. Then, we have extended the "HW_Hybrid" stereotype by outlined hybrid techniques which is illustrated in Fig. 8.

3.7.1 Solar_Thermal

In this hybrid Solar Thermal cell [16], we integrated a dye sensitized solar cell (DSSC) and thermal-electric cell (TG) in the same device. Solar energy is converted

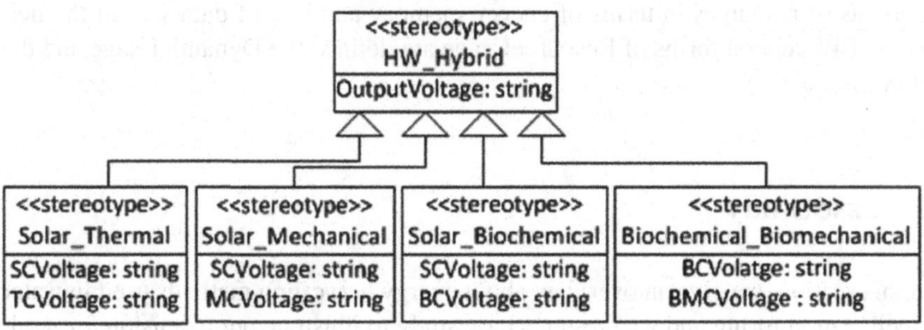

Fig. 8 Proposed MARTE extension for hybrid energy harvesting modeling

into electrical energy and heat. Then, the heat is transmitted to the TC used for thermoelectric conversion. Every cell is characterized by its voltage.

3.7.2 Solar_Mechanical

In this hybrid Solar Mechanical cell [17], we integrated a DSSC and piezoelectric nanogenerator (NG) in the same device. This type is used specially for small electronics devices. Every cell is characterized by its voltage.

3.7.3 Solar_Biochemical

In this hybrid Solar Biochemical cell [18], we integrated a DSSC and Biofuel cell in the same device. Every cell is characterized by its voltage.

3.7.4 Biochemical_Biomechanical

In this hybrid cell [18], we integrated a Biofuel cell and NG in the same device. This type of cell is used for devices implemented in the body (for biomedical applications). Every cell is characterized by its voltage.

4 Behavioral View of the Proposed Extensions

To describe the behavioral view of the new proposed stereotypes, we propose to use of the "ResourceUsage" stereotype which is a class from the GRM profile. The ResourceUsage package is used for representing the consumption of a resource. The HW_Vibration, HW_Thermal, HW_Kinetic, HW_Acoustic, HW_RF, HW_Biochemical and HW_Hybrid stereotypes are regarded as resources such as they produce energy and power to supply the wireless network. The ResourceUsage package provides the UsageTypedAmount package which represents different types of amounts of resources in terms of energy memory and size of data sent in the network. Two general forms of ResourceUsage are defined the DynamicUsage and the StaticUsage.

5 Case Study

In this section, we give an overview about energy harvesting methods used in water pipeline monitoring and we present a case study to illustrate our extension for modeling energy harvesting devices.

5.1 Energy Harvesting Methods in Water Pipeline Monitoring

An interesting application domain of WSNs is monitoring the water distribution systems where WSNs are a potential alternative to maintain a continuous monitoring of water infrastructure by detecting different failure mechanisms such as leaks in the pipeline, change of water quality and other anomalies. Harvesting energy from ambient environment would be an attractive option in water pipeline as sensor nodes are currently supplied by unmanageable batteries in inaccessible environment such as buried underground water pipelines. Several possibilities exist to harvest energy from the water pipe and its environment [19]. A direct way to generate power from water is harvesting energy from the water flow using turbines. In [20], authors proposed this technique as a solution to power sensor nodes. This technique called kinetic energy from water flow. It allows an enough amount of power, but it suffers from the big size of turbines, their hard installation and water quality disturbance. Another technique is the vibration energy harvesting. This alternative is feasible to power water pipeline network [21]. Flow-induced vibration is one of the potential energies can be harvested from water for powering sensor nodes. Piezoelectric energy harvester [22], electromagnetic energy harvester and electrostatic energy harvester are used in this field. Another renewable energy source is the thermal energy produced by the heat flow caused by the temperature gradient between the water and the ambient air. In [23], authors described thermal energy harvesting between the Air/Water Interface for powering sensor nodes.

5.2 A Case Study on Water Pipeline Monitoring

In order to illustrate our extension for the modeling of the energy harvesting devices in WSNs systems, we consider, in case study form, a wireless sensor node used in water distribution monitoring to detect and localize leakages in the pipe. A node consists basically of four units: sensing unit, processing unit, transceiver unit and a power unit. The proposed system consists of a pressure sensor to measure the water pressure in order to localize leaks in the pipe. This pressure sensor is powered by a AA Lithium battery which provides 3.9 V with an operating frequency equal to 2.4 GHz and its life time vary from 2 months to 1 year. To solve the energy supply problem in water monitoring pipeline and extend the network lifetime, we integrate energy harvesting module to the sensor node to harvest energy from the water flow in the pipe and then recharged the battery with the harvested energy. The potential solution is using a vibration harvester based on the vibration induced by the water flow. The best mechanism to harvest energy from flow induced vibration is the piezoelectric technique which offers the higher voltage level compared to electromagnetic and electrostatic mechanisms [24]. We consider a piezoelectric energy harvester used for harvesting energy from water flow

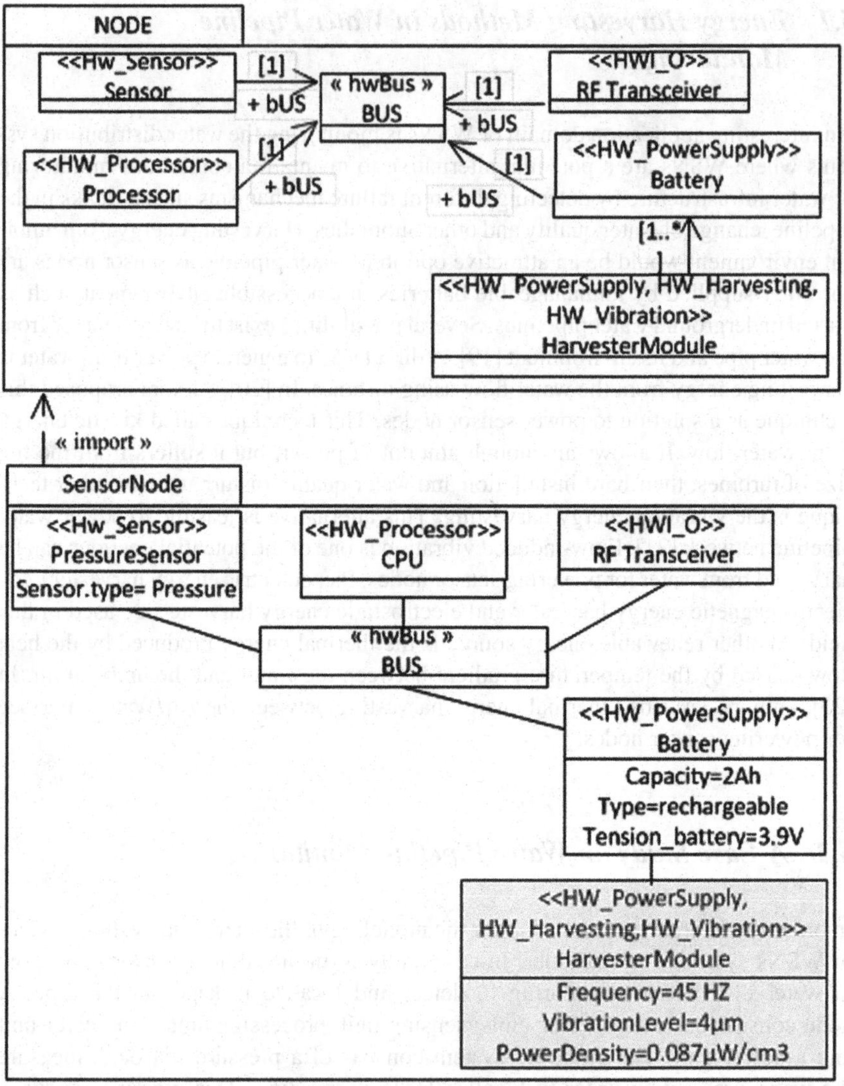

Fig. 9 Class diagram annotated by MARTE/HRM stereotypes describing the proposed sensor node structure

described in [25]. The sensor node is considered as a Hardware (HW) component in MARTE standard. To model the wireless sensor node, MARTE uses the HRM) sub-profile. Each sensor node is specified by the "HW_Device" stereotype. Figure 9 specifies a class diagram annotated by MARTE/HRM stereotypes describing the sensor node structure. The processor is specified by the stereotype "Hw_Processor". The sensor is modeled as a "Hw_Sensor" component. The RF transceiver is specified by the stereotype "HwI_O". Finally, the battery is modeled by the stereotype "HW_PowerSupply" and the energy harvesting module is modeled by the proposed stereotype "HW_Vibration" for modeling the vibration energy harvesting device.

6 Conclusion

This paper presented a high level methodology for designing different energy harvesting types for a WSNs node. We provided an overview of the UML/MARTE standard capabilities for modeling the power section of such system. We then presented our contribution to the modeling of energy harvesting devices by extending MARTE profile with new semantics to design harvesting devices. We are interested on seven energy sources: vibration, thermal, kinetic, acoustic, RF, biochemical and hybrid. This contribution makes UML/MARTE standard able to support energy harvesting devices which present a potential alternative to extend the network lifetime in WSN applications. As an example, we have illustrated our proposed extension with a water pipeline monitoring application which aims to detect and localize leaks and failures in the pipe. We plan in future works to deal with reconfigurable wireless sensor node and we aim to define a high level modeling of this system by extending MARTE standard in order to minimize the energy consumed by the process of reconfiguring sensor nodes.

Acknowledgments This work was supported by King Abdulaziz City for Science and Technology (KACST) and Digital Research Centre of Sfax (CRNS).

References

1. Al Ameen, M., Liu, J., Kwak, K.: Security and privacy issues in wireless sensor networks for healthcare applications. J. Med. Syst. **36**(1), 93–101 (2012)
2. de Lima, G.H.E.L., e Silva, L.C., Neto, P.F.R.: Wsn as a tool for supporting agriculture in the precision irrigation. In: 2010 Sixth International Conference on Networking and Services (ICNS), pp. 137–142 (2010)
3. Akbari, S.: Energy harvesting for wireless sensor networks review. In: 2014 Federated Conference on Computer Science and Information Systems (FedCSIS), pp. 987–992 (2014)
4. Schmidt, Douglas C.: Guest editor's introduction: model-driven engineering. Computer **39**(2), 25–31 (2006)
5. OMG Object Management Group: A UML Profile for MARTE: Modeling and Analysis of Real-Time Embedded systems, ptc/2011-06-02. Object Management Group, June 2011
6. Argyris, I., Mura, M., Prevostini, M.: Using marte for designing power supply section of wsns. In: M-BED 2010: Proceedings of the 1st Workshop on Model Based Engineering for Embedded Systems Design (a DATE 2010 Workshop), Germany (2010)
7. Roundy, S., Wright, P.K., Rabaey, J.: A study of low level vibrations as a power source for wireless sensor nodes. Comput. Commun. **26**(11), 1131–1144 (2003)

8. Yoon, Y.-J., Park, W.-T., Li, K.H.H., Ng, Y.Q., Song, Y.: A study of piezoelectric harvesters for low-level vibrations in wireless sensor networks. Int. J. Precis. Eng. Manuf. **14**(7), 1257–1262 (2013)
9. Naruse, Y., Matsubara, N., Mabuchi, K., Izumi, M., Suzuki, S.: Electrostatic micro power generation from low-frequency vibration such as human motion. J. Micromech. Microeng. **19**(9), 094002 (2009)
10. Beeby, S.P., Torah, R.N., Tudor, M.J., Glynne-Jones, P., O'Donnell, T., Saha, C.R., Roy, S.: A micro electromagnetic generator for vibration energy harvesting. J. Micromech. Microeng. **17**(7), 1257 (2007)
11. Lu, X., Yang, S.-H.: Thermal energy harvesting for wsns. In: 2010 IEEE International Conference on System Man Cybernetics (SMC), pp. 3045–3052 (2010)
12. Azevedo, J.A.R., Santos, F.E.S.: Energy harvesting from wind and water for autonomous wireless sensor nodes. Circuits, Devices Syst. IET **6**(6), 413–420 (2012)
13. Pillai, M.A., Deenadayalan, E.: A review of acoustic energy harvesting. Int. J. Precis. Eng. Manuf. **15**(5), 949–965 (2014)
14. Sim, Z.W., Shuttleworth, R., Alexander, M.J., Grieve, B.D.: Compact patch antenna design for outdoor rf energy harvesting in wireless sensor networks. Prog. Electromagn. Res. **105**, 273–294 (2010)
15. Niu, P., Chapman, P., DiBerardino, L., Hsiao-Wecksler, E.: Design and optimization of a biomechanical energy harvesting device. In: IEEE Power Electronics Specialists Conference, PESC 2008, pp. 4062–4069, June 2008
16. Tan, Y.K., Panda, S.K.: Energy harvesting from hybrid indoor ambient light and thermal energy sources for enhanced performance of wireless sensor nodes. Ind. Electron. IEEE Trans. **58**(9), 4424–4435 (2011)
17. Georgiadis, A., Collado, A., Via, S., Meneses, C.: Flexible hybrid solar/em energy harvester for autonomous sensors. In: 2011 IEEE MTT-S International Microwave Symposium Digest (MTT), pp. 1–4, June 2011
18. Chen, X., Pan, C., Liu, Y., Wang, Z.L.: Hybrid cells for simultaneously harvesting multi-type energies for self-powered micro/nanosystems. Nano Energy **1**(2), 259–272 (2012)
19. Ye, G., Soga, K.: Energy harvesting from water distribution systems. J. Energ. Eng. (2011)
20. Kokossalakis, G.: Acoustic data communication system for in-pipe wireless sensor networks. PhD thesis, Massachusetts Institute of Technology (2006)
21. Mohamed, M.I., Wu, W.Y., Moniri, M.: Power harvesting for smart sensor networks in monitoring water distribution system. In: 2011 IEEE International Conference on Networking, Sensing and Control (ICNSC), pp. 393–398, Apr 2011
22. Xie, J., Yang, J., Hongping, H., Yuantai, H., Chen, X.: A piezoelectric energy harvester based on flow-induced flexural vibration of a circular cylinder. J. Intell. Mater. Syst. Struct **23**(2), 135–139 (2012)
23. Davidson, J., Collins, M., Behrens, S.: Thermal energy harvesting between the air/water interface for powering wireless sensor nodes. In: SPIE Smart Structures and Materials + Nondestructive Evaluation and Health Monitoring, pp. 728–814. International Society for Optics and Photonics (2009)
24. Walton, R., Sadeghioon, A.M., Metje, N., Chapman, D., Ward, M.:Smart pipes: the future for proactive asset management. In: Proceedings of the International Conference on Pipelines and Trenchless Technology, vol. 2629, pp. 1512–1523, Beijing, China (2011)
25. Wang, D.-A., Liu, N.-Z.: A shear mode piezoelectric energy harvester based on a pressurized water flow. Sens. Actuators, A **167**(2), 449–458 (2011)

A Survey on Web Service Mining Using QoS and Recommendation Based on Multidimensional Approach

Ilhem Feddaoui, Faîçal Felhi, Imran Hassan Bergi and Jalel Akaichi

Abstract The process of web service mining intends to discover required services so as to provide the users with the services that are important and desired. While as the system that has been proposed has an important role in the recommendation of services to the users. Multiple techniques have been projected to execute the proposed actions, the collaborative filtering technique is mostly used for the recommended system here, we will describe different approaches which make use of collaborative filtering and also QOS, (a technical notation that is applied to the Web service mining). We will also discuss some methodologies of recommended system which use the multidimensional approach.

Keywords Web service mining · Recommender system · Collaborative filtering · QoS · Data warehouse

1 Introduction

In beginning era of internet, the use of web was just to store information and that too only text and images. The Web has evolved since then to a great extent into host of multimedia information as well as the service-providing applications. The service applications include map-finding, weather-reporting, e-commerce and many more.

I. Feddaoui (✉) · F. Felhi · J. Akaichi
BESTMOD Laboratory, High Institute of Management,
Tunis University, Tunis, Tunisia
e-mail: ilhemfd@yahoo.fr

F. Felhi
e-mail: felhi_fayssal@yahoo.fr

J. Akaichi
e-mail: jalel.akaichi@isg.rnu.tn

I.H. Bergi
King Khalid University, Abha, Saudi Arabia
e-mail: imhassan@kku.edu.sa

G. De Pietro et al. (eds.), *Intelligent Interactive Multimedia Systems and Services 2016*, Smart Innovation, Systems and Technologies 55,
DOI 10.1007/978-3-319-39345-2_38

There are also real time applications using hardware devices like temperature sensor systems and the traffic monitor cameras. The Businesses and the government organizations realized the fact of integrating the existing Web applications, so as to provide the value-added services which were unavailable in the past. Unfortunately, customized interfaces that are required for accessing applications and deficiency of semantics in data they consume and produce have made unified integration and assimilation of the applications and the fast deployment of the value-added services very much challenging. To address all of these problems, the two qualifying technologies can be seen. The first one is initiative of the Web service that was announced in year 2000 by the IBM, HP, Sun and the Microsoft. The given initiatives included the Web services of IBM, HP's e-speak Sun's Open Network Environment (ONE) and Microsoft's.NET. The joint effort among such big companies and the other institutions resulted in World Wide Web Consortium (W3C) publishing specifications of the Web service [1]. Giving to W3C specification, Web service is Web application the functionalities of which can be accessed programmatically via set of the homogeneous XML interfaces. As a transition of web from Web of data to the web of data as well as services, where the Web services can be seen as first-class objects, an increase in opportunities to combine the potentially interesting as well as gainful Web services from the existing services can be seen. Since collective opportunities in composing the services can exceed anyone's imagination, lot many of such opportunities would be hiding in Web of the available services, which would be unexpected to most of the people. Sometimes we may not have the particular queries needed for searching them, but unrevealing of these opportunities in early stages in would be seen very advantageous in today's competitive business environment. For government organizations, doing so will means that the citizens in advance would receive the useful and the potential services. It is therefore important to actively attempt in discovering the useful services even if goals are not specified at moment, or are difficult to imagine or are unknown. Lot like easy access of surplus data which has provided the fertile ground for research in data mining, we do expect that increase in availability of Web services would also spur the need as well as opportunities giving the new grounds to Web service mining.

There is a strong belief that the Web service mining will be a key in realizing Semantic Web services in full potential by leveraging big investments in the applications which have been operated till now as the non-interoperable silos.

With augmentation the number of web services and the task to discover the desired service is becoming increasingly difficult. Firstly, it is complicated for the user to find the desired service, because the number of services recovered compared to research done can be enormous. On the other hand two concepts semantically different could have an identical representation, which will further lead to feeble precision. Consequently, services can be recovered unrelated with the need for their consumers. That is why we need to establish an efficient and reliable process of web services mining. The proposed system can be defined as a system that can help user to discover useful services according to their customized preferences, past behavior or based on their similar tastes with other users. Therefore, the recommended system has an important role in their ability to recommend a service to the customer. It provides a ranking of services in predicting what services are the most necessary for users, based on the history, preferences and user constraints. Since many Web services have same functionality, another parameter should be introduced to be defined as a deciding factor. The quality of service (QoS) is the decisive factor appropriate, a set of information such as Accessibility [2], Accuracy, Availability, Cost, Execution time, Latency, Performance, Reliability, Response time, Scalability: Scalability [2], Success ability [3], Throughput [3], Reputation, Self-adaptability [4]. A recommendation can be done from the data warehouse. It is a database object-oriented, integrated, time-variant, contains a collection of non-volatile data that is used mainly in organizational decision making. Generally, the performance indicators here are more related to query throughput and response time. The technology well suited to this is OLAP (On Line Analytical Processing), it is a software technology that enables analysts, managers and executives to gain insight into the data through a coherent and easy access to a wide variety of information that has been processed and stored in data warehouses. A first step towards solving the problems of Web service mining is to study the current methods for the extraction of useful Web services. The main objective of this document is to provide a complete classification of Web services mining approaches using recommendation systems based on multidimensional and QoS techniques. Our classification is based on the following lines: Web service mining approaches by QoS, Recommended system based on the multidimensional and web service mining by recommended systems (collaborative filtering) and QoS.

This document is organized as following. Section 2 presents an overview of Web service concepts and presents a comparative study of Web service mining techniques. Section 3 presents a comparison of various techniques of recommendation system and show the usefulness of the data warehouse. Section 4 presents various Web services mining techniques based on collaborative filtering and QoS. While in Sect. 5 we will give our opinion on approaches studied during our work. Finally, Sect. 6 concludes the paper.

2 Web Service Mining

2.1 Definition

Web services mining is a new research discipline, which allows the discovery of the desired Web services, it has become a hot topic in those years. Web service mining is based on particular criteria given by the user. Search the right service that can meet the needs of users is still a problem. We can categorize the Web services mining into two classes that are syntactic based approach and semantic based approach.

2.2 Web Service Mining by QoS

QoS using rank Web services and select the best Web service from a list of services with similar features; it is used to categorize the Web services. The service that has the value of the highest quality of service is first selected. In this part, we studied the existing work of Web services mining by QoS, we'll take a little analysis on this work. Our study is presented in Table 1.

3 Proposed System

3.1 Definition

The proposed system plays an important role in the extraction of Web services. The proposed system can be classified as shown in Fig. 1 Content-based, Collaborative Filtering, Knowledge-based, Demographic-based Hybrid and Recommendations. In this paper we will focus on Collaborative Filtering.

3.2 Recommender System Using a Multidimensional

In this section, we will mention some approaches of literature that use the multi-dimensional in recommender system.

In [5] the authors propose a new approach for evaluation of recommender system. This approach is built on multidimensional analysis, allowing the consideration of various important aspects to judge the quality of proposed system in terms of applications in real time. They proposed a multidimensional framework to integrate OLAP technology.

Whereas in [6] the authors proposed a multidimensional approach to recommender system which can provide recommendations based on additional background information and more information about users and the typical elements used in most existing recommendation systems. This system supports multiple dimensions, additional information and hierarchical aggregation of recommendation. This recommender system could simultaneously acquire advantage of basic content-based recommendation, knowledge-based recommendation and collaborative filtering recommendation.

The authors in [7] introduced a generic recommendation system, which is able to provide advice for different types of applications. This system provides capabilities of multidimensionality of two-dimensional recommendation system. This upgrade allows generating more specific recommendations.

The authors present a new extension to the traditional approaches of recommendation systems, to be able to support the capabilities of data warehouse. They

Table 1 Comparative study of Web service mining based on QoS approaches

Refs.	Authors	Used technique	Advantages	The negative points
[27]	Makhlughian et al.	It uses data mining algorithm for classification	Can manipulate changes in dynamic environments	This approach takes time
			Satisfy user requests	Algorithm used in this approach is not optimal
[28]	Sachan, Kumar Dixit, and Kumar	Based on the multi-agent system	Efficiency and validity of this model	Only some QoS parameters are considered
			The agent is used to facilitate access to the registry service	
			The agent made the interaction with the UDDI	The approach based on agents is not secure, it's easy to replace them with malicious users
[29]	Ran	A new model based on UDDI register	The functional and nonfunctional requirements are used for the discovery service	Problem to determine the matching algorithms between desired and provided QoS
			A New role is introduced in this context the certifier (s)	
			They verify claims with Web service providers	
[30]	Yu and Lin	Two approaches were used—Multiple Choice Knapsack Problem (MCKP)—Graphical approach	Performance and effective approaches	Defeat for large problems
				Complexity
[31]	Khutade and Phalnikar	Using the matchmaker algorithm	Two steps: matchmaker and selection of Web services	Matchmaker algorithm used in this approach is unable to take a correct decision
			Simplicity of this approach	
[32]	D'Mello and Ananthanarayana	Use a semantic broker system. Search the Web services based on Quality of Service (QoS) and commercial offers (BO)	Effective	Only some QoS parameters are considered
			The broker also reads requirements from the requester and finds the best (profitable) Web service by matching and ranking the advertised services	

(continued)

Table 1 (continued)

Refs.	Authors	Used technique	Advantages	The negative points
[33]	Alrifa, Risse, Dolog and Nejdl	Based on a heuristic algorithm, which decomposes the optimization problem into smaller subproblems can be solved more efficiently than the original problem	Fast and scalable	Based on a very limited set of architectural requirements
			The optimality of the results	
[34]	Kritikos and Plexousakis	OWL-Q based approach	The functional and nonfunctional requirements are used for the discovery service	This approach takes time
				Only some QoS parameters are considered
[35]	Shi, Jinan, Zhang, Liu and Cui	It used a Clustering and Regression Algorithm	Web service selection which can provide the approximate QoS value for users	The clustering based on user data is not a good solution
			It clusters the users based on location and network condition	Only some QoS parameters are considered

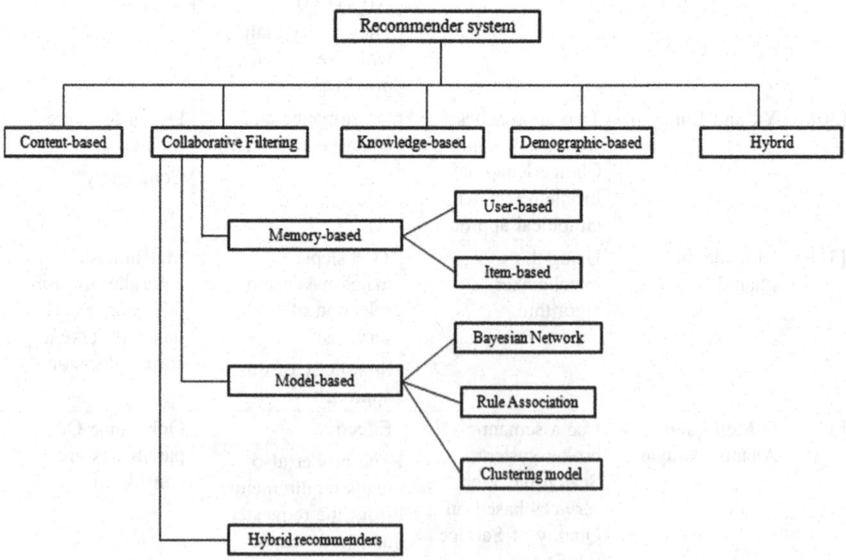

Fig. 1 Classification of recommender system technique

propose a recommendation system that works in multidimensional with the ability to support multiple dimensions. This approach is also supported on the OLAP technology [8].

The author present AWESOME (Adaptive website recommendations), a new data warehouse based on recommendation system capturing and evaluating user feedback on presented recommendations. It allows the use of a large number of prescribers generated automatically by websites recommendations. The recommendations are dynamically determined by a flexible approach based on specific rules [9].

The author present a multidimensional approach to the recommendation in the commerce. This approach has three phases. In the first phase it is possible to represent users, items, context information and the relationship between them in a multidimensional space. The second then determines each user's usage patterns in different contextual situations. While the third phase will create a new space of the recommendation in two dimensions. The final recommendation is made in this space [10].

The authors present an approach to recommender systems which can provide multidimensional recommendations which are based on additional background information besides the information of users and items used in most current recommendation systems. This approach supports multiple dimensions, extensive profiling and aggregation of hierarchical recommendations. This article also presents a method for estimating the multidimensional able to select two dimensions of segments [11].

Here they proposed to use OLAP technology during their exploration as a basis for recommending requests to the user. They proposed a generic framework to recommend queries based on OLAP server. This framework is generic in the sense that the change of its parameters changes how the recommendations are calculated. They showed how to use this framework to recommend simple MDX (Multi-Dimensional Expressions) Query [12].

In this article the authors proposed to apply approach based on collaborative filtering that exploits the ancient explorations of cube for recommending OLAP queries. This approach has been implemented with OLAP technology for recommend MDX queries [13].

The author proposed a generic framework of recommendations MDX queries and four instantiations of the framework. This framework is generic in the sense that it can be instantiated in different ways to change the method of recommendation calculations. This Framework consists of three steps:

- All registered queries pretreatment.
- Candidate recommendations generation.
- Candidate recommendations scheduling.

3.3 Collaborative Filtering

We can define collaborative filtering (CF) as a process that selects the desired services in relation to a user request from the preferences of other users. The collaborative filtering is classified into two classes as shown below:

Memory based Collaborative Filtering. In this type of collaborative filtering, a set of items or users is generated based on the relevance of the user or item. This system works on implicit or explicit evaluations. The user-item ratings are stored in the system and allow generating the list of items or user to be recommended. There are two types of memory based Collaborative Filtering: item-based and user-based [14]. The collaborative filtering based on user evaluates the interest of a user for an item using the ratings from other users for this item. On the other hand, item-based predicted rating of an item to a user, based on similar products, enjoyed by the user.

Model-based Collaborative Filtering. The basic idea of this approach is to model the interactions of user-item with the main characteristics of the extraction. This model is then trained by using these data. Several algorithms are used in this approach as Bayesian clustering.

4 Web Service Mining by Collaborative Filtering and QoS

We explored various research approaches to study the Web services mining by collaborative filtering techniques and the quality of service and present their limitations.

The proposed work in [15] provides a method for the selection of Web services based on the hybrid Collaborative Filtering and QoS. In [15] the selection of Web services is based on two phases: (i) the selection of Web services using the Collaborative filtering based on memory and QoS (ii) selection of Web services using the Collaborative filtering based on model and QoS.

While the authors in [16] provide a mechanism for Web service discovery for two phases based on the collaborative filtering and QoS. The first phase is used to filter the services with QoS values is incorrect. While the second phase is used to recommend services to users based on their non-functional requirements, which describe by the QoS provided by the consumer.

In [17], the authors propose a new recommendation algorithm of Web services, this approach does not require additional Web services invocations. It is based on the predicted values of QoS, Web services recommendations by QoS can be produced to help users select the optimal service. To improve the accuracy of prediction, they propose in this article a system of recommendation geolocation Web service that uses the quality of Web services and user locations for wished service prediction.

In [18], the authors propose a framework for the efficient discovery of Web services that provides customers with service quality information and reduces the probability of failure by analyzing the statistics of previous service recommendations. This proposed framework allows the selection of Web services based on QoS properties implicitly collected by an agent-based distributed system.

In [19], the authors propose a new Web service recommendation system based on collaborative filtering, which can help users find specific services using QoS. The recommendation system uses location information and service quality values for users and recommends a personalized service for users based on the results of the grouping. Different to other studies, this approach uses the characteristic of QoS and achieves a considerable improvement in the recommendation accuracy.

In [20], the authors present in this article a new collaborative filtering system with QoS to recommend large-scale Web services. The proposed scheme automates the Web service selection process using the quality of service, it facilitates the selection of Web services. Most existing service selection approaches ignore the great diversity in the service environment and assume that individual users receive identical QoS of the same service provider. This can lead to inappropriate selection decisions.

In [21], the authors propose a Web service mining based on collaborative filtering and QoS, which classifies Web services to the recommendations. In particular, the similarity between two Web services is measured by the correlation coefficient. In this approach, we noticed that the authors also improve NDCG measure (Normalized Discounted Cumulative Gain) to assess the accuracy of the recommendations returned with the order of ranking.

In [22], the authors present a recommendation of Web services based on QoS. It will help users choose the correct optimal solution. This approach is based on collaborative filtering of Web services, the recommendation works by collecting the QoS records of users and the correspondence with users that share the same information or the same likes.

The work presented in [23] is a Web service recommendation system that allows discovering and proactively managing Web services. The authors focus on the underlying search and classification algorithms of services for active recommendations. The basic idea in this approch is to involve the user in their recommendation of Web services system. Users are encouraged to share their observation for Web services quality with others in the recommendation system. This can be effectively used by users for a better selection of Web services. The experimental part in this article demonstrates the effectiveness of this technique.

In [24], the authors propose a method for predicting the quality of personalized service. It not only considers the impact of the network but also the Web server environment, in particular the needs of individual users. It analyzes the behavior of previous users. When there is no information on the target motif, the system uses collaborative filtering to recover data on other motifs. Experimental results show that the proposed method makes it possible to significantly improve the accuracy of the QoS prediction.

In [25], the authors propose algorithms for semantic building of recommendation Web services using collaborative filtering. This work enables the use of semantic markup for Web services to increase the accuracy of the recommendations based on collaborative filtering algorithms when user-item matrix is sparse.

In [26], the authors proposed a system based on collaborative filtering of Web services to help users to find certain services with optimal performance QoS. The basic idea of this system is to predict the quality of Web services and recommend the best values for active users on the basis of historical records Web services quality. This proposed method uses NLP protocols to obtain enriched recommendation focusing on user feedback. The system successfully merges the ranking of Web services and user feedback to provide a hybrid solution for an appropriate recommendation.

5 Discussions

The first part of our research focuses on Web service mining by QoS. Here, the approaches we have studied, used the QoS information as classification means, they allow us to get that optimal services. These approaches are used to discover services wanted but they present several problems such as:

- Only some QoS parameters are considered.
- Lack semantic modeling of QoS parameter.
- Security problem especially for approaches that are based on the agents.
- Defeat for large problems.

While the second part of our research is on the recommendations systems that have a significant impact on the discovery of Web services. The recommendation from data warehouse plays an important role in the Web service mining, since the multidimensional allows us to make a reliable and accurate recommendation especially using OLAP technology.

The final part of our research is on the Web service mining by collaborative filtering and QoS, in this part we have shown the utility of QoS in the selection of Web services, according to the QoS parameters we can classify Web services and choose the best service from a list of services that have the same functionality. Current approaches to Web services mining present several weaknesses among which:

- Don't take into account the behavior of the user.
- Are not adaptable to change.
- Lack of precision in selection of the desired service.
- With the increasing number of services, the Web service mining system crashes every time.

6 Conclusion

This paper presents a classification of Web service mining approaches. Our classification is based on the proposed system technique and the QoS. So we talk about the recommendations is done from the data warehouse. The purpose of Web services mining is to select the best Web service for a particular task. The QoS has an important role in the discovery of Web services. We conducted an analysis of these approaches and highlighted some of their merits and shortcomings.

References

1. Sharma, V., Kumar, M.: Web service discovery research: a study of existing approaches. Proc. Int. J. Recent Trends in Eng. Technol. **5**(1) (2011)
2. Understanding quality of service for Web services. http://www.ibm.com/developerworks/library/ws-quality/index.html
3. Summary of Quality Model for Web Services. https://www.oasisopen.org/committees/download.php/15444/Comparsion.WSQMFE.doc
4. Felhi, F., Akaichi, J.: Real time self adaptable web services to the context: case study and performance evaluation. Int. J. Web Appl. (IJWA) **7**(1), 1–9 (2015)
5. Grimberghe, A.K., Nanopoulos, A., Thieme, L.S.: A novel multidimensional framework for evaluating recommender systems. In: Proceedings of the ACM RecSys 2010 Workshop on User-Centric Evaluation of Recommender Systems and Their Interfaces (UCERSTI)
6. Rahman, M.M.: Contextual recommender systems using a multidimensional approach. Int. J. Emerg. Technol. Adv. Eng. (2013)
7. Uzun, A., Rack, C.: Using a Semantic Multidimensional Approach to Create a Contextual Recommender System
8. Adomavicius, G., Tuzhilin, A.: Extending Recommender Systems: A Multidimensional Approach
9. Thor, A., Rahm, E.: AWESOME A Data Warehouse-based System for Adaptive Website Recommendations
10. Pozveh, M.H., Nematbakhsh, M., Movahhedinia, N.: A multidimensional approach for context-aware recommendation in mobile commerce. (IJCSIS) Int. J. Comput. Sci. Inf. Secur. **3**(1) (2009)
11. Adomavicius, G., Sankaranarayanan, R.: Incorporating Contextual Information in Recommender Systems Using a Multidimensional Approach
12. Giacometti, A., Marcel, P., Negre, E.: A Framework for Recommending OLAP Queries
13. Giacometti, A., Marcel, P., Negre, E.: Recommending Multidimensional Queries
14. Su, X., Khoshgoftaar, T.M.: A survey of collaborative filtering techniques. In: Advances in Artificial Intelligence, vol. 2009 (2009)
15. Urmela, S., Joseph, K.S.: An effective web service selection based on hybrid collaborative filtering and QoS-Trust evaluation. Int. J. Adv. Res. Comput. Eng. Technol. (2015)
16. Lina, S.Y., Lai, C.H., Wu, C.H., Lo, C.C.: A trustworthy QoS-based collaborative filteting approach for Webservice discovery. J. Syst. Soft. **93** (2014)
17. Patil, N.K., Pawar, P., More, S., Tupe, B.: Web Service recommendation using qos parameters and users location. Int. J. Adv. Res. Comput. Commun. Eng. **4**(3) (2015)
18. Kokash, N.: Web Service Discovery with Implicit QoS Filtering
19. Gurjar, N.R., Rode, S.V.: Personalized QoS-aware Web service recommendation via exploiting location and collaborative filtering. Int. J. Adv. Res. Comput. Sci. Softw. Eng. **5** (1) (2015)

20. Yu, Q.: QoS-aware service selection via collaborative QoS evaluation
21. Chen, M., Ma, Y.: A Hybrid Approach to Web Service Recommendation Based on QoS-Aware Rating and Ranking
22. Chen, X., Zheng, Z., Liu, X., Huang, Z., Sun, H.: Personalized QoS-Aware Web Service Recommendation and Visualization
23. Suria, S., Palanivel, K.: An enhanced web service recommendation system with ranking QoS information. Int. J. Emerg. Trends Technol. Comput. Sci. (IJETTCS)
24. Zhang, L., Zhang, B., Pahl, C., Xu, L., Zhu, Z.: Personalized Quality Prediction for Dynamic Service Management based on Invocation Patterns
25. Coello, J.M.A., Tobar, C.M., Yuming, Y.: Improving the performance of Web service recommenders using semantic similarity. JCS&T **14**(2), 80 (2014)
26. Mohalkar, R., Fadtare, K., Todkar, S.: Web service recommendation via quality of service information. Int. J. Comput. Sci. Inf. Technology Research **3**(2), 517–524 (2015)
27. Makhlughian, M., Hashemi, S., Rastegari, Y., Pejman, E.: Web service selection based on ranking of qos using associative classification. Int. J. Web Serv. Comput. (IJWSC), **3**(1) (2012)
28. Sachan, D., Dixit, S., Kumar, S.: QoS aware formalized model for semantic Web service selection. Int. J. Web Semant. Technol. (IJWesT) **5**(4) (2014)
29. Ran, S.: A Model for Web Services Discovery with QoS
30. Yu, T., Lin, K.: Service selection algorithms for Web services with end-to-end QoS constraints. IseB **3**, 103–126 (2005). doi:10.1007/s10257-005-0052-z.19
31. Khutade, P., Phalnikar, R.: QOS based Web service discovery using oo concepts. Int. J. Adv. Technol. Eng. Res. (IJATER)
32. D'Mello, D.A., Ananthanarayana, V.S.: Semantic Web service selection based on service provider's business offerings. In: IJSSST, vol. 10, no. 2
33. Alrifai, M., Risse, T., Dolog, P., Nejdl, W.: A scalable approach for QoS-based Web service selection
34. Kritikos, K., Plexousakis, D.: Requirements for QoS-based Web service description and discovery
35. Shi, Y., Zhang, J.K., Liu, B., Cui, L.: A new QoS prediction approach based on user clustering and regression algorithms. In: IEEE International Conference on Web Services (ICWS), Washington, DC, 4–9 July 2011. ISBN: 978-0-7695-4463-2
36. Negre, E.: Exploration collaborative de cubes de données. Doctoral thesis, University François Rabelais Tours (2009)

Mapping and Pocketing Techniques for Laser Marking of 2D Shapes on 3D Curved Surfaces

Federico Devigili, Davide Lotto and Raffaele de Amicis

Abstract Laser marking has been used since the invention of lasers but it is only in the last decade that it started evolving into 3D surface marking. The problem of defining the toolpath for a 3 axis laser marking machine can seem to be the same as the definition of the toolpath for the CNC milling machines but this is not completely true. In the case of laser marking is not only the last pass that will affect surface finish but every pass made. This implies that to obtain the desired final effect on the material it is crucial to define different pocketing and filling patterns together with the laser parameters. Defining new patterns that meet the requirements for the laser marking on 3D curved surface is a non-trivial problem; the toolpaths, depending on the application, may need to have different properties such as constant distance or density between path lines, non-crossing of path lines or defined angle of intersection. When trying to mark non flat surfaces with 2D images or paths, in certain cases, distortion of the 2D space cannot be avoided. This paper will analyze different proposed techniques for mapping and marking 3D solids with a 3 axes, mirror based, laser marking CNC machine analyzing advantages and disadvantages of each one from the software development point of view.

Keywords Laser marking · Pocketing

F. Devigili (✉) · D. Lotto · R. de Amicis
Fondazione Graphitech, Trento, Italy
e-mail: federico.devigili@graphitech.it

D. Lotto
e-mail: davide.lotto@graphitech.it

R. de Amicis
e-mail: Raffaele.de.Amicis@gmail.com

G. De Pietro et al. (eds.), *Intelligent Interactive Multimedia Systems and Services 2016*, Smart Innovation, Systems and Technologies 55,
DOI 10.1007/978-3-319-39345-2_39

1 Introduction

Laser marking has been around since the invention of lasers in 1960; the amount of materials which can be worked is countless, metals, plastics and even transparent materials. On the other hand, only in the last decade, software for laser marking on curved and 3D surfaces started to be developed. Regardless of the type of laser marking machine, marking 3D surfaces require adoption of special focus lenses to change the focal point. Changing the focal point or changing distance of the emitter enables the laser to focus at different distances thus giving a third axis of control. The problem of defining the toolpath for a 3 axis laser marking machine can seem to be the same as the definition of the toolpath for the CNC milling machines but this is not completely true. Milling machines toolpaths have different requirements: Laser toolpath does not require checking for tool collision and the piece can be considered static since very little material is removed. In the case of milling, only the last pass will affect surface finish, while in laser marking every pass is important for the final result. This implies that, to obtain the desired final effect on the material, it is crucial to define different pocketing and filling patterns together with laser parameters. In the paper we will explore techniques and challenges faced during the development of a 3D marking software.

In Sect. 2, we give an overview of the issues and challenges encountered. In Sect. 3 we describe: (1) a mapping approach based on mesh description and UVW mapping; (2) a technique that allows the generation of the toolpath without the need of geometry data; (3) a technique that guarantees constant distance between toolpaths' lines which deeply affect marking results. Finally, in Sect. 4, we compare the surveyed techniques and explain when to use them based on toolpath requirement and available data.

2 Overview and Challenges

The process of defining a 3D laser toolpath can be summarized in two steps.

The first step is the definition of the mapping between the 2D marking and the 3D surface. This mapping will define the position, dimension and eventual distortion of the 2D shape on the 3D surface. The shape can be a simple contour that needs to be filled or a completely ready 2D toolpath. If the 2D toolpath does not need to be filled, only the transformation of the 2D coordinates to the 3D coordinates of the surface is required. In most cases the input to the marking software is a shape that needs to be filled.

Filling of the shape is the second step of the process. Filling or pocketing of the areas can be done in 2D or 3D dimensions. Filling or pocketing those areas in the

Fig. 1 2D cross filled text mapped on 3D surface

bidimensional rather than tridimensional space leads to different results and problems. For instance, Fig. 1 shows a preview of a toolpath generated from the text 'graphitech' on the Utah teapot.

Defining new patterns that meet the requirements for the laser marking on 3D curved surface is a non-trivial problem, the toolpaths, based on the application, must have different properties such as constant distance or density between path lines, non-crossing of path lines or defined angle of intersection.

Another problem which arises when trying to mark non flat surfaces with 2D images or paths is that in certain cases distortion of the 2D space cannot be avoided. When the solid, or the section of the solid being marked, is not unfoldable, the mapping between the 2D space and the 3D space will create a distortion which can be acceptable or not. It does not matter what technique is used to define the laser path, you will either have an already defined 3D path or you will need to define a transformation from a 2D path to the 3D surface.

3 Surveyed Techniques

The techniques described in this section have been designed during the development of commercial software for laser marking. Each technique addresses specific issues faced during practical use of laser marking machines.

3.1 Mesh-Based Surface Mapping

Mesh based surface mapping is a classic approach that defines the 2D to 3D transformation of the marking with a standard mesh UV mapping. This technique requires the 3D model of the physical object complete of a set of UV coordinates. The UV coordinates should be created for the specific purpose of mapping the area that will be marked. This approach requires good matching between the physical position of the real object and the position of the mesh in the virtual 3D space, failing to align the two reference frames will inevitably lead to areas where the laser will go out of focus. This is especially relevant when the area being marked has a high slope since a small lateral misalignment will lead to a greater vertical offset.

Assuming the real object is aligned with the virtual reference frame the computational steps required to achieve mapping are the following:

1. Define the position of the marking in UV space thus defining an actual 2D to 3D transformation.
2. Intersect marking lines with 2D UV mapping lines.
3. Project the intersected segments defined in 3D to obtain the toolpath.

Note that this process is valid both when the marking is only a contour that requires filling or pocketing or when the marking is a 2D toolpath complete of area fillings. In the case of area-only marking, filling can be done either on the 2D UV space as for classical laser marking and then transformed into 3D surface space or can be done directly in the 3D surface space as described in Sect. 3.

This technique presents advantages and disadvantages.

Among the advantages we can count the speed, the procedure does not, in fact, require a vast amount of time, even with complex meshes; it allows also to define precise pocketing paths, by specifying the line density and angle. This, as stated before, is of utmost importance when trying to create effects on the marked surface. The limitation of this approach is the unfolding of the object. In fact, not every object has a clear unfolding, or it may have a distorted one. Taking the sphere as an example, we know that it is not possible to unfold it in a plane without distorting it, the only thing that we can control is to choose where the unfolding will be less distorted.

The other problem of this technique is that usually, when an object is unfolded to create an uv map, some of the triangles of the map reference to more than one triangle in the object; thus trying to map a line from the 2D space to the 3D space will result in a duplicated line, and a wrong toolpath.

Clearly the advantages listed above are deeply related to the quality of the unfolding, while a good unfolding produces good and expected result, a bad one will reduce the quality of the toolpath caused by excessive avoidable distortion.

3.2 Depth-Based Surface Mapping

Depth based surface mapping approach is based on the idea that if the laser machine has a way to detect the objects in the marking area and pass this information to the control software, then the control software can use this information to modify and create the surface transformation required to place the marking image on top of the 3D model surface. To be able to detect the object on the laser work area any depth sensing technique can be used: photogrammetry, structured light, etc. One advantage of using depth based mapping, regardless of the technology used for the implementation is that the physical position of the object is intrinsically included in the detected depth information; this means that the user do not need to accurately place the object before marking [1]. Previously discussed techniques need both precise positioning and 3D information of the object including UV mapping of the surface. Even if 3D model of the marked object is available UV mapping is usually not and it requires specific knowledge to be created. On top of that when marking, only a specific part of the object needs to be mapped. This leads to the consideration that if the user were able to map in a simple and intuitive way the area to be marked, or even better to directly apply the image on the three-dimensional model without worrying about the coordinates, the ease of use of the marking software would increase significantly.

In the case of a mirror based laser CNC the origin of the marking laser is fixed and the field-of-view is constant so most of the mapping needs can be reduced to 2 fundamental cases. The mapping of nearly flat surfaces mapped with planar projection and mapping mapped curved surfaces with an appropriately distorted cylindrical projection. Any other cases with more complex surfaces, such as a sphere, no longer would benefit from special automatic mappings as there are no projections able to cancel the intrinsic distortion. This means that if we need to map a spherical object even if we use a spherical UV mapping the distortion would be the same as using a cylindrical mapping.

Regardless of the technology used; to implement this technique a depth sensor working in parallel with the laser machine is required. The 3D object surface data and position needs to be detected with enough accuracy to keep the laser focused on the surface of the object. The software should show the detected surface on a 3D view and the user should be able to place the 2D marking image on the detected surface abstracting mapping data.

3.3 3D Pocketing with Isolines

When the filling toolpath is calculated from the object unfolding, distortion problems can arise, causing a wide range of problems. Among them is the pocketing line distance, which varies depending on the distortion of the given area. In Fig. 2 we can see a visual representation of the spatial distortion between 2D and 3D of a

Fig. 2 Distortion in 2D to 3D mapping on a curved surface

curved surface (Utah teapot side). The distortion is one of the most concerning problems, because the line distance in the pocketing toolpath deeply affects the resulting marking.

To overcome this issue, it is possible to calculate the filling toolpaths with other techniques that do not require an UV map and work directly with the 3D shape. However, the technique does not remove the need of the unfolding because the image to be marked still needs to be mapped to the 3D surface.

One approach that allows computing pocketing toolpaths directly on the mesh consists in calculating the contour lines, that is, the set of points at a constant distance from a source on the object surface (isolines). Figure 3 shows an example of isolines calculated using the Saddle Vertex Graph algorithm. Calculating isolines of a discrete 3D mesh is a non-trivial task, and requires quite a lot of time. The discrete geodesic problem was successfully addressed for the first time in 1987 by Mitchell et al. [2]. They described an algorithm that allowed the computation of minimum shortest paths in a discrete 3D object. Recently, Xiang et al. [3] described how to reduce considerably computational time needed to calculate isolines by using a data structure called saddle vertex graph.

After having computed the isolines of a given object it is easy to create the output toolpath. First, the border of the image needs to be mapped from 2D to 3D, using its unfolding, and then the border is intersected with the isolines, only keeping the contour lines inside the image border.

The utilization of isolines open different possibilities regarding the patterns of the filling toolpaths, since it is possible to choose more than a vertex as source point it could be possible, for example, to highlight the features of the 3D object by choosing its edge as source points.

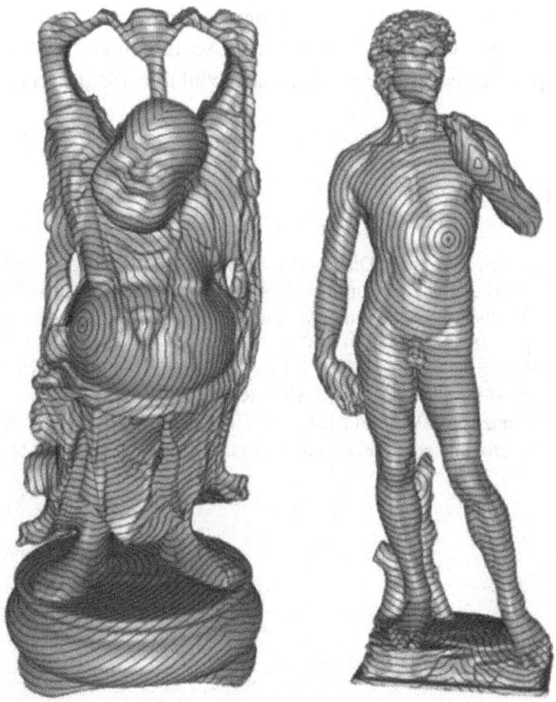

Fig. 3 Example of isolines on polyhedra meshes. Surazhsky et al. [4]

4 Conclusion and Techniques Comparison

To decide which technique use to generate the toolpaths for laser marking, it is essential to know the availability of the mesh's geometry and unfolding.

If it is difficult to obtain the unfolding or geometry of the object that needs to be marked, a depth based approach should be adopted, since this technique automatically obtains the data by analyzing the object itself. The downside of this approach is the extra hardware needed, which increases the cost and complexity of the laser marking machine.

As opposite, if the mesh's data is available, it is important to take in consideration the error of its unfolding. Since each mapping from 2D to 3D introduces a distortion, if the toolpaths need a very low tolerance, the right choice would be to adopt the isolines to calculate the pocketing, as this technique works directly on the mesh's surface, reducing considerably the error introduced.

In conclusion the mesh based approach should be followed when the mesh's unfolding is available and there are no particular requirements regarding the toolpath tolerance. This technique is useful because of its speed and straightforwardness.

A laser marking suite, which includes both software and hardware, should give the user the possibility to choose the most appropriate technique based on the specific requirements and data available for the given task.

References

1. Diaci, J., Bračun, D., Gorkič, A., Možina, J.: Rapid and flexible laser marking and engraving of tilted and curved surfaces. Opt. Lasers Eng. **49**(2), 195–199 (2011)
2. Mitchell, J.S., Mount, D.M., Papadimitriou, C.H.: The discrete geodesic problem. SIAM J. Comput. **16**(4), 647–668 (1987)
3. Xiang, Y., Xiaoning, W., Ying, H.: Saddle vertex graph (SVG): a novel solution to the discrete geodesic problem. ACM SIGGRAPH Asia **32**(6) (2013)
4. Surazhsky, V., Surazhsky, T., Kirsanov, D., Gortler, S. J., Hoppe, H.: Fast exact and approximate geodesics on meshes. ACM Trans. Graph. **24**(3), 553–560 (2005)

3DHOG for Geometric Similarity Measurement and Retrieval on Digital Cultural Heritage Archives

Reimar Tausch, Hendrik Schmedt, Pedro Santos, Martin Schröttner and Dieter W. Fellner

Abstract With projects such as CultLab3D, 3D Digital preservation of cultural heritage will become more affordable and with this, the number of 3D-models representing scanned artefacts will dramatically increase. However, once mass digitization is possible, the subsequent bottleneck to overcome is the annotation of cultural heritage artefacts with provenance data. Current annotation tools are mostly based on textual input, eventually being able to link an artefact to documents, pictures, videos and only some tools already support 3D models. Therefore, we envisage the need to aid curators by allowing for fast, web-based, semi-automatic, 3D-centered annotation of artefacts with metadata. In this paper we give an overview of various technologies we are currently developing to address this issue. On one hand we want to store 3D

R. Tausch (✉) · H. Schmedt · P. Santos · D.W. Fellner
Fraunhofer Institute for Computer Graphics Research IGD, Fraunhoferstr. 5,
64283 Darmstadt, Germany
e-mail: reimar.tausch@igd.fraunhofer.de
URL: http://www.igd.fraunhofer.de

H. Schmedt
e-mail: hendrik.schmedt@igd.fraunhofer.de

P. Santos
e-mail: pedro.santos@igd.fraunhofer.de

D.W. Fellner
Interactive Graphics Systems Group, TU Darmstadt, Fraunhoferstr. 5,
64283 Darmstadt, Germany
e-mail: cgv@cgv.tugraz.at
URL: http://www.gris.tu-darmstadt.de

D.W. Fellner
Institut für ComputerGraphik and Wissensvisualisierung, TU Graz, Inffeldgasse 16c,
8010 Graz, Austria
URL: http://www.cgv.tugraz.at

M. Schröttner
Fraunhofer Austria Research GmbH, Visual Computing, Inffeldgasse 16c,
8010 Graz, Austria
e-mail: martin.schroettner@vc.fraunhofer.at
URL: http://www.fraunhofer.at

G. De Pietro et al. (eds.), *Intelligent Interactive Multimedia Systems and Services 2016*, Smart Innovation, Systems and Technologies 55,
DOI 10.1007/978-3-319-39345-2_40

models with similarity descriptors which are applicable independently of different 3D model quality levels of the same artefact. The goal is to retrieve and suggest to the curator metadata of already annotated similar artefacts for a new artefact to be annotated, so he can eventually reuse and adapt it to the current case. In addition we describe our web-based, 3D-centered annotation tool with meta- and object repositories supporting various databases and ontologies such as CIDOC-CRM.

Keywords 3D object retrieval · Classification · Descriptor

1 Introduction

When cultural heritage objects are captured by 3D scanning methods, they become available as a digital mesh. Subsequently they are often stored in a database with all their corresponding provenance data. In times of approaches like CultLab3D [1] targeting affordable 3D mass digitization, the sheer number of digitized artefacts will overwhelm the curators if measures to simplify and automate the annotation process are not equally developed and explored. One approach which has the potential to help curators reduce their efforts, is to generate similarity descriptors with each digitized 3D model and store them alongside the object. It is our assumption that when searching for similar objects during the annotation process, similar objects to the one being annotated might yield similar links to provenance data. Those results can be conveyed to experts through expectation lists where they decide if they are going to be adopted for the current object. An additional advantage of this approach lies in preserving uniform terminology for the metadata and reducing the margin for errors, which could otherwise be dependent of the author's gender, age, religion or other background characteristics. Finally such an approach would be embedded in a larger context, which is our web-based, 3D-centered Annotation tool developed, inspired by initial efforts in 3D-COFORM which we subsequently describe.

2 Related Work

There are multiple ways to describe a 3D model. An overview can be seen in the survey from Tangelder and Veltkamp [2] whose classification of methods are borrowed for the following section. First of all there are **feature based methods**. These methods use features to describe the model with an n-dimensional vector. There are three kinds of feature based methods. First there are global features like the work of Osada et al. [3]. Next there are spatial maps which try to save the spatial position of a feature in a map such as Ankerst et al. [4]. Another possibility are local features like in the work by Knopp, Prasad and Gool [5] combined with the bag of words principle. Another type of methods are **graph based methods**. Additionally to the geometry they try to describe the topology as well. One way to do this is by using model graphs like the work from El-Mehalawi and Miller [6]. Another possibility are skeleton graphs that represent the object's basic structure like in the work of Sundar et al. [7]. Alternatively reeb graphs are used to represent an object whose origin lie in the Morse theory. This kind of graphs were used by Mohamed and Ben Hamza [8] by using an algorithm for skeleton graphs in combination with a mixed distance function. The last type are **geometry based methods**. Most popular are the view based approaches who use images of the 3D model from different angles and describe these images with 2D Descriptors like the works from Yoon and Kuijper [9]. Another kind of approaches are based on the volumetric error. Sanchez-Crus and Bribiesca [10] combine the volumetric error with a transportation distance measuring. This gives indication of how much effort is necessary to transform one object into another.

In our work we use the 3D Histogram of Oriented Gradient descriptor (3D HOG) which is a global feature based method. The decision to use this descriptor was made because the original 2D approach for images is still a state-of-the-art approach and previous evaluation of 3D HOG [11, 12] showed promising results for 3D models with low resolution. Also in comparison with other approaches 3D HOG offers many parameters to optimize the performance for the data. Additionally we wanted to apply 3D HOG to models in high resolutions and verify that 3D HOG might work even better with more information about the model. Cultural heritage artefacts digitized by CultLab3D [1] shall at this point be stored with their 3D similarity descriptor in a knowledge based database. Existing and used standards like the LIDO which combines elements of CIDOC CRM and Museumdat[1] for the minimum description of digital data used in virtual exhibition environments will be taken into account. Current content management systems used in the museum domain are very text-centric, such as Adlib Museum [13], MuseumPlus [14] or Museumindex [15]. They provide support for a variety of metadata schemata such as CIDOC CRM, LIDO, METS and a range of document and image and media formats. However, native support of 3D formats to annotate, store and display virtual 3D/4D models is poor or non-existent, which is even more true for the visualization and analysis of 3D/4D information. The workaround mostly used, is to store the 3D data and then link to external tools to open, visualize and work with it, which means the data is not natively integrated in the museum content management system and that there are no native tools to further explore, annotate or work with the virtual 3D/4D models. A good overview of current content management systems can be found with the Collectionstrust, UK [16]. There have been some projects focussing on 3D centered interaction such as 3DSA [17] from the University of Queensland in which annotations could be placed directly on the surface of objects and connected to metadata or measurements could be taken. Also some scanner companies such as Artec [18] are starting to include basic annotation functionality to work with 3D models in their products. Other examples include the Smithsonian explorer in collaboration with Autodesk [19] using proprietary technology to showcase items on the web in 3D but not integrated into their database backend. It allows to explore, measure and light objects

[1]http://www.museumdat.org/index.php?ln=en&t=home.

and present narrative stories centered on the 3D artefact. ENCoUNTeRS wants to build on a paradigm shift initiated by EU-funded 3D-COFORM [20], which created a diverse assortment of tools [21] with the objective of capturing, annotating and storing three-dimensional artefacts to virtually preserve them. With the advent of 3D mass digitization, annotation with provenance and metadata is centered around the 3D Model with the possibility to create annotations using drag and drop connecting them to the surface of an object, or marking surface areas or regions of interest and connecting them to documents, pictures or other 3D models. The frontend consists of the IVB (Integrated Viewer Browser) [22] and the backend consisted of a CIDOC-CRM [23] and CRMdig [24] conform metadata repository [25] and object data repositories at the corresponding museums.

3 Approach

The Object Retrieval Module is embedded in the CHR (Cultural Heritage Repository *CHR*), our web-based, 3D centered annotation system. When an object is first stored in the repository a descriptor is calculated representing the geometric appearance of the 3D model. Based on these attached compact descriptors a fast geometric similarity comparison can be achieved using a simple distance function.

3.1 Object Normalization

Before calculating the descriptor the polygonal mesh is normalized in terms of size, position and orientation. The object normalization relies on the following three steps: The first step unifies the position by translating the center of mass of the polygonal mesh into the origin of the coordinate system. The next step rotates the object by a weighted Principal Component Analysis (wPCA) [26]. Finally the object is scaled to fit inside a unified cube.

3.2 Descriptor Generation

To compare the polygonal meshes it is important to get a numerical representation. For this purpose we used and improved a method named 3D Histogram of Oriented Gradients (3D HOG) [11, 12] which consists of seven steps: voxelization, distance field computation, gradient computation, cell aggregation, histogram computation, block normalization, description vector construction. 3D HOG is a global description method which is originally based on the 2D HOG [27]. Each part of the resulting feature vector then represents the magnitude of the gradient in a certain direction of a certain volumetric cell capturing a part of the model.

3.3 Distance Calculation

Because of its histogram characteristics the 3D HOG descriptor vectors can be compared using a simple distance function such as the euclidean distance. It complies with the distances axioms of Filler et al. [28] and led to the best results in our tests when compared to the Manhattan and angle distance.

3.4 Applied Changes

The 3D HOG method originally introduced by Scherer et al. [11] and Walter [12] had only been available as a Java implementation. To speed up calculation time we ported it to C++ and improved the memory management. We exchanged the original implementation of object normalization. The construction of the rotation matrix was altered by entering the Eigenvectors line-by-line into the matrix which led to a more consistent alignment of the objects. In addition we improved the scaling to prevent information loss at the border of the bounding volume. Furthermore we found that a non-linear distance field led to more precise comparisons by improving the differentiation between occupied and empty cells.

4 Evaluation Results and Integration

To evaluate our Object Retrieval System we used 32 objects scanned with a lateral resolution of 130 μm and a depth resolution of 16 μm. The dataset (Fig. 1) consisted of four series (Busts, Conch instruments, Statues and Woodwind instruments) each with 8 objects from the State Museums of Berlin.

4.1 Experiments

We examined different parameter settings and a higher gradient field resolution than Scherer et al. [11] or Walter [12] due to our fast C++ implementation. Parameters tested were the gradient field resolution, the block normalization (BN), the number of cells and the polygonal mesh quality. To evaluate the different parameters we calculated the R-Precision and used a result visualization. The R-Precision is defined as $\frac{r}{R}$ by the number of objects r similar to the query object from the number of stored and eligible objects R for this query (similarity would not necessarily have to be the only attribute searched for). In addition we were interested in calculating similarity descriptors for a variety of resolutions of each object and test their stability. We therefore decimated the meshes from 100 to 10, 1, 0.1 and 0.01 % (Fig. 2) and

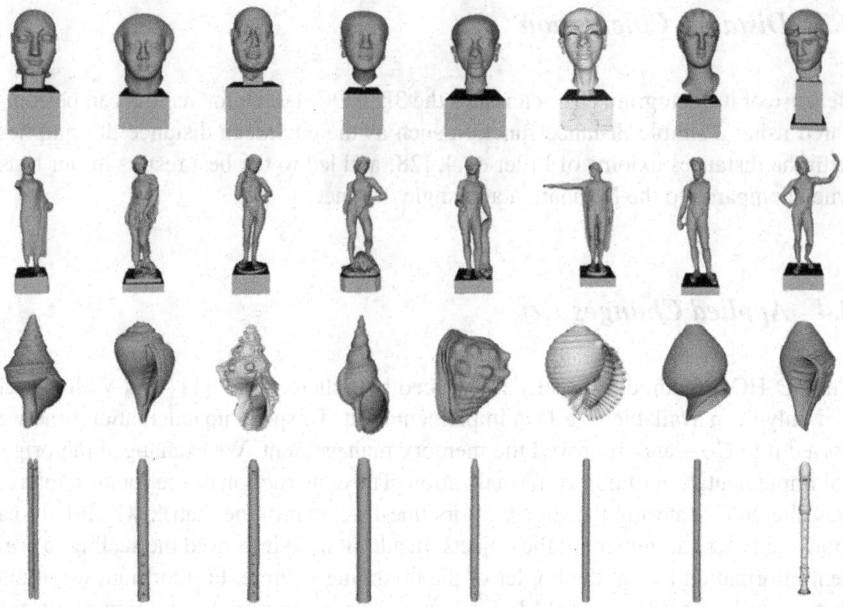

Fig. 1 Dataset overview: The *rows* represent busts, statues, conches and woodwind instruments

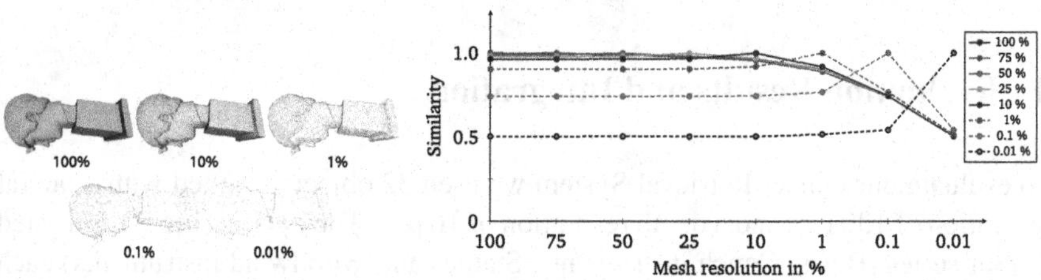

Fig. 2 *Left*: Overview of the polygonal mesh of one object with different resolutions while 100 % equals 1.3 mio. vertices. *Right*: Similarity between one object under different resolutions

calculated the R-Precision for each scenario. Additionally the produced similarity descriptors of high resolution meshes (100 %) and low resolution meshes (0.01 %) were compared. This was done using the distance between each n-dimensional vector and applying MDS (multidimensional scaling).

4.2 Results

An overview of the results for the tested parameters is given in Table 1. It is notable that the use of a block normalization delivers better results similar to the 2D HOG. Also the number of cells is important for a high accuracy. A higher number of cells

Table 1 Listing of the average R-Precision for each series and the overall value depending on the stated parameter

	Busts	Statues	Conches	Woodwind	Overall
Block normalization					
No	0.75	0.875	0.607	0.964	**0.799**
Yes	0.75	0.911	0.661	1.0	**0.83**
Number of cells					
6 (8)	0.75	0.911	0.661	1.0	**0.83**
4 (12)	0.732	0.946	0.786	1.0	**0.866**
2 (24)	0.714	0.821	0.857	1.0	**0.848**
Gadient field resolution					
$48 \times 48 \times 48$	0.732	0.946	0.786	1.0	**0.866**
$96 \times 96 \times 96$	0.821	0.875	0.857	1.0	**0.888**
Mesh resolution					
100 %	0.821	0.875	0.857	1.0	**0.888**
10 %	0.821	0.875	0.875	1.0	**0.893**
1 %	0.821	0.875	0.875	1.0	**0.893**
0.1 %	0.821	0.893	0.839	1.0	**0.888**
0.01 %	0.804	0.857	0.839	1.0	**0.875**

leads to a very narrow description which is not always useful. Using too little cells makes the description too broad. For the data sets used our result indicate a number of 4 cells is best. Apparently the number of cells working best depends on the specific data set. Another important parameter is the gradient resolution. The higher the resolution the higher the received R-Precision. But since the computation time rises with a larger gradient field, we determined that $96 \times 96 \times 96$ gradients yield the best tradeoff for our purpose. The last examination was the quality of the polygonal meshes and if higher resolution leads to higher accuracy. The results show that a reduction of the number of vertices can yield better results. However, if the mesh quality is overly reduced results become worse. Furthermore, we took a great interest in knowing how much a 3D model can be decimated while the similarity descriptor calculated remains largely the same (Fig. 2). This is extremely interesting, because the performance of our geometric search in a database can benefit very much from returning the same results with greatly decimated models. Therefore we reduced the number of vertices for each object from the high resolutional meshes from 100 to 10 %, 1 %, 0.1 % and 0.01 % and the similarity was calculated. The results showed, that similarity results would only start deteriorating when artefacts scanned at 16 µm reached less than 1 % of the initial mesh resolution.

4.3 Integration in the Cultural Heritage Repository

New and improved technologies for mass digitization like Cultlab3D lead to a rapid increase of digitized cultural heritage objects. Once the digitized cultural heritage assets and their metadata are available, an appropriate storage solution is needed. But only storing the data is not enough, the system should include a complete digital library service handling, e.g., indexing, retrieval and permission management. The Repository Infrastructure (RI) [25] developed as part of the 3D-COFORM project is able to accomplish the demands on a storage system for cultural heritage assets. Many different tools for users and specific use-cases were developed, but an integrated, flexible and easy to use graphical user interface (GUI) is still missing. In contrast to state-of-the-art museum annotation databases, the Cultural Heritage Repository *CHR*, which is based on the ideas of the RI, puts the 3D consolidated model at the center of all activities. It uses X3D by the Web3D consortium which is integral part of all current browsers supporting at least HTML5. The latest state of the art 3D Web Technology enables platform independent single page web applications with the opportunities for presenting, processing and annotating 3D-content, allows reduction of the hardware requirements for museums and curators to almost any web-capable device able to access the Internet from almost anywhere at any time and open the door to community participation in the process of annotating artefacts democratizing access to cultural heritage. In addition, the 3D-HOG technology feature geometric search over already existing 3D consolidated models in the database as well as over available web databases to simplify annotation of similar provenance data and identification of similar artwork across museums and databases. The idea behind is to allow curators to semi-automatically annotate newly scanned artefacts with CultLab3D which are stored in the repository database in various mesh resolutions with their corresponding similarity descriptors ready to be queried. Once a query is successful the curator will be given an expectation list of possible corresponding artefacts and their annotations linked to provenance data that he may or may not adopt to annotate the current artefact.

Figure 3 shows a schematic representation of the *CHR*. The *CHR* unifies and integrates the user tools in one web application, it is called *CHRWebApp*. In Fig. 4 you can find two examples of the web frontend: The query interface and the annotation tool.

The user tools are implemented with HTML5, JavaScript and CSS3. For communication with the backend a REST based common server interface (*CHRBackendAPI*) is used. As the *CHRBackendAPI* a flexible solution supporting any kind of ontology and database is put in place therefore allowing to support CIDOC-CRM/CRDig as much as Dublin Core and EDM (Europeana Database Model) and Databases such as MySQL, MSSQL and others as needed. It includes user, permission, data and metadata management and service handlers for data services. The integration of the stand-alone 3D HOG tool takes in place via a service and is connected to the service handler. In this way the C++ based 3D format transcoding, the descriptor calculation or the Distance calculation can be integrate in our web based *CHR*. The used metadata ontology like CIDOC-CRM or METS is applied to

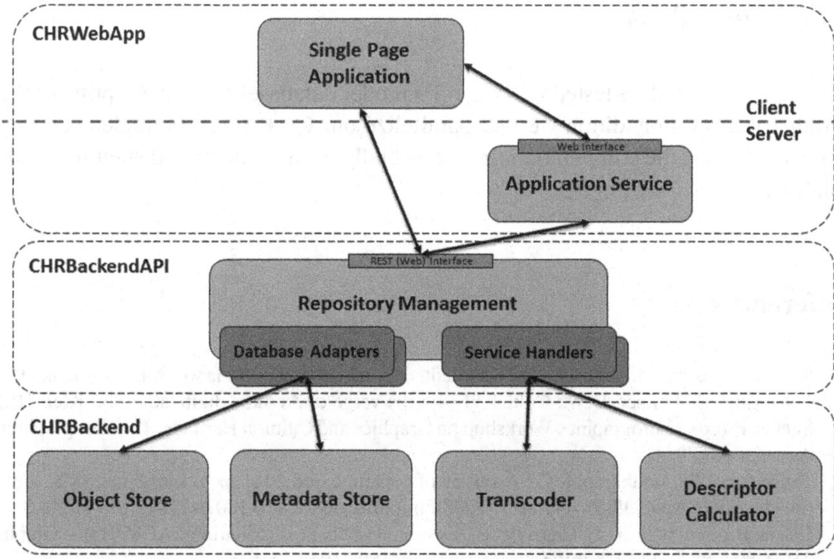

Fig. 3 Schematic representation of the CHR

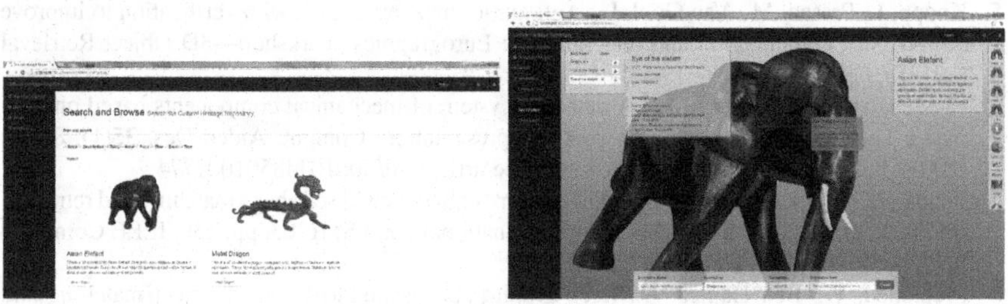

Fig. 4 Web-based 3D-centered annotation browser: Query and annotation tool

the metadata by the appropriate adapter for the metadata store. Results from services like a calculated descriptor, is automatically linked to the original 3d object by adding to the object metadata in the used schema. The database abstraction enables the opportunity for exchanging of the used databases with minimal adjustment, if requirements for storing binary data, metadata and semantic search are satisfied.

5 Conclusions

Our tests showed that the selection of parameters is critical for the accuracy of 3D HOG descriptor. Mostly the gradient field size number of cells influence the results of the introduced approach. It was also shown that mesh models can be decimated or can come in different resolutions and still produce comparable similarity descriptors.

5.1 Future Work

The algorithms will be tested on larger 3D model databases to further optimize the parametrization. In addition we will parallelize our C++ 3DHOG implementation to further reduce the computation time, especially of the expensive distant field calculation.

References

1. Santos, P., Ritz, M., Tausch, R., Schmedt, H., Monroy, R., Stefano, A.D., Posniak, O., Fuhrmann, C., Fellner, D.W.: Cultlab3d—on the verge of 3d mass digitization. In: Klein, R., Santos, P. (eds.) Eurographics Workshop on Graphics and Cultural Heritage. The Eurographics Association (2014)
2. Tangelder, J.W., Veltkamp, R.C.: A survey of content based 3d shape retrieval methods. Multimedia Tools Appl. **39**(3), 441–471 (2008). http://dx.doi.org/10.1007/s11042-007-0181-0
3. Osada, R., Funkhouser, T., Chazelle, B., Dobkin, D.: Shape distributions. ACM Trans. Graph. **21**(4), 807–832 (2002). http://doi.acm.org/10.1145/571647.571648
4. Ankerst, M., Kastenmller, G., Kriegel, H.P., Seidl, T.: 3d shape histograms for similarity search and classification in spatial databases. In: SSD'99, pp. 207–226. Springer (1999)
5. Knopp, J., Prasad, M., Van Gool, L.: Automatic shape expansion with verification to improve 3d retrieval, classification and matching. In: Eurographics Workshop—3D Object Retrieval (2013)
6. El-Mehalawi, M., Miller, R.A.: A database system of mechanical components based on geometric and topological similarity. part i: representation. Comput. Aided Des. **35**(1), 83–94 (2003). http://www.sciencedirect.com/science/article/pii/S0010448501001774
7. Sundar, H., Silver, D., Gagvani, N., Dickinson, S.: Skeleton based shape matching and retrieval. In: Proceedings of the Shape Modeling International 2003 SMI '03, pp. 130. IEEE Computer Society, Washington, DC, USA (2003)
8. Mohamed, W., Ben Hamza, A.: Reeb graph path dissimilarity for 3d object matching and retrieval. Visual Comput. **28**(3), 305–318 (2012)
9. Yoon, S.M., Kuijper, A.: 3d model retrieval using the histogram of orientation of suggestive contours. In: Proceedings of the 7th international conference on Advances in visual computing—Volume Part II, ISVC'11, pp. 367–376. Springer-Verlag, Berlin, Heidelberg (2011). http://dl.acm.org/citation.cfm?id=2045195.2045238
10. Snchez-Cruz, H., Bribiesca, E.: A method of optimum transformation of 3d objects used as a measure of shape dissimilarity. Image Vision Comput. **21**(12), 1027–1036 (2003)
11. Scherer, M., Walter, M., Schreck, T.: Histograms of oriented gradients for 3D model retrieval. In: Proceedings of International Conference in Central Europe on Computer Graphics, Visualization and Computer Vision, pp. 41–48. University of West Bohemia, Plzen (2010)
12. Walter, M.: Implementierung und Evaluierung einer gradientbasierten Methode für das 3D Model Retrieval. Bachelor thesis, Darmstadt, TU (2009)
13. Adlib museum, http://www.adlibsoft.com (2016). Accessed 1 Feb 2016
14. zetcom ltd, http://www.zetcom.com (2016). Accessed 02 Feb 2016
15. System simulation, http://www.ssl.co.uk (2016). Accessed 01 Feb 2016
16. Collections trust uk, http://www.collectionstrust.org.uk/collections-link/collections-management/spectrum/choose-a-cms (2016). Accessed 01 Feb 2016
17. Hunter, J., Yu, C.H.: Assessing the value of semantic annotation services for 3d museum artefacts. In: Sustainable Data from Digital Research Conference (SDDR 2011) (2011)
18. Artec, http://www.artec3d.com/software/artec-studio (2016). Accessed 03 Feb 2016

19. Smithsonian explorer, http://3d.si.edu/ (2016). Accessed 01 Feb 2016
20. Arnold, D.: 3d-coform: Tools and expertise for 3d collection formation. In: Proceedings of Electronic Information, the Visual Arts and Beyond, vol. 21, pp. 94–99 (2009)
21. Arnold, D.: Computer graphics and cultural heritage: continuing inspiration for future tools. Comput. Graphics Appl. **34**, 70–79 (2014)
22. Serna, S.P., Schmedt, H., Ritz, M., Stork, A.: Interactive semantic enrichment of 3D cultural heritage collections. In: Arnold, D., Kaminski, J., Niccolucci, F., Stork, A. (eds.) VAST: International Symposium on Virtual Reality, Archaeology and Intelligent Cultural Heritage. The Eurographics Association (2012)
23. Crofts, N., Doerr, M., Gill, T., Stead, S., Stiff, M.: Definition of the cidoc conceptual reference model. In: ICOM/CIDOC CRM Special Interest Group (2006)
24. Doerr, M., Theodoridou, M.: Crmdig: A generic digital provenance model for scientific observation. In: 3rd Workshop on the Theory and Practice of Provenance, TaPP'11, Heraklion, Crete, Greece, 20–21 Jun 2011. https://www.usenix.org/conference/tapp11/crmdig-generic-digital-provenance-model-scientific-observation
25. Pan, X., Schröttner, M., Havemann, S., Schiffer, T., Berndt, R., Hecher, M., Fellner, D.W.: A repository infrastructure for working with 3d assets in cultural heritage. Int. J. Herit. Digit. Era **2**(1), 144–166 (2013)
26. Paquet, E., Rioux, M., Murching, A.M., Naveen, T., Tabatabai, A.J.: Description of shape information for 2-d and 3-d objects. Sig. Proc. Image Comm. **16**(1–2), 103–122 (2000)
27. Dalal, N., Triggs, B.: Histograms of oriented gradients for human detection. In: CVPR, pp. 886–893 (2005)
28. Filler, A.: Euklidische und nichteuklidische Geometrie. BI-Wiss.-Verlag, Mathematische Texte (1993)

Computer Aided Process as 3D Data Provider in Footwear Industry

Bita Ture Savadkoohi and Raffaele De Amicis

Abstract One of the new emerging of ubiquitous manufacturing is Mass Customization. The combination of 3D scanning systems with mathematical technique makes possible the development of CAD system which can help the selection of good footwear for a given customer. During the surface reconstruction process, a mesh is calculated from points cloud. This mesh may have holes corresponding to deficiencies in the original point data. Although these data are given in an arbitrary position and orientation in 3D space. Moreover, there is a need to mesh segmentation in order to shape retrieval form shoe last data base. Thus to apply sophisticated modeling operation on data set, substantial preprocessing is usually required. In this paper first, we describe an algorithm for filling hole. Then all of the models are aligned and at the end the models are segmented in order to use for further analysis.

Keywords Mass customization · Filling hole · Alignment · Segmentation

1 Introduction

Given the changing characteristics of consumer's foot and industrial competition, it is imperative to adapt mass customization. The design of new shoes starts with the design of the new shoe last which is a mechanical form of human foot model. In such industry the shoe last should be designed rapidly from the individual foot model [1, 2]. Due to both of the object complexity and scanning process, some areas of the object outer surface may never be accessible, thus obtained data typically contains

B. Ture Savadkoohi (✉)
Department of Computer and Electrical Engineering,
Seraj Higher Education Institute, Next to Municipal Museum,
Maghsoodieh Avenue, Nowbar Alley, Tabriz, Iran
e-mail: bita.turesavadkoohi@gmail.com

R. De Amicis
Fondazione GraphiTech, Via Alla Cascata 56/c, 38123 Povo, Trento, Italy
e-mail: Raffaele.de.Amicis@gmail.com

G. De Pietro et al. (eds.), *Intelligent Interactive Multimedia Systems and Services 2016*, Smart Innovation, Systems and Technologies 55,
DOI 10.1007/978-3-319-39345-2_41

missing pieces and holes. However this deficiency is not acceptable, where the geometric models are using in design processing in manufacturing. Although data from 3D scanning are given in arbitrary position in 3D scape and moreover there is need meaningful sub-meshes in order to best fit shape retrieval. Thus reverse engineering (RE), automatic alignment and segmentation have become an important step in the design, manufacture of new product and shape analysis [1]. An approach for preprocessing of 3D data for manufacture of new product and shape analysis is proposed in this paper. This paper in structures as follows: Sect. 2 is presented the filling hole in triangle mesh for building complete 3D model. An alignment of 3D foot model with shoe last data base is described in Sect. 3 while segmentation for shape retrieval is presented in Sect. 4. Finally, conclusion and remarks are summarized in Sect. 5.

2 Hole Filling Algorithm

Since several factors such as occlusion, low reflectance or even missing pieces in the original geometry can lead to incomplete data and this deficiency is not acceptable when the 3D model is taking into actual application. Thus certain repairs must be done before taking these models into application. Existing approaches to fill holes in meshes can be distinguished in two main categories: the geometric and non-geometric approaches [3, 4]. A hole filling process that is applied here [5] is summarized as it follows: first, identify the holes automatically by looking close loop of boundary edges. Next, cover the holes with Advanced Front Mesh technique [6]. Then, estimate desirable normal instead of relocating them directly for modification of triangles in initial patch mesh. Finally, rotate triangles by local rotation and reposition these coordinate by solving the Poisson equation according to desirable normal and boundary vertices of the hole in order to make algorithm more accurate.

As shown in Fig. 1a, all triangles that share one common vertex are called 1-ring triangle of vertex, all edges that share one common vertex are called 1-ring edges of the vertex and all vertices on 1-ring edges of a vertex (except itself) are called 1-ring vertex. Two triangles that share a common edge, they called adjacent triangle.

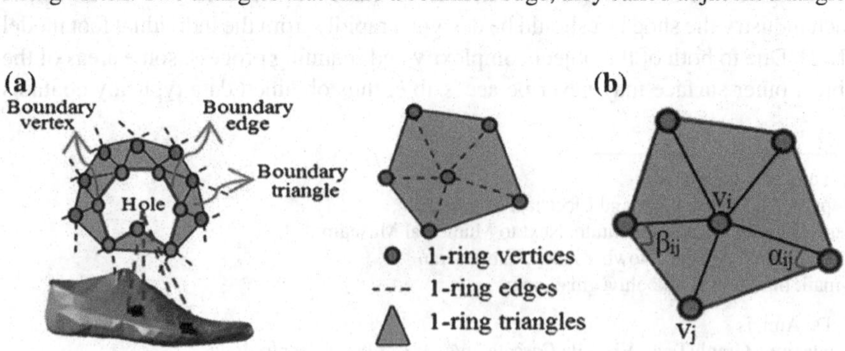

Fig. 1 **a** Preliminaries related to *triangle* mesh and hole. **b** Angles opposite to edge v_i, v_j

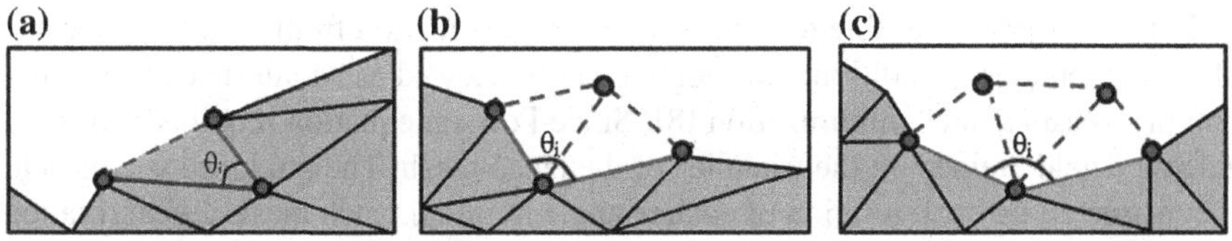

Fig. 2 Initial patch mesh generation: $\theta \leq 75°$, $75° \leq \theta \leq 135°$, $\theta \geq 135°$

A boundary edge is adjacent to exactly one triangle and a boundary vertex is a vertex used to define a boundary edge. Thus, a closed cycle of boundary edges defies a hole. A boundary triangle is a triangle that owns one or two boundary vertices.

The Advanced Front Mesh technique is applied over the hole to generate an initial patch mesh as it follows: First, for each vertex v_i on the front calculate the angle θ between two adjacent boundary edges. Next, depending on the angle between e_i and e_{i+1} build the new triangles on the plane, see Fig. 2. Then, calculate the distance between new vertex and related boundary vertices, if distance between them is less than given threshold, then they should merge.

Then, the triangles involved in the initial patch mesh are modified by estimating their desirable normals instead of relocating them directly. The most important task of discrete **harmonic function** is to map a given disk-like S_T onto a plane S^* which is introduced by Eck et al. [7]. The goal is to find a suitable piecewise linear mapping such as $f : S_T \rightarrow S^*$ to minimize Dirichlet energy as follows:

$$E = \frac{1}{2} \int_{S_T} \|grad_{S_T} f\|^2 ds \tag{1}$$

Consider one triangle $T = \{v_1, v_2, v_3\}$ in the surface S_T. The Dirichlet energy can be expressed as follow:

$$2 \int_{S_T} \|grad_{S_T} f\|^2 = \frac{1}{2}(\cot \theta_3 \|f(v_1) - f(v_2)\|^2 + \cot \theta_2 \|f(v_1) - f(v_3)\|^2 + \cot \theta_1 \|f(v_2) - f(v_3)\|^2) \tag{2}$$

where θ_3, θ_2 and θ_1 are the angles between edges (v_3v_1, v_3v_2), (v_2v_1, v_2v_3) and (v_1v_3, v_1v_2). The normal equation for the minimization problem can be defined as the following linear system of equations:

$$\sum_{v_j \epsilon N_i} w_{i,j}(f(v_j) - f(v_i)) = 0 \qquad v_i \in V_I \tag{3}$$

where $w_{i,j} = \cot \alpha_{i,j} + \cot \beta_{i,j}$ and angles are showed in Fig. 1b. The associated matrix is symmetric and positive, so the linear system is uniquely solvable with iterative methods such a conjugate gradient method. Moreover that system has to be solved three time, once for x-, once for y- and once for z-coordinate.

Poisson equation is able to reconstruct a scalar function from a guidance vector filed and boundary condition. Although it can be viewed as an alternative formulation of a least squares minimization [8]. Since Poisson equation requires a discrete guidance field defined on the triangles of the patch mesh. The guidance vector field is constructed by local rotation of each triangle in initial patch mesh. Local rotation is applied to each triangle of initial patch mesh.

Let n be the original normal of triangle and n' be the new normal of triangle that is calculated with desirable normal of vertices of triangle and c be the center of triangle. The rotation can obtain by rotating n to n' around c.

After rotation, the original patch mesh is torn apart and triangles are not connected anymore and these torn triangles are used to construct a guidance vector filed for Poisson equation. Finally, the disconnected triangles are stitched by solving Poisson equation. Consider an unknown scalar function, f, the Poisson equation with Dirichlet boundary condition is defined by:

$$\nabla^2 f = \nabla . w \quad over\ \Omega, \quad with\ \ f|_{\partial\Omega} = f^*|_{\partial\Omega} \tag{4}$$

where w is Guidance Vector Filed, $\nabla . w = \frac{\partial w_x}{\partial x} + \frac{\partial_{w_y}}{\partial y} + \frac{\partial_{w_z}}{\partial z}$ is the divergence of $w = (w_x, w_y, w_z)$, f^* provides the desirable values on the boundary $\partial\Omega$, $\nabla^2 = (\frac{\partial^2}{\partial x^2} + \frac{\partial^2}{\partial y^2} + \frac{\partial^2}{\partial z^2})$ is Laplacian operator. Thus it can define as least-squares minimization problem:

$$\min_f \int_\Omega |\nabla^2 f - \nabla w|^2 \quad with\ \ f|_{\partial\Omega} = f^*|_{\partial\Omega} \tag{5}$$

A discrete vector field on a triangle mesh is defined to be a piecewise constant vector function whose domain is the set of point on the mesh surface. A constant vector is defined for each triangle, and this vector is coplanar with the triangle. For a discrete vector filed w on the mesh, its divergence at vertex v_i can be defined with:

$$(div\, w)(v_i) = \sum_{T_k \in N_i} \nabla B_{ik}.w|T_k| \tag{6}$$

where $|T_k|$ is the area of triangle T_k, $\nabla B_{i,k}$ is the gradient vector of the vertices within T_k, N_i is the 1-ring vertices of v_i. The discrete gradient of the scalar function f on a discrete mesh is expressed as:

$$\nabla f(v) = \sum_i f_i \nabla_{\emptyset_i}(v) \tag{7}$$

where $\emptyset_i(.)$ piecewise linear basis function valued 1 at vertex v_i and 0 at all other vertices and f_i is a scalar vector value attached to vertex v_i and it is one of the coordinate of v_i. The discrete Laplacian operator can determine as:

$$\Delta f(v_i) = \frac{1}{2} \sum_{v_j \in N_i} (\cot_{\alpha_{i,j}} + \cot_{\beta_{i,j}})(f_i - f_j) \tag{8}$$

where $\alpha_{i,j}$ and $\beta_{i,j}$ are the two angles opposite to edge in the two triangles sharing edge (v_i and v_j) and N_i is the set of the 1-ring vertexes of vertex v_i. Finally discrete Poisson equation is expressed as follows: $\nabla^2 f = div(\nabla f) = \nabla w$. Discrete Poisson equation with known boundary condition can be defined by linear system such as: $Ax = b$. So that the coefficients matrix A is determined by Eq. 8 and the vector b is determined by Eq. 6 and unknown vector x is the coordinate of all vertices on the patch mesh. The smooth and accurate patch mesh is constructed as follow: First, compute the gradient of each new vertex on the adjacent triangle by using Eq. 7. Next, calculate the divergence of every boundary vertex by using Eq. 6. Then, determine the coefficient matrix A by Eq. 8 and vector b in this equation is determined by using divergence of all boundary vertices. Finally, solve the Poisson equation and obtain the new coordinate of all vertices of the patch mesh.

3 Alignment of 3D Foot with Shoe Last Data Base

Principle Component Analysis (PCA) is a method that has been extensively used for analysis, modeling and recognition [9]. In analysis, instead of applying the PCA in classical way (sets of 3D point-clouds), for taking into account different size of triangles **W**eighted **P**rinciple **C**omponent **A**nalysis(WPCA) is applied [1]. To achieve the alignment we describe the main steps and details of WPCA in the next steps. First, we apply step 1 through step 6 for the first models in shoe last data base, see Fig. 3. Then, for alignment of another models with the first model we apply step 1 through step 5 and steps 7 and 8.

Let A be the total sum of the areas of all triangles in the mesh, A_i be the area of triangle i within the mesh, T_{c_i} be barycenter of each triangle, T_c be the total sum of barycenter of all triangles in mesh, matrix *OME* be **O**rigin **M**atrix of

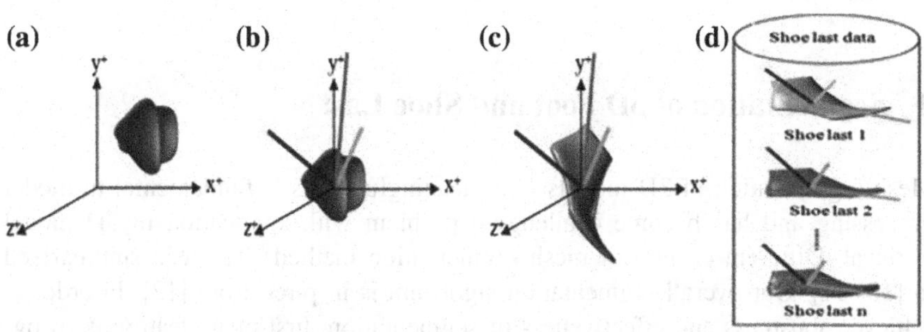

Fig. 3 **a** Input 3D model. **b** Translated barycenter of the 3D model to the origin. The *red*, *green* and *blue lines* are eigenvectors. **c** Rotated 3D model with it's eigenvectors. **d** Target 3D models

Eigenvectors that is included the position of eigenvectors of model as column and *CM* be **C**ovariance **M**atrix type (3 × 3). The main steps of WPCA are as it follows:

Step 1. Accomplish the translation invariance by finding the barycenter of the model as:

$$x_c = \frac{1}{n}\sum_{i=0}^{n} x_i, \quad y_c = \frac{1}{n}\sum_{i=0}^{n} y_i, \quad z_c = \frac{1}{n}\sum_{i=0}^{n} z_i \tag{9}$$

Step 2. Translate 3D model to origin. That is to say, for each coordinate, a corresponding transformation is done by $v_i^{'} = \{x_i - x_c, y_i - y_c, z_i - z_c\}$ and define the new vertex set $v^{'} = \{v_1^{'}, v_2^{'}, ...v_n^{'}\}$.
Step 3. Calculate matrix *CM* as follows:

$$CM = \begin{bmatrix} Cov_{xx} & Cov_{xy} & Cov_{xz} \\ Cov_{yx} & Cov_{yy} & Cov_{yz} \\ Cov_{zx} & Cov_{zy} & Cov_{zz} \end{bmatrix} \tag{10}$$

where $T_c = \frac{\sum_{i=0}^{n} T_{c_i}}{n}$ and $Cov_{xx} = \frac{\sum_{i=0}^{n} A_i(T_{c_i}x - T_c x)(T_{c_i}x - T_c x)}{A}$. Obviously the matrix *CM* is a symmetric real matrix, there for its eigenvalues are non negative real numbers and orthogonal.
Step 4. Sort the eigenvectors in a decreasing order and find the corresponding eigenvectors. The eigenvectors are scaled to Euclidean unit length and form the rotation matrix *R* which has scaled eigenvectors as rows.
Step 5. Apply matrix *R* to all of the vertices of a triangle and form a new vertex sets called, $v'' = \{R \times v_1^{'}, ..., R \times v_n^{'}\}$.
Step 6. Rotate first shoe last with its eigenvectors in Fig. 3c up to a position where the 3D model becomes parallel with x-y plane, see Fig. 3d and build matrix *OME* with the new position of these three eigenvectors as Columns.
Step 7. Set **T**ranspose of a **M**atrix *R* as *TM*. The alignment is accomplished by constructing a rotation matrix through $R^{'} = OME \times TM$.
Step 8. Get the matrix $R^{'}$ and apply it to all $v''$and calculate new vertex sets V_A as the alignment of the model $V_A = \{R^{'} \times v_1'', ..., R^{'} \times v_n''\}$.

4 Segmentation of 3D Foot and Shoe Last

Mesh segmentation of 3D models into meaningful parts is fundamental to model processing and has become challenging problem with application in 3D model retrieval. Different criteria of mesh segmentation methods has been summarized in [10, 11]. Our overall segmentation algorithm is inspired from [12]. In order to achieve robustness and effectiveness of segmentation, first plane feature is recognized by applying normal based initial decomposing. Then further segmentation based on curvature criteria and Gauss mapping, followed by the detection of quadric surfaces is performed. Finally, the segmentation is refined by B-spline surface fitting technology.

Let $F = \{f_1, f_2, ..., f_n\}$ be the surrounding triangular faces, $N = \{n_1, n_2, ..., n_n\}$ and $AN = \{a_1, a_2, ..., a_n\}$ be normal vector and the area of each subset of F, $d_{i,j}$ be the control points, $N_{i,3}(u)$ and $N_{j,3}(v)$ be stand for the univariate B-spline basis functions of 3-order in the u- and v-direction, defined over the knot vector vectors $U = \{u_0, u_1, ..., u_{m+4}\}$ and $V\{v_0, v_1, ..., v_{n+4}\}$.

For **planer segmentation** the initial normal vectors of all vertices of the mesh is calculated as:

$$n_i = \frac{\sum_{j=1}^{n} a_j . n_{f_j}}{\sum_{j=1}^{n} a_j} \tag{11}$$

Then the normal filtering method is applied in order to obtain more accurate normal vector [13]. So that instead of moving the original vertices of the mesh, the normal vectors of the vertices are re-computed. As the first seed, the region is initialized with an arbitrary, unsegmented triangular face. Next, when the normal vector angle between the added face and the seed face is smallest, (less than a pre-defined threshold), three neighboring faces of the seed is added and the region is growing. Then, the seed is updated with the latest added face. Finally, the region growing is iterated until no more faces can be added into the current region. So, this region is referred to as a segment. The above process is repeated until all faces of mesh are segmented. Thus, the initial segmentation is achieved. The first seed face is stored for each segment and the normal vector angles between this seed face and all other faces within the segment are calculated. When all those vector angles are less than a small value, this segment is remarked as a plane, otherwise, further curvature-based segmentation is performed.

Curvature-based segmentation is applied for all non-plane segments. In order to determine the curvature for each vertex, the quadric surface fitting technology is applied. As shown in Fig. 4, a local surface patch around a vertex is approximated with a quadric surface: $S(u, v) = (u, v, h(u, v))$. This principle is applied to a discrete mesh and the Local Coordinate System (LGS) of each vertex is constructed to fit a quadric surface on the neighboring vertices of the vertex.

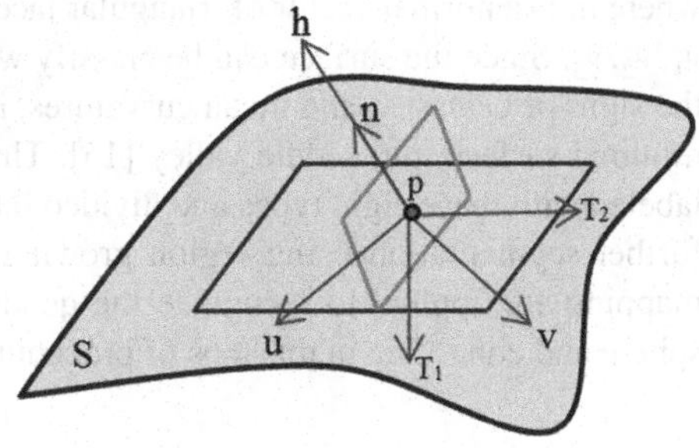

Fig. 4 An illustration of the local coordinate system on a surface: v is the origin, h-axis is along the normal vector n at p on S, and u-, v-axes are two arbitrary, orthogonal unit vectors in the tangent plane at v

Relevant to the surface theory [14], a boundary of a model is the intersection of adjacent surfaces, which mentions that the geometric property of each vertex around a boundary should be calculated from its surrounding neighborhood within one surface. On the other hand the anisotropic neighborhood is applied in order to eliminate the error. Moreover based on the initial segmentation result, the neighborhood of a vertex are exclusively searched from the segment of the vertex, i.e., the neighborhood of a vertex will never cross the segment of this vertex. At the same time, in order to guarantee that neither the local sharp edges nor the global sharp edges would be included in the neighborhood of the vertex such as the underlying surface of the vertex, the constraint that the normal vectors of all triangular faces of the neighborhood are closed to the normal vector of the vertex is applied. Thus more accurate curvature would be obtained. Then the quadric surface is fitted by transforming the neighboring vertices from the **G**lobal **C**oordinate **S**ystem(GCS) to **L**ocal **C**oordinate **S**ystem(LCS) as:

$$h(u, v) = au^2 + buv + cv^2 + eu + fv + g \tag{12}$$

The least squares methods is utilized here. The Gaussian and mean curvature of a vertex are determined by

$$\begin{cases} K = \frac{4ac-b^2}{(e^2+f^2+1)^2} \\ H = \frac{c+ce^2+a+af^2-bef}{(e^2+f^2)+1)^{\frac{3}{2}}} \end{cases} \tag{13}$$

The normal-weighted averaging method is applied to compute the curvature of a triangular face as follows:

$$\begin{cases} K_f = \frac{\sum_{i=1}^{3} K_{vi}.e^{\frac{-\|n_{ni}-n_f\|^2}{2\sigma^2}}}{\sum_{i=1}^{3} e^{\frac{-\|n_{ni}-n_f\|^2}{2\sigma^2}}} \\ H_f = \frac{\sum_{i=1}^{3} H_{vi}.e^{\frac{-\|n_{ni}-n_f\|^2}{2\sigma^2}}}{\sum_{i=1}^{3} e^{\frac{-\|n_{ni}-n_f\|^2}{2\sigma^2}}} \end{cases} \tag{14}$$

where n_f is a normal vector of triangular face f, $n_{v_1}, n_{v_2}, n_{v_3}$ are the normal vectors of v_1, v_2, v_3. Since the surface can be classify with eight fundamental types in terms of the signs of Gaussian and mean curvatures, including peak, ridge, saddle ridge, flat, minimal surface, pit, saddle valley [15]. Thus the triangular faces of the mesh are labeled with these eight types and divided into different patches. In order to perform further segmentation using region growing method statistics methods and Gauss mapping are applied to recognize the quadric surface feature, including cylinder, sphere and cone. The main steps of procedure to detect cylinder include:

Step 1. As a seed face, select a triangular face and initialize the current region with this seed.
Step 2. Add each connected neighboring faces of the seed face into region when the minimal average and variances of the minimal and maximal curvature are less than a small positive value. Otherwise, skip this face and end the growing at this point by setting a flag on the face.
Step 3. Stop the growing when all neighboring faces of the seed face are skipped. Otherwise, add this face into the region and update the seed face with this face and go to step 2. In the case of sphere surface the Gaussian and mean curvature are constant. Since, a cone surface has no particular property in term of curvature, so Gauss mapping method is applied to segment and consist of the following three steps:

Step 1. As a seed face, select a triangular face and initialize the current region with the seed face and map its normal vector onto the Gauss sphere.
Step 2. Add each connected neighboring faces of the seed face into region when that face's normal vector mapping is located around a small circle of the Gauss sphere, thus that face is on the cone surface.
Step 3. Stop the growing when all neighboring faces of the seed face are skipped. Otherwise, update the seed face with an added neighboring face of the seed face within the current region and go to step 2.

The Gauss mapping of the normal vectors of faces on the cone surface form a small circle of the Gauss sphere. Due to the noises and computational errors the mapping can not exactly located on a small circle. Thus, the mappings are fitted with a circle and the fitting error is calculated. The mapping are considered to be around a small circle and the region is seen as a cone surface segment when the error is less than a small value and the radius of the fitted circle is less than the radius of Gauss sphere, i.e., 1.

In the neither planer nor quadric surface, **B-spline surface fitting method** is performed to refine the segmentation. A bicubic B-spline surface $S(u, v)$ is determined by

$$S(u, v) = \sum_{i=0}^{m} \sum_{j=0}^{n} d_{i,j} N_{i,3}(u) N_{j,3}(v), \quad 0 \le u, v \le 1 \tag{15}$$

The surface is fitted when the control points $d_{i,j}$ are found, so that the distance of the data points P to $S(u, v)$ is minimized:

$$E = \sum_{k=0}^{N} dist^2(v_k, S) \tag{16}$$

A few connected triangular faces within the surface segment are chosen as a seed region and its vertices are fitted with a bicubic B-spline surface. Taking the fitted B-spline surface as an input the vertices which are not included in but connected with the current region are used for growing the region. For each added vertex, three condition should be fulfilled: **(i)**, the projection distance of the vertex onto the fitted face is less than a small value. **(ii)** the normal vector of the projection point on the fitted surface should be close to that of the vertex. **(iii)** the principle curvature of the projection point should be close to those of the vertex. Followed by fitting a new fitted B-spline surface on the region, when no more vertices can be added into region, the region growing process terminates. Then all faces in the current region are cleared and updated with origin seed region. By taking new fitted B-spline surface as input the region growing procedure above is applied again. This is repeated until the maximal region is achieved and a final segment is obtained.

5 Conclusions

In this paper preprocessing algorithms of 3D scanning data before taking 3D models to the real application is proposed. For this aim, the hole in triangle mesh is filled by generating initial patch mesh. Then the initial patch mesh is modified with calculation the desirable normal based on Harmonic function. Finally accurate and smooth triangle mesh are obtained by solving the Poisson equation according to desirable normals and the boundary vertices of the hole. After obtaining complete 3D models, we applied the **W**eighted **P**rinciple **C**omponent **A**nalysis technique for alignment of 3D foot with shoe last data base. Finally, in order to shape retrieval within the shoe last data base, segmentation method based is demonstrated. At first, to recognize the plane feature, normal based initial decomposition is applied. Then, by detecting the quadric surface feature, curvature criteria and Gauss mapping is performed. Finally, the segmentation is refined by using B-spline surface fitting technology.

References

1. Ture Savadkoohi, B., De Amicis, R.: Mass customization strategy in footwear industry. J. Adv. Comput. Sci. Appl. IJCSIA **5**(2), 2250–3765 (2015)
2. Ture Savadkoohi, B., De Amicis, R.: 3D foot scanning in flexible manufacturing: comparisation and deformation solution. In: International Conference on Computer Graphics, Visualization, Computer Vision and Image processing. Las Palmas de Gran Canaria, Spain, 22–24 July 2015
3. Liepa, P.: Filling holes in meshes. In: Proceedings of the Eurographics/ACM SIGGRAPH Symposium on Geometry Processing, pp. 200–205. Granada, Spain, 1–6 Sept 2003
4. Nooruddin, F.S., Turk, G.: Simplification and repair of polygonal models using volumetric techniques. J. IEEE Trans. Visual. Comput. Graph. **9**(2), 191–205 (2003)
5. Zhao, W., Gao, S., lin, H.: A robust hole-filling algorithm for triangular mesh. J. Visual Comput. **23**(12), 987–997 (2007)
6. George, L.P., Seveno, E.: The advancing-front mesh generation method revisited. J. Numer. Methods Eng. **37**(7), 3605–3619 (1994)
7. Eck, M., DeRose, T.D., Kuchamp, T., Hoope, H., Lounsbery, M., Stuetzle, W.: Multi-resolution analysis of arbitrary meshes. In: Processing of SIGGRAPH, pp.173–182. Los Angeles, CA, USA, 6–11 Aug 1995

8. Perez, P., Gangnet, P., Blake, A.: Poisson image editing. In: Processing of SIGGRAPH, pp. 313–318. San Diego, California, USA, 27–31 July 2003
9. Paquet, E., Rioux, M.: Nefertiti: a query by content system for three-dimensional model and image databases management. J. Image Vis. Comput. **17**(2), 157–166 (1999)
10. Shamir, A.: A survey on mesh segmentation techniques. J. Comput. Graph. Forum **27**(6), 1539–1556 (2008)
11. Chen, X., Golovinskiy, A., Funkhouser, T.: A benchmark for 3D mesh segmentation. J. ACM Trans. Graph. **28**(3) (2009)
12. Wang, J., Yu, Z.: Surface feature based mesh segmentation. J. Comput. Graph. **35**(3), 661–667 (2011)
13. Sun, X., Rosin, P., Martin, R., Langbein, F.: Fast and effective feature-preserving mesh denoising. J. IEEE Trans. Visual. Computer Graph. **13**(5), 925–938 (2007)
14. do Carmo, M.: Differential Geometry of Curves and Surfaces. Prentice-Hall (1976)
15. Besl, P.J., Jain, R.: Segmentation through variable-order surface fitting. J. IEEE Trans. Pattern Anal. Mach. Intell. **10**(2), 167–192 (1998)

CoolTour: VR and AR Authoring Tool to Create Cultural Experiences

Nagore Barrena, Andrés Navarro, Sara García and David Oyarzun

Abstract This paper presents a platform to create complex VR and AR cultural experiences in an easy way. Each experience is composed by steps that contain a virtual scene and that are connected among them via end-user interactions. The platform presents three pillars that allow the author to configure the experience: the authoring tool, the experiences server and a mobile app. A use case consisting in a treasure hunt in a city has been developed to validate the platform.

Keywords Virtual reality · Augmented reality · Authoring tools · User experiences

1 Introduction

Traditional forms of printed content are increasingly being replaced by multimedia experiences in a variety of fields such as learning, health, tourism or architecture. Motivated by the demands of highly interconnected users accustomed to smartphones, multimedia experiences have deeply spread into popular culture and have become a central part of how we consume information.

However, massive growth and access to multimedia content can become overwhelming and difficult the engagement of users who have habituated to short and immediate information. Augmented Reality (AR) or Virtual Reality (VR) tech-

N. Barrena (✉) · A. Navarro (✉) · S. García (✉) · D. Oyarzun (✉)
Vicomtech-IK4, Paseo Mikeletegi 57, 2009 Donostia/San Sebastián, Spain
e-mail: nbarrena@vicomtech.org

A. Navarro
e-mail: anavarro@vicomtech.org

S. García
e-mail: sgarcia@vicomtech.org

D. Oyarzun
e-mail: doyarzun@vicomtech.org

G. De Pietro et al. (eds.), *Intelligent Interactive Multimedia Systems and Services 2016*, Smart Innovation, Systems and Technologies 55,
DOI 10.1007/978-3-319-39345-2_42

nologies can enhance this engagement by creating the perception of being present in the multimedia experience. In addition, these technologies can use elements such as simulators, glasses and other virtual gadgets that lead to a better immersion into that experience.

VR and AR technologies unveil a new palette of tools to creators. It is vital to select the appropriate ones to accomplish the goal proposed in the user experience. For instance, the goal to achieve should be clear, the message direct and interaction should be perceived as natural, keeping any technical detail hidden to the user. All of this will attract more users, increase their bond, improve the effectiveness of our message and ultimately, better communicate to an increasingly demanding public.

To achieve this goal, it is important to pay attention not only to hardware, usability or graphics. It is also essential to consider the reactions, emotions and thoughts of those that are using the system. Therefore, it is crucial to design taking into account the reactions of the users, making the experience full and satisfying.

We are at a stage where there is a need to develop an experience that requires a mixture of such technologies and knowledge of the field of design and communication. The type of user-creators varies to a non-expert profile, with less technical knowledge. They need a simple tool, but at the same time versatile, allowing them to create rich and engaging content.

To fulfil this need, an easy-to-use tool that meets all the requirements detailed previously has been created. This tool facilitates the creation of rich VR and AR experiences that are easily adaptable to any platform and that can incorporate different multimedia content for creating very attractive combinations.

The platform should fulfill two key features:

- It allows the creation of complex multimedia content.
- It is enough flexible to apply several engagement strategies (i.e.: gamification techniques) that motivate the end user.

This is a contribution to the field of marketing and to show all its qualities, a use case with this tool has been designed. Specifically, a tourist experience was created that uses the folklore of a city to show the most characteristic places of it. Moreover, in the visit, key points of the environment are linked to the cultural heritage, showing its various film festivals and popularly photographed landscapes.

This paper is organized as follows. Section 2 presents the related work. Section 3 introduces the platform, describing its features and architecture. Section 4 describes the use case. Finally, Sect. 5 presents conclusions and future work.

2 Related Work

The use of authoring tools has been a common practice in VR and multiple examples can be found including tools for videogames or simulations. As opposed to VR, where facing the creation of an experience without an authoring tool would not be considered, AR authoring tools are not that frequent. However, several tools have emerged to create AR experiences, ranging from high to low level approaches and requiring less or more technical skills.

Wang et al. [1] propose to classify those tools between programmers and non-programmers oriented tools, low and high abstraction level and for desktop or mobile platforms. Authoring tools for programmers consist of software libraries that expose certain interface that require programming knowledge to create an application, while tools for non-programmers do not require to write code. Although abstraction level is related to programmers categorization, it represents the abstraction level required to generate content. Hence in the lower abstraction level specific logic should be specified through a programming or visual interface, while in the higher one only the content should be configured through a simple set of steps [2].

Focusing on AR authoring tools for non-programmers, PowerSpace [3] is an authoring tool that allows to create simple 3D scenes—composed by 3D models, primitives and text—and reference them to the tracking marker. DART [4] is an authoring tool built on top of Macromedia Director that extends scene creation with interactivity through Director behaviours and actors, although complex interactions need some scripting requiring certain programming skills. AMIRE [5] on the other hand presents a visual programming framework that allows to create interactive AR content. MARS [6] authoring tool is oriented to develop experiences for their mobile client, allowing creators to generate time based AR documentaries that contain audio, video, images, text and 3D models associated to a location. ComposAR [7] is an authoring tool focused on associating virtual content to real objects and define interactions on those objects, by a trigger-action paradigm where triggers are the real objects, the markers, and the actions update the virtual content, a 3D scene.

AR commercial solutions have also offered authoring tools to generate content for their software. Although no longer available, Metaio[1] provided a graphical authoring tool that allowed to define 3D scenes and animations and associate them with markers. Interaction with multimedia content was possible through overlaying buttons.

3 Platform Overview

The platform has been designed pursuing not only the easiness of use but also the deployment of an integral solution, which is able to give support to all the stages of the authoring process.

It is composed by three main pillars as shown in Fig. 1: the *authoring tool*, the *experiences server* and the *final app*.

[1]https://www.metaio.com/.

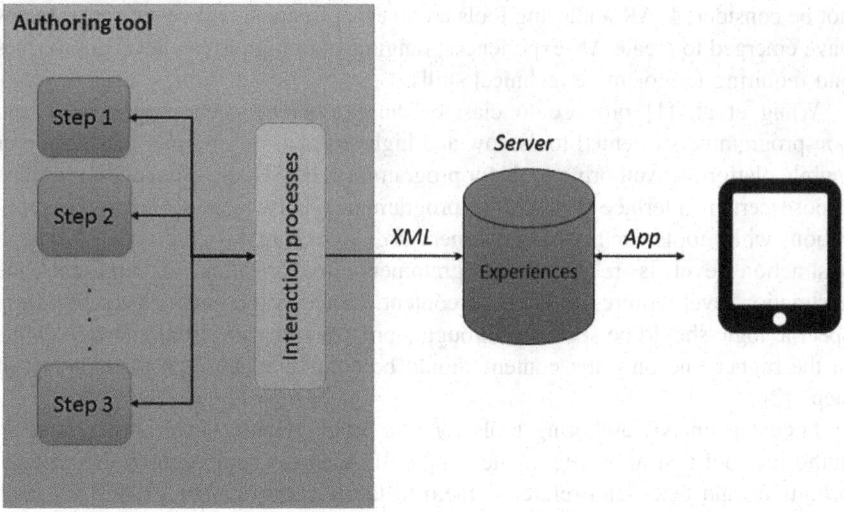

Fig. 1 Conceptual structure of the CoolTour platform

The *authoring tool* is able to create VR and AR experiences with any level of complexity. Therefore, tools for configuring contents and application interaction processes have been integrated into a common and easy-to-use visual interface.

Process. Each AR experience, independently of its level of complexity, is divided into steps that are connected among them by user actions. The author of the experience is the responsible of defining the granularity of each of these steps.

A step is composed by a real marker and a virtual scene. Author-defined interactions in the virtual scene will transport the user to a new step, providing in this way the application flow.

Interactions can be as easy as a scene timeout or more user-dependent, as clicking or moving virtual objects.

Content. The variety of content is a critical aspect to provide enough flexibility to the system to create complex VR and AR experiences. Therefore, any kind of media can be loaded into a virtual scene in a transparent way for the author: text, audio, video, 3D content (including animations).

The *experiences server* stores the configuration of each VR or AR experience in a XML-compliant format.

The final mobile *app*, developed for iOS, Android and Windows devices, is able to read the contents in the experiences server and to show them to the end user.

4 Use Case

The presented tool allows an easy workflow from the initial idea and its development to the final utilization by the user. It is not necessary to have technical or programming knowledge to obtain a functional outcome. The interface (Fig. 2) is simple and intuitive. For the user it is as creating the plot of a story.

To develop the application logic, it is necessary to create a sequence of actions and conditions that is named diagram. The diagram is a method commonly used in programming. Basically it encodes schematically what you want to do in each moment of the game. In every action, a content type is determined and the condition that will trigger the change to another action is defined. The conditions may differ from a timer in the displaying of the material to a button to perform the state change. This allows great versatility to tell any kind of story.

The tool supports a variety of audiovisual material, mainly video, image, audio, text or links to content on the web. In addition, virtual elements in three dimensions can be included. These elements go together with a marker. This way, the content is displayed when the user points the camera of a device to that image.

The tool is as intuitive for the user to enjoy the experience as for the creator to generate it. Once the application is opened on the mobile device the user have to follow the instructions that are given in each section. The result is visually rich in details, and it becomes a product that engages and motivates the public to continue with the experience until the end.

The validation of this process is done through a treasure hunt prepared for a child audience that makes a tour through the city. Children reach a destination and in that place they will be offered audiovisual information related to the environment. The

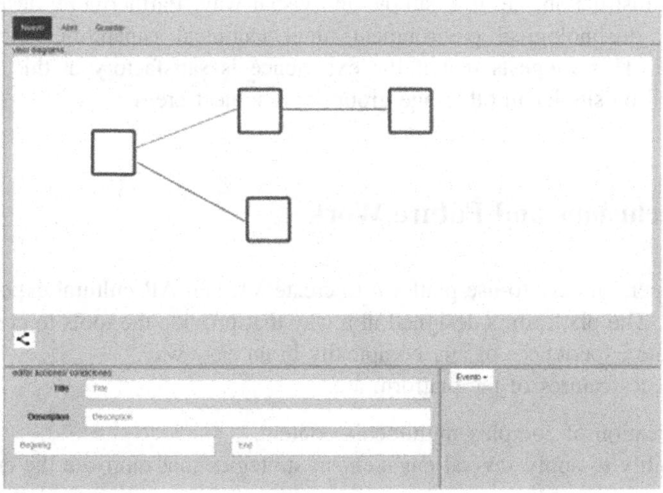

Fig. 2 Creation of the interaction processes of the use case

Fig. 3 Composition of the use case showing the use of the mobile app

way to advance will be watching the material and answering the questions proposed. They must observe the actual environment in which each section is performed, because that information will be key to solve future challenges. If they finish all the tasks they will receive a prize.

A conceptual composition describing the vision of the use case application is shown in Fig. 3.

Making a test with young audience is the best way to observe possible shortcomings or difficulties that the tool may have. This is an audience that needs well explained instruction, i.e. in a simple and visual way. Furthermore, despite being born in a technological environment, their technical capabilities are not yet developed. This suggests that if the experience is satisfactory at this level, the results will be similar in other age groups and subject areas.

5 Conclusions and Future Work

In this paper, an easy-to-use platform to create VR and AR cultural experiences is presented. The platform is designed in a way that provide the tools to create, store and execute experiences of any complexity in an easy way.

The basic features of the platform are:

- The creation of complex multimedia content.
- Flexibility to apply several engagement strategies that motivate the end user.

A whole experience is composed by a set of steps connected among them via user interactions. The platform that is presented is able to create, configure and execute the experience. The platform is composed by three main pillars: a visual and intuitive authoring tool, a server containing the created experiences and a mobile app.

The authoring tool allows the author to create the steps that compose the experience by defining contents in the virtual scene and the interaction processes that cause the change of step.

An AR experience consisting on a treasure hunt around a city has been developed to validate the platform. This treasure hunt is focused on children and primary school students.

As future work, several tests will be carried on with groups of students to validated usability and interaction aspects.

Moreover, a user profiling module will be integrated to adapt the content to the user expectations and preferences.

References

1. Wang, Y., Langlotz, T., Billinghurst, M., Bell, T.: An authoring tool for mobile phone AR environments. Proc. New Zeal. Comput. Sci. Res. Student Conf. 1–4 (2009)
2. Jee, H.-K., Lim, S., Youn, J., Lee, J.: An augmented reality-based authoring tool for E-learning applications. Multimed. Tools Appl. **68**, 225–235 (2014)
3. Haringer, M., Regenbrecht, H.T.: A pragmatic approach to augmented reality authoring. Proc. Int. Symp. Mix. Augment. Real. 237–245 (2002)
4. MacIntyre, B., Gandy, M., Dow, S., Bolter, J.D.: DART: a toolkit for rapid design exploration of augmented reality experiences. UIST Proc. Annu. ACM Symp. User Interface Software Technol. 197–206 (2004)
5. Grimm, P., Haller, M., Paelke, V., Reinhold, S., Reimann, C., Zauner, R.: AMIRE—authoring mixed reality. First IEEE Int. Work. Augmented Real. Toolkit 2 (2002)
6. Guven, S., Feiner, S.: Authoring 3D hypermedia for wearable augmented and virtual reality. In: Proceedings of the Seventh IEEE International Symposium on Wearable Computers, pp. 118–126 (2003)
7. Seichter, H., Looser, J., Billinghurst, M.: ComposAR: an intuitive tool for authoring AR applications. In: 2008 7th IEEE/ACM International Symposium on Mixed and Augmented Reality, pp. 177–178 (2008)

Emotional Platform for Marketing Research

Andres Navarro, Catherine Delevoye and David Oyarzun

Abstract In order to face new social and mobile realities, Marketing has advanced towards the reinforcement of customers' relationship based on understanding consumers' behavior. Emotions, having such an important role in our lives, are a key aspect on this behavior and should be incorporated in its analysis. We present in this paper a platform based on Affective Computing technology that allows marketers to gain insights of consumers' emotional response, facilitating them with a great tool to study and design their marketing strategy, and enabling the generation of better consumers' experiences. In this paper, we describe the design and implementation of the platform, as well as the uses cases that will allow the validation of the platform.

Keywords Affective computing · Marketing · Emotional recognition · Facial expression

1 Introduction

Marketing is continuously evolving and adapting to the new realities of e-commerce, social media and mobile expansion. Understanding consumers' behavior is a key on designing a successful marketing strategy. Emotions play a very important role in our decision-making process or how memories are store or retrieve. Hence, a decisive factor on the behavior of the consumers are emotions

A. Navarro (✉) · D. Oyarzun (✉)
Vicomtech-IK4, Paseo Mikeletegi 57, 2009 Donostia/San Sebastián, Spain
e-mail: anavarro@vicomtech.org

D. Oyarzun
e-mail: doyarzun@vicomtech.org

C. Delevoye (✉)
Technoport SA, 9, avenue des Hauts-Fourneaux, 4362 Esch-sur-Alzette, Luxembourg
e-mail: catherine.delevoye@technoport.lu

G. De Pietro et al. (eds.), *Intelligent Interactive Multimedia Systems and Services 2016*, Smart Innovation, Systems and Technologies 55,
DOI 10.1007/978-3-319-39345-2_43

and getting insights of the related emotions to our marketing strategy are very valuable.

Traditionally, the assessment of consumers' response towards a topic or a product has been performed from their self-assessment using questionnaires or following an interview. However, technology can provide valuable and objective information of this response, enhancing previous assessments or even discovering a new vision of those reactions.

In this paper we present a platform to analyze consumers' response to a product, an advertising campaign or a marketing outcome applying Affective Computing technology. From the recorded image of the user through the frontal camera of a mobile device or a webcam, user's facial expression is recognized and interpreted into an emotional state that provides valid information for the marketing analysis.

The platform will be evaluated on different scenarios that show the value of the gathered emotional information to analyze different aspects of a marketing strategy.

The presented work has been developed as part of the c-Space project,[1] which aims to develop a new generation of creative tools to create unique experiences.

This paper is organized as follows. Section 2 presents the related work on the field. Section 3 details the platform including its architecture, the emotional recognition process and the developed tools. Section 4 describes several uses cases. Finally conclusions and future work are presented.

2 Related Work

Marketing has evolved changing its focus from the product itself to the relationship with its customers [1], increasing the interest on analyzing users' attitude towards a certain topic or product. As a result, marketing performance can no longer be only analyzed in terms of sales and should incorporate metrics of consumers' satisfaction, communication and reactions. Designing a successful marketing strategy requires to predict and model the consumers' behavior, which is greatly influenced by their emotional state [2].

From the Human Computer Interaction perspective, Affective Computing aims to develop systems and devices that can recognize, interpret and simulate human affects. Allowing the development of systems that are aware of users' emotions, interaction with computers as we currently know can be highly improved [3]. Under this emotional vision of the interaction, several technologies have been developed to estimate the users' emotions. Such recognition can be performed from the multiple modalities that humans express emotions. Our facial expression is driven by our emotional state and there is universal evidence of its correlation [4]. The text that we write or the content of our speech contain emotional attributes, and in the latter, our voice presents several paralinguistic features related to our emotional

[1]http://c-spaceproject.eu/.

state. Our body language and posture also express our affective state and several physiological signals haven been studied to detect emotions, such as monitoring cardiovascular, electrodermal or brain electrical activities [5].

Focusing on the marketing field, Affective computing brings plenty of possibilities to understand consumers' behavior because of the effect of emotions on the different consumption-related processes [6]. Affective state is used as a metric to determine the subjective opinion of the consumer with respect to a topic, as part of what is referred as sentiment analysis. Examples can be found of linguistic-based approaches [7] as well as including facial expressions and voice analysis [8]. Moreover, affective response to marketing stimuli is analyzed in the in-lab study of consumers' behavior, as part of the research field of neuromarketing [9], using several biometrics such as the heart rate, the galvanic skin response or the brain electrical activity.

3 Platform Overview

3.1 System Architecture

The developed emotional analysis platform is integrated in the c-Space project architecture. Figure 1 shows the complete c-Space architecture where involved modules and services in the presented case are highlighted. In this way, a client mobile application was developed that makes use of the affective recognition module to estimate the user's emotional state and communication library to upload results to the server. The server offers its functionality through the c-Space REST API. In this case, the Content Access Infrastructure is used to retrieve and store multimedia content being presented as well as emotional information. Finally, a web application was developed that using the affective analysis tool allows the assessment of the experiments.

3.2 Emotion Recognition from Facial Expression

The developed facial expression recognition solution is based on Dornaika and Davoine [10] proposal for a simultaneous facial action tracking and expression recognition. This approach follows a two-step process. In the first step, the 3D head pose is estimated using an Online Appearance Model deterministic approach. In the second step, the facial expression features are estimated using a stochastic approach based on a particle filter.

As a deformable model of the face, a modified version of the Candide-3 model [11] is used for better performance in low-end devices. This model defines a set of faces and vertices, as well as different modifiers of those vertices based on the

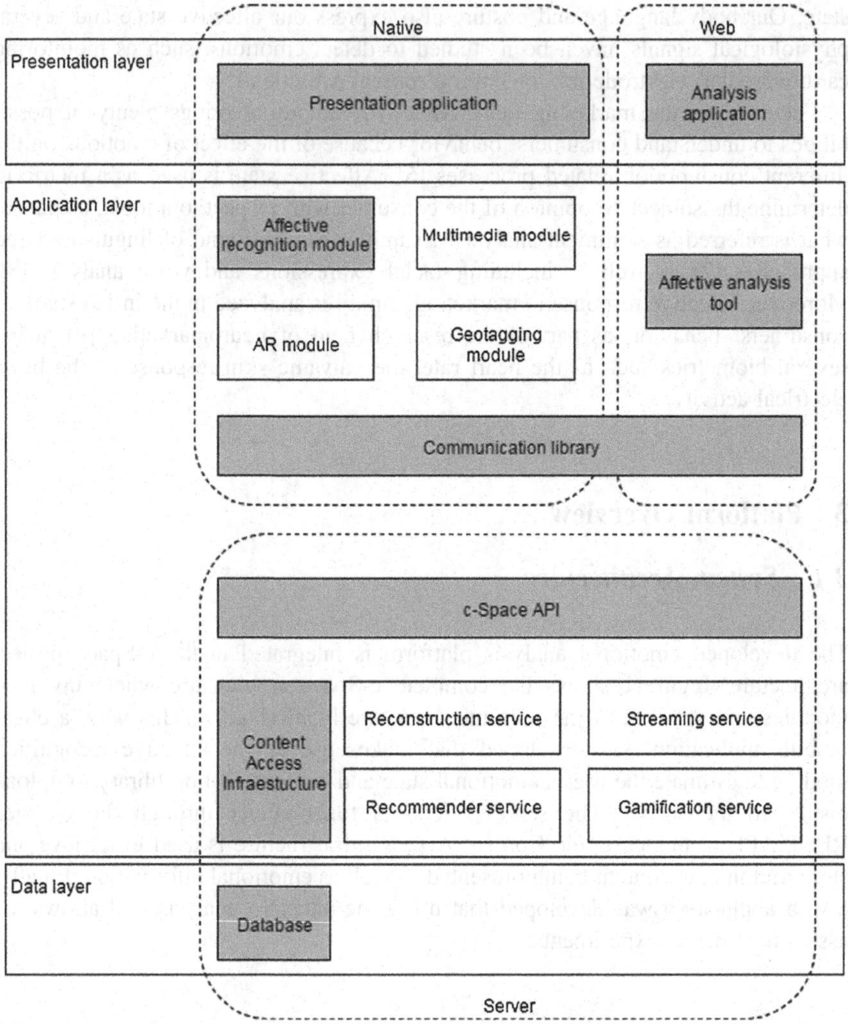

Fig. 1 c-Space architecture

characteristics of each person's face (i.e. head size, mouth width, eyes width, position and separation) and the current expression (primarily involving mouth and eyebrow movements).

In order to calculate the 3D head pose, an incremental tracking guided by a region-based registration technique is applied. Given that, the current image is warped with previous pose and it is compared against the reference appearance model, which is based on a multivariate Gaussian with a diagonal covariance matrix of the pixel intensities of the corresponding region.

Thus, the current solution is obtained by minimizing the Mahalanobis distance between the warped region and the current appearance mean, which can be solved using an iterative gradient descent method, setting the previous solution as a starting point. It is also noteworthy that the parameters of the reference appearance model are updated with the current observation, i.e., they evolve dynamically (Online Appearance Model).

At this point the current 3D head pose and the updated appearance model are completed with the facial expression detected in the previous frame, which is considered relevant because it is assumed that for each basic expression class there is a stochastic dynamic model describing the temporal evolution of the facial actions. Thus, given all this information, the goal of this second step is to simultaneously infer the facial actions as well as the expression label associated with the current frame. In order to combine all this information, a particle filtering approach is used.

The key idea of this technique is to represent the posterior distribution of facial actions and expression state given the appearance model by a set of random samples (called particles) with associated weights. This technique is divided into two stages: particle generation and particle evaluation. In the first step, several samples of facial actions and expression are generated to simulate all possible solutions (each sample corresponds to a different particle).

Facial actions are perturbed using a second-order auto-regressive model, while expression is perturbed using a manually predefined transition matrix that defines the probability of changing from one expression to another. The particle evaluation step, in turn, is responsible of assigning a weight to each particle and selecting the correct one.

In this case, the weight of each particle is calculated by warping the current image according to the facial actions defined in the corresponding particle and comparing this warped image region to the reference appearance model. Due to the recursive nature of the particle filter approach, particles of the current frame are propagated using the particles of the previous frames.

3.3 Analysis Tool

In order to inspect the emotional response of the user towards the marketing stimulus a visual analysis tool was developed. To allow gathering a richer insight of the consumer's emotional response different components of the affective state of the consumer are presented: (i) a continuous affect represented in a bi-dimensional space of valence and arousal, (ii) recognized prototypical emotional episodes categorized in 7 basic emotions (happiness, anger, disgust, sadness, fear, surprise and neutral), (iii) mood, understood as a diffuse affective state of low intensity but relative long duration. Figure 2 presents an image of the real-time monitoring of the consumer's emotional state where continuous core affect is represented by a line

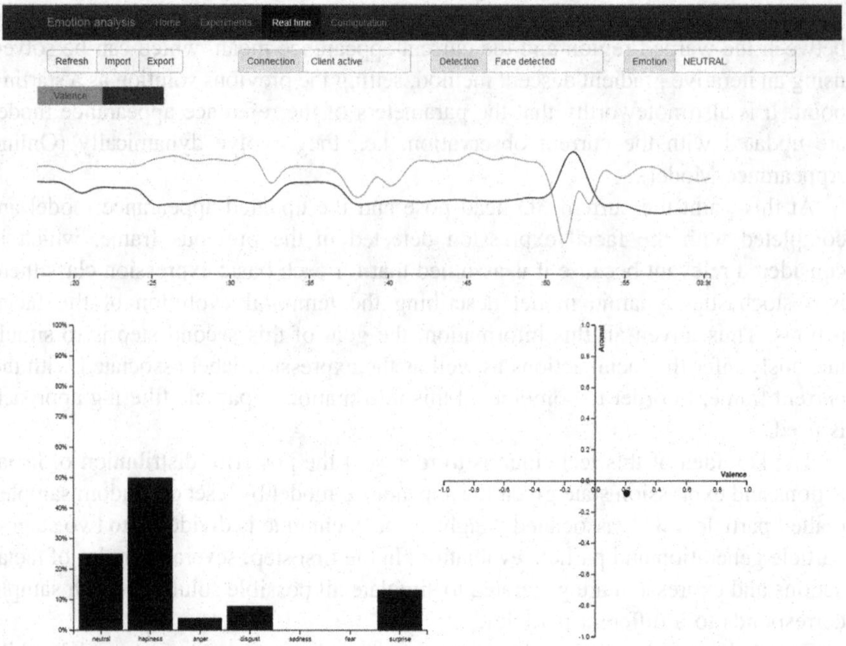

Fig. 2 Emotional analysis tool

graph on the top, prototypical episodes by an histogram on the left and mood by its position on the VA dimensions on the right.

The analysis tool allows to define experiments where a single content can be evaluated or comparisons can be established between different products or scenarios. Having performed the experiment with multiple consumers, the tool allows to visualize their combined response during the duration of the marketing content or as an accumulated value. Moreover, results are aggregated based on consumers' demographical information, allowing the user of the tool to study the differences between consumers' profiles.

4 Use Cases

The effectiveness of marketing is related to a two-fold process: first the ability to create a specific feeling or opinion about a topic, then the ability to use that opinion to induce a specific behavior. The scope of the uses cases is to further explore and research the role of human affects in a purchasing decision and the extent to which emotions are affected or influenced when the audience is being immersed in a given environment or is being exposed to specific pictures or scenes: what is the emotional response to a given stimuli? How does this response evolve over time? Is the opinion globally the same inside the target group or are there different responses to a similar scene? To which extent is the purchasing process rational versus emotional? What is the impact of emotions on the purchasing behavior? What are the requirements for the entire decisional process to be completed and which external factors can influence the decision?

4.1 Advertising Campaign

Marketing is an industry where novelty and innovation are playing a key role. Affective computing allows to investigate the point when viewers begin to tire of an advertising campaign, that is to say the point when further displays of the same marketing content will not only raise less attention and interest but also start inducing negative emotions. In some cases, the fatigue is related not only to the campaign but to the product itself, the design and branding of which should then be modified. Marketers and designers are therefore continuously trying the find the right balance between getting the most of their work by making it last and avoiding to offer something not enjoyable anymore to the customer, who will start feeling bored or even annoyed.

By measuring emotions expressed by a same panel of viewers over time, the platform supports decision makers in determining the timing after which a given visual should be changed as well as the effectiveness of the rebranding exercise which is being undertaken. The metrics provided by the system are feeding the decision to change something which may be well-known and efficient but also perceived as wear-out or not in line anymore with the expectations and the global "mood" of the targeted end users. The path followed by the Luxembourgish company Airboxlab illustrates this necessary process. After launching their connected indoor air pollution monitoring device, thanks to a focus group experiment, they could see that the name and design of the product were not considered appealing anymore in the long run. Moreover, the red lights triggering on the device when pollution was detected, as well as the global promotional campaign, proved to be deterrent in terms of product acquisition. They were anxiety-generating, when the emotion expected to be derived from the product in contrast was an increased feeling of security, inducing comfort and well-being. In the perception of the product, many details appeared to play a key role including the dominant colors being chosen for the campaign. A blue background had a calming effect while red or black tended to be associated with anger or fear, contributing to generate such emotions Fig. 3.

Fig. 3 Initial versus redesigned product campaign

4.2 *Price Negativity*

Beyond comparing the impact of different campaigns and designs over time, the emotion analysis platform can be used to compare the perceptions and feelings induced by a specific campaign or offer among different communities of stakeholders. Groups of end users from different countries are invited to interact with the same online marketing content. Reactions are being monitored and the results of the tests are communicated to the entrepreneurs. In some cases, a specific step during the purchase procedure is being perceived by the potential client as a negative "event" arising in the process. That is reflected as a peak or change in the graphs generated by the emotion analysis system. This is often related to the disclosure of the price to the customer. That emotion is not utterly out of the control of online marketers, who can make modifications not only to the pricing itself, but also to the way it is being displayed to the customer.

With the support of the emotional platform, marketers can test different processes online and optimize their product and price information, as well as the way those are being disclosed to the final client. Comments and feedback are being sought in case of divergent views among the communities of clients to better understand the context and the reasons why an offer is not leading to a sale and the tests are typically being performed on different geographical markets. The results allow to detect cultural specificities and differences of opinion in terms of product's perception. The value and features of a specific product can thus be adapted to expectations on a specific foreign market if the local clients tend to express surprise, sadness, disgust or anger when a specific offer is showcased, while the national consumers tend to be happier or neutral.

4.3 Technical Products

Advertising is at the crossroad between the provision of information about a product (news) and the induction of pure emotions and desires (art). Where to put the cursor is not an easy decision for marketing departments, especially when having to promote highly technical products. In order to efficiently promote some wood scanning machines for instance, to which extent should the emphasis be put on the functionalities of the machine and to which extent on the global industrial process, including the quality of the final output? Which combination between the two is likely to produce the best results in terms of marketing? Recording of emotions while performing technical demonstrations to clients and when performing less product-oriented campaigns allows to compare the two approaches and find the right balance. Tests performed with video material from LuxScan Technologies and the marketing department of its mother company Weinig demonstrated that industrial engineers, though highly interested in the technical performance of a given machinery, were also touched by a broader message, highlighting the value of their work and the beauty of the final products. It is therefore relevant for industrial marketing departments to produce a very wide range of marketing material, promoting brands in a generic way to create a positive image about their firm but highlighting as well the technical innovations embedded in their products and their enhanced performance versus competition. The emotional analysis platform can provide a very valuable measure to find the correct balance between both approaches.

4.4 Novel Interfaces

Last but not least, affective computing allows to understand to which extent the emergence of new displays in particular is having an impact on the perception of a campaign by the final end users. Marketing is fast changing, in line with the emergence of new media and consumer-oriented ICT technologies. The monitoring of emotions when the people are interacting with street furniture, augmented content and immersive environments allows to understand to which extent those new tools are enhancing the impact when being introduced in a campaign and to which extent they are perceived as complex or intrusive by the consumers, creating new challenges to overcome for the advertising sector. Beyond marketing and advertising, the emotion recognition system provides information on the way people react to a changing environment and more or less easily adapt to the emergence of new technologies. In a number of market segments, such as smart cities, smart homes, educational tools and entertainment, the use of emotion recognition in combination with new hardware offers a strong testbed for emerging services and consumer applications. Affective computing thus supports and enhances innovation and transformation in a number of industries, paving the way towards a world where the frontier between virtual and real environments is expected to be more and more blurred, and providing key information about the market readiness of emerging concepts in terms of consumer acceptance.

5 Conclusions

Understanding consumers' behaviors, attitudes and sentiments can provide a relevant tool for marketing, and what is more important from the consumer point of view, lead to better customer experiences.

In this paper we presented a platform based on Affective Computing technology, that being easily incorporated in the marketing evaluation procedures can provide valuable information of the consumers' emotional response. From the image of the consumer's face recorded for example by the frontal camera of a smartphone, the platform estimates a rich emotional state of the customer, allowing the marketer to gathered information of how a product or an advertising campaign is perceived.

Four different scenarios have been envisioned where the developed platform can provide very valuable insights of the consumers' behavior: (i) evaluation of an advertising campaign, (ii) study of perceived price negativity, (iii) balance of technical content and (iv) analysis of novel interfaces effect. The evaluation showed the capability of the platform to provide marketers with an effective tool on analyzing consumers' behavior.

The on-going project is starting its evaluation phase, where the presented platform will be validated in the context of the presented uses cases. Future work will be focused on extending emotional recognition from other modalities of emotional expression and on incorporating further analysis tools to allow marketers to share and compare collected information, enhancing their knowledge on consumers' behavior.

Acknowledgments The research leading to these results has received funding from the European Community's Seventh Framework Program (FP7/2007–2013) under the Grant Agreement number 611040. The authors are solely responsible for the information reported in this paper. It does not represent the opinion of the Community. The Community is not responsible for any use that might be made of the information contained in this paper.

References

1. Kotler, P., Keller, K.L.: Marketing Management. Pearson Prentice Hall (2006)
2. Fong, B., Westerink, J.: Affective computing in consumer electronics. IEEE Trans. Affect. Comput. 3, 129–131 (2012)
3. Picard, R.W.: Affective Computing (1995)
4. Ekman, P.: Basic emotions. In: Handbook of Cognition and Emotion. pp. 45–60. Wiley (1999)
5. van den Broek, E.L.: Affective Signal Processing (ASP): Unraveling the mystery of emotions. http://doc.utwente.nl/78025/ (2011)

6. Watson, L., Spence, M.T.: Causes and consequences of emotions on consumer behaviour. Eur. J. Mark. **41**, 487–511 (2007)
7. Ren, F., Quan, C.: Linguistic-based emotion analysis and recognition for measuring consumer satisfaction: an application of affective computing. Inf. Technol. Manag. **13**, 321–332 (2012)
8. Poria, S., Cambria, E., Howard, N., Huang, G.-B., Hussain, A.: Fusing audio, visual and textual clues for sentiment analysis from multimodal content. Neurocomputing **174**, 50–59 (2015)
9. Lee, N., Broderick, A.J., Chamberlain, L.: What is "neuromarketing"? A discussion and agenda for future research. Int. J. Psychophysiol. **63**, 199–204 (2007)
10. Dornaika, F., Davoine, F.: Simultaneous Facial Action Tracking and Expression Recognition in the Presence of Head Motion. Int. J. Comput. Vis. **76**, 257–281 (2007)
11. Ahlberg, J.: Candide-3. An updated parameterised face (2001)

Gamification as a Key Enabling Technology for Image Sensing and Content Tagging

Bruno Simões and Raffaele De Amicis

Abstract The advent of mobile phone with a multi-megapixel camera and autouploaders has democratised photography. Taking pictures and acquiring annotations is no longer an expensive task as it used to be. Yet performing these tasks in a systematically way is still very cumbersome for most users. In this paper, we outline two game mechanics that can be exploited for the purpose of large-scale image sensing and content annotation. Our first mechanic allows for better control over when, how and where people should acquire images. The problem with existent image providers is that their services usually do not cover the entire area of interest, are inaccurate or very expensive. Our second mechanism aims at making the annotation of crowd-sourcing images more engaging. It leverage on large end-user communities to annotate images while avoiding the pitfall of using annotations that are meaningful only to domain experts. Annotations that are not relevant to users' interests cannot be directly leveraged to enable search and discovery. A drawback of using crowdsourced annotations is that they have low agreement rates. Our approach aims at a finding a balanced agreement rate between pre-established annotations and those defined by users.

1 Introduction

The ubiquity of full-fledged sensing, computing, and communication devices like smartphones is paving the wave for new perspectives on how to accomplish large-scale sensing. The phenomenon of large-scale sensing, which is better known in the literature as participatory sensing, is an approach to data collection and analysis in which individuals and communities use their personal devices to acquire and explore specific aspects of their surroundings. The number of applications is quite vast and

B. Simões (✉) · R. De Amicis
Graphitech, Trento, Italy
e-mail: bruno.simoes@graphitech.it

R. De Amicis
e-mail: Raffaele.de.Amicis@gmail.com

G. De Pietro et al. (eds.), *Intelligent Interactive Multimedia Systems and Services 2016*, Smart Innovation, Systems and Technologies 55,
DOI 10.1007/978-3-319-39345-2_44

can range from environmental issues to culture. For example, we could easily digitalise objects (e.g. animals, buildings, etc.) and even entire cities over time if we can collect images and videos that cover every single corner of a town, at different times of the day and year [1, 2]. The impact of 4D reconstructions would go much beyond a common digitalization process. It would facilitate the creation of ubiquitous virtual worlds and pave the way to new ways of user experience. It would also contribute to the preservation and experience of daily life moments like when a building is torn down, repaired or constructed, outdoor festivals, etc. This data could also feed by domain-specific applications to support users in monitoring and analysing, for example, the infrastructure of a building, or simply for the benefit of individual digital content creators.

A number of on-line services that aim at providing access to large structured (e.g., Bing Maps, Google Street View, etc.) and unstructured collections of images (Flickr, Picasa, Panoramio, Facebook) already exist. However, these repositories have several drawbacks if used for the aforementioned scope. One the one hand, acquiring imagery from structured repositories can be financially expensive—structured repositories are often the core business of commercial enterprises. Additionally, data is only available within some areas or it captures only certain features of the environment, e.g. streets. These repositories are also updated very sporadically, thus, these image collections become obsolete quite easily; similarly, these image collections might target a limited number of moments of day or of the year, e.g. satellite imagery. On the other hand, we have various unstructured repositories that are maintained by users. One flaw in these services is that only attractive landmarks or points of view are well-represented, against everything else that is very sparsely captured by users.

For a 4D reconstruction to be successful we need to find a sustainable approach that facilitates the capture of all those 'missing' view angles, along the time span we want to reconstruct, and without requiring special equipment [1, 2]. There are many successful cases of crowdsourcing applications. Still, we do not have a good understanding of how to motivate massive crowds to participate. Well accepted strategies that influence user behaviour towards doing a specific action are individual motivation, serious games, and gamified activities. Serious games are "games that do not have entertainment, enjoyment or fun as their primary purpose" [3, 4]. Gamification is primarily characterised by the fact that the game, if defined, is always secondary to the tasks that have to be performed, e.g. adding a points system, peer pressure, leaderboards, and things that normally would not be considered. In all scenarios, effective gamification exploits the user context to provide motivation specific to the situation, instead of simply displaying badges and leaderboards [5, 6].

In this paper, we introduce a game platform to crowdsource the gathering of videos and pictures. Our key contribution is the game design that is proposed. In our game, the simple action of taking pictures serves two purposes: players perceive it as a mechanism to capture magical creatures, collect artefacts, and compete against other users; from our perspective it represents a strategy to gather pictures and collect metadata. The outcome of the game, which is transparent to the gameplay, is a set of images spanning over large areas, from multiple angles, and a set of 4D models reconstructed from those images. Crowdsourced annotations are equally supported.

2 Key Research Challenges

In this section we describe research challenges that we took into consideration while designing our game.

1. Dealing with Sparse Sensing. Data collected with mobile phones is typically randomly distributed over time and space, and therefore in most of the cases it is incomplete in terms of coverage. The true challenge, thus, is to maximise the amount of data that can be collected for a given spatio-temporal phenomenon. With this objective, we have implemented an autonomous algorithm that uses a spatial grid and users' activity to compute the probability of generating valuable artefacts. Indirectly, this strategy aims at rewarding players that take pictures of places that are less active or attractive.
2. Evaluating Data Trustworthiness. There is also a strong need for verifying and confirming that any data uploaded is simply not fabricated, i.e. it was collected at the claimed location. Hence, mechanisms to detect malicious users have to be investigated. In our framework, we have implemented a few algorithms to verify the authenticity of user-generated content. Analysis of GPS user activity is the first among several. The user is allowed to take pictures faster if the GPS is active. A second strategy is the use of computer vision. First we try to match features of that image with features of other images. This can help us to identify the possible location of the picture. The second method is to identify elements (e.g. monuments, architectural style, etc.) in the picture. This latter approach gives us 14 % chances of spotting fake pictures for known landmarks.
3. Maximising User Engagement. Participatory sensing is an activity that might not provide enough motivation per-se. In order to maintain the players engaged in this kind of prolonged activity it is necessary to integrate additional mechanisms [7]. In the next section we explain the mechanisms that were implemented to maximise user engagement.

3 Related Work

One of the objectives behind the creation of Serious Games (SG) is to keep players entertained while they pursue a greater purpose. The purpose of the game may vary from simple tasks like data collection to advanced topics where the objective is to teach a concept. The design of many serious games relies on the work of von Ahn, who developed together with other authors, Peekaboom [8] and ESP game [9]. But, there are other good examples [10, 11]. The ESP game was designed for the purpose of labelling random Internet pictures. The game is grounded on the idea that whenever two players visualise the same image they should come up with textual tags that match. In this game, players are paired randomly without any means of communication; therefore to earn points, they must find tags that most people would associate to the image. Peekaboom, on the other hand, involves defining the location of objects.

Foldit [12] is an experiment that serves to exemplify the relevance and efficiency of using games that take advantage of human' innate spatial reasoning abilities, to solve problems that computers fail to resolve. In this game, people can help research scientists to solve a protein-folding problem that had baffled them for more than a decade. The objective of the game is to find, for each digital 3D protein structure, the most tightly packed configuration.

Astro Drone [13] is another crowdsourcing effort based on data collection through a game. The aim of the game is to play a spaceship simulation game in which players have to control a spacecraft that has to fly close to a comet and then release a lander on it. The game sends the data extracted from the camera of the drone to ESA to be then used in the study of new automatic control systems (e.g. obstacles avoidance and docking).

PhotoCity [14] is a game with the purpose of collecting photos. This game relies on small groups of expert photographers highly motivated to acquire pictures that are useful to the reconstruction of buildings. Players are rewarded with castles and flags as they contribute to 3D replicas. The authors used vision techniques to reward only those players with useful input. The drawback of this strategy is that this game is designed to reward only those players with exceptional skills in photography, as a way to compensate for a limitation in their 3D reconstruction framework. Consequently, the game fails to implement a mechanism that can attract large communities and that can maintain them engaged independently of their skills or interests.

Finally, EyeSpy is another relevant game that proposed by Bell et al. [15]. The objective of this game is to collect pictures and tags that can be functional in a navigational context, for example, to give directions to someone based on landmarks that can be easily recognisable. To play the game, we just need to walk around, take pictures, and insert tags. Additionally, the player can geolocate and tag pictures collected by other players.

4 Methodology

Our game platform is designed to provide four types of game experience: acquiring artefacts, taking quizzes, exchanging artefacts and competing against other players. Artefacts are collectable cards that are either generated automatically by our platform or manually defined by an expert user. Optionally, these artefacts might have an educational scope. Capturing artefacts requires players to go outdoors and use their smartphones in AR mode, see Sect. 4.5. There are two strategies for capturing artefacts: users take a picture of an augmented artefact (visible through the display) or they take a "blind" picture to unveil hidden gems. An artefact is unlocked only when the user uploads the picture to our backend infrastructure—an action that within the game corresponds to clone the creature from its DNA artefact. We implemented this strategy of unlocking creatures at home or via WIFI to avoid exceeding their data plans. At the backend, we assess the veracity as well as the added value provided by the picture, which is a parameter in the computation of the set of abilities that is assigned to the creature.

In this game, artefacts might be abundant in quantity but they are not in quality. As we will explain later, these artefacts can be used in competitions that give valuable artefacts better chances of wining or evolving. This mechanic was integrated to push players into taking pictures even if they do not need more artefacts. It also generates dynamics that appeal to explorers user-type whom are looking to complete their list of artefacts, and to killer user-type that want to gain the upper hand against others, e.g. influence points.

Fighting pits two players one against another for the supremacy of having more influence points or gaining control over a place. Players will have to use artefacts they have captured in order to win. There are two different ways of engaging in a battle: (1) the player attacks directly the other player using the online game platform or (2) the game automatically initiates a fight between two players when a new artefact is collected. The latter, called proximity battle, forces players to fight each other if they are capturing artefacts in the same zone. Players can withdraw by sacrificing a certain number of artefacts. Another way of competing against other players is by taking quizzes. Quizzes can be unlocked by when two players tag an image with a common keywords. To facilitate the process the game provide players with clues of other players.

In the next sections, we describe the most crucial gameplay elements in conjunction with the effect they should generate on the collection of pictures during the intervention.

4.1 Distribution and Elemental Affinity

In our game, artefacts are aligned to one or more fictional elements that have been selected for the game, see Fig. 1. Environmental affinities change the probability of finding creatures of those elements, making it easier for elements with a stronger affinity and harder for elements with a weaker affinity.

However, this property does not change the fact that some elements are incompatible with certain environments. For example, fire-based artefacts are unlikely to be spotted inside lakes, see Sect. 4.3 for more details. Hence, affinities is a tools that game designers can use to control the distribution of artefacts within a given environment.

4.2 Distribution and Content Value

To further push the interaction between players and to balance the way players can earn influence points, cities (physical space) and towns are automatically sub-divided into zones.

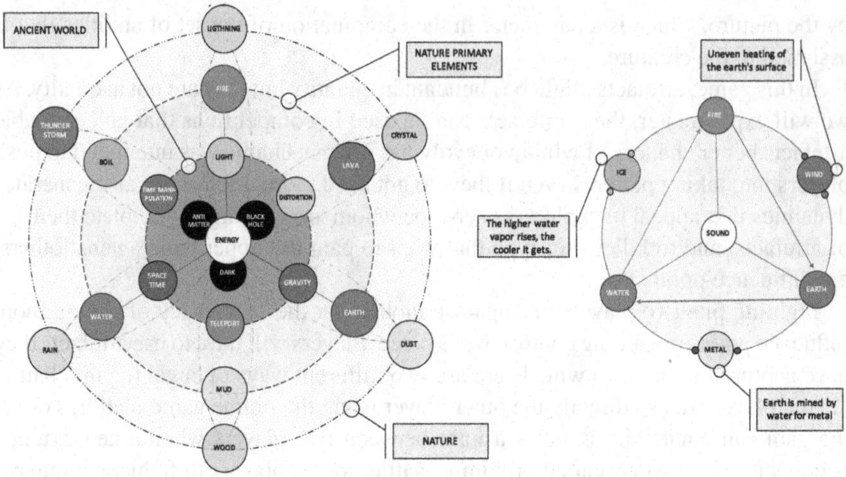

Fig. 1 List of possible elements/strengths and weakness

The size of each zone in the spatial grid is inversely proportional to the number of pictures acquired for the area, following the concept of a quad-tree. The purpose of this mechanism is to balance the spatial distribution of pictures and consequently to control the way players can earn points. However, players shall perceive it differently: when a player captures an artefact the control of that player in that specific zone is increased (influence points). This new probability of finding an artefact is presented as a decrease in the population of artefacts (localised). The population is automatically restored over time, based on players behaviour or on external factors (e.g. an happening, a festival, etc.). Restoration over time has designed for continuous player involvement: newbies can always find a place to go for hunting, while veteran players are required to return regularly to their zones in order to maintain their local influence points. This strategy attracts not only the attention of killers and achievers player types [16, 17], but also of achievers and explorers that are interested in finding rare gems where elements have not been depleted. The location and bearing of a picture within a zone are two of most relevant variables used to compute the probability of finding valuable artefacts, see Section. An internal algorithm was implemented to balance the scores by comparing these properties at local and global level—players always compete with someone but preference is given to other local players.

4.3 Habitat and Elemental Affinity

To maximise the realism of our game and to attract explorer-type players, we have implemented a service that constrains the probability of finding artefacts creatures in certain areas. Hence, marine-related artefacts are most likely to be found nearby the sea or near to the habitat of specific plants species, e.g. useful for biology-related games. The game procedure is fully automatic, however, at the moment it supports only the European territory.

The habitat mapping service is engineered as a two layers data service. The first data layer exposes the Corine Land Cover (CLC) of Europe as a service. The CLC nomenclature aggregates 44 land cover classes in a three-level hierarchy. Five main categories are "artificial surfaces", "agricultural areas", "forest and semi-natural areas", "wetlands" and "water bodies". The second data layer requests, to the first layer, the CLC code of a given location and then computes the list of compatible elements to that nomenclature.

The number of new artefacts that is generated depends on the relevance of the area. Our system is capable of generating thousands of different artefacts and descriptions without human intervention. During the creation process the system takes into account the strategies described above. Every time a new card is generated, we have to update the database of the living artefacts. See Sect. 4.5 for a few examples of automatically generated cards. Limited and thematic editions will also be considered.

4.4 Artefacts Lifecycle

Given that many of our mechanics are based on probabilistic algorithms, both for generating and combining artefacts, we have implemented an ageing effect—artefacts will age and become inactive—to prevent the use of singularities in warfare activities. Artefacts will age 1 year (artefact's calendar) for each week (player's calendar). After a certain number of years ($\exists n \in \mathbb{N}, 2^n = i$, where i is the age of the card in weeks) the artefact goes through an elemental evolution. At each evolution, the player can decide which element of the artefact to level-up. An artefact can live up to 110 years in average. Players can capture artefacts with ages raging from 10 to 40 years old. Under special conditions, e.g. legendary artefacts that correspond to high-value pictures, artefacts can live up to 390 years. The maximum lifespan of an artefact is also shortened after each battle by 1 year. However, as a recompense, artefacts evolve whenever the condition $\gcd(n, F_{n+1}) > 2$ is satisfied. The sum of all battles is represented by n; F is the Fibonacci Sequence and gcd is the greatest common divisor.

4.5 Game Platform

Our online game platform uses artefacts captured by the user as soldiers that can conquer places owned by other players. Players initially have to designate where

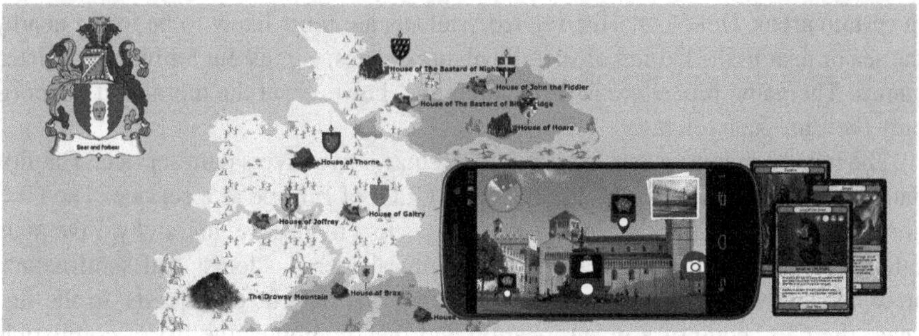

Fig. 2 Online game platform, augmented reality game, and collectible artefacts

their physical home is located (latitude, longitude). A house and a coat-of-arms is automatically assigned to the player, see Fig. 2.

Although maps are a 1 to 1 match with the geography of real countries they are re-styled to look more like a game. When users click in the icon representing they house, they access all the information on the artefacts they own. Direct attacks on other players can inflict injuries on both sides or even casualties. Injuries is a feature game feature that attempts to support offline players during attacks. Artefacts can recover from injuries after a given amount of time. Hence, battles can take place both when the two players are online or when just one of them is online. If the defender is not online, then the odds of defending the house are much greater.

Given that competition can be intensive, a virtual stock market with automatically generated news was also implemented. The value of each player stock in the market depends on many factors like market feeling, if the user is progressing, losing assets, and so fort. Players can use this mechanism to earn money which can also be used to buy artefacts.

5 Evaluation

Our evaluation of the framework relies on three experiments. The objective of the first experiment was to collect images for reconstruction purposes. The goal of this experiment was to compare the user engagement when using our serious game approach and a simple crowdsourcing application that simply takes and uploads pictures. Two groups of 120 contacts were randomly invited to participate in the experiment. The number of users that accepted to play our game were 36 (60 % of the number of users invited) against 20 users who accepted to use the crowdsourcing application (33 %). Hence, users were clearly more willing to participate if the activity provides some kind of entertainment.

After one week, users collected a total of 14.056 images. Users using our serious game collected an average of 48 images a day and those using the alternative

Table 1 Descriptive data of two tagging datasets

	Not gamified	Gamified
Number of images tagged	6984	432
Number of unique tags	3744	352
Average tagging time per image (in s)	9.1980	17.97
Average number of images per tag	1.95	1.12
Mean (median) tag intuitiveness	4.1	4.8
Average number of images (per user session)	18.8	26.2
Number of users	20	36

approach an average of 19 images. Although, our data confirmed that users playing in a game environment were more willing to take pictures as well as pictures from less conventional angles (as suggested by the game), the results of both crowdsourcing activities were not enough for a successful large-scale reconstruction.

In our second experiment we examined how games can influence user-generated tags. In this experiment we asked users to tag random images from our dataset of 14.056 images. Table 1 summarises the results of our two experiments. As the data shows, the taggers who were playing spent significantly more time to decide the image tag. This also provides an evidence that tags created in a game environment were more diverse than those assigned under normal tagging conditions. Another aspect that was investigated was how often tags were reused across different images. Our result demonstrate that tags obtained using gamified applications are reused less. This is probably a consequence of our game mechanics that incentive players to use more diverse tags. This diversity indicator also implies that the number of images that is retrievable with a single tag is smaller, thus, making the process of finding particular images more pecular. Hence, we need to investigate better the implications of tag diversity by analysing if tag diversity and tag intuitiveness have some kind of correlation.

For that purpose, we conducted a last experiment were users were asked to rate image tags from 0 to 5 based on their intuitiveness, see “Mean (median) tag intuitiveness”. We concluded that tags produced in a game context were more meaningful than those created under normal conditions. This is partially justified by the fact in the game environment users have to pick tags that can be easily guessed by other users.

6 Conclusion

In this paper we describe two distinct gameplay designs. Our first gameplay design aims at motivate players to take pictures, so they can be used for digital reconstruction purposes. The game was designed to attract different types of players' personality. We archived this by integrating the use of different game play strategies, e.g. engaging in combats, collecting cards, exploring physical spaces, including social interactions, etc. The design of mechanisms to deal with sparse sensing and to evaluating data trustworthiness were among our biggest considerations. Our second game aims at motivate users to tag content and keep them engaged for longer periods of time.

In addition to our objective of image sensing, our game platform can be used to create educational games that have strong links to physical locations. For example, payers can collect educational cards on specific types of plants when walking near to areas where those plants are common, see Sect. 4.3. This is a feature that has been appreciated by some educational institutions.

As future work, we need to investigate how to improve the security and reliability of the framework and how to handle intellectual property and privacy issues, e.g. reconstruction a person without the proper permission.

References

1. Simões, B., De Amicis, R.: Digital earth in a user-centric perspective. In: 2014 Fifth International Conference on Computing for Geospatial Research and Application (COM. Geo), pp. 47–48. IEEE (2014)
2. Simões, B., Aksenov, P., Santos, P., Arentze, T., Amicis, R.D.: c-Space: fostering new creative paradigms based on recording and sharing "casual" videos through the internet. In: 2015 IEEE International Conference on Multimedia Expo Workshops (ICMEW), pp. 1–4, June 2015
3. Michael, D.R., Chen, S.L.: Serious Games: Games That Educate, Train, and Inform. Muska & Lipman/Premier-Trade (2005)
4. Djaouti, D., Alvarez, J., Jessel, J.-P., Rampnoux, O.: Origins of serious games. In: Serious Games and Edutainment Applications, pp. 25–43. Springer (2011)
5. Marczewski, A.: Gamification: a simple introduction. Andrzej Marczewski (2012)
6. Zichermann, G., Linder, J.: The gamification revolution (2013)
7. Ueyama, Y., Tamai, M., Arakawa, Y., Yasumoto, K.: Gamification-based incentive mechanism for participatory sensing. In: 2014 IEEE International Conference on Pervasive Computing and Communications Workshops (PERCOM Workshops), pp. 98–103. IEEE (2014)
8. Von Ahn, L., Liu, R., Blum, M.: Peekaboom: a game for locating objects in images. In: Proceedings of the SIGCHI Conference on Human Factors in Computing Systems, pp. 55–64. ACM (2006)
9. Von Ahn, L., Dabbish, L.: Labeling images with a computer game. In: Proceedings of the SIGCHI Conference on Human factors in Computing Systems, pp. 319–326. ACM (2004)
10. Von Ahn, L., Ginosar, S., Kedia, M., Blum, M.: Improving image search with phetch. In: IEEE International Conference on Acoustics, Speech and Signal Processing, 2007, ICASSP 2007, vol. 4, pp. IV–1209. IEEE (2007)

11. Eickhoff, C., Harris, C.G., de Vries, A.P., Srinivasan, P.: Quality through flow and immersion: gamifying crowdsourced relevance assessments. In: Proceedings of the 35th International ACM SIGIR Conference on Research and Development in Information Retrieval, pp. 871–880. ACM (2012)
12. Cooper, S., Khatib, F., Treuille, A., Barbero, J., Lee, J., Beenen, M., Leaver-Fay, A., Baker, D., Popović, Z., et al.: Predicting protein structures with a multiplayer online game. Nature **466**(7307), 756–760 (2010)
13. de Croon, G., Gerke, P.K., Sprinkhuizen-Kuyper, I.: Crowdsourcing as a methodology to obtain large and varied robotic data sets. In: 2014 IEEE/RSJ International Conference on Intelligent Robots and Systems (IROS 2014), pp. 1595–1600. IEEE (2014)
14. Tuite, K., Snavely, N., Hsiao, D.-Y., Tabing, N., Popovic, Z.: Photocity: training experts at large-scale image acquisition through a competitive game. In: Proceedings of the SIGCHI Conference on Human Factors in Computing Systems, pp. 1383–1392, ser. CHI '11. ACM, New York, NY, USA. http://doi.acm.org/10.1145/1978942.1979146 (2011)
15. Bell, M., Reeves, S., Brown, B., Sherwood, S., MacMillan, D., Ferguson, J., Chalmers, M.: Eyespy: supporting navigation through play. In: Proceedings of the SIGCHI Conference on Human Factors in Computing Systems, pp. 123–132. ACM (2009)
16. Bartle, R.: Hearts, clubs, diamonds, spades: players who suit muds. J. MUD Res. **1**(1), 19 (1996)
17. Stewart, B.: Personality and play styles: a unified model. Gamasutra. http://www.gamasutra.com/view/feature/6474/personality_and_play_styles_a_.php (2011)

Bridging Heritage and Tourist UX: A Socially-Driven Perspective

Paola La Scala, Bruno Simões and Raffaele De Amicis

Abstract The paper illustrates the potential of smart-phones as a medium of exchange of memories and experiences. Our application aims at providing diverse types of cultural user experience: to enable tourists to explore new places from a social-driven perceptive; to support new forms of connection and interaction between users and information (data exchange, contents sharing, feedback); to compose interactive narrative conveying the richness of information of interest to the user; to allow users to experience the narrative and underlying physical environment as a mixed-reality experience while allowing for deeper, context-specific exploration at any time through AR system.

1 Introduction

Social transformations conveyed by technological evolution have been in the origin of intense changes in mental attitude and in the ability to interpret cultural heritage. Advanced technologies for data exploration are increasingly becoming part of our reality. Interaction and multimedia are instruments that can facilitate user involvement; if well designed allow the audience to involve into a complex experience of learning and research.

Applications for portable devices such as smart-phones can be of great value in exploratory experiences of places. In effect, diverse context-aware applications can be combined in order to find an effective methodology for exchange and visualization

P. La Scala
University of Palermo, Palermo, Italy
e-mail: paola.lascala@unipa.it

B. Simões (✉) · R. De Amicis
Graphitech, Trento, Italy
e-mail: bruno.simoes@graphitech.it

R. De Amicis
e-mail: Raffaele.de.Amicis@gmail.com

G. De Pietro et al. (eds.), *Intelligent Interactive Multimedia Systems and Services 2016*, Smart Innovation, Systems and Technologies 55,
DOI 10.1007/978-3-319-39345-2_45

of visitors' memories, e.g. through mapping techniques. Additionally, the combination of different tools and technologies can also be used as method for stimulating the interactions between users and cultural heritage [1].

Due to proliferation of technologies that are able to overcome the obstacles of time and space in the cultural heritage storytelling, one would expect these tools to support them during their visit to a city, to gain an understanding of its history, and to communicate and share effectively their experiences with other users. However, some technological advances cause people to be overly distracted and increasingly isolated. Surely, technology has a significant impact on what it means to be social and interactive [2].

In recent years, the use of mobile application guides has become quite popular [3]. Despite the many technologies that exist for this purpose, there are still open challenges with respect to user's interaction (both with people and predefined content) and visitor experience. Hence, we need another user-perspective that actively involves visitors physically, intellectually, emotionally, and socially.

A considerable part of the literature is focused on the implications of digital media in our contemporary society, while a few researches are more interested in the way new technologies influence user interaction. In this sense the challenge is the realization of a mobile application that is user-friendly and that is capable of being a learning stimulus to its visitor. In fact, the multimedia should create or straighten a real emotional connection between visitors and different points of interests underlying in the narrative [4].

The paper illustrates the potential use of smart-phone technologies, as system for the exchange of memories and cultural experiences. The c-Space cultural heritage application aims at providing four different types of experience on cultural paths: to enable tourists to explore and learn about new places from a social perceptive; to support new forms of connection and interaction between users and information (data exchange, contents sharing, and feedback); to compose interactive narrative conveying the richness of information that is of interest to the user; and to allow users to experience the narrative and underlying physical environment as a mixed-reality experience while allowing for deeper, context-specific exploration at any time through AR system.

2 Related Work

The concept of 'interaction' is still not easy to define despite the various attempts made by authors from many fields. The interaction, widely understood, should have a specific informational and educational purpose and it must be conducted on a simple communication process that is affected by what the visitor does, what he learns and even what he feels. Designing an interactive system, means to integrate communication purposes like 'what visitors want to learn', with behavioral aims, 'what visitors want to do', and, finally, the emotional purposes, 'what visitors want to feel': the Mclean triad learn-do-feel [5]. In this paper, we use the term interaction when we refer to interactive experiences as those which involve visitors into an activity that is simultaneously physical and conceptual. Much of the most important learning happens through social interaction including forms of communication which take place with various technologies. Interaction between people and content, insofar as they address learning, are also relevant.

Then, what is social interaction? We recognise Rummel's definition according to which "social interactions are the actions or practices of two or more people mutually oriented towards each other's selves, that is, any behavior that tries to affect or take account of each other's subjective experiences or intentions. This means that the parties to the social interaction must be aware of each other -have each other's self in mind. Social interaction is not defined by type of physical relation or behavior, or by physical distance. It is a matter of a mutual subjective orientation towards each other?" [6].

In the field of tourism, considering the strong host-guest relationship, residents play a fundamental role in enhancing city heritage [7].

In this sense, Armenski et al. state that the promotion of travel for pleasure between countries contributes not only to economic growth but to the interchange between citizens which helps to achieve understanding and co-operation [8]. Mutual social and cultural interaction between resident and tourists is inevitable. Thus, the quality of interaction between tourists and residents contributes to both tourists experience and perception of the visited destination.

3 Methodology

The understanding of the overall narrative when following an exploratory experience can be facilitated by the interaction with others individuals. A mobile application could help different type of users to experience the city and its history, and to forge connections between tourists, citizens and touristic employers. It also means that the social interaction could happen at different interaction levels, and between people and content.

As Garcia-Crespo et al. argue the tourism industry needs to expand its comprehension systems with dynamic technologies, which can offer interactivity and entertainment to visitors [9]. Sophisticated applications and Augmented Reality can change the way people explore cities allowing tourists and inhabitants to choose the most effective and emotionally stimulating code of communication. Thus, the simultaneous use of multiple languages and tailored content helps to better explore the city and to absorb diverse information about it. Figure 1 depicts an example of an application that augments physical spaces with historical images.

Our research work started from an initial assumption that there are two different types of interaction involving visitors: the interaction between users and the interaction with content. Hence, the application should allow a multilayer learning mechanism. In fact, the different levels of interaction, being based on an idea of progressive knowledge, are complementary but at the same time can be exploited individually.

Fig. 1 Augmented history

3.1 Shared Tours Experiences

One of the most important aspects when experiencing a cultural path are: personalisation of information and localisation of information. Existent visitor applications for city tours now follow one of two well established conventions. Either they provide visitors with a particular path through the city, a directed tour, introducing the points of interest in a fixed, but logical order; or they allow visitors to move freely, by selecting the order they view particular points of interest. In both cases, the information provided to all visitors is the same, regardless of their particular interests and their personal capabilities [10]. Therefore pre-established tours might seem static contrary to the concept itself of city discovering. However, if we look at visitors as an active community then we might get better insights on the meaning of dynamic tours.

Advances in technology mean that both of these limitations in interaction with multimedia guides can now begin to change radically. Personalisation in particular becomes important because visitors may feel overwhelmed by the amount of information provided to them by the guided tour. The personalization concept is not limited to the content of the application, but can also includes personalisation of the system that is shared with other users. In fact one method that tourists use to answer their questions is to request help or information in specialized online forums. Tourism is a influential social activity and it might be an added value if tourists can share their path experiences with others users that have similar interests [11].

The first feature of our framework allow users to easily access and explore the city using routes that were suggested and shared by other users. When a user decides for a specific cultural, he/she becomes part of a sub-community that can communicate anonymously in real-time. Possibilities for this type of interaction are huge. For example, a visitor traveling alone could easily communicate to users with similar plans and eventually join them. Undoubtedly it could not left out the aspect of social sharing. Then at the end of the tour, the visitor can share content, for instance, recorded comments on POIs on social platform which are “popular infrastructures for communication, interaction, and information sharing on the Internet" [12]. In the emergent social web, interaction and content has become an immersive experience.

3.2 Bridging Interest-Based Communities

A big part of what makes a place special is its people. Confronting oneself with other persons, even if they do not speak the same language, can show to visitors the real character of the city they are traveling through [13]. It is well known that, despite the amount of information that is provided by an application, the user can have doubts about the city he/she is visiting. A system that can bring together inhabitants, tourism professionals, and visitors can support and create awareness of the local heritage in a sustainable way because it can also be used as a tool for development and economic growth [14]. Nevertheless, it is important to honour the relationship between tourists, local businesses and residents, whilst also seeking to improve the tourist experience. In fact, a truly innovative system would allow regular users and professionals (e.g. a touristic guide) to help visitors personalizing their tours, as well as to help citizens to plan a culturally-coherent visit along hidden corners of the city. In our heritage platform, users can launch the "Live user support" application and gives them access to an interface that is personalized to their specific profile, e.g. visitor, tourism profession, citizen, etc. Both user-driven interaction and content-driven features encourage users to socialize through both the technology of the system and the content of the conversation. Figure 2 provides a glimpse on the user interface.

Fig. 2 Connecting interest-based communities

3.3 Augmented Cultural Experiences

Augmented Cultural Experiences is the third layer of interaction of our cultural heritage platform. Tourism and cultural heritage understanding often have an awkward relationship [15]. It is not always easy for visitors which do not know the history of a city to interpret it, in particular for those parts are no longer recognizable. Nevertheless, the easiest way to narrate a specific storytelling about the city is through historical images.

3.3.1 Image-Based Narratives

If narratives include information on physical events then it is also important to contextualize the images of the story in a temporal geographical background. The interaction between pictures and storytelling can facilitate the narrative and visualization of the history itself, but it can also facilitate the association of historical facts with places [16]. However, the visualization of historical images per-se does not bring any added value if they are not inline with the interests of the visitor. A desirable interactive system should be able to present storylines in a way that is meaningful to the user, in order to maintain engagement. A valuable input on visitor's interests is the list of POIs that were selected by the visitor during the tour creation. These points of interest are elements that we know for sure that are of interest, but this information by itself is not enough to create storylines that break with traditional ways of creating storylines. For this purpose, our mobile application makes use of the dynamic needs of the user that, for example, take into account factors like saturation and cultural themes, see Simões et al. [17] for further details. Storylines are then presented to the user as an animation of historical images that were extracted from a semantic storyline network that was created specifically for the visitor, see Fig. 3.

Fig. 3 Immersive storytelling system

Interaction between the device and the user is also used to refine and propose new storylines when conditions are ideal, e.g. physical location has a deep connection to the proposed storyline. Hence, this approach of presenting historical images is a break-through for traditional systems because not only images are selected accordingly to visitor interests but they are also ordered according to an historical logic.

3.3.2 VR/AR-based Narratives

VR/AR systems are an excellent medium for an interpretative and experiential learning experience. Hence, we decided to introduce augmented reality (AR) into the storytelling not only as an interactive technology but also as a way to engage wider audiences. AR is not a new concept, however, factors such as increased smart-phone ownership and constant use of mobile devices to experience cultural heritage have contributed to the recent interest and growth of AR [18]. In a heritage scenario, AR can be used "to help tourists in accessing valuable information and improving their knowledge regarding a touristic attraction or a destination, while enhancing the tourist experience and offering increased levels of entertainment throughout the process" [4]. AR can also be used to create interactive spaces, which Hornecker defines as "interactive spaces rely on combining physical space and objects with digital displays" [21].

Our mobile framework uses the AR framework described in [19, 20] to design and provide tourists with immersive storylines experiences. Figure 3 shows an example of how digital content (medieval reconstruction of the city) can be used to overlap the real view of the city. This is just an example of a multimodal interaction that can play an important role in the augmentation of the storylines created from sets of images, as it allow users to freely look around instead of being limited to a static perspective.

3.3.3 Post-visiting Experiences

Lastly, we investigated the importance of post-visiting experiences. Post-visiting experiences play a fundamental role in tourist's overall experience. In effect, tourists, especially those who travel in groups, like to share their experiences at the end of their journey with those that were not present. The use of photographs represent is therefore a very important medium because of their capability to illustrate stories and to engage those who were and were not present. Hence, by providing post-visiting experiences we extend the enjoyment of a tourist visit way beyond the visit itself.

Our application supports the creation of post-visiting experiences. When users arrive to predefined POI's, they become automatically eligible to post-visiting bonus. Post-visiting consist of historical pictures that were prepared for this specific scope, and that allow users to insert themselves into the picture (users becomes part of story).

4 Conclusion

In this paper, we discuss the use of smart-phone applications as a medium to promote the interaction between tourists and citizens, and as a tool to create community groups that preserve and promote the culture of a city. For this purpose, we have created an application that exploits digital content in many different ways and for many different kinds of interaction and learning: to describe the territory, to provide information on some specific place, to promote interaction between users and cultural heritage, to inform tourists about cultural places, to create storytelling experiences, and to keep under control the impacts of tourism (social, cultural, environmental) on territory.

As future work, we plan to evaluation the following aspects: user satisfaction—the attitude of users toward using the application; simplicity—the comfort with which users find a way to accomplish tasks; comprehensibility—how easily users can understand content presented on the mobile device; perceived usefulness—to what extend the application has met its implementation objectives; system adaptability—how the system adapts to the user requirement. In our future evaluation, we will consider the involvement of different actors (e.g. tourists, cultural heritage institutions and employers) as an important aspect in the understanding of the user experiences produced.

Acknowledgments This research has been supported by the European Commission (EC) under the project c-Space. The authors are solely responsible for the content of the paper. It does not represent the opinion of the European Community. The European Community is not responsible for any use that might be made of information contained herein.

References

1. Rolando, A., Scandifflo, A.: Mobile applications as tool for exploiting cultural heritage in the region of turin and milan. ISPRS-Int. Arch. Photogrammetry, Remote Sens. Spat. Inf. Sci. **1**(2), 525–529 (2013)
2. C. DEIS-CSITE: Spatio-temporal approaches for the fruition of the cultural heritage
3. Kenteris, M., Gavalas, D., Economou, D.: An innovative mobile electronic tourist guide application. Pers. Ubiquit. Comput. **13**(2), 103–118 (2009)
4. Kounavis, C.D., Kasimati, A.E., Zamani, E.D., Giaglis, G.M.: Enhancing the tourism experience through mobile augmented reality: challenges and prospects. Int. J. Eng. Bus. Manage. **4**(10), 1–6 (2012)
5. McLean, K.M.: Planning for people in museum exhibitions, vol. 1 (1993)
6. Rummel, R.J.: The Conict Helix. Sage Publications (1976)
7. Skipper, T.L.: Understanding tourist-host interactions and their inuence on quality tourism experiences (2009)
8. Armenski, T., Dragicević, V., Pejović, L., Lukić, T., Djurdjev, B.: Interaction between tourists and residents: Inuence on tourism development. In: Polish Socio-logical Review, pp. 107–118 (2011)
9. García-Crespo, A., Chamizo, J., Rivera, I., Mencke, M., Colomo-Palacios, R., Góomez-Berbbís, J.M.: Speta: social pervasive e-tourism advisor. Telematics Inform. **26**(3), 306–315 (2009)

10. Ardissono, L., Aroyo, L., Bordoni, L., Kay, J., Kuik, T.: Patch 2013: Personal access to cultural heritage
11. Yovcheva, Z., Buhalis, D., Gatzidis, C.: Smartphone augmented reality applications for tourism. e-Rev. Tourism Res. (eRTR) **10**(2), 63–66 (2012)
12. Abraham, A., Hassanien, A.-E.: Computational Social Networks: Tools, Perspectives and Applications. Springer Science & Business Media (2012)
13. Brown, B., Chalmers, M.: Tourism and mobile technology. In: ECSCW 2003, pp. 335–354. Springer (2003)
14. Stylidis, D., Biran, A., Sit, J., Szivas, E.M.: Residents' support for tourism development: the role of residents, place image and perceived tourism impacts. Tourism Manage. **45**, 260–274 (2014)
15. Johnson, N.C.: Framing the past: time, space and the politics of heritage tourism in ireland. Polit. Geogr. **18**(2), 187–207 (1999)
16. Balabanović, M., Chu, L.L., Wolff, G.J.: Storytelling with digital photographs. In: Proceedings of the SIGCHI Conference on Human Factors in Computing Systems, pp. 564–571. ACM (2000)
17. Simoes, B., Aksenov, P., Santos, P., Arentze, T., De Amicis, R.: c-Space: fostering new creative paradigms based on recording and sharing casual videos through the internet. In: 2015 IEEE International Conference on Multimedia & Expo Workshops (ICMEW), pp. 1–4. IEEE (2015)
18. Poppe, E., Brown, R., Johnson, D., Recker, J.: Preliminary evaluation of an augmented reality collaborative process modelling system. In: 2012 International Conference on Cyberworlds (CW), pp. 77–84. IEEE (2012)
19. Simões, B., Prandi, F., De Amicis, R.: Creativity support in projection-based augmented environments. In: Augmented and Virtual Reality, pp. 168–187. Springer (2015)
20. Simões, B., De Amicis, R.: Experience-driven framework for technologicallyenhanced environments: key challenges and potential solutions. In: Intelligent In-teractive Multimedia Systems and Services. Springer (2016)
21. Hornecker, E., Buur, J.: Getting a grip on tangible interaction: a framework on physical space and social interaction. In: Proceedings of the SIGCHI Conference on Human Factors in Computing Systems, pp. 437–446. ACM (2006)
22. Rubidge, S., MacDonald, A.: Sensuous geographies: a multi-user interactive/responsive installation. Digit. Creativity **15**(4), 245–252 (2004)
23. Draper, J., Woosnam, K.M., Norman, W.C.: Tourism use history: exploring a new framework for understanding residents attitudes toward tourism. J. Travel Res. (2009)

A Personalised Recommender System for Tourists on City Trips: Concepts and Implementation

Petr Aksenov, Astrid Kemperman and Theo Arentze

Abstract In this paper we introduce a new recommender system for urban tourists. The goal of the system is to enrich tourists' experience by offering them personalised tour recommendations tailored to their dynamic user profiles. Particular attention in the proposed approach is paid to the influence of basic leisure needs of an individual, which include new experiences, entertainment, being in open area, relaxation, physical exercise, and socialising, on the tour composition. These needs tend to be dynamic and give rise to saturation effects and variety seeking behaviour. The system is developed as part of the larger c-Space framework, in which a number of technologies, such as projective augmented reality, a newly proposed near real-time 4D dynamic scene reconstruction, and affective computing, are brought together and used to enrich experiences of users in their interactions with built environments. The paper describes the main concepts of the recommender system and its implementation in the specified context of the city of Trento, Italy.

Keywords Recommender system · Tourism · Leisure needs · Dynamic preferences

1 Introduction

Personalised recommender systems for tourists have received a strong impetus from the proliferation of smartphones/tablets and the mobile internet. The new technologies have made it easier to access, share, and exchange the vast amounts of

P. Aksenov · A. Kemperman · T. Arentze (✉)
Urban Systems and Real Estate Group, Eindhoven University of Technology, Eindhoven, The Netherlands
e-mail: t.a.arentze@tue.nl

P. Aksenov
e-mail: p.aksenov@tue.nl

A. Kemperman
e-mail: a.d.a.m.kemperman@tue.nl

G. De Pietro et al. (eds.), *Intelligent Interactive Multimedia Systems and Services 2016*, Smart Innovation, Systems and Technologies 55,
DOI 10.1007/978-3-319-39345-2_46

information on which sites to visit and activities to participate in. As a consequence, the number of choice options in terms of which places to visit and how to allocate the available time across a range of activities may be overwhelming. In order to benefit from all this information, a sophisticated filtering approach is required that would meet and respond to the interests and needs of an individual tourist and would account for the available time and money budget.

Personalised recommender systems offer their users an opportunity to utilise recent advancements in information technology and to choose activities in ways that are better adapted to their interests and needs [1, 2]. A number of approaches to derive highly-relevant and personalised information from users' activity patterns or from their use of social media for subsequently feeding it into a recommender system have been proposed in literature [3–5]. A potential issue with these solutions is that they are often not detailed enough in the sense that they do not fully take into account a tourist's quickly changing individual context, which stretches beyond their demographics and whereabouts. Preferences for certain activities at any given moment in time are often dynamic (e.g., they depend on the tourist's current motivational and affective state). Focusing on leisure activities, Nijland et al. [6] identified six need dimensions that generally play a role in leisure activity choice: (1) new experiences/information, (2) entertainment, (3) relaxation, (4) being in open air/green environment, (5) physical exercise, and (6) social contact. A common characteristic of these needs is that they tend to be dynamic and give rise to saturation and variety seeking behaviour [7]. Moreover, individuals differ in terms of the relative weights they assign to these needs, which leads to considerable heterogeneity among tourists with respect to their preferences for activities [8]. As a consequence, in determining an activity plan for a trip, it is important to find a good balance between the goal of satisfying leisure needs (dynamic) on the one hand, and meeting more stable personal interests (static) on the other hand.

In this paper, we describe the concept and the implementation of a new recommender system for tourists (in short RecSys) that as an innovative feature takes dynamic (leisure) needs into account in recommending activities. The system is part of the larger c-Space framework, where a number of technologies (such as the projective augmented reality paradigm, affective computing, and a newly developed 4D dynamic scene reconstruction from videos and still images) are combined into a personal information system with the aim to stimulate creativity and enhance individuals' experiences in built environments [9]. RecSys focuses on cultural city trips, i.e. it offers recommendations for tourists visiting a (historical) city for a (few) day(s). The recommendations include: (1) activity programming—the selection of POIs (points of interest) to visit, with an indication of time to be spent at each of them; (2) activity scheduling—arranging the selected POIs into a suitable and meaningful sequence, and (3) travel planning—determining a plan (route and transport mode) for traveling needed to reach the sites. In the recommendations, the influence of dynamic leisure needs as well as the more static interests are taken into account altogether.

In the sections that follow we will describe the main concepts of the system, the algorithms for generating activity plans, and the structure of and the relevant considerations about the POI database. We conclude with a summary of the proposed contribution and give an outline of future steps.

2 Concept and Architecture of the System

RecSys considers a tourist who is visiting a historical city for the purpose of making a cultural tour. The time for the tour may vary from a few hours (part of a day) to a few days, depending on the tourist's time-budget. The specific interests and the monetary budget the tourist has may also vary. For using the RecSys the user is asked to provide some basic data regarding their preferences and available (time and money) budgets. In terms of preferences and constraints, the user indicates:

- specific themes and activities that are of interest;
- relative importance of specific needs (e.g., new information, entertainment, etc.);
- "base" location during the trip (e.g., a hotel);
- available time (via the arrival and departure times) and finances (amount of EUR);
- available transport modes as well as the preferred maximum-walking-distance.

The interaction is supported through a graphical user interface (UI) on the client side, where the user can specify these personal preferences and other data for the city trip. Besides, the user can pre-define certain activities in the program. For this the user can search through the set of all available POIs at a destination and select those which are to be included in the program regardless of their actual (i.e. algorithmical) fitness. This so-called wish-list of POIs defines an initial program for the tour.

The collected data is passed on to the RecSys service running on the server side. The service processes the received input and responds with a list of POIs recommended to be included in the tour's program (in the case of a multi-day trip, each day receives its own program). Using the UI, the user can confirm, reject, or make changes to this initially suggested program: insert new POIs, remove suggested POIs, or use drag-and-drop to prioritise items. Once the program is confirmed, the RecSys arranges the selected POIs into an ordered sequence of stops on the tour. For traveling between the sites, the schedule includes a travel plan—the chosen transport mode (walking, cycling, driving, or using public transport) and the route. The schedule is presented to the user, with a possibility to explore the routes' details.

During the trip, the UI might suggest content that matches with the user's interests specified during the information collection step, and that can be particularly explored at the current location. We call these packages of content *storylines*. The UI also provides a list of similar alternatives that can be explored by the user. This list is represented as a set of movie films that are sorted vertically by relevance and that provide a glimpse of content that will be presented.

The RecSys is a component within the larger c-Space framework, and it interacts with the latter's various other parts. In particular, it uses a central database (in turn, available through a larger content access infrastructure system, the CAI) for storing and retrieving data. Data communication concerns the locations and attributes of POIs, data about transportation networks and (registered) users. The RecSys' engine responsible for generating recommendations (that is, identifying suitable POIs and determining corresponding travel plans) runs on the server side, and the user's current state (i.e. needs and saturation) is kept and continuously updated on the client side.

3 The Engine of the Recommender System

The engine of the recommender system consists of two building blocks:

1. A comprehensive model for compiling activity programs for city tours;
2. OpenTripPlanner (OTP)—a multimodal route planner for generating complementary travel plans.

The aforementioned compilation model is a new technique specifically developed for the c-Space recommender system. It takes into account a range of diverse relevant criteria and offers advice tailored to the user's personal preferences and constraints. The second component's core element, OpenTripPlanner, is an existing open source software that implements the latest technology in the area of routing in integrated multimodal transport networks (called supernetworks) [10].

In the remainder of this section, we will describe the model, briefly introduce the so called LATUS concept, which is at the core of the model's dynamic component, and describe the main activity-selection algorithm followed by a short consideration on the model's parameterisation used for fine-tuning.

3.1 POI Utility Model

The model compiles a program by making a well-balanced selection of POIs, where each POI is assigned an individual score it has toward the tourist in question. The score is determined taking into account the personal preferences and interests of the tourist, the impacts on a full range of (dynamic) needs, existing matches with storylines, and (multimodal) travel opportunities and travel costs. In general, the score is calculated based on the following utility function:

$$U_{in}^{t} = V_{in}^{interest} + V_{in}^{attract} + V_{in}^{story} + V_{in}^{t} + V_{in}^{travel} + V_{in}^{cost} \qquad (1)$$

where:

U_{in}^{t} is the total utility (score) assigned to POI i for individual n at time t;

$V_{in}^{interest}$ is the utility component related to the POI's match with the tourist's interests;

$V_{in}^{attract}$ is the utility component related to the POI's attraction value;

V_{in}^{story} is the utility component related to the POI's role in a storyline of interest;

V_{in}^{t} is the utility component related to the POI's match with current needs;

V_{in}^{travel} is the (negative) utility component related to the travel effort and costs required to reach the location of the POI;

V_{in}^{cost} is the (negative) utility component related to the monetary costs involved in visiting the POI (e.g., the entrance fee).

The computed scores represent personalised preference values. Individuals usually differ in terms of how they weight the different factors (utility components) in a POI's score. These weights are part of the user profile, and in order to take them into account, the utility components are defined in the following form:

$$V_{in}^{f} = \gamma_{n}^{f} \cdot X_{i}^{f} \qquad (2)$$

where:

f stands for a particular performance factor ($f \in$ {interest, attract, story, needs, travel, cost});

X_{i}^{f} is the performance of POI i regarding factor f;

γ_{n}^{f} is the weight that individual n assigns to factor f

Equation (2) takes into account the fact that the preference for visiting a particular POI is generally a trade-off between multiple aspects including the attraction value, the travel costs, an entrance fee, a match with the current needs, a match with personal interests, and, where available, a possible fit into a storyline of interest.

The time superscript, t, in Eq. (1) indicates that a score depends on the tourist's current state, i.e. the state the tourist is in at the (expected) moment of the actual visit to the POI under evaluation. This reflects the notion that the utility obtained from fulfilling a person's needs depends on their size and hence on the activities previously conducted on the tour (which may decrease or increase existing needs). The corresponding needs-based dynamic component, V_{in}^{t}, is drawn from c-Space's earlier achievement addressing leisure activity travel utility simulation (LATUS) as described in [11], that has been further adapted to the context of a city visit. This component is an essential part of the proposed recommendation model and the process behind it, therefore we briefly outline its major concepts in Sect. 3.2 below.

3.2 LATUS Model for the Dynamic Utility Component

In accordance with the LATUS model, an activity (visiting a POI) may have an influence on multiple needs at the same time. For example, visiting a botanic garden may satisfy partly or completely the need for being in the open air and the need for relaxation. All six need dimensions (Nijland et al. [6]) are taken into account in the RecSys. The impacts of activities on the need dimensions are represented by parameters of activities stored in the POI database (see Sect. 4). The utility depends on the size of the current need, the weight associated with the need and the degree of satisfaction of the need—which are then summed across all the needs affected by the activity. A second complementary process that is modelled in LATUS is saturation. Saturation refers to a loss in the ability of an activity to satisfy a particular need if this activity has been implemented recently—for example, a second-time visit to a museum during the same trip would have less impact than a first-time museum visit (that is, to any other museum) even if the need for new information has not been fully satisfied yet. Satisfaction and saturation both decay with time, so that the ability of an activity to generate satisfaction (joy, pleasure, etc.) tends to recover gradually. For a complete description of the LATUS logic the reader is referred to [11].

3.3 Activity (POI) Selection Algorithm

Besides the utility function in Eq. (1), the model includes a new algorithm for selecting POIs. Since activities may overlap in terms of the needs they have an impact on, the utility (score) of any given activity (POI) depends on the other activities (POIs) already conducted (or planned). The decision process assumed in the recommendation model is based on principles of optimal time allocation. In this process, activities are selected sequentially from an exhaustive list of activities.[1] The duration (*D*) of a candidate activity is determined simultaneously with the selection decision. Because the available time on a day is limited, it is rational to select an activity only if the utility per unit time under optimal duration choice exceeds a threshold value [12]. In every step of the process, the activity that maximises the utility and meets the threshold constraint is selected. The threshold constraint is defined (see [11, 12]) as:

$$\frac{U_i^t(D)}{D + D_i^{\text{trav}}} \geq \mu \tag{3}$$

where μ is a threshold utility per unit time representing the scarcity of time available on the day under concern, and D_i^{trav} is the travel time required to get to the activity's location. Because the marginal utility decreases monotonously with duration, the optimal duration is the one where the value of the lastly added unit is larger than or equal to the threshold. If no solution exists then the activity fails to meet the

[1]Note that the sequence in which activities are selected does not necessarily define the sequence in which they are going to be conducted.

threshold constraint and can be discarded. It is also worth noting that when a certain activity is selected, the utilities of the other activities are updated accordingly.

Using this process the model generates a recommended program of activities (POIs) for each day of the trip. The tourist's (dynamic) state variables include the satisfaction level (symbol *B*) for each of the six need dimensions and the degree of saturation, represented by sensitivity levels (symbol *s*) for each combination of activity (POI) and need dimension. Given these state variables, the process for generating a recommended program for each day can be written as follows:

1. At the start of the day, retrieve the tourist's satisfaction levels, *B*, for each of the six need dimensions
2. At the start of the day, retrieve the tourist's sensitivity levels, *s*, for each POI regarding the six need dimensions
3. Initialise a program for the first day of the trip:
   (a) Start with an empty program
   (b) Retrieve the tourist's wish-list of activities for the day; add them to the program
   (c) Update satisfaction levels, *B* (needs)
   (d) Update sensitivity levels, *s* (POIs)
4. Determine the recommendation for (additional) activities:
   (a) Identify the activities that meet the threshold constraint
   (b) If no activity meets the threshold level then stop (go to 5)
   (c) Among the activities that meet the threshold find the one with the highest utility
   (d) Add the activity to the program:
      (i) update satisfaction levels, *B* (needs)
      (ii) update sensitivity levels, *s* (POIs)
   (e) Repeat from 4(a)
5. Go to the next day:
   (a) implement the decay of satisfaction levels: $B - \beta$
   (b) implement the recovery of activities: $s - \alpha$

The model's parameters related to the personal preferences of individual tourists include two sets of weighing factors: (1) relative weights of the main utility components, γ (Eq. 2), and (2) relative weights ω_g of the six need dimensions. Without any information about an individual user, the parameters are set to population averages. To estimate these averages, preference measurements were conducted based on choice experiments administered in a survey in which a large sample of individuals participated. In fact, the initial setting will be somewhat more refined and based on a segmentation of tourists that we were able to identify in the empirical study as well. The set-up and the results of the survey are discussed in

[13]. The settings are stored as part of the user profile and can be changed by the user when needed.

4 The Database and Test Site

An application of RecSys considers a specific city or city region as the cultural tour's destination. The routing component (OTP) uses OpenStreetMap data to compile a navigation graph of the street network and GTFS data to compile a navigation graph for public transport connections. Routes are then calculated into a matrix of possible origins and destinations for each transport mode. The travel time information is used by the RecSys for recommending tours including transport modes to be used for trips and the corresponding travel plans.

All information about the POIs available for cultural tours is stored in the CAI database. The information includes the geographical location, images, references and the relevant attributes of the POIs. Table 1 shows the structure of the POI-database. The *URLs* provide links to websites, as well as (still) images and videos of and about the POIs that are also stored on CAI as part of the c-Space framework and, therefore, can be fetched by the client upon request. Furthermore, data on *opening hours* and *ticket costs* (entrance fee) are stored. *Duration* represents

Table 1 Attributes of POIs (fields in the POI database)

Label	Description
Name	Name of the POI
Latitude	Latitude coordinate
Longitude	Longitude coordinate
WebSite	URL to website where information on the POI can be found
PicURL	URL to image of the POI
VidURL	URL to video of the POI
TicketCost	Entrance fee (Euro per person)
Duration	Recommended duration of the visit to the POI (in minutes)
OpeningHours	Opening hours of the POI
AttractionValue	Attraction value (5-point scale)
Typology	Type of POI (as detailed further in Table 2)
Theme	Theme related to the POI
Storyline	Set of Ids of story lines the POI is part of (possibly none)
SFPHYS	Satisfaction value, λ, Exercise need (%)
SFSOC	Satisfaction value, λ, Socialising need (%)
SFRELAX	Satisfaction value, λ, Relaxation need (%)
SFOUT	Satisfaction value, λ, Open air/green environment need (%)
SFEXP	Satisfaction value, λ, New experiences need (%)
SFENTT	Satisfaction value, λ, Entertainment need (%)

the recommended duration of a visit to the POI (in minutes). *Typology* and *theme* refer to two complementary classifications of POIs (the involved typologies are presented in Table 2; the themes are organised in a similar structure). *AttractionValue*, *Typology* + *Theme* pair and *Storyline* contain, respectively, information about the attraction value (popularity), the theme (subject), and the storylines the POI is part of (if any). The theme categories correspond to the themes presented in the dialogue for specifying the user's personal interests. Finally, the database stores information about the likely influence of the POI on specific needs. To determine this impact, the profile of each POI *i* is defined as a needs-vector of stimuli intensities, λ_{ig} (g = 1, …, 6), and for each of the six identified dimensions this field indicates the extent to which the POI is able to satisfy the corresponding need as a percentage of a total value. These settings are fine-tuned based on the data collected through a survey of individuals' judgments of these impacts.

As a use-case to validate the recommended tours and programs with users within the scope of the c-Space project, an application was developed for the historical part of the city of Trento, Italy. Hence, the OpenStreetMap data, public transport data and POI data were collected for the aforementioned area to fill the database. In terms of the POI database, the specific, culture- and tourism-related, collections (i.e. typology, theme, and storyline) were provided by the authorities involved in the process during the data collection stage for the actual POI set of the use case (Table 2). The ratings for the λ-parameters were conducted per class of POIs and based on expert judgment.

Table 2 Typology classification of POIs

Main	Sub	Main	Sub
Archive		Services	Coffee bar
Library			Picnic Area
Museum	Blind people		Accommodation
	Open Air		Restaurant
	Indoor		Mountain Refuges
Heritage	Building		Camping
	Church		Agritour
	Clock		Parking
	Garden		Bike Parking
	Monument		Shop
	Fountain		View Point
	Fort		Nature
	Castle	Entertainment	Cinema
	Street		Music
Sport	Indoor		Theatre
	Outdoor	Presentation	
		Unclassified	

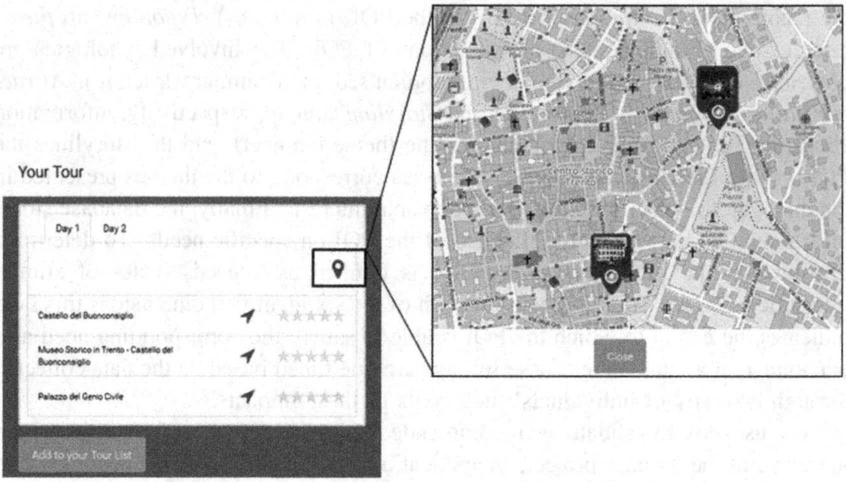

Fig. 1 Example of a recommended tour and its visualisation on a map

Figure 1 shows an example of the recommendation that the RecSys service has provided on a certain user input, and the arrangement of the suggested activities on a map (OpenStreetMap). At the moment of writing, the model has been implemented and functionally verified as a component of the larger c-Space system, ensuring its successful integration into the entire c-Space framework. Both its validation and field user testing are scheduled as the project's next steps.

5 Conclusion

In this paper we have described a new recommender system that was developed as part of the c-Space framework. The system offers personalised recommendations to tourists who visit a (historic) city for one or multiple days and are interested in conducting tours for sightseeing. The proposed innovation is that the system generates activity plans for a tour that are well-balanced in terms of the satisfaction of basic leisure needs on the one hand, and specific interests in certain POIs on the other. The preference parameters used in (dynamic) utility functions for evaluating options are estimated empirically based on data collected through choice experiments conducted as part of developing the system. A multimodal routing component is integrated in the system for the purpose of generating travel plans (transport mode and route) for reaching locations to be visited. A database for the city of Trento, Italy, was produced for the system's validation use case. The RecSys service has been implemented and functionally verified as a component of the larger c-Space system, with its integration into the c-Space framework. The system's validation with the actual users and its extensive field testing are scheduled as the next steps in the project.

Acknowledgments The research leading to these results has received funding from the European Community's Seventh Framework Program (FP7/2007-2013) under the Grant Agreement number 611040. The authors are solely responsible for the information reported in this paper. It does not represent the opinion of the Community. The Community is not responsible for any use that might be made of the information contained in this paper.

References

1. Kabassi, K.: Review: personalizing Recommendations for Tourists. Telematics Inform. **27**(1), 51–66 (2010)
2. Felfernig, A., Gordea, S., Kannach, D., Teppan, E., Zanker, M.: A short survey of recommendation technologies in travel and tourism. OEGAI J. **25**(7), 17–22 (2007)
3. Tsai, C.-Y., Chung, S.-H.: A personalized route recommendation service for theme parks using RFID information and tourist behavior. Decis. Support Syst. **52**(2), 514–527 (2012)
4. Batet, M., Moreno, A., Sánchez, D., Isern, D., Valls, A.: Turist@: agent-based personalised recommendation of tourist activities. Expert Syst. Appl. **39**(8), 7319–7329 (2012)
5. Liu, L., Xu, J., Liao, S.S., Chen, H.: A real-time personalized route recommendation system for self-drive tourists based on vehicle to vehicle communication. Expert Syst. Appl. **41**(7), 3409–3417 (2014)
6. Nijland, L., Arentze, T.A., Timmermans, H.J.P.: Eliciting the needs that underlie activity-travel patterns and their covariance structure. Transp. Res. Rec. **2157**, 54–62 (2010)
7. Arentze, T., Timmermans, H.: A need-based model of multi-day, multi-person activity generation. Transp. Res. Part B: Methodological **43**(2), 251–265 (2009)
8. Dellaert, B.G.C., Arentze, T.A., Horeni, O.: Tourists' mental representations of complex travel decision problems. J. Travel Res. **53**, 3–11 (2014)
9. Simoes, B., Aksenov, P., Santos, P., Arentze, T.: C-Space: fostering new creative paradigms based on recording and sharing "casual" videos through the internet. In: 2015 IEEE International Conference on Multimedia & Expo Workshops (ICMEW), p. 4, June 29 2015–July 3 2015
10. OpenTripPlanner. http://www.opentripplanner.org. Accessed 07 Mar 2016
11. Arentze, T.A.: LATUS: a dynamic model for leisure activity-travel utility simulation. In: Proceedings the 94th Annual TRB Meeting, p. 19 (2015)
12. Arentze, T.A., Ettema, D., Timmermans, H.J.P.: Incorporating time and income constraints in dynamic agent-based models of activity generation and time use: Approach and illustration. Transp. Res. Part C **18**, 71–83 (2010)
13. Kemperman, A.D.A.M., Aksenov, P., Arentze, T.A.: Measuring dynamic needs and preferences for recommending personalized tourist routes. In: Proceedings 22nd Recent Advances in Retailing and Consumer Science Conference. Montreal, Canada (2015)

Experience-Driven Framework for Technologically-Enhanced Environments: Key Challenges and Potential Solutions

Bruno Simões and Raffaele De Amicis

Abstract Technologically-enhanced physical environments are gradually replacing a natural part of our working and living environments. Although their use in everyday life is still rare, there has been a good effort towards a better understanding of their implications in our society, and the type of user experiences that can be provided. In this paper, we discuss various issues related to user-centric designs of technologically-enhanced environments based on our results and previous literature. We ground our argument on the hypotheses that these environments are becoming an indispensable part of people's daily lives. The integration of technologically-enhanced environments into the fabric of everyday life requires, therefore, to take into consideration a set of human experience dimensions (e.g. emotional, instrumental value, etc.) when shaping the interaction between people and the environment they live and act in. Understanding people expectations, the experience they seek in technologically-enhanced environments, and enabling them to easily influence and shape their self-crafted environments according to their needs and ever-changing expectations, in a trustworthy and controllable, can help us shape a more acceptable and meaningful vision of technologically-enhanced environments for all those living and acting in them.

1 Introduction

Technologically-enhanced physical environments (TPE) have been envisioned for more than 30 years. In initial research studies, they were envisioned as a seamless integration between Information and Communication Technology (ICT) enabled objects and a physical environment and expected to be pro-actively capable of identifying its inhabitants and responding properly to their presence, in a seamless way

B. Simões (✉) · R. De Amicis
Graphitech, Trento, Italy
e-mail: bruno.simoes@graphitech.it

R. De Amicis
e-mail: Raffaele.de.Amicis@gmail.com

G. De Pietro et al. (eds.), *Intelligent Interactive Multimedia Systems and Services 2016*, Smart Innovation, Systems and Technologies 55,
DOI 10.1007/978-3-319-39345-2_47

and irrespective of space and time [1]. Therefore, people would gain new possibilities to act, to interact with each other, and receive situation-aware assistance in their daily lives activities.

In the past few years, this vision finally found its way into our reality, to some extent, because of an on-going societal shift towards an "experience economy" [2], of great advances in the field of mobile technologies and computing, artificial intelligence, sensor networks, robotics, social media, and because of new interaction tools capable of acquire and infer knowledge about their users and environment and then use it to enhance or focus their experience in particular details. A deep understanding of people's expectations for technology-enhanced environments, in an experience economy, has therefore become an indispensable consideration because the expectations of a user are a clear reflection of an anticipated behaviour and therefore can influence the process of forming the actual user experience of the product [3]. Karapanos et al. takes an additional step by claiming that "often, anticipating our experiences with a product becomes even more important, emotional, and memorable than the experiences per se" [4].

In this paper we argue that it is indispensable to understand to whom these enhanced environments are designed and what are the user expectations. We investigate the design process of these environments from three distinct perspectives: do-it-yourself design experience, participatory user experience, and user acceptance. In the next section, we discuss the instrumental value that can be created by such environments and then we analyse what users see as a threat and which factors affect the user's willingness to adopt them. Afterwards, we devote a last sub-section to the understanding of aspects of user experience resulting from the actual user participation in a technologically-enhanced physical environment. In Sect. 3, we introduce a novel approach that aims at extending the role of users in pre-designed enhanced environments into a more proactive one where they can become co-crafters of environment itself, that is, we draft a new vision of technologically-enhanced environments that can be shaped and personalised by its inhabitants as theirs needs change. We conclude with a short description of our experience-driven design platform and brief discussion of its advantages and implications.

2 Background

The relationship between people and technologically-enhanced environments is not longer a one-to-one relationship, and consequently it requires now a more holistic understanding of user experience. We have to understand the expectations of those people using the services embed into the environment and we need to consider the experience of those who simply cohabit or share the environment [5]. Therefore, we should start by making a distinction between "user experience" and "experience". Roto [6] defines "user experience" as a two-way interaction which requires the possibility of users being able to manipulate or control the system. Experience should be defined as a much broader concept which covers all the other forms of interaction, i.e. some environments may not require any form of human interaction but still can they can create experiences that affect the way people live.

It is also indispensable to obtain a proper insight the properties that intertwine all three types of enhanced environments: living, service and production environments [5]. Living environments are the personal spaces where people manage their lives. They are not defined as a specific physical space, but rather as a space that changes over time, e.g. people commuting between home and work. Service enhanced environments integrate different types of infrastructures to support users, e.g. traffic lights control. In production environments the objective is to support people in their activities and usually involves some sort of collaboration between actors towards a common goal, e.g. industrial production process in a factory, offices, etc.

2.1 Expectations and Acceptance Factors

The willingness of a user to embrace a new technology is defined by personal expectations of a probable experience when using it [7]. Personal expectations arise from how technology and information is perceived socially and from former experiences with similar technologies. In literature, the user's perceptions and expectations are categorised into factors of user acceptance [8]. Next, we describe some of the factors that have larger impact on user acceptance of TPE.

Anticipated Usefulness and Value. The anticipated usefulness and the added value of a new technology are two of the factors that can make a big impact on the user acceptance of a technology [7, 8]. The anticipated usefulness comprises mostly the motivation to use the technology. The value it creates is based on the actual needs and future benefits that can be obtained from its use. The meaning of these factors depends on the type of user and scenario envisioned. Lee and Yoon [9] conducted an interview study to understand, in particular, the value that users expected from ubiquitous technology. Overall, participants have expressed the desire for the technology to increase their personal and families safety, to help them with effective decision-making, to increase their personal freedom, and to decrease the living cost. Röcker [10] and Niemelä et al. [11] investigated the usefulness and user-friendliness expected in several application scenarios from technology-enhanced environments. Their results indicate that users do not expect technologically-enhanced smart offices to significantly help them to become more efficient at work. Elderly people consider the value of technologically-enhanced home environments to be in the increase of their own safety and and in the creation of effortless health care activities. In the scenario of a museum, participants found the value to be in the enrichment of the user experience with supplementary multi-modal content.

Ease to Integrate, Use and Control. The hypothesis that autonomous environments based on algorithmic intelligence are the most desirable can be proven wrong if built on incorrect assumptions. Previous research studies confirmed that if a user is not able understand the logic behind a system behaviour, then it can result in a loss of user control and therefore in a decrease of user satisfaction, trust and acceptance [12]. However, it is also known that in some cases users prefer to give up on control in exchange for usefulness and benefits. In a study conducted by Barkhuus et al. [13], users reported to be willing to exchange tasks that require manual personalisation by automated ones even if such trade would erode they feeling of being in control of the system. In a different study by Cheverst et al. [14], users expressed similar opinions as well as the desire to have the possibility to switch between automatic mode or by prompt. Another study, Zaad and Allouch [15] compared the intention of use, by elderly people, of fully automatic and by prompt motion monitoring systems. Participants who perceived some sort of control over systems, and that could be easily integrated as part of their existing practices, had a greater intention of adopting it due to the fact that they perceived more control over their well-being [11]. Multi-modal interaction is another factor that enables a more personal use of enhanced environments, and that makes them more flexible and easier to control. In the design of multi-modal interactions, we should take into consideration that interaction may become less intuitive if embed as hidden and as natural as possible because of a lack of awareness [16].

Trust and Privacy. Intelligent, context-aware and pro-active services can not escape the privacy dilemma: their intelligence is derived from personal and historical data of the user [17]. Trust can be perceived as a consequence of how intelligent systems control, monitor and secure sensitive data [18]. In other words, the user acceptance decreases if users do not believe to be fully in control of the system or if they do not trust it to protect their privacy. In an experiment of Koskela et al. [19], authors concluded that the biggest challenge in gaining user's trust is to help them overcome the fear of losing their privacy. Moran and Nakata [20] investigated several the effects of monitoring (unsupervised collection of data by the system) on user behaviour. In their paper, they suggested that the awareness of such activities has a negative effect on the behaviour of the user.

Cultural and Individual Differences. The willingness of a user to embrace a new technology is influenced by contextual and individual factors, e.g. social, cultural and physical aspects. Examples of factors that may matter to the design of the system include region of residence, education, gender, job, age, wealth, and social status [21]. As a matter of fact, research studies have shown that different cultures might even expect different functionalities from the system [22]. Additionally, the social pressure, the impact expected on user's social status, and the characteristics of the user's personality such as neuroticism, extraversion, conscientiousness and agreeableness can as well affect the intention to use and act in such enhanced environments [23]. Rogers classified different types of individuals into five categories of adopters that always exist: laggards, early majority, late majority, early adopters and innovators [24]. Hence, it is important to ensure that the design of these environments is suitable for a variety of different types of users and not only to those who are ready and willing to use the system. It may well happen that system will never be unanimously accepted [17], but it should be always designed to avoid the grow of any permanent digital divide is between "laggards" and "innovators" [24].

2.2 *Factors Influencing the User Experience*

A widely accepted assumption about user experience is that it generally depends on cultural and social factors and on the individual (i.e., subjective factors) using it. These factors are known to evolve temporally and dynamic over the instances of use [25]. Another widely accepted assumption is that user experience depends on actual user's expectations. Thus, the understand of actual experiences and its manifestations can pinpoint aspects that have an important role in users' expectations. In general, user experience can have manifestations consisting of behavioural reactions such as memories, expressive reactions (e.g. postural, vocal and facial expressions), subjective feelings, inaction and avoidance, and physiological reactions [26]. Memory itself often becomes the novel quality attribute of a physical product [27]. Next, we focus on user experience provided by a number of novel interaction tools that intertwine virtual and physical elements.

Expectations of Tangible, Tactile and Haptic Interaction. Previous studies on tangible interaction suggest that they can enhance communication between people and creativity. Works on tangible interaction by Zuckerman et al. [28] and Ryokai et al. [29] indicate that tangible interfaces provide effective means to promote the communication and collaboration among children in learning environments. Hornecker and Buur [30] conducted similar studies on shared experiences resulting from the use of multi-user applications. Results confirmed that the same is true for adults. Known advantage of tactile communication is that information can be converted into a different sense modality to convey richer and more immerse experience, e.g. transformed into by certain smell, tactile feeling, or type of sound [31]. Haptic technology can be effective when users need to select and manipulate with high accuracy (without direct contact) objects in the real world [32]. Overall expectations include the interaction to be enjoyable, socially acceptable, practical, efficient and intuitive [31].

Expectations of augmented reality. The next generation of AR headsets and ultra-portable hand-held projectors can change people's perception of the physical and digital world into a unified world [33]. Simöes et al. conducted an experiment that used hand-held projectors and smart-phones to augment the physical world, as an alternative to he use of traditional AR headsets. Experimental results indicate that interpersonal communication and collaboration is more effective with the use of hand-held projectors than with hand-held AR devices due, partially, due to its tactile nature. Hornecker [30] conducted a comparative study to investigate the type of user experience provided by a telescope device and a multi-touch interactive table. The authors concluded that telescope devices can provide more immersive experiences, but multi-touch table were more efficient in supporting social and shared experiences.

3 Users as Co-crafters of Smart Environments

In this section, we describe our approach to empower people with the possibility to influence or actively take part in the construction process of their own technology-enhanced environments. Our visual programming approach guarantees at any moment of the user experience the four principles: (1) Users can access information on running processes and re-design in real-time system actions, thus preventing the threat of a system taking control over the user. (2) The possibility of imposing feedback routines and system confirmation actions (3) Access to mechanisms that enforce user authentication in particular when actions involve sharing sensitive information, and (4) Ways of controlling physical devices so they provide the user experience that is expected.

The objective of our visual programming platform is not to facilitate the development of software applications but to provide a high-level interface that enables ordinary people to create a technologically-enhanced environment in a do-it-yourself fashion. Hence, advantages of our approach include providing greater control to the user, lowering deployment costs, and the facilitation of ad-hoc personalisations [34]. User engagement is expected to be greater in domestic environments [34], and is considered an important asset in the creation of evolutionary environments that are truly in-line with user's expectations [35].

3.1 Event-Driven Design

An important aspect in the design of a user experience or learning process is: if users take an action then the user already accepted the consequences of that action. In our framework we define an "event" as a condition that has a clearly defined consequence. Hence, we can use an event-driven approach to shape the behaviour of technology towards a specific user experience, e.g. if (event) users is sad, then (consequence) change the colour of retro-projected walls into blue.

In our framework, users have the possibility to model six groups of events: events that are driven by a location, emotion, time, people involved, hardware available and content. A middleware software that can be installed in any device that runs Android is then used to monitor and control events. Location-driven events require system access to location sensors (e.g. GPS sensors that are normally embed smartphone) and can be combined with instances of smart geofences to help users minimising energy consumption. The use of location-driven events facilitates the design of user experiences that are tied to a physical location (e.g. use projectors to project the instructions of a recipe near to the user only if he/she is cooking in the kitchen, or use projectors to show relaxing animations if kids are in their rooms). Location-driven events can be combined at free will with another group of events, e.g. emotion-driven triggers. Emotion-driven events require the access to at least one camera to make possible the detection and analysis of facial emotions. We can use these events to design a user experience that can react to emotional responses of inhabitants of the environment. For example, a technology-enhanced environment can introduce or readjust ambience music based on user's actual feelings. Time-driven events provide a further level of flexibility to our user experience design framework. Time-driven events facilitate the design of user experiences along a temporal axis. Therefore, they play a key role in the design of experiences that consider, for instance, user saturation as an important variable. User-driven events are a fundamental support to the design of authentication and collaborative tasks. However, the design of authentication and collaborative tasks requires more then just a subject. It requires also a context. The last two groups of events are functional to the definition of an activity context: data-driven events and hardware-driven events. Data-driven events are designed to provide increased control over the data, e.g. who is accessing the data? where? and how?. The answer to these three questions sets the ground to the proper design of collaborative experiences. Hardware-driven events are the last group and are intended to support adaptive user experiences based that are hardware-dependent. For example, an image can be projected but as an alternative it can also be visualized in a normal LCD display.

3.2 Hardware-Driven Design

The design of user experiences in technologically-enhanced environments clearly depends on the hardware that is physically installed. In this section, we discuss how we integrate hardware that supports augmented reality experiences. For more information, see [36].

Experiences based on augmentation of the reality can happen at two different levels: as an individual experience (e.g. physical world is augmented using see-through displays) or as a shared experience (e.g. pico-projectors and computer vision are used to project and track content around users). Our AR prototype uses pico-projectors connected to smart-phones with cameras as a medium to augment and track objects in the physical space, see Fig. 1. Our framework supports also wearable, see-through displays and multi-touch interactive tables as alternative means of interaction. Our hardware-driven design introduces a set of functionalities that facilitate the design of an information flow through the list of devices available.

3.3 Data-Driven Design

Event and hardware-driven designs provide the basic means to manipulate data. Our data-driven design focus primary activities like finding, managing and transforming data. For this purpose, our platform integrates, in a seamless way, a visual programming interface that implements mechanics that support data collection, access and

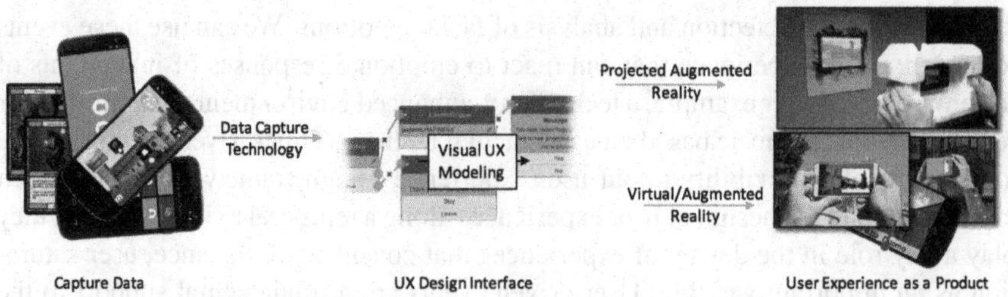

Fig. 1 User experience design platform (UXP)

presentation. The first group of features enable users to gamify activities with the purpose of engaging users in large data collection processes. For more details, see [37]. A second set of features was designed to support data transformation processes, e.g. structure-from-motion techniques to transform videos and pictures into animated 3D models. For more details, see [38]. A last group of features was projected to give users the possibility of interacting and experiencing multi-modal content. For example, 3D models that were reconstructed from videos can be streamed to AR headsets or projected into physical surfaces in a way that are susceptible to user interaction.

Figure 1 depicts a complete data work-flow in our platform, from data collection and ingestion to user interaction. At the core of the user experience design platform there is our visual programming interface.

4 Conclusions

In this paper, we investigate user expectations from three different perspectives: the willingness of a user to adopt technology-enhanced physical environments and their services, actual use and types of interaction with such systems, and finally the user-centric perspective of users as active crafters of their own environments. Additionally, we describe a web-based visual programming service that can be used in real-time to design and control existent experiences embed in technology-enhanced physical environments. An advantage of our platform, entitled c-Space UXP, is that it presents unlimited potential for self-crafted experiences if creativity and craftsmanship is provided from the part of its users.

Our c-Space UXP is still an immature technology, therefore, there are plenty of problems that were left unsolved. As a future work, we shall focus our efforts on creating a more scalable strategy to runtime management of UXP framework.

References

1. Plomp, J., Heinilä, J., Ikonen, V., Kaasinen, E., Välkkynen, P.: Sharing content and experiences in smart environments. In: Handbook of Ambient Intelligence and Smart Environments, pp. 511–533. Springer (2010)
2. Pine, B.J., Gilmore, J.H.: Welcome to the experience economy. Harvard Bus. Rev. **76**, 97–105 (1998)
3. Arhippainen, L., Tähti, M.: Empirical evaluation of user experience in two adaptive mobile application prototypes. In: Proceedings of the 2nd International Conference on Mobile and Ubiquitous Multimedia, pp. 27–34 (2003)
4. Karapanos, E.: User experience over time. In: Modeling Users' Experiences with Interactive Systems, pp. 57–83. Springer (2013)
5. Kaasinen, E., Kymäläinen, T., Niemelä, M., Olsson, T., Kanerva, M., Ikonen, V.: A user-centric view of intelligent environments: user expectations, user experience and user role in building intelligent environments. Computers **2**(1), 1–33 (2012)
6. Roto, V., et al.: Web browsing on mobile phones: characteristics of user experience. Helsinki University of Technology (2006)
7. Davis, F.D.: Perceived usefulness, perceived ease of use, and user acceptance of information technology. MIS Q. 319–340 (1989)
8. Shin, D.-H.: Ubiquitous computing acceptance model: end user concern about security, privacy and risk. Int. J. Mobile Commun. **8**(2), 169–186 (2010)
9. Lee, J., Yoon, J.: Exploring users' perspectives on ubiquitous computing. In: Proceedings of the 4th International Conference on Ubiquitous Information Technologies & Applications, 2009, ICUT'09, pp. 1–6. IEEE (2009)
10. Röcker, C.: Perceived usefulness and perceived ease-of-use of ambient intelligence applications in office environments. In: Human Centered Design, pp. 1052–1061. Springer (2009)
11. Niemelä, M., Fuentetaja, R.G., Kaasinen, E., Gallardo, J.L.: Supporting independent living of the elderly with mobile-centric ambient intelligence: user evaluation of three scenarios. In: Ambient intelligence, pp. 91–107. Springer (2007)
12. Lim, B.Y., Dey, A.K., Avrahami, D.: Why and why not explanations improve the intelligibility of context-aware intelligent systems. In: Proceedings of the SIGCHI Conference on Human Factors in Computing Systems, pp. 2119–2128. ACM (2009)
13. Barkhuus, L., Dey, A.: Is context-aware computing taking control away from the user? three levels of interactivity examined. In: UbiComp 2003: Ubiquitous Computing, pp. 149–156. Springer (2003)
14. Cheverst, K., Byun, H.E., Fitton, D., Sas, C., Kray, C., Villar, N.: Exploring issues of user model transparency and proactive behaviour in an office environment control system. User Model. User-Adap. Inter. **15**(3–4), 235–273 (2005)
15. Zaad, L., Allouch, S.B.: The influence of control on the acceptance of ambient intelligence by elderly people: an explorative study. In: Ambient Intelligence, pp. 58–74. Springer (2008)
16. Valjamae, A., Manzi, F., Bernardet, U., Mura, A., Manzolli, J., Verschure, P.F., et al.: The effects of explicit and implicit interaction on user experiences in a mixed reality installation: the synthetic oracle. Presence **18**(4), 277–285 (2009)
17. Punie, Y.: A social and technological view of ambient intelligence in everyday life: What bends the trend. Key Deliverable, Eur. Media Technol. Everyday Life Netw. (EMTEL) (2003)
18. Aarts, E., Markopoulos, P., de Ruyter, B.: The persuasiveness of ambient intelligence. In: Security, Privacy, and Trust in Modern Data Management, pp. 367–381. Springer (2007)
19. Koskela, T., Väänänen-Vainio-Mattila, K.: Evolution towards smart home environments: empirical evaluation of three user interfaces. Pers. Ubiquit. Comput. **8**(3–4), 234–240 (2004)
20. Moran, S., Nakata, K.: Analysing the factors affecting users in intelligent pervasive spaces. Intell. Build. Int. **2**(1), 57–71 (2010)
21. Noh, M.J., Kim, J.S.: Factors influencing the user acceptance of digital home services. Telecommun. Policy **34**(11), 672–682 (2010)

22. Forest, F., Arhippainen, L.: Social acceptance of proactive mobile services: observing and anticipating cultural aspects by a sociology of user experience method. In: Proceedings of the 2005 Joint Conference on Smart Objects and Ambient Intelligence: Innovative Context-Aware Services: Usages and Technologies, pp. 117–122. ACM (2005)
23. Devaraj, S., Easley, R.F., Crant, J.M.: Research note-how does personality matter? relating the five-factor model to technology acceptance and use. Inf. Syst. Res. **19**(1), 93–105 (2008)
24. Rogers, E.M.: Diffusion of Innovations. Simon and Schuster (2010)
25. Law, E.L.-C., Roto, V., Hassenzahl, M., Vermeeren, A.P., Kort, J.: Understanding, scoping and defining user experience: a survey approach. In: Proceedings of the SIGCHI Conference on Human Factors in Computing Systems, pp. 719–728. ACM (2009)
26. Desmet, P., Hekkert, P.: Framework of product experience. Int. J. Des. **1**(1) (2007)
27. Pine, B.J., Gilmore, J.H.: The Experience Economy. Harvard Business Press (2011)
28. Zuckerman, O., Arida, S., Resnick, M.: Extending tangible interfaces for education: digital montessori-inspired manipulatives. In: Proceedings of the SIGCHI Conference on Human Factors in Computing Systems, pp. 859–868. ACM (2005)
29. Ryokai, K., Marti, S., Ishii, H.: I/O brush: drawing with everyday objects as ink. In: Proceedings of the SIGCHI Conference on Human Factors in Computing Systems, pp. 303–310. ACM (2004)
30. Hornecker, E., Buur, J.: Getting a grip on tangible interaction: a framework on physical space and social interaction. In: Proceedings of the SIGCHI Conference on Human Factors in Computing Systems, pp. 437–446. ACM (2006)
31. Heikkinen, J., Olsson, T., Väänänen-Vainio-Mattila, K.: Expectations for user experience in haptic communication with mobile devices. In: Proceedings of the 11th International Conference on Human-Computer Interaction with Mobile Devices and Services, p. 28. ACM (2009)
32. Kaasinen, E., Niemelä, M., Tuomisto, T., Välkkynen, P., Ermolov, V.: Identifying user requirements for a mobile terminal centric ubiquitous computing architecture. In: International Workshop on System Support for Future Mobile Computing Applications, 2006, FUMCA'06, pp. 9–16. IEEE (2006)
33. Azuma, R.: Tracking requirements for augmented reality. Commun. ACM **36**(7), 50–51 (1993)
34. Kawsar, F., Nakajima, T., Fujinami, K.: Deploy spontaneously: supporting end-users in building and enhancing a smart home. In: Proceedings of the 10th International Conference on Ubiquitous Computing, pp. 282–291. ACM (2008)
35. Roelands, M., Claeys, L., Godon, M., Geerts, M., Feki, M.A., Trappeniers, L.: Enabling the masses to become creative in smart spaces. In: Architecting the Internet of things, pp. 37–64. Springer (2011)
36. Simões, B., Prandi, F., De Amicis, R.: Creativity support in projection-based augmented environments. In: Augmented and Virtual Reality, pp. 168–187. Springer (2015)
37. Simões, B., De Amicis, R.: Gamification as a key enabling technology for image sensing and content tagging. In: Intelligent Interactive Multimedia Systems and Services. Springer (2016)
38. Simoes, B., Aksenov, P., Santos, P., Arentze, T., De Amicis, R.: c-Space: fostering new creative paradigms based on recording and sharing casual videos through the internet. In: 2015 IEEE International Conference on Multimedia & Expo Workshops (ICMEW), pp. 1–4. IEEE (2015)

Touchless Disambiguation Techniques for Wearable Augmented Reality Systems

Giuseppe Caggianese, Luigi Gallo and Pietro Neroni

Abstract The paper concerns target disambiguation techniques in egocentric vision for wearable augmented reality systems. In particular, the paper focuses on two of the most commonly used selection techniques in immersive environments: Depth Ray and SQUAD. The design and implementation of such techniques in a touchless augmented reality interface, together with the results of a preliminary usability evaluation carried out with inexpert users, are discussed. The user study provides insights on users' preferences when dealing with the precision-velocity trade-off in selection tasks, carried out in an augmented reality scenario.

Keywords Touchless interface · Freehand interaction · Wearable augmented reality · Depth ray · SQUAD

1 Introduction

The recent availability of low cost wearable augmented reality (WAR) technologies and the consequent possibility of showing computer-generated information directly in the field of view (FOV) of the user are leveraging the design of applications in many different domains in order to support users in their daily activities. Together with such applications, in recent years, we have been witnessing a growing exploration of interaction techniques which aim at reducing the users' efforts and, especially, at simplifying the way in which they can interact with the virtual information. This trend has led to interaction modalities mainly tailored to specific applications,

G. Caggianese (✉) · L. Gallo · P. Neroni
Institute for High Performance Computing and Networking National Research Council of Italy (ICAR-CNR), Naples, Italy
e-mail: giuseppe.caggianese@na.icar.cnr.it

L. Gallo
e-mail: luigi.gallo@na.icar.cnr.it

P. Neroni
e-mail: pietro.neroni@na.icar.cnr.it

G. De Pietro et al. (eds.), *Intelligent Interactive Multimedia Systems and Services 2016*, Smart Innovation, Systems and Technologies 55,
DOI 10.1007/978-3-319-39345-2_48

most of which exploit vocal control, touch control or a combination of both (e.g., the Google Glass project [1]).

However, the possibility of mixing real objects with virtual information both in the same space, has triggered the demand for more natural interaction approaches. In fact, users who are wearing a pair of glasses in order to exploit an AR application as a support for their daily activities are not likely to use specific devices to control the system. The direct manipulation of computer-generated information through the point-and-click paradigm also has the potential to reduce the gap between intention and action [1].

Moreover, another important issue to deal with in the WAR systems is the limited FOV of the actual device, which is often the cause of a cluttered virtual environment. In this case the information easily becomes difficult to access and direct manipulation proves to be more practical than using vocal control or touchpads.

In this paper we address the selection problem in a dense virtual environment by presenting two disambiguation mechanisms associated with a touchless selection technique specifically implemented for a WAR system. The chosen disambiguation mechanisms are Depth Ray and SQUAD, a single-step and two-step process selection, respectively. Both of these mechanisms have been described and evaluated in a WAR context associated with a ray-casting technique used to point at virtual objects in the scene. In this paper, we detail the design of these techniques, adapted to a specific application in which the depth sensor is worn by the user, in egocentric vision, and where the interaction is based on a markerless hand tracking system. In this way, the two mechanisms realized become *Touchless Depth Ray* and *Touchless SQUAD*. All the encountered problems during their implementation are described and the results of a preliminary qualitative evaluation aimed at verifying the preferences of users between the two disambiguation mechanisms in dense environments are discussed.

The rest of the paper is structured as follows. In the next section, Sect. 2, we show related work present in the literature. In Sect. 3, we present the WAR system. In Sect. 4, we describe the two disambiguation techniques and the main problems encountered in implementing them in the WAR system. Next, Sect. 5 focuses on the qualitative evaluation test performed, providing a discussion of the results. Finally, in Sect. 6 we present our conclusions.

2 Related Work

Selection represents a fundamental task in the interaction of the user with computer-generated information. It allows the user to specify a single object in the surrounding virtual environment on which to execute subsequent actions [2].

Among all the selection techniques, ray-casting is perhaps the most frequently applied solution in virtual environments due to its simplicity, generality and applicability to any distance. However, the selection becomes more demanding when the visual size of the target becomes small due to the object size, the distance from the user's point of view or occlusion in a cluttered environment. In these situations the ray-casting technique proves to be slow and error-prone because it does not provide high-precision pointing at a small object at a distance and proves to be limited to simply selecting the first target which is intersected [3].

In order to improve the selection performance, by reducing the required precision and user effort, a first attempt relies on employing volumetric selection tools such as cones and spheres [3, 4]. However, these tools do not allow you to indicate a single object at once, especially in a dense environment [5]. In these conditions, the user needs some possibilities to disambiguate between the indicated objects. The common solution is to split the selection into two different phases: selection and disambiguation [6]. In the selection phase, the user indicates with a low accuracy a group of objects containing the target, while the disambiguation phase allows the user to discriminate between the objects of the previous phase, so selecting the one desired. The disambiguation phase can be explicit or implicit depending on whether an additional cognitive load on the user is required. Argelaguet et al. [7] classify the approaches in which the disambiguation phase is explicit as manual whereas the approaches characterized by an implicit disambiguation phase are indicated as heuristic and behavioural.

Among the manual approaches, the simplest solution provides a button which allows the user to cycle among all previously indicated objects [8]. However, this solution does not scale well with the number of indicated objects. Other approaches propose displaying the indicated objects of the selection phase, so allowing the user to perform a disambiguation phase for a subsequent refinement. Grossman et al. [6] propose Flow Ray in which all the objects intersected with the ray-casting are displayed in a marking menu selection. Similarly, Kopper et al. [9] present an approach called SQUAD in which the disambiguation phase is composed of progressive refinement and in [10] this approach has been used to develop two interactive variations for dense dynamic environments called Zoom and Expand. Finally, a different approach relies on the fact that techniques like ray-casting require only two degrees of freedom (DoF) to work, so that the extra DoF can be used to disambiguate the selection. Depth Ray [6] provides a 3D cursor constrained along the virtual ray controlled by the user by moving her/his hand forwards and backwards.

3 The Wearable AR System

In recent years, researchers have designed several mobile systems with the aim of transforming a cultural space into a smart cultural environment to enhance the enjoyment and satisfaction of the people involved [11–14]. The WAR system used to test the proposed implementation of the disambiguation techniques is a system realized for a fully featured enjoyment of the cultural heritage. To realize the system we used one of the first integrated wearable devices specifically devoted to AR and available off-the-shelf (Fig. 1). The META 1 Developer Kit [15] integrates in a single device a WAR display, a low consumption 3-axis inertial sensor and a RGB-D camera. In

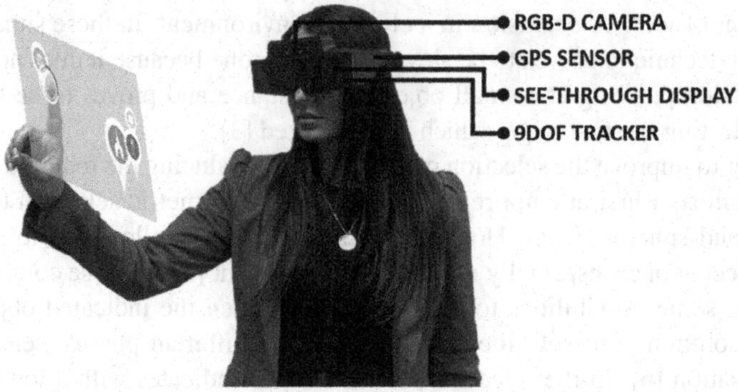

Fig. 1 The WAR system, hardware and application. The user's FOV is augmented with a set of POIs, which can be selected via a touchless point-and-click interface

more detail, the WAR device has been realized by integrating an augmented reality eye-wear equipped with a binocular see-through display with a FOV of 23°, together with an RGB-D sensor that allows you to capture RGB images using a standard CMOS sensor. The user's head movements are tracked by using the inertial sensors embedded in the device with 3 degrees of freedom (DoF) whereas an external GPS sensor is used to localize the user.

The WAR application allows an individual to visualize georegistered points of interest (POI) relative to cultural, historical and tourist information, overlaid onto the real world view. The user is allowed to visualize this information organized in eight different categories (e.g. churches, museums and archaeological sites). The different categories are represented by using different icons and colours in order to facilitate POI visualization and recognition. This classification can be used together with a parameter of distance indicating the areas of interest in the surrounding environment, to filter the visualized POIs.

The system allows the user to interact with the POIs by using our touchless point-and-click interface based on a hand tracking system that exploits the RGB-D sensor camera integrated in the wearable device [16]. By selecting one of the displayed POIs, the user visualizes additional information about it or navigation aids to reach it. When the user executes the pointing gesture (all the fingers closed except for the index finger), the system visualizes a virtual pointer over the fingertip position. To use the pointing metaphor in a virtual space, which is bigger than the working space of the user, defined as the area reachable with an extended arm, we have used the ray-casting technique [17].

However, because of the reduced FOV in WAR displays, as the number of POIs to visualize increases, the computer-generated environment easily becomes dense requiring the user to interact very carefully to accomplish a single precise selection (Fig. 2). In this highly cluttered environment, which is very frequent in the cultural heritage domain, a further technique to reduce the required precision in disambiguating the selection, is needed.

Fig. 2 Example of cluttered environment issues encountered when considering the real world in relation to the specific context of the cultural heritage

For this purpose, two different disambiguation techniques have been designed and developed for our WAR application and associated with our point-and-click interface, namely Touchless Depth Ray and Touchless SQUAD, described in the next section.

4 Disambiguation Techniques

4.1 Touchless Depth Ray

The touchless depth ray disambiguation mechanism exploits the extra DoF, unused in the ray-casting techniques, to disambiguate between all the indicated synthetic objects. In this mechanism the selection and the disambiguation phase are executed at the same time in a single-step process. In the selection phase the user points toward the target and, in the case of a dense environment, all the intersected objects become semi-transparent allowing the user to quickly verify if her/his target is in the indicated group (Fig. 3a).

If the target is already in the group the user can start to execute the disambiguation phase. Continuing to point toward the set of objects, the user can bring her/his hand forward and backward of the sensor by exploiting the dimensionality that does not affect the position of the pointer. This movement controls the depth cursor that moves along the ray selecting the POI that lies in its correspondence. The closest POI to the cursor will be the only one totally coloured in the indicated set (Fig. 3b, c). As soon as the depth cursor highlights the desired POI, the user can execute the trigger action to confirm the selection.

Although remaining distinct the two phases can be executed at the same time. In fact, by moving the pointer, the user controls the position of the ray, modifying the indicated set of objects. At the same time, by acting simultaneously on the extra DoF, she/he can disambiguate among the multiple objects. While this may allow for a fast selection, the two phases could potentially interfere with each other. The fine-grained adjustments of the pointer could cause the depth cursor to move and vise-versa.

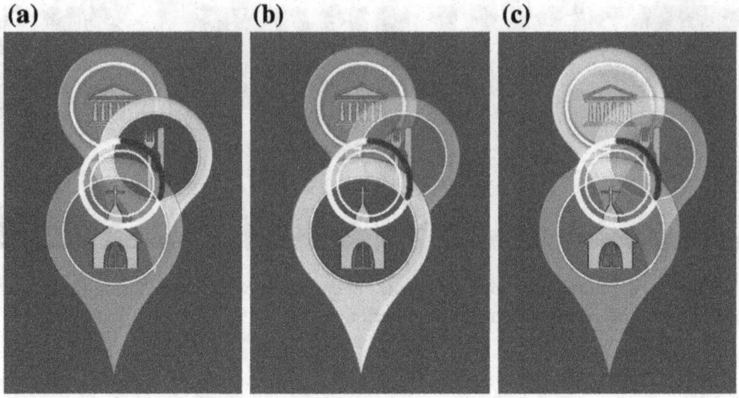

Fig. 3 Touchless Depth Ray. **a** The user moves the pointer performing the selection phase with the result that all the POIs become semi-transparent except the one in the middle which is closer to the position of the depth cursor. **b** The user performs the disambiguation phase by moving her/his fingertip backward in order to select the closest POI. **c** The user performs the disambiguation phase by moving her/his fingertip forward in order to select the farthest POI

The first problem we dealt with in implementing this disambiguation mechanism for the WAR system was relative to the extra DoF used to control the disambiguation phase. Effectively, the user has a limited range of movement along the axes outgoing from the camera, and thus her/his motor space proves to be much smaller than the virtual space with the result that it becomes necessary to map between the two spaces. We decided to discretely cycle from one target to the next, as suggested by Hinckley et al. [8], and fix the initial position of the depth cursor, considering a physical interaction area of 80 cm lying 20 cm far from the sensor up to the maximum extension of the arm from the sensor. However, the physical movement needed to switch between consecutive POIs depends on the number of POIs in the selection phase. Therefore, a dense list of POIs requires also a careful pulling of the hand along the z axis, an aspect that further complicates the user interaction.

Finally, a velocity-based pointing enhancing technique was been used [18] to smooth the pointer movements and to improve the precision of the ray and so simplify the control of the selection phase for the user.

4.2 Touchless SQUAD

The second mechanism we considered exploits a totally different approach. In the touchless SQUAD mechanism, in fact, the selection and disambiguation phases are carried out sequentially, in a two-step process.

In this case we used a modified version of ray-casting that casts a cylinder that starts from the nearest intersecting surface and moves forward. This volume, used to determine which POIs will be selectable, remains constant although the dimension of the POIs in the WAR application changes in accordance with their distance from the user's point of view. This solution allows the user to select fewer POIs when they are close to the point of view, but more when they are far from it.

With the touchless SQUAD mechanism, each time the user moves the pointer in the virtual space crossing a set of POIs all the POIs that lie in the volume of the cylinder are highlighted. In this way, the user can immediately verify if the target object is in the highlighted group. If so, always by pointing at the same group, the user executes the click action so finalizing the selection phase (Fig. 4a). Alternatively, the user can continue to move the pointer until her/his target object is highlighted.

After performing the selection phase, all the indicated objects animate within the user's point of view covering the environment in the background. In this way, the user loses visibility of the virtual environment acquiring a better visibility on the selected objects organized in four quadrants on the available FOV of the WAR display (Fig. 4b). At this point, the user can easily refine her/his selection by pointing at the area of the FOV which contains the target object and then at the target object among those that appear in the grid (Fig. 4c).

In the case of a wrong selection, this technique requires the mechanism to move back to the selection step. In our implementation, the user simply needs to move her/his hand out of the FOV of the RGB-D sensor.

As we well know, the SQUAD mechanism presents two main limitation: first, the indicated objects lose their spatial context during the refinement steps and, secondly, the mechanism does not perform well if the objects are not sufficiently different from each other. In our application the different categories of POIs prove to be visually different but it often happens that multiple POIs of the same category can be in the same virtual scene. Therefore, in order to preserve the spatial context we decided to distribute the items in the quad menu according to their distance from the user's point of view in order to reduce the probability of having many POIs of the same type in the same quadrant (Fig. 4a). We decided, once a quadrant is selected, to show the POIs belonging to that quadrant in a grid menu inspired by the work of Cashion et al. [10] (Fig. 4b). In this way, the maximum number of steps to select a POI is three.

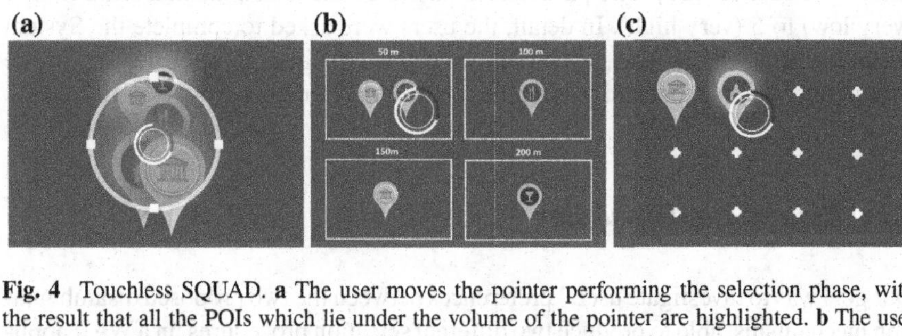

Fig. 4 Touchless SQUAD. **a** The user moves the pointer performing the selection phase, with the result that all the POIs which lie under the volume of the pointer are highlighted. **b** The user performs the disambiguation phase, which corresponds to the first refinement, and selects the area with the desired POI. **c** In a subsequent level of refinement, the user selects the target from the set of POIs in the grid

5 User Study

5.1 Experimental Procedure

In order to perform the study we recruited 10 unpaid volunteers. The participants' ages ranged from 26 to 36 years old, with an average age of 30.7. Three of the participants were female and all of them were right-handed.

The experiment was performed by using a single pair of smart glasses and the same simulated scenario for all the users, guaranteeing the same set-up and conditions for each tester. The simulated scenario was populated by different POIs having different sizes in order to simulate different distances from the user's point of view.

Before starting, a facilitator showed, to each volunteer, the functionalities of the system and how to use the two proposed disambiguation techniques.

After, in a practice session, designed to make the participants more comfortable and relaxed, the volunteers were left free to familiarize themselves with the interface without any limitations. They were able to switch between the two proposed techniques, so improving their confidence with both mechanisms of disambiguation.

Next, the test session started with the facilitator issuing precise instructions, requiring the testers to select a specific object in the virtual scene by using one technique at a time. In order to avoid any learning effect, the two techniques were counterbalanced by using a balanced Latin square and, systematically after each session, the tester was asked to the tester to evaluate the technique by using a post-hoc questionnaire. No time limit was imposed on the participants during the performance of each technique and each user's interaction was observed to collect all the possible impressions of her/his experience. Whit the aim of better collecting information, they testers were asked to think aloud, describing their intentions and any possible difficulties. In the same way, no time limit was imposed on the completion of the proposed questionnaire. However, the testers were asked to record an immediate answer to each question, rather than spending a long time thinking about the items.

The questionnaires used to measure both the usability and the user's experience were a five point Likert-scale questionnaires [19] structured to fulfil all of the criteria listed by Uebersax [20]. The participants answered the questions in a scale from 1 (very low) to 5 (very high). In detail, the users were asked to complete the System Usability Scale (SUS) [21], which allows you to obtain a rapid evaluation of the system interaction expressed as a single number which ranges from 0 to 100.

5.2 Results and Discussion

Our goal was to investigate users' preferences between the two proposed disambiguation mechanisms, able to produce two different selection procedures, in a WAR application. In fact, when using a touchless depth ray approach, the selection proves to be faster but, at the same time, requires more precision during the pointing. On the contrary, the touchless SQUAD, based on progressive refinement, results in a slower selection that requires less precision.

The majority of the participants complained about fatigue when using the Touchless Depth Ray techniques, mainly due to the requirement not to move the pointer during the disambiguation phase, an aspect also known as the Gorilla arm effect [22]. This inconvenience caused a high rate of wrong selections, forcing the tester to continuously correct the pointer position. Moreover, most of the users expressed dissatisfaction with the visual feedback proposed in the touchless depth ray. The semi-transparency used to mark all the indicated POIs in the selection phase was considered not very comprehensible. Nevertheless, after a few minutes in the practice session, all the testers understood how to use the technique and were able to select the desired objects.

The touchless SQUAD technique was the one more appreciated by the testers, mainly because of the low level of precision required in the selection phase. This technique, which is supposed to require more time to accomplish the selection because of the consecutive refinements, was in fact considered faster then the touchless depth ray by some participants. Actually, during the test we observed that this was not the case, the users being faster in selecting the group of POIs containing the target but disoriented during the refinement steps. In fact, many of testers complained about difficulties in identify the target POI in the refinement steps because it was out of their spatial context. Moreover, after the practice session, the criteria used to distribute the POIs in the quad menu was considered useful in order to identify the target POI, so reducing the sense of disorientation.

Finally, these observations were also confirmed by the results of the SUS questionnaire. The average SUS score, whose values range from 0 to 100, was 50.83 for the Touchless Depth Ray and 59.58 for the Touchless SQUAD.

6 Conclusions

Nowadays, when WAR systems are generally characterized by a huge amount of information resulting in a cluttered virtual environment a significant challenge consist in the design of a truly natural technique to reduce the users' effort and simplify the way in which they can interact with the virtual information.

In this paper, we have presented two disambiguation mechanisms designed for a WAR system with a dense virtual environment and have also enumerated the problems encountered during their implementation. Moreover, we have compared them in a preliminary usability evaluation performed with ten volunteers, describing the results obtained.

The results show that fatigue is the main difficulty in touchless, mid-air interaction. Our future work will focus on enhancing the interface by addressing the users' comments collected during the tests. Moreover, we will perform a quantitative evaluation on a larger group of users.

References

1. Starner, T.: Project glass: an extension of the self. IEEE Pervasive Comput. **12**(2), 14–16 (2013)
2. Steed, A.: Towards a general model for selection in virtual environments. In: IEEE Symposium on 3D User Interfaces, 2006, 3DUI 2006, pp. 103–110. IEEE (2006)
3. Steed, A., Parker, C.: 3D selection strategies for head tracked and non-head tracked operation of spatially immersive displays. In: 8th International Immersive Projection Technology Workshop, pp. 13–14 (2004)
4. Liang, J., Green, M.: JDCAD: a highly interactive 3D modeling system. Comput. Graph. **18**(4), 499–506 (1994)
5. Vanacken, L., Grossman, T., Coninx, K.: Exploring the effects of environment density and target visibility on object selection in 3D virtual environments. In: IEEE Symposium on 3D User Interfaces, 2007, 3DUI'07. IEEE (2007)
6. Grossman, T., Balakrishnan, R.: The design and evaluation of selection techniques for 3D volumetric displays. In: Proceedings of the 19th Annual ACM symposium on User Interface Software and Technology. pp. 3–12. ACM (2006)
7. Argelaguet, F., Andujar, C.: A survey of 3D object selection techniques for virtual environments. Comput. Graph. **37**(3), 121–136 (2013)
8. Hinckley, K., Pausch, R., Goble, J.C., Kassell, N.F.: A survey of design issues in spatial input. In: Proceedings of the 7th Annual ACM Symposium on User Interface Software and Technology, pp. 213–222. ACM (1994)
9. Kopper, R., Bacim, F., Bowman, D., et al.: Rapid and accurate 3D selection by progressive refinement. In: 2011 IEEE Symposium on 3D User Interfaces (3DUI), pp. 67–74. IEEE (2011)
10. Cashion, J., Wingrave, C., LaViola Jr., J.J.: Dense and dynamic 3D selection for game-based virtual environments. IEEE Transa. Visual. Comput. Graph. **18**(4), 634–642 (2012)
11. Amato, F., Mazzeo, A., Moscato, V., Picariello, A.: Exploiting cloud technologies and context information for recommending touristic paths. Stud. Comput. Intell. **511**, 281–287 (2014)
12. Amato, F., Chianese, A., Moscato, V., Picariello, A., Sperli, G.: Snops: a smart environment for cultural heritage applications. In: Proceedings of the Twelfth International Workshop on Web Information and Data Management, WIDM '12, pp. 49–56. ACM, New York, NY, USA (2012). http://doi.acm.org/10.1145/2389936.2389947
13. Chianese, A., Piccialli, F.: Improving user experience of cultural environment through IoT: the beauty or the truth case study. Smart Innovation, Syst. Technol. **40**, 11–20 (2015)
14. Chianese, A., Piccialli, F., Valente, I.: Smart environments and cultural heritage: a novel approach to create intelligent cultural spaces. J. Location Based Serv. **9**(3), 209–234 (2015)
15. Meta spaceglasses: https://www.metavision.com/
16. Caggianese, G., Neroni, P., Gallo, L.: Natural interaction and wearable augmented reality for the enjoyment of the cultural heritage in outdoor conditions. In: De Paolis, L.T., Mongelli, A. (eds.) Augmented and Virtual Reality, Lecture Notes in Computer Science, pp. 267–282. Springer International Publishing (2014)
17. Mine, M.R.: Virtual environment interaction techniques. Technical report, University of North Carolina at Chapel Hill, Chapel Hill, NC, USA (1995)
18. Gallo, L., Minutolo, A.: Design and comparative evaluation of smoothed pointing: a velocity-oriented remote pointing enhancement technique. Int. J. Hum. Comput. Stud. **70**(4), 287–300 (2012)
19. Likert, R.: A technique for the measurement of attitudes. Arch. Psychol. **22** (1932)
20. Uebersax, J.S.: Likert scales: Dispelling the confusion. http://www.john-uebersax.com/stat/likert.htm
21. Brooke, J.: SUS: a quick and dirty usability scale. In: Jordan, P.W., Weerdmeester, B., Thomas, A., Mclelland, I.L. (eds.) Usability Evaluation in Industry. Taylor and Francis (1996)
22. Boring, S., Jurmu, M., Butz, A.: Scroll, tilt or move it: using mobile phones to continuously control pointers on large public displays. In: Proceedings of the 21st Annual Conference of the Australian Computer-Human Interaction Special Interest Group: Design: Open 24/7, OZCHI '09, pp. 161–168. ACM, New York, NY, USA (2009)

System Architecture and Functions of Internet-Based Production of Tailor-Made Clothes

Petra Perner

Abstract Smart spaces are currently developed for different kinds of applications. We want to describe a smart space for the production of tailor-made clothes. In the paper, we review the production process for tailor-made clothes. Then, we describe the architecture for the smart production space. We describe the advantages of such a model for different commercial services that can be linked to the online production process. We also describe the kind of sensors and the functions of the sensors that are connected online to each other. In addition the software is described that allows controlling the smart space. Finally, we summarize our work.

Keywords Smart spaces · Internet of things · Tailor-made cloth · Image processing for tailor-made cloth

1 Introduction

Smart spaces are currently developed for different kinds of applications, such as kids care [1], home care [2] or for intelligent cultural spaces [3, 4]. We want to describe a smart space for tailor-made clothes.

Micro, small and medium-sized enterprises have been identified by scholars as the backbone of the global economy [5]. Shops for tailor-made clothes are usually micro or small-sized enterprises. An investigation on the Milan economy [6] showed that fashion is a serious business [7] and can have a big economic impact on an area. Especially rural areas can profit from this fact. The introduction of computer technology [8, 9] and the use of the internet can substantially help to increase this business and to market the clothes worldwide [10]. It can help to improve the demand for tailor made clothes and the production process [11]. It will allow different locally dispersed services to work together.

P. Perner (✉)
Institute of Computer Vision and Applied Computer Sciences, IBaI, Kohlenstr. 2, Leipzig, Germany
e-mail: pperner@ibai-institut.de

G. De Pietro et al. (eds.), *Intelligent Interactive Multimedia Systems and Services 2016*, Smart Innovation, Systems and Technologies 55,
DOI 10.1007/978-3-319-39345-2_49

Here, custom-made clothes are investigated [12]. We will study tailor-made clothes. Normally the production of tailor-made clothes is a time-consuming process. It requires a lot of time from the customer, ashe has to choose the design and the fabric, and has also to go several times to the shop for allowing the tailor to fit the clothes to his body measures. When the clothes have been completed, the customer has to consider that the clothes have to be treated in the right way. Dry cleaning is a problem, since special materials might have been used and not all dry-cleaning shops are cleaning specialists for those particular materials. If the clothes are damaged, a special repair service has to be provided.

Marketing of tailor-made clothes over the internet [13] can help to get new customers who might not be living in the area where the tailor is located. Research is going on that investigates how customer search for special clothes can take place over the internet [14]. If the service is well presented, it might attract customers all over the world [15]. That requires a smart production process over the internet, so that the customer does not need to go to the tailor's shop, and it requires a good logistic system [11] in order to ship the clothes worldwide, thus offering anew business to small parcel shops.

A webshop also offers further opportunities for marketing the clothes. E-mail marketing can improve the awareness of the service of the tailor shop [16] and can reduce the costs of marketing measures. The use of a web presence will give an insight into the customer's behavior. While visiting the shop and searching for special products and services, the customer is leaving traces in the log-file [17] or information in the database of the shop. This information can be used to learn the customer's profile and web usage of the shop [18]. The latter can result in an improved web-presence setup based on the customer's needs, while the customer profile can help to improve the service of the shop. Knowing the customer's profile, the marketing actions can be set up in a more efficient way. Moving from a web presence to e-commerce for a tailor shop will result in a new and higher quality of his business [19]. That requires studying the new business model under the prerequisite of an e-commerce tailor shop [20].

In this paper, we present our study on "How can we carry out the production of tailor-made clothes via the internet?" We review the tailor-made production process in Sect. 2 and study the framework for a smart production space of tailor-made clothes. We identify the typical function of a tailor-made clothes production process. Then we describe possible new tasks in such an e-commerce framework that can help to improve the design and the quality of tailor made clothes. We also identify services that can be included in an e-commerce model. These services would help the customer to handle and to maintain the clothes often made from special and valuable or high-quality material.

In Sect. 3, we describe the overall architecture of the smart production space. Not only do we take into account the customer/tailor site, we also include the service sites into this model. With data-mining methods, one can investigate the obtained common database of log files and other information. By doing so new insight about the customer, his web usage and the offered services can be observed.

Section 4 introduces the communication channel between the customer and the tailor. Special hardware based on cameras and VoIP is necessary in order to obtain all the functions that are necessary for the fitting of the clothes and the communication. We explain the communication process [21] based on that we describe the functions of the special tailor software.

Finally, we give conclusions in Sect. 5.

2 A Framework for a Smart Production Space for Tailor-Made Clothes

First we want to review the production process of tailor-made clothes. We are only focussing on the important work stages. Figure 1 shows the production of tailor-made clothes from the time the customer enters the shop until he is handed over the final clothes.

Different work stages are necessary. First the tailor makes a briefing with the customer on the design and fabric of the clothes. Often the customer will have to bring the fabric, as the tailor will not always offer this service. Then the tailor will take the body measurements and make the cut of the clothes. Afterwards the fabric gets cut into parts for the clothes.

Then, the tailor makes a first draft of the clothes. The first fitting is necessary to see if the dress fits the customer's measurements. Then the tailor makes the necessary modifications on the clothes to better fit the customer. Afterwards a second fitting is necessary to make sure that the clothes really fit the customer. Now the

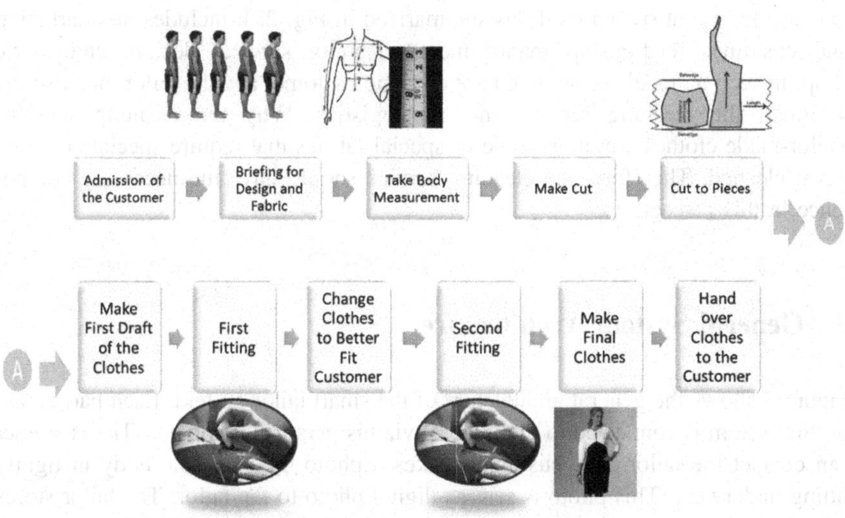

Fig. 1 The production process of tailor-made clothes

tailor proceeds to make the final version of the clothes. The customer comes to the shop and examines the tailor-made clothes. If everything is in order he takes the clothes and leaves the shop.

This brief review of the production process shows that producing tailor-made clothes is not a simple process neither for the tailor nor for the customer. It also is a time-consuming process. The customer has to come several times to the tailor—to conclude the contract, twice for fitting, and afterwards for fitting and inspecting the final version of the clothes.

A tailor is not a designer. He can help the customer to find a design for his clothes, however, if something special is wanted, a designer should be involved in the process.

The customer is not involved in all of the work stages. The tailor carries out the cut and tailoring of the clothes all by himself. Therefore, we have two different functions in a smart production space: first, the interaction with the customer and second the tailoring.

The typical interactive functions in a tailor shop are the following:

- Customer acquisition
- Transfer customer image to different sites to the tailor and the designer
- Briefing with the customer concerning design and fabric
- Take body measurements
- First, second, and final fitting.

If these functions could be realized within a smart space it would save time for the customer and could help to increase the request for tailor-made clothes.

Around all the above-described functions, we can identify more functions that could be realized in a smart space. This can lead to a new model for the production and marketing of tailor-made clothes. This new model should be realized on-line.

Our idea about such a model is summarized in Fig. 2. It includes the marketing and customer relationship management, cleaning service, design, and fabric shop. In such a model are involved not only the customer and the tailor, but also the designer, the cleaning service and the logistics. Why the cleaning service? Tailor-made clothes are often made of special fabrics that require special attention when cleaned. Therefore, the cleaning service should have the necessary competence in this respect.

3 General-System Architecture

Figure 3 shows the general architecture of the smart tailor system. Each participant in this system is connected to the internet via his personal computer. The customer can contact the tailor. The customer makes a photo of his whole body in tightly fitting underwear. This photo is sent as digital photo to the tailor. The tailor stores this photo with the customer-relevant data in the customer database. The designer,

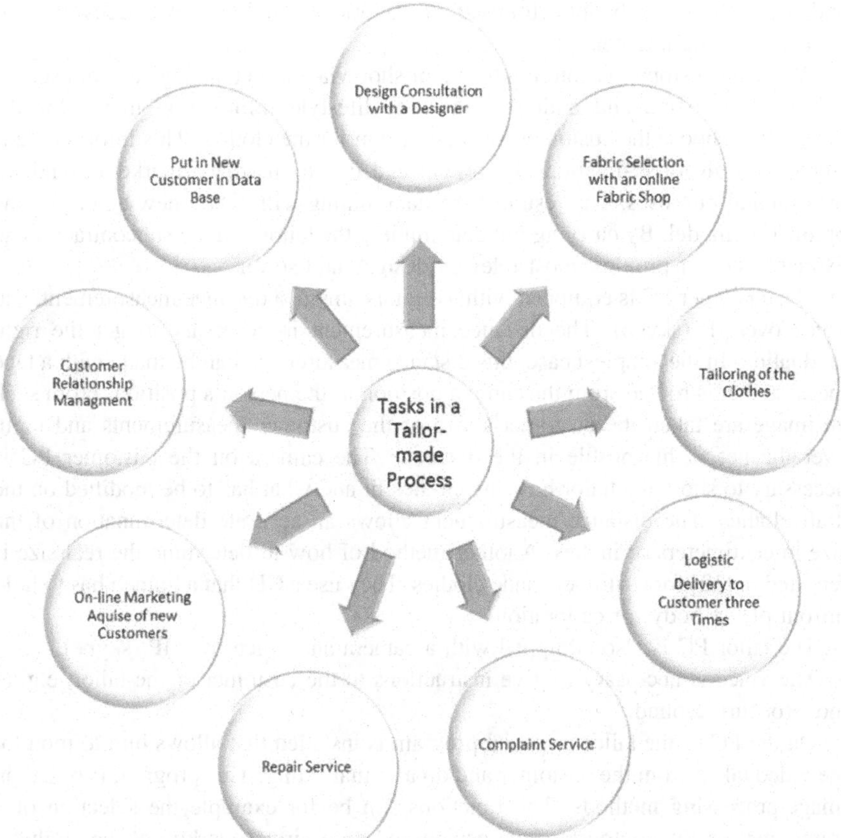

Fig. 2 New tasks in a tailor-made process

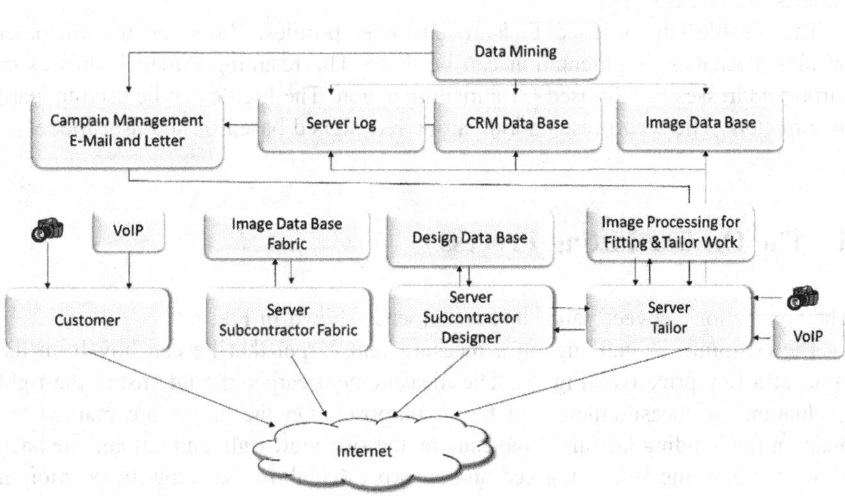

Fig. 3 General architecture of smart tailor production space

and, possibly the subcontractor with the fabric collection, have access to this database via the internet.

When the customer is entering the tailor shop via internet, he registers himself by inputting his name and address and some lifestyle information that helps the designer to choose the customer-relevant design for the clothes. This information is stored in a customer-relationship system and can be used for marketing and for data-mining purposes. The result of the data mining will give a new quality to the production model. By carrying out data mining, the tailor and the subcontractor can extract customer profiles, most relevant designs and so on.

The customer PC is equipped with a digital camera, a distance measurement, and voice over IP (skype). The distance measurement is necessary to get the right graduation. In the simplest case, this distance measurement can be made with a tape measure that is routed from the camera position to the person's position. From such an image are taken the customer's size or the customer measurements and input over the net to his profile in the database. The camera on the customer PC is necessary to show the tailor how the clothes fit and what has to be modified on the draft clothes. The distance measurement allows an accurate determination of the size in centimeters or inches. Another method of how to determine the real size is reported in [12] for customer-made clothes. They use a CD that a human has to hold infront of the body for calibration.

The tailor PC is also equipped with a camera and voice over IP (skype).

The voice is necessary to give instructions to the customer or the tailor, e.g. to move or turn around.

On the PC of the tailor a special program is installed that allows him to monitor the video taken from the customer and do a virtual fitting. This program is based on image-processing methods. The functions can be for example the selection of a frame, getting the contours of the person and do a virtual markup of the clothes.

The internet-based architecture also opens possibilities for further functions, such as marketing and redesigning of the web presentation according to the customer's preferences [18].

The log-file can be used to learn customer profiles. Based on the customer profiles a database segmentation can be done. The resulting e-mail addresses or surface addresses can be used for a mailing action. The logfile can be used to learn user models. The web presentation can be redesigned based on the user model.

4 The On-line Fitting Process

The connection between tailor and customer is shown in Fig. 4.

The customer is standing on a measurement carpet that he can buy from the tailor at a low price (see Fig. 5). The measurement carpet should ensure the right graduation for measurement and fitting purposes. On the carpet are marked the position for standing infront of the camera, the side view right and left and the back view. A measuring tape is marked on the carpet that allows viewing the position in

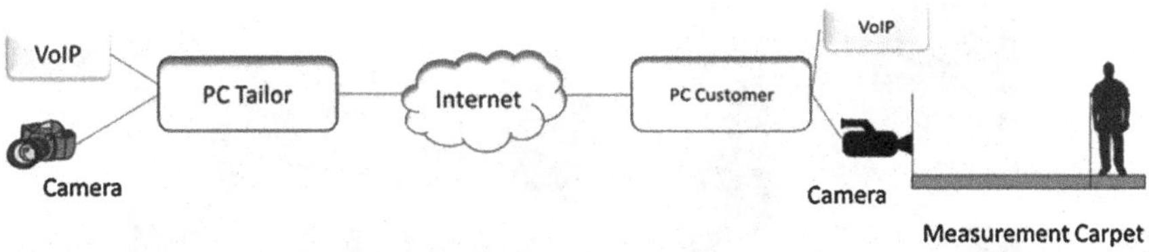

Fig. 4 Connection between tailor and customer

centimeters. In the same way as for the measurement process in [12] a disc is used to get the right graduation. The measurement and fitting software does the calibration.

The tailor controls the camera on the customer site. He can zoom in and out of the scene. A video is recorded, but if necessary the tailor can stop the video and display the single image on the PC screen. By doing so he can have a closer look to the different? Parts of the clothes and see how they fit the customer's body (see Fig. 6). What he is doing in real life with a piece of chalk, he is doing in the smart space by marking things out in the image. The measurement scale can be superimposed to the image by the software. Besides that, the tailor can take notes about the fit and the agreement with the customer about the design and annotate them in the image. The possibility to annotate the images by his notes will result in a quality improvement of the production process.

The tailor communicates with the customer over VoIP. He can give instructions for moving left or right or turning around and for controlling the position of the customer.

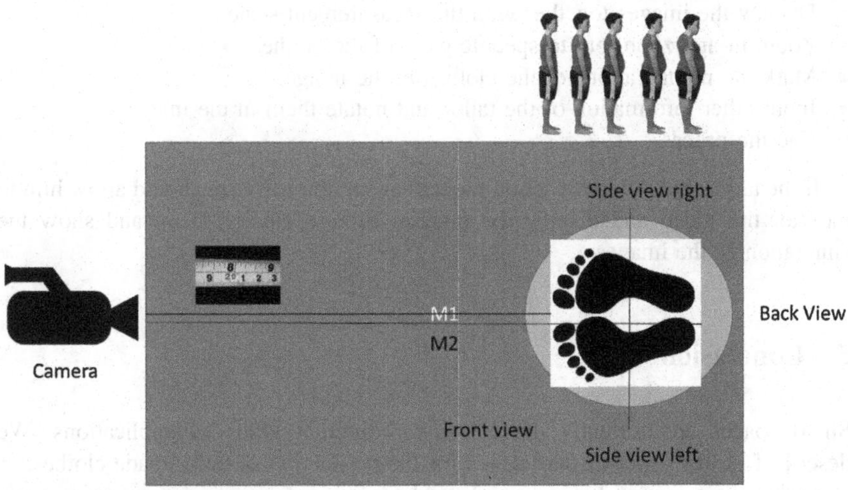

Fig. 5 Measurement and fitting process

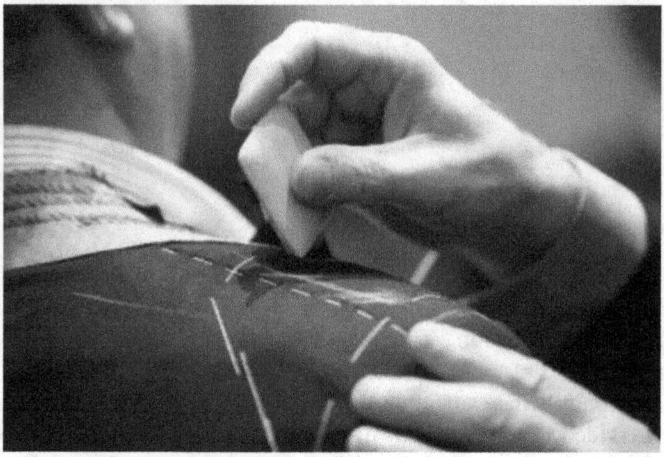

Fig. 6 The fitting process

The special functions, which the tailor-made production software should have, are as follows:

- Establishing connection between the customer and the tailor
- Record name of the customer, date and location, the name of the clothes under production, and the production steps?
- Store photo of the customer from frontal, left and right side view, back view
- Start measurement or fitting process
- Calculate the graduation and store the information
- Record the video and stop for a single image the tailor wants
- Display the images together with the measurement scale
- Zoom in and zoom out to specific parts of the clothes
- Mark the modifications of the clothes in the image
- Input other information of the tailor and notate them in the image
- End the process.

If the tailor starts his production process again, the software should allow him to monitor the video and display the images, browse among them, and show the annotation in the images.

5 Conclusion

Smart spaces are currently developed for different kinds of applications. We described in this paper a smart space for the production of tailor-made clothes.

Fashion is a serious business and can have a big economic impact on an area. Especially rural areas can profit from this fact. The services enclosed in this business model can be located at different places far away from each other.

Here, custom-made clothes are investigated. We studied the tailor made clothes-production process. Normally the production of tailor made clothes is a time-consuming process. It requires a lot of time from the customer and the tailor. The usage of computer technology and internet can help to reduce this time for both the customer and the tailor.

We described the tailor-made clothes-production process and identified other services that should be combined with the business model of tailor-made clothes. Starting on that we explained the functions a tailor-made clothes e-commerce shop should have.

We developed the architecture for such a business model and described the sensors that are necessary for the communication between the customer and the tailor.

The usage of a web presence does not only result in a new production model, it also has an impact on the marketing of the service of a tailor and other services. He can identify the customer profiles by using the log file and other information input in a database by the customer by using data mining methods. These customer profiles will help to improve the tailor's service to correspond better to the customer's needs. He can also use these customer profiles for e-mail marketing.

The communication between the customer and the tailor takes place via the internet. The hardware that is necessary on both sites consists of a camera and a microphone for VoIP. The customer should have a measurement carpet that allows fixing the position and the graduation. We adopted the idea with the CD for calculating the graduation. The tailor should control the camera on the customer's site. A special software is necessary in order to do the measurement and fitting. We described the functions of this software.

Acknowledgments This work has been done under the grant "A Study for the Business Model of Tailor-Made Clothes based on New Computer Technology and the Internet-of-Things", ETailor, IT 2014-6.

References

1. Ma, J., Yang, L.T., Apduhan, B.O., Huang, R., Barolli, L., Takizawa, M.: Towards a smart world and ubiquitous intelligence: a walkthrough from smart things to smart hyperspaces and UbicKids. J. Pervasive Comput. Commun. **1**, 53–68 (2005)
2. Evans, C., Mount, S., Xhafa, F., Chapman, C.: Smart Care Spaces. Int. J. Ad Hoc Ubiquit. Comput. **16**(4), 268–282 (2014)
3. Cuomo, S., De Michele, P., Galletti, A., Ponti, G.: Visiting styles in an art exhibition supported by a digital fruition system. In: International Conference on Signal-Image Technology and Internet-Based Systems, SITIS 2015 (2015)
4. Chianese, A., Piccialli, F., Valente, I.: Smart environments and cultural heritage: a novel approach to create intelligent cultural spaces. vol. 9, pp. 209–234 (2015)
5. Kisato, J.: Utilisation of E-Marketing Tools and Influencing Forces on the Performance of Micro and Small Fashion Enterprises in Nairobi County, Kenya, Ph.D. thesis, Kenyatta University, Oct 2014

6. Merlo, E., Polese, F.: Turning Fashion into Business: The Emergence of Milan as an International Fashion Hub, Business History Review 80 (Autumn 2006): 415Ð447. © 2006 by The President and Fellows of Harvard College
7. Habte Selassie, S.: Behaviours and attitudes towards a sustainable consumption of fashion, Master's degree in Fashion Management, The Swedish School of Textiles. Accessed 21 June 2010
8. Choi, S., Xinran, Y., Oleary, J.: What does the consumer want from a DMO website? a study of US and Canadian tourists' perspectives. Int. J Tourism Results **9**, 59–72 (2007)
9. Choi, S., Morrison, A.: Website effectiveness for bricks and mortar travel retailers. Anatolia: Int. J. Tourism Hospitality Res. **16**(1), 63–78 (2006)
10. Booth, M.: The future of fashion: the fashion industry's old business model is out of style. Los Angeles Times. http://articles.latimes.com/2009/sep/13/entertainment/et-future (2009). Fashion 13. Accessed 12 Mar 2012
11. Brun, A., Caniato, F., Caridi, M., Castelli, C., Miragliotta, G., Ronchi, S., Sianesi, A., Spina, G.: Logistics and supply chain management in luxury fashion retail: empirical investigation of Italian firms. Int. J. Prod. Econ. **114**, 554–570 (2008)
12. Knoke Felix 2011: Berlin Start-Up Aims to Revolutionize Clothes Shopping. Spiegel Online International. http://www.spiegel.de/international/business/tailored-for-success-berlin-start-up-aims-to-revolutionize-clothes-shopping-a-798250.html. Accessed 17 Nov 2001
13. Costa, M.: Online shopping: online fashion is extension of the high street. Market. Week 24–26 (2010)
14. Cillo, P., Verona, G.: Search styles in style searching: exploring innovation strategies in fashion firms, long range planning. Int. J. Strateg. Manag. **41**(6), 650–671 (2008)
15. El-Gohary, H.: New digital marketing and micro businesses. Int. J. Market. (2011)
16. Chaffey, D.: Total E-mail Marketing, Second Edition: Maximizing Your Results from Integrated E-Marketing, 2nd edn. Butterworth-Heinemann, London (2007)
17. Reichle, M., Perner, P., Althoff, K.-D.: Data preparation of web log files for marketing aspects analyses. In: Petra Perner (ed.): Advances in Data Mining, Applications in Medicine, Web Mining, Marketing, Image and Signal Mining, lncs 4065, p. 131–145. Springer (2006)
18. Perner, P., Fiss, G.: Intelligent e-marketing with web mining, personalization and user-adpated interfaces. In: Perner, P. (ed.) Advances in Data Mining, Applications in E-Commerce, Medicine, and Knowledge Management, LNAI 2394, pp. 39–57. Springer (2002)
19. Fisher, J.: Moving from a web presence to e-commerce: the importance of a business—web strategy for small-business owners. J. Electron. Markets (2007)
20. Scheer, C., Hansen, T., Loos, P.: Business models to offer customized output in electronic commerce. Integrat. Comput. Aided Eng. **10**(163), 163–175 (2003)
21. Perner, P.: Einrichtung zur automatischen Bestimmung von Änderungen bei der Anprobe inviduell bereitgestellter Kleidungsstücke, Patent DE 10 2013 003 438 A1

Influence of Some Parameters on Visiting Style Classification in a Cultural Heritage Case Study

Salvatore Cuomo, Pasquale De Michele, Ardelio Galletti and Giovanni Ponti

Abstract A smart system for a cultural exhibition, generally, has the ability to infer interests of users and to track the propagation of the information into the event. We are interested in analysing and studying the visiting styles of users in a real cultural heritage exhibition, named *The Beauty or the Truth*. Starting from data that was collected during this exhibition, the interaction between visitors and artworks and the influences of the available technology on their behaviours are studied. Finally, we analyse how the tuning of some parameters on a classification strategy influences the users' visiting styles. The obtained results have revealed interesting issues also to understand hidden aspects in the data and unattended in the analysis.

Keywords Clustering · Data mining · User profiling · Cultural heritage

1 Introduction

In the cultural heritage environment, visiting style dynamics and classification of users are very interesting research topics. Generally, digital technologies and innovative fruition strategies are combined in order to increase the quality of services and

---

S. Cuomo (✉) · P. De Michele
Department of Mathematics and Applications, University of Naples Federico II,
Strada Vicinale Cupa Cintia 26, 80126 Naples, Italy
e-mail: salvatore.cuomo@unina.it

P. De Michele
e-mail: pasquale.demichele@unina.it

A. Galletti
Department of Computer Science, University of Naples Parthenope,
Centro Direzionale, Isola C3, 80100 Naples, Italy
e-mail: ardelio.galletti@uniparthenope.it

G. Ponti
DTE-ICT-HPC, ENEA Portici Research Center, Piazzale Enrico Fermi 1,
80055 Portici, Naples, Italy
e-mail: giovanni.ponti@enea.it

G. De Pietro et al. (eds.), *Intelligent Interactive Multimedia Systems and Services 2016*, Smart Innovation, Systems and Technologies 55,
DOI 10.1007/978-3-319-39345-2_50

to enhance the spectator experiences. Several works are usually devoted to develop suitable technologies for adapting the event contents to single visitors or groups of these. Interesting approaches based on observing the behaviour of visitors and on profiling them with *a priori* data in order to provide non-intrusive bootstrapping of user models, have been proposed in [13]. In [11] the supervised data groupings by means of a Bayesian classifier is exploited. In [9] a new classification approach based on clustering techniques, in order to discover a reliable strategy to find data groupings in an unsupervised way. Moreover, in [7] a biologically inspired mathematical model simulating social network behaviours is presented. A more detailed analysis of these approaches is presented in [10]. Starting from these works, in [5, 6, 8] we have classified the visiting styles of spectators by means of a mathematical model and some heuristics, and have studied how visitors of a cultural heritage event interact with available technologies. In this works we focus on the analysis and the classification of visiting styles, starting form data collected by the overall technology available in a real exhibit. Deeply investigating these aspects, in this paper we discuss how the tuning of some parameters could influence the classification results. Moreover, we perform a clustering task, which emploies the well-known K-means algorithm [12], in order to achieve data groups that can reflect the users' classification obtained with our classifier. The art event we have analysed was named "*The Beauty or the Truth*",[1] and took place in the monumental complex of *San Domenico Maggiore* in the historical centre of Naples. The cultural objects, 271 sculptures divided into 7 thematic sections, were provided of the capability to interact with people, environments and other objects, and to transmit the related knowledge to users through multimedia facilities. Moreover, the visitors were active players to which have been offered the pleasure of the perception and the charm of the discovery of a new knowledge. A wireless sensor network, using bluetooth technology and able to sense the surrounding area for detecting user devices presence in a museum, has been a part of the ICT framework used in the exhibit and discussed in [1–4].

The paper is organized as follows. In Sect. 2 we formalize the visiting style and useful parameters for user classification. In Sect. 3 we shown how a quality value (i.e., a score) is assigned to a visitor's path in the exhibit. Section 4 is devoted to the experiments on the classification and the clustering task. Finally, the conclusions are drawn in Sect. 5.

2 Visiting Styles Definition

In order to classify visiting styles in art exhibition, we start from work shown in [13], where personalized information presentation in the context of mobile museum guides are reported and visitor movements are classified comparing these to the behaviours of four typical animals. In our work, we adapt this classification to find how

[1]http://www.ilbellooilvero.it.

Table 1 Characterization of the visiting styles' classification

	(a)	(b)	(c)
Animal	Viewed artworks	Average time	Path
A	high	–	high
B	high	–	low
F	low	low	–
G	low	high	–

visitors interact with the ICT technology. Accordingly, the visitor's behaviour can be compared to that of:

- an ANT (**A**), if this tends to follow a specific path in the exhibit and intensively enjoys the furnished technology;
- a FISH (**F**), if this moves around in the centre of the room and usually avoids looking at media content details;
- a BUTTERFLY (**B**), if this does not follow a specific path but rather is guided by the physical orientation of the exhibits and stops frequently to look for more media contents;
- a GRASSHOPPER (**G**), if this seems to have a specific preference for some preselected artworks and spends a lot of time observing the related media contents.

The classification of the visiting styles is characterized by three different parameters related to the visitor: (a) the number of artworks viewed, (b) the average time spent by interacting with the viewed artworks, and (c) the path determining the order of visit of the exhibit sections.

As we can observe in Table 1, high values for the parameter (a) characterize both **A**s and **B**s, while low values are related to **F**s and **G**s. Moreover, the parameter (b) does not influence the classification of **A**s and **B**s, while low values are typical for **F**s and high values are inherent in **G**s. Finally, the parameter (c) does not influence the classification of **F**s and **G**s, whereas high values are related to **A**s and low values characterize **B**s. Each parameter is associated with a numerical value normalized between 0 and 1. The values of the parameters (a) and (b) correspond to percentages of viewed artworks and average time spent for these, respectively. In Sect. 3 we will show how to assign a value to the parameter (c).

3 Assigning a Quality Value to the Parameter "Path"

In this Section, we are interested in defining a set of rules to assign a score between 0 and 1 to the parameter (c). We recall the notation used in a previous work [8], where we introduced a first way to assign this score. Assuming that the N sections (rooms) of a museum are organized in an increasing order as follows:

$$\boxed{1} \rightarrow \boxed{2} \rightarrow \cdots \rightarrow \boxed{\text{N-1}} \rightarrow \boxed{\text{N}},$$

we have denoted by

$$p: \quad S_1 \rightarrow S_2 \rightarrow \cdots \rightarrow S_{M-1} \rightarrow S_M$$

the path p of a visitor. Optionally, we can assume the entrance in 1 and the exit in N, so that it results $S_1 = 1$ and $S_M = N$. Moreover, because a visitor could visit each room more than once, or never, the length M of the path is independent on the number N of rooms. Furthermore, we have assigned a score $v(p)$ to the path p in the following way. Considering the $M - 1$ movements

$$S_{j-1} \rightarrow S_j, \qquad \forall\, j = 2, \ldots, M,$$

we have assigned at each of them a value m_j by setting

$$m_j = \begin{cases} 1, & \text{if } S_j = S_{j-1} + 1 \\ 0.5, & \text{if } S_j > S_{j-1} + 1 \\ S_j - S_{j-1}, & \text{if } S_j < S_{j-1} \end{cases}$$

Finally, in this way, the score of the path was

$$v(p) = \max \left\{ 0, \frac{1}{N-1} \sum_{j=2}^{M} m_j \right\}.$$

The idea underlying is to consider correct the forward movements ($m_j = +1$ for $S_j = S_{j-1}+1$, or $m_j = +0.5$ for $S_j > S_{j-1}+1$) and to penalize the backward ones ($m_j = -(S_{j-1} - S_j)$). It is easy to prove that to divide by $N - 1$ gives the sum of m_j's less than 1. A further *max* condition has been imposed to avoid negative values. Notice that the *optimal path*

$$p_1: \quad 1 \rightarrow 2 \rightarrow 3 \rightarrow \cdots N - 1 \rightarrow N,$$

receives the maximum value $v(p_1) = 1$. However, we remark that this way to assign the v value seems to be too penalizing in several cases where we believe that the quality of the path is not so bad. For instance, the path

$$p_2: \quad 1 \rightarrow 2 \rightarrow 3 \rightarrow \cdots N - 1 \rightarrow N \rightarrow 1,$$

is quasi-optimal, in the sense that it *contains* the optimal path and an unique repeated room. But this path receives the score $v(p_1) = 0$, while we would reserve the worse value 0 to the empty path only (i.e. to the path where no room has been visited). This phenomenon is due to the presence of a (too) penalized backward step. Observe that the same thing also happens to a path as

$$p_3 : \quad \underbrace{1 \to 2 \to \cdots \to N}_{\text{1-st cycle } (N-1 \text{ movements})} \;\to\; \underbrace{1 \to 2 \to \cdots \to N}_{\text{2-nd cycle } (N-1 \text{ movements})} \;\to\; \underbrace{1 \to N}_{\text{3-rd cycle (1 movement)}},$$

containing two optimal *cycles*, which receives a small score $v(p_3) = 1/(N-1)$. With the term cycle we mean to define each sub-path consisting in increasing steps only. Of course, a new cycle begins, along the path, every time a backward movement occurs. These drawbacks in assigning the score are probably one of the main causes in some troubles of our classification scheme in [8] (measured with respect to the k-means results). Then, to enhance these classification results, here we propose a new way to assign the score of a path which does not excessively penalize paths containing optimal sub-paths (cycles). To this aim, we denote the generic path consisting in $L \geq 1$ cycles, with the notation

$$p : \quad \underbrace{S_{c_0+1} \to \cdots \to S_{c_1}}_{\text{1-st cycle}} \to \underbrace{S_{c_1+1} \to \cdots \to S_{c_2}}_{\text{2-st cycle}} \to \cdots \to \underbrace{S_{c_{L-1}+1} \to \cdots \to S_{c_L}}_{L\text{-th cycle}},$$

where $c_0 = 0$ and $c_L = M$. As before, we give a value $\bar{m}_j$ to each movement $S_{j-1} \to S_j$ but in this case no negative values are assigned. We consider the steps in the first cycle better than the steps in the other cycles, i.e. we set the $\bar{m}_j$ values decreasing along the cycles. More in detail, we set

$$\bar{m}_j = 0.9^{c-1} \qquad \text{if } S_j \text{ is in the } c\text{-th cycle}$$

The basis 0.9 is chosen to weight less and less (90 %) the steps in the new cycles. A mean of all $\bar{m}_j$ values, that is their sum divided by $L \cdot N$, gives a possible score where a path with few optimal cycles is not bad evaluated but, again, cases as p_2 will receive a small score. Our idea to take also into account the presence of optimal sub-paths, is to assign the score with the following rule:

$$\bar{v}(p) = \alpha \max_{i=1,\ldots,L} \frac{c_i - c_{i-1}}{N} + (1-\alpha)\frac{1}{N \cdot L}\sum_{j=1}^{M} \bar{m}_j \qquad (0 \leq \alpha \leq 1).$$

Using this formula, the score depends on the weights α and $1-\alpha$ and on the maximum length of the sub-paths. The parameter α is crucial: a high value (close to 1) preserves path with at least an optimal sub-path, while a small value (close to 0) would penalize *chaotic* paths regardless of they may contain good sub-paths. For instance, setting $\alpha = 0.7$ and recalling previous paths p_1, p_2, p_3, we get:

$$\bar{v}(p_1) = 0.7 \cdot \frac{N}{N} + 0.3 \cdot \frac{1}{N}\sum_{j=1}^{N} 1 = 1,$$

$$\bar{v}(p_2) = 0.7 \cdot \frac{N}{N} + 0.3 \cdot \frac{1}{2N} \left(0.9 + \sum_{j=1}^{N} 1 \right) \approx 0.85,$$

$$\bar{v}(p_3) = 0.7 \cdot \frac{N}{N} + 0.3 \cdot \frac{1}{3N} \left(\sum_{j=1}^{N} 1 + \sum_{j=1}^{N} 0.9 + 0.81 \right) \approx 0.89.$$

In the following Section, we apply this rule to assign the path score and use the results to perform the visiting style classification.

4 Experiments on Data Classification

The experiments described in this Section were carried out from a dataset of 253 log files, one for each visitor of the exhibition. We have tracked the visitor behaviour by using a suitable Extrapolation Algorithm (EA), which has a JSON file as input data. Such files are particularly suitable to identify visitors' behaviours not only regarding their interactions with the artworks, but also with respect to the whole exhibition. In fact, properly looking at the JSON files, for each visitor it is possible to determine the sequence of visited sections and, with respect to each artwork, the time visitor spent to enjoy audio and image contents, and if he has displayed or not text information. Starting from the data collected in the exhibit, we deployed a classifier in order to characterize the visiting behaviours by means of some heuristics and to investigate how visitors interact with the supporting technology. According with this and with the characterization of the classification illustrated in Table 1, in a preliminary phase, we fixed 10 % for viewed artwork (a), 60 % for average time (b), and 70 % for path (c). With these thresholds, we obtained the results shown in Fig. 1. We observe that there are a 5.5 % of **As**, a 17 % of **Bs**, a 26.9 % of **Fs** and a 50.6 % of **Gs**. The distribution of the "animals" at the end of the exhibit is suitable to understand how visitors interact with cultural objects through the technology. On the other hand, in

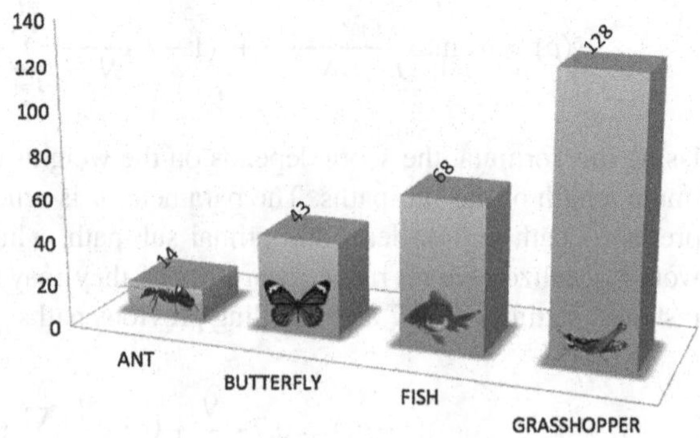

Fig. 1 Classification results obtained choosing (a) = 10 %, (b) = 60 % and (c) = 70 %

order to accomplish this task, it is necessary to validate the classification results, resorting to unsupervised approaches, as shown in the next Subsection.

4.1 Clustering

We performed a further experimental phase based on clustering algorithm. The aim of this step consists in executing an unsupervised data mining algorithm in order to achieve data groups that can reflect the user classification obtained with our classifier. In this direction, we propose a data structure that reflects the above mentioned observations about the path and the media content fruition. The dataset is built from the log files and is structured in the ARFF Weka format, as shown in the following.

```
@RELATION ARTWORKS
@ATTRIBUTE viewed NUMERIC [0..1]
@ATTRIBUTE avg_time NUMERIC [0..1]
@ATTRIBUTE path NUMERIC [0..1]
@ATTRIBUTE class {A,B,F,G}
@DATA
...
0.0836653386454,0.846588116217,?,G
0.0478087649402,0.317966675258,?,F
0.119521912351,?,0.714285714286,A
0.175298804781,?,0.470303571429,B
...
```

In particular, the dataset contains fields regarding the percentage of viewed artworks (i.e., `viewed`), the average time spent interacting with an artwork (i.e., `avg_time`), and the percentage of path followed during the visit (i.e., `path`). Note that these three attributes reflect the parameters (i.e., (a), (b) and (c)) introduced in Sect. 2, which characterize the visitors' behaviours. We have resorted to the well-known K-means partitional clustering algorithm [12] and set the number of classes to $K = 4$. In Table 2 we report the results of the clustering (with $K = 4$) for the entire exhibit. Note that **Cluster0** corresponds to **G**, **Cluster1** is **F**, **Cluster2** represents **B** and **Cluster3** is **A**, as this is a typical majority voting based cluster assignment. This clustering

Table 2 Results of the clustering ($K = 4$) for the entire exhibit

Animals	Cluster0	Cluster1	Cluster2	Cluster3
A	0	0	0	14 [**14**]
B	0	0	32 [**43**]	11
F	7	61 [**68**]	0	0
G	128 [**128**]	0	0	0

In brackets the classification results

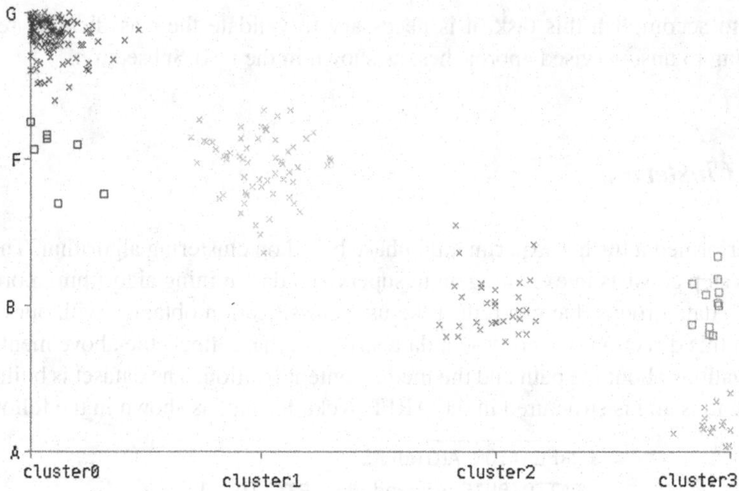

Fig. 2 *K*-means cluster assignment ($K = 4$) with respect to the classification results choosing (a) = 10 %, (b) = 60 % and (c) = 70 %

session provides very interesting results that can be seen in the Table. In fact, the four categories in our data have been correctly identified by our classifier with an accuracy very close to the K-means results, with a percentage of incorrectly clustered instances equal to about 7.11 %. Figure 2 shows the cluster assignments for tuples in the dataset (the square points indicate misclassified instances). These are coloured by following the classes' attribute, whereas on the axes there are class-IDs (i.e., x-axis) and cluster-IDs (i.e., y-axis). Clustering results confirmed what we addressed in the preliminary classification task, as the algorithm correctly identify classes **A** and **G**, achieving a very high accuracy results. The poor clustering error is due to the classes **B** and **F**, with the error of 11 tuples for **B** and 7 tuples for **F**. This is not a surprising result, since we yet noticed in the previous classification task that **B** and **F** have very similar trends.

4.2 Tuning of the Parameters

In the previous experimental phase, we obtained very interesting results in terms of accuracy, with a very low clustering error (i.e., 7.11 %). However, such results came from an our intuition in setting proper thresholds in classifying visitor behaviours. In this Section, we focus on the tuning of the classification parameters, in order to show how they bias the cluster accuracy. We described in Sect. 2 how setting thresholds in order to classify visiting styles reflecting the ethnographic behaviour. Based on a possible range of values for each parameter, we defined a set of experiments in which the classification changes and the clustering strategy/setting remains the same. The

Table 3 Tuning of the classification parameters

Quality	Parameters (%)			Accuracy error (%)	# Misclassified
	(a)	(b)	(c)		
Best	10	50	58	0	0
Medium	10	6	7	7.11	18
Worst	20	60	70	35.57	90

Table 4 Results of the clustering ($K = 4$) for the entire exhibit with (a) = 10 %, (b) = 50 % and (c) = 58 %. In brackets the classification results

Animals	Cluster0	Cluster1	Cluster2	Cluster3
A	0	0	0	25 [**25**]
B	0	0	32 [**32**]	0
F	0	61 [**61**]	0	0
G	135 [**135**]	0	0	0

goal here is to discover the best setting for the classification that produces the highest accuracy results in the clustering phase for the entire exhibit. We found several clustering accuracy results. In Table 3 we show only three most relevant settings, which produce the best, the medium and the worst quality results. Here, the first column expresses the quality evaluation of the tuning and the following three columns represent the settings for the three classification parameters, that are "viewed" artworks (a), "average time" (b), and "path" (c). We can notice that the first setting produces the best accuracy in clustering (0 % error). However, this result is very close to the one we obtained without the tuning (7.11 % error), and this suggests that our preliminary intuitions were right.

The worst setting produces an error of about 35.57 %, which is anyway acceptable in clustering especially if we consider that our dataset is not so big (i.e., 253 tuples). The classification with (a) = 10 %, (b) = 50 % and (c) = 58 %, at the end of the exhibit, produces 25 **A**s, 32 **B**s, 61 **F**s and 135 **G**s. Table 4 compares these with the clustering results in which we can observe that there are no misclassified instances.

5 Conclusions

In this paper we deal with the characterization of the visiting style classification starting from real datasets. The results strongly depend on the heuristic metrics related to the digital fruition system. We have presented a strategy to fine set important parameter of the proposed classification approach. To validate the results, we resorted to the K-means clustering algorithm to discover data groups reflecting visitors' behaviours in all the sections of the exhibit. In our experiments, to perform the classification and the clustering tasks, we defined a structured database based on real log files. Experimental results shown that the clustering task correctly identifies the visitors' behaviours, providing high accuracy clusters that reflect the classification results.

Acknowledgments Authors thank DATABENC, a High Technology District for Cultural Heritage management of Regione Campania (Italy) and the CHIS project that has financial supported this work.

References

1. Amato, F., Chianese, A., Mazzeo, A., Moscato, V., Picariello, A., Piccialli, F.: The talking museum project. Procedia Comput. Sci. **21**, 114–121 (2013)
2. Chianese, A., Marulli, F., Moscato, V., Piccialli, F.: Smartweet: a location-based smart application for exhibits and museums. Signal-Image Technol. Internet-Based Syst. (SITIS) **2013**, 408–415 (2013)
3. Chianese, A., Piccialli, F., Riccio, G.: Designing a smart multisensor framework based on beaglebone black board. Comput. Sci. Appl. **330**, 391–397 (2015)
4. Chianese, A., Piccialli, F., Valente, I.: Smart environments and cultural heritage: a novel approach to create intelligent cultural spaces **9**, 209–234 (2015)
5. Cuomo, S., De Michele, P., Galletti, A., Pane, F., Ponti, G.: Visitor dynamics in a cultural heritage scenario. In: DATA 2015—Proceedings of 4th International Conference on Data Management Technologies and Applications, pp. 337–343. Colmar, Alsace, France, 20–22 July 2015
6. Cuomo, S., De Michele, P., Galletti, A., Piccialli, F.: A cultural heritage case study of visitor experiences shared on a social network. In: 10th International Conference on P2P, Parallel, Grid, Cloud and Internet Computing, 3PGCIC 2015, pp. 539–544. Krakow, Poland, 4–6 Nov 2015
7. Cuomo, S., De Michele, P., Galletti, A., Ponti, G.: A biologically inspired model for analyzing behaviours in social network community and cultural heritage scenario. In: Tenth International Conference on Signal-Image Technology and Internet-Based Systems, SITIS 2014, pp. 485–492. Marrakech, Morocco, 23–27 Nov 2014
8. Cuomo, S., De Michele, P., Galletti, A., Ponti, G.: Visiting styles in an art exhibition supported by a digital fruition system. In: Eleventh International Conference on Signal-Image Technology and Internet-Based Systems, SITIS 2015. Bangkok, Thailand, 23–27 Nov 2015
9. Cuomo, S., De Michele, P., Ponti, G., Posteraro, M.R.: A clustering-based approach for a finest biological model generation describing visitor behaviours in a cultural heritage scenario. In: DATA 2014—Proceedings of 3rd International Conference on Data Management Technologies and Applications, pp. 427–433. Vienna, Austria, 29–31 Aug 2014
10. Cuomo, S., De Michele, P., Ponti, G., Posteraro, M.R.: Data management technologies and applications. In: Third International Conference, DATA 2014. Vienna, Austria, 29–31 Aug 2014. Validation Approaches for a Biological Model Generation Describing Visitor Behaviours in a Cultural Heritage Scenario, pp. 154–168. Springer International Publishing, Cham (2015)
11. Cuomo, S., De Michele, P., Posteraro, M.R.: A biologically inspired model for describing the user behaviors in a cultural heritage environment. In: 22nd Italian Symposium on Advanced Database Systems, SEBD 2014, pp. 292–302. Sorrento Coast, Italy, 16–18 June 2014
12. Jain, A., Dubes, R.: Algorithms for Clustering Data. Prentice-Hall (1988)
13. Zancanaro, M., Kuflik, T., Boger, Z., Goren-Bar, D., Goldwasser, D.: Analyzing museum visitors' behavior patterns. User Modeling 2007. Lecture Note in Computer Science **4511**, 238–246 (2007)

Opinions Analysis in Social Networks for Cultural Heritage Applications

Flora Amato, Giovanni Cozzolino, Sergio Di Martino, Antonino Mazzeo, Vincenzo Moscato, Antonio Picariello, Sara Romano and Giancarlo Sperlí

Abstract Social media provide a great amount of valuable information in the form of messages posted by users. Information extracted from posts can be considered like features giving insights about the preferences of users towards certain events. These features can be used to generate recommendations looking forward for upcoming events they might find interesting. In this work we present system for opinion analysis from tweets and recommendation of cultural heritage events. At this aim, we detect the events of interest from Tweets and propose a methodology for associating a sentiment degree with a tweet using NLP techniques.

F. Amato (✉) · G. Cozzolino · S. Di Martino · A. Mazzeo · V. Moscato · A. Picariello · S. Romano · G. Sperlí
DIETI - Department of Electrical Engineering and Information Technology, University of Naples “Federico II”, Naples, Italy
e-mail: flora.amato@unina.it

G. Cozzolino
e-mail: giovanni.cozzolino@unina.it

S. Di Martino
e-mail: sergio.dimartino@unina.it

A. Mazzeo
e-mail: antonino.mazzeo@unina.it

V. Moscato
e-mail: vincenzo.moscato@unina.it

A. Picariello
e-mail: antonio.picariello@unina.it

S. Romano
e-mail: sara.romano@unina.it

G. Sperlí
e-mail: giancarlo.sperli@unina.it

S. Romano
CeRICT scrl - Centro Regionale Information Communication Technology, Complesso Universitario di Monte Sant’Angelo, Naples, Italy

G. De Pietro et al. (eds.), *Intelligent Interactive Multimedia Systems and Services 2016*, Smart Innovation, Systems and Technologies 55,
DOI 10.1007/978-3-319-39345-2_51

1 Introduction

In the field of recommender systems, in recent years a lot of research efforts have been devoted to explore the possibility of using social networks to get insights on public opinion, with the goal to suggest the most relevant "facts" to a user according to the user learned or stated preferences. Indeed, thanks also to the spreading of mobile technologies allowing users to "geotag" locations of interest on Social Networks, it is now possible not only to detect events where multiple users congregate, but also to get an insight on their "hype". This is clearly a rich feature space to understand user behaviours and preferences towards certain events, and that can be exploited to generate new recommendations for upcoming events of interest for users.

Among existing Social Networks, Twitter has become a reference point for worldwide events and discussions, thanks to the immediateness of its model, enabling its users to post/read *tweets*, i.e. short messages up to 140 characters. Twitter messages provide valuable information about events of different types, reflecting the users perspectives and interests [1–6]. In the literature, many works have shown that Twitter users spread news about events faster than the traditional news media [5, 7] and also in anticipation of a planned event, which can lead to early identification of interest in these events. Moreover, Twitter users reflect their interests on local events where traditional news coverage is low or nonexistent.

Due to these interesting features, several approaches to event recommendation have been proposed in the literature leveraging Twitter data. In [8] the authors propose an event recommender system for Twitter users, which identifies twitter activity co-located with previous events, and uses it to drive geographic recommendations via item-based collaborative filtering. Another aspect is to provide recommendation for twitter users for the creation of relationships between users [9]. In [10] is proposed a news recommender system suggesting articles based on user interests rather than presenting articles in order of their occurrence. In [11] sentiment analysis is exploited for social recommender system to get insights on user interest similarities.

Nevertheless, the full exploitation of the knowledge in the Twitter messages is still an open and challenging issue. On one hand, event detection from Twitter is a problem due to the heterogeneity of messages, which can contain personal updates and re-tweets not related to any real-word event [12]. On the other hand, the limitation of 140 characters per Twitter message, and sometimes the low quality of the textual data makes difficult to associate a tweet to the events.

In this work we present a system for topic detection and opinion analysis from tweets having the aim of suggesting upcoming cultural heritage events. The system aims at suggesting events occurring to user proximity according to the user preferences extracted automatically from the posted tweets. At this purpose, we analyse the user past behaviour on Twitter for determining his preferences on the base of extracted topics and sentiments. On the other hand our system provides facilities to discover all the tweets for upcoming events in a given area. The events happening in the user's neighborhood are ranked and recommended according to their automatically extracted preferences.

The reminder of the paper is structured as follows: in Sect. 2 we present the proposed system for opinion analysis from tweets and recommendation of cultural heritage events; in Sect. 3 we describe the methodology for associating a *sentiment degree* with a tweet and for detecting the related most important *topics*; in Sect. 4 we provide the description of the methodology adopted to detect the most popular events; in Sect. 5 we provide a case study highlighting the system functionalities; in the end, in Sect. 6 we present some conclusions and future work.

1.1 Motivating Example

In this section, we describe an application example in the Cultural Heritage domain that helps to better understand our approach. In particular, we consider a scenario in which a user interested in history and archeology wants to participate to cultural events such as guided tours, exhibitions or shows.

Likely, an archeology enthusiast visitor has participated to discussion groups related to cultural heritage topics. Thus, the system can collect users preferences analysing the Twitter stream in order to identify the text and the tags of their tweets, and can rank which event can be more interesting according to the user's preferences [13].

The system has to provide a set of functionalities for recommending personalized cultural thematic events. It should consider user preferences (both declared and acquired analysing tweets), context information (current location, weather forecasts, crowding at the entrance to other buildings, etc.), opinions and behaviors of other people who have similar experiences. In the same time, the system monitors Twitter stream in order to detect the upcoming events, and organize them for area and topic. In this way, the system can alert an user of the presence of interesting events in a certain area.

As an example, we assume that in the historical center of Naples a guided tour has been organized to show the masterpieces of Caravaggio (Michelangelo Merisi) located in some churches. Many users like people from the Naples Metropolitan area and people interested in historical artifacts, post several tweets regarding the tour. In addition, mass media, institutional agency and etc. promote the events on many channels of communication, as social networks including Twitter.

Once the system has collected these tweets, it can infer the details of the event useful for providing recommendations. If the user is walking through the streets of the historical district, when the system detects the position of the user by means of GPS, it alerts her/him in order to report events. In this way the user is aware of an event of potential interest, without having actively searched for information about it.

2 System Overview

In this section, we provide a brief description of the proposed system. Figure 1 depicts a functional overview in terms of the main system components that we are detailing as in the following.

The *Tweet Analyzer* component has as main task the automatic extraction of topics and sentiments from tweets through a deep analysis of the related texts realized by a set of specific modules.

A *Meta-Noise Filter* module aims at rejecting, and thus removing, inaccurate or irrelevant words in the input text, called *meta-noise* words, because they have not attached useful meaning. Once this task has been performed, typical noise words coming from tags are rejected. This activity exploits some linguistic restrictions over the considered vocabulary, such as *stop-word lists*, and different *text filters* based on the maximum word length and on several syntactic anomalies. Moreover, a particular filter is used to reject *emoticons* and *http links*.

In addition, a *Named Entity Filter* module is used to find particular words (*named entities*) with their related semantic classes (person, organization, location, date, etc.) in the input text and filter them on the base of some available *dictionaries* [14, 15]. Only, named entities related to the Cultural Heritage domain are then considered for the successive elaboration steps [16, 17].

The *Linguistic Normalization & Tagging* module applies the proper NLP pipeline on the pre-processed text. In particular, classical *sentence splitting*, *stemming*, *lemmatization* and *tokenization* activities are performed in order to avoid influence in

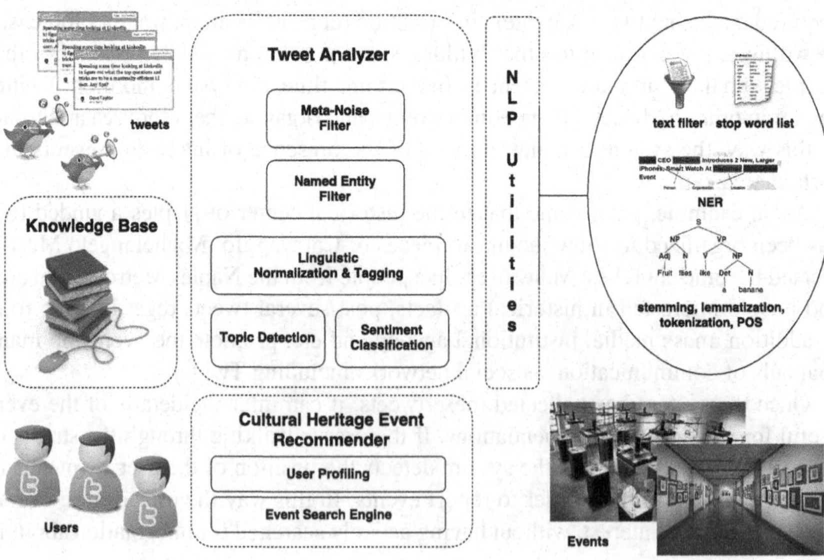

Fig. 1 System overview

text processing of word inflections, capital characters, plurals, punctuation, etc. Successively, a *part of speech tagging* associates each extracted token with the related grammatical class (noun, adjective, verb, etc.). Note that the hashtags are subject to further processing in order to split them in the composing words.

The *Topic Detection* and *Sentiment Classification* modules have the goal of discovering the main *topics* and *sentiment degree* related to a given tweet, respectively. Each module exploits a part of extracted tokens and uses specific linguistic corpora (as training sets) together with machine learning facilities to accomplish the related task, as described in the following sections.

The *Cultural Heritage Event Recommender* has the objective of automatically suggesting upcoming cultural heritage events (e.g. art exhibitions in museums or historical buildings, availability of special offers for visiting museums, archeological sites, etc.) on the base of user *cultural preferences* (e.g. paintings of Baroque genre).

The *User Profiling* module examines user past behavior (and behavior of users belonging to the same Twitter groups) and determines the related preferences on the base of topics and sentiments extracted from his/her published tweets, using a *rule-based* approach. From the other hand, the *Event Search Engine* provides facilities to discover all the tweets for upcoming events in a given area.

The *Knowledge Base* component contains and manages all the information necessary to the system operation. It consists of all linguistic resources (corpora, vocabularies, dictionaries, word lists, etc.), user tweets, preferences and logs (timestamped actions of Twitter users), and maintains all the results of the text analysis.

Finally, the *NLP Utilities* component collects all the described NLP functionalities.

Concerning implementation details, all the modules have been realized using JAVA technologies. In addition, the NLP Utilities are provided as web services and leverage the *Stanford NLP libraries*,[1] while the Knowledge Base is built on the top of *MongoDB*[2] document database and *Cassandra*[3] column-oriented repository.

3 Sentiment Analysis and Topic Detection for CH Tweets

In this section we describe the proposed methodology for associating a *sentiment degree* with a tweet and for detecting the related most important *topics*.

In a first preliminary stage, each tweet is processed exploiting the NLP pipeline described in Sect. 2.

[1]http://nlp.stanford.edu/software/.

[2]http://www.mongodb.org.

[3]http://cassandra.apache.org/.

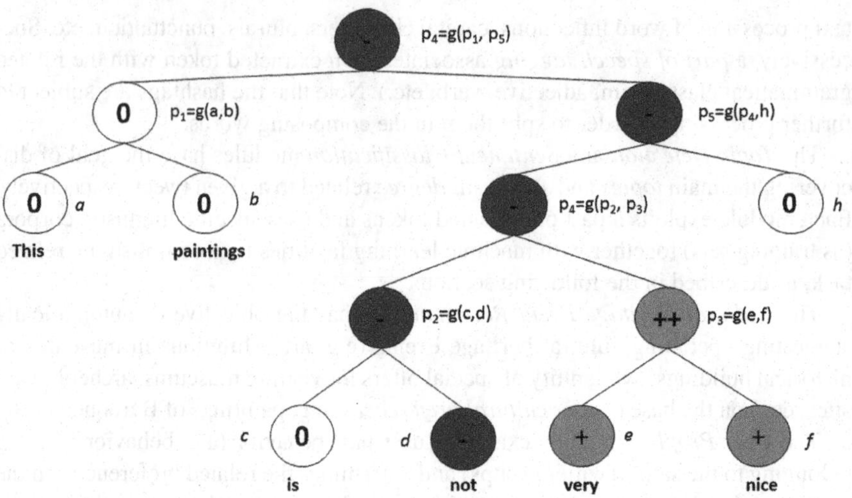

Fig. 2 Example of RNN approach

From one hand, the sentiment analysis phase allows to compute a sentiment score for each tweet using machine earning approaches [18]; in particular, we classify each tweet in one of the following five classes: *very negative*, *negative*, *neutral*, *positive* and *very positive*. We firstly build a tree which contains semantic annotations of each tweet sentence and successively employ a *Recursive Neural Network (RNN)* to compute the score for each sentence in order to classify the related text in one of the five classes leveraging *compositional vector representations* for phrases of variable length and syntactic type.

In this model, we split the text of a tweet into sentences and for each one building a binary tree: each leaf, corresponding to a single word, is a dimension of the vector space model. Then, we use RNN in order to compute parent vectors in a bottom up way using a compositionality function g as shown in Fig. 2.

Such vectors constitute the input features of a *softmax* classifier that computes the a-posteriori probability over labels, in order to classify tweets into one of the five classes. The aim of the approach is thus to minimize the *cross entropy* error between the vector representation at the root of each parsetree and the corresponding target vector distribution.

From the other hand, the topic detection activity is performed using a LDA-based approach that some of the authors have proposed in [19].

4 Discovery of Events of Interest from Tweets

One of the main challenges in the definition of the proposed system is to obtain meaningful information about upcoming cultural events in a given area. While it is pretty easy to get a list of all the on-going and upcoming exhibitions for a venue (for example downloading it from the venue web site), it is still hard to predict the "hype" for each event given the current state of the technology. Indeed, to date it is hard to understand, given a list of exhibitions, which of them will attract a lot of visitors, and which of them will be not be successful.

The idea we propose is to extract this knowledge by mining tweets, looking for the most popular exhibitions, i.e. those that obtain the highest number of tweets with positive sentiments. To this aim, we need to integrate a burst detection algorithm and sentiment analysis subsystem.

More in details, the solution we developed is to integrate a surveillance algorithm, intended as a burst detector, in the pipeline discussed in the previous Section, with the goal to analyze the Cultural Heritage-related Twitter data streams. In particular, we used some algorithms included in *R* within the free package *Surveillance* from CRAN project repository, which implements multiple statistical methods for the Spatio-Temporal Modeling and Monitoring of data series [20]. We employed the algorithms from the `Early Aberration Reporting System` (EARS) family [21], which compute a test statistic on day t as follows:

$$S_t = \max(0, (X_t - (\mu_t + k\,\sigma_t))/\sigma_t) \quad (1)$$

where X_t is the count of episodes on day t, k is the shift from the mean to be detected, and μ_t and σ_t are the mean and standard deviation of the counts during the baseline period.

EARS uses three baseline aberration detection methods:

- `C1-Mild`, where the baseline is determined on the average count from the past 7 days;
- `C2-Medium`, where the baseline is determined on the average count from the 7 days in the 10 days prior to 3 days prior to measurement.
- `C3-High`, that uses the same baseline as C2, but takes a 3 day average of events to determine the measure.

The rationale behind the selection of the EARS family is that it requires a very limited number of previous data to provide an alert, thus being potentially suitable for the burst detection of new topics regarding events. Moreover, it has basically no parameters to tune, make its applicability straightforward.

The final outcome of this module is a ranked list of all the most popular exhibitions for a given geographical area. Thus, tweets coming from the monitored area, and subjected to topic detection and sentiment classification activities, will be then used by the developed system for suggesting cultural events on the base of the similar users that have been located in the surrounding areas.

5 Cultural Heritage Applications: A Case Study

Based on the proposed system depicted in Sect. 2, we did a first exploratory study with two-fold focus: (1) to test with real users the effectiveness of the sentiment analysis methodology for associating a sentiment degree and for related topics; and (2) to gain insight in current context in order to elicit sites suitability thus enhancing user experience of Cultural Heritage [22–24].

We consider as real case study the *Historic Centre of Naples*, the largest historic centre in Europe, UNESCO World Heritage Site in 1995. A great number of cultural and artistic resources are located in this area.

The users interact with our system using—in the present release—an *Android* client application. The application collects user's preferences analysing the information coming from tweets or provided by the personal profile.

Once the user location is detected, the core process of Cultural Heritage Event Recommender module proposes the set of candidates by considering the set of ranked upcoming events that are closer to the current user's position, whose topics match with the user's preferences and the average sentiment is positive.

Successively, the system sends a notification to the mobile application to inform the user of potentially interesting events, properly visualized on the area map.

Tracking the GPS coordinates of user mobile phone, the system can get an implicit feedback about the suggestions (Fig. 3). In addition, the user can rate the proposed events so evaluating her/his experience.

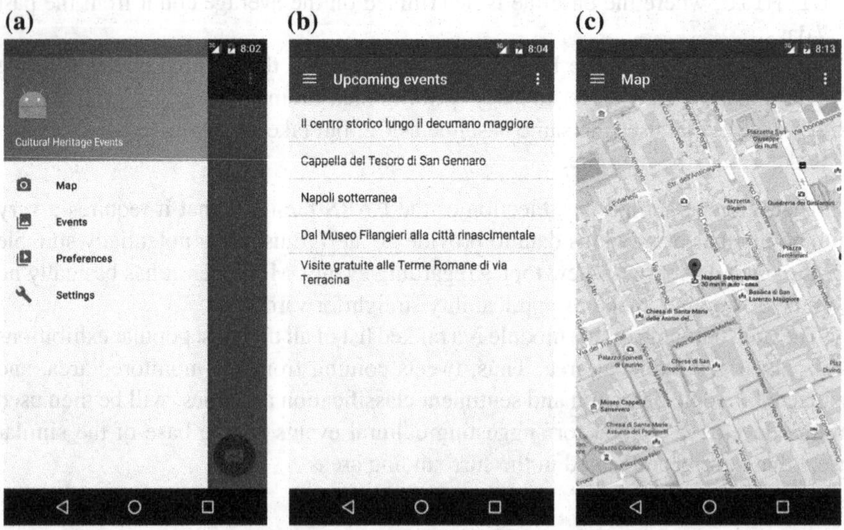

Fig. 3 Android client application of the proposed recommender system. **a** Client Application menu. **b** List of upcoming events. **c** Map of the proposed event

6 Conclusions and Future Work

In this work we presented a system for topic detection and opinion analysis from tweets and recommendation of cultural heritage events. At this aim, we developed a detection method for recognizing events of interest from tweets, implementing a methodology to associate a sentiment degree with a tweet using NLP techniques. It is a common belief, indeed, that social media provide a great amount of valuable information in the form of messages posted by users. Information extracted from posts can then be considered like features, giving insights about the preferences of users towards certain events. The idea behind this paper is that these features can be used to generate recommendations, looking forward for upcoming events they might find interesting.

Future works will be devoted to validate the proposed approach by an exhaustive experimental campaign in an attempt to understand the relationships among the user posts and his preferences in the specialist domain of the cultural heritage.

References

1. Becker, H., Naaman, M., Gravano, L.: Beyond trending topics: real-world event identification on twitter (2011)
2. Diakopoulos, N., Naaman, M., Kivran-Swaine, F.: Diamonds in the rough: social media visual analytics for journalistic inquiry. In: 2010 IEEE Symposium on Visual Analytics Science and Technology (VAST), Oct 2010, pp. 115–122
3. Yardi, S., Boyd, D.: Tweeting from the town square: measuring geographic local networks. In: International Conference on Weblogs and Social Media. American Association for Artificial Intelligence, May 2010. http://research.microsoft.com/apps/pubs/default.aspx?id=122433
4. Becker, H., Naaman, M., Gravano, L.: Learning similarity metrics for event identification in social media. In: Proceedings of the Third ACM International Conference on Web Search and Data Mining, ser. WSDM'10, pp. 291–300. ACM, New York, NY, USA (2010)
5. Sakaki, T., Okazaki, M., Matsuo, Y.: Earthquake shakes twitter users: real-time event detection by social sensors. In: Proceedings of the 19th International Conference on World Wide Web, ser. WWW'10, pp. 851–860. ACM, New York, NY, USA (2010). http://doi.acm.org/10.1145/1772690.1772777
6. Sankaranarayanan, J., Samet, H., Teitler, B.E., Lieberman, M.D., Sperling, J.: Twitterstand: news in tweets. In: Proceedings of the 17th ACM SIGSPATIAL International Conference on Advances in Geographic Information Systems, ser. GIS'09, pp. 42–51. ACM, New York, NY, USA (2009)
7. Kwak, H., Lee, C., Park, H., Moon, S.: What is twitter, a social network or a news media? In: Proceedings of the 19th International Conference on World Wide Web, ser. WWW'10, pp. 591–600. ACM, New York, NY, USA (2010). http://doi.acm.org/10.1145/1772690.1772751
8. Magnuson, A., Dialani, V., Mallela, D.: Event recommendation using twitter activity. In: Proceedings of the 9th ACM Conference on Recommender Systems, ser. RecSys'15, pp. 331–332. ACM, New York, NY, USA (2015)
9. Hannon, J., Bennett, M., Smyth, B.: Recommending twitter users to follow using content and collaborative filtering approaches. In: Proceedings of the Fourth ACM Conference on Recommender Systems, ser. RecSys'10, pp. 199–206. ACM, New York, NY, USA (2010)
10. Jonnalagedda, N., Gauch, S.: Personalized news recommendation using twitter. In: Proceedings of the 2013 IEEE/WIC/ACM International Joint Conferences on Web Intelligence (WI) and

Intelligent Agent Technologies (IAT), ser. WI-IAT'13, vol. 03, pp. 21–25. IEEE Computer Society, Washington, DC, USA (2013)
11. Pham, T.-N., Vuong, T.-H., Thai, T.-H., Tran, M.-V., Ha, Q.-T.: An experimental study, sentiment analysis and user similarity for social recommender system. In: Science, Information, Applications (ICISA), pp. 1147–1156. Springer Singapore, Singapore (2016)
12. Naaman, M., Boase, J., Lai, C.-H.: Is it really about me? message content in social awareness streams. In: Proceedings of the 2010 ACM Conference on Computer Supported Cooperative Work, ser. CSCW'10, pp. 189–192. ACM, New York, NY, USA (2010)
13. Essmaeel, K., Gallo, L., Damiani, E., De Pietro, G., Dipanda, A.: Comparative evaluation of methods for filtering kinect depth data. Multimedia Tools Appl. **74**(17), 7331–7354 (2015)
14. Wang, X., Huang, D., Akturk, I., Balman, M., Allen, G., Kosar, T.: Semantic enabled metadata management in petashare. Int. J. Grid Util. Comput. **1**(4), 275–286 (2009)
15. Lai, C., Moulin, C.: Semantic indexing modelling of resources within a distributed system. Int. J. Grid Util. Comput. **4**(1), 21–39 (2013)
16. Eftychiou, A., Vrusias, B., Antonopoulos, N.: A dynamically semantic platform for efficient information retrieval in P2P networks. Int. J. Grid Util. Comput. **3**(4), 271–283 (2012)
17. Cha, B., Kim, J.: Handling and analysis of fake multimedia contents threats with collective intelligence in P2P file sharing environments. Int. J. Grid Util. Comput. **4**(1), 1–9 (2013)
18. Socher, R., Perelygin, A., Wu, J.Y., Chuang, J., Manning, C.D., Ng, A.Y., Potts, C.: Recursive deep models for semantic compositionality over a sentiment treebank. In: Proceedings of the Conference on Empirical Methods in Natural Language Processing (EMNLP), vol. 1631, p. 1642 (2013)
19. Colace, F., Santo, M.D., Greco, L., Amato, F., Moscato, V., Picariello, A.: Terminological ontology learning and population using latent dirichlet allocation. J. Vis. Lang. Comput. **25**(6), 818–826 (2014). http://dx.doi.org/10.1016/j.jvlc.2014.11.001
20. Hhle, M., Meyer, S., Paul, M.: Surveillance: Temporal and Spatio-Temporal Modeling and Monitoring of Epidemic Phenomena, 2015, r package version 1.8-3. http://CRAN.R-project.org/package=surveillance
21. Hutwagner, M.L., Thompson, M.W., Seeman, G.M., Treadwell, T.: The bioterrorism preparedness and response early aberration reporting system (ears). J. Urban Health **80**(1), i89–i96 (2003)
22. Chianese, A., Piccialli, F., Valente, I.: Smart environments and cultural heritage: a novel approach to create intelligent cultural spaces. J. Location Based Serv. **9**(3), 209–234 (2015)
23. Chianese, A., Piccialli, F., Riccio, G.: Designing a smart multisensor framework based on beaglebone black board. Lecture Notes in Electrical Engineering, vol. 330, pp. 391–397 (2015)
24. Caggianese, G., Neroni, P., Gallo, L.: Natural interaction and wearable augmented reality for the enjoyment of the cultural heritage in outdoor conditions. In: Augmented and Virtual Reality, pp. 267–282. Springer (2014)

A Forward-Selection Algorithm for SVM-Based Question Classification in Cognitive Systems

Marco Pota, Massimo Esposito and Giuseppe De Pietro

Abstract Cognitive Systems have attracted attention in last years, especially regarding high interactivity of Question Answering systems. In this context, Question Classification plays an important role for individuation of answer type. It involves the use of Natural Language Processing of the question, the extraction of a broad variety of features, and the use of machine learning algorithms to map features with a given taxonomy of question classes. In this work, a novel learning approach is proposed, based on the use of Support Vector Machines, for building a set of classifiers, each one to use for different questions and comprising the respective features, chosen through a particular forward-selection procedure. This approach aims at decreasing the total number of features, by avoiding those giving scarce information and/or noise. A Question Classification framework is implemented, comprising new sets of features with low numerosity. The application on a benchmark dataset shows classification accuracy competitive with the state-of-the-art, by considering a lower number of features.

Keywords Cognitive systems · NLP · Question answering · Question classification · Feature extraction

1 Introduction

Cognitive Computing refers to the development of computer systems modelled after the human brain, which has natural language processing capability, learns from experience, interacts with humans in a natural way, and helps in making

M. Pota (✉) · M. Esposito · G. De Pietro
Institute for High Performance Computing and Networking (ICAR),
National Research Council of Italy (CNR), Naples, Italy
e-mail: marco.pota@na.icar.cnr.it

M. Esposito
e-mail: massimo.esposito@na.icar.cnr.it

G. De Pietro
e-mail: giuseppe.depietro@na.icar.cnr.it

G. De Pietro et al. (eds.), *Intelligent Interactive Multimedia Systems and Services 2016*, Smart Innovation, Systems and Technologies 55,
DOI 10.1007/978-3-319-39345-2_52

decisions based on what it learns. Cognitive Systems have attracted attention in last years, especially regarding Question-Answering (QA) systems, which present high interactivity, responding directly and precisely to natural language prompts, with relevant responses and associated confidence scores, by accessing both structured and unstructured information.

In order to individuate a concise answer to be returned, a QA system should properly understand the expectation of a question. Question Classification (QC) plays an important role to achieve this goal. Indeed, QC is aimed at analysing a question and assigning it to an appropriate category representing its expected answer type, thus it enables to limit the search for possible answers only to specific information sources, to enact the proper strategy to find a response, and to narrow down the number of possible candidate answers. In practice, given a finite set of possible categories, arranged in a question taxonomy, the goal of a QC module is to learn the mapping from questions to answer types. For example, for the question "*What is the clinical consequence of pneumococcus?*", the task of QC is to assign it to the category "disease", since the answer to this question (*pneumonia*) is the name of a disease.

QC approaches based on machine-learning have been devised over the last few years, with very promising results. In order to build a classifier, Natural Language Processing (NLP) methods [1] could be used to extract features from a dataset made of classified questions, then a machine-learning algorithm can be applied to build a model, mapping question features to classes. The trained classifier can be used to predict the answer type of further questions. The UIUC benchmark dataset of classified questions [2] is widely used in literature to train and test QC modules. On the other hand, while NLP methods are quite well established, a broad variety of features, usually divided into lexical, syntactic and semantic [3, 4], can be extracted from questions [5]. Support Vector Machines (SVMs) were revealed as the best machine learning method for this context [4–7], primarily for their feasibility to manage many features, since good results were gained as long as features were employed in a very high number [3, 4, 6–8].

In this work, some new sets of features are proposed, presenting low dimensionality, for summarizing the information of larger sets.

In addition, a novel learning approach for QC is proposed. It is based on the use of SVMs for building a number of classifiers, to use for different cases, each one comprising the respective features, chosen based on a particular forward-selection procedure. This approach aims at improving two aspects, typical of QC: (i) decreasing the total number of features, and (ii) avoiding, in some cases, to consider features that for such cases contribute with scarce information and/or even with noise.

The proposed approach is applied on the benchmark dataset built by Li and Roth [2]. In order to do this, a framework enabling QC is implemented, which is able to extract from questions a broad variety of lexical and syntactic features, to train different SVM-based classifiers, and to classify each question by the proper classifier.

2 Background and Related Work

In order to characterize each question, its analysis is performed by the following modules of NLP: tokenization divides the question into elementary fragments, which can be words, or punctuation, numbers, acronyms, dates, and so on; tagging assigns to each token the corresponding morphologic-lexical label; stemming substitutes each token with its radical form; finally, parsing allows representing the syntactic structure of the question, in the form of a parse tree. An example is given in Table 1.

The questions can be divided into classes, corresponding to the semantic categories of possible answers, which are arranged in the form of a taxonomy. At the same time, a collection of questions associated with the same classes is needed to train the classifier. Different taxonomies have been proposed in literature. The earliest one [9] divides questions into two hierarchical levels: the first one consists of 9 classes including 8 "wh-words" plus a class labelled "Name", while the second level defines subclasses. A more complex classification is made by the ISI taxonomy, which divides questions into 140 categories [10]. A well-known taxonomy, introduced in [11], defines two hierarchical levels, constituted by 6 "coarse" and 50 "fine" classes. More recent studies [12, 13] used the Bloom's Taxonomy, arranged in six levels: knowledge, comprehension, application, analysis, synthesis, and evaluation.

Different supervised machine learning algorithms have been used in literature for question classification purposes and successfully applied on the UIUC dataset [11]. In particular, the Sparse Network of Winnows algorithm [14] have been specifically tailored for employing a high number of features. Later, the Decision Trees, compared with other algorithms, outperformed their performances [7]. However, SVMs showed the best feasibility to high dimensional data, and often the best results [6–8, 15].

Table 1 Example of NLP

Input	"*What is the clinical consequence of pneumococcus?*"
Tokenization	{*what, is, the, clinical, consequence, of, pneumococcus, ?*}
Tagging	{*what*(WP), *is*(VBZ), *the*(DT), *clinical*(JJ), *consequence*(NN), *of*(IN), *pneumococcus*(NN), *?*(.)}
Stemming	{*what, be, the, clinical, consequence, of, pneumococcus, ?*}
Parsing	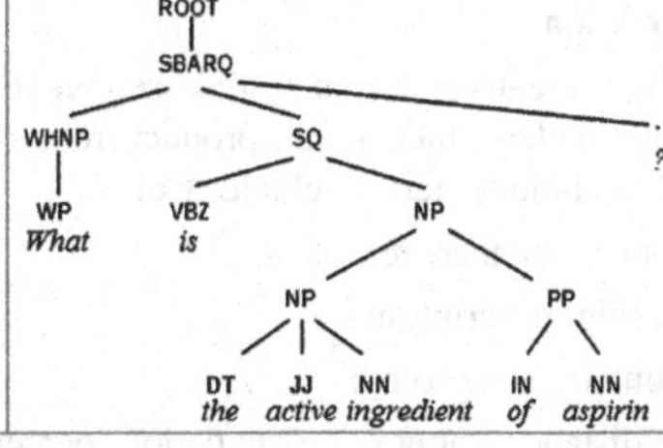

For a complete list of lexical and syntactic tags, see: http://web.mit.edu/6.863/www/PennTreebankTags.html

3 The Proposed Approach

In this section, the proposed approach for building a QC module is described. It enables to classify questions according to a certain taxonomy of classes, therefore it is applied on a coherent question dataset, as reported in Sect. 3.1. Various features, chosen among those commonly used and some others proposed here, are extracted from questions by NLP techniques, and are listed in Sect. 3.2. The learning procedure of each classifier is based on SVMs algorithm, as reported in Sect. 3.3. Finally, in Sect. 3.4 the proposed forward-selection approach is described, which enables to choose the best classifier for each question.

3.1 Taxonomy and Dataset

Since the taxonomy introduced in [11] is the most used in literature, many results exist to make comparisons. Moreover, a benchmark dataset [2] is available, classified according to the same taxonomy. Therefore, in this work, the Li and Roth taxonomy [11], reported in Table 2, is used. Since the chosen taxonomy comprises two hierarchical levels, two different classification results are required for each run: one regarding coarse classes and the other regarding fine classes. Moreover, the UIUC dataset [2] is used, made of 5452 questions for training and 500 for testing.

3.2 Features Extraction and Representation

In order to characterize questions and extract from them the features reported below, each of the NLP modules described in Sect. 2 was implemented by using available Java libraries. The best tokenization in our opinion was taken from the Stanford coreNLP tool [16], since some problems were found in other implementations, regarding particular types of tokens. The GATE Developer 8.1 tool [17]

Table 2 Two levels taxonomy developed by Li and Roth [11]

Coarse classes	Fine classes
ABBREVIATION	Abbreviation, expression
ENTITY	Animal, body, colour, creative, currency, disease, event, food, instrument, language, letter, other, plant, product, religion, sport, substance, symbol, technique, term, vehicle, word
DESCRIPTION	Definition, description, manner, reason
HUMAN	Group, individual, title, description
LOCATION	City, country, mountain, state, other
NUMERIC	Code, count, date, distance, money, order, period, percent, speed, temperature, size, weight, other

was taken for the implementation of both tagging and stemming phases. Finally, in order to execute the parsing of the question, the Stanford PCFG parser [18] was chosen, since it results the best one in our opinion, even if it presents some bugs.

Once the question is analysed, different sets of features can be extracted from it and then used for the classification. Usually, features are divided into lexical, syntactic and semantic. In particular, here the focus is on lexical and syntactic features, since the use of semantic features like hypernyms greatly contribute to increase the dimensionality. In the following, the features extracted in this work are detailed. Among them, some are proposed here for the first time. An example is given in Table 3.

- **Stemmed Unigrams (SU)**. This is the mostly used set of features, since it usually allows obtaining the best results [3, 4, 6, 8]. Unigrams are obtained from the set of tagged tokens of the question, by eliminating tokens with some tags, like DT, IN, and punctuation. Then, these unigrams are stemmed into their radical form.
- **Principal Wh-Word (PWW)**. Question words, if present, are recognized in the tagging phase with a WP label. These words are naturally helpful for question classification. Moreover, they are in a very limited number, since the only question words in English language (and some less frequent with similar meanings) are: what (and whatever, which, whichever), how, when, where (and whence, whither, wherever), why (and wherefore), who (and whom, whose, whoever, whomever, whosoever). The principal wh-word is obtained here by considering the parse tree and applying a set of rules specifically tailored for searching, if any, the wh-word of the principal phrase of the question. For more information, see [19]. Moreover, here, wh-words with similar meanings are grouped, so that different wh-words of each group correspond to the mostly used one.
- **Bigrams**. These are n-grams with $n = 2$, where n-grams are sequences of n consecutive words in the question. Since the number of bigrams is very high, and usually they do not help much to improve accuracy [5], here only some are proposed:
  - **After-How (AH)**: This bigrams are made by the word "*how*", in case it is the Principal-Wh-Word, plus the following word ("*how can*", "*how much*", …).
  - **Adjective-Noun (AN)**: Bigrams consisting of an adjective plus the following (non-proper) noun are proposed here to be considered.

Table 3 Example of feature extraction

Question	"*What is the clinical consequence of pneumococcus?*"
Stemmed Unigrams	{*what, be, clinical, consequence, pneumococcus*}
Principal Wh-Word	{*what*}
After-How	{}
Adjective-Noun	{*clinical consequence*}
Head-Word	{*consequence*}
Head-Verb	{*is*}
Multiple-Head-Words	{*what, is, consequence*}

- **Head-Word (HW)**. This feature aims at individuating the most informative word of the question for classification purposes. Introduced in [8], it is widely used and recognized to be useful [3, 4, 6, 8, 20]. It is extracted here from the parse tree, by using redefined rules, similar to those proposed by Collins [21], already modified in other works [6, 8]. In particular, while Collins' rules prefer verbs, here nouns are favoured, and rules are defined in different manners, depending on whether the Principal-Wh-Word exists. For more information, see [19].
- **Head-Verb (V)**. This feature, already proposed in [20], is considered here in order to avoid losing verb information. It is extracted from the parse tree, by using rules similar to Collins' rules [21]. In particular, while Collins' rules prefer auxiliary verbs, here non-auxiliary ones are chosen if present. For more information, see [19].
- **Multiple-Head-Words (MHW)**. These features are proposed here, with the aim of extracting the most informative words of a question, directly from the parse tree. They can be viewed as a midway between the very numerous unigrams and the single informative head-word. They are extracted by considering all the third-generation children starting from the root, then obtaining the corresponding sub-trees and extracting the head-word of each sub-tree.

For each type of features, the union of those extracted from different questions of the training dataset constitutes the set of features of that type. The dataset can thus be represented as a matrix, whose rows are the questions and whose columns are the features. Each position is fulfilled with an element equal to 1 if the corresponding feature is extracted from the corresponding question, otherwise the position is empty (sparse representation).

3.3 SVM Algorithm

The SVM algorithm for learning classifiers is used here, since it is the most suitable to be used with large numbers of features.

In previous research on the application of SVMs for QC, $C = 1$ (where the constant C is a fixed trade-off between the two objectives of having maximum margin separating classes and of minimizing the number of wrongly classified samples), and linear kernel functions have been widely adopted, and results supported these choices [22]. More than two classes have been suitably separated by the one-against-all strategy [23]. In this work, the previous established settings are adopted.

An available Java library, i.e. LIBSVM [24], is used for the SVM implementation.

3.4 Forward-Selection Approach

The approach proposed here for QC is inspired by the forward-selection algorithms for feature selection. For this type of problem, the number of features is particularly high, and they depend on the training dataset, thus the selection of one feature at a time for building the best unique classifier is not effective. Therefore, a generalized forward-selection is made here by considering and selecting together the features of each set described in Sect. 3.2.

The hypothesis is made that the PWW is the most important features set for QC purposes, for the following reasons: (i) it is the less numerous, since the number of features is limited by the total number of wh-words of English language, i.e. 6 (how, what, when, where, who, why), without considering those with similar meanings; (ii) typically, results obtained by using wh-words are particularly accurate, taking into account the low features number; (iii) this set of features does not depend on the questions dataset, therefore an approach based on it can be considered scalable. Therefore, the questions of the training dataset are divided into seven subsets, six comprising questions presenting different PWW, and one comprising questions without PWW.

Each subset of the training data is used to train an SVM-based classifier. Each classifier comprises sets of features chosen by a generalized forward-selection approach. The selection of further sets of features stops when a maximum of predictability is reached, i.e. when the accuracy calculated on the test set does not increase anymore. Clearly, the selected sets of features can be different, depending on the considered subset of training questions. As a result, seven SVM-based classifiers are obtained, and for each question, only one of them is used, depending on the PWW.

4 Results and Discussion

In Table 4, the results obtained on the whole test set by SVM classifiers comprising different single sets of features are reported.

The results reported in Table 4 show that PWW is actually the best set of features, if both the accuracies and the numerosities are considered. It can also be noticed that the MHW set, proposed here, is very effective, since it allows to reach the same accuracy for coarse classes and slightly lower accuracy for fine classes, with respect to the SU set, which is the best one, but much more numerous.

For each subset of training questions, corresponding to some PWW or no PWW, a PWW-specific classifier employing SVM algorithm is obtained by adding sets of features by the forward-selection procedure. Results on the corresponding test subset of questions are reported in Table 5.

The best sets of features, highlighted in Table 5, are selected by considering firstly the accuracy of coarse classes, then that of fine classes, and, if accuracies are

Table 4 Classifiers comprising single sets of features

Features sets	Features number	Coarse classes (%)	Fine classes (%)
{PWW}	6	45.6	46.8
{AH}	29	25.2	16.4
{V}	864	48.2	39.6
{AN}	1845	21.4	12.4
{HW}	2445	63.4	40.2
{MHW}	4431	86.6	77.2
{SU}	8324	86.6	80.8

Table 5 PWW-specific classifiers

PWW	Features sets	Coarse classes (%)	Fine classes (%)	Features sets	Coarse classes (%)	Fine classes (%)
HOW	{}	94.1	26.5			
	{AH}	100.0	82.4			
	{V}	100.0	47.0	{AH + V}	100.0	88.2
	{AN}	94.1	14.7	{AH + AN}	100.0	79.4
	{HW}	91.2	35.3	{AH + HW}	100.0	82.3
	{MHW}	94.1	88.2	{AH + MHW}	100.0	91.1
	{SU}	94.1	85.3	{AH + SU}	100.0	91.1
WHAT	{}	26.1	35.2			
	{V}	49.7	45.8	{SU + V}	82.2	74.2
	{AN}	26.7	35.8	{SU + AN}	82.2	75.3
	{HW}	55.3	69.7	{SU + HW}	85.6	78.3
	{MHW}	82.2	71.4	{SU + MHW}	85.6	77.2
	{SU}	82.8	76.1			
WHEN	{}	100.0	100.0			
WHERE	{}	100.0	92.6			
WHO	{}	100.0	93.6			
WHY	{}	100.0	100.0			
none	{}	0.0	0.0			

equal, the numerosity. More comprehensive classifiers that do not improve accuracies are not reported. From Table 5, it can be noticed that:

- if the PWW is "how", then AH is the best set of features, and MHW set is the best coupled with AH;
- if the PWW is "what", then SU is the best set, and HW is the best coupled with SU;
- in case PWW is "when" or "why", perfect classification is obtained without any other feature, just by assigning to those questions the most frequent class of the

training subsets (this happens because the only class of each test subset is the most frequent in the corresponding training subset);

- in case PWW is "where" or "who", no features set is useful, since the best accuracies are obtained just by assigning to those questions the most frequent class of the training subsets (this happens because the wrongly classified questions of each test subset are very different from those of the corresponding training subset);
- none of the considered features set allows to accurately classify questions without PWW (this happens because the questions in the test set without PWW are completely different from those of the training set), thus the classifier without any other feature is chosen, which assigns to the questions the most frequent class of the training subset.

As a result, the following algorithm is individuated for classifying different types of questions depending on the PWW:

```
Extract PWW
Switch
  Case PWW="HOW", then
    Extract AH and MHW
    Apply "HOW"-specific SVM classifier based on {AH,MHW}
  Case PWW="WHAT", then
    Extract SU and HW
    Apply "WHAT"-specific SVM classifier based on {SU,HW}
  Case PWW="WHEN", then class is "NUMERIC:date"
  Case PWW="WHERE", then class is "LOCATION:other"
  Case PWW="WHO", then class is "HUMAN:individual"
  Case PWW="WHY", then class is "DESCRIPTION:reason"
  Case no PWW, then class is "HUMAN:individual"
End Switch
```

The accuracy values calculated on the whole test set, resulting by applying the described method, are reported in Table 6, together with other results of previous works on the same data, using lexical and syntactic features and employing linear SVMs.

From results shown in Table 6, it can be noticed that accuracies obtained by the presented approach are higher than those obtained by using any single features set,

Table 6 Accuracy gained on the whole dataset, compared with best literature results

	[4]	[6]	[7]	[19]	This work
Coarse classes (%)	89.0	91.8	87.4	89.6	89.2
Fine classes (%)	84.0	84.2	80.2	82.0	82.4

and are competitive with the state-of-the-art. On the other hand, the proposed approach allows to extract a total number of features that is very lower, with respect to the others.

5 Conclusion

The problem of Question Classification, which is particularly important for Question Answering purposes in the ambit of Cognitive Computing, was faced in this paper. Question Classification is typically performed by applying Natural Language Processing methods on the question, then extracting some features characterizing it, and finally mapping these features with a given taxonomy of classes.

In this work, some new types of features were proposed, in order to characterize questions by more compact sets of features, summarizing information of more numerous sets. In particular, the features named Multiple-Head-Words result very effective, since they allow to reach accuracies similar to those obtained by all the unigrams, but are very less numerous.

Moreover, a novel approach was proposed, for learning the mapping between features and classes. Questions are divided depending on the Principal-Wh-Word, then for each type of question an SVM-based classifier is trained. The learning procedure is based on the proposed generalized forward-selection procedure, by means of which whole sets of features are considered and selected together. As a result, each question is classified as follows: if the Principal-Wh-Word is "how", an SVM-based classifier can be used, employing as features After-How bigrams and Multiple-Head-Words; if the Principal-Wh-Word is "what", an SVM-based classifier can be used, employing as features Stemmed Unigrams and Head-Words; if the Principal-Wh-Word is "when", "where", "who", or "why", trivial classifiers are found. The accuracy obtained by the described method, are higher than those obtained by using any single features set, and are competitive with the state-of-the-art on the same dataset. On the other hand, the proposed approach allows to extract a total number of features that is very lower, with respect to the other competitive approaches.

Surely, this method for QC can be improved or adapted, in order to use it on different datasets. In particular, it suffers from some limits, like the poor discrimination in case of some Principal-Wh-Word, or the dependence of specific classifiers on the training dataset. These limits depend fundamentally on the biased hand-made benchmark dataset used as a proof of concept, and on its separation between training and testing sets. Using a more numerous training questions dataset, and choosing the testing set according to a cross-validation technique, would help to improve scalability and to obtain even better predictability.

References

1. Dale, R.: Classical approaches to natural language processing. In: Indurkhya, N., Damerau, F. J. (eds.) Handbook of Natural Language Processing, Chap. 1. Chapman & Hall/CRC (2010)
2. http://cogcomp.cs.illinois.edu/Data/QA/QC
3. Mishra, M., Mishra, V.K., Sharma, H.R.: Question classification using semantic, syntactic and lexical features. Int. J. Web Semant. Technol. **4**(3) (2013)
4. Loni, B., van Tulder, G., Wiggers, P., Tax, D.M.J., Loog, M.: Question classification by weighted combination of lexical, syntactic and semantic features. LNAI, vol. 6836. Springer, Berlin, pp. 243–250 (2011)
5. Loni, B.: A survey of state-of-the-art methods on question classification. Delft University of Technology, Technical report (2011)
6. Silva, J., Coheur, L., Mendes, A.C., Wichert, A.: From symbolic to sub-symbolic information in question classification. Artif. Intell. Rev. **35**, 137–154 (2011)
7. Zhang, D., Lee, W.S.: Question classification using support vector machines. In: Proceedings of the 26th Annual International ACM SIGIR Conference on Research and Development in Information Retrieval, NY USA, pp. 26–32 (2003)
8. Huang, Z., Thint, M., Qin, Z.: Question classification using head words and their hypernyms. In: Proceedings of the 2008 Conference on Empirical Methods in Natural Language Processing, Honolulu, pp. 927–936 (2008)
9. Moldovan, D., Harabagiu, S., Pasca, M., Mihalcea, R., Goodrum, R., Girju, R., Rus, V.: LASSO: a tool for surfing the answer net. In: Eighth Text Retrieval Conference, National Institute of a Standards and Technology, vol, pp. 175–184. 500–246. NIST Special Publication, Gaithersburg, MD (1999)
10. Hovy, E., Hermjakob, U., Ravichandran, D.: A question/answer typology with surface text patterns. In: Proceedings of the DARPA Human Language Technology Conference (HLT), San Diego, CA (2002)
11. Li, X., Roth, D.: Learning question classifiers. In: Proceedings of the 19th International Conference on Computational Linguistics (COLING'02), Morristown, NJ, USA, pp. 1–7 (2002)
12. Haris, S.S., Omar, N.: A rule-based approach in Bloom's Taxonomy question classification through natural language processing. In: Proceedings of 7th International Conference on Computing and Convergence Technology (ICCCT), pp. 410–414 (2012)
13. Yahya, A.A., Osman, A.: Automatic classification of questions into Bloom's cognitive levels using support vector machines. In: Proceedings of the International Arab Conference on Information Technology, pp. 1–6. Naif Arab University for Security Science (NAUSS) (2011)
14. Roth, D.: Learning to resolve natural language ambiguities: a unified approach. In: Proceedings of the 15th National/10th Conference on Artificial Intelligence/Innovativer Applications of Artificial Intelligence, Menlo Park, USA, pp. 806–813 (1998)
15. Metzler, D., Croft, W.B.: Analysis of statistical question classification for fact-based questions. Inf. Retrieval **8**(3), 481–504 (2005)
16. Manning, C.D., Surdeanu, M., Bauer, J., Finkel, J., Bethard, S.J., McClosky, D.: The Stanford CoreNLP natural language processing toolkit. In: Proceedings of 52nd Annual Meeting of the Association for Computational Linguistics: System Demonstrations, pp. 55–60 (2014)
17. Cunningham, H., et al.: Text Processing with GATE. University of Sheffield (2011)
18. Klein, D., Manning, C.D.: Accurate unlexicalized parsing. In: Proceedings of the 41st Meeting of the Association for Computational Linguistics, pp. 423–430 (2003)
19. Pota, M., Fuggi, A., Esposito, M., De Pietro, G.: Extracting compact sets of features for question classification in cognitive systems: a comparative study. In: Proceedings of 10th International Conference on P2P, Parallel, Grid, Cloud and Internet Computing, November 4–6 2015, Krakow, Poland

20. Li, F., Zhang, X., Yuan, J., Zhu, X.: Classifying what-type questions by head-noun tagging. In: Proceedings of the 22nd International Conference on Computational Linguistics, Manchester, pp. 481–488 (2008)
21. Collins, M.: Head-driven statistical models for natural language parsing, Ph.D. thesis, University of Pennsylvania (1999)
22. Loni, B.: Enhanced question classification with optimal combination of features, Master's thesis (2011)
23. Hsu, C.-W., Lin, C.-J.: A comparison of methods for multiclass support vector machines. IEEE Trans. Neural Netw. **13**(2), 415–425 (2002)
24. Chang, C.-J., Lin, C.-J.: LIBSVM: a library for support vector machines. ACM Trans. Intell. Syst. Technol. **2**(3), 27 (2011)

Supporting Autonomy in Agent Oriented Methodologies

Valeria Seidita and Massimo Cossentino

Abstract Designing a software solution for a complex systems is always a demanding task, it becomes much more complex if we consider to design a multi agent system where agents have to exhibit autonomy; which abstractions and which concepts to take into consideration when using a design methodology we would like to support autonomy? In this paper, we answer this question by studying and analyzing literature on the concept of agents in order to establish the basic set of concepts an agent oriented methodology has to deal with.

Keywords Autonomy · Multiagent systems · Design process

1 Introduction

So far, researchers have faced the problem of understanding, from the point of view of multiagent system (MAS) development and implementation, how to construct interacting or communicating agents, how to let agents reason, how to ensure they take useful decisions and so on. Several good solutions, in terms of techniques and methods, have been identified but however we find they mainly focus on an implementation perspective and on the features agents must possess, not how these features have to be abstracted for providing means for carrying on good analysis phases. This fact does not create so many problems when dealing with concrete features, like for instance goal, but become much more awkward when dealing with non tangible features, like for instance *autonomy*.

V. Seidita (✉)
Dip. di Ingegneria Chimica, Gestionale, Informatica, Meccanica,
University of Palermo, Palermo, Italy
e-mail: valeria.seidita@unipa.it

V. Seidita · M. Cossentino
Istituto di Reti e Calcolo ad Alte Prestazioni,
Consiglio Nazionale delle Ricerche, Palermo, Italy
e-mail: cossentino@pa.icar.cnr.it

G. De Pietro et al. (eds.), *Intelligent Interactive Multimedia Systems and Services 2016*, Smart Innovation, Systems and Technologies 55,
DOI 10.1007/978-3-319-39345-2_53

What we want to do now is moving the problem towards the design process and identifying which abstractions do have to, or do not, be present in the analysis phase of an agent oriented methodology for supporting agent autonomy.

The objective of this work is to explore agents features in order to identify which elements (or abstraction) an agent based methodology have to deal with for being countered among the methodologies implementing autonomy. In doing this, in the next section, we explore building definitions of agent, multiagent system and agent based methodology; among the three, *agent* is the concept that meets smaller consensus among researchers, however this does not affect our analysis indeed we explore literature on agent definition from the point of view of the designer that has to identify and highlight which elements he has to deal with mainly during analysis and then during the design in order to implement agents to be useful for solving real world problems. For doing this we start from the features an agent has to own and we report some of the most popular definitions from the plainest to the most structured.

Within the end of Sect. 3 we provide a reasonable list of elements that have to be present in a methodology for supporting autonomy design whereas in Sect. 4 we show two agent oriented methodologies supporting autonomy in step with our evidences. Finally, some discussions and conclusions are provided.

2 Agents Definitions

In this section we overview literature in order to examine the notion of agent and autonomy in agency.

Russell and Norvig define the agent as follows:

> An agent is anything that can be viewed as perceiving its environment through sensors and acting upon that environment through effectors [15].

In this sense, we may venture that everything that uses inputs from an environment and provides output may be considered an agent.

According to Maes:

> Autonomous agents are computational systems that inhabit some complex dynamic environment, sense and act autonomously in this environment, and by doing so realize a set of goals or tasks for which they are designed [13, 14].

Maes directly defines autonomous agents, she does not separate the term agent from the term autonomous, they are tightly interrelated, and, in addition to Russel and Norvig's idea, autonomy is also in the action and sensing and a new important element is considered, the goal. This definition is more restrictive, it is not sufficient to act and sense but autonomous agent has to autonomously act and sense in a dynamic environment in order to pursue its own goals.

Haes-Roth defines:

> Intelligent agents continuously perform three functions: perception of dynamic conditions in the environment; action to affect conditions in the environment, and reasoning to interpret perceptions, solve problems, draw inferences, and determine actions [11].

Hayes adds for agent the reasoning and introduces the concept of affecting the environment with actions and determining which action to perform.

In a white paper by IBM:

> Intelligent agents are software entities that carry out some set of operations on behalf of a user or another program with some degree of independence or autonomy, and in so doing, employ some knowledge or representation of the user's goals or desires [10].

The definition of IBM instills the presence of users and the fact that the agent acts on the behalf of them, and also gives suggestion for understanding autonomy by means of the word "independence". Moreover, the need for user's goals and desires representation arises.

Brustoloni gives the following definition of autonomous agents.

> Autonomous agents are systems capable of autonomous, purposeful action in the real world [2].

A very concise definition that highlights three elements, the ability of acting, the environment rather the real world, the presence of driven or committed behavior suggested by the word *purposeful*.

Franklin and Graesser state that:

> An autonomous agent is a system situated within and a part of an environment that senses that environment and acts on it, over time, in pursuit of its own agenda and so as to effect what it senses in the future [9].

Franklin and Graesser embrace all the previous definitions and restate the fact that the environment is continuously changing also for the effect of agents' actions.

It is worth to note that in all these definitions the concept of reaction to the environment changes seems to be hidden in the concept of action.

One of Wooldridge's definitions of agent reports that:

> An agent is a computer system that is situated in some environment, and that is capable of autonomous action in this environment in order to meet its delegated objectives [17].

The previous definition summarizes the more complete one from [16, 17], it is the *weak notion of agent* for Wooldridge and Jennings but however complete enough for the purposes of this paper:

> An agent is a system enjoying the following properties:

- autonomy: agents encapsulate some state (that is not accessible to other agents), and make decisions about what to do based on this state, without the direct intervention of humans or others;
- reactivity: agents are situated in an environment, (which may be the physical world, a user via a graphical user interface, a collection of other agents, the INTERNET, or perhaps many of these combined), are able to perceive this environment (through the use of potentially imperfect sensors), and are able to respond in a timely fashion to changes that occur in it;

- pro-activeness: agents do not simply act in response to their environment, they are able to exhibit goal-directed behavior by taking the initiative;
- social ability: agents interact with other agents (and possibly humans) via some kind of agent-communication language, and typically have the ability to engage in social activities (such as cooperative problem solving or negotiation) in order to achieve their goals.

So, Wooldridge, too, highlights the goal directed behavior of agents in an environment that changes while the procedures for pursuing goals are running. This leads to a kind of reactivity that is different from the one of the previous definitions, the agent has to react in a timely fashion to the changes. Agents reach autonomy if there is not the intervention of humans and if they are engaged in social activities.

Moreover, in [17], Wooldridge says "An agent takes sensory input from the environment, and produces as output, actions that affect it. The interaction is usually an ongoing, non-terminating one." An agent continuously looks at the environment and gains information from it, on the basis of this information it decides what to do next, what action to perform next in order to pursue its own agenda or goals, each action typically affects and changes the environment; once acting the agent senses again the environment and so on in a continuos loop *sense-decide-act-sense-decide*. During this loop, decisions depend on agent's (mental) state, it will take different decisions basing on the particular interaction with the environment (hence its own state and the environment state at a specific time).

In our opinion, one of the most comprehensive definition found in literature is the one given by Jacques Ferber in [8]:

An agent is a physical or virtual entity:

- which is capable of acting in and perceiving its environment, can communicate directly in an environment and possesses resources of its own;
- which has only a partial representation of this environment (and perhaps none at all), is driven by a set of tendencies (in the form of individual objectives) and possesses skills and can offer services;
- whose behavior tends towards satisfying its objectives, taking account of the resources and skills available to it and depending on its perception, its representation and te communication it receives.

Ferber's definition points out that agent is able of acting and not only reasoning (different viewpoint from AI field), by simply acting agent modifies its environment and its future interaction with it and its decision making process. An agent communicates with other agents that are part of the environment (hence environment includes not only the static object but also other autonomous agents). Finally, autonomy is meant in the sense that agents are not directed by commands from users, but by a set of tendencies, i.e. goals or some other kind of desire the agent wants to realize.

In the following section we explain how all these definitions, which we know contribute in a different way to the concepts of agent and autonomy, led us to identify the elements an agent design methodology has to present in order to support autonomy.

3 Towards Implementing Autonomy in Agent Oriented Methodologies

Several agent design methodologies are reported in literature and are used for solving different problems (see [1, 7, 12] for a wide overview), but how may we affirm that a design methodology supports analysis and design of autonomous agents given that, as we have already seen, the concept of agent and of autonomy in a multiagent system includes a variety of aspects to be taken into consideration and that are sometimes not coherent?

Besides, too many methodologies force us to think in terms of objects whereas for an agent oriented methodology we have to think in terms of agents being led by the sentence "on the behalf of humans" seen in some previous definitions. When we perform object oriented design we tend to represent, to model, the users' interactions with the system (with the software objects representing or implementing the objects in the real world) in a functional way, hence conducting analysis on the base that some inputs have to be elaborated to give some outputs. When we perform agent oriented design (it is ascertained that objects and agents are truly different) we have to think of the agents of the system and to model the software under the point of view of agents, of their interaction with the world and with other agents that populate the world; at the same time we have to think that environment changes and all this is away from the classical software engineering perspective of functional design. But what does it mean and imply? From all the definitions of the previous section we may abstract that: an agent is a purposefully originator of actions; agent's actions are chosen in the interest of the agent itself; agent senses the environment; actions modify or shape the environment; agent takes decisions in an independent fashion on the actions it may do. Let us now further analyze these sentences with the aid of the Table 1.

From a methodological point of view, a designer has to analyze what an agent has to do in order to accomplish the user design objectives hence its goals, a list of actions it is able to do and a list of actions it is not allowed to do; the first depends on the kind of agent situated in an environment whereas the second on the kind of environment it lives in; the environment includes other agents. In other words, it is important to analyze and to establish the boundaries within which agents may act, thus not deciding how it behaves but only modeling agent decision process that is realized by dynamic, not fixed, plans for pursuing goals and may be influenced by several factors (an important one is the set of non-functional requirements of the systems, we don't detail this in the present paper).

From the study of the autonomy definitions and the survey of many methodologies, we identified the minimum set of elements a design methodology has to encompass (see Fig. 1). In the following section we try to examine two methodologies on the base of these three levels in order to say if they support autonomy.

Table 1 Design abstractions

Agent autonomy feature	Consequences in the design abstractions
"agent is purposefully originator of actions"	An agent has to be endowed with *knowledge* on its agenda and *goals* and on the *action* it may perform **or not** in order to pursue them
"actions are chosen on the interest of agent"	Agent has to be aware of the action it is allowed (*constraint*) to perform in order to pursue an objective
"agent senses the environment"	Agent has to be endowed with *knowledge* about its *environment*
"actions modify or shape the environment"	There is a relation between *action* and *environment*, the list of actions an agent may do are related to its environment
"agent takes decision in an independent fashion on the actions"	There must not be someone who says to an agent what it has to do, hence *no static plan* but autonomous *decision process*

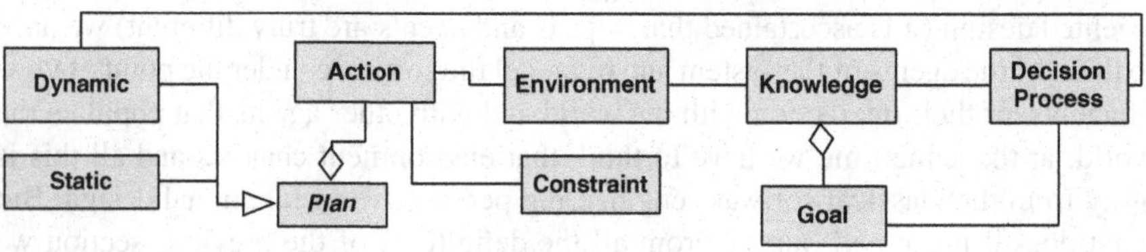

Fig. 1 The minimum set of design abstractions for autonomy

4 Analyzing How Gaia and ASPECS Support Autonomy

In this section we use the results illustrated in Sect. 3 and in Fig. 1 in order to examine two well know agent oriented methodologies and to give an example of how to support autonomy.

Gaia [3, 7, 18, 19] is a complete agent oriented methodology allowing the designer to develop multiagent systems going from analysis to code and, so as a lot of other methodologies in literature, considers the requirement elicitation as an independent phase. During analysis phase Gaia moves from abstract concepts to concrete ones creating models detailed in an incremental fashion. Gaia, in its latest version [19], focuses on organizational abstraction in order to analyze and design MASs that work in complex and open environment.

Gaia is composed of three main phases: analysis, architectural design and detailed design; the first phase is mainly concerned with the identification of organizations of the MAS and their roles, the second phase deals with organizational structure and the third specifies the AgentType concept for modeling what agent plays one or more roles and all the activities, agents may perform, in terms of services (see the provided references for a deeper review of Gaia).

ASPECS [4–6] is an agent design methodology for developing Holonic MASs; it covers all the activities from requirement analysis to code and allows to design open, dynamic and complex systems. ASPECS combines the holonic and agency concepts for completely modeling the whole structured organizational aspect of entities composing a complex system, thus an organizational approach is rife in all the methodology.

ASPECS is composed of three main phases: System Requirements, Agent Society and Implementation and Deployment. The first phase deals with the definition of system requirements and the identification of organizations, the second handles the definition of roles, communications, agents and the holonic architecture and finally during the third phase a solution is complemented by using platform dependent concepts.

In Tables 2 and 3 we illustrate how each abstraction of Fig. 1 is dealt with in Gaia and in ASPECS. Both Gaia and ASPECS support autonomy, indeed for both of them we found ways of dealing with the abstractions in the table. Gaia explicates knowledge on both actions and environment also providing a model for this whereas ASPECS does not specifically model environment; knowledge of the domain and on the allowed actions is supplied respectively by the ontology model and capacity description. The concept of capacity is very important because different aspects of

Table 2 Autonomy supported in Gaia

GAIA	
Environment	In GAIA the environment is modeled by determining all the entities and resources that MAS may exploit; the environment is considered both physical and virtual so providing a complete and wide representation of all the entities, agents or not, living in it
Action	Entities and resources identified in the environment model are used for modeling what actions according to specific permission may be performed by roles
Knowledge	Gaia provides knowledge on the resources of the environment described in terms of symbolic names (associated with the type of action an agent may do on it); also, knowledge on how to achieve goals by means of *skills*
Goal	Goals are identified and modeled during the analysis phase, Gaia principally focusses on the goals of organizations, one or more agents into an organization accomplish goals
Constraints	Rules expressing constraints the organization has to obey in its behavior while executing an activity. In particular during the *Identify Protocol Dependencies* and *Identify Environmental Constraints*, Gaia provides norms and constraints for the interaction among roles and with the environment
Plan	There are not static plan, and this is favorable for well supporting autonomy, and how to take decisions is helped by the identification of Safety and Liveness properties, the first guarantees to prevent undesirable behavior whereas on the contrary the second enables desirable behavior
Decision process	The identification of Responsibility together with Safety and Liveness properties model the decision process in terms of expected behavior of a Role

Table 3 Autonomy supported in ASPECS

ASPECS	
Environment	A representation of the environment may be extracted from two different perspectives, the knowledge domain represented through the ontology and the set of resources that may be manipulated by roles through capacities
Action	During the System Analysis phase the concept of *capacity* is used for describing the "know-how" of each agent, a sort of behavioral building block and also the specification of the system and environment transformation under certain constraints
Knowledge	An ontology is used for conceptualizing knowledge domain, hence concepts/objects of the world, assertions on concepts using predicates, and actions an entity may perform for changing properties of one or more concepts
Goal	ASPECS models individual and collective goals, the first kind relates to the self-interest goal of each agent, whereas the second to the goals shared among other agents. Goals are identified during the design phase
Constraints	Social rules and norms are represented in the ontology and are used by roles that represent expected behavior and a set of rights and permissions
Plan	Both static and dynamic plans are modeled, the second through Interactions that is a not a priori known sequence of events that may trigger effects on the system
Decision process	It is modeled in the recruitment process that allows to establish and assigning capacities to a holon

autonomy are supplied, however they are not clearly made explicit as it is in Gaia. It is not in the scope of this paper to compare methodologies or provide metrics for autonomy, we used these two methodologies for showing that, although deeply different, they both support autonomy; if only one of the above concepts missed we could not affirm the same.

5 Discussion and Conclusions

Agent based systems technology is a paradigm born for conceiving, analyzing, designing and implementing complex systems by means of a very powerful abstraction: *agent*. It is recognized that the most of today's complex systems may be managed through organized societies of agents that communicate and interact in order to pursue their own goals or the ones of the society they live in. In doing this, agents strongly interact with their environments.

All the researchers, working in this field, cannot disagree that using the agent paradigm goes beyond providing a system with problem solving or social interaction capabilities; there must be much more. Autonomous behavior, hence *autonomy*, of agents or of the whole multiagent system is a key concept in agency. So we wondered: since an agent is mostly used as a design paradigm, how may we say that an agent methodology supports autonomy in building MASs? In order to answer these questions we firstly analyzed literature definitions of agent and we conducted a study for catching all the elements that could be used as design abstractions for guaranteeing autonomy.

The result was that, for being autonomous, an agent has to be endowed with the ability to be reactive towards the environment and then to have a representation of it; reactive in the sense that, while pursuing their objectives, agents (living) in a dynamic and uncertain environment change the environment itself and react to that in a useful fashion applying decision process and not static plans imposed by the designers. Moreover, an agent has to be provided of knowledge on the actions it can do, all its abilities, and the rules (or constraints) preventing it to do what is not allowed in a specific environment while pursuing objectives.

Thus the minimum set of elements to be managed during (mainly) analysis phase is: environment, action, knowledge, goal, constraint, decision process and plan; this latter has to be dynamically created by a decision process for realizing autonomy.

This representation, however, does not consider the fact that, see Wooldridge's definition, agents interact with other agents in order to achieve their goals, hence they are engaged in social actions; indeed, generally speaking, few goals may be achieved without the interaction with other people, without getting organized with others, negotiating and cooperating. Thus, we may say, the different levels of autonomy in realizing complex function pass through establishing social actions and creating organizations. Discussing this is out of the scope of this paper and will be argued in the future.

The paper concludes with an example of how two well known agent methodologies support autonomy; this work does not want to be a framework for comparing methodologies on the autonomy aspects, but is a starting point for conducting studies for identifying which are the best practices for constructing methodologies, or for modifying existing ones, well supporting all the aspects of agency, from autonomy to pro-activeness and social ability. So in the future we plan to extend this study to all other peculiar aspects of agency.

References

1. Bergenti, F., Gleizes, M.-P., Zambonelli, F.: Methodologies and Software Engineering for Agent Systems: The Agent-Oriented Software Engineering Handbook, vol. 11. Springer (2006)
2. Brustoloni, J.C.: Autonomous Agents: Characterization and Requirements (1991)
3. Cernuzzi, L., Juan, T., Sterling, L., Zambonelli, F.: The gaia methodology. In: Methodologies and Software Engineering for Agent Systems, pp. 69–88. Springer (2004)
4. Cossentino, M., Galland, S., Gaud, N., Hilaire, V., Koukam, A.: An organisational approach to engineer emergence within holarchies. Int. J. Agent-Oriented Softw. Eng. **4**(3), 304–329 (2010)
5. Cossentino, M., Gaud, N., Hilaire, V., Galland, S., Koukam, A.: Aspecs: an agent-oriented software process for engineering complex systems. Auton. Agents Multi-Agent Syst. **20**(2), 260–304 (2010)

6. Cossentino, M., Hilaire, V., Gaud, N., Galland, S., Koukam, A.: The ASPECS process. In: Handbook on Agent-Oriented Design Processes, pp. 65–114. Springer, Berlin (2014)
7. Cossentino, M., Hilaire, V., Ambra Molesini, and Valeria Seidita. *Handbook on Agent-Oriented Design Processes*. Springer (2014)
8. Ferber, J.: Multi-agent Systems: An Introduction to Distributed Artificial Intelligence, vol. 1. Addison-Wesley, Reading (1999)
9. Franklin, S., Graesser, A.: Is it an agent, or just a program? A taxonomy for autonomous agents. In: Intelligent Agents III Agent Theories, Architectures, and Languages, pp. 21–35. Springer (1996)
10. Gilbert, D.: Intelligent Agent: The Right Information at the Right Time
11. Hayes-Roth, B.: An architecture for adaptive intelligent systems. Artif. Intell. **72**(1), 329–365 (1995)
12. Henderson-Sellers, B., Giorgini, P.: Agent-Oriented Methodologies. Information Science Reference (2005)
13. Maes, P.: Designing Autonomous Agents: Theory and Practice from Biology to Engineering and Back. MIT Press (1990)
14. Maes, P.: Artificial life meets entertainment: lifelike autonomous agents. Commun. ACM **38**(11), 108–114 (1995)
15. Russell, S., Norvig, P.: Artificial Intelligence. A Modern Approach. Artificial Intelligence. Prentice-Hall, Egnlewood Cliffs, vol. 25, p. 27 (1995)
16. Wooldridge, M., Jennings, N.R.: Intelligent agents: theory and practice. Knowl. Eng. Rev. **10**(2), 115–152 (1995)
17. Wooldridge, M.: An Introduction to Multiagent Systems. Wiley (2009)
18. Wooldridge, M., Jennings, N.R., Kinny, D.: The gaia methodology for agent-oriented analysis and design. Auton. Agents Multi-agent Syst. **3**(3), 285–312 (2000)
19. Zambonelli, F., Jennings, N.R., Wooldridge, M.: Developing multiagent systems: the gaia methodology. ACM Trans. Softw. Eng. Methodol. (TOSEM), **12**(3), 317–370 (2003)

A Data-Driven Approach to Dynamically Learn Focused Lexicons for Recognizing Emotions in Social Network Streams

Diego Frias and Giovanni Pilato

Abstract Opinion Mining aims at identifying and classifying subjective information in a collection of documents. A variety of approach exists in literature, ranging from Supervised Learning to Unsupervised Learning. Currently, one of the biggest opinion resource of opinionated texts existing on the Web is represented by Social Networks. Networks are not only a vast collection of documents but they also represent a dynamic evolving resource as the users keep posting their own opinions. We based our work relying on this idea of dynamicity, building an evolving model that updates itself in real time as users submit their posts. This is done through a set of supervised techniques based on a Lexicon of emotionally-tagged terms (i.e. anger, disgust, fear, joy, sadness and surprise) that expands accordingly to user's dynamic content.

Keywords Social networks • Emotion analysis • Data-driven models

1 Introduction

Social Networks represent a great place, maybe the best, to gather information about people's opinions since they are generally used to express personal thoughts and to discuss with other people about a subject [1, 10]. These opinions are really useful to understand and classify the emotion of an event (a social event, a product, a person, etc.) and analyze his trend [6, 8]. The identification of the event is also made easier in Social Networks by the use of hashtags a metadata tag that users use to group messages relating to the same subject, for example #USA, #Traffic, #business and so on.

D. Frias
Scuola delle Scienze di Base e Applicate, University of Palermo, Palermo, Italy

G. Pilato (✉)
Istituto di Calcolo e Reti ad Alte Prestazioni - ICAR (CNR), Palermo, Italy
e-mail: giovanni.pilato@cnr.it

G. De Pietro et al. (eds.), *Intelligent Interactive Multimedia Systems and Services 2016*, Smart Innovation, Systems and Technologies 55,
DOI 10.1007/978-3-319-39345-2_54

User's posts are divided into streams, usually private streams or public streams. Public streams are generally populated by posts about actuality, world events most of the times, where users confront each other supporting their opinions on the subject.

The basic idea of the work presented in this paper is to have an evolving method to pragmatically analyze the emotion, or the sentiment, of a Social Network stream. We achieved this goal by a learning procedure that somehow simulates human cognitive processes capable to update a basic lexicon used for the analysis of the stream. The approach is useful in social network streams, where new terms are continuously coined and where a fast and adaptive approach for recognizing useful features is desirable. The approach makes use of Class Association Rules (CARs) and proper weighting formulas in order to expand and improve the effectiveness of a lexicon specifically suited to recognize emotion in texts.

2 The Proposed Approach

The basic idea of the work illustrated in this paper is to have an evolving method to pragmatically analyze a Social Network public stream in order to detect the emotions, or the sentiment orientations expressed in the posts. The approach consists in a set of supervised learning techniques aimed at concurrently classifying the posts on the stream and using the tagged posts to update a set of features.

Supervised learning is widely used in Sentiment classification. Many classifiers, like Support Vector Machine, Term Frequency Inverse Document Frequency and Naïve Bayesian classifiers, have been used to accomplish this task [2, 4, 7, 11].

All these classifiers need to be trained in order to learn the model with which they will classify the documents. This training is done through a set of already tagged documents, which constitute a training set. This set is usually represented as a set of vectors, and the performance of the classifiers heavily depends on the quality of such training datasets, and on the features that are exploited by the classifiers.

2.1 The Model

Our model is based on a starting training dataset, a lexicon, made of words already tagged with an emotion. This lexicon is the core of the model and it is also automatically expanded by the proposed approach through an iterative procedure. The goal is to tackle new terms (or even sequences of characters that at first sight may not have a sense outside the context in which they are used) that are continuously created and used in social network posts. We used Carlo Strapparava and Alessandro Valituttis emotions lexicon [12] as a starting dataset. The lexicon consists of approximately 1,500 words that have been manually tagged into 6 classes, representing the six fundamental emotions given by Ekman [3]: namely *anger*, *disgust*, *fear*, *joy*, *sadness* and *surprise*. Further information about the lexicon can be found in [12].

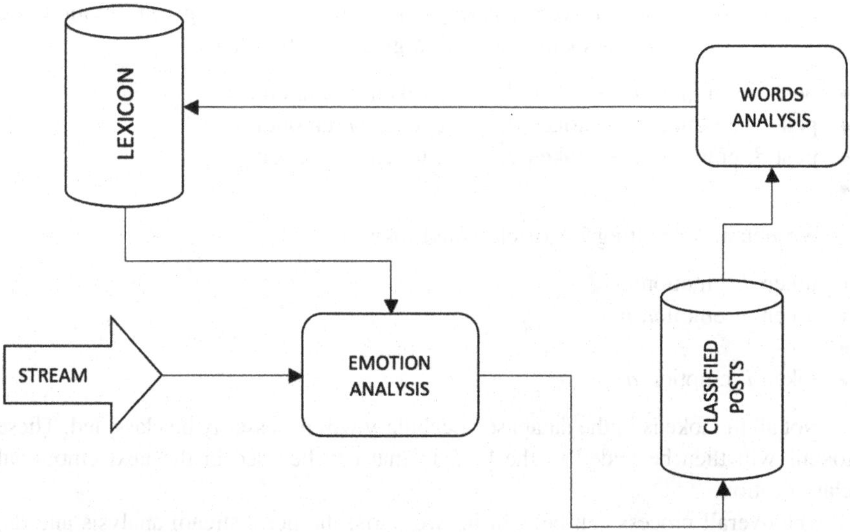

Fig. 1 Schema of the iterative approach

Our approach is iterative: at the beginning we classify the posts by using a supervised approach trained with the starting lexicon, then we use this dataset of tagged posts to classify the words contained in them in order to find new terms, or generally speaking, items, to add in the lexicon. Afterwards, a new cycle begins and we proceed with the classification of the posts by using the extended core lexicon and so on, like a distant supervised learning approach.

We can see a schematization of the idea in Fig. 1. This cyclic approach works as follows:

1. Connect and listen to a Social Network stream;
2. Every time a post is returned, let us perform the sentiment/emotional analysis by exploiting a Lexicon *L* and let us add the single analyzed post to a Dataset *D*;
3. When *D* has reached a given number of entries, let us perform a token analysis and let us add the resulting tokens, labeled with an emotion tag, to *L*.
4. Perform optional cleanup on the lexicon *L*.

The sentiment or emotional analysis in step (2) can be performed by using any method that uses a lexicon of terms. This point is obviously mandatory with our approach and a starting seed lexicon is needed at the beginning.

Once we have classified enough posts to achieve a reliable classification we proceed with the analysis in step (3). Any kind of analysis can be used: Association Rules, Naïve Bayesian, Support Vector Machine, Term Frequency, etc.

This analysis will parse every word in every database entry and it will correlate it with the resulting emotion. In particular, we have tested our approach by means of Association Rules and Naïve Bayesian classifier.

Hence, from a set of posts automatically labelled with a given emotion, we extract the tokens and associate them to a given emotion label:

- post *1*, emotion a → token *1*, · · ·, token n, emotion a
- post *2*, emotion b → token *3*, · · ·, token q, emotion n
- post *3*, emotion a → token *1*, · · ·, token i, emotion a
- . . .

We obtain a resulting list of classified tokens:

- token *1*, emotion a
- token *3*, emotion n
- . . .
- token i, emotion a

Not all the tokens in the database vocabulary will necessarily be classified. These tokens will then be added to the lexicon that can be used in the next emotional classification.

The overall process can be split in two parts: the posts stream analysis and the words analysis.

2.2 Stream Analysis

We begin analyzing the posts in the public stream using one of the many Supervised techniques that exploit our lexicon. Here the posts are seen as documents in a collection (i.e. the dataset to classify) and the lexicon is the training set. In our experiments we used Timothy P. Jurka Naïve-Bayes technique [5]. It first creates the Term Frequency Matrix (TFM) of the posts. TFM is a matrix where the rows represent the documents in the collection, the columns represent the words in the global dictionary and the elements are integer values representing the number of occurrences of the word in the document. Every row is also a vector representation of the relative document in a term-vector model. So the documents will be in the form:

$$d_j = (w_{1,j}, w_{2,j}, w_{3,j}, \ldots, w_{i,j}) \tag{1}$$

where d_j stands for the jth document and $w_{i,j}$ is the integer representing the number of times the word i occurs in document j.

The vectors (1) are then used to compute a probabilistic model, i.e. the Naïve Bayes classifier. Jurka's probabilistic model calculates the likelihood of a term-vector for each emotional class (e.g. anger, disgust, fear, joy, sadness and surprise) assigning a score based on the matches of the words with the core lexicon. In this kind of analysis, each document is independent from another one.

The classified posts are then stored in a database ready to be exploited in the words analysis phase.

2.3 Words Analysis

The goal of this step is to find and correctly classify the words in the posts in order to expand and/or renew our lexicon. To do so we have to build a classifier which, given a collection of tagged documents, recognizes and labels the words of such a collection. In order to reach this goal, we used two different algorithms: an algorithm inspired to the Term Frequency-Inverted Document Frequency (TF-IDF) approach and Class Association Rules (CAR).

The used TF-IDF model was inspired by Ghag and Shah's SentiTF-IDF [4]. It uses the same TFM used by the Naïve Bayes probabilistic model in the Stream analysis, together with the Term Presence Matrix (TPM), which has the same structure of the TFM but the elements are binary numbers representing the presence, 1, or the absence, 0, of the words in the document. Obviously the rows in the TPM are also term-vectors representations of each document with the same model as (1). SentiTF-IDF uses TFM and TPM to compute the logarithmic differential TF-IDF of a word in positive and negative tagged documents, so we had to adapt it to our 6-class model for it work with our lexicon.

In another experiment we used CAR in place of SentiTF-IDF. CAR is a particular kind of mining model used to find all the correlations between the words in a collection of text documents and each topic (class). A CAR is an implication in the form:

$$\mathrm{D} \rightarrow c, \text{ where } \mathrm{D} \subseteq V, \text{ and } c \subseteq C \tag{2}$$

With V being the vocabulary of the dataset and C the set of class labels c. CAR finds all the rules of the form [2] that satisfy a given minimum support and a given minimum confidence. Support and confidence are two measures of the rule's strength: the former estimates the probability $Pr(D \cup c)$, and the latter the conditional probability $Pr(\mathrm{D} \mid \mathrm{c})$. Since we searched for single word associations, we did not go further than $|\mathrm{D}| = 1$, so we stopped at the first level of the association rules mining and we did not analyzed subsets whose cardinality was $|\mathrm{D}| > 1$.

A comparative of the two approaches is shown in the next chapter.

As opposed to the Stream analysis, in Words analysis the quality of the output relies on the whole collection of text, i.e. the documents are not independent. For good results the tagged posts database should be sufficiently big.

The output of this analysis is a database formed by a set of pairs of the form (T, E), where T is a given term and E is the associated emotion. These pairs of the database can be directly added to the starting (or the current) lexicon.

3 Experimental Results

In our experiment, we focused on analyzing the daily emotional trends of hashtags on Twitter. We used the Python language during all the experiment.

For the catch phase, we used twitter APIs in order to catch the Twitter live stream and get all the tweets throughout the month of December 2015 with at least one hashtag. During this time slot, we created a database of about 4,000,000 tweets. Eventually we calculated the top fifty most used tags and focused on their emotional trend. Surprisingly the *#MerryChristmas* hashtag was only 8th on December 25, and 143th during the whole period.

For the Stream analysis we made a Python porting of a sentiment and emotional text analysis package in R [5]. It classifies the six fundamental emotions given by Eckman (i.e. *anger, disgust, fear, joy, sadness, surprise*) of a set of texts by using a Naïve Bayes classifier trained on an emotions lexicon. The lexicon is the crucial point during this phase. The starting lexicon is based on Strapparava and Valitutti's emotions lexicon, and it is expanded every cycle as shown in Fig. 1. So the system computes the 6 fundamental emotions expressed by the ~4,000,000 tweets in order to proceed with the analysis of the words in each tweet.

In the subsequent subsections we illustrate four methodologies of construction of the lexicon, i.e. the words analysis phase, and we illustrate at the end a comparison of the results, together with the analysis of the *joy* emotion associated to the *#MTVSTARS* hashtag.

3.1 Experiment 1

In our first experiment we implemented a Python version of the SentiTF-IDF [4]. SentiTF-IDF uses the Term Frequency Matrix and the Term Presence Matrix to calculate the principle logarithmic proportion of TFIDF of a term across positively tagged documents and negatively tagged documents (in other words, the polarity of the term), so we had to slightly modify the algorithm for it to work with any number of classes in order to use it with our 6 classes emotional analysis.

The only pre-processing tasks performed were the stop-words removal and the use of words long at least 3 characters. Moreover, we applied a POS tagger [9] before the token analysis so we could better focus our analysis on nouns, adjectives and adverbs.

In order to give a good precision to the TF-IDF, the system stores 500 documents before proceeding with the token analysis.

For the final phase we re-applied the emotion analysis given by [5] and trained with the resulting lexicon, generated in the previous phase, to all the tweets in the database containing at least one hashtag from our 100 most used hashtags December list.

3.2 Experiment 2

In this experiment we used an aPriori-CAR algorithm in place of the SentiTF-IDF for the token classification during the learning phase (i.e. the words analysis). The support and confidence values were 15 % and 50 % respectively.

Here we applied a pre-processing similar to the one in the first experiment: stop-words and punctuation removal and a minimum length term of 3. No POS-tagging was applied. The clusters were of 500 documents too.

In order to speed up the process and since we need to tag only single terms, the algorithm was stopped at the first level (i.e. only subset whose cardinality was 1).

3.3 Experiment 3

We used a lexicon generated by the intersection (AND) of the above lexicons, the TF-IDF one (L_1) and the CAR one (L_2). The resulting Lexicon (L_3) has terms derived from the intersection and the combined emotions carried by both the lexicons. For example, if both L_1 and L_2 had the term *'music'* but the classified emotion was different, let's say *'joy'* for L_1 and *'surprise'* for L_2, L_3 would have two entry of the term *'music'*, one labeled with *'joy'* and one with *'surprise'*.

3.4 Experiment 4

Same as the Experiment 3 but using the union (OR) of the two lexicons to generate the new one (L_4).

3.5 Comparison of Results

The resulting lexicons of the four above illustrated experiments were quite different. TF-IDF classified many more terms than CAR (almost 7 times more terms), and also the resulting emotions were different enough. **Joy** is the uppermost emotion in L_1 representing more than 95 % of the whole lexicon. The emotions in L_2 are prevailed by **disgust**, taking around 45 % of the lexicon, followed by **joy** (21 %). The remaining 35 % is almost fairly divided by the other 4 emotions.

We then used these lexicons to calculate the emotional feelings of the fifty most used hashtags in December (Table 1). Most of the tweets resulted associated to the **joy** emotion by using L_1.

Table 1 Top 20 retrieved hashtags in December 2015

Hashtag	Occurrences
MTVSTARS	527,193
F4F	262,239
MGWV	262,196
RT2GAIN	258,131
VIDEOMTV2015	226,784
FOLLOWTRICK	174,487
E	53,622
SOUNDCLOUD	47,161
NP	40,721
TVPERSONALITY2015	34,184
JOBS	34,172
VIVIANDSENA	33,979
NOWPLAYING	31,155
MTVSTAROF2015	27,795
RT	23,404
YGTPB	23,135
USA	22,190
TEAMFOLLOWBACK	21,062
MARKETING	20,901
SOCIALMEDIA	20,263

With L_2, 80 % of the tweets were tagged as **surprise** (**surprise** represented <10 % of the emotional classified terms in L_2). L_3 and L_4 performed almost identical as L_1.

In Fig. 2 we can observe a chart representing the **joy** trend of the most used hashtag (#MTVSTARS) during the whole period.

We expected L_4 to give a result much closer to the one of L_1 than L_2. This was because L_1 is much bigger than L_2, so most of the terms in L_4 came from L_1. However, L_3 is made-up from the two lexicons' intersection (L_3 is almost the same size of L_2 and its emotional percentage distribution is close to L_2), but its classification is still almost identical to L_1 and L_4's one. Comparing it with Fig. 2 we observed that the emotional trend in **joy** of *#MTVSTARS* was virtually identical.

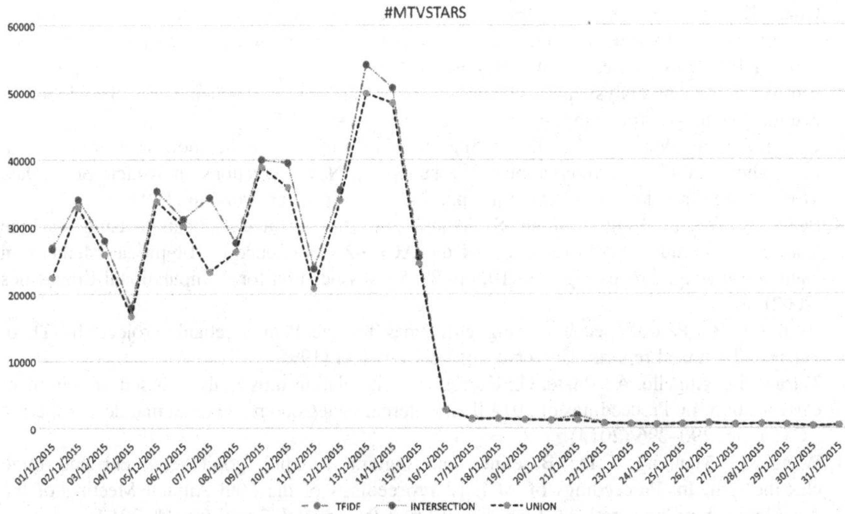

Fig. 2 Chart of the daily trend of #MTVSTARS in JOY using L_1, L_2 and L_3

4 Conclusions and Future Work

This paper proposes a model that dynamically learns new terms to add to lexicons in order to provide better accuracy in the classification of the emotion of opinionated texts. This is made possible thanks to the existence of Social Networks, which provide a non-stop stream of opinionated documents. Our model evolves with the evolution of the emotions of the users and becomes more accurate every time that a user posts a comment.

We are doing further experiments to provide the best approach to our model. In particular we are going to try different classifier in our Term analysis (such as SVM or Artificial Neural Networks) and compare with our current results, furthermore we will investigate more in-depth the CAR approach by using different values of Support and Confidence thresholds.

References

1. D'Avanzo, E., Pilato, G.: Mining social network users opinions' to aid buyers' shopping decisions. Comput. Hum. Behav. **51**, 1284–1294 (2014)
2. Dinu, L., Iuga, I.: The naive bayes classifier in opinion mining: in search of the best feature set. In: Computational Linguistics and Intelligent Text Processing: 13th International Conference, CICLing 2012, New Delhi, India, 11–17 Mar 2012, Proceedings, Part I, pp. 556–567. Springer, Berlin (2012)
3. Eckman, P.: An argument for basic emotions. Cogn. Emot. **6**(3/4), 169–200 (1992)

4. Ghag, K., Shah, K.: SentiTFIDF—sentiment classification using relative term frequency inverse document frequency. Int. J. Adv. Comput. Sci. Appl. Sci. Inf. Organ.
5. Jurka, T.P.: Tools for Sentiment Analysis, 01 Aug 2012
6. Liu, B.: Sentiment analysis and subjectivity. In: Indurkhya, N., Damerau, F.J.: Handbook of Natural Language Processing, pp. 627–665. CRC Press (2010)
7. Liu, B., Hsu, W., Ma, Y.: Integrating classification and association rule mining. In: Proceedings of the 7th International Workshop on New Directions in Rough Sets, Data Mining, and Granular-Soft Computing, pp. 443–447. Springer, London (1999)
8. Pang, B., Lee, L., Vaithyanathan, S.: Thumbs up? Sentiment classification using machine learning Techniques. In: Proceedings of the ACL-02 Conference on Empirical Methods in Natural Language Processing, vol. 10, pp. 79–86. Association for Computational Linguistics (2002)
9. Santorini, B.: Part-of-speech tagging guidelines for the Penn treebank project. In: D. o. Science, Technical reports. University of Pennsylvania (1995)
10. Terrana, D., Augello, A., Pilato, G.: Facebook users relationships analysis based on sentiment classification. In: Proceedings of 2014 IEEE International Conference on Semantic Computing (ICSC), pp. 290–296 (2014)
11. Wang, S., Manning, C.D.: Baselines and bigrams: simple, good sentiment and topic classification. In: Proceedings of ACL'12 Proceedings of the 50th Annual Meeting of the Association for Computational Linguistics: Short Papers, vol. 2, pp. 90–94 (2012)
12. Strapparava, C., Valitutti, A.: WordNet-affect: an affective extension of WordNet. In: Proceedings of the 4th International Conference on Language Resources and Evaluation (LREC 2004), Lisbon, pp. 1083–1086 (2004)

Disaster Prevention Virtual Advisors Through Soft Sensor Paradigm

Agnese Augello, Umberto Maniscalco, Giovanni Pilato and Filippo Vella

Abstract In this paper we illustrate the architecture of an intelligent advisor agent aimed at limiting, or as far as possible preventing, the damages caused by catastrophic events, such as floods and landslides. The agent models the domain and makes forecasting by exploiting both ontology models and belief network models. Furthermore, it uses a monitoring network to recommend preventive measures and giving alerts, if necessary, before that the event happens. The monitoring network can be implemented through both physical and soft sensors: this choice makes the measurements more adequate and available also in case of failure of some of the physical sensors. The front-end of the agent is made by a chat-bot, capable to interact with human users using natural language.

Keywords Decision support systems · Intelligent conversational agents · Soft sensors

1 Introduction

Decisional processes require the capability to analyze complex and dynamic scenarios, monitoring different strategic variables, and taking into account the consequences of the choices. Moreover, administrative decisions are important and even critical to prevent, or limit, the damages caused by catastrophic events such as floods and landslides. For these reasons, it is necessary to monitor variables and extreme situations and then make the right choices to avoid as much as possible heavy direct or indirected consequences. As an example, although the main cause of a pluviometric emergency is a natural event, a wrong urban planning choice can indirectly contribute to enlarge its consequences and amplify the damages due to intense rain.

A. Augello (✉) · U. Maniscalco · G. Pilato · F. Vella
Istituto di Calcolo e Reti ad Alte Prestazioni - ICAR (CNR), Palermo, Italy
e-mail: augello@pa.icar.cnr.it
URL: http://www.icar.cnr.it/en/node/466

G. De Pietro et al. (eds.), *Intelligent Interactive Multimedia Systems and Services 2016*, Smart Innovation, Systems and Technologies 55,
DOI 10.1007/978-3-319-39345-2_55

In this scenario, decision support systems [19] represent a set of useful tools that can be profitably exploited by decision makers and city administrators.

In [10] spatio-temporal bayesian models have been introduced to evaluate complex risk scenarios. In particular the system has been used in the context of forest fire emergencies management, to model temporal evolution of forest fires, taking into account climatic variables and territorial features. Models based on bayesian networks have been proposed also in [20] for fast disaster assessment in unconventional emergences, and in a context of emergencies produced by river floods [17].

Symbolic models such as ontologies can be useful to integrate and formalize information useful for disasters management. Information exchange between different organizations is crucial. However, it is very difficult since different vocabularies and representations are used by the involved organizations. For this reason, the authors in [4] proposed an ontology aimed at the description of both disasters damages and the resources needed for fighting them.

Hybrid models, considering both probabilistic and ontological models can be useful to build agent-based decision support systems. In [2] a distributed expert system based on a multi-agent-architecture has been proposed. The system is composed of a community of intelligent conversational agents playing the role of specialized advisors for the government of a virtual town. They are able to interact in natural language, and give suggestions about the best strategies to apply in order to manage the town. In [11] a conversational agent able to query a semantic Bayesian networks (SeBN) has been proposed. Data mining can be used to generate knowledge models that can be dynamically embedded into the agents; this allow the agents to adapt their behaviour to different scenarios, enhancing the adaptability, re-usability and versatility of these systems. In [21] a methodology and a tool for transferring extracted knowledge into newly created agents has been presented.

In this work we propose the exploitation and the integration of different methodologies in order to model an agent acting as an intelligent virtual advisor for city administrators. The agent could use forecasting models connected to a monitoring network in order to recommend preventive measures and giving alerts if necessary, before that the event happens. The agent can analyse environmental observations achieved by a monitoring network, composed of sensors dislocated in strategic locations. Its knowledge base can be formalized into ontological and probabilistic models of knowledge, allowing him to reason about the available data, making also estimation about uncertainty or not directly observable variables. This choice allows the agent to analyze the consequences of the selection of a sequence chain of actions among several alternatives.

In our model the agent can interact with the administrator and the citizens by means of an interactive natural language module. Through the conversation with the user it can acquire useful information about the current state of the place and information about user preferences. As a consequence, it can give suggestions about the best strategies to apply, forecasting the consequences of determined choices. The knowledge models can be used also to analyze past data for detecting the causes due to a bad administration, that had contributed to catastrophic events. The agent can interact also with the citizens, in order to inform them about the protocol to adopt in case of emergency.

Here we describe the proposed model according to an hypothetical case of study, regarding the creation of a advisor agent for disaster prevention into a virtual city.

2 Proposal for a Virtual Advisor

In this work we propose a model for a virtual advisor, able to provide support to human beings in decision-making processes.

The architecture, shown in Fig. 1, is composed of three modules named *measurement, knowledge management*, and *dialogue*.

The agent monitors strategic variables of a domain of interest in order to detect critical situations and reasoning about the best policies to propose to the user. Information about elements of interest for the analyzed domain can be acquired through a dialogue with the user and through observations of the environment. The interaction with the user is obtained by means of a dialogue module, and it can be used by the agent to communicate with decision makers and citizens.

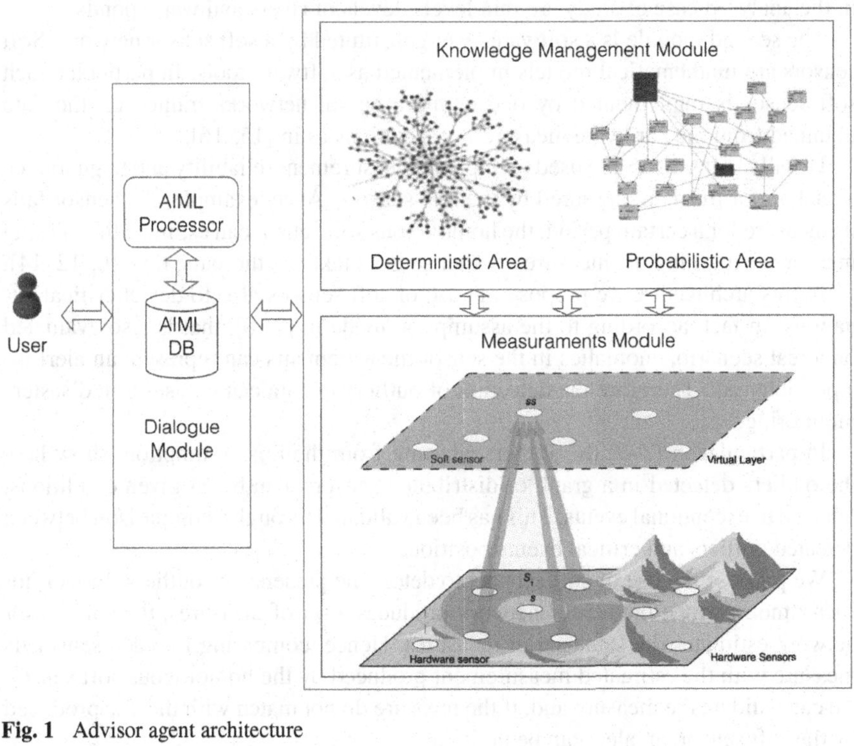

Fig. 1 Advisor agent architecture

The measurement module is used to collect and process sensorial data. In particular it is composed of two layers, a *physical* layer, represented by a sensors network geographically distributed into an area of interest, and a soft layer, represented by mathematical models named in literature soft sensors.

The knowledge on the domain of interest is formalized into proper models inside the knowledge management module, that are updated by the agent according to the data achieved by the measurements module and the information acquired during the conversation with the users. This module is therefore used by the agent to perform different types of reasoning in order to infer useful information and get the best strategies.

Measurements Module This module is aimed at the collection and analysis of data coming from multiple heterogeneous sources, data that are then used to populate the knowledge models of the virtual advisor. This allows the agent to reason about the extracted information, and in particular to identify critical or exceptional events.

The measurements module is composed of two layers. The first one is a physical layer, represented by a networks of measurements stations dislocated in the urban area that must be monitored, connected to a remote station devoted to the collection of all the data by mean of a communication infrastructure. It typically is a wireless sensor network (WSN) [1, 5]. The measurements are obtained by means of proper sensors located in each measurement station. Example of measurements of interest in the analysed case of study are rain levels, levels of rivers and water ponds.

The second module is a software layer constituted by a soft sensor network. Soft sensors are mathematical models implemented as software tools. In particular each soft sensor is implemented by one or more neural networks trained on the data obtained from the hardware measurements stations as in [15, 16].

Usually soft sensors are used to improve measurement reliability and to guarantee a backup for the data measured by the real sensors. As an example, if a sensor fails its measure for a certain period, the homologous soft sensor can estimate the missing measures exploiting the measure of other sensors taken at the same time [6, 12–14].

In this architecture we propose the use of soft sensors also to detect critical situations. In fact, according to the assumption made in [7, 8], that is also evaluated on a real scenario, anomalies in the sensor measurements can represent an alert for a possible risk. Therefore the detection of outliers in data can be useful in disasters monitoring.

In particular in [7–9] the authors, starting from the this assumption, show how the outliers detected in a graph of distributed sensors, can be, at given conditions, referred to exceptional events. This has been validated through a comparison between detected outliers and critical events positions.

We propose the use of soft sensors to detect the presence of outliers. In fact, for each time in which the sensor network produces a set of measures, the soft sensor network estimates the same set of measures. Hence, comparing for each sensor its measure with the estimated measurement produced by the homologous soft sensor, we can validate the measure and, if the measure do not match with the one produced by the soft sensor, an alert can be notified.

The historical evolution of the collected data correlated with occurrences of real catastrophic events can be also used to train the models of the knowledge management module.

Knowledge Management Module The knowledge management model of the agent is a hybrid model, constituted by the integration of a traditional ontology-based reasoning system (deterministic area in Fig. 1) with a probabilistic one (probabilistic area in Fig. 1). This choice is due to the requirement of managing both deterministic information and information characterized by uncertainty, which cannot be treated only by ontology-based systems. The integration of ontologies with bayesian decision networks allows the agent to exploit the advantages of both models.

The deterministic area gives a detailed description of the domain in an ontological model. The choice of the ontology will be made considering the specific case of study, the available data and the type of reasoning that will be necessary to perform [3].

The ontology is used to describe the main concepts, taxonomy, properties and rules of the domain of interest. In particular it allows for a description of the strategic variables and their properties and the definition of rules that can help the agent to take decisions that reflect the user preferences. As an example, for the case of study illustrated in this paper, the ontology can contain the definition of the concept *hydrogeological emergency* and its possible instances such as such as: *overflow, cloud burst, deluge, flooding, flood, severe storm, violent downpour*.

An inferential deterministic engine allows the agent to reason about the knowledge formalized in the ontology. It gives support to the user by inferring facts and making forecasting about the domain.

In particular, this type of reasoning is used by the agent to give the user information related to the domain, and answering to his requests. For example, citizens can query the system to deepen concepts related with the domain of interest, e.g.: "what are rain bands?", "What are the mechanisms that trigger floods?", "What is the Transitional Plan for Water Improvement?" or to learn how to behave in case of an hydrological emergency. "What do I do if a flood occurs?", "What do I do if I find myself on the street during a flood?".

An agent can exploit also a probabilistic area to manage information and situations characterized by uncertainty. In particular, it can take advantages by the exploiting of bayesian decision networks properly modelled and populated with the support of domain experts, or trough adequate supervised learning algorithms based on historical data. The observations of the domain variables can be processed in order to determine causal relationships between the variables. The network contain also decision and utility nodes that allows to reason about the decision that can lead to desired output with a certain utility. Utility nodes are defined and filled according to the preferences of the administrator interacting with the virtual advisor [18, 22].

Thanks to the decision network, the agent can reason about uncertain information, and suggest to the user the decisions that it believes to be the best to take in order to reach a specific goal. The agent can estimate the benefits and the potential risks resulting from the adoption of different strategies, and, as a consequence, to evaluate the effects over the variables of interest of the analyzed domain. The decisions are suggested taking into account also the preferences of the user. The decisional analysis is conducted through a direct comparison among the utility values corresponding to all possible choices.

For example, the agent can query this model to estimate the flooding risk given the evidences related to river level data when he does not clear meteorological information, or he can query it in order to estimate the possible risks due to the building of a mall in a certain area, considering at the same time its own utility in sing this choice would bring him many votes by a great amount of citizen that could benefit from this operation.

Dialogue Module This module allows the agent to interact in a natural way with decision makers to inform them about the current state of the domain and providing a support for decisional processes. The dialogue is implemented by exploiting the ALICE chatbot technology [23]. This choice is due to the fact that this simple technology allows to easily set up a dialogue engine. The rules for the understanding of user sentences and for the choice of the corresponding answers are defined by means of the mark-up language called AIML (Artificial Intelligence Mark-up Language). The chatbot KB is written in AIML and it is represented by question answer modules, named *categories*, that are compared time to time with the user questions according to a pattern matching algorithm. The user request is caught and processed. If it is an information strategic for the domain of interest, it is stored in the agent knowledge models and used by the reasoner, that then actuates the most proper strategy to suggest. Therefore a sentence is consequently composed *ad hoc* in order to answer to the user. User queries could include support request for a given decision policy, information about the domain, a refinement or revision of information represented in the deterministic and decisional areas. Ad hoc AIML tags allows the agent to query the ontology modelled in the deterministic area, or the decision networks belonging to the probabilistic area, in order to respectively provide users with useful information about the domain and to support them in decisional processes. Let us suppose that the model has been properly modelled into a Cyc ontology. Then, if the user asked for an information, the agent can exploit the CYD module [3] to query the ontology and give an appropriate response. As an example the following AIML category:

```
<category>
  <pattern>WHAT I HAVE TO DO IN CASE OF * </pattern>
  <template>
    This is the suggested protocol:
    <cycsystem>
      (cyc-query'(#$comment<cycterm><star/></cycterm>))
    </cycsystem>
  </template>
</category>
```

allows the system to manage the following interaction:

User: *What i have to do in case of hydrogeological emergency?*

Agent: *This is the suggested protocol: Do not rest in basements, cellars or garages; do not stay where a landslide may occur. Avoid transit or stopping on the banks of watercourses, bridges or footbridges. Avoid using your car unless strictly necessary...*

If the user asked for the possible consequences of an action, the agent, by exploiting the reasoning on the decision networks inform the user about the possible repercussions of a move. For example by the implementation of ad-hoc module for bayesian networks querying, introducing specific AIML tags, it would be possible to manage the following interaction:

User: *I want to build a mall*

Agent: *Write the exact location.*

User: *X-Avenue*

Agent: *If you build a mall in X-Avenue the impact on a possible hydrogeological emergency is high.*

The interaction can be obtained by writing categories such the following:

```
<category>
  <pattern>I WANT TO BUILD A MALL</pattern>
  <template>
    Write the exact location.
  </template>
</category>

<category>
  <pattern>*</pattern>
  <that>write the exact location.</that>
  <template>
    If you build a mall in <star/> the impact on a possible
    hydrogeological emergency is
     <bayesquery network="HYDRO_DN" node="risk">
       building mall location <star/>
     </baysquery>
  </template>
</category>
```

Off course this scenario requires the introduction of an hypothetical AIML tag *bayesquery* that needs information about the decisional network to query, the node on which make the forecasting, and a list of evidences on a set of nodes (in this case the value *mall* for the *building* node and the value *X-Avenue* for the *location* node).

3 Conclusion

In this paper we have presented a work-in-progress system, with embedded knowledge representation and probabilistic reasoning capabilities, aimed at supporting decisions in critical scenarios. The use of such an agent that suggests the best strategies to adopt, even in uncertainty situations, in order to manage an emergency is a useful test-bench for using knowledge representation and reasoning in real life. The agent makes use of both deterministic and probabilistic models of the world and furthermore it exploits advanced sensor networks capable to give accurate measurements and detecting possible high-risk situations. The chat-bot interface makes the interaction with users very friendly and effective.

Acknowledgments We would like to thank Emanuele Cipolla and Dario Stabile for their work in the set up of the visualization system and the collection of the data inside the activities for the systems able to filter data and process information for environmental multi risk analysis.

References

1. Akyildiz, I.F., et al.: Wireless sensor networks: a survey. Comput. Netw. **38**(4), 393–422 (2002)
2. Augello, A., Pilato, G., Gaglio, S.: Intelligent advisor agents in distributed environments. In: Information Retrieval and Mining in Distributed Environments, pp. 109–124. Springer, Berlin (2010)
3. Augello, A., Pilato, G., Vassallo, G., Gaglio, S.: Chatbots as interface to ontologies. In: Advances onto the Internet of Things, pp. 285–299. Springer (2014)
4. Babitski, G., Probst, F., Hoffmann, J., Oberle, D.: Ontology design for information integration in disaster management. GI Jahrestagung **154**, 3120–3134 (2009)
5. Chong, C.-Y., Srikanta, P.: Sensor networks: evolution, opportunities, and challenges. Proc. IEEE **91**(8), 1247–1256 (2003)
6. Ciarlini, P., Maniscalco, U., Regoliosi, G.: Validation of soft sensors in monitoring ambient parameters. In: Advanced Mathematical and Computational Tools in Metrology and Testing VII, vol. 72, p. 142 (2006)
7. Cipolla, E., Maniscalco, U., Rizzo, R., Stabile, D., Vella, F.: Analysis and visualization of meteorological emergencies. J. Ambient Intell. Hum. Comput. (2016)
8. Cipolla, E., Vella, F.: Boosting of association rules for robust emergency detection. In: 2015 Eleventh International Conference on Signal-Image Technology and Internet-Based Systems (SITIS), pp. 248–255. IEEE (2015)
9. Cipolla, E., Vella, F.: Identification of spatio-temporal outliers through minimum spanning tree. In: 2014 Tenth International Conference on Signal-Image Technology and Internet-Based Systems (SITIS), pp. 248–255. IEEE (2014)
10. Giretti, A., Carbonari, A., Naticchia, B.: A spatio-temporal Bayesian network for adaptive risk management in territorial emergency response operations. INTECH Open Access Publisher (2012)
11. Kim, K.-M., Hong, J.-H., Cho, S.-B.: A semantic Bayesian network approach to retrieving information with intelligent conversational agents. Inf. Process. Manag. **43** (2007)
12. Maniscalco, U., Pilato, G., Vassallo, G.: Soft Sensor based on E-αNETs. In: Apolloni, B., Bassis, S., Morabito, C.F. (eds.) Frontiers in Artificial Intelligence and Applications, vol. 226, pp. 172–179 (2010). ISSN: 0922-6389
13. Maniscalco, U., Rizzo, R.: A virtual layer of measure based on soft sensors. J. Ambient Intell. Hum. Comput. pp. 1–10 (2016)

14. Maniscalco, U., Rizzo, R.: Adding a virtual layer in a sensor network to improve measurement reliability. In: Advanced Mathematical and Computational Tools in Metrology and Testing X. World Scientific Publishing Co., Singapore, pp. 260–264 (2015)
15. Maniscalco, U.: Virtual sensors to support the monitoring of cultural heritage damage. In: Biological and Artificial Intelligence Environments, pp. 343–350 (2005)
16. Maniscalco, U., Pilato, G.: Multi soft-sensors data fusion in spatial forecasting of environmental parameters. Adv. Math. Comput. Tools Metrol. Test. IX **84**, 252–259 (2012)
17. Molina, M., Fuentetaja, R., Garrote, L.: Hydrologic models for emergency decision support using Bayesian networks. In: Symbolic and Quantitative Approaches to Reasoning with Uncertainty. Springer, Berlin, pp. 88–99 (2005)
18. Nielsen, T.D., Jensen, F.V.: Bayesian Networks and Decision Graphs. Information Science and Statistics, 2nd ed. vol. XVI, 448 p. (2007). ISBN: 978-0-387-68281-5
19. Power, D.J., Sharda, R., Burstein, F.: Decision Support Systems. Wiley (2015)
20. Song, L., Jie, W., Hui, Y., He-ping, Z.: Bayesian network model for fast disaster assessment in unconventional emergencies management. In: 2011 International Conference on Information Systems for Crisis Response and Management (ISCRAM), pp. 375–381. IEEE (2011)
21. Symeonidis, A.L., Kyriakos, C.C., Athanasiadis, I.N., Mitkas, P.A.: Data mining for agent reasoning: a synergy for training intelligent agents. Eng. Appl. Artif. Intell. **20**(8), pp. 1097–1111 (2007). doi:10.1016/j.engappai.2007.02.009. ISSN: 0952-1976
22. Williamson, J.: Bayesian Nets and Causality: Philosophical and Computational Foundations (2004)
23. www.alice.com

A Personal Intelligent Coach for Smart Embodied Learning Environments

Agnese Augello, Ignazio Infantino, Adriano Manfré, Giovanni Pilato, Filippo Vella, Manuel Gentile, Giuseppe Città, Giulia Crifaci, Rossella Raso and Mario Allegra

Abstract Within a Smart Learning Environment (SLE) learners are involved in a new learning process tailored to create a continuum of education by extending the current educational formal settings to real-life informal learning context. The goal of this paper is to describe the Cognitive Architecture (CA) of a Personal Intelligent Coach able to manage learning tasks and interactions within a complex Smart Learning Environment (SLE). PICo has two possible embodiments: humanoid robot, and an avatar on mobile device. We argue that the proposed intelligent coach can adapt to the contents, to the students needs and can evolve its strategies according the learning process.

Keywords Cognitive architecture · Embodied cognition · Smart learning environment

1 Introduction

The concept of Smart Learning Environment (SLE) has a recent foundation [26]; it emerges from a multidisciplinary interaction amongst different areas (mainly epistemology, psychology and ICT) that connect three core concepts: learning, environment and smartness [26]. Learning concerns stable and persisting changes in what a person knows and can do [25]. Cognition and learning are linked to the body and to the physical world (environment). That is, the embeddedness of cognition in a physical body and in a physical world means that not all knowledge needs to be put into the head; it can reside in a specific relation between body and world [8, 24].

A. Augello (✉) · I. Infantino · A. Manfré · G. Pilato · F. Vella
ICAR - National Research Council of Italy, Viale delle Scienze - Edificio 11, 90128 Palermo, Italy
e-mail: augello@pa.icar.cnr.it; agnese.augello@gmail.com

M. Gentile · G. Città · G. Crifaci · R. Raso · M. Allegra
ITD - National Research Council of Italy, Via Ugo La Malfa 153, 90146 Palermo, Italy
e-mail: manuel.gentile@itd.cnr.it

G. De Pietro et al. (eds.), *Intelligent Interactive Multimedia Systems and Services 2016*, Smart Innovation, Systems and Technologies 55,
DOI 10.1007/978-3-319-39345-2_56

Accordingly, the principles of embodied cognition [10, 14, 22] are taken into account in order to create a context within the learning process that involves the core relation between body and world. Furthermore they allow a multi-modal and multi-sensory cognition [12]. A learning environment is smart if it is able to support end-users with different levels of knowledge and with different backgrounds [11]. A SLE can adapt itself to learners learning profiles and allows action planning and several learning strategies both for learners and teachers.

Within a SLE learners are involved in a new learning process tailored to create a continuum of education by extending the current educational formal settings to real-life informal learning context. This process allows to unlock ubiquitous learning opportunities to students in a more engaging and entertaining manner, within an inclusive edutainment (educational entertainment) paradigm.

At the same time, in the past, cognitive architectures (such as ACT-R, SOAR, and so on), have been successfully employed to develop learning tutors [1, 2]. Such systems include learning and memory theories, and other learning and cognitive science models. Recent findings in this research field tried to explore the potentiality of biologically inspired architectures [23], and to include complex human-like cognitive aspects such as emotions [13], motivations, and social capabilities.

In this paper, we propose a model of cognitive architecture for a Personal Intelligent Coach named PICo, able to manage learning tasks and interactions within a complex Smart Learning Environment (SLE). According to two foundational theoretical concepts mentioned above (embodied cognition and ubiquitous learning) [6, 20], PICo has two possible embodiments: an humanoid robot, and an avatar on mobile device. Both play the role of catalysts of learning processes, and realize the Personal Intelligent Coach (PICo) of the SLE.

The paper is structured as follows: Sect. 2 describes the main desirable goals and functionalities that may characterize a personal intelligent coach, Sect. 3 describes the modules of the PICo architecture and its embodiments in the smart learning environment, Sect. 4 explicates the PICo functionalities by means of a concept example and finally, Sect. 5 contains conclusions and discussion on future works.

2 Personal Intelligent Coach: Aims and Functionalities

The main goal of Personal Intelligent Coach is to manage learning activities in order to create personalized learning paths according to the user needs and the different contexts in which the learning activities could take place.

By means of storyteller robot/avatar PICo provides students with multi-modal and multimedia learning contents and paths within three different connected learning contexts: laboratory, school, household (Fig. 1).

The mentioned SLE integrates robots and virtual robots as storytellers and catalyst of learning processes (Fig. 2).

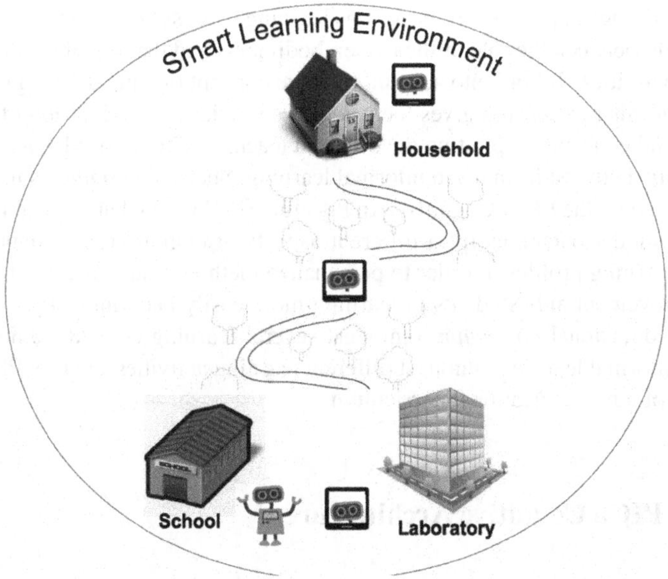

Fig. 1 PICo framework: intelligent learning coach in smart learning environments (SLEs)

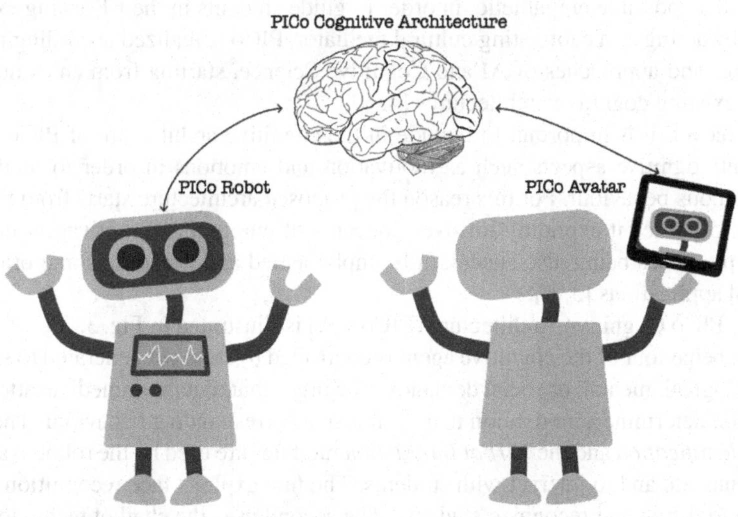

Fig. 2 PICo possible embodiements: humanoid and avatar on smart device

When the robot/avatar plays the active role of principal actor in the learning activities, the teacher has the role of a facilitator. When the robot/avatar has a passive role the teacher guides directly the learning activities.

The key activity of PICo is storytelling. PICo within the SLE upstages the current meaning of storytelling [18] or digital storytelling since it creates a new version of this methodological practice. It makes this learning process new, establishing a close relationship between the role of the two embodiments and the role of the contents of learning activities. Taking into account the core concept of embodied cognition, the creation of this relationship gives the students an easier comprehension of learning contents and concepts, exploiting the different teaching sources of SLE and creating a *continuum* between formal and informal learning. Such a *continuum* is maintained and managed by the PICo-Cognitive Architecture (PICo-CA) that is capable to evaluate multimodal outcomes from different aspects of cognition and adapt itself to students learning profiles in order to personalize methods and contents of teaching. This improvement aids students in learning more easily and efficiently. Creating a complex educational *continuum* (amongst several learning contexts, amongst formal and informal learning, amongst different learning activities), makes PICo-CA a promoter of a new educational curriculum.

3 The PICo Cognitive Architecture

A PICo can be considered as an artificial cognitive agent endowed with knowledge management and reasoning mechanisms. A key purpose is also to make the agent as much as possible empathetic, in order to guide students in their learning experiences by acting as a captivating cultural mediator. PICo is realized according to key concepts and approaches of AI and Cognitive Science, starting from an evaluation of the existing cognitive architectures [4, 9].

Moreover, it is important to include in the cognitive architecture of PICo some relevant cognitive aspects such as motivation and emotions in order to model an autonomous behaviour. For this reason the proposed architecture starts from the Psi model [5], since it explicitly involves concepts of emotion and motivation in cognitive processes being also successfully implemented and tested on some practical control applications [3, 9].

The PICo Cognitive Architecture (PICo-CA) is illustrated in Fig. 3.

The behaviour of the cognitive agent is controlled by its *urges* generated to satisfy physiological, mental, or social demands. The urges that require immediate attention or action determine a motivation to accomplish a corresponding behaviour. The *Student Identification* and the *Verbal Interaction* modules are used by the robot/avatar to communicate and to interact with students. The first exploits face recognition algorithms to detect and recognize students. The second uses the chatbot technology to implement a conversational agent. Traditional chatbot architectures can be improved by considering semantic analysis and reasoning methodologies in order to make the chatbot answers as coherent as possible with the requests of the user. The chatbot exploits the planning module in order to be able to assume a proactive behaviour, by stimulating the conversation according to the current context, extrapolated from the perceived data and according to the user profile. The *Personalized learning* module is based on the recognition of the student, and the recall of the learning plan created by an expert teacher, taking into account age, gender, and linguistic behaviour [7].

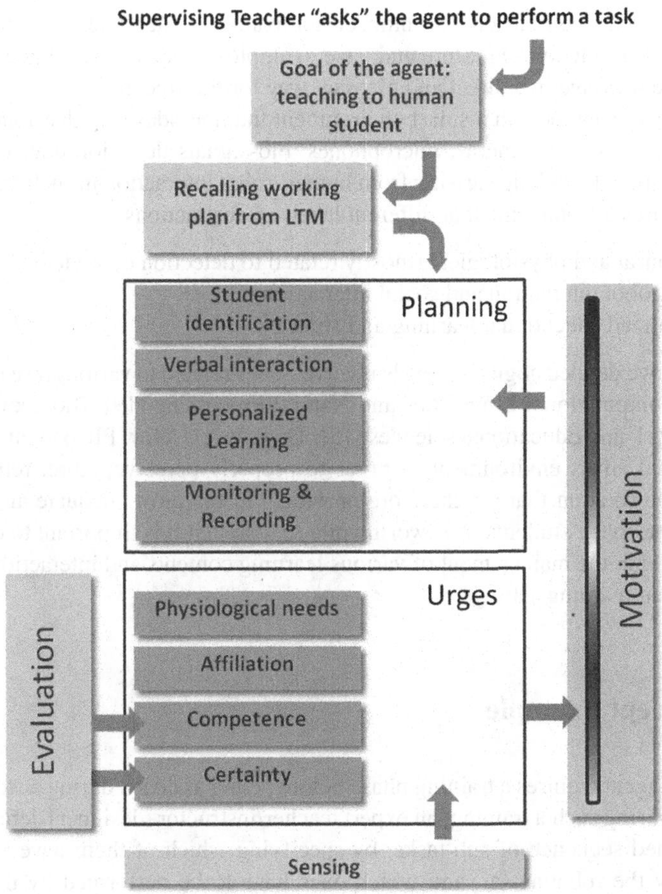

Fig. 3 PICo architecture based on PSI cognitive model

The learning plan could consider some variations activated during plan execution by monitoring the human-robot interaction e.g. affective state detection, evaluation of learning parameters, and so on. During the execution of the plan, the information and parameters are monitored and recorded by means of the *Monitoring and Recording* module, so that these data are useful to improve the next learning session. Evaluation mechanisms will be also considered in order to test the learning process. They affect Competence and Certainty Urges of the PICo Agent, and influence its motivation that in turn affects its behaviour during execution of a plan.

PICo has its own knowledge base suited to the topic to be treated. Furthermore each teacher (planner) can set the parameters of the PICo-model in order to define the more appropriate agent's personality according to the psychological profile of the student. Moreover, if the student has new needs, or some of the previous ones were not satisfied, PICo reschedules the session to try to fulfil all needs of the student.

Moreover, by means of teacher training, customization on the student and storytelling skills, PICo is able to use gesture and gaze to employee persuasive strategies and to convey the message in a direct and pleasant way for the student.

The PICo agent acts in a smart environment that includes also distributed smart sensors (other video cameras, microphones, bio-signals detection devices, and so on). The global knowledge arising from human-robot interaction in such an environment requires a management at different levels of abstractions:

- behavioural and physiological (mostly related to detection of student reactions);
- human-robot interaction and social interaction;
- personalized educational learning and strategy.

The above defined cognitive levels involve issues related to various research fields such as Computational Linguistics and Natural Processing [19], Biomedical Engineering [21], and Educational sciences [15]. These levels allow PICo to interact with the external smart environment in order to properly perceive, store, retrieve and process information that are therefore provided, in the most adequate and appealing manner, to the students. It is worthwhile to note that it is important to ensure an adaptability in the management of various learning contents and interactions within the different learning settings.

4 Concept Example

The PICo agent requires a training phase before acting as coach during a storytelling session. During such a training, an expert teacher/instructor (designer) defines a pre-programmed sequence of sub-tasks, by specifying which of them have to be performed by the robot/avatar, and which ones have to be performed by the human co-teacher. The teacher defines also possible substitutes to such sub-tasks in order to manage possible replanning according to the evolution of the interaction. The following scenario takes inspiration from a work of Breazel [17]. At the beginning the co-teacher introduces the PICo robot to the student, explaining the rules of the learning environment, describing the robot/avatar sensory capabilities and its interaction skills. Similarly he/she introduces the student to the robot, by specifying his/her name, gender, age, and other features useful to choose a suitable planned session of learning.

PICo recalls from its LTM a planning schema, and starts to tell a first story, monitoring the students reactions and collecting, at the same time, the data coming from the smart environment. Each storytelling subtask ends with few questions directed to the student. The student's responses are the processed by speech recognition and NLP modules; subsequently the results are stored for further evaluation.

Following a pre-programmed sequence, the robot tells another story, or waits that the human co-teacher executes his/her sub-task. By a physical touch of the robot or a vocal command, human can communicate to PICo the end of his/her intervention.

Moreover, in order to engage as much as possible the student and to avoid boredom, PICo can introduce a sort of variability exploiting computational creativity mechanisms. Its openness to creativity is strongly influenced by the motivation parameter. Therefore, during the storytelling sub-session, the PICo agent can have creative behaviour that determines the choice of different movements, postures, sentences, and other backchanneling behaviors (e.g., smiles, face appearances, sounds, gaze [16]). All the detected values of the whole session are stored together with the final judgements of the student and the co-teacher about their satisfaction about learning and interaction.

5 Discussion and Future Works

In this work a proposal for an architecture for the implementation of a personal intelligent coach has been discussed. The PICo cognitive architecture is a high performing architecture designed considering other cognitive architectures but improving them with regard to several aspects.

The proposed architecture has an high degree of adaptability, and it could be deployed on an artificial (embodied) agent that represents a powerful learning tool for teachers. Robotics, and artificial intelligence findings have been often used in learning contexts, and cognitive architectures could be an added value: active artificial agents can be in fact useful to evaluate both learning evolution and social interactions by exploiting their sophisticate sensing capabilities, and cognitive processing modules. At the moment, we are developing the storytelling modules that will be integrated in the cognitive architecture. Meanwhile, we are planning some experimentations involving students and teachers of the primary school, in order to determine the best settings of parameters of PICo architecture.

References

1. Anderson, J.R., Corbett, A.T., Koedinger, K.R., Pelletier, R.: Cognitive tutors: lessons learned **4**(2), 167–207
2. Anderson, J.R., Matessa, M., Lebiere, C.: ACT-R: a theory of higher level cognition and its relation to visual attention **4**(12), 439–462 (1997)
3. Augello, A., Infantino, I., Pilato, G., Rizzo, R., Vella, F.: Introducing a creative process on a cognitive architecture. Biol. Inspired Cogn. Archit. **6**, 131–139 (2013)
4. Bach, J.: Principles of Synthetic Intelligence PSI: An Architecture of Motivated Cognition, vol. 4. Oxford University Press (2009)
5. Bartl, C., Drrner, D.: Psi: a theory of the integration of cognition, emotion and motivation. In: Proceedings of the 2nd European Conference on Cognitive Modelling, pp. 66–73. DTIC Document (1998)
6. Benitti, F.B.V.: Exploring the educational potential of robotics in schools: a systematic review. Comput. Educ. **58**(3), 978–988 (2012)
7. Bowers, E.P., Vasilyeva, M.: The relation between teacher input and lexical growth of preschoolers. Appl. Psycholinguist. **32**(01), 221–241 (2011)

8. Byrge, L., Sporns, O., Smith, L.B.: Developmental process emerges from extended brainbody-behavior networks. Trends Cogn. Sci. **18**(8), 395–403 (2014)
9. Cai, Z., Goertzel, B., Geisweiller, N.: OpenPsi: Realizing dörners "Psi" cognitive model in the OpenCog integrative AGI architecture. In: Artificial General Intelligence, pp. 212–221. Springer, Berlin (2011)
10. Chemero, T.: Radical Embodied Cognitive Science. MIT Press, Cambridge, MA (2009)
11. Città, G., Crifaci, G., Prenjasi, E., Raso, R., Gentile, M.: Designing a new smart, adaptive and embodied learning environment. In: Bruni, D., Carapezza, M., Cruciani, M., Lo Bosco, G., Plebe, A., Perconti, P., Tabacchi, M.E. (eds.) Il futuro prossimo della scienza cognitiva. NEA-SCIENCE—Giornale Italiano di Neuroscienze, Psicologia e Riabilitazione, vol. 7, pp. 35-37 (2015). ISSN: 2282-6009
12. Crifaci, G., Città, G., Raso, R., Gentile, M., Allegra, M.: Neuroeducation in the light of embodied cognition: an innovative perspective. In: Proceedings of the 2015 International Conference on Education and Modern Educational Technologies (EMET2015), pp. 21–24 (2015)
13. Faghihi, U., Fournier-Viger, P., Nkambou, R.: CELTS: a cognitive tutoring agent with human-like learning capabilities and emotions. Intelligent and Adaptive Educational-Learning Systems, pp. 339–365. Springer, Berlin (2013)
14. Gallagher, S.: Philosophical antecedents of situated cognition. In: Robbins, P., Aydede, M. (eds.) The Cambridge Handbook of Situated Cognition, pp. 3551. Cambridge University Press (2009)
15. Garber-Barron, M., Si, M.: Adaptive storytelling through user understanding. In: Ninth Artificial Intelligence and Interactive Digital Entertainment Conference (2013)
16. Ham, J., Cuijpers, R.H., Cabibihan, J.J.: Combining robotic persuasive strategies: the persuasive power of a storytelling robot that uses gazing and gestures. Int. J. Soc. Robot. **7**(4), 479–487 (2015)
17. Kory, J., Breazeal, C.: Storytelling with robots: Learning companions for preschool children's language development. In: The 23rd IEEE International Symposium on Robot and Human Interactive Communication, 2014 RO-MAN, pp. 643–648. IEEE (2014)
18. Leite, I., McCoy, M., Lohani, M., Ullman, D., Salomons, N., Stokes, C.K., Scassellati, B.: Emotional storytelling in the classroom: individual versus group interaction between children and robots. In: HRI, pp. 75–82 (2015)
19. MacWhinney, B.: The CHILDES Project Tools for Analyzing TalkElectronic Edition Part 2: The CLAN Programs (2016)
20. Mubin, O., Stevens, C.J., Shahid, S., Al Mahmud, A., Dong, J.J.: A review of the applicability of robots in education. J. Technol. Educ. Learn. **1** (2013)
21. Nacke, L.E., Kalyn, M., Lough, C., Mandryk, R.L.: Biofeedback game design: using direct and indirect physiological control to enhance game interaction. In: Proceedings of the SIGCHI Conference on Human Factors in Computing Systems, pp. 103–112. ACM (2011)
22. Noë, A.: Out of Our Heads: Why You Are Not Your Brain, and Other Lessons from the Biology of Consciousness. Macmillan (2009)
23. Samsonovich, A.V., De Jong, K.A., Kitsantas, A., Peters, E.E., Dabbagh, N., Kalbfleisch, M.L.: Cognitive constructor: an intelligent tutoring system based on a biologically inspired cognitive architecture (BICA), 311–325 (2008)
24. Smith, L.B.: Cognition as a dynamic system: principles from embodiment. Dev. Rev. **25**(3), 278–298 (2005)
25. Spector, J.M.: Foundations of Educational Technology: Integrative Approaches and Interdisciplinary Perspectives. Routledge (2012)
26. Spector, J.M.: Conceptualizing the emerging field of smart learning environments. Smart Learn. Environ. **1**(1), 1–10 (2014)

A Model of a Social Chatbot

Agnese Augello, Manuel Gentile, Lucas Weideveld and Frank Dignum

Abstract Traditional chatbots lack the capability to correctly manage conversations according to the social context. However a dialogue is a joint activity that must consider both individual and social processes. In this work we propose a model of a social chatbot able to choose the most suitable dialogue plans according to what in sociological literature is called a "social practice". The proposed model is discussed considering a case study of a work in progress aimed at the development of a serious game for communicative skills learning.

Keywords Chatbots · Social practice · Serious games

1 Introduction

Conversational agents are becoming more and more present in everyday life. Thanks to the widespread chatbot technology, it is easy to add a conversational interface to a large set of applications with various purposes [1–4]. Chatbots are based on simple pattern matching rules, therefore it is possible to avoid several issues concerning natural language processing, and to quickly set up *simple* dialogue scenarios [5, 6].

Obviously, when dealing with more complex situations, requiring a thorough analysis of the conversation and the generation of adequate answers, it becomes necessary to increase the flexibility of these systems, providing them with linguistic

A. Augello (✉)
ICAR - National Research Council of Italy,
Viale Delle Scienze - Edificio 11, 90128 Palermo, Italy
e-mail: augello@pa.icar.cnr.it

M. Gentile
ITD - National Research Council of Italy, Via Ugo La Malfa 153, 90146 Palermo, Italy
e-mail: manuel.gentile@itd.cnr.it

L. Weideveld · F. Dignum
Utrecht University, The Netherlands Princetonplein 5, De Uithof,
3584 CC Utrecht, The Netherlands
e-mail: F.P.M.Dignum@uu.nl

G. De Pietro et al. (eds.), *Intelligent Interactive Multimedia Systems and Services 2016*, Smart Innovation, Systems and Technologies 55,
DOI 10.1007/978-3-319-39345-2_57

analysis and reasoning capabilities and extending their knowledge with external repositories, such as ontologies and semantic spaces [7, 8]. Moreover, chatbots lack the capability to correctly manage social conversational practices. A dialogue is a joint activity that must consider both individual and social processes [9]. Different communication strategies and then different utterances can be used according to the specific social context, and each conversation has a social effect because it contributes to changes in people, their social relations, and the material world [10].

Although there is an increased attention to the development of "more social" conversational agents [11], having a social identity [12] and being able to recognize the social attitude of their interlocutors [13], to the best of our knowledge, little has been done to put social context at the basis of the deliberation in conversation. Including social context could lead to adding excessive complexity to the chatbots implementation, especially considering that currently they lack a proper dialogue management: all the conditions that characterize the state of the conversation, and the effects of the choices on different variables related to the state of the dialogue and the interlocutor must be handled at every dialogue move, further complicating the writing of their knowledge base.

Moreover, the interpretation of the social context does not only influence the understanding of the current sentence and the issuance of a certain response. It leads the agent to assign specific roles and meanings to the actors and the elements involved in the conversation and to pursue specific courses of actions driven by specific expectations. It is very important to consider a theoretical model that can help to formalize how the social context influences the deliberation process and therefore the communication choices of a conversational agent.

In this work we use the Social Practices theory [14] and its formalization into an agent architecture as proposed by Dignum et al. [15], where a social practice refers to a routinized type of behaviour typically and habitually performed in a society. The aim is to take a first a step towards the implementation of what we called a "social" chatbot, a conversational agent that is not simple reactive, but that is aware of the current social practice and consequently is able to pro-actively choose the more suitable dialogue plans. The proposed architecture extends chatbots with proper modules for the recognition and management of social practices in conversations. In particular, the work focuses on the deliberation of such an agent after the recognition of the current social practice. As a case study we propose an application for a serious game aimed to improve communicative skills.

2 Theoretical Background

2.1 The Alice Chatbot

Alice, like other chatbots, relies on pattern matching mechanisms to manage the conversation with the user. Its knowledge base (KB) is described by a mark-up language named AIML (Artificial Intelligence Mark-up Language) [16]. Specific AIML tags are defined to properly interpret the meaningful elements of the sentence written by the user and to properly generate the answers. The tag *category* represents a question-answers module, composed in turn of a tag *pattern* enclosing a sentence that will be compared to the user question, and a *template* tag which encloses the rules to generate the chatbot answer when the matching with the related pattern is found. The template can contain other AIML tags; in this case the answer is dynamically composed by properly analyzing them or recursively calling other categories.

As claimed in the introduction, chatbots have a limited management of the dialogue, but this is an essential feature for the development of social conversational agents. In AIML the dialogue is managed keeping track of the last conversation exchange and setting conversation topics. These mechanisms are accomplished by means of the special tags *that* and *topic*. Categories could be organized in topic-related collection by putting them inside a specific tag *topic*. The AIML processor allows the designer to manage simple text based variables in order to dinamically customize the template. The following category shows an example of matching that keeps into account the chatbot's last answer:

```
<category>
      <pattern>MY NAME IS *</pattern>
      <that>HELLO THERE WHAT IS YOUR NAME</that>
      <template>Nice to meet you <star /></template>
</category>
```

2.2 Social Practice Model

A social practice refers to a routinized type of behaviour typically and habitually performed in a society. Social Practice theory is a sociological theory that studies the behaviour of groups of people and their interactions according to social practices [14]. In [15], this theory is analyzed considering an individual perspective in order to formalize it into an agent architecture. The proposal for a model of social practice, provides a representation scheme that allows the implementation of cognitive agents able to use the social practice as a first-class construct in the agent deliberation process [15]. According to this model a social practice is characterized by the following elements:

- *Physical Context*: the set of environment elements, that are both the physical objects with a meaningful role in the practice (*Resources*), the agents (human beings or autonomous systems) involved in the practice (*Actors*) and the locations of objects and actors (*Places*).
- *Social Context*: the social interpretation of the elements sensed in the environment, and in particular the *Roles*, describing specific behaviours that can be expected from specific actors, the interpretation of all the other elements, called *Social*

Interpretation, and the *Norms* identifying the rules that are expected inside a social practice.

- *Activities*: possible courses of actions that agents can perform. In a very fine grained level we can tie them to speech acts [17].
- *Plan Patterns*: ordered sets of scenes with a specific sub-goal that restrict the type of plans that can be used.
- *Meaning*: social meanings for the agent's activities and plans.
- *Competences*: the abilities that the agent should have to perform the activities of the social practice.

3 Toward a Social Chatbot Architecture

The architecture of the social chatbot is shown in Fig. 1 and it is composed of three main modules named *Social Practice Selection*, *Identity* and *Deliberation Engine*. In this work we describe the last two modules, while a description of the *Social Practice Selection* module is reported in [18].

According to the proposed architecture, the identification of the current social practice influences the agent's identity and drives its deliberation process, influencing its conversational choices. According to the selected social practice, the agent updates its beliefs about social interpretation of the elements with which it interacts, and starts a deliberation process in order to manage the conversation within the context of that practice.

The deliberation module is composed of a rules engine that allows to properly reason on the agent's rules, and by a Social AIML (S-AIML) processor used to manage the conversation. The interaction of these two modules allows the agent to manage the dialogue according to the social practice model. The conversation allows the agent to further discriminate the context and if necessary to re-evaluate the current practice. This happens also if some norms bound to that practice are violated. In the following sections the architecture is explained according to a specific case study.

3.1 A Case Study

Effective communicative skills are important in different fields, and are the basis of every social relation. Serious games can be useful in this context, as a valid approach to train people to properly carry out conversations by means of simulation of dialogues with virtual characters [19–21]. In particular the case study, taking inspiration from a game named Communicate! [22], deals with the simulation of medical consultation in order to train physicians to communicate better with their patients. Actually we are defining a conversational agent for such a game playing the role of a virtual patient.

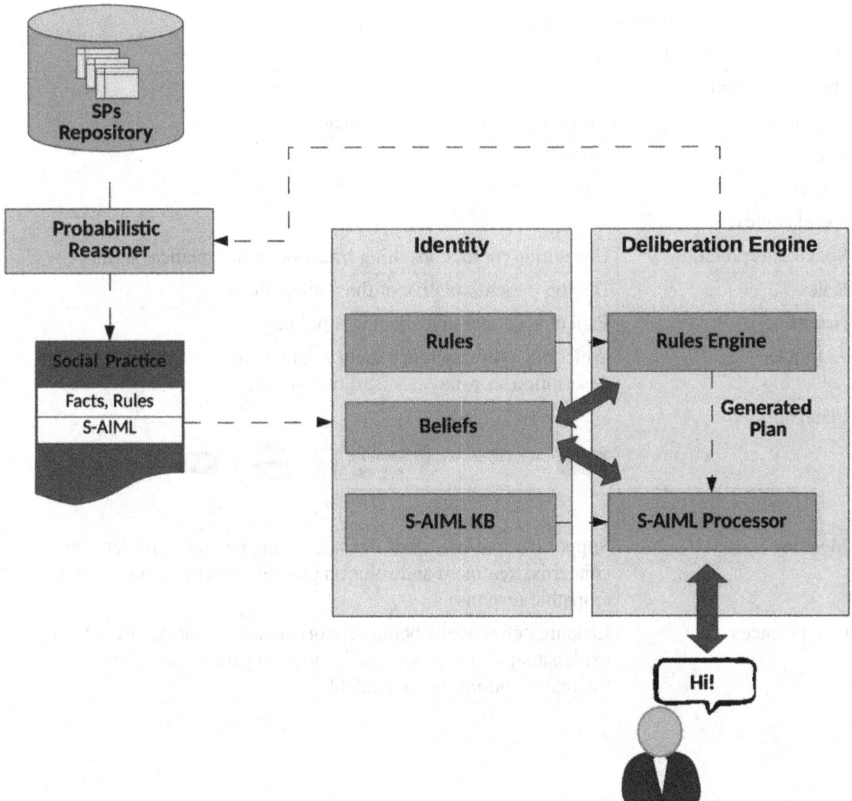

Fig. 1 Social chatbot architecture

3.2 An Example of Social Practice

In this section the components of social practices are described by referring to the concrete example concerning a consultation with an unknown doctor. Table 1 summarizes the components of the social practice under investigation.

The consultation takes place in a hospital, inside a room and at a specific time. The two actors involved in the practice assume respectively the roles of doctor and patient. Another possible role of such a practice could be a relative of the patient. The doctor-patient dialogue generally takes place during consultation hours in a special room used for the purpose. The plan pattern associated with the social practice is based on a classical medical protocol and it is organized in scenes. A scene is normally characterized by a specific goal. For example, the goal of the first scene of the practice "Welcome and Presentation" is to start a doctor/patient relation in order to proceed to the consultation. Moreover, the scenes are linked to the set of activities

Table 1 The doctor-patient dialogue as a social practice

Abstract social practice	Doctor patient dialogue
Physical context	
Resources	Current time, medical instruments
Places	Hospital, office
Actors	User, agent
Social context	
Social interpretation	Consulting room, consulting time,doctor has medical skills
Roles	Doctor, patient, relative of the patient, nurse
Norms	Patient is cooperative, doctor is polite
Activities	Welcome, presentation, patient's data gathering, patient symptom description, constative acts, directive acts
Plan patterns	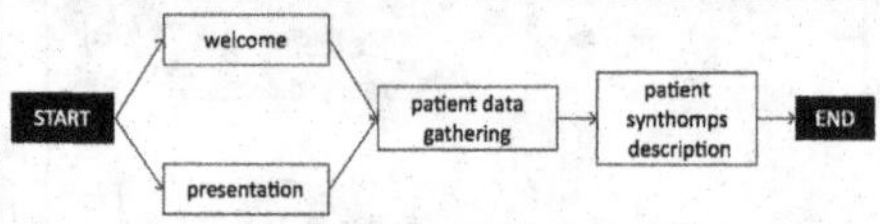
Meaning	Support the patient, create trust, eliciting patient's problems and concerns, treatment and solution possible, empathic opportunity, empathic response
Competences	Listening effectively, being supportive and empathic, use effective explanatory skills, adapt conversation to patient, discussing treatment options understandable

they take could take place in, for example communicative acts used by the agents to interact. The meaning of communicative acts includes its intended effects [9].

During the execution, the situation can evolve and lead to quit the social practice, for example an agent could quit from a social practice also in the case of a norm violation.

3.3 *Identity*

The identity of the agent formalizes his beliefs, the information related to the possible social practices, the state of the dialogue and the rules for the generation of plans, the analysis of possible norm violations and the state variables updating.

A formalization based on the concept of social practice allows the agent to interpret the context from a social point of view and to perform the more suitable plan pattern. Moreover the identity of the agents includes a set of categories, representing its linguistic knowledge required to manage the conversation with the user playing the game. In particular the categories are described according to S-AIML, an extension of the AIML language, that allows to bind the categories to specific practices and their activities.

The S-AIML language and its relative processor, are enhancements of the traditional dialogue engine of Alice that allows the chatbot to manage the dialogue according to a specific social practice. In particular specific tags have been introduced to contextualize the dialogue inside a social practice (*social practice* tag), according to a specific activity (*activity* tag), when specific precondition are satisfied (*precondition* tag) and to have more freedom in the effect management (*inc* and *dec* tags have been introduced to increase or decrease dialogue variables). The enhancement of the AIML language was required because the AIML dialogue designer cannot use only the variables/parameters to control the evolution of the dialogue. This is better explained in the next section.

3.4 Deliberation Engine

This module allows the agent to analyse the state of the dialogue and deliberate and act according to the social practice recognized by the *social practice selection* module. It is composed of two main components. The first is a reasoner that exploits facts, data and rules to infer conclusions that can lead to the accomplishment of an action. The action can be the updating of a fact, the execution of a plan or a specific activity inside the plan. The plan can be executed following specific conversational activities.

The other component is the S-AIML processor that manages the dialogue with the user processing the S-AIML KB. The rules engine and the S-AIML processor allows for a dynamic activation of a set of S-AIML categories related to the current social context. It is important to notice that the reduction of the number of possible S-AIML categories that can match user sentences simplifies the creation of the S-AIML knowledge base, leading to the possibility to write categories having more generic patterns.

The plan pattern associated to a social practice triggers a set of activities. The course of a particular activity is conditional on the fulfilment of a certain rule described in the rules engine. For example, the *presentation* activity can be performed only within certain social practices and for determined state preconditions. The agent can update the value of the dialogue state variables through the reasoning process but he can also update by means of the S-AIML processor. The update of a variable can also be indirectly caused by the updating of other variables in the course of the dialogue, or by switching to other social practices.

Depending on the actual situation the expectations of the social practice are confirmed and the practice is continued or the social practice is re-evaluated. In fact, thanks to a reasoning on the rules formalized in the *Identity* module, the agent continuously monitors possible violations of social practice norms. In case of a violation the agent will act in a proper manner, stopping the execution of that practice if necessary. A violations of one of the strict norms related to the current practice would lead him to re-analyze the situation and to select a new practice.

3.5 *An Example of S-AIML*

In what follows, a traditional AIML category is compared to a S-AIML one in order to show some advantages of a social-practice oriented approach. The example shows how modelling the chatbot knowledge base becomes more complex when the same user question must be analyzed according to different social contexts and conversation states.

Let us suppose that the doctor wants to reduce patient worries but he uses a wrong phrasing. The effect of such a communicative action depends largely on the context in which it is uttered. If the context is that of a generic conversation with an unknown interlocutor, the sentence should be interpreted as an insult. Particularly, if the interlocutor is a just met doctor, the tone is unprofessional and the patient could react interrupting the consultation. Instead, if the interlocutor is the patient personal doctor, he has different expectations, since he knows that the level of confidence they have, can allow the doctor to friendly scold him. The current dialogue regards the discussion of a patient's pathology, therefore, if the doctor temporises in answering the patient about his worries, he could become even more worried, with a negative effect on his level of trust.

The following category shows how this scenario should be managed without a model of social practice, using the traditional AIML.

```
 <category>
   <pattern>your worries are stupid</pattern>
   <template>
      <condition name="interlocutor" >
        <li value="unknown_doctor">
         Are you insulting me? I don't think this is a professional tone!
         <think><set name="emotion">angry</set>
                <set name="trusting">low</set>
         </think>
         </li>
        <li value="mydoctor">
             <condition name="answer_time"
                <li value="low">
                 Ok, I am more relieved!
                </li>
                <li value="high">
                Are you hiding something from me?
                <think><set name="emotion">surprise</set>
                       <set name="trusting">medium</set>
                </think>
                </li>
             </condition>
        <li> Are you insulting me?
         <think><set name="emotion">angry</set> </think>
         </li>
      </condition>
  </template>
</category>
```

The dialogue is managed by checking the values of specific state variables in the template, which is done after matching the user sentence with the AIML pattern. Instead, if the agent knows the current social practice, it already knows the values of other variables, for example the role of its interlocutor, and checking it is no needed anymore. Distributing the knowledge partly to the social practice, partly to the state and partly to the AIML rules can simplify the dialogue modelling.

It is worthwhile to note that, even if we manage the social practice in traditional AIML as a variable that implies the setting of other variables, it is important that the check is done before the pattern matching. In fact if the check is done after the pattern matching, different conditions have to be inserted on a single category as shown in the previous example. The specification of a dialogue rule would be very large and complex. Moreover, at the moment we have considered only the differentiation for a social practice, but in some cases it would be necessary to differentiate the agent's reactions to the same question, considering different activities inside a single practice. The following S-AIML category shows that managing dialogue according to social practices and putting the control before the deliberation simplifies the dialogue modelling and management. New agent behaviours can be added in a simple manner, leading to a more controllable and scalable process of modelling.

```
<social practice name="mydoctor_consultation">
  <category>
    <precondition>
      <condition> answer_time > 5 </condition>
    </precondition>
    <pattern>your worries are stupid</pattern>
    <template>
     Are you hiding something from me?
      <think>
        <set name="emotion">surprise</set>
        <dec name="trusting">3</dec>
      </think>
    </template>
  </category>
</social practice>
```

4 Conclusion

In this work a model for a social chatbot, based on the social practice theory, has been discussed. The architecture puts social practice at the heart of the deliberative process of an agent, allowing for a more accurate interpretation of user sentences. The introduction of S-AIML as an enhancement of the well known AIML language and interaction between a rules engine and the S-AIML processor, allows for a dynamic activation of the chabot KB, depending on the current social practice, the pursued plan, the ongoing activity, and finally, at the lowest level the agent's identity. This allows for a great flexibility in the conversation while at the same time simplifying the formalization of the chatbot KB. In our work we presented a case study related to the implementation of such a conversational agent in the Communicate! serious game. Future work will contain a more developed implementation of the agent with particular attention to the semantic understanding of the conversation.

Acknowledgments We are grateful to Prof. J. Jeuring, for his useful explanation of the Communicate! project and for providing us the scenarios of the game.

References

1. Augello, A., Santangelo, A., Sorce, S., Pilato, G., Gentile, A., Genco, A., Gaglio, S.: A multimodal interaction guide for pervasive services access. In: IEEE International Conference on Pervasive Services, pp. 250–256 (2007)
2. Pilato, G., Vassallo, G., Gentile, M., Augello, A., Gaglio, S.: LSA for intuitive chat-agents tutoring system. In: G., Allegra, M., Chifari, A., Ottaviano, S.(eds.) Methods and Technologies for Learning Chiazzese (2005)
3. Huang, J., Li, Q., Xue, Y., Cheng, T., Xu, S., Jia, J., Feng, L.: Teenchat: a chatterbot system for sensing and releasing adolescents stress. In: Health Information Science, pp. 133–145. Springer (2015)
4. Comendador, B.E.V., Francisco, B.M.B., Medenilla, J.S., Nacion, S.M.T., Serac, T.B.E.: Pharmabot: a pediatric generic medicine consultant chatbot. J. Autom. Control Eng. **3**(2), 137–140
5. AbuShawar, B., Atwell, E.: ALICE chatbot: trials and outputs. Computacin y Sistemas **19**(4) (2015)
6. Abdul-Kader, S.A., Woods, J.: Survey on chatbot design techniques in speech conversation systems. Int. J. Adv. Comput. Sci. Appl. (IJACSA) **6**(7) (2015)
7. Augello, A., Pilato, G., Machi, A., Gaglio, S.: An approach to enhance chatbot semantic power and maintainability: experiences within the FRASI project. In: 2012 IEEE Sixth International Conference on Semantic Computing (ICSC), pp. 186–193. IEEE (2012)
8. Agostaro, F., Augello, A., Pilato, G., Vassallo, G., Gaglio, S.: A conversational agent based on a conceptual interpretation of a data driven semantic space. In: AI*IA 2005: Advances in Artificial Intelligence, pp. 381–392. Springer, Berlin (2005)
9. Clark, H.H.: Using Language. Cambridge University Press, Cambridge, MA (1996)
10. Fairclough, N.: Analysing Discourse: Textual Analysis for Social Research. Psychology Press (2003)
11. Klwer, T.: I like your shirt—dialogue acts for enabling social talk in conversational agents. In: Intelligent Virtual Agents. Springer, Berlin (2011)
12. Cassell, J.: Social practice: becoming enculturated in human-computer interaction. Universal Access in Human-Computer Interaction. Applications and Services, pp. 303–313. Springer, Berlin (2009)
13. Carofiglio, V., De Carolis, B., Mazzotta, I., Novielli, N., Pizzutilo, S.: Towards a Socially Intelligent ECA. IxD&A **5**, 99–106 (2009)
14. Reckwitz, A.: Toward a theory of social practices a development in culturalist theorizing. Eur. J. Soc. Theor. **5**(2), 243–263 (2002)
15. Dignum, V., Dignum, F.: Contextualized planning using social practices. In: Coordination, Organizations, Institutions, and Norms in Agent Systems X, pp. 36–52. Springer (2014)
16. http://www.alice.com
17. Searle, J.R.: Speech Acts: An Essay in the Philosophy of Language. Cambridge University Press, Cambridge (1969)
18. Augello, A., Gentile, M., Dignum, F.: Social practices for social driven conversations in serious games. In: Proceedings of the Games and Learning Alliance Conference, GaLA 2015, Roma, 9–11 Dec 2015 (in press)

19. Swartout, W., Artstein, R., Forbell, E., Foutz, S., Lane, H.C., Lange, B., Morie, J., Noren, D., Rizzo, S., Traum, D.: Virtual humans for learning. AI Mag. **34**(4), 13–30 (2013)
20. Babu, S.V., Suma, E., Hodges ,L.F., Barnes, T.: Learning cultural conversational protocols with immersive interactive virtual humans. Int. J. Virtual Real. **10**(4), 25–35 (2011)
21. Dargis, M., Koncevics, R., Silamikelis, A., Kirikova, M.: Game-based training of communication skills in requirements engineering. In: REFSQ Workshops 2014, pp. 147–148
22. Jeuring, J., Grosfeld, F., Heeren, B., Hulsbergen, M., IJntema, R., Jonker, V., Mastenbroek, N., Van Der Smagt, M., Wijmans, F., Wolters, M., Van Zeijts, H.: Demo: Communicate!—a serious game for communication skills. In: Proceedings EC-TEL 2015: 10th European Conference on Technology Enhanced Learning (2015)

An Experience of Engineering of MAS for Smart Environments: Extension of ASPECS

Philippe Descamps, Vincent Hilaire, Olivier Lamotte and Sebastian Rodriguez

Abstract This paper presents a methodological approach for the engineering of Smart Environments based upon Multi-Agent Systems. This approach is an extension of an existing MAS methodology, namely ASPECS. The extension of ASPECS is allowed by the Situational Method Engineering principles underlying ASPECS and takes the form of several existing modified activities and corresponding meta-model elements. The key elements that are targeted by the contribution are: the identification of goals hierarchy, the expression of detailed requirements and associations of goals to sensors/effectors, different levels of ontologies to describe on the one hand, the problem conceptualization, and, on the other hand, the several involved expertness. The remaining, existing activities, refine the models in order to identify organizational structures/behaviours and theirs agentification. The approach is illustrated through a case study that consists in a platform dedicated to the Monitoring of patients with heart failures.

1 Introduction

The emergence of technological devices in our everyday lives has given birth to what may be called Smart Environments. The common feature of these Smart Environments is to add sensors and effectors of many sorts combined with an information system in order to build a new generation of information systems able to improve our everyday life. The applications covered by this kind of systems range from health and safety to power efficiency and comfort.

Such systems may be difficult to analyze and design, as they are composed of numerous interacting entities immersed in a dynamic and partially predictable environment. Moreover, the information systems part of these environments have to

P. Descamps · V. Hilaire (✉) · O. Lamotte
Univ. Bourgogne Franche-Comté, UTBM, IRTES EA7274, 90100 Belfort, France
e-mail: vincent.hilaire@utbm.fr

S. Rodriguez
GITIA, UTN-FRT, Rivadavia 1050, San Miguel de Tucuman, Argentina

G. De Pietro et al. (eds.), *Intelligent Interactive Multimedia Systems and Services 2016*, Smart Innovation, Systems and Technologies 55,
DOI 10.1007/978-3-319-39345-2_58

exhibit specific capabilities such as: autonomy, reactivity, and pro-activeness. In this paper we propose an approach based on Multi-Agent Systems (MAS) for the realization of the information system part. More specifically, the contribution consists in a methodological process fitted for engineering MAS dedicated to Smart Environments. This methodological process is based on an existing MAS process, namely ASPECS [1]. The principle is to adapt the original process by modifying the previously defined activities. The approach is illustrated through a case study that consists in a platform dedicated to the Monitoring of patients with heart failures [2]. Due to space reasons a part of the methodology and of the illustration example is presented. All implementation details are omitted. However, using ASPECS allows an easy deployment with it deployment language SARL [3].

The first activity consists in a goal identification for the system. The goal identification relies on the S-Moise+ notation [4] and is integrated within ASPECS. From this goal analysis our proposition is to refine goals by a problem ontology description. This first ontology aims at conceptualizing the problem and define the context for the realization of the MAS. The third activity refines the goals hierarchy, produced by the first activity, and the problem ontology, produced by the second activity, in order to define a model of the refined goals with association to sensors/effectors involved in their fulfillment and eventual added requirements. This latter activity uses as notation an extended version of the UML automation profile [5]. Starting from the model of requirements the next activity, which may be iterated, produces an ontology per specific expertise involved. The rest of the existing activities, refine the models in order to identify organizational structures/behaviours and theirs agentification.

This paper is organized as follows: Sect. 2 presents the methodology principles and models. Section 3 illustrates the methodology with a simplified (due to space concerns) analysis of the e-Health SE. Section 4 presents related works and concludes.

2 Models and Methodology

2.1 *Methodology*

The principle proposed in this paper in order to engineer SE consists in applying the activities described by the Fig. 1. The first activity, Goals Identification, aims at the definition of the global goals of the system. This activity produces a goal hierarchy using the S-Moise+ model and notation [4]. The next activity, Domain Ontology Description consists in conceptualizing the problem domain knowledge. In this conceptualization, the pertinent elements are those that are of interest for the SE. Each of these elements are represented by concepts. Each concept is then detailed in terms of attributes or features. Among these concepts some are able to act and produce modifications in the SE environment. Such acting concepts are associated to their possible action(s). Each action is also associated to the others concepts that are

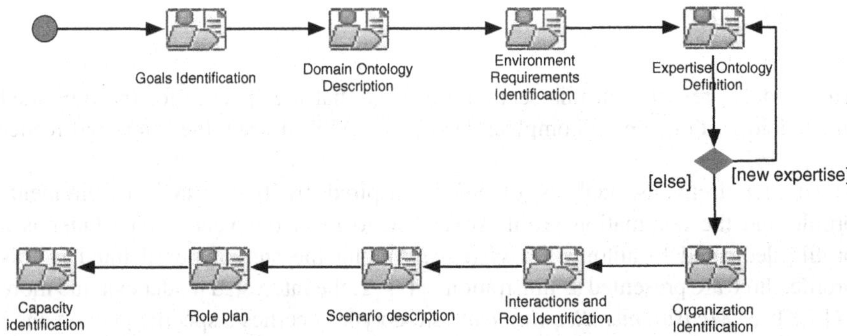

Fig. 1 Overview of the methodology

possibly modified. A concept may also be characterized by predicates that establish their possible states and relationships. Among the predicates some are constraints. Environment Requirements Identification, the third activity, aims at the identification of the problem requirements. The notation used is inspired by the UML automation profile. The resulting model should identify: the Human user's interaction with the SE, the different phenomenon that are monitored or controlled within the SE and the different capabilities of the SE in terms of control. Moreover, non functional requirement can be introduced. These latter requirements, as opposed to functional such as Measurement and Actuation do not describe systems functions but the acceptable states of the system. In this sense an analyst may express quality or safety/security features for example. This activity also focus on an analysis of what the SE should do, in what type of environments the SE will be situated and what are the properties that should be provided by the SE. What the SE should do are expressed by control requirements. The environment of the SE is specified by two categories of elements. First, the measurement and actuation requirements are, respectively, concepts related to the inputs and outputs of the SE by way of sensors and effectors. Second, user instrumentation requirements specify the possible means of Human-SE interactions. Once such a model is produced the SE analyst can have an initial view of the SE to be. The Expertise Ontology Definition deals with the definition of applications ontologies. By application ontology, we mean ontologies that are specific for a given application domain. For this activity, each specific application expertise should give birth to a specific ontology that will be used by agents to communicate and within reasoning mechanism. These ontologies thus must have a sufficient level of details to handle SE current situations.

The rest of the methodology activities described in Fig. 1 are similar to the previous Domain Analysis phase of ASPECS and result from the whole set of metamodels produced in an organizational structure. They result in the identification of organizations, that are logically grouped collective behaviors in charge of goals fulfillment, and the assignment of requirements to organization that will thus be in charge of their satisfaction. Once organizations are identified they can be refined into smaller interacting behaviors called roles.

2.2 Models

The models presented in this section are those that are specific for the presented methodology. For a more complete description of S-Moise+ the interested reader can see [4].

The requirements analysis of SE is inspired by the SysML requirements profile and the automation profile (extended to meet our needs). The latter is a profile dedicated to automation systems. Among the stereotypes defined in this profile, three are presented as illustration in Fig. 2, the interested reader can find more in [5]. The "*MeasurementRequirement*" stereotype specifies a specific phenomenon to be monitored. This kind of requirement can be further refined by a specific sensor. The "*ActuationRequirement*" stereotype specifies a specific phenomenon that is influenced by the SE. This requirement can be further refined by a specific actuator. Both "*MeasurementRequirement*" and "*ActuationRequirement*" are characterized by the following attributes: a textual description, a field parameter (what is measured or acted upon) and a set of data named, typed and directed (input or output). Eventually, the "*ControlRequirement*" stereotype in interaction with "*MeasurementRequirement*" and "*ActuationRequirement*" specifies an autonomous control mechanism based upon perception (resp. action) inputs (resp. outputs) information. Each "*ControlRequirement*" also contains a textual description and the data specifications coming from measurement or going to actuation. Concerning the "*NonFunctionalRequirements*", a description states the constraint to be fulfilled by the system.

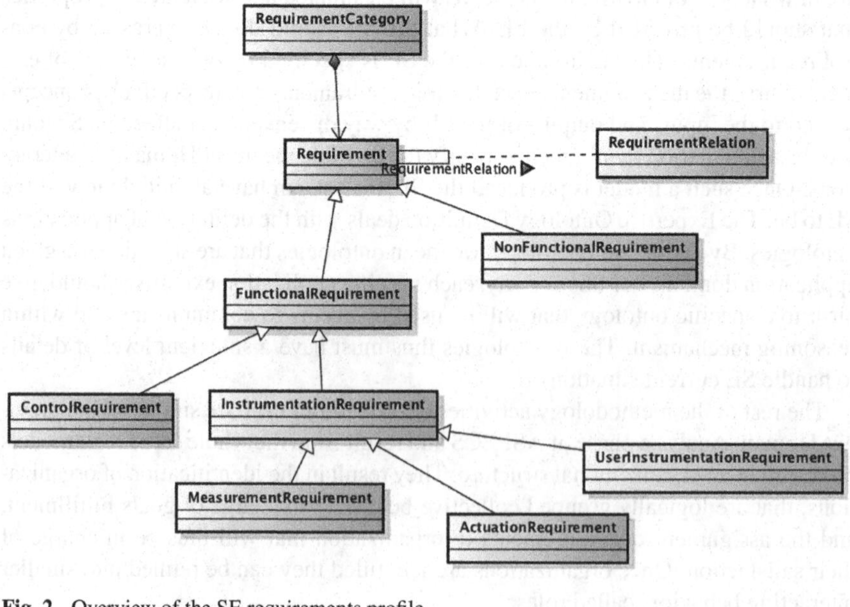

Fig. 2 Overview of the SE requirements profile

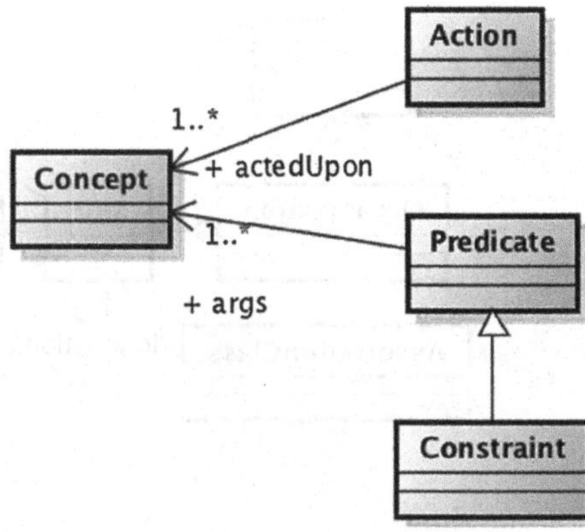

Fig. 3 Overview of the SE problem ontology profile

The stereotypes defined by the Problem Ontology profile are "*Concept*", "*Action*", "*Predicate*" and "*Constraint*". These stereotypes are represented by the Fig. 3.

Each "*Concept*" refers to an entity in the domain of interest during the analysis phase. These entities can represent: resources, actors, manipulated objects, etc. An "*Action*" represents a treatment or processing in the field of interest. This treatment or this transformation can also be specified by attributes and operations. Classes representing actions can possibly be bound by an association to a "*Concept*" performing this action. This association is named "*Actor*". A second association, also optional, defines a link between an action and a set of "*Concepts*" used as parameters for this action. This association is named "*Argument*". A "*Predicate*" is used to represent domain knowledge in the form of a property. This property is expressed by a first-order logic predicate. A predicate can be combined with a set of "*concepts*" which are variables of the predicate. "*Constraint*" are specific kind of "*Predicate*". It is described by a set of equations that constraint elements of a (or several) "concept". Using the Domain Requirement Description, a skeleton of Problem Ontology can be generated. For each "*MeasurementRequirement*", a specific "*Concept*" representing the field value is generated and linked to a isMeasurable "*Predicate*". The same is true for each "*ActuationRequirement*" and isControllable "*Predicate*". The created concept allow to refine the field value by adding semantics and properties. Each "*ControlRequirement*" generates an "*Action*" in the ontology as it is the case for "*UserInstrumentationRequirement*". For each concept representing field values the analyst should try to associate a "*Concept*" that represents the object, existing in the SE environment and characterized by the field value.

For the organization identification activity, the profile defines a specific diagram that allow the definition of "*Organization*" represented as a specific kind of Package grouping identified "*Requirements*". Possible strategies for organization identification and requirements grouping are, for example, based on the coherence of "*ControlRequirement*" or based on the structure in terms of concept aggregation proposed by the ontology. Once identified, organizations can be refined in terms of roles, interactions and capacities. The dedicated steretotypes are presented in Fig. 4.

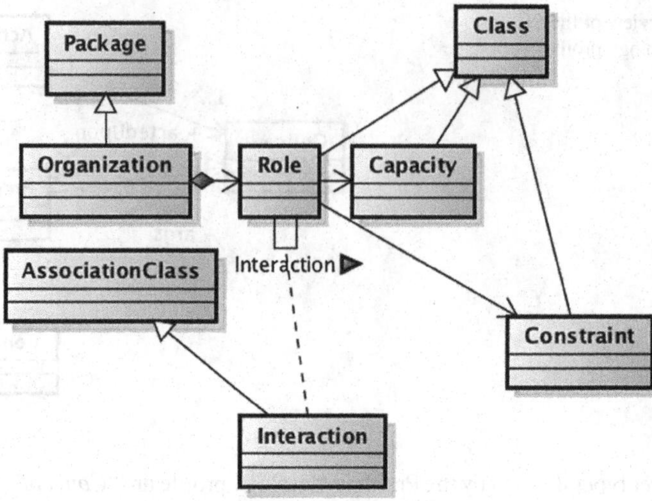

Fig. 4 Overview of the CRIO profile

Organizations are still represented by specific kind of Package that contain, for this diagram, "*Role*" that are specific kind of class, "*Interaction*" between roles that are represented by association classes, "*capacities*" and "*constraints*" are other specific kinds of classes. A skeleton of this kind of diagram can be generated from the previous ones. Indeed, each "*ControlRequirement*" generates at least one "*Role*". Each "*MeasurementRequirement*" and "*ActuationRequirement*" generates capacities (one per measurement or actuation).

3 Case Study

This section intends to illustrate the proposed approach with a simplified example (due to space concerns only a limited part of the analysis is presented). This example is drawn from an eHealth project named ECare [2]. It aimed at monitoring patients with heart failure disease. The goal hierarchy presented in Fig. 5 is the result of the first activity. The "Monitor patient" goal is the global goal of the system. This decomposed in two simultaneous goals: "Store data" and "Send warning". Data storage allows future analysis and warnings are sent to targeted persons if a potential danger is detected. The "Send warning" goal is in turn decomposed in two simultaneous goals: "Collect data" and "Detect abnormal situations". The "Collect data" goal. The latter is satisfied by a sequential goals: "Collect" for sensors data capture, "Control" for verification of captured data and "Transmit" for broadcasting the data to the allowed receivers. The "Detect abnormal situations" goal is decomposed in "Filtering" and "Reasoning". These goals respectively ensure the choice of pertinent data

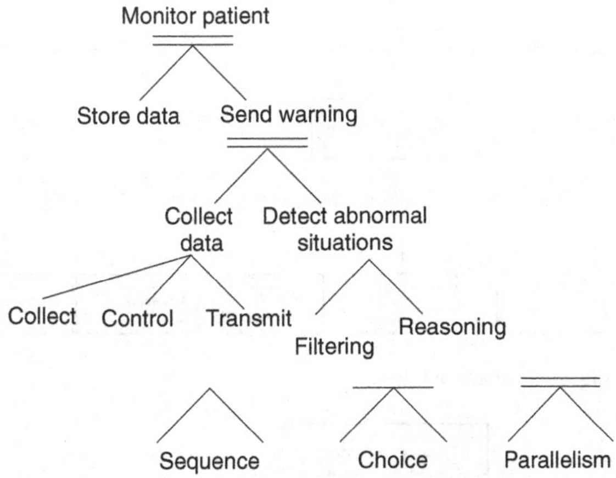

Fig. 5 Goal Analysis

in the continuous flow send by sensors and produce alerts with a corresponding level of severity.

The Domain ontology of Fig. 6 conceptualizes the domain of the targeted problem. The aim is to deploy *Environment* that are composed of *Sensor*. In order to fulfill the monitoring goal we define a constraint *isEquiped*. These sensors are in charge of producing a specific data related to their type. This production is specified by the action *Measure*that is related to a *Patient* a special kind of *User*. The other kind of *User* is *Medic*that denotes medical specialists that may be informed by the system.

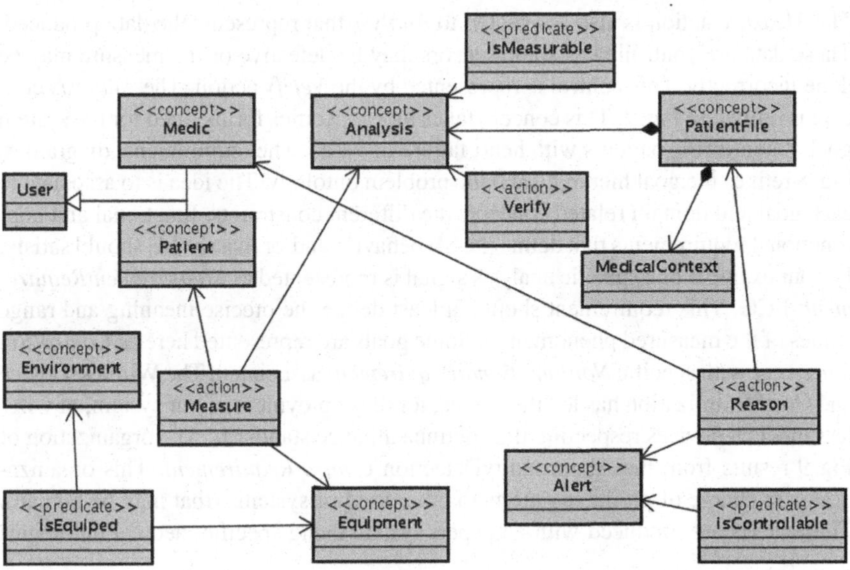

Fig. 6 Problem ontology

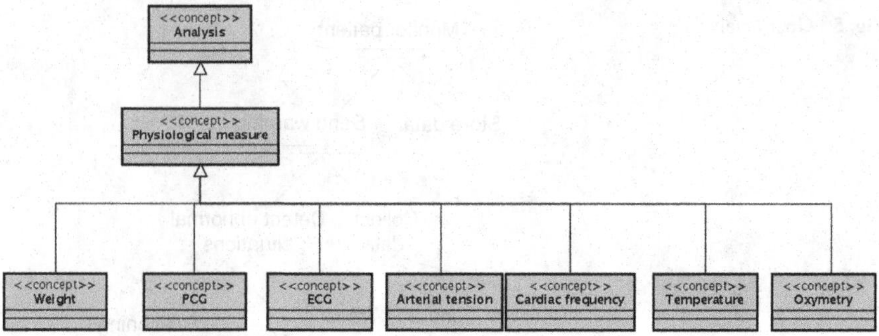

Fig. 7 Domain ontology

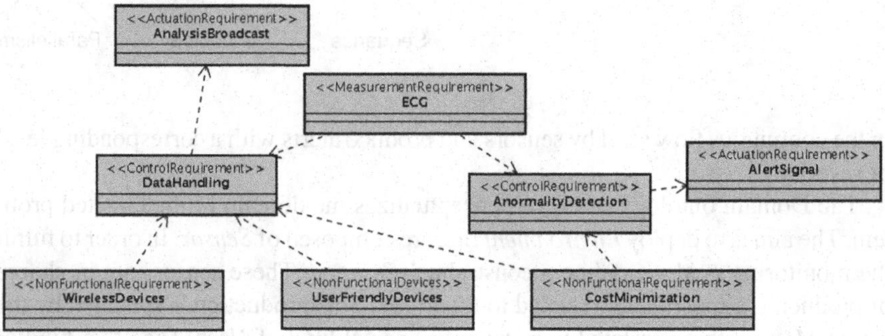

Fig. 8 Requirements

The *Measure* action is also associated to *Analysis*that represents the data produced. These data are controlled as some sensors may be defective or the measure may be done incorrectly. This control is represented by the *Verify* action. The *Analysis* concept is refined in Fig. 7. This concept takes more specifics forms fitted for the system goal: "monitoring patients with heart failure disease". The requirements diagram of Fig. 8 refines the goal hierarchy and the problem ontology. The idea is to associate to each goal and domain related concepts the different constraints, functional and non-functional requirements that define the SE behavior and/or that the SE should satisfy. For our example one specific analysis signal is represented as *MeasurementRequirement*: ECG. This requirement should at least define the precise meaning and range values of the measured phenomenon. Some goals are represented here as *ControlRequirement* with specific *NonFunctionalRequirement* associated. The WirelessDevice and CostMinimization has led the choice, for the deployment of our system, of wireless medical devices respecting the continua alliance standard.[1] The organization of Fig. 9 results from the AbnormalityDetection *ControlRequirement*. This organization is in charge of producing alerts to other kind of systems (that may be human). These alerts are produced with an expert system using specific medical ontologies

[1]http://www.continuaalliance.org/.

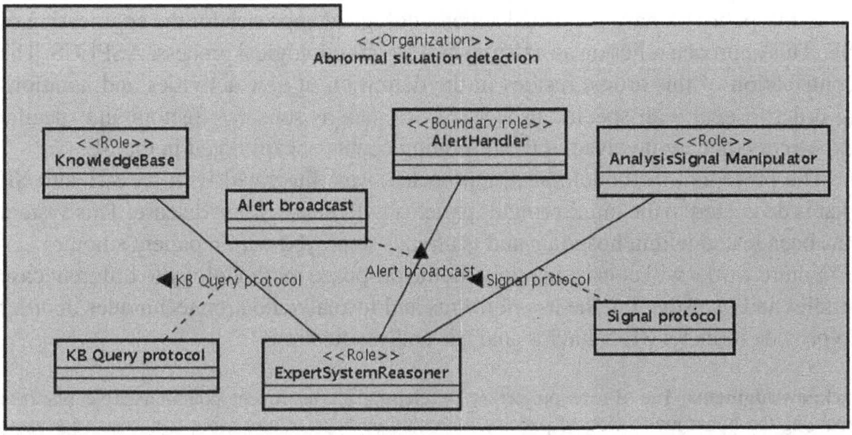

Fig. 9 CRIO

that are represented and encapsulated by the KnowledgeBase *Role*. The results are send to the AlertHandler *BoundaryRole* that represent these external systems. This expert system takes as input the different verified signals coming from the medical devices and sent by the AnalysisSignal Manipulator *Role*.

4 Related Works and Conclusion

There already exist several methodologies for MAS. Referencing all existing ones is out of the scope of this paper. The interested reader can check the following sources [6, 7]. These methodologies have proven that they can be applied for the analysis/design/implementation of MAS each assuming a specific set of hypothesis. However, the concept of smart environment that is the target of this paper is not taken as a first class concept. The chosen stance is to use dedicated metamodel elements for describing sensors/effectors and goals as soon as the early analysis activities.

For the e-Health domain we are interested in, there are several existing systems that have proven some efficiency. For example, in [8] the authors propose a system that monitor pregnant women suffering from Gestational Diabetes Mellitus. The underlying technique used is based on agents and logical reasoning models. However, SE that require an intelligent software part may be difficult to design. Methodologies trying to define guidelines and activities in order to ease the analysis and development of such systems are very few. In [9] the authors propose a methodology for e-Health system that is oriented towards e-Health interacting information and that do not take into account SE as discussed in this paper. Approaches such as [10] or [11] are important contributions to the SE methodological challenge. The differences with the approach proposed in this paper are the focus on sensors/effectors and monitored/controlled phenomenon and the integration of multiple expertness.

In this paper we have presented a methodological approach for the engineering of SE. This approach relies on an existing MAS methodological process, ASPECS. The contribution of this article resides in the definition of new activities and notations in order to deal with specific aspects of SE such as sensors/effectors and specific expertness that are involved in the reasoning agents encapsulated in the SE.

The presented methodological approaches was illustrated with an e-Health SE that is dedicated to the monitoring of patients with heart failure disease. This system has been tested within hospitals and is already deployed within patient's home.

Future works will consist in applying the proposed methodology to different case studies and investigating the use of norms and formal validation techniques in order to provide tools for SE behavior analysis and verification.

Acknowledgments The E-care project is developing an intelligent communicative platform enabling the home monitoring of patients with chronic heart failure using non-invasive sensors and was funded by the Grant: "Investissements d'avenir" from the French government.

References

1. Cossentino, M., Hilaire, V., Molesini, A., Seidita, V. (eds.) The ASPECS Process, Chapter 4, 1 edn., pp. 216–236. Springer (2013)
2. Benyahia, A.A., el Hassani, A.H., Hilaire, V., Hajjam, M.: Ontological Architecture for Management of Telemonitoring System and Alerts Detection, Chapter 5, pp. 85–96. InTech (2012)
3. Rodriguez, S., Gaud, N., Galland, S.: SARL: a general-purpose agent-oriented programming language. In: International Conference on Intelligent Agent Technology (IAT14), Warsaw, Poland, pp. 103–110. IEEE Computer Science (2014)
4. Hubner, J., Sichman, J., Boissier, O.: S-moise+: a middleware for developing organised multi-agent systems. In: Proceedings of International Workshop on Organizations in Multi-Agent Systems, from Organizations to Organization Oriented Programming in MAS. LNCS, vol. 3913, pp. 64–78 (2006)
5. Ritala, T., Kuikka, S.: Uml automation profile: enhancing the efficiency of software development in the automation industry. In: 2007 5th IEEE International Conference on Industrial Informatics, vol. 2, pp. 885–890 (2007)
6. Cossentino, M., Hilaire, V., Molesini, A., Seidita, V.: Handbook on Agent-Oriented Design Processes. Springer (2014)
7. Gómez-Sanz, J.J., Fuentes-Fernández, R.: Understanding agent-oriented software engineering methodologies. Knowl. Eng. Rev. **30**(4), 375–393 (2015)
8. Schumann, R., Bromuri, S., Krampf, J., Schumacher, M.I.: Agent based monitoring of gestational diabetes mellitus (demonstration). In: van der Hoek, W., Padgham, L., Conitzer, V., Winikoff, M., (eds.) AAMAS. IFAAMAS, pp. 1487–1488 (2012)
9. Garcia, E., Tyson, G., Miles, S., Luck, M., Taweel, A., Van Staa, T., Delaney, B.: Analysing the suitability of multiagent methodologies for e-health systems. In: Müller, J.P., Cossentino, M. (eds.) AOSE. Lecture Notes in Computer Science, vol. 7852, pp. 134–150. Springer (2012)
10. Pavón, J., Gómez-Sanz, J.J., Fernández-Caballero, A., Valencia-Jiménez, J.J.: Development of intelligent multisensor surveillance systems with agents. Robot. Auton. Syst. **55**(12), 892–903 (2007)
11. Ayari, N., Chibani, A., Amirat, Y., Matson, E.T.: A novel approach based on commonsense knowledge representation and reasoning in open world for intelligent ambient assisted living services. In: IROS, pp. 6007–6013. IEEE (2015)

A Norm-Based Approach for Personalising Smart Environments

Patrizia Ribino, Carmelo Lodato, Antonella Cavaleri and Massimo Cossentino

Abstract People have a great variety in their needs. There is a great demand for personalized services, especially those interacting with their environment. In this paper, we propose a norm based approach for personalizing smart environments that constraint user requirements by means of non functional requirements expressed in terms of permissions, obligations or prohibitions.

Keywords Smart systems · Norms · Smart environment

1 Introduction

In recent years, a growing trend is the development of smart systems to improve well-being of individuals in their environment by making everyday activities more convenient and enjoyable. Smart systems aim at augmenting real environments to create smart spaces where users are provided with pervasive electronic devices. Usually each device can provide a set of services and functionalities. A smart system connects such electronic devices into a network and control them by using advanced ICT technologies in such away the devices satisfy user requirements. A common architecture for smart environment is sketched in Fig. 1a. The highest layer is related to the interaction with users. The second layer provides intelligence of the environment,

P. Ribino (✉) · C. Lodato · A. Cavaleri · M. Cossentino
Istituto di Calcolo e Reti ad alte Prestazioni, Consiglio Nazionale delle Ricerche, Palermo, Italy
e-mail: ribino@pa.icar.cnr.it
URL: http://ecos.pa.icar.cnr.it

C. Lodato
e-mail: c.lodato@pa.icar.cnr.it

A. Cavaleri
e-mail: a.cavaleri@pa.icar.cnr.it

M. Cossentino
e-mail: cossentino@pa.icar.cnr.it

G. De Pietro et al. (eds.), *Intelligent Interactive Multimedia Systems and Services 2016*, Smart Innovation, Systems and Technologies 55,
DOI 10.1007/978-3-319-39345-2_59

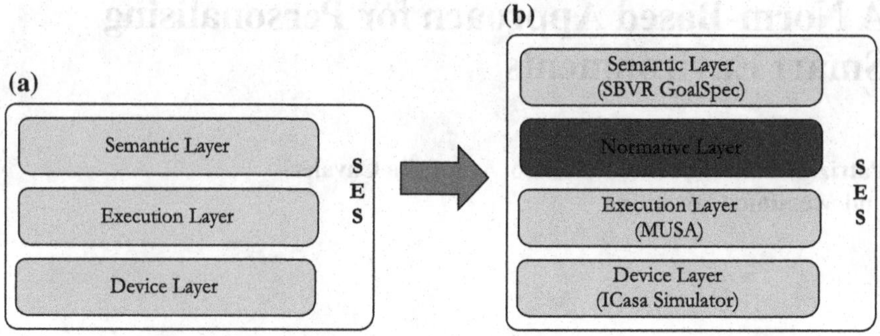

Fig. 1 The proposed layered smart environment system

often using artificial intelligence capabilities. Finally, the lowest layer is related to the electronic devices based on current technologies used for smart systems.

The use of a smart environment is variable from a user to another. Designing ad-hoc systems for each individual is not realistic economically. As well as, it is not possible to hard-wire all possible user scenarios. Hence, how to personalise smart environments? The most reasonable answer is that users have to be able to define their own requirements, thus defining their own usage scenarios. There are several kinds of systems that try to address this question [1]:

- *Predefined Scenario Systems* include centralized systems based on predefined scenarios. They only provide the users the possibility to choose the scenarios they want to execute. A new scenario definition generally consists in assembling existing components [2, 3].
- *Service Control Systems* include systems that allow users to control available services by providing automatic detection of devices in their environment. A user may interact with the system for triggering service executions, but he cannot define complex scenarios [4, 5].
- *Scenario Definition Systems* enable users to define their own scenarios. Only some of them allow runtime scenario definition and in some cases such scenarios are sequences of service calls [1].

Such kind of systems do not allow to manage non functional requirements[1] during a predefined scenario execution. In this paper, we propose a norm based approach for the definition of highly personalised scenarios during the normal execution of the system. The proposed approach allows for constraining user requirements by means of non functional requirements (i.e.: variable user personalizations) expressed in terms of norms (i.e.: permissions, obligations or prohibitions). In order to give the users the capability to define their own initial requirements and to modify the

[1]For the scope of this paper, we consider only user constraints. Thus, hereafter we use non functional requirements, user constraints or user personalization as synonymous.

behaviour of the system during its normal execution according to user personalizations, we introduced a Normative Layer in the classical smart environment architecture (see Fig. 1a).

The main contribution of this paper is an algorithm implemented in the Normative Layer, which integrates non functional requirements expressed by means of norms (i.e.: permissions, obligations and prohibitions) into user requirements (expressed in terms of goals) in order to introduce personalizations into predefined user scenarios.

In order to show some practical examples, in this paper, we use GoalSpec [6] and SBVR [7] for modelling user requirements and personalizations in the semantic layer and MUSA [8] for the composition and the orchestration of services in the Execution Layer. GoalSpec is a language designed for specifying user-goals and enabling at the same time goal injection and software agent reasoning. The Semantics of Business Vocabulary and Business Rules (SBVR) is an adopted standard of the Object Management Group (OMG). It is designed for formalizing complex business rules. MUSA (Middleware for User-driven Service Adaptation) is a holonic multi-agent system for the composition and the orchestration of services in a distributed and open environment that is founded on the BDI paradigm [9].

The remaining of the paper is organized as follows. Section 2 presents the proposed norm-based approach along with some definitions is based on. Section 3 shows some application scenarios. Some related works are illustrated in Sect. 4. Finally, in Sect. 5 some conclusions are drawn.

2 The Normative Layer

Many studies have been conducted in order to develop systems that satisfy user requirements in smart environments. These systems generally provide predefined scenarios corresponding to general requirements and enable users to select those he/she wants to trigger. Such hard-wired behaviours limit the personalization degree of such systems. The main purpose of this work is to introduce more flexibility in the scenario definition for personalising smart environments. In order to reach this aim, we conceived a Normative Layer independent of the Execution Layer (see Fig. 1b) and we propose a norm based approach in order to introduce personalizations in such smart environment systems.

The proposed approach is based on three key concepts: state of the world, goal and norm.

Definition 1 (*State of the world*)

Let $\mathcal{D}$ the set of concepts defining a business domain. Let $\mathcal{L}$ be a first-order logic defined on $\mathcal{D}$ with $\top$ a tautology and $\perp$ a logical contradiction, where an atomic formula $p(t_1, t_2..., t_n) \in \mathcal{L}$ is represented by a predicate applied to a tuple of terms $(t_1, t_2..., t_n) \in \mathcal{D}$ and the predicate is a property of or relation between such terms that can be true or false.

A *state of the world* in a given time t ($\mathcal{W}^t$) is a subset of atomic formulae whose values are true at the time t:

$$\mathcal{W}^t = [p_1(t_1, t_2, ..., t_h), ..., p_n(t_1, t_2, ..., t_m)]$$

The *state of the world* represents a set of declarative information concerning events occurred within the environment and relations among events at a specific time. An event can be defined as the occurrence of some fact that can be perceived by or be communicated to the smart system. Events can be used to represent any information that can characterize the situation of an interacting user as well as a set of circumstances in which the smart system operates at a specific time. Definition 1 is based on the close world hypothesis that assumes all facts that are not in the state of the world are considered false.

Definition 2 (*Goal*)
Let $\mathcal{D}$, $\mathcal{L}$ and $p(t_1, t_2..., t_n) \in \mathcal{L}$ as previously introduced in the Definition 1. Let $t_c \in \mathcal{L}$ and $f_s \in \mathcal{L}$ formulae that may be composed of atomic formulae by means of logic connectives AND($\wedge$), OR ($\vee$) and NOT ($\neg$).

A *Goal* is a pair $\langle t_c, f_s \rangle$ where t_c (*trigger condition*) is a condition to evaluate over a state of the world $\mathcal{W}^t$ when the goal may be actively pursued and f_s (*final state*) is a condition to evaluate over a state of the world $W^{t+\Delta t}$ when it is eventually addressed:

- *a goal is active iff* $t_c(\mathcal{W}^t) \wedge \neg f_s(\mathcal{W}^t) = true$
- *a goal is addressed iff* $f_s(\mathcal{W}^{t+\Delta t}) = true$

Goals express what is the desired state of the world the system has to result in. Conversely, norms denote the way the system has to operate in order to achieve the desired state of the world in compliance with the normative context in which that process takes place.

Definition 3 (*Norm*)
Let $\mathcal{D}$, $\mathcal{L}$ and $p(t_1, t_2..., t_n) \in \mathcal{L}$ and let a *state of the world* in a given time t ($\mathcal{W}^t$) as previously introduced in the Definition 1. Let $\phi \in \mathcal{L}$ and $\rho \in \mathcal{L}$ formulae composed of atomic formula by means of logic connectives AND($\wedge$), OR ($\vee$) and NOT ($\neg$). Moreover, let $D_{op} = \{permission, obligation, prohibition\}$ the set of deontic operators. A *Norm* is defined by the elements of the following tuple:

$$n = \langle \leqq r, g, \rho, \phi, d \rangle \quad (1)$$

where

- $r \in \mathcal{R}$ is the *Role* the norm refers to. The special character "_" indicates that the norm refers any role.

- $g \in \mathcal{G}$ is the *Goal* the norm refers to. The special character "_" indicates that the norm refers to any goal.
- $\rho \in \mathcal{L}$ is a formula expressing the set of actions and state of affairs that the norm disciplines.
- $\phi \in \mathcal{L}$ is a logic condition (to evaluate over a state of the world $\mathcal{W}^t$) under which the norm is applicable;
- $d \in D_{op}$ is the deontic operator applied to ρ that the norm prescribes to the couple $(r, g) \in \mathcal{R} \times \mathcal{G}$.

In particular $d(\rho) = \begin{cases} \rho & \text{iff } d = \textit{obligation} \\ \neg\rho, & \text{iff } d = \textit{prohibition} \\ \rho \vee \neg\rho & \text{iff } d = \textit{permission} \end{cases}$

In other words, let a state of the world $\mathcal{W}^t$ a norm prescribes to a couple (r, g) the deontic operator d applied to ρ if ϕ is true in $\mathcal{W}^t$.[2] *Norms* represent system regulations by specifying obligations, permissions or prohibitions to be followed during system activities in case of certain conditions occur, thus relaxing or restricting a process.

Definition 4 (*State of Norm*) Let a norm $n = \langle r, g, \rho, \phi, d\rangle$ where $g = \langle t_c, f_s\rangle$ and let a state of the world in a given time t ($\mathcal{W}^t$)

A norm can assume the following states:

- n is *applicable at time t* if $\phi(\mathcal{W}^t) = true \vee \phi = \top$
- n is *active at time t* if n is applicable and $t_c(\mathcal{W}^t) = true$
- n is *logically contradictory* if ϕ *is* $\bot$
- n is *in opposition to goal* if $f_s \wedge d(\rho)$ *is* $\bot$

Moreover, let a state of the world (W^t) and let two norms $n_1 = \langle r_1, g_1, \rho_1, \phi_1, d_1\rangle$ and $n_2 = \langle r_2, g_2, \rho_2, \phi_2, d_2\rangle$ where $r_1 = r_2$, $g_1 = g_2$, $\rho_1 = \rho_2$

- n_1 and n_2 are *deontically contradictory* iff $\begin{cases} \phi_1(\mathcal{W}^t) \wedge \phi_2(\mathcal{W}^t) = true \\ d_1 \neq d_2 \end{cases}$

It is worth noting that we talk about *logically contradictory* when the contradiction concerns the logical conditions ($\phi \in \mathcal{L}$) under which the norms are applicable. On the contrary, we talk about *deontically contradictory* when the contradiction concerns the semantic meaning of the deontic operator ($d \in D_{op}$) the norms apply.

Algorithms: The aim of the Normative Layer is to provide some mechanisms that allow to modify the behaviour of the smart environment (that is based on some

---

[2]It is worth noting that in order to be compliant in $\mathcal{W}^t$ with (1) an obligation ρ must be true, (2) a prohibition $\neg\rho$ must be true (3) a permission ρ or $\neg\rho$ may be true. In the context of this paper, we assume that the system does not violate norms.

predefined user scenarios) in order to adapt it to user personalization expressed by means of SBVR Rules in the Semantic Layer.[3]

Algorithm 1: Norms into Requirements

Data: a set of goals $\mathcal{G}$ and a set of norms $\mathcal{N}$
Result: $\mathcal{G}_{\mathcal{N}}$
$\mathcal{G}_{\mathcal{N}} \leftarrow \emptyset$;
create a list *NormList*, *size*(*NormList*) = *card*($\mathcal{N}$);
create a list *GoalList*, *size*(*GoalList*) = *card*(*G*);
`// Loop for detecting logically contradictory norms`
① **for** $j \leftarrow 1$ **to** *card*($\mathcal{N}$) **do**
  $\langle r, g, \rho, \phi, d \rangle \leftarrow n_j$;
  if n_j *is not logically contradictory* **then**
    add $\langle r, g, \rho, \phi, d \rangle$ to *NormList*;

GoalList ← *G*;
`// Loop for encapsulating norms into goals`
② **for** $i \leftarrow 1$ **to** *size*(*GoalList*) **do**
  `// see Algorithm 2`
  $(\phi_{mergedOR}, \phi_{mergedAND}) \leftarrow$ *compose_norm*(*GoalList*[*i*], *NormList*);
  `// Goal composition`
  $\langle t_c, f_s \rangle \leftarrow$ *GoalList*[*i*];
  $t_c \leftarrow$ *OR_composition*$(t_c, \phi_{mergedOR})$;
  $t_c \leftarrow$ *AND_composition*$(t_c, \phi_{mergedAND})$;
  add $\langle t_c, f_s \rangle$ to $\mathcal{G}_{\mathcal{N}}$;

Algorithm 1 is the core of the Normative Layer. It allows to modify goals, making them norm compliant. By encapsulating the condition expressed by the norms inside the goal they refer, it is possible to modify the activation of that goal thus making it compliant with the norms. Algorithm 1 consists of an initial pre-filtering (Step ①) of logically contradictory norms (see Definition 4). Then norms are encapsulated into goals (see Step ②) by composing new trigger conditions for goals from the norm conditions (see Algorithm 2).

Such composition takes into consideration different types of norm and addresses the following question: when norms regulate a goal, in what cases that goal is activated? The activation table shown in Fig. 2 shows all the possible cases in which the system can pursue a goal: (i) when the trigger condition is true and the norm is not applicable or (ii) when the norm is applicable and its deontic operator is permission; (iii) when the norm is active and its deontic operator is an obligation. As we can see in Fig. 2, when an obligation is introduced in the system, it does no effective change to the original activation of the goal. Because in any case the original trigger condition has to be satisfied.

[3]It is out of the scope of the paper the algorithms that convert SBVR and GoalSpec in the formalisms managed by the Normative Layer.

Algorithm 2: Compose_Norm

```
Data: a goal g_current, a list of norms NormList
Result: a couple (φ_mergedOR, φ_mergedAND)
Listφ_OR ← ∅;
Listφ_AND ← ∅;
// Identification of norm types
①for j ← 1 to size(NormList) do
    ⟨r, g, ρ, φ, d⟩ ← NormList[j];
    ⟨t_c, f_s⟩ ← g;
    // Choose among norms of the current goal, which are
       directly linked to the goal final state and which are
       not in opposition to the same goal
    if (g = g_current) ∧ (f_s = ρ) ∧ ((f_s ∧ d(ρ)) ≠ ⊥) then
        switch d do
            case Obligation
                break;
            case Prohibition
                add ¬φ to Listφ_AND;
            case Permission
                add φ to Listφ_OR;
// Permissions give alternatives (OR)
②if Size(Listφ_OR) ≠ 0 then
    φ_mergedOR ← Listφ_OR[1];
    for h ← 2 to Size(Listφ_OR) do
        φ_mergedOR ← OR_composition(φ_mergedOR, Listφ_OR[h]);
// Prohibition are mandatory (AND)
③if Size(Listφ_AND) ≠ 0 then
    φ_mergedAND ← Listφ_AND[1];
    for h ← 2 to Size(Listφ_AND) do
        φ_mergedAND ← AND_composition(φ_mergedAND, Listφ_AND[h]);
```

In common smart environments all the parameters that trigger the execution of some scenarios have to be defined a priori for each of them. Our approach allows to define prohibitions or permissions (that express conditions under which something is allowed or forbidden) without a specific relation with a scenario. In our approach norms are transversal to several scenarios, thus decoupling what we want the system do and the boundary in which it has to move.

In the following, we show some simple application scenarios.

$n = \langle \_, g, \rho, \varphi, d \rangle \; g = \langle t_c, f_s \rangle$

(a) Permission

$t_c \vee \varphi$	t_c	φ
1	1	0
1	0	1
0	0	0
1	1	1

(b) Obligation

$t_c \vee (\varphi \wedge t_c)$	t_c	φ
1	1	0
0	0	1
0	0	0
1	1	1

(c) Prohibition

$t_c \wedge \neg\varphi$	t_c	φ
1	1	0
0	0	1
0	0	0
0	1	1

Fig. 2 Goal activation regulated by norms. **a** Permission. **b** Obligation. **c** Prohibition

3 Application Scenarios

In this section we provide some examples of application scenarios. We simulate such scenarios in ICasa simulator [10]. The purpose is to show as the same smart environment behaves according to different user personalizations. Let us suppose to install the same smart home system in two different apartments. Thus, each apartment is furnished with sensors for the detection of individuals, with digital cooking stove, air conditioner, electronic shutters and so on. The first apartment is occupied by two adults, Sara and John, and a teenager, Luke. Sara and John establish some norms they want their smart apartment follows. In particular, they forbid the system to activate any cooking device if there are no adults at home. Thus, supposing the system owns an appropriate knowledge, such norm could be expressed in SBVR as follows:

N1: *It is prohibited turning on the cooking device if Sara is not at home or John is not at home.*

In the second apartment live Mike and his wife. Mike forbids the system to turn off the exterior lights of the apartment during the night if the house is empty or if only one person is at home. The corresponding norm could be:

N2: *It is prohibited turning off exterior lights if it is night and house habitant are not at home.*

Let us suppose the smart system has two predefined scenarios: *good night* and *have a lunch*. In the *good night* scenario, the exterior lights of home turn off and the curtains and shutters of the bedroom will be automatically closed. In the *have a lunch* scenario, the TV switches on, the conditioner turns on at the desired temperature and the cooking stove turns on if there is a pot on. Such scenarios could be expressed in GoalSpec as follows:

Good night:
WHEN on(10:00 pm) THE system SHALL ADDRESS closed(shutter) AND closed (curtains) AND exterior_lights(off).

Have a lunch:
WHEN on(1:00 pm) THE system SHALL ADDRESS television(on) AND conditioner(on,T)
WHEN on(1:00 pm) AND pot_on(cooking_stove) THE system SHALL ADDRESS cooking_stove(on)

Although in each apartment both the scenarios have been activated, the system behaves differently. In the first apartment the good night scenario will be executed as well as it is defined. In Fig. 3a, Sara and John are coming back at home later in the night. Luke is at home. The system at 10 o'clock closes shutters and curtains and turns off exterior lights. In the second apartment the *good night* scenario will be differently executed. In Fig. 3b, Mike went to walk after the dinner, his wife is at home. At 10 o'clock the system closes shutters and curtains. The system does not turn off exterior lights, until Mike comes back at home. Similarly, in the *have a lunch* scenario, if both families are at home the system behaves in each apartment in the same way. TV and air conditioner are turned on and cooking device starts to cook the lunch (see Fig. 4a). A day, Luke unusually comes back at home for lunch before of his parents. At 1 o'clock the TV and air conditioner turn on while cooking device continues to be off, although there is a pot on the stove (see Fig. 4b). It is worth noting that norms defined by Mike and Sara are not directly related to *good night* or *have a lunch* scenario. They may also act in other scenarios that include the same actions the norms discipline.

These are simple examples we can simulate. But the proposed approach goes beyond the smart home systems. We can apply our approach also in other contexts.

(a) **(b)**

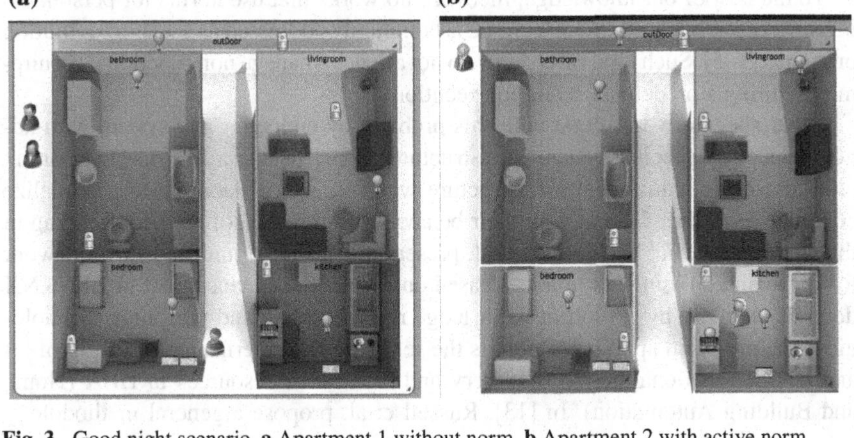

Fig. 3 Good night scenario. **a** Apartment 1 without norm. **b** Apartment 2 with active norm

(a) (b)

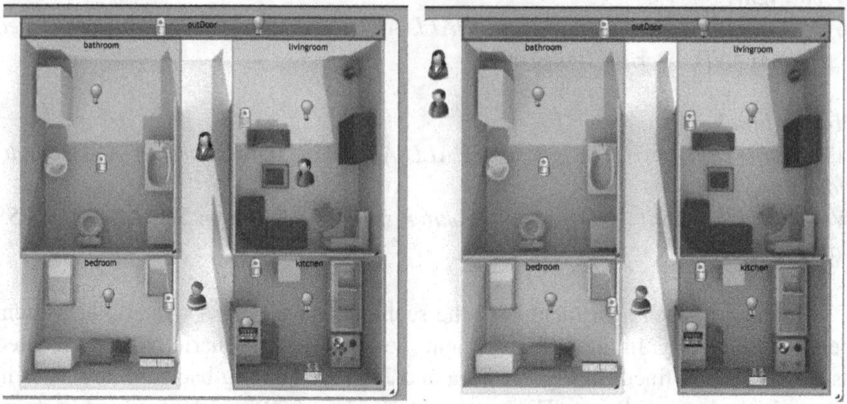

Fig. 4 Have a lunch scenario. **a** Apartment 1. **b** Apartment 1 with active norm

For example in smart traffic and transportation contexts, we can conceive a smart system that allows to automatically drive a vehicle. Such systems have to follow different traffic regulations, according to the country they will operate.

4 Related Works

Personalization is a key issue to be addressed in order to widely diffuse the use of smart environments. The approach we propose is quite simple, but at the same time effective. It works to a higher abstraction level by introducing norms for personalizing user requirements, thus not modifying the lower execution layer of a smart environment system.

To the best of our knowledge, there are no works that use norms for personalising smart environments. Such an issue is commonly addressed as scenario definition problems [1–5]. Such kind of systems do not allow to manage non functional requirements during a predefined scenario execution.

Other specific works cope with this problem by endowing the system with ad-hoc devices or specific system infrastructures. Only to cite a few, in [11] Giroux et al. present a multi-agent infrastructure where agents collaborate to personalize cognitive assistance by adapting their behaviour to the need of the people living in the smart home. In [12], Loseto et al. present a distributed multi-agent framework for home and building automation, based on a semantic enhancement of EIB/KNX domotic standard by exploiting knowledge representation and reasoning technologies. The proposed approach supports the semantic characterization of user profiles and device functionalities to discovery and orchestrate resources in HBA (Home and Building Automation). In [13], Russell et al. propose a general methodology that defines how to use unobtrusive sensors within smart environments to provide physical context and metadata for system personalization. In [14], Bergesio et al. propose a method for personalizing the behaviour of smart environments by means of the configuration of smart objects through mobile devices. With respect to the previous ones, our approach introduces more flexibility in the scenario definition by means of a Normative Layer that is independent from the Execution Layer.

5 Conclusions

The paper proposes an approach for personalising user requirements by expressing non functional requirements (i.e.: variable user personalizations) by means of norms (i.e.: permissions, obligations or prohibitions). The proposed norm-based approach is able to support the specification of the variable parts of a scenario execution in a smart environment. The approach we propose facilitates the customization of a smart environment to a particular usage context by decoupling the variable parts (i.e.: user personalization) from standard user requirements. Then an algorithm implemented in the Normative Layer combines system requirements with user personalizations. Norms allow to specify user personalizations in a way that is understandable by the user, but also executable by the Execution Layer, thus bridging the gap between user and technology. At the moment we are working on the integration in the Normative Layer of an algorithm to allow runtime modification of the system. We are also developing a norm editor that allows to easily compose norms according to the domain the system will operate. The work proposed in this paper is part of a larger one that aims at creating normative smart environments that can be compliant with the legislative environments they are plugged in.

References

1. Hamoui, F., Huchard, M., Urtado, C., Vauttier, S.: Specification of a component-based domotic system to support user-defined scenarios. In: SEKE 2009: 21st International Conference on Software Engineering and Knowledge Engineering, pp. 597–602. Knowledge Systems Institute Graduate School (2009)
2. Bottaro, A., Gérodolle, A., Lalanda, P.: Pervasive service composition in the home network. In: 21st International Conference on Advanced Information Networking and Applications, 2007. AINA'07, pp. 596–603. IEEE (2007)
3. Grondin, G., Bouraqadi, N., Vercouter, L.: Madcar: an abstract model for dynamic and automatic (re-) assembling of component-based applications. In: Component-Based Software Engineering, pp. 360–367. Springer (2006)
4. Ishikawa, H., Ogata, Y., Adachi, K., Nakajima, T.: Building smart appliance integration middleware on the OSGI framework. In: Proceedings. Seventh IEEE International Symposium on Object-Oriented Real-Time Distributed Computing, 2004, pp. 139–146. IEEE (2004)
5. Chao-Lin, W., Liao, C.-F., Li-Chen, F.: Service-oriented smart-home architecture based on osgi and mobile-agent technology. IEEE Trans. Syst. Man Cybern. Part C Appl. Rev. **37**(2), 193–205 (2007)

6. Sabatucci, L., Ribino, P., Lodato, C., Lopes, S., Cossentino, M.: Goalspec: a goal specification language supporting adaptivity and evolution. In: Engineering Multi-Agent Systems, pp. 235–254. Springer (2013)
7. Object Management Group. Semantics of business vocabulary and business rules (sbvr). version 1.3 (2015)
8. Cossentino, M., Lodato, C., Lopes, S., Sabatucci, L.: Musa: a middleware for user-driven service adaptation. In: Proceedings of XVI WORKSHOP "DAGLI OGGETTI AGLI AGENTI", Napoli, June, 17–19 2015, vol. 1382 (2015)
9. Rao, A.S., Georgeff, M.P.: Bdi agents: from theory to practice. In: Proceedings of the First International Conference on Multi-agent Systems (ICMAS-95), pp. 312–319, San Francisco (1995)
10. iCASA—a dynamic pervasive environment simulator. http://adele.imag.fr/icasa-a-dynamic-pervasive-environment-simulator
11. Giroux, S., Castebrunet, M., Boissier, O., Rialle, V.: A multiagent approach to personalization and assistance to multiple persons in a smart home. In: Workshops at the Twenty-Eighth AAAI Conference on Artificial Intelligence (AAAI-14), p. 11 (2014)
12. Loseto, G., Scioscia, F., Ruta, M., Di Sciascio, E.: Semantic-based smart homes: a multi-agent approach. In: 13th Workshop on Objects and Agents (WOA 2012), vol. 892, pp. 49–55 (2012)
13. Russell, L., Goubran, R., Kwamena, F.: Personalization using sensors for preliminary human detection in an iot environment. In: 2015 International Conference on Distributed Computing in Sensor Systems (DCOSS), pp. 236–241. IEEE (2015)
14. Bergesio, L., Marquinez, I., Bernardos, A.M., Besada, J.A., Casar, J.R.: Perseo: a system to personalize the environment response through smart phones and objects. In: 2013 IEEE International Conference on Pervasive Computing and Communications Workshops (PERCOM Workshops), pp. 640–645. IEEE (2013)

Adopting a Middleware for Self-adaptation in the Development of a Smart Travel System

L. Sabatucci, A. Cavaleri and M. Cossentino

Abstract A smart travel system is a complex distributed system acting as a tour operator for organizing holiday packages and supporting travelers on-the-run. A couple of key characteristics of such a system are the ability of self-configuring a set of heterogeneous services and self-adapting to unexpected circumstances. This paper reports an experience of developing a smart travel system by adopting MUSA, *a Middleware for User-driven Service Adaptation*. The prototype supports users in organizing their time by the specification of goals: this triggers the automatic composition and dynamic orchestration of touristic services. The chosen middleware has played a fundamental role by simplifying the development process thus to speed up the time-to-complete.

1 Introduction

Traditionally, the orchestration of services [4] uses approaches based on static models that provide little support for allowing self-configuration and adaptation of activities at run-time. Among these approaches, BPEL [14]—the main standard for implementing the orchestration of services—does not support advanced features for facing mutable and dynamic operative environment. Some of the well-known weakness are: (1) the flow of activities can not be changed when the execution context changes; (2) incorporating dynamic user preferences complicates the modeling activity; (3) revising the whole workflow is necessary every time a new services is introduced in the model; (4) even if it is possible to include service failures, the process is not robust enough to react to unexpected situations.

L. Sabatucci (✉) · A. Cavaleri · M. Cossentino
ICAR-CNR, Palermo, Italy
e-mail: sabatucci@pa.icar.cnr.it

A. Cavaleri
e-mail: a.cavaleri@pa.icar.cnr.it

M. Cossentino
e-mail: cossentino@pa.icar.cnr.it

G. De Pietro et al. (eds.), *Intelligent Interactive Multimedia Systems and Services 2016*, Smart Innovation, Systems and Technologies 55,
DOI 10.1007/978-3-319-39345-2_60

This work reports the development of a smart travel system as a customization of MUSA [2, 11, 13] (a Middleware for User-driven Service Adaptation). The aim is to aggregate heterogeneous services on-demand and to orchestrate touristic services in a dynamic, open and geographically distributed environment. Creating new travel experiences grounds on putting traveler at the center of the process. Firstly users may express travel preferences supported by a flexible language and a specific interface to convey her goals about: places to visit, activities to do and—in general—the kind of vacation. It is worth noting that this language deals with undefined (or better not completely defined) requirements, thus leaving users the choice to not specify something, and delegating the system to propose alternative ways to complete travel itineraries. The second point is to allow the system to act as a local guide for traveler—running on personal device—by providing contextual information, monitoring the state of reserved resources and proposing viable alternatives on the run. Indeed, when traveler deviates from the planned route (willingly or not) the smart travel system will re-arrange services and resources to meet new emerging needs. So far, the developed prototype simulates this latter point.

The paper is organized as follows. Section 2 presents the main features of the smart travel system and motivates the choice of MUSA for the development. Section 3 focuses on two of the main characteristics provided by the middleware: self-configuration and self-adaptation. Therefore, the steps for implementing the system are briefly described in Sect. 4. Concluding remarks are drawn in Sect. 5.

2 The Smart Travel System

The smart travel system is a complex software designed to act as a tour operator, combining travel components on-demand to create a holiday package. The user may serve of the smart travel system to organize a vacation by specifying set of preferences about the kind of desired vacation including the geographic area of interest, places of interest, activities to perform, budget and so on. The system arranges alternative solutions to these specifications by composing a set of travel services: flights, transfers, hotels and tickets for museums, for the opera and for other local events. Moreover, when a user selects and pay for the package, the system works as a local assistant, providing contextual information, warning about train delays, checking flight cancellations, and re-organizing the vacation on the need. Such a software may be classified as a socio-technical system in which tourists and touristic services must be orchestrated in order to satisfy the former (improving the whole experience) and to maximize the use of the latter (increasing incomes for providers).

It is worth to underline the nature of the involved services. From the one hand services are *real* [5], i.e. each of them is composed of an electronic interface (e.g. the web protocol for booking a flight) and of an actual service the user will directly consume (e.g. flying). On the other hand, these services are heterogeneous–providing different benefits—and are geographically distributed in the territory.

Other significant features of the system are: (i) user preferences: these must be flexible enough to allow configuring many aspects of the expected travel but also supporting user indecision, by allowing to specify only partial information; (ii) self-configuration: the system must be able to check available touristic services and to arrange one o more solutions (travel packages) that address all the user preferences; (iii) monitoring: as a local assistant, the system has to check that travel proceeds correctly; (iv) self-adaptation: during the travel, the system must assure valid on-the-run alternatives, when something changes. It is possible that traveler changes his desires, or that a booked service is no more available. The system must be able to re-configure the vacation, respecting new contextual constraints.

2.1 A Smart Travel Scenario

In this section we illustate a scenario for the smart travel system.

Herbert from Munich wants to travel to Sicily for a week with his family. By the smart_travel_system he plans a vacation of 7 days and set the following preferences: to stay 2–3 days in Palermo to look around the old city; at least 1 day at the beach to please his son; to watch a performance in the Greek theater in Syracuse; and finally at least 1 day in Catania city. He decides to leave the system free to suggest something for the remaining 2 or 3 days.
As response, the smart_travel_system suggests to flight to Palermo, where to stay 2 days, than 1 relaxing day at the Cefalù beach, 1 day in Catania, 2 days in Syracuse, (tickets for assist at the Greek tragedy are available only on day 5th), then 1 day at Agrigento and finally back to Palermo for the return flight.
Herbert confirms the travel plan and the smart_travel_system books the two flights, buys the ticket for the Greek tragedy, reserves hotels and buy tickets for transfers.

The whole package is shown on the left side of Fig. 1. The second part of this scenario illustrates the ability of the system to adapt.

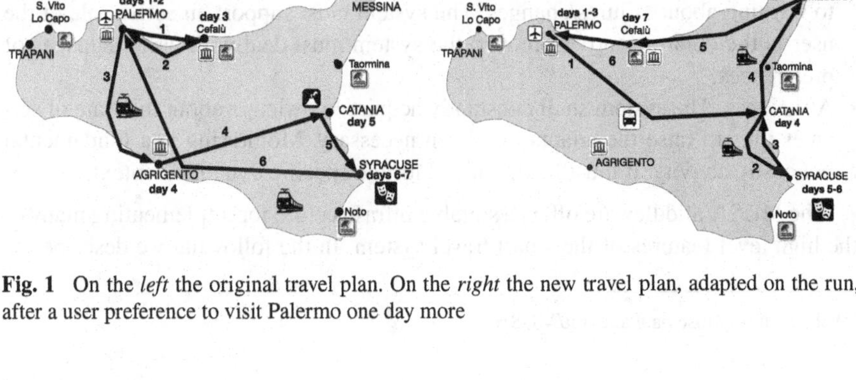

Fig. 1 On the *left* the original travel plan. On the *right* the new travel plan, adapted on the run, after a user preference to visit Palermo one day more

Herbert and his family are enjoying their vacation. They have been visiting Palermo for two days and decide to extend staying in the city by 1 day (variation to the plan).
Therefore the smart_travel_system proposes to variate the vacation, trying to maintain the tickets for the theater that are not reimbursable. The new plan comprises to stay another day in Palermo (addressing the new desire), and to skip the day at Agrigento and delaying the beach day (Cefalù) at the end. The new package is shown on the right side of Fig. 1.
Herbert confirms the new travel plan, so the smart_travel_system cancels trains and hotels reservations that are no more necessary according the new plan.

We conducted a traditional study of the domain, followed by a requirement engineering analysis. A subset of the requirements are detailed in the next section.

3 Self-composition and Adaptation

This section analyzes some of the Smart Travel System's requirements, in order to drive strategic choices for the implementation phase. In this context we present MUSA [2], a Middleware for User-driven Service Adaptation.[1]

A summary of the main characteristics of the system to implement are listed and described hereafter.

- Heterogeneity. The same result may be addressed by properly composing different categories of services and resources. The design phase requires new design abstractions for describing services and the corresponding resources.
- Proactivity. The smart travel system holds a degree of freedom in taking decisions about how and when achieving user's goals. This requires a flexible description language to convey user's preferences that drive the system decision making.
- Dynamism. When the travel package is formed, the system must deal with a situation in which the operative context may change: services could fail, resources may be unavailable or even user's goals may change.
- Mobility. The system shall run on personal mobile devices with the aim of monitoring traveler's position and service failures. The system control loop must be flexible enough to support distributed data.
- Human Interaction. During the travel, users may either change preferences or react to warning about failures/changes. The system must support an active role of the user in the control loop. Moreover, the system must deal with user's changes of preferences.
- Awareness. The system shall constantly acquire knowledge about the state of services thus to raise the adaptation when necessary. Monitoring is a fundamental but costly activity. It must be optimized for the specific dynamic context.

The MUSA middleware offers a suitable infrastructure for implementing many of the high level features of the smart travel system. In the following we describe the

[1]Website: http://aose.pa.icar.cnr.it/MUSA/.

main characteristics of the middleware, highlighting between brackets the impact on a category of requirements. The MUSA backbone is the decoupling of the dimensions of what and how. The GoalSPEC language [13] allows run-time specification of users requirements (*human interaction*) whereas the system adopts a descriptive logic for supporting the high-level reasoning on service semantic (*heterogeneity*).

Specific abstractions [12] are provided for representing what the system knows being able to do (*awareness*). The main function of MUSA is the ability of automatically associating system functions to user's goals thus to address the desired requirements (*proactivity*). This approach allows for implementing an architecture for self-configuration and self-healing (*dynamism*).

In MUSA the working key is configuring a solution as a response to the users request. It is implemented as a multi-agent system where entities are autonomous and driven by a proactive goal-directed behavior. Agents guarantee knowledge acquisition, distributed coordination and robustness. When a solution becomes operative, the system executes special monitoring activities that capture the deviations between expected results and runtime performance, thus to adjust its behavior accordingly.

3.1 Main Concepts of the System

We informally introduce some important concepts used in the middleware. For the sake of clarity formal definitions and the reasoning framework are described with more details in [11, 12].

A ***Capability*** is a semantic wrapper for services that allows the developer to specify i) how to invoke the specified functionality (which data must be passed and which data will be returned) and ii) which effect is expected by executing the encapsulated application or service. The capability also has the advantage of being composable in order to address a complex result.

A ***User-Goal*** is "a desired *change* in a state of affair the user wants to achieve". The concept of goal is often used in the context of business process for representing enterprise strategic interests that motivate the execution of business processes [15].

A ***Configuration*** is a set of capabilities that address a set of user-goals. The main advantage of MUSA is the ability to self-configure, i.e. to automatically discover and aggregate capabilities to address dynamic user-goals.

A ***User-Norm*** is the description of a constraint the system must hold when addressing user-goals. It is described as a function in the domain of configurations that returns a boolean indicating admissible/non-admissible aggregation of capabilities.

A ***User-Metric*** is a user defined quality, associated to how the system will address user-goals. It is described as a function in the domain of configurations that returns a real number. This number may be used to compare two different aggregation of capabilities.

3.2 The Three Layered Architecture

The MUSA's core architecture for self-configuration and self-adaptation is composed of three communicating levels.

The uppermost layer of this architecture is the ***Goal Layer*** responsible for arranging system evolution. The user may specify the expected behavior of the system in terms of high level goals, norms and metrics. Whereas goals are dynamic entities specified through a language, norms and metrics are, so far, hard coded at design time. At run time, the goal injection phase allows free specification of user-goals; conversely norms and metrics are selected from pre-set lists. Injecting goals triggers a change in the system behavior. The user may be involved also as supervisor of the process, in order to take decisions about alternative cases of adaptation.

The second layer is the ***Capability Layer***, responsible of managing the strategic deliberation. The objective is selecting, aggregating and configuring available capabilities [11] as the response to (i) self-evolution events (generated from the upper level) and (ii) self-adaptive events (generated from the lower level). This activity may be very costly when the number of services (and their dependencies) grows. To reduce the complexity, the implemented algorithm reasons about services and environment through abstract data rather than concrete data [7, 8], so to be more affordable and scalable. The consequent output is one or more configurations of abstract services.

The third layer is the ***Service Layer***, responsible of translating from abstract services to concrete ones by adding the coordination logic necessary for enacting the corresponding business process. This layer provides atomic blocks of computation for acquiring and analyzing real data from the environment and producing the desired result. The proposed implementation adopts a distributed MAPE-K model [1, 9] that comprises: (i) a *Monitor component* that acquires information from the environment, and updates the system knowledge accordingly; (ii) an *Analyze component* that uses the knowledge to determine the need for adaptation with respect to expected goals and capabilities failure; (iii) a *Plan component* that synchronizes the available capabilities according the goals to address and, finally, (iv) an *Execute component* that modifies the environment by activating the appropriate capability.

4 Implementing the Smart Travel System

The middleware we have decided to adopt for implementing the advanced features of the smart travel system is a general purpose one: it has been recently used for implementing a document management system, a cloud mashup application process and a emergency management scenario. The strength is the easiness in customizing self-configuration and self-adaptation features for the specific domain of interest. The purpose of this section is to illustrate the steps for this customization.

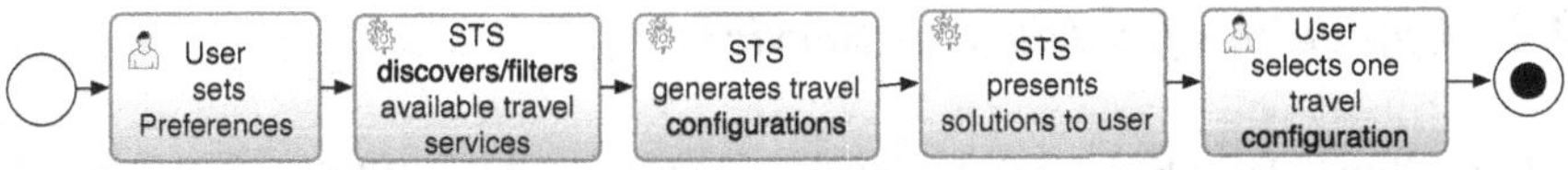

Fig. 2 Flow of activities for generating a travel plan

4.1 Implementing Self-configuration

In the smart travel system, self-configuration is mainly intended for automatically generating a number of alternative travel plans that are suitable for the user's goals. The challenge relies on the fact that travel services are heterogeneous, and they are not designed to be composed. Aggregating them means evaluating a great number of combinations. Figure 2 illustrates, in practice, the business process behind this activity.

1. The user will set his own preferences via either a web interface or a personal device.
2. The smart travel system discovers available services, and filters those may be useful for addressing the user request.
3. The self-configuration algorithm arranges travel services, considering user's goals, uncertainty and domain constraints.
4. The result is a set of possible configurations (i.e. travel plans) to present via a user interface.
5. Finally, the user selects the travel plan he prefers.

The MUSA system allows to easily generate such behavior by associating user's preferences to user-goals/norms/metrics and by implementing travel services as capabilities.

The first step is a preliminary study of the domain for building the reference ontology. Ontologies provide a shared understanding of a domain of interest to support communication among human and computer agents. An ontology will be the base for matching design-time elements (norms, metrics and capabilities) and run-time elements (goals). An ontology for the travel domain has to cover (i) geographical places (ex: Sicily, Palermo, Syracuse), (ii) activities (visiting, swimming, watching opera), (iii) transportation (by train, by car), and (iv) immaterial qualities (cost, hotel rating). A plethora of web ontologies already exist, ready to be reused. In our system we adopted a subset of the Knublauch's OWL ontology [3] for a Semantic Web of tourism.[2]

[2]Available via the Protegé website: http://protege.cim3.net/file/pub/ontologies/travel/travel.owl.

4.2 Capabilites for Touristic Services

A Capability is described as a self-contained autonomic entity. Whereas web-services are typically passive entities that act when receive the control [6], Capabilities are based on software agents able of self-managing, sensing the environment and making decisions on their own [11]. Despite their intrinsic complexity, developing a capability is not such a hard work, because MUSA provides basic facilities for self-awareness, self-adaptation, social interactions and self-configuration. The capability-developer has to specify two additional aspects.

- *The abstract description*: a specification of why, when and how the related web-service may be used. This specification exploits a description-logic language based on the reference ontology. Typical fields are: pre/post conditions, information for simulating the aggregation, data input and data output.
- *The concrete implementation*: contains the entry points for invoking the specific web-service and a reference to the protocol to use (examples are HTTP, SOAP or REST).

In the context of the smart travel system we have defined 6 capabilities for touristic services: flight, hotel, train, bus, theater and museum. Each of them is able of gathering information and making/canceling reservation. An additional capability—visit_city—has been conceived for arranging time for freely visiting a place. During the visit, it has the special purpose of providing useful information via personal device and to monitor the traveler's position.

4.3 Norms and Metrics for Traveling Preferences

Besides the capabilities, norms and metrics are fundamental mechanism to increase user's flexibility in expressing their desires. Despite the intrinsic simplicity of the presented norms and metrics—more complex one could be added without changing the model—they represent a powerful instruments to dynamically refine the set of requirements.

A norm is a rule that describes the configurations of capabilities that produce undesired situations. For the smart travel system we have defined four norms the user may attach to goals:

- budget(MAX): for ensuring total expenses will not exceed a given budget.
- km(MAX): for ensuring the total number of kilometers will not exceed a max number.
- noTrain, noBus, a couple of norms for preventing respectively the use of either trains or buses for the transportation.

On the other side, metrics are high level elements that measure the quality associated to a configuration, thus to allow the system to compare and sort them. They are generally used in MUSA for enriching user's goals for specifying non-functional preferences. For the smart travel system we have defined the following metrics:

- artCityTour, that rewards spending time in famous cities of art.
- beachTour, that rewards spending time in seaside places.
- trainLovingTour, that rewards trips with continuous movements by train.
- wineFoodTour, that rewards visiting farmhouses and wineries.

4.4 A Web Application for Facilitating the Definition of Goals

MUSA provides a high-level language, GoalSPEC [13], to specify requirements as set of goals. For increasing flexibility and user-friendliness, it has been conceived as a controlled natural language. However such a language can not be directly used by smart travel users because it requires specific skills [10].

To this aim a web-based interface has been designed to mediate between users and their goals. It is based on a set of templates for specifying goals. Each template is a fixed structure in which some elements must be replaced by data coming from the forms.

This approach reduces the flexibility of the language, but it allows an unskilled user to produce a valid set of goals without supervision. He also may select metrics and norms from pre-filled combo-boxes.

When the user confirms the whole set of preferences, then the web-portal generates the corresponding set of goals (step 1 of Fig. 3) and injects them in the platform thus to activate the self-configuration phase (step 2 of Fig. 3). The self-configuration phase consists in aggregating available services according to user preferences (goals, norms, metrics) and contextual availability of resources. The result is presented in geotagged maps where the travel plan is detailed step-by-step.

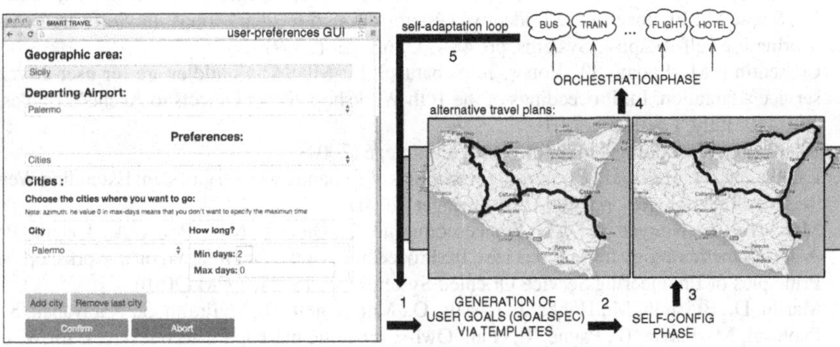

Fig. 3 An example of the MUSA approach to the case study of the smart travel

So far, we have not implemented the payment module necessary to buy each service of the travel package, but we have rather built a simulator engine that replaces the orchestration phase (step 2 of Fig. 3). It simulates the progression of the stages of the vacation, by updating the current state as if the user is going to benefit of the acquired services. The evolution of the trip is represented directly in a geotagged map via an animation. During this simulation the user interface also allows to interactively produce events for adaptation. For instance, it is possible to mark a resource (e.g.: a train route) as unavailable. This triggers a new cycle of self-adaptation, as shown in step 5 of Fig. 3, for proposing an alternative plan for the remaining part of the trip.

5 Conclusions

We presented a practical experience of customizing the general-purpose MUSA middleware in the context of the smart travel. This represents a novel approach for the development of smart and complex system where most of the complexity is inherited by reusing the underlying platform. After a traditional requirement engineering phase, the adoption of MUSA has facilitated the overall effort by limiting the development to three fundamental steps: (1) building (or reusing) an ontology of the domain, (2) developing and deploying a repository of autonomous functions that incorporate the available services (capabilities) and finally (3) designing and developing the interface for supporting user participation in the loop. By following this approach, fundamental characteristics such as self-configuration, self-evolution and self-adaptation are automatically integrated and do not impact the time-to-complete.

References

1. Brun, Y., Serugendo, G.D.M., Gacek, C., Giese, H., Kienle, H., Litoiu, M., Müller, H., Pezzè, M., Shaw, M.: Engineering self-adaptive systems through feedback loops. In: Software Engineering for Self-Adaptive Systems, pp. 48–70. Springer (2009)
2. Cossentino, M., Lodato, C., Lopes, S., Sabatucci, L.: MUSA: a middleware for user-driven service adaptation. In: Proceedings of the 16th Workshop "From Objects to Agents", Naples, pp. 1–10, Italy, 17–19 June 2015
3. Knublauch, H.: Editing owl ontologies with protégé (2004)
4. Laukkanen, M., Helin, H.: Composing workflows of semantic web services. In: Extending Web Services Technologies, pp. 209–228. Springer (2004)
5. Marchetto, A., Nguyen, C.D., Di Francescomarino, C., Qureshi, N.A., Perini, A., Tonella, P.: A design methodology for real services. In: Proceedings of the 2nd International Workshop on Principles of Engineering Service-Oriented Systems, pp. 15–21. ACM (2010)
6. Martin, D., Burstein, M., Hobbs, J., Lassila, O., McDermott, D., McIlraith, S., Narayanan, S., Paolucci, M., Parsia, B., Payne, T., et al.: Owl-s: semantic markup for web services. In: W3C Member Submission, vol. 22 (2004)
7. Moore. R.C.: Reasoning about knowledge and action. Ph.D. thesis, Massachusetts Institute of Technology (1979)

8. Newell, A.: The knowledge level. Artif. Intell. **18**(1), 87–127 (1982)
9. Patikirikorala, T., Colman, A., Han, J., Wang, L.: A systematic survey on the design of self-adaptive software systems using control engineering approaches. In: 2012 ICSE Workshop on Software Engineering for Adaptive and Self-Managing Systems (SEAMS), pp. 33–42 (2012)
10. Ryan, K., Maiden, N., Glinz, M.: If you want innovative re, never ask the users; a formal debate. In: 18th IEEE RE, pp. 388–388. IEEE (2010)
11. Sabatucci, L., Cossentino, M.: From means-end analysis to proactive means-end reasoning. In: Proceedings of 10th International Symposium on Software Engineering for Adaptive and Self-Managing Systems, Florence, Italy, 18–19 May 2015
12. Sabatucci, L., Lodato, C., Lopes, S., Cossentino, M.: Highly customizable service composition and orchestration. In: Service Oriented and Cloud Computing. LNCS, vol. 9306, pp. 156–170. Springer (2015)
13. Sabatucci, L., Ribino, P., Lodato, C., Lopes, S., Cossentino. M.: Goalspec: a goal specification language supporting adaptivity and evolution. In: Engineering Multi-Agent Systems, pp. 235–254. Springer (2013)
14. Weerawarana, S., Curbera, F., Leymann, F., Storey, T., Ferguson, D.F.: Web services platform architecture: SOAP, WSDL, WS-policy, WS-addressing, WS-BPEL. WS-reliable messaging and more. Prentice Hall PTR (2005)
15. Yu, E., Mylopoulos, J.: Why goal-oriented requirements engineering. In: Proceedings of the 4th International Workshop on Requirements Engineering: Foundations of Software Quality, p. 15 (1998)

Pentas: Using Satellites for Smart Sensing

Lorena Otero-Cerdeira, Alma Gómez-Rodríguez, Francisco J. Rodríguez-Martínez, Juan Carlos González-Moreno and Arno Formella

Abstract Smart environments are an important field of study and have suffered from an important evolution in the last decade. Smart technology makes an intensive use of wireless technology as the way of communicating all the elements of the system, including sensors. This paper tries to show how the use of satellites and radio waves may provide smart characteristics to environments that lack the common wireless technologies usually utilized. A system using Agent Oriented Software Engineering and based on that technology has been constructed and shows the suitability and advantages of the proposal.

1 Introduction

Last decade constitutes a period of extensive research in the field of Smart environments. The field has suffered from an important evolution and has been used in a wide range of applications from health monitoring or assistance to city management. Some examples of the variety of applications can be found in [2, 4, 12, 21, 23].

L. Otero-Cerdeira · A. Gómez-Rodríguez (✉) · F.J. Rodríguez-Martínez · J.C. González-Moreno · A. Formella
Departamento de Informática, Ed. Politécnico, Campus As Lagoas, Universidad de Vigo, Ourense, Spain
e-mail: alma@uvigo.es

L. Otero-Cerdeira
e-mail: locerdeira@uvigo.es

F.J. Rodríguez-Martínez
e-mail: franjrm@uvigo.es

J.C. González-Moreno
e-mail: jcmoreno@uvigo.es

A. Formella
e-mail: formella@uvigo.es

G. De Pietro et al. (eds.), *Intelligent Interactive Multimedia Systems and Services 2016*, Smart Innovation, Systems and Technologies 55,
DOI 10.1007/978-3-319-39345-2_61

Nowadays, the basis for smart spaces relies on the data obtained by sensors which are at the core of the system's functionality. Data are the center of the system and its relevance is twofold. The acquisition of data is fundamental, and therefore systems pay attention to get accurate data on time. Secondly, the obtained data must be stored and adequately treated.

This paper focuses mainly on data acquisition. Currently, Smart technology makes an intensive use of network connection resources, which are not always available, specially in certain geographical areas. Trying to overcome this limitation, this paper introduces space technology, namely pico-satellites for data acquisition. In this way, the system is able to obtaining data even when there is no reception coverage. This feature makes the system suitable for a specific kind of environments. Namely, those where there is a noticeable lack of network connection. This circumstance may be rather common in remote areas, perhaps places with singular climate conditions, like deserts, wild environments, etc.

Despite these problems, these areas are of high interest for establishing a network of sensors in order to monitor environmental values of interest. The main purpose being the prevention and forecasting of potentially dangerous situations. The system proposed in this paper may be used in such situations for providing data for analysis and prevention.

The remainder of the paper is organized as follows. Section 2, introduces other works in the field and compares them with the proposal presented in this paper. In addition, Sect. 3 describes the problem that the system tries to solve. Next, in Sect. 4, the system is detailed and its architecture is provided. Section 5 introduces an example of use of the architecture proposed in a simple system. Finally, in Sect. 6 we present the concluding remarks and directions for future work.

2 Related Work

As stated before, Smart environments constitute an important field of research and have may applications. In [2] a broad view of the applications of smart technology is provided. Among the most relevant applications of smart environments, smart homes is one of the most common ones, as [4] shows. Nevertheless, much work has been done in the use of smart technology for health monitoring or elderly people support, as introduced in [9, 12, 21, 23]. In all the previous proposals, sensors provide the data to the system using network based communications and therefore need that infrastructure. As stated previously, Smart technology is used in a very different way in Pentas system.

Other works show the use of sensors in space missions for aerospace-based monitoring and diagnosis, such as [1]. The work addresses the definition of the network architecture for a set of picosatellites and the algorithm to control that network. Its main contribution is the capability of the system to reconfigure the network. The focus in Pentas, the system proposed in this paper, is far from that, as its main goal is to provide a new way of communication and data transmission using picosatellites. In this case, it is assumed that the satellites have a correct function and communication and Pentas system takes profit of such communication.

In literature there are some works which relate Multiagent systems and space applications or even picosatellites [15, 20]. Nevertheless in these works, the goal is in the use of Multiagent systems on-board, as a way of optimizing the picosatel- lites constellation, regarding different criteria. As it was introduced before, the goal of the present paper is different and considers the picosatellite just a tool for guaranteeing communication.

Finally, other authors address the relationship between communication and multiagent systems [13, 14]. These works show the suitability of considering multiagent approach for its use in communications. Despite its drawbacks, these works conclude that agent technologies are excellent mechanisms for communication because agents provide benefits such as: flexibility, maintainability, etc. These reasons are also in the foundations of choosing agent orientation for Pentas proposal.

3 Problem Description

The system proposed in this paper (Pentas) is a part of a bigger project, HumSAT 2.0 [7, 8, 22]. HumSAT (Humanitary Satellite Network Project) is a project led by the University of Vigo with the participation of universities of the different states of the ESA (European Space Agency), Japan and the United States, and also is sponsored by the UNOOSA (United Nations—Ofice for Outer Space Afairs). The project aims at building a constellation of cubesat satellites to connect a set of users with a network of sensors distributed throughout the world.

The satellites are located in a non-geostationary orbit which implies that they move at a different speed than the Earth and so they won't always cover the same zone. The sensors set on Earth are responsible for the acquisition of data and their transmission to the satellites via radio waves. There is a great diversity of sensors to be integrated within the platform to measure and monitor different types of parameters, such as water temperature or wind speed. However monitoring capabilities are not only restricted to the measurement of environmental parameters. System features allow the users to define their own sensors and integrate them into the network.

This paper tries to show how the use of these technologies may provide smart characteristics to environments that lack wireless technologies usually used in Smart environments. In populated areas, the use of different kinds of sensors to obtain the data from the environment relies on the use of wireless technologies [10, 19] provided by commercial providers. This makes the communication between sensors and smart systems very easy and facilitates that smart spaces provide information to the users. Nevertheless, there can be situations or environments where such kind of technologies are not available. For instance in developing countries or in zones of the planet which are not populated (Amazon rainforest, deserts or ice areas, etc.).

The main aim of this paper is to show how space technology may be used for treating relevant data obtained from sensors installed in such remote areas. It is important to notice that information from remote areas can be of great relevance. For instance, the upstream volume of water in a river may be a fundamental data for preventing river flows in populated areas down river. In these situations, the acquisition of data from sensors may not be done using traditional technologies, therefore these cases could benefit from the proposal introduced in this paper.

4 System Description

The system described in this paper allows to obtain data from sensors using satellites for sending the data and receiving them in the earth stations following the satellite. The system is hence integrated by highly heterogeneous hardware devices that should seamlessly interact. These hardware devices range from small satellites and sensors, to earth stations. The most outstanding feature of the system is the dispersion of the components, since those on earth are globally spread, but the system also integrates those in space. In a situation like this, the software component turns crucial for the success at all levels.

The main purpose of the system is to seamlessly achieve the transmission of the data measured by the sensors on earth to the constellation of satellites and from these back to earth to the earth stations. The success of such functioning is highly dependable on the integrating software infrastructure.

The software layer, which is basic to every space related project, gains even more importance in this project. The software infrastructure is designed to not only give support to the hardware devices, but also to provide the whole system with intelligent features, such as an enhanced performance, the ability to self adapt to potential requirements from users and an expanded the set of functionalities beyond the basic ones.

This software system may be seen as complementary to the software embedded in the different devices that are integrated in the system, as well as to be used for the sensor-satellite and satellite-earth station communication.

A general view of the system is shown in Fig. 1. As it can be seen, the system obtains data from sensors and sends them to satellite, which in turn, send these data to earth. Sensors sending the data may be placed in remote locations such as a river flow, a forest, a desert, etc. As stated before, the whole communication is made via radio waves.

The main goal may be decomposed into the following smaller ones:

1. Definition of the mission's software, that is, the on board system that will be embedded in the satellites once put into orbit. The system shall include the necessary functionalities to support the satellite and to allow the communication with the **Smart Nodes**, as well as the later transmission of the data down to earth. This cycle of communications is a strategic feature of the success of the mission and

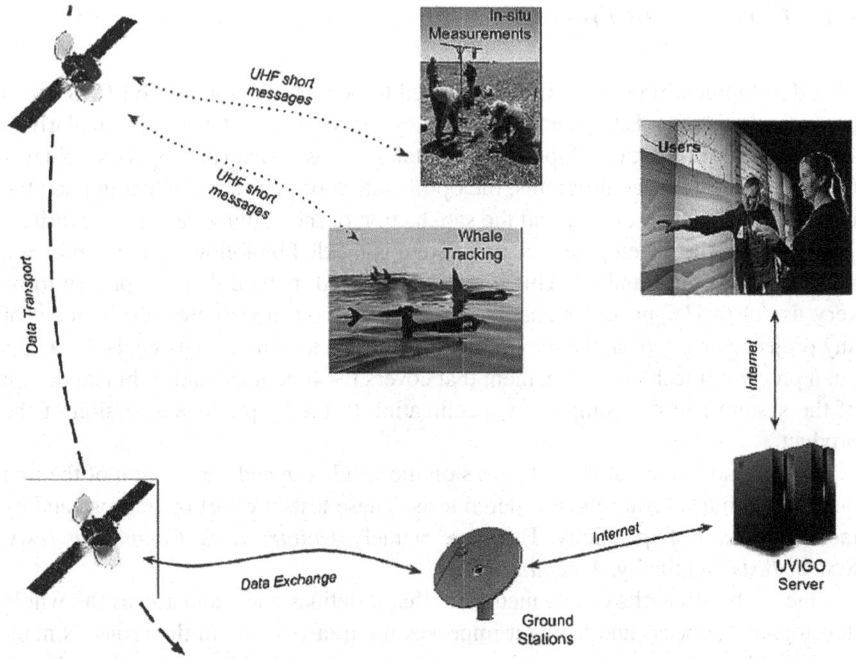

Fig. 1 Humsat 2.0: overview of the whole system architecture

hence of the system. This system is in charge of the optimization of the bandwidth, which can not be done statically since it must respond to the changing environmental circumstances and therefore it must be provided with some degree of adaptability and dynamic flexibility. This key feature relies on time windows adjusted to the visibility of the earth station or the link with an specific satellite.
2. Definition and development of the software support systems for the **Smart Nodes**. By developing this goal, the necessary infrastructure to support the motes (hardware device that includes one or several sensors) is achieved. In addition, the software allows the communication of the measured values and the optimization of such communication.
3. Construction of a simulation framework of the real environment that may allow to run the test as closer to the real environment as possible. This is the most ambitious goal of this system from a developing point of view. The software included in the satellite must grant at every time high levels of reliability, robustness and adequacy, considering the tremendous costs that would imply a poor functioning.

This software solution defines the foundation of the project, since it provides support for different parts of the whole system, and becomes a binding point that joins all the different subsystems.

4.1 Development Process

This development process has been selected to be compliant with the ECSS European standard by the European Space Agency for space activities [5], particularly in what refers to software development. This standard has as main purposes the achievement of effective space programs, the optimization of their economical impact, the improvement of their quality and the satisfaction of the required levels of security.

Therefore, the development of the system is tackled by following the *V* process, accordingly to the standard. This is an standardized method that has proven to be very useful in ITC project management. This method models the development of any project starting from the requirement definition according to its goals. It carries out a process of technical refinement that covers the functional and technical design of the system and the component specification to finally produce a version of the product.

The second branch of the *V* focuses on the validation and verification of the first one by running several sets of ordered tests. These tests are sort by purpose and by increasing level of specificity. These are, namely, *Unitary Tests, Component Tests, System Tests* and finally, *Acceptance Tests.*

One of the strengths of this method is that it defines and standardizes the whole development process and hence it improves the transparency in the project's management. As a consequence of this specification, it is possible to identify earlier in the development or even prevent, the possible risks of the project as well as minimize their consequences.

By turning the development process into a standardized one several advantages are provide, such as guaranteeing the completeness and uniformity of the expected results in the different milestones. It may also help in unifying the different elements that integrate the process in order to ease the communication among the different entities that integrate a collaborative project as this one. Finally, the standardization eases the calculation of the effort required to achieve the different milestones and hence favors their optimization and even reduction.

Following the general process proposed by the standard, the development has also been compliant with the formal definition process for Multiagent Systems proposed in [3, 6]. This standard provides a formal sequence of activities, tasks and work products to obtain within the general overview of a Scrum like process, using the INGENIAS methodology for development [16, 17].

4.2 Multi Agent System Architecture

The system has been constructed using the Agent Oriented Software Engineering principles. The software architecture of the system is constituted of several agents with different capabilities. These agents are:

- Each one of the end devices or *Motes* has an associated mobile agent which is responsible for obtaining the data from the hardware and communicate them to the satellite.
- The "on board" software, on the satellite, is modeled as a single agent, with the goal of receiving data and send it to the earth stations.
- Earth stations contain the main part of the system agents which receive the communications and provide these data for analysis.

The system has several constraints regarding the limitations of the environment. In particular, the transmission cost must be minimized, fitting these restraints:

- The size of the data packet to be transmitted. The greater the amount of data to be transmitted, the greater the time consumed in transmission.
- Transmission time. Each node has a time interval in which the transmission will be effective (data reach the satellite). The total time available to each node corresponds to the size of the transmission window. The time that will consume the total transmission of the data packet, must be calculated and check its feasibility in relation to the transmission window.
- Weather conditions. In cases of unfavorable weather conditions, an extra increment of transmission power may be required to ensure the integrity of the data received at the satellite.

The optimization of data is particularly importante. Mobile agents and the software on board make an efficient storage of data, trying to minimize time and space. In particular, data interchange is coded in JSON [11, 18].

Following this architecture a simple prototype has been constructed and used to obtain data of flow in the local river. Next section introduces the main characteristics of such prototype.

5 Case Study

As starting point, a prototype of the system previously described has been developed. At this moment, there are some cubesat satellites orbiting, and they have been used to prove the suitability of the system previously described. This infrastructure has been used by a system with the architecture described which monitors the local river, and in particular its flow, in order to provide data for avoiding river overflows. The purpose of constructing this prototype is to test the results obtained by this experimentation system against the data obtained by local authorities using the traditional methods.

A general overview of the system is shown in Fig. 2. As shown, the system consists of some sensors which have been put in the river and measure the level of flow. A picture of the hardware and the environment is shown in the photographs in the upper part of figure. The data taken by sensors is sent using the satellite infrastructure and reaches the control office (photographs at the bottom of the figure).

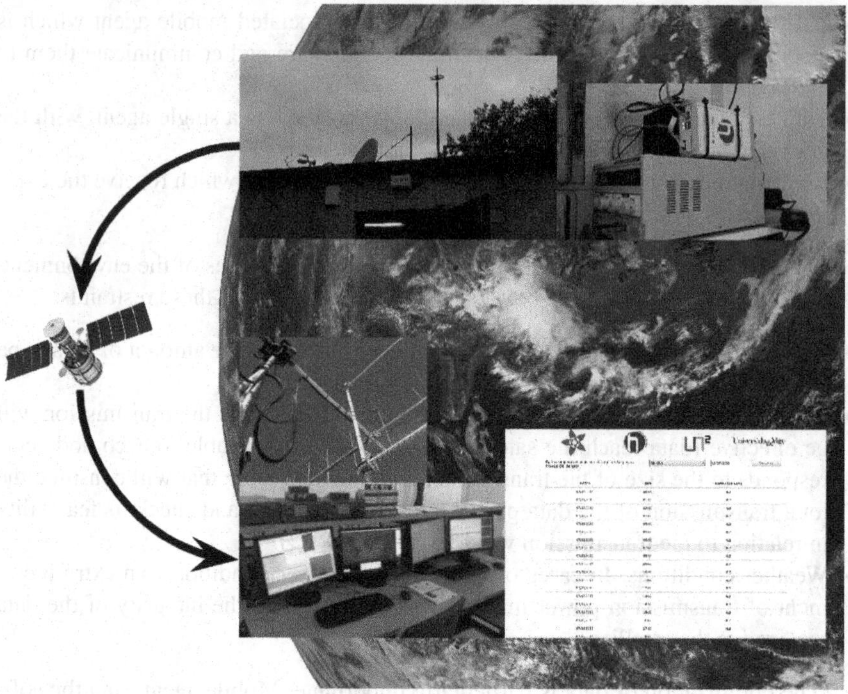

Fig. 2 Pentas: system overview

At this moment the system provides the following functionalities:

1. Data transmission. The system consists physically of a set of sensors, which provide data hourly. Afterwards, data are processed by the mobile agent associated to the sensors. These data are packed and made fit within the restrictions. The size of the package is optimized to minimizing bandwidth and time consumption. Nevertheless, the particular size depends on the sensor hardware and the variables measured. These packed data are transmitted to satellite when the transmission window is on. For each particular area, the window has a duration of approximately five minutes every hour.
2. Data reception. The earth node obtains data from the satellite within its transmission window, by means of the on board software provided. Data are stored and processed by an earth node. In addition, the data are published online and also may be obtained by a mobile app.

The system has worked correctly and obtained the results showed in Fig. 3. The figure shows the table which visualizes the data obtained at a particular date and time from the infrastructure previously described. The table contains three columns which provide respectively the Date, Hour and Flow river level. The data were compared with those provided by authorities and they was proven to be accurate and correct.

Universidade Vigo

Datos proporcionados por el satélite Serpens a través del sensor: AINRIO1 12/07/2015 Ver gráfica

Fecha	Hora	Nivel del rio (m)
07/12/2015	00:00	0.2
07/12/2015	01:00	0.2
07/12/2015	02:00	0.2
07/12/2015	03:00	0.2
07/12/2015	04:00	0.2
07/12/2015	05:00	0.2
07/12/2015	06:00	0.2
07/12/2015	07:00	0.2
07/12/2015	08:00	0.2
07/12/2015	09:00	0.2
07/12/2015	10:00	0.2
07/12/2015	11:00	0.2
07/12/2015	12:00	0.2
07/12/2015	13:00	0.2
07/12/2015	14:00	0.2
07/12/2015	15:00	0.2

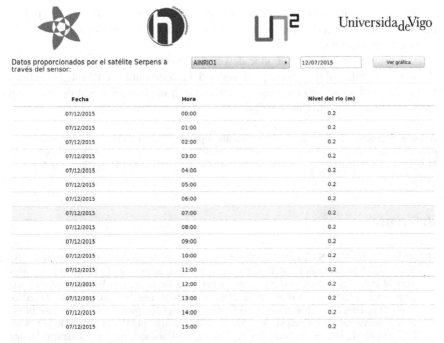

Fig. 3 Data obtained using the prototype

6 Conclusions and Future Work

By means of this work we have provided a general overview of how satellites and radio waves may be used for transmitting data for smart sensing when no wireless communication are available. The system for achieving this transmission takes profit from a constellation of satellites which have previously been launched for academic uses by the University of Vigo and other collaborators.

Pentas project proposes a Multiagent architecture which allows to transmit the data, such system is compliant with all the restrictions defined. The main applications that arise for the project may be:

- Autonomous Monitoring of Climate Change: Sensor networks deployed will control environmental parameters related to climate change in scattered locations throughout the globe, such as Arctic and Atlantic Oceans.
- Emergency beacons location: The system can be used to locate priority emergency signals, giving support to the intervention in emergency or humanitarian initiatives.
- Communications support in remote areas: The satellites constellation configuration will allow to monitor different locations of the planet using sensors since these places, for their unique characteristics, lack the infrastructure necessary to maintain communication through other channels.

As a toy example, the system constructed has been successfully used for river flow monitoring. Besides, the data have been publicly available through a Web and an mobile application.

In the future, we plan to continue developing more complex systems which will show the suitability of the system proposed and the communication through satellites for any kind of smart environment. In particular, it is planned to address the main areas of application introduced before.

Acknowledgments This work has been partially supported by grant *PentaS: Smart Software to provide global service to sensors using satellites* [ESP2013- 47935-C4-2-R] with funding Bodies: Ministerio de Economa y Competitividad (Spanish Government) and FEDER (European Union).

References

1. Arslan, T., Haridas, N., Yang, E., Erdogan, A.T., Barton, N., Walton, A.J., Thompson, J.S., Stoica, A., Vladimirova, T., McDonald-Maier, K.D., Howells, W.G.J.: Espacenet: a framework of evolvable and reconfigurable sensor networks for aerospace? based monitoring and diagnostics. NASA/ESA Conference on Adaptive Hardware and Systems, pp. 323–329 (2006)
2. Cook, D.J., Das, S.K.: How smart are our environments? an updated look at the state of the art. Pervasive and Mob. Comput. **3**(2), 53–73 (2007). Design and Use of Smart Environments
3. Cossentino, M., Hilaire, V., Molesini, A., Seidita, V.: Handbook on Agent-Oriented Design Processes. Springer (2014)
4. De Silva, L.C., Morikawa, C., Petra, I.M.: State of the art of smart homes. Eng. Appl. Artif. Intell. **25**(7). 1313–1321, (2012). Advanced issues in Artificial Intelligence and Pattern Recognition for Intelligent Surveillance System in Smart Home Environment
5. ECSS european standard. http://www.ecss.nl/ (2016). Accessed 10 Feb 2016
6. González-Moreno, J.C., Gómez-Rodríguez, A., Fuentes-Fernandez, R., Ramos-Valcárcel, D.: Ingenias-scrum. In: Handbook on Agent-Oriented Design Processes, pp. 219–251. Springer (2014)
7. Guerra, A.G.C., Francisco, F., Villate, J., Agelet, F.A., Bertolami, O., Rajan, K.: On small satellites for oceanography: a survey (2015). arXiv preprint arXiv:1512.07442
8. Humsat satellite homepage. http://www.humsat.org/ (2016). Accessed 10 Feb 2016
9. Isern, D., Moreno, A.: A systematic literature review of agents applied in healthcare. J. Med. Syst. **40**(2), 43:1–43:14 (2016)
10. Jauregui-Ortiz, S., Siller, M., Ramos, F., Scalabrin, E.: Smart environmental architecture for node localization in a wireless sensor network. In: 2012 8th International Conference on Intelligent Environments (IE), pp. 222–227, June 2012
11. JSON home page. http://www.json.org/ (2016). Accessed 10 Feb 2016
12. Kevin, I.K., Wang, W., Abdulla, H., Salcic, Z.: Fast track article: ambient intelligence platform using multi-agent system and mobile ubiquitous hardware. Pervasive Mob. Comput. **5**(5), 558–573 (2009)
13. Manvi, S.S., Kakkasageri, M.S., Pitt, J.: Multiagent based information dissemination in vehicular ad hoc networks. Mob. Inf. Syst. **5**(4), 363–389 (2009). December
14. Manvi, S.S., Venkataram, P.: Applications of agent technology in communications: a review. Comput. Commun. **27**(15), 1493–1508 (2004)
15. Ocon, J., Rivero, E., Montero, A.S., Cesta, A., Rasconi, R.: Multi-agent frameworks for space applications. In: SpaceOps-10. Proceedings of the 11th International Conference on Space Operations. Huntsville, Alabama (2010)
16. Pavón, J., Gómez-Sanz, J.J., Fuentes, R.: The ingenias methodology and tools. Agent-oriented Methodol. **9**, 236–276 (2005)

17. Pavón, J., Gómez-Sanz, J.: Agent oriented software engineering with ingenias. In: Multi-Agent Systems and Applications III, pp. 394–403. Springer (2003)
18. Pohls, H.C.: Json sensor signatures (jss): end-to-end integrity protection from constrained device to iot application. In: 2015 9th International Conference on Innovative Mobile and Internet Services in Ubiquitous Computing (IMIS), pp. 306–312, July 2015
19. Polastre, J., Szewczyk, R., Mainwaring, A., Culler, D., Anderson, J.: Analysis of wireless sensor networks for habitat monitoring (2004)
20. Schetter, T., Campbell, M., Surka, D.: Multiple agent-based autonomy for satellite constellations. Artif. Intell. **145**(1–2):147–180 (2003)
21. Touati, F., Tabish, R.: U-healthcare system: state-of-the-art review and challenges. J. Med. Syst. **37**(3), 1–20 (2013)
22. Tubío-Pardavila, R., Vigil, S.A., Puig-Suari, J., Agelet, F.A.: The humsat system: a cubesat-based constellation for in-situ and inexpensive environmental measurements. In: AGU Fall Meeting Abstracts, vol. 1, p. 3365 (2014)
23. Vallee, M., Ramparany, F., Vercouter, L.: A Multi-agent system for dynamic service composition in ambient intelligence environments. In: The 3rd International Conference on Pervasive Computing (PERVASIVE 2005) (2005)

A Multiple Data Stream Management Framework for Ambient Assisted Living Emulation

Jorge J. Gómez-Sanz and Pablo Campillo Sánchez

Abstract The development of Ambient Assisted Living systems would be facilitated if there was a development environment that allowed to simulate in a computer the physical environment, its inhabitants, as well as the Ambient Assisted Living system. This requires, on the one hand, an infrastructure for simulating the physical environment and, on the other hand, an infrastructure for emulating the Ambient Assisted Living devices. Both can be interconnected through data streams that allow emulated devices to behave as if they were connected to the real world, since they get similar sensor input. This paper introduces advances on a simulation framework for ambient intelligence so that it becomes capable of producing such data streams.

Keywords Ambient intelligence · AAL emulation · AAL development

1 Introduction

The research in Ambient Assisted Living (AAL) is inherently expensive. Renting facilities, buying specialised hardware, and involving volunteers, are only a few of the activities a researcher has to assume. These maybe some of the reasons why AAL systems are not as affordable by end users as they ought to be. To change this situation, previous work [3, 8] has shown how a simulation of a 3D environment can facilitate the development of Ambient Assisted Living (AAL) solutions in an inexpensive way. The proposed solution, now called Ambient Intelligence Development Environment (AIDE, http://grasia.fdi.ucm.es/aide), was made of 3D environments where simulated characters interacted with several emulated Android devices representing the AAL solution.

J.J. Gómez-Sanz (✉) · P.C. Sánchez
Facultad de Informática, Departamento de Ingeniería del Software e Inteligencia Artificial, Universidad Complutense de Madrid, 28040 Madrid, Spain
e-mail: jjgomezm@ucm.es

P.C. Sánchez
e-mail: pabcampi@ucm.es

G. De Pietro et al. (eds.), *Intelligent Interactive Multimedia Systems and Services 2016*, Smart Innovation, Systems and Technologies 55,
DOI 10.1007/978-3-319-39345-2_62

Despite its effectiveness, the use of Android constrains developers to a single technology that, despite being widespread, its not well prepared for commercial AAL. Among others, it lacks of arrays of sensors/actuators at the same level as other hardware platforms, such as Arduino, Beagle Bone, or RaspBerry Pi; and does not have the necessary means for real time operation, unlike real time Operating Systems.

At the same time, AIDE is not considering AAL specific middleware created precisely to deal with this hardware diversity, such as UniversAAL or OpenAAL. These two middleware frameworks do not regard the emulation of the AAL infrastructure and, least of all, the simulation of the physical environment. To achieve integration between AIDE simulation capabilities and any of these middleware solutions, a platform independent sensor data stream approach would be needed so that custom platform specific clients can adapt such data streams to what each middleware expects.

It could be said that the solution to the extension of AIDE services to other non-Android hardware and the standard middleware integration problem could be solved in part with the specification and implementation of sensor data stream management services. This paper discusses how these sensor data streams can be integrated with AIDE and shows evidence of integration into other emulated linux based platforms. The sensor data stream management requires a discovery mechanism, and a way of accessing the different sensor/actuator infrastructure. Both are introduced in this paper.

Software developed for this paper is part of the AIDE, which it is distributed as free software under GPLv3. The advances are situated of the context of previous results [3, 8] HAIS [4]. These previous results will be discussed too along the paper.

The paper is structured as follows. First, Sect. 2 introduces limitations of previous works and what improvements this contribution is bringing. Then, Sect. 3 introduces the elements that aim to overcome problems identified in Sect. 2. In particular, the text focuses on data obtained from accelerometers, but explains how this could be generalised to other relevant data streams. Section 4 presents a practical application of multiple devices getting input from the same information model, the accelerometer signals. These results are compared with the current literature in Sect. 5. The paper finishes with the conclusions in Sect. 6.

2 Previous Work

Initial work [3, 4], described a framework, now called AIDE, where the developer simulated a house environment together with the house inhabitants. AIDE considered as well the interplay between a situated and emulated Ambient Assisted Living solution and the house inhabitants. The main advantage of this formula was the software-in-the-loop approach that facilitated reusing the software controlling the emulated AAL system in a real AAL. An example of what AIDE provides is shown in Fig. 1. This figure shows a scenario with two people and three emulated Android devices, two of them attached to each actor's wrist and one to the bathroom. The first

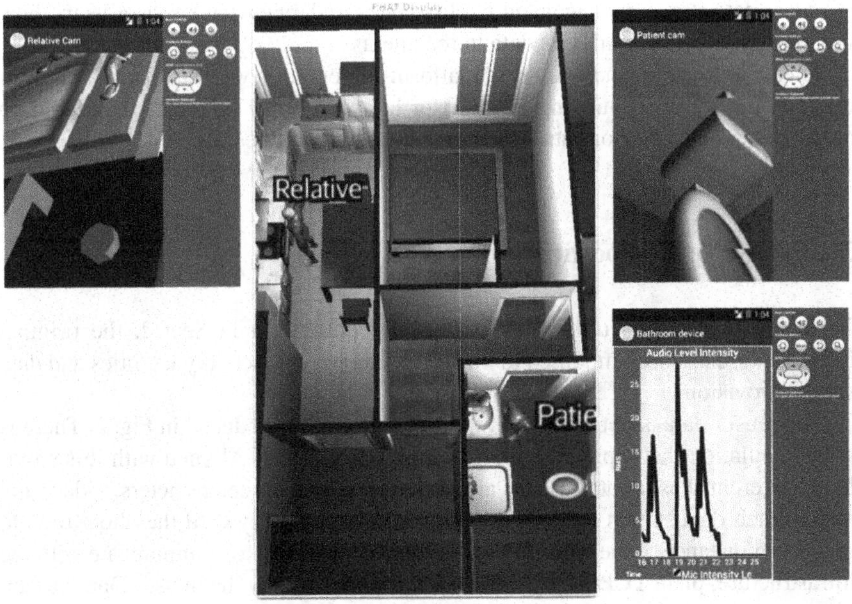

Fig. 1 An Android deployment of an AAL solution interacting with a 3D simulation

two are connected to video data streams obtained from the simulation, whereas the second is connected to audio data streams. The AAL solution is a monitoring one whose software is running inside the Android emulators.

However, in AIDE, the simulation engine provided sensor data streams to the emulated device only once. If the developer needed to try out different clients or stop and restart whatever emulated device, the simulation had to be started from scratch. Besides, AIDE, as it was before, was unable to deal with the Internet of Things (IoT) configurations, which requires going beyond Android. It should be possible to connect different clients, perhaps hosted in remote computers, connected through standard middleware to the simulation, and implemented using something different from Java.

The simulation also was dealing mainly with physical models supported by the open source game engine JMonkey. JMonkey has a physics engine called JBullet which allows to infer the result of object collisions, among other kinds of physical interactions. However, JBullet is limited. It does not address the propagation of smoke in the ambient, for instance. Freely attaching and developing models that run together with the simulation, is something needed to go beyond the 3D animation. Examples of relevant additional data streams are biostats of the character such as pulse rate, electroencephalography, or sweat. Since excessive sweating and rapid heart rate can be symptoms of a heart attack [7], the presence of such models could be used to test health monitors and the algorithm they use to foresee this risk. A better sensor data stream management facilities can help to attain this goal, too.

A sensor data stream management facility needs to define what a sensor is and how this sensor service provides its data to the clients.

Then, deploying on non-Android platforms, integration with standard AAL middleware, as well as simulating new sensor hardware, may be issues that could be aided with a better sensor data stream management facilities.

3 Framework Modification

To overcome the difficulties and limitations pointed out in Sect. 2, the original infrastructure has been modified to incorporate services discovery facilities and data stream provision.

The sensor data is obtained from the simulation, as introduced in Fig. 2. There is a 3D simulation that represents the physical environment. Aligned with this simulator, different sensor data streams are obtained, such as accelerometers, video, and audio. Each data stream is assigned a tag, which is the keyword the client uses to request an instance of the data stream. To enable any client to communicate with the infrastructure, plain TCP/IP sockets are used. Clients query the Sensor Data Stream manager which data streams are available and how to connect with them. Afterwards the information is received, the client can connect to the particular sensor data stream and start processing the data.

Each data stream has to remain consistent with the situation depicted in the 3D simulation. The accelerometer signal is a most basic example. The simulation defines a number of accelerometers attached to different parts of one simulated character or other objects of the environment. The movements of the character are analysed by an

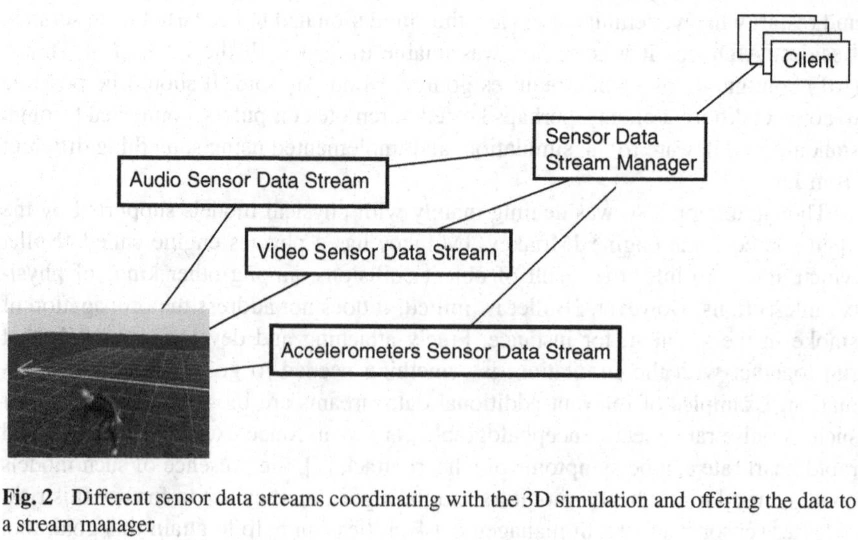

Fig. 2 Different sensor data streams coordinating with the 3D simulation and offering the data to a stream manager

accelerometer model to generate a stream of tuples $[a_x, a_y, a_z]$ where each component represents the acceleration at that instant of time, including the gravity, in a specific axis. These tuples conform the data stream that external clients are interested into.

The number of sensor data streams an external client wants is undetermined. One client may require a couple of sensors, while others can require a handful of them. Such situations are frequent when analysing the performance or effectiveness of different activity recognition solutions based on accelerometer data. Identifying the number and location of sensors is an important task in the development of an AAL solution and it has to be thoroughly experimented. This takes time. Therefore, instead of creating different real prototypes, it is worth evaluating first at the simulation level which configuration is more suitable for the intended activity recognition task.

Access to the data stream is hidden to the clients that operate remotely. There is an API that provides access to the sensors. This API is the same for local or remote clients. The API is implemented by mediators that provide a wrapper to the real sensor, or a proxy to the data stream. One case or the other is identified automatically by the infrastructure, so the developer can ignore this and proceed the same way in the real devices or emulated devices.

Internally, the adequate data stream has to be created. In the case of accelerometer, video, and audio data streams, the information is already integrated in the simulation, so the corresponding sensor data stream can be directly delivered. The creation process can compromise the simulation if, for instance, the necessary data cannot be implied from the 3D model, as in the previous examples. In those cases, additional elements are needed that produce such data stream in a consistent way, i.e., coordinated with the 3D simulation progress. Hence, if there is a new CO_2 sensor to be incorporated and the 3D simulation does not include data on CO_2 generation and propagation, a new CO_2 data stream needs to be created. Such stream ought to be consistent with the simulation, so, it should show low levels when the character is idle or absent from the room where the sensor is.

4 Case Study

The case study is a character that walks through a plain terrain and repeatedly falls. The goal is to analyse the accelerometer signals and infer patterns that permit, later on, to determine if the character has fallen or not. To check the multiple data stream management capabilities, data streams are transferred to local, remote, and clients within virtual machines. The data stream is generated using a sensor situated in the chest of the actor. In the Fig. 3, the accelerometer is drawn as a black box. This accelerometer generates acceleration data across x,y, and z axes. The data is offered to three different clients. The two first, local and remote, are java based, while the third remote one is python based. The first local client is taking data directly from the simulation, i.e., using no sensor data stream. The second remote java client is taking the information from a socket connection. The third remote client is implemented in

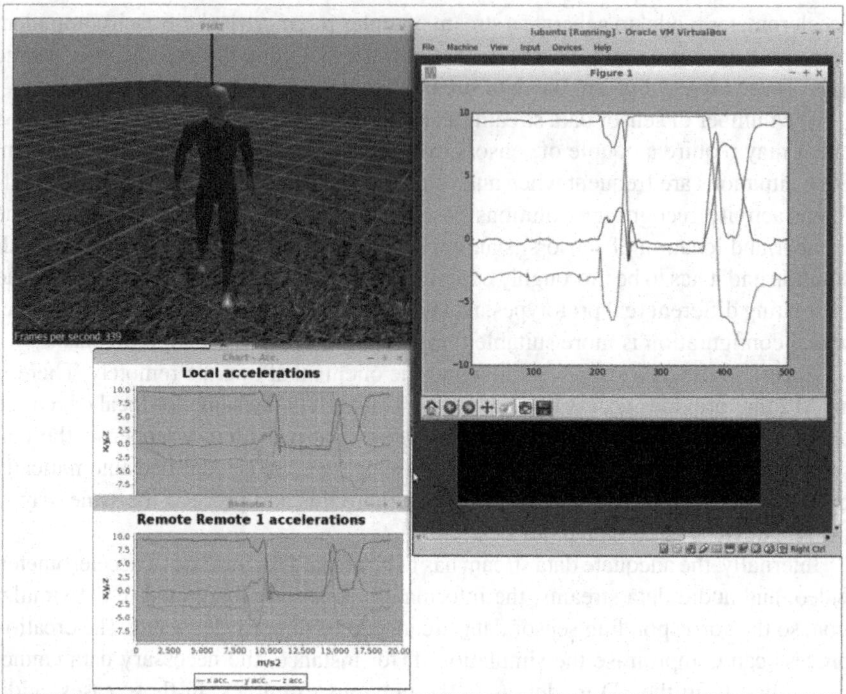

Fig. 3 Chart representing the signals obtained from accelerometer situated at the chest of the actor and delivered to different clients

python and works inside of a virtual machine. It uses the same socket connection to get the sensor data stream.

Data presented across all variants is the same. The java local and java remote clients were added to prove fast prototyping was possible. The virtual machine variant demonstrated that, at the same time, an emulation of the target platform could access the same data. This virtual machine could represent a non-Android device running some linux distribution.

Since all clients are sharing the same information, it is possible to compare the different implementations running them in parallel and compare the results. This step is important, since it is not ensured that initial prototyping stages will use the same technologies as later stages of the development, where platforms are closer to the production ones and the technology may not be that developer friendly.

Along the case study execution, each remote client could be switched off and on independently. As long as the simulation was running, the client could re-attach to the sensor data stream and resume the information analysis. This feature is important in order to not constrain the developer when preparing experiments.

5 Related Work

The work presented in this paper is compared with current literature about two research lines: the generation of sensor data and the emulation of the AAL infrastructure.

In the first case, there is literature considering high-fidelity simulations and virtual humans. In both, a simulation produces realistic sensor data streams.

Virtual humans are studied in [5]. A virtual human reproduces symptoms to recognise heart attacks. The work uses game engines to represent the situation and to modify the appearance and status of the character to simulate the symptoms, such as paleness of the patient. When compared to this contribution, it can be said the simulation is used to feed devices which take decisions and perform actions, whereas the work from [5] is devised mainly for training purposes.

High-fidelity patient simulators are mentioned in [6], though applied to dogs. It proposes simulating in a 3D environment different situations to let trainees discover/decide what to do. When compared to this contribution, it can be said the purpose, though it aims at reproducing the actor in a high-fidelity way, is not limited to training. The main goal is to obtain data streams that could cheat AAL devices to think they are receiving real data, instead.

With respect the emulation of AAL infrastructure inside of a 3D simulation, literature cites several initiatives.

Activity recognition is addressed in SIMACT [1], which is a 3D smart home simulator. The researcher defines scenarios with a scripting language. Sensor data comes from values stored in a database. When compared to this contribution, the complexity of simulations is potentially lower than the ones introduced in this paper. Characters in this contribution can show a variety of behaviours which generate sensors data accordingly to the scene currently rendered. Besides, the data is fed to the emulated devices so that the behaviour of the AAL solution can be evaluated.

In the project http://hbms-ainf.aau.at/en/HBMS.html, the developer has to define activities and then sensor data is gathered. Activities definition includes valid variations of the same activity. However, that project focuses on producing a knowledge base, rather than connecting the simulation to emulated devices, as in this work.

On the other hand, Naranjo et al. [9] structures a process for the co-creation of simulations in the context of AAL. Nevertheless, that work is oriented towards the production of immersive virtual representations for end-users, not reproducing the AAL solution behaviour, as in this contribution. This contribution tries to address every sensor data stream to feed this data to emulated devices. The work [9], instead, provides with video and/or audio, not other kind of signals, such as accelerometers.

Sala et al. [10] continues the work of [9]. Both works present the results of VAALID project in developing this approach of creating tools for design and simulation of AAL Solutions using VR and Mixed Reality, supporting the early involvements of beneficiaries in the process. However, project is centred in VR using InstantReality[1] where the definition of the scenarios is closed and deterministic due

[1] http://www.instantreality.org/.

to, in part, lack of a physics engine. Besides, the obtained sensor data is limited. This paper tries to give access to different kinds of sensors, such as accelerometers, audio, and video sensors.

Work in UbikSim is related too. Campillo-Sanchez et al. [2] extends UbikSim framework for testing smart phone applications using a 3D simulation. Campillo-Sanchez et al. [2] was using a simple 3D engine and no physics. The absence of physics makes harder to track the movement of characters and provide convincing sensory output towards the AAL devices. This work, however, addresses this issue by using the physics engine JBullet, which is included in the underlying game engine JMonkey. This allows to model more realistic falling situations or to explore the result of pushing objects and interferring the activity of the characters.

Lastly, Shoulson et al. [11] introduce ADAPT, which is an authorware platform for creating rich and expressive human characters in a virtual world. ADAPT is not interfacing with emulated AAL devices, or intending to do so. Besides, ADAPT is a closed platform, what makes difficult to address this functionality and produce something similar to the work introduced in this paper.

6 Conclusions

The paper has introduced an architecture for enacting the interconnection of multiple heterogeneous devices to a simulated platform to facilitate the development of software that control the AAL solution. The architecture has been illustrated with a proof of concept development where different accelerometer sensors attached to a walking character. The resulting data has been streamed to a virtual machine running a light linux distribution at the same time it was transferred to other clients. Also, python and Java were used indistinctly in the clients, enabling a more flexible development.

Acknowledgments We acknowledge support from the project "SOCIAL AMBIENT ASSISTING LIVING - METHODS (SociAAL)", project "Collaborative Ambient Assisted Living Design (ColoSAAL)", and mobility grant grant EEBB-I-15-10097, supported by Spanish Ministry for Economy and Competitiveness, with grant TIN2011-28335-C02-01 and TIN2014-57028-R respectively; and MOSI-AGIL-CM (S2013/ICE-3019) co-funded by Madrid Government, EU Structural Funds FSE, and FEDER. We mobility from the ministry for a short stay in TU Delft.

References

1. Bouchard, K., Ajroud, A., Bouchard, B., Bouzouane, A.: SIMACT: A 3D open source smart home simulator for activity recognition with open database and visual editor. Int. J. Hybrid Inf. Technol. (IJHIT) **5**, 13–32 (2012)
2. Campillo-Sanchez, P., Botia, J.: Simulation based software development for smart phones. In: Novais, P., Hallenborg, K., Tapia, D.I., Rodrguez, J.M.C. (eds.) Ambient Intelligence—Software and Applications, Advances in Intelligent and Soft Computing, vol. 153,

pp. 243–250. Springer, Berlin, Heidelberg (2012). http://dx.doi.org/10.1007/978-3-642-28783-1_31
3. Campillo-Sanchez, P., Gómez-Sanz, J.J.: Agent based simulation for creating ambient assisted living solutions. In: Advances in Practical Applications of Heterogeneous Multi-Agent Systems. The PAAMS Collection—12th International Conference, PAAMS 2014, Salamanca, Spain, 4–6 June 2014. Proceedings, pp. 319–322 (2014)
4. Campillo-Sanchez, P., Gómez-Sanz, J.J., Botía, J.A.: PHAT: physical human activity tester. In: Hybrid Artificial Intelligent Systems—8th International Conference, HAIS 2013, Salamanca, Spain, 11–13 Sept 2013. Proceedings, pp. 41–50 (2013)
5. Cavazza, M., Simo, A.: A virtual patient based on qualitative simulation. In: Proceedings of the 8th International Conference on Intelligent User Interfacesm IUI'03, pp. 19–25. ACM, New York, NY, USA (2003). http://doi.acm.org/10.1145/604045.604053
6. Fletcher, D.J., Militello, R., Schoeffler, G.L., Rogers, C.L.: Development and evaluation of a high-fidelity canine patient simulator for veterinary clinical training. J. Vet. Med. Educ. **39**(1), 7–12 (2012)
7. Furie, K.L., Kasner, S.E., Adams, R.J., Albers, G.W., Bush, R.L., Fagan, S.C., Halperin, J.L., Johnston, S.C., Katzan, I., Kernan, W.N., et al.: Guidelines for the prevention of stroke in patients with stroke or transient ischemic attack a guideline for healthcare professionals from the American Heart Association/American Stroke Association. Stroke **42**(1), 227–276 (2011)
8. Gómez-Sanz, J.J., Sanchez, P.C.: Achieving Parkinson's disease patient autonomy through regulative norms. In: Trends in Practical Applications of Agents, Multi-Agent Systems and Sustainability—The PAAMS Collection, 13th International Conference, PAAMS 2015, Salamanca, Spain, 3–4 June 2015, Special Sessions, pp. 97–104 (2015)
9. Naranjo, J.C., Fernandez, C., Sala, P., Hellenschmidt, M., Mercalli, F.: A modelling framework for ambient assisted living validation. In: Proceedings of the 5th International on Conference Universal Access in Human-Computer Interaction. Part II: Intelligent and Ubiquitous Interaction Environments, UAHCI'09, pp. 228–237. Springer-Verlag, Berlin, Heidelberg (2009). http://dx.doi.org/10.1007/978-3-642-02710-9_26
10. Sala, P., Kamieth, F., Mocholí, J.B., Naranjo, J.C.: Virtual reality for AAL services interaction design and evaluation. In: Proceedings of the 6th international conference on Universal Access in Human-Computer Interaction: Context Diversity—Volume Part III, UAHCI'11, pp. 220–229. Springer-Verlag, Berlin, Heidelberg (2011). http://dl.acm.org/citation.cfm?id=2022539.2022566
11. Shoulson, A., Marshak, N., Kapadia, M., Badler, N.I.: ADAPT: the agent development and prototyping testbed. In: Proceedings of the ACM SIGGRAPH Symposium on Interactive 3D Graphics and Games, I3D'13, pp. 9–18. ACM, New York, NY, USA (2013). http://doi.acm.org/10.1145/2448196.2448198

Soft Sensor Network for Environmental Monitoring

Umberto Maniscalco, Giovanni Pilato and Filippo Vella

Abstract This paper shows the application of a soft sensor network for the detection of meteorological events. A set of hard (real) sensor are placed in a territory, where they measure heterogeneous quantities. Starting from their measurements, a soft sensor network provides useful information coming from the data. In this contribution we show how prediction and validation of data can be done through machine learning approach by collecting data from the historical series. Furthermore, we show how the cluster based on correlation analysis among the data achieved by the sensors can be sensibly different from the ones simply drawn on geographical distance.

Keywords Soft sensor · Monitoring · Multivariate regression · Environmental risk · Wireless sensor network

1 Introduction

The capability to monitor large areas is an essential step for the preservation of the environment and a deep control of the places we live. Nowadays a detailed monitoring is possible through the adoption of a set of sensors that measure heterogeneous quantities and transmit the samples measurements to a central unit. Starting from a set of sampled values it is possible to extract higher level information and apply business intelligence algorithm to highlight important events and analyze the evolution of occurring phenomena [1, 2]. In general, an architecture based on these

U. Maniscalco (✉) · G. Pilato · F. Vella
Istituto di Calcolo e Reti ad Alte Prestazioni - C.N.R, Viale delle Scienze ed. 11, 90128 Palermo, Italy
e-mail: umberto.maniscalco@cnr.it
URL: http://www.icar.cnr.it

G. Pilato
e-mail: giovanni.pilato@cnr.it

F. Vella
e-mail: filippo.vella@cnr.it

G. De Pietro et al. (eds.), *Intelligent Interactive Multimedia Systems and Services 2016*, Smart Innovation, Systems and Technologies 55,
DOI 10.1007/978-3-319-39345-2_63

principles can be used to monitor domestic houses, cities and different kind of environments [3]. Recent technologies have enabled the possibility to deploy sensors on ground, in air, under water or even on people to capture relevant data for multiple tasks [3]. Environmental and habitat monitoring is a natural candidate for sensor network application, since the measured variables are distributed over a large region. The network is endowed with a connection infrastructure that connects sensor with central node. Typically high speed network are used for satellite and aircraft sensors while ground based sensors are typically connected through low speed networks. The connection infrastructure is beyond the scope of this paper and we will consider the sensors are connected with a time delay equal to zero.

Some applications for the environment monitoring include the measurements of physical quantities and the tracking of movements of animals, insects and livestocks in a given area. Some of them can provide an important impact on the environment are Fire detection, Biocomplexity mapping and flood detection [4]. In particular a set of interesting information and evaluations for flood detection is presented in the Alert System.[1] Multiple quantities are measured and when a particular phenomenon has to be detected, all the physical events that concur to the formation of the exceptional case must be recorded and investigated. The aim is to detect the exceptional event that can cause damages to people and things and try to forecast what is happening or just monitoring the evolution of a given situation.

This information can be used for detecting the occurrence of exceptional events and can therefore improve the environment preservation and management. The cost that is asked is the sampling of quantities, the registration of multiple parameters and the storage of the recorded values. In this context, a very positive initiative is the opening of the databases making the citizens and organizations able to download and process the data giving their contribution, through the analysis of the data, or just monitoring the environment. But indeed, the usability of these data can be limited by some difficulties. For example the format is rarely standard, quantities are presented and acquired with different standards and protocols, and furthermore, different kind error can be present. The measurement station, or in general the infrastructure, can have some failure and don't present data for some time that can be as short as days or, in the worse cases, longer than months. Finally, the detection of exceptional events intrinsically shows a problem for the presence of values in the data. The anomaly is typically present in a local area that does not affect all the stations but few of them (Cloudburst, Forest Fire, Floods). A simple threshold mechanism is not enough because making the station too sensible in raising alerts could generate too many alerts that would be false positives. Also in the temporal scale the anomalous events are limited in time and they do not have a long temporal extension. For all these reasons an additional overlay layer is needed to capture the data from the measurement stations and extract information from the lower layers. We argue that this layer can be implemented with a set of virtual soft sensors trained with all the historic measurement

[1] http://www.alertsystems.org.

of the stations and can provide interpolation and, in some case, extrapolation starting from the measured data.

In the current paper we consider a spatial connection of the data deriving which station are more correlated with the other and showing that a mere distance criterion can be not enough to create coherent clusters showing the geographic areas that are affected by similar events.

2 Soft Sensors

Soft sensors are software tools capable to calculate quantities that are either *difficult* to measure or that *cannot be measured*. In this context, *difficult* means that the "cost" of the hardware architecture to measure a specific parameter is too much high. Furthermore, "*cannot be measured*" means that the value of the parameter that we wish to measure could be out of range for any hardware instrument or, it could be a non-physical parameter.

The main characteristic of a soft sensor is the capability to learn the functional relations among the measures and to use this acquired knowledge to perform some kind of forecasting. In other terms, a soft sensor can be viewed as an associative model that links an input pattern with an output pattern.

As shown in Fig. 1, a soft sensor can be described as a black-box capable to estimate a set of output parameters $\widetilde{Y}_n$ starting from an input set of parameters Y_m, with U representing the command signals.

Everyone can devise many approaches that can be adopted to implement the model MSS_n. More in detail, it is possible to distinguish between two different classes of methods, the first one, based on the first principles models and, the second one, based on the data. The former approach, is often used when the process under investigation is well known and, as a consequence, it is possible to build a sound mathematical model. This is the case of biological models [5], fluid dynamic models or some kind of control processes [6]. The second kind of approach is necessary when the model MSS_n is completely unknown but many data, regarding the phenomenon under investigation, are available. In particular, when series of couple of input-output pairs of data have been acquired. In this latter case, input-output data can be employed in a machine learning paradigm to build the MSS_n model. Typically, this kind of approach results in neural networks models or statistical models. We have a particular

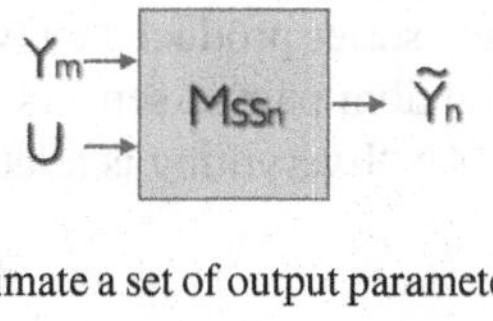

Fig. 1 A soft sensor capable to estimate a set of output parameters $\widetilde{Y}_n$ exploiting a set of parameters Y_m. U is a command signal

experience in designing system based on soft sensors for noninvasive monitoring in the field of cultural heritage [7–10] and in risk prevention applications [11, 12].

Soft sensors are often used to substitute hardware sensors. They can replace hardware sensors at all or they can work as a backup copy of real sensors. Anyhow, when soft sensors are used to replace some real sensor, their *metrological* performances should be made explicit.

The classical procedures for evaluating the performances of a system based either on machine learning paradigm or based on statistical model, can be applied also to soft sensors. These procedures usually make use of statistical evaluators capable to assess the forecasting performance on test set of data.

Since such evaluators, mostly compute overall values, may hide some particular behavior, like the response in case of a rare event. For this reason such procedures do not give an acceptable description of the metrological behaviors of soft sensor. To obtain an in-depth description of a soft sensor behaviors similarly to what happens for the real sensors, we have adopted an evaluation procedure divided in two phases: the first is the the statistical evaluation, based on specific estimators and, the second is the validation by comparison [13]. This procedure uses a set of overall evaluators jointly to a set of analytic evaluators in order to compose a "data sheet" for a soft sensor under investigation. Moreover, it describes a methodology to compare the soft sensors permanences both among them and with respect to a particular soft sensor called "null sensor".

3 The Virtual Layer of Measure

Our case of study regards a wireless sensor network (later referred to as WSN) of pluviometers placed in different locations of Tuscany, Italy. Thus, we have designed a soft sensor network (later referred to as SSN) that fits this WSN of pluviometers. In this paragraph, a methodology to design a SSN is introduced. In such network there is a one-to-one correspondence between a soft sensor and a hardware sensor of the WSN, as shown in Fig. 2. This SSN is capable to backup the WSN by exploiting the functional relations among the measures of the same parameter produced by the hardware sensors of the WSN.

Different ways to capture this functional relation among the measures in a WSN are possible. Looking at Fig. 1, the output parameter (here we are considering a single parameter) $\widetilde{Y}$, of a given hardware sensor s, could be estimated by exploiting the measures Y_m of a cluster[2] of sensors that are, in some way, correlated with s. It is easy to suppose that the chronological series produced by two neighbors sensors are more correlated than the ones of a random pair of sensors. Such a concept is formalized by the first law of geography [14]: "Everything is related to everything else, but near

[2]In this manuscript we use the term "cluster" to indicate the set of hardware sensors used to forecast the measures of another hardware sensor.

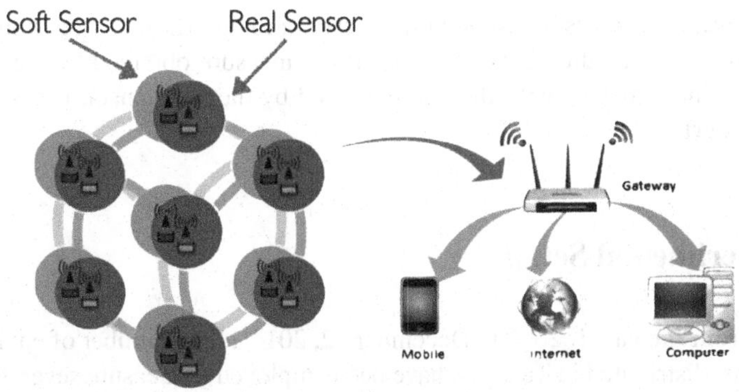

Fig. 2 A graphical representation the SSN that is a mirror of the WSN

things are more related than distant things". Then, according to this law, we might estimate the measures of a certain sensor *s* by the using of the measures of sensors in the neighborhood. But, in some cases, this "general" law is not true. We think that an analysis of the correlation among the measures of all sensors that constitute the WSN could be useful to form the best clusters, in terms of both elements and size.

As we will show in Sect. 4, clusters obtained by considering the geographical closeness and clusters formed by using the correlation analysis could be deeply different, as shown in Fig. 4.

In order to implement the MSS_n models, we have chosen to use a multivariate regression technique. To this aim, we have composed, for each sensor, clusters of different sizes based both on geographical closeness and correlation analysis [12]. According to Eq. 1 for each sensor *s*, the β coefficients and the residue *u* have been computed by the using of a training set of measure $\bar{s_1}, \bar{s_2}, \ldots, \bar{s_n}$ obtained from the sensors $\{s_1, s_2, \ldots, s_n\}$. In doing so, for each hardware sensor, several different models MSS_n and then several soft sensors have been obtained. In fact, varying the size of the clusters and choosing a different criterion to compose them we get different β coefficients and different residues *u*.

$$\bar{s} = \beta_0 + \beta_1 * \bar{s_1} + \beta_1 * \bar{s_2} + \cdots + \beta_n * \bar{s_n} + u \tag{1}$$

In order to evaluate the performances of each soft sensor, we applied a procedure based on specific estimators and a validation by comparison introduced in [13]. Thus, for each pluviometer, we have have chosen the soft sensor that has the best behavior, from a metrological point of view.

This virtual layer of measure can be used in to different modes. It can be considered a backup for some pluviometer that fails its measurements for a certain period of time. In this case, the homologous soft sensor can estimate the missing measures exploiting the measure of other pluviometer registered at the same time. Another mode can be considered a "control mode". For each time *t* in which the WSN produce a set of measures, the SSN estimates this same set of measures. Thus, comparing, for each hardware sensor its measure with the estimation produced by the homologous soft sensor, we can validate the measure. If the measure obtained by the hardware sensor does not "match" with the one produced by the soft sensor, the SSN might notify an alert.

4 Experimental Setup

Starting from January 1, 2000 to December 12, 2014 a high number of environmental stations, distributed in Tuscany, have been employed to measure several parameters. We have taken into account only the pluviometric registrations. The set of the pluviometers counts 279 units. However, during the 15 years of observations, some of these units have never worked whilst other ones have worked for a shorter time. Moreover, several units produced corrupted time series.

By the analysis of the measures distribution, we have decided to involve in our experiments only the pluviometers that have produced a number of valid measures greater then the 80 % of the number of measurements. In this way, we have considered only 160 pluviometers instead 279 (see Fig. 3).

The experiments have been conducted by the using of MATLAB®, thus all registrations have been stored in a matrix of 160×5479 elements labeled *Pluvio*. The 160 rows, of the Pluvio matrix, contain the different pluviometers and its 5479 columns correspond to the number of days of the considered time interval. They correspond also to all the dates taken into account. This matrix contains also several "not a number" (NaN), because some pluviometer for some days has produced no valid values.

Fig. 3 All the environmental units (*right*). The environmental units that have produced a number of valid measures exceeding 4500 (*left*)

Starting from this matrix we have computed other 2 symmetric matrices of 160×160 elements. The first one, is the correlation matrix that contains the correlation coefficients computed between a station and all the other ones. The second matrix is the distances matrix. Each element of this latter matrix contains the distance between the environmental station indicated by the row index and the environmental station indicated by the column index. The distance is computed using the geographical coordinates, in terms of latitude and longitude, of each environmental station.

Considering the correlation matrix and the distance matrix, we sort them in this way: for each row, the values of correlation matrix have been sorted in descending order and the values of distance matrix have been sorted in ascending order. In doing so, we obtained, the list of the best correlated stations and the list of the neighbors stations.

These two sorted matrices make we able to get clusters of environmental stations of different size just selecting the first n elements form a ith row. We can get clusters based on the geographical closeness criterion by the using the distance matrix, or we can get clusters based on the correlation criterion considering the correlation matrix.

For example, the Fig. 4, represents two different clusters of environmental stations referred to the station labeled "P1001189". We have got them by selecting the first five elements both of the correlations sorted matrix and of the distance sorted matrix. The figure also shows how these 2 clusters are overlapped only for one environmental station. This is an evidence of the first law of geography, can have some exception as previously argued.

Considering the *Pluvio* matrix, The training set and the test set are formed for each soft sensor selecting the pair rows of the *Pluvio* matrix to form the training set and selecting the odd rows to form the test set. In doing so, all the measures are divided in two equal parts, and both sets cover the same period.

According to equation , by using the training set, β coefficients and the residue u for each sensor s for each real sensor have been computed obtaining the

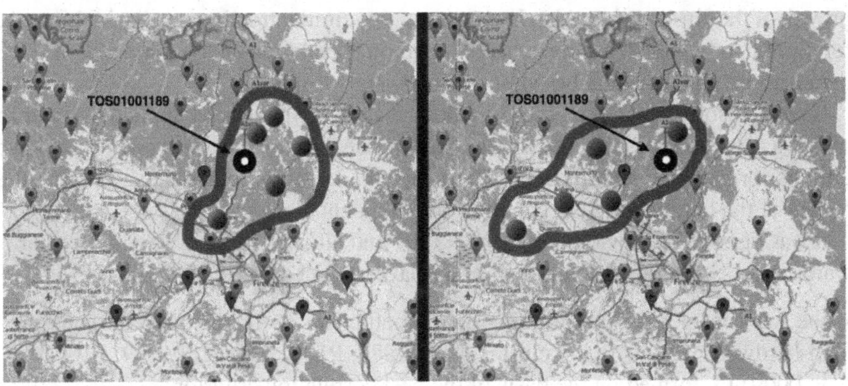

Fig. 4 The most five correlated stations to the station P1001189 (*right*). The nearest five stations to the station P1001189 (*left*)

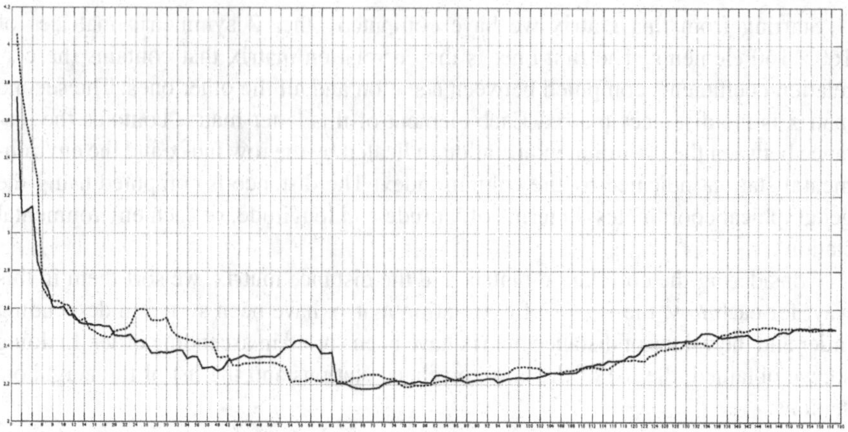

Fig. 5 The SDVs as a function of the clusters size. The *solid line* and *dotted line* for clusters based on the correlation analysis and the geographical closeness respectively

corresponding soft sensor. Thus, by using the test set we have measured the performances of each soft sensor. In order to find the best size of each cluster for each sensor, we have varied the size of the both kind of clusters, starting from 1 to 159.

The Fig. 5 shows the trends of the standard deviation of the error vectors, as a function of the growth of the clusters size, for the soft sensor "P1001189". Both trends are similar, but considering clusters based on the correlation analysis, a minimum in the standard deviation function is reached employing a lower number of stations than considering the clusters based on geographical closeness. That is we need to involve less hardware sensors to build the model MSS_n. Moreover, for clusters of small sizes, the clusters based on correlation analysis always produce better results than clusters based on geographical closeness.

The same kind of results have been obtained by considering the mean, over all the soft sensors, of the standard deviation of the error vectors, as a function of the growth of the clusters size. Also Fig. 6 shows two similar trends, and also in this case, for clusters of small sizes, the clusters based on correlation analysis always produce better results than clusters based on geographical closeness.

For each soft sensor, the procedure described in [13] has been applied in order to obtain a sort of data sheet. According to this procedure, for each soft sensor, we have built its correspondent *null-sensor*. Here, the *null-sensor* is constituted averaging the output of the hardware sensors in the correspondent clusters.

Considering the pluviometer "P1001189" and clusters formed by only five hardware sensors, as represented in Fig. 4, we have obtained:

- the soft sensor $SC_{P1001189}$ finding the β coefficients by the using of the measures of hardware sensors in the cluster {P1001273, P1001205, P1001269, P1001634, P1000911}, formed on the basis of the best correlation criterion;

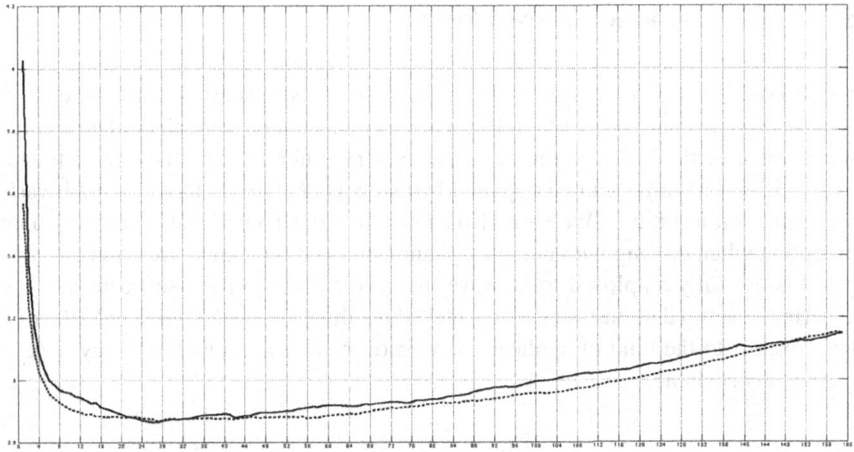

Fig. 6 The trends of the mean of standard deviation as a function of the growth of the clusters size

- the soft sensor $SD_{P1001189}$ finding the β coefficients by the using of the measures of hardware sensors in the cluster {P1000911, P1000941, P1000926, P1004621, P1004784}, formed considering the geographical closeness criterion;
- the null-sensor $Null_{P1001189}$ by the averaging the output values of test set of hardware sensors in the cluster {P1000911, P1000941, P1000926, P1004621, P1004784}, formed considering the geographical closeness criterion.

For these three kinds of soft sensors we have computed the following evaluators

- ratio RR between the ranges of the estimates and the test set target values;
- minimum, maximum, mean and standard deviation values for error vector;
- CC, correlation coefficient between the estimates and the test set values;
- the percentage R_ϵ of estimates lower than a fixed value ϵ;
- the percentage R_E of estimates greater than a fixed value E;

and here, we have reported the results in the following Table 1.

We have computed similar data sheets for all the soft sensors taken into account in this manuscript, finding similar results. Table 1 shows how all the considered evaluators confirm a better behavior for soft sensors based on the correlation analysis.

Table 1 Statistical validation and comparison for the soft sensor P1001189

Evaluator	$SC_{P1001189}$	$SD_{P1001189}$	$Null_{P1001189}$
RR	0.8752	0.9949	0.8330
$\|Mean\|$	1.0216	1.1108	1.1652
Std	2.8479	3.2831	3.3366
CC	0.9249	0.9005	0.8998
$R_\epsilon(\epsilon = 1\text{ mm})$ (%)	96.74	96.83	94.78
$R_E(E = 20\text{ mm})$ (%)	0.25	0.38	0.50

5 Conclusion and Future Works

We showed how a virtual layer of measures, based on soft sensors, can be used as a backup for real sensors in a WSN. It can also be used to notify an alert when a hardware sensor of the WSN produces measures very different from the ones estimate by the SSN. This layer can be obtained by using MSS_n models based on multivariate regression technique. We have also shown how the Tobler's law can have some exception when dealing with complex phenomena and how the correlation analysis could be usefully employed to improve the performances of the soft sensors. The experiment setup demonstrates that achieved result seem overall good and effective with respect to the kind of application. Moreover, the shown methodology is geographically scalable.

References

1. Cipolla, E., Vella, F.: Identification of spatio-temporal outliers through Minimum Spanning Tree. In: 2014 Tenth IEEE International Conference on Signal-Image Technology and Internet-Based Systems (SITIS), pp. 248–255 (2014)
2. Cipolla, E., Maniscalco, U., Rizzo, R., Stabile, D., Vella, F.: Analysis and visualization of meteorological emergencies. J. Ambient Intell. Hum. Comput. (2016). ISSN: 1868-5137
3. Chee-Yee, C., Kumar, S.P.: Sensor networks: evolution, opportunities, and challenges. Proc. IEEE **91**(8), 1247–1256 (2003)
4. Akyildiz, I.F., et al.: Wireless sensor networks: a survey. Comput. Netw. **38**(4), 393–422 (2002)
5. Paulsson, D., Gustavsson, R., Mandenius, C.F.: A soft sensor for bioprocess control based on sequential filtering of metabolic heat signals. Sensors **14**(10), 17864 (2014)
6. Lee, S.D., Zahrani A.J.: Employing first principles model-based soft sensors for superior process control and optimization. In: IPTC 2013: International Petroleum Technology Conference (2013)
7. Maniscalco, U., Pilato, G., Vassallo, G.: Soft sensor based on E-αNETs. Front. Artif. Intell. Appl. **226**, pp. 172–179 (2010). ISSN: 0922-6389
8. Maniscalco, U., Pilato, G.: Multi soft-sensors data fusion in spatial forecasting of environmental parameters. In: Advanced Mathematical and Computational Tools in Metrology and Testing IX, vol. 84, pp. 252–259 (2012)
9. Maniscalco, U.: Virtual sensors to support the monitoring of cultural heritage damage. Biol. Artif. Intell. Environ. 343–350 (2005)
10. Ciarlini, P., Maniscalco, U.: Mixture of soft sensors for monitoring air ambient parameters. In: Proceedings of the XVIII IMEKO World Congress (2006)
11. Maniscalco, U., Rizzo, R.: Adding a virtual layer in a sensor network to improve measurement reliability. In: Advanced Mathematical and Computational Tools in Metrology and Testing X, pp. 260–264. World Scientific Publishing Co, Singapore (2015)
12. Maniscalco, U., Rizzo, R.: A virtual layer of measure based on soft sensors. J. Ambient Intell. Hum. Comput. (2016). ISSN: 1868-5137
13. Ciarlini, P., Maniscalco, U., Regoliosi, G.: Validation of soft sensors in monitoring ambient parameters. In: Advanced Mathematical and Computational Tools in Metrology and Testing VII, vol. 72, p. 142 (2006)
14. Tobler, W.R.: A computer movie simulating urban growth in the Detroit region. Econ. Geogr. 234–240 (1970)

The Use of Eye Tracking (ET) in Targeting Sports: A Review of the Studies on Quiet Eye (QE)

Dario Fegatelli, Francesco Giancamilli, Luca Mallia, Andrea Chirico and Fabio Lucidi

Abstract The Quiet Eye (QE) consists in the final visual fixation before the initiation of a critical phase of the movement, and functionally represent the time needed for the precise control of movements. The aims of the manuscript is provide a mini-review of the studies analyzing through Eye Tracking (ET) the Quiet Eye phenomena in ecological sport settings in the last decade. Using Scopus database was performed a search (January 2005–December 2015) including a combination of "Eye Track*" with "Quiet Eye", and with "Sport" as keywords, and extracting only original research including adult athletes and focused on targeting sports (e.g. shooting, golf, etc). Overall, 30 studies were reviewed, confirming that ET was a useful instrument to address different research issues within sport domain. However, new studies need to confirm these results, and to combine ET with other instruments in order to understand deeply the processes underpinning successful performance in sport.

Keywords Eye tracking · Targeting sports · Quiet eye · Ecological settings

1 Introduction

Eye tracking is the process of measuring the point of gaze (where one is looking) and the motion of the eyes (typically divided into fixations and saccades). Eye Tracker (ET) is a device able to register eye positions and eye movement. The ET has been successfully applied in a wide variety of research domains (e.g. marketing, medicine, etc) including psychology. The research in psychology has been used ET

D. Fegatelli (✉) · F. Giancamilli · A. Chirico · F. Lucidi
Department of Psychology of Development and Socialization Processes,
Sapienza University of Rome, Rome, Italy
e-mail: dario.fegatelli@gmail.com

L. Mallia
Department of Movement, Human and Health Sciences,
University of Rome "Foro Italico", Rome, Italy

G. De Pietro et al. (eds.), *Intelligent Interactive Multimedia Systems and Services 2016*, Smart Innovation, Systems and Technologies 55,
DOI 10.1007/978-3-319-39345-2_64

to investigate how eye movements are related to cognitive processes during different tasks [1].

In the past years the use of ET was limited, since some of them were very complex (e.g. manual calibration, image optimization), time consuming both for researchers and participants, and quite expensive. Additionally, ET was typically used in laboratory settings, limiting the possibility to study the phenomena in ecological settings. Over the last two decades, however, remarkable improvements in ET's technology were made. A new generation of portable ETs with more automatic (and user-friendly) settings (e.g. calibration) and analysis procedures (e.g. eye movement track, fixations duration, etc) allows nowadays researchers to set up a quick and portable eye analysis both in laboratory and in ecological settings, studying participants' eye movements directly in real-life situations.

At the same time, different studies [2] have dealt with understanding the role of visual strategy in sport. Athletes need to be able to pick up some important information in a very complex and changing environment. The best athletes know the best way to collect this information from the surrounding environment. For an excellent performance athletes need to know "*where*" and "*when*" look, since this awareness enables a more efficient use of their time. The ETs allowed researchers to explore and analyze the visual strategies of athletes during their sport performances. Within this research line, empirical evidences with ET suggested that gaze control was critical for skills requiring precise cue selection, optimal timing and the ability to focus for long durations under extreme performance conditions [3]. Scholars suggested that the final fixation made by a performer must not only be located on the target, but must also be of a long enough duration to ensure accuracy [4, 5]. Vickers [6] analyzing with ET the athletes' performance of targeting sports (e.g. basket shooting) evidenced the presence of a typical visual fixation strategy, the so called Quiet Eye (QE). The QE is defined, for a given motor task, as the final fixation or tracking gaze directed to a single location or object in the visual motor work-plan within 3° of visual angle (or less) for a minimum of 100 ms [6].

Here, we provide a mini-review of the studies analyzing through ET the Quiet Eye phenomena in ecological sport settings.

2 Materials and Methods

This study was performed using the guidelines set out by the preferred reporting items for systematic reviews [7]. A broad search was performed using Scopus database in the date range from January 2005 to December 2015, and including a combination of "Eye Track*" with "Quiet Eye", and with "Sport" as keywords. Once the main articles were identified, a second search was carried out using the citations within each article, to supplement the already mentioned search terms.

We considered only original research articles, including adult athletes and focused on targeting sports (e.g. shooting, golf, etc). Non-English-language studies and reviews were excluded. After a screening based on full-text review, 30 articles were included. The main data (e.g. sample, sport, experimental conditions, ET and QE parameter used and results) from each study were reported in Table 1.

3 Results

The studies extracted and reviewed focused on different sports such as golf/putting (n = 13), basket/free throws (n = 4), shooting (n = 3), Soccer/penalty kick (n = 4), Ice-Hockey (n = 2), bowling (n = 1), dart (n = 2) and biathlon (n = 1). Overall, the studies reported in Table 1 evidenced some crucial research issues that will be briefly summarized below. It is important to notice that many studies focused of more than one research issue at the same time.

3.1 The Effects of Expertise on QE

The first research issue is the link between the QE and the expertise of the athletes. Overall, 7 studies [8, 11, 13–15, 23, 24] compared the QE of expert with one of no-expert athletes, across different sports, training or experimental conditions (e.g. difficulty, pressure). The results of five [8, 13–15, 24] of these studies confirmed the evidence of pioneristic studies [5, 6], showing that experts have longer quiet eye durations (QED), early quiet eye onset (QEO), and/or later QE offset. Furthermore, some studies [e.g. 8, 24] showed a positive correlation between QED and or QEO and performance. Only two studies did not register differences in QE parameters [11, 23].

3.2 The Effect of Task Difficulty on QE

A second issue is the analysis of the link between the difficulty of the sport task and the QE. Few studies focused on this issue, and many of them did not confirm the past results [6] showing that QED increase with task difficulty [13, 28, 33]. However, a study focused on dart players [12] showed that QED increase with task difficulty.

Table 1 Studies selected and data extraction

First author	Sport	Sample	Experimental Conditions and Design	ET used and QE parameters measured	Other measures	Main results related to QE
Causer [8]	Shooting	24 elite 24 sub-lite shooters	No conditions	Mobile Eye II (ASL, Waltham, MA) • QED and QEO	• Gun barrel kinematics • Performance	• Elite shooters had both earlier QEO and longer QED • Longer QED and earlier QEO during successful trials for both groups
Causer [9]	Shooting (skeet)	20 international level skeet shooters	• Perceptual training (PT) group • Control group Pretests and posttests along with an 8 wk training	ASL Mobile Eye II (ASL, Waltham, MA) • QED and QEO	• Gun barrel kinematics • Performance	• Athletes of the PT group significantly increased their QED, used an earlier QEO and recorded higher shooting accuracy scores from pretest to posttest
Causer [10]	Shooting (Skeet)	16 elite level shooters (skeet)	2 counterbalanced conditions • Low anxiety (practice round) • High anxiety (competition round).	ASL Mobile Eye II (ASL, Waltham, MA) • QED and QEO	• Anxiety (MRF-3) • Mental Effort (RSME) • Gun barrel kinematics • Performance	Athletes under the high anxiety condition showed shorter QED, less efficient gun motion, along with a decreased performance
Fischer [11]	Basket (free throws) and dart	14 medium-age 7 older-aged less skilled. 15 medium-age 15 older-aged	2 different tasks: • Basketball free throws • Throwing darts (transfer task)	Lightweight head-mounted ET system (Arrington Research, Inc., Scottsdale, AZ) • QED	• Throwing accuracy	No significant differences in QED across the skill or age groups in either task, indicating that expertise in a

(continued)

Table 1 (continued)

First author	Sport	Sample	Experimental Conditions and Design	ET used and QE parameters measured	Other measures	Main results related to QE
		skilled players				perceptual motor task (such as the basketball free throw) can be retained in older athletes
Horn [12]	Dart	24 novices players	Assignment to 2 conditions: • Random targets • Blocked targets	ASL 5000(ASL, Bedford, MA) • QED	• Throwing Accuracy: Radial error	• Longer QED during trials of the random targets group • No relations between QED and accuracy
Chia [13]	Bowling	6 expert 6 novice players	2 Conditions: • Easy (1 pin) • Hard (10 pins)	Mobile Eye-XG movement system. (ASL, Waltham, MA) • QED	• Perceived task difficulty and confidence	• Expert had significantly longer QED in both conditions • No relation between QED, accuracy and task condition
Klostermann [14]	Golf (Putting)	12 expert 12 near-expert golfers	2 focus-of-attention instructions: • Movement-related instruction (MI) • Effect-related instruction (EI)	VICON ET system (EyeSeeCam, 220 Hz). • QEO, QED and QE offset	• Performance • Head movement initiation	• Experts showed an overall later QE offset, longer QED, than near-experts • Larger QE offset differences between experts and near-experts in the MI compared with the EI condition

(continued)

Table 1 (continued)

First author	Sport	Sample	Experimental Conditions and Design	ET used and QE parameters measured	Other measures	Main results related to QE
Mann [15]	Golf (Putting)	10 expert 10 near expert players	No conditions	BIOPAC EOG 100B; (BIOPAC Systems, Santa Barbara, CA) • QED	• EEG and EMG activities • Performance	• Experts had longer QED and greater cortical activation in the right-central region compared with non experts • Association between cortical activation and QE duration
Moore [16]	Golf (Putting)	30 novice participants	Random assignment to: • QE training group • Control group (technical training) Baseline, training, retention, and pressure putts	ASL Mobile Eye II (ASL, Waltham, MA) • QED	• Anxiety (MRF-3) • Cognitive Appraisal • Performance errors	The quiet eye trained group performed more accurately, displayed more effective gaze control (longer QED), and appraised the pressure test more favorably
Moore [17]	Golf (Putting)	127 novice golfers	Randomly assignment to 2 instruction groups: • Challenge group • Threat group	Mobile Eye tracker (ASL; Bedford, MA) • QED	• Cognitive Appraisal • Immediate Anxiety Measurement Scale • Performance Errors • Cardiac and EMG activity • Putting Kinematics	The challenge group performed more accurately, reported more favorable emotions, and displayed more effective gaze (longer QED), putting kinematics, and muscle activity than the threat group

(continued)

Table 1 (continued)

First author	Sport	Sample	Experimental Conditions and Design	ET used and QE parameters measured	Other measures	Main results related to QE
Moore [18]	Golf (Putting)	40 novice golfers	Random assignment to: • QE training group • Control group Pre-training, practice, training, follow up retention task and pressure test	Mobile Eye Tracker (ASL; Bedford, MA) • QED	• Anxiety (MRF-3) • Perfomance • Cardiac and EMG activity • Putting kinematics	The quiet eye trained group performed more accurately, displayed more effective gaze control (longer QED), and appraised the pressure test more favorably
Panchuk [19]	Golf (Putting)	29 amateur golfers	3 intervention groups: • Marker under the ball • Hole-focus instruction • Novel putting device (PBoS). • Control group Pre/ post test.	Mobile Eye tracker (ASL; Bedford, MA) • QED • QE dwell time	• Putting accuracy	• Longer QED on hits than on misses • The control and PBoS groups did not change the QED while the hole-focus group had a decrease, and the marker group an increase in QED. • QE dwell time increased only for marker group
Panchuk [20]	Ice-Hockey	8 elite goaltenders	2 experimental conditions: • 5 m shots • 10 m shots	501 Eye Tracker (ASL) • QED, QEO and QE Offset		• Longer QED and earlier QEO for saves compared to goals • QE offset occurred later in 5 m trials compared to 10 m trials

(continued)

Table 1 (continued)

First author	Sport	Sample	Experimental Conditions and Design	ET used and QE parameters measured	Other measures	Main results related to QE
Panchuk [21]	Ice-Hockey	8 experienced goaltenders 9 experienced shooters	4 occlusion conditions: • Head and upper body • Lower body, • Puck/stick, • All but puck flight	Mobile eye-tracking system (ASL) • QED, QEO, QE offset	• Performance (save percentage)	• QEO earlier on saves than goals . • Relative QED longer on saves than goal
Piras [22]	Soccer (penalty kick)	7 intermediate level goalkeeper	2 different type of kicks: • Instep kick • Inside kick	Mobile ET system (ASL MA, USA) • QED		• Participants had longer QED for saves compared to goals
Rienhoff [23]	Dart	13 skilled 16 less skilled players	3 different viewing conditions: • Baseline • Central • Peripheral	Head-mounted eye-tracking system was used (Eyelink II, SR Research) • QED	• Performance	• For Baseline condition, significant correlations between QED and both throwing accuracy and consistency for the less skilled players • QED and throwing accuracy in central vision conditions were significantly positively correlated for less skilled players
Rienhoff [24]	Basket (free throws)	9 expert, 9 advanced 9 novice players	3 manipulations of attention focus: • On the ball (external) • On their hands (internal) • No instruction (control)	Light-weight, head-mounted, ET (Arrington Research BS007, Scottsdale, AZ) • QEO and QED	• Shooting performance	• External focus of attention lead to a decrease in performance and reduction QED • Better performance was associated with longer QED across all skill levels

(continued)

Table 1 (continued)

First author	Sport	Sample	Experimental Conditions and Design	ET used and QE parameters measured	Other measures	Main results related to QE
Vickers [4]	Biathlon	10 elite biathlon shooters	2 different pressure condition: • Low pressure (LP) • High pressure (HP)	Model 501 mobile eye tracker (ASL; Bedford, MA) • QED	• Cognitive Anxiety (CSAI-2) • Physiological arousal (RPE and HR) • Performance accuracy • Monark cycle ergometer power output	• QED longer on hit than misses • In 100 % Power Output condition, QED decreased • Increasing QED on highest workload prevent choking during HP condition • In lower workload condition QED could be shorter with no effects to performance
Vine [25]	Golf (Putting)	14 novice players	Random assignment to: • QE training group • Control group Pre-test, acquisition phase, retention-transfer (pressure)-retention test putts	ASL Mobile Eye II (ASL, Waltham, MA) • QED	• Cognitive state anxiety • Performance errors	• The QE-trained group maintained more effective attentional control and performed better in the pressure test • Longer QED were associated with better performance across all test putts
Vine [26]	Golf (Putting)	22 elite golfers	Assignment to: • QE training group • Control group	ASL Mobile Eye II (ASL, Waltham, MA) • QED	• Cognitive state anxiety • Competitive and	• QED predicted 43% of the variance in putting performance

(continued)

Table 1 (continued)

First author	Sport	Sample	Experimental Conditions and Design	ET used and QE parameters measured	Other measures	Main results related to QE
			Pre-training, practice, training, follow up retention task and pressure test		experimental Performance	• The QE-trained group maintained their optimal QE under pressure conditions, and had better performance
Vine [27]	Basket (free throws)	20 novice athletes	Assignment to 2 groups: • QE Training group • Control group Pre test, retention test 1, pressure test, retention test 2	Mobile ET ASL (ASL; Bedford, MA) • QED	• Anxiety (MRF-3) • Performance	• QE trained group had significantly longer QE periods and better performance under heightened levels of cognitive anxiety
Vine (In press) [28]	Golf (Putting)	27 skilled golfers	3 conditions of vision occlusion: • Early (prior to backswing), • Late (during putter stroke), • No (control)	Mobile ET(ASL; Bedford, MA). • QED	• Performance • Putting Kinematics	• No significant differences in QED across conditions, • Performance decrease in late occlusion condition and not change in early occlusion condition
Vine [29]	Golf (Putting)	45 novice golfers	Assignment to 3 instruction groups: • QE • Analogy • Explicit Baseline-Retention-Pressure-Retention putts	ASL Mobile ET (ASL, Waltham, MA) • QED	• Anxiety (MRF-3) • Movement Specific Reinvestment Scale (MSRS)	• All the groups increased QED following training • QE-trained group outperformed the Analogy group in the Retention tests, and

(continued)

Table 1 (continued)

First author	Sport	Sample	Experimental Conditions and Design	ET used and QE parameters measured	Other measures	Main results related to QE
					• Performance errors	both other groups in the Pressure test; underpinned by superior visualattentional control (longer QED)
Vine [30]	Golf (Putting)	50 expert golfers	Participants performed putts under pressure until they missed ("shootout")	ASL Mobile ET (ASL, Waltham, MA) • Total QED • QED-pre and QED-online • QED- Dwell	• Anxiety (MRF-3) • Movement phase durations.	• Total QED was shorter for the final (missed) putt compared with the first and penultimate (successful) putts • QED-pre was similar across the three putts, while QED-online and QED-dwell were shorter on the missed putt
Wilson [31]	Basket (free throws)	10 experienced players	2 counterbalanced conditions: • Low anxiety condition • High anxiety condition	Mobile ET (ASL; Bedford, MA). • QED e QEO	• Anxiety (MRF-L) • Performance	• Participants had longer QED and earlier QEO for successful shots (hits) compared with misses- • High anxiety resulted in significant reductions QED and free throw success rate

(continued)

Table 1 (continued)

First author	Sport	Sample	Experimental Conditions and Design	ET used and QE parameters measured	Other measures	Main results related to QE
Wilson [32]	Golf (Putting)	6 university team golfers	Player performed 25 putts in a randomized order; five on each slope of flat, –0.9° left-to-right and right-to-left –1.8° left to-right and right-to-left	ASL Mobil Eye gaze-registration system (Bedford, Massachusetts) • number of fixations • QED	• Performance (holed or missed and error)	Shorter QED on missed than on holed putts
Wood [33]	Soccer (penalty kick)	18 experienced footballer	2 counterbalanced conditions: • Threat (low vs. high) • Goalkeeper movements (stationary vs. waving arms)	Mobile ET (ASL; Bedford, MA) • QED	• The anxiety thermometer • Performance	No significant main effects were found for threat, or goalkeeper movement on QED
[34]	Soccer (penalty kick)	20 university level soccer players	Randomly assignment to 2 groups: • QE training group • Control group Baseline, retention test, pressure test (penalty shootouts)	Mobile Eye gaze registration System (ASL; Bedford, MA) • QED aiming phase (A-QE) • QED execution phase (B-QE)	• Anxiety (MRF-3) • Performance	• The QE training group showed longer A-QED and B-QED respect to control group, • No differences on performance were registered between the two groups
Wood [35]	Soccer (penalty kick)	20 experienced footballer	Randomly assignment to 2 groups: • QE training program • Practice program Baseline, retention, transfer (penalty shootouts)	Mobile Eye tracker (ASL; Bedford, MA,USA) • QED	• Anxiety (MRF-3) • Control beliefs • Performance	• The QE training group showed longer QED and better performance in all the conditions and increased their perceptions of shooting ability (competence) and ability to score and cope with the pressure (control)

(continued)

Table 1 (continued)

First author	Sport	Sample	Experimental Conditions and Design	ET used and QE parameters measured	Other measures	Main results related to QE
Ziv [36]	Golf (Putting)	72 novice players	Assignment to 3 groups: • Internal focus group (24) • External focus group (24) • Control group (24) 2 experimental conditions: • Non Distraction • Distraction/Non distraction	Mobile Eye tetherless ET system (ASL, Bedford, MA, USA) • QED	• Performance Errors	• Under non-distracted conditions QED were longer in the EXT participants than other groups while their performance did not improve • Under distracted conditions, higher performance was observed in both the INT and EXT attentional focus participants than in the Controls

3.3 The Effect of Focus of Attention

A third issue is the link between the focus of the attention during the sport task and the QE. Studies on this topic emerged recently in literature, these studies focused on the effects of different attentional focus on QE and performance [14, 24, 36]. It seems that an external focus provides a better performance and a longer QED [36] and different focus location in an external condition of attention brings variation on QE parameters as offset [14]. However, other study provided opposite effects [24] especially related to QED and performance when external attentional focus occurred.

3.4 Competitive Anxiety and QE

Six studies deal specifically with the issue of the relationship between athletes' competitive anxiety/pressure and the QED. Four studies [10, 17, 30, 31] showed that the athletes that perform under high anxiety conditions in different sports (shotgun, basket and golf) have a shorter QED along with a decreased performance. Only one study focused on soccer [33] didn't find any relation between "high threat" condition and QED while a relevant study focused on biathlon [4] stressed that an increasing of QED prevent choking during high pressure conditions.

3.5 QE Training

A final issue is the possibility to improve QED and consequently the performance through specific training. Overall, seven studies [18, 25–27, 29, 34, 35] analyzed the effect of specific QE training protocols on the QE parameters during the performance. The results showed that training focused on QE were efficacious in different sports (golf putting, basket free throw, soccer penalty kick): the athletes significantly increased their QED and recorded higher performance. Moreover, these studies showed that the athletes trained with specific QE protocols were able to maintain a longer QED and provide a better performance [with exception of 34] than the other athletes under high anxiety conditions.

4 Discussion

The present studies identified 30 original studies that in the last decade used the ET systems to study the Quiet Eye in different sport settings. Overall, from the data extracted by these studies emerged that ET was a useful instrument to address different research issues within sport domains. However, some inconsistencies in results were emerged across the different issues, reflecting mainly differences in data analysis as well as in definition and operationalization of task difficulty and task-specific skills. Consequently, new studies using comparable methodologies need to confirm the results emerged in this mini-review in other/ new sport specialties (e.g. combined event in pentathlon). Furthermore, according with some studies included in this review [15, 17, 18], additional future studies needs to combine ET with other instruments (e.g. EEG, EMG) in order to jointly analyze the different processes involved in the athletes' best performance.

References

1. Mele, M.L., Federici, S.: Gaze and eye-tracking solutions for psychological research. Cogn. Process. **13**(1), S261–S265 (2012)
2. Mann, D.Y., Williams, A.M., Ward, P., Janelle, C.M.: Perceptual-cognitive expertise in sport: a meta-analysis. J. Sport. Exerc. Psychol. **29**(4), 457–478 (2007)
3. Vickers, J.N.: Mind over muscle: the role of gaze control, spatial cognition, and the quiet eye in motor expertise. Cogn. Process. **12**(3), 219–222 (2011)
4. Vickers, J.N., Williams, A.M.: Performing under pressure: the effects of physiological arousal, cognitive anxiety, and gaze control in biathlon. J. Mot. Behav. **39**, 381–394 (2007)
5. Williams, A.M., Singer, R.N., Frehlich, S.G.: Quiet eye duration, expertise, and task complexity in near and far aiming tasks. J. Mot. Behav. **34**(2), 197–207 (2002)
6. Vickers, J.N.: Visual control when aiming at a far target. J. Exp. Psychol. Hum. Percept. Perform. **22**, 342–354 (1996)
7. Moher, D., Liberati, A., Tetzlaff, J., Altman, D.G.: PRISMA group: preferred reporting items for systematic reviews and metanalysis: the prisma statement. Ann. Intern. Med. **151**, 264–269 (2009)
8. Causer, J., Bennett, S.J., Holmes, P.S., Janelle, C.M., Williams, A.M.: Quiet eye duration and gun motion in elite shotgun shooting. Med. Sci. Sport. Exerc. **42**(8), 1599–1608 (2010)
9. Causer, J., Holmes, P.S., Williams, A.M.: Quiet eye training in a visuomotor control task. Med. Sci. Sport. Exerc. **43**(6), 1042–1049 (2011)
10. Causer, J., Holmes, P.S., Smith, N.C., Williams, A.M.: Anxiety, movement kinematics, and visual attention in elite-level performers. Emotion **11**(3), 595–602 (2011)
11. Fischer, L., Rienhoff, R., Tirp, J., Baker, J., Strauss, B., Schorer, J.: Retention of quiet eye in older skilled basketball players. J. Mot. Behav. **47**(5), 407–414 (2015)
12. Horn, R.R., Okumura, M.S., Alexander, M.G.F., Gardin, F.A. Sylvester, C.T: Quiet eye duration is responsive to variability of practice and to the axis of target changes. Res. Q. Exerc. Sport. **83**(2), 204–211 (2012)
13. Chia, S.J., Chow, J.Y., Kawabata, M., Dicks, M., Lee, M.: An exploratory analysis of variations in quiet eye duration within and between levels of expertise. Int. J. Sport Exerc. Psychol. 1–15 (2016). doi:10.1080/17461391.2015.1122093
14. Klostermann, A., Kredel, R., Hossner, E.-J.: On the interaction of attentional focus and gaze: the quiet eye inhibits focus-related performance decrements. J. Sport Exerc. Psychol. **36**(4), 392–400 (2014)
15. Mann, D.T.Y., Coombes, S.A., Mousseau, M.B., Janelle, C.M.: Quiet eye and the Bereitschaftspotential: visuomotor mechanisms of expert motor performance. Cogn. Process. **12**(3), 223–234 (2011)
16. Moore, L.J., Vine, S.J., Freeman, P., Wilson, M.R.: Quiet eye training promotes challenge appraisals and aids performance under elevated anxiety. Int. J. Sport Psychol. **11**(2), 169–183 (2013)

17. Moore, L.J., Vine, S.J., Wilson, M.R., Freeman, P.: The effect of challenge and threat states on performance: an examination of potential mechanisms. Psychophysiology **49**(10), 1417–1425 (2012)
18. Moore, L.J., Vine, S.J., Cooke, A., Ring, C., Wilson, M.R.: Quiet eye training expedites motor learning and aids performance under heightened anxiety: The roles of response programming and external attention. Psychophysiology **49**(7), 1005–1015 (2012)
19. Panchuk, D., Farrow, D., Meyer, T.: How can novel task constraints be used to induce acute changes in gaze behaviour? J. Sports Sci. **32**(12), 1196–1201 (2014)
20. Panchuk, D., Vickers, J.N.: Gaze behaviors of goaltenders under spatial-temporal costraints Hum. Mov. Sci **25**(6), 733–752 (2006)
21. Panchuk, D., Vickers, J.N.: Using spatial occlusion to explore the control strategies used in rapid interceptive actions: predictive or prospective control? J. Sports Sci. **27**(12), 1249–1260 (2009)
22. Piras, A., Vickers, J.N.: The effect of fixation transitions on quiet eye duration and performance in the soccer penalty kick: instep versus inside kicks. Cogn. Process. **12**(3), 245–255 (2011)
23. Rienhoff, R., Baker, J., Fischer, L., Strauss, B., Schorer, J.: Field of vision influences sensory-motor control of skilled and less-skilled dart players. J Sports Sci Med. **11**(3), 542–550 (2012)
24. Rienhoff, R., Fischer, L., Strauss, B., Baker, J., Schorer, J.: Focus of attention influences quiet-eye behavior: an exploratory investigation of different skill levels in female basketball players. Sport Exerc. Perform. Psychol. **4**(1), 62–74 (2015)
25. Vine, S.J., Wilson, M.R.: Quiet eye training: Effects on learning and performance under pressure. J. Appl. Sport Psychol. **22**(4), 361–376 (2010)
26. Vine, S.J., Moore, L.J., Wilson, M.R.: Quiet eye training facilitates competitive putting performance in elite golfers. Front. Psychol. **2**, 8 (2011)
27. Vine, S.J., Wilson, M.R.: The influence of quiet eye training and pressure on attention and visuo-motor control. Acta Psychol. **136**, 340–346 (2011)
28. Vine, S.J., Lee, D.H., Walters-Symons, R., Wilson, M.R.: An occlusion paradigm to assess the importance of the timing of the quiet eye fixation. Eur. J. Sport Sci. (2015). doi:10.1080/17461391.2015.1073363
29. Vine, S.J., Moore, L.J., Cooke, A., Ring, C., Wilson, M.R.: Quiet eye training: a means to implicit motor learning. Int. J. Sport Psychol. **44**, 367–386 (2013)
30. Vine, S.J., Lee, D., Moore, L.J., Wilson, M.R.: Quiet eye and choking: online control breaks down at the point of performance failure. Med. Sci. Sports Exerc. **45**(10), 1988–1994 (2013)
31. Wilson, M.R., Vine, S.J., Wood, G.: The influence of anxiety on visual attentional control in basketball free throw shooting. J. Sport Exerc. Psychol. **31**(2), 152–168 (2009)
32. Wilson, M.R., Pearcy, R.C.: Visuomotor control of straight and breaking golf putts. Percept. Mot. Skills **109**(2), 555–562 (2009)
33. Wood, G., Wilson, M.R.: A moving goalkeeper distracts penalty takers and impairs shooting accuracy. J. Sport Sci. **28**(9), 937–946 (2010)
34. Wood, G., Wilson, M.R.: Quiet- eye training for soccer penalty kicks. Cogn. Process. **12**(3), 257–266 (2011)
35. Wood, G., Wilson, M.R.: Quiet-eye training, perceived control and performing under pressure. Psychol. Sport Exerc. **13**(6), 721–728 (2012)
36. Ziv, G., Lidor, R.: Focusing attention instructions, accuracy, and quiet eye in a self-paced task —an exploratory study. Int. J. Sport Exerc. Psychol. **13**(2), 104–120 (2015)

The Elapsed Time During a Virtual Reality Treatment for Stressful Procedures. A Pool Analysis on Breast Cancer Patients During Chemotherapy

A. Chirico, M. D'Aiuto, M. Pinto, C. Milanese, A. Napoli, F. Avino, G. Iodice, G. Russo, M. De Laurentiis, G. Ciliberto, A. Giordano and F. Lucidi

Abstract Virtual reality (VR) during chemotherapy has resulted in an elapsed time compression effect, validating the use of VR in the treatment of some stressful conditions. In the past literature the framework of the pacemaker–accumulator cognitive model of time perception resulted very reliable to explain this effect. This pilot-study explored the efficacy of Virtual Reality in reducing the perception of time during receipt of intravenous chemotherapy. Patient's retrospective estimates of time elapsed during this treatment were evaluated versus patient's treated with Music-Therapy. Materials and methods 47 breast cancer patients were randomly assigned to 20 min of VR treatment (N = 24) or 20 min of Music-therapy (N = 23) during chemotherapy infusion. Difference between actual and perceived elapsed time during chemotherapy with VR and with MT were evaluated with

A. Chirico (✉) · F. Lucidi
Department of Psychology of Development and Socialization Processes, Sapienza University of Rome, Rome, Italy
e-mail: andrea.chirico@uniroma1.it

A. Chirico · C. Milanese · G. Russo
Sbarro Health Research Organization, Temple University, 1900 12th street, Philadelphia, PA 19122, USA

M. D'Aiuto · F. Avino · G. Iodice · M. De Laurentiis
Breast Department, INT "Pascale" IRCCS, Naples, Italy

M. Pinto
Quality of Life Department, INT "Pascale" IRCCS, Naples, Italy

A. Napoli
Department of Electrical and Computer Engineering, Temple University, Philadelphia, PA, USA

G. Russo · G. Ciliberto
Scientific Direction, INT "Pascale" IRCCS, Naples, Italy

A. Giordano
College of Science and Technology, Temple University, Philadelphia, PA, USA

G. De Pietro et al. (eds.), *Intelligent Interactive Multimedia Systems and Services 2016*, Smart Innovation, Systems and Technologies 55,
DOI 10.1007/978-3-319-39345-2_65

ANOVA analysis. Results: The VR group underestimated the time spent with VR treatment, instead MT treatment group overestimated it. In one step anova model, the VR versus MT treatment showed a significant difference in terms of altered time perception ($F = 5.06$, $p = 0.0008$). Further analysis on the same panel of patients will be conducted to explain also the role of some possible mediator of this effect.

1 Introduction

The burden of cancer mortality is improving in the last 5 years in Italy, and in other developed countries. [1] A Major factor of the improvement in survival are the innovative treatments that involve different kind of chemotherapies [2]. Chemotherapy has significantly extended life in patients with many forms of cancer, but these outcomes have been achieved at the cost of heavy collateral damage to patients from adverse side effects and impaired quality of life, also in breast cancer patients [3]. Despite considerable improvement in specificity and side effect profiles of chemotherapeutic regimens, prevalence of adverse treatment-related symptoms remains high [4]. Anxiety and negative expectations are both significantly associated with worse treatment-related adverse effects, which in turn may trigger the development of conditioned responses to chemotherapy administration reducing quality of life. [5] High levels of treatment-related distress increase the risk of noncompliance, decreased or delayed dosing, and interruption or discontinuance of chemotherapy [6–8], which can reduce the likelihood of remission or cure and jeopardize survival [7]. In view of these considerations, early delivery of interventions to improve tolerability of initial chemotherapy sessions is a key to achieving better patient outcomes.

It is well-established that music produces emotional effects and changes due to many variables, such as physiological appraisal, memories evoked, cognitive changes, physiological arousal [9]. Considered a support to the traditional medical practice, MS relieving effects on cancer patients going through chemotherapy are well known since long time [10].

In recent years, we have been experiencing a real changing world owing to the technological advances which allowed the development of new user-friendly devices in different settings and for different functions. Virtual Reality devices are increasing their possibilities by using in health contexts. A recent review of literature showed significant effects of Virtual Reality treatment for reducing distress of patients in different oncologic settings [11].

2 Time Perception, Pace Accumulator Model

Occupation of treatment time has been identified as an important process-related concern of chemotherapy recipients, but patients' experience of the time spent during treatment has not been extensively explored [12]. Pacemaker-accumulator model is one of the framework used to explain the perception of time from a cognitive point of view [13–16]. This cognitive model, a dominant paradigm in time perception research, has been applied to studies of both prospective and retrospective duration estimation for intervals ranging from fractions of a second to many minutes [17]. The PA model has different processing stages: clock, memory, and decision. The clock includes a pacemaker which emits regular pulses and an accumulator which collects them, connected by an on/off switch controlling pulse transfer. The memory stage includes working (short-term) and reference (longterm) memory. Working memory receives the accumulator output and packages both for transfer to reference memory (which adds it to duration reference data compiled from earlier experiences) and to the comparator. The comparator (the locus of the decision stage) compares the duration representation from working memory with significant durations previously stored in reference memory; the outcome of the comparison determines a behavioral response, such as the estimate given by an individual when asked to report the duration of an interval. Figure 1 shows an ideal model of PA accumulator in neutral environment. However chemotherapy setting is not a neutral environment and aversive stimuli distort time perception by promoting negative physiological arousal, which accelerates the rate of pulse generation in the pacemaker; they also focus patients' attention on the passage of time during chemotherapy, which keeps the switch between pacemaker and accumulator in the "on" position [14–16] throughout treatment, allowing all generated pulses to reach the accumulator. The accumulator therefore collects more pulses during a treatment interval than during an equivalent period of time in a neutral environment, so the experienced duration of treatment exceeds its actual duration, yielding an overestimate of total time elapsed.

Schneider et al. in her studies showed significant differences ($p < 0.001$) between perceived and actual time spent with VR device [12, 18, 19].

The aim of this study is to evaluate the effect of VR on the time perception of breast cancer patients during chemotherapy infusion.

Fig. 1 Means plots

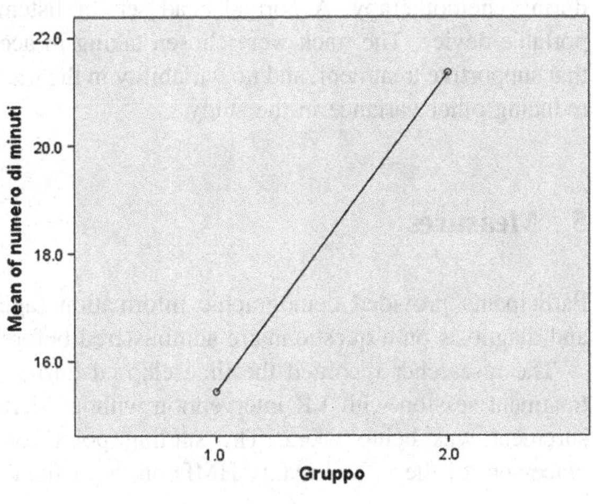

3 Materials and Methods

This is a secondary analysis of pooled data collected from three trials evaluating effectiveness of VR in reducing chemotherapy-related symptom distress in cancer patients.

47 Female patients with breast cancer diagnosis were recruited at the National Cancer Institute of Napoli "Fondazione G. Pascale", Italy. Inclusion criteria comprehended patients between 18 and 70 age with breast cancer diagnosis, receiving chemotherapy treatments. Exclusion criteria comprehended subjects with epilepsy, diabetes, addictions, metastasis, wearing glasses, wearing port.

Each participant was randomly assigned to a VR treatment or to a Music therapy treatment during their second chemotherapy infusion.

Each participant has been informed by their doctor of possibility of virtual reality immersion and/or music-therapy treatment during chemotherapy as distraction intervention.

Each study was approved by the human subjects review board and scientific direction review committee of the participating cancer center. Informed consent was obtained from all participants.

4 Procedure

The administration of the intervention (Music or VR) last totally 20 min.

The VR was administered through a head-mounted glasses (Vuzix Wrap 1200 VR) interfacing a computer capable of interactive three-dimensional (3D) visualization sensing the position/orientation of the user. Each patient had a joystick consenting the interaction with the virtual environment. Virtual reality and animation, represented relaxing landscapes environment, created on Second Life® platform (Linden Lab).

Music-therapy was realized through the listening of recorded relaxing music during chemotherapy. A normal head set for listening music was attached to a portable device. The track were chosen taking in account other studies [20] using that supportive treatment, and no variability in the tracks were offered to patients for reducing other variance in the study.

5 Measures

Participants provided demographic information (age, gender, and race/ethnicity) and diagnosis on a questionnaire administered before the chemotherapy treatment.

The researcher recorded the time elapsed during each patient's chemotherapy treatment session with VR intervention without alerting the patient that this measurement was being taken. The starting point for time measurement was the placement of the virtual reality HMD on the patient's head as infusion began; the end point was the removal of the HMD as infusion keep continuing (waiting time in the chemotherapy suite and time required for IV/port access were not included). Thus, for the purposes of this study, the actual time elapsed during the chemotherapy session with VR intervention is defined as the full duration of the treatment. Patients were not informed beforehand that they would be asked any questions about the passage of time during chemotherapy. Instead, as the HMD was being removed at the end of the treatment session, the researcher asked the patient to retrospectively estimate the amount of time that had elapsed (in minutes) while the patient was using the virtual reality intervention. The estimate provided by the patient in response to this request is defined as the estimated time elapsed.

6 Data Analysis

Each participant's estimate of the number of minutes elapsed during the chemotherapy session with VR distraction was subtracted from the actual number of minutes elapsed as recorded by the researcher for that treatment session. This time difference variable (which takes a positive value when participants underestimate time elapsed and a negative value when participants overestimate time elapsed) was evaluated. Independent variables of interest included age, gender, and diagnosis. After initial screening of data for errors and outliers, descriptive statistics were calculated for all variables.

A one-way Analysis Of Variance (ANOVA) was performed in order to analyze differences in the perceived time spent during the intervention between the two groups. Data were analyzed with SPSS Inc version 16.

7 Results

7.1 Descriptive Statistics

Sample demographics are presented in Table 1. Actual time elapsed during chemotherapy treatment sessions with VR intervention as recorded by the researcher averaged 20 min for the pooled sample.

Table 1 Demographic information of the sample

Status	miss	1
	Married	39
	Separate	2
	Single	2
	Widowed	3
Education	Low	3
	College Degree	7
	Middle	10
	High	27
N		47

Table 2 Descriptive Statistics of time perception

	N	Mean	Std. deviation	Std. error	Minimum	Maximum
VR	23	15.435	4.9802	1.0384	10.0	25.0
MT	22	21.364	5.6023	1.1944	15.0	30.0
Total	45	18.333	6.0302	0.8989	10.0	30.0

VR: Virtual Reality
MT: Music Therapy

Table 3 ANOVA

	Sum of squares	Df	Mean square	F	Sig.
Between groups	395.257	1	395.257	14.108	0.001
Within groups	1204.743	43	28.017		
Total	1600.000	44			

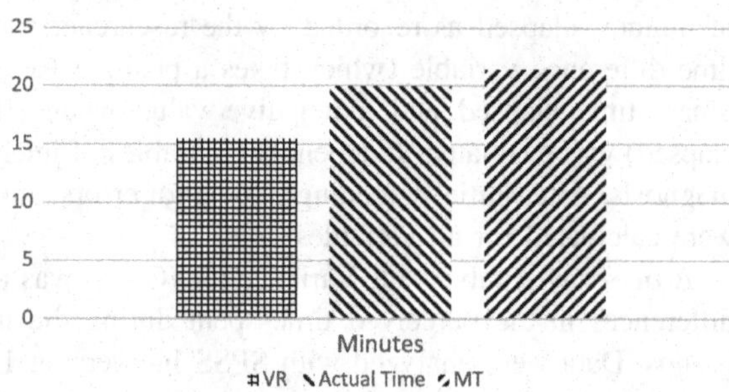

Fig. 2 Actual time elapsed during chemotherapy treatment with VR immersion versus patient estimates of time elapsed during the VR versus MT treatment

Most of participants (56 %) underestimated the duration of VR intervention during chemotherapy sessions, therefore, considerable amount of participants (31.8 %) of MT intervention overestimated the time spent with the equipment (Table 2).

7.2 *Analysis of Variance*

The results of the ANOVA showed significant differences between the VR and the MT group in the perceived time spent with the equipment ($p < 0.001$) (Table 3).

On average patients treated with VR interventions during chemotherapy infusion underestimate the time actual spent with the equipment and, therefore, the patients treated with MT intervention overestimate the actual time spent (Fig. 2).

8 Discussion

The goal of the study was to evaluate the efficacy of VR intervention as distraction intervention during chemotherapy for breast cancer patients. Other study found similar results, specifically in breast cancer patients and also with a large number of subjects [12] The analysis of descriptive statistics confirmed the result found in previous literature, also with greater differences. 62.5 % of participants in the VR intervention group underestimate the time spent and no one overestimate it, therefore 69.6 % of participants in the MT group overestimated the time spent and no one underestimated it. The ANOVA analysis confirmed the significant differences between the two groups. With respect to the study of Schneider et al. [12], in which were treated different diagnosis with different symptoms severities, in our study we enrolled only breast cancer patients at the second chemotherapy treatment after surgery, this allow us to reduce the influencing bias due to the symptoms severity on the perception of time. In the PA model, more severe symptoms may preferentially direct patients' attention to negative stimuli (accelerating pulse generation in the pacemaker) and inhibit their ability to immerse themselves in VR (increasing awareness of the passage of time during chemotherapy). Such patients would thus be less likely to underestimate (and more likely to overestimate) time elapsed during treatment.

The study is part of a broader project using VR in oncological setting taking in account also other measurements (biological, neurological and physiological). However the result showed here could be only considered as a descriptive analysis that need more other analysis to fully understand the time perception mechanism during VR treatment. Different variables could play an important role and need to be assessed and correlate to this analysis, anxiety, mood, self-efficacy, symptoms severity could influence the perception of the patients. Future studies should focus on the possibilities of explore factors that can influence the intervention in order to bring tailored "distressing tools" that meet specific needs of each patient.

Acknowledgment Andrea Chirico and Antonio Giordano were funded by Sbarro Health Research Organization (www.shro.org) and the Commonwealth of Pennsylvania, Department of Health, Biotechnology Research Program.

References

1. Stephens, P.A.: Breast Disease. In: Breast Disease Comprehensive Management, pp. 515–522 (2015)
2. DeVita, V.T., Chu, E.: A history of cancer chemotherapy. Cancer Res. **68**(21), 8643–8653 (2008)
3. Langford, D.J., Paul, S.M., Cooper, B., Kober, K.M., Mastick, J., Melisko, M., Levine, J.D., Wright, F., Hammer, M.J., Cartwright, F., Lee, K.A., Aouizerat, B.E., Miaskowski, C.: Comparison of subgroups of breast cancer patients on pain and co-occurring symptoms following chemotherapy. Support. Care Cancer **24**(2), 605–614 (2016)

4. Cohen, L., de Moor, C.A., Eisenberg, P., Ming, E.E., Hu, H.: Chemotherapy-induced nausea and vomiting: incidence and impact on patient quality of life at community oncology settings. Support. Care Cancer **15**(5), 497–503 (2007)
5. Hofman, M., Morrow, G.R., Roscoe, J.A., Hickok, J.A., Mustian, K.M., Moore, D.F., Wade, J.L., Fitch, T.R.: Cancer patients' expectations of experiencing treatment-related side effects: a University of Rochester Cancer Center–Community Clinical Oncology Program study of 938 patients from community practices. Cancer **101**(4), 851–857 (2004)
6. Abetz, L., Coombs, J.H., Keininger, D.L., Earle, C.C., Wade, C., Bury-Maynard, D., Copley-Merriman, K., Hsu, M.-A.: Development of the cancer therapy satisfaction questionnaire: item generation and content validity testing. Value Health **8**(Suppl 1), S41–S53 (2005)
7. Goldspiel, B.R.: Chemotherapy dose density in early-stage breast cancer and non-hodgkin's lymphoma. Pharmacotherapy **24**(10), 1347–1357 (2004)
8. Greer, J.A., Pirl, W.F., Park, E.R., Lynch, T.J., Temel, J.S.: Behavioral and psychological predictors of chemotherapy adherence in patients with advanced non-small cell lung cancer. J. Psychosom. Res. **65**(6), 549–552 (2008)
9. Juslin, P.N., Sloboda, J.A.: Music and Emotion: Theory and Research. Oxford University Press, Oxford UK (2011)
10. Bailey, L.M.: The effects of live music versus tape-recorded music on hospitalized cancer patients. Music Ther. **3**(1), 17–28 (1983)
11. Chirico, A., Lucidi, F., De Laurentiis, M., Milanese, C., Napoli, A., Giordano, A.: Virtual reality in health system: beyond entertainment. A mini-review on the efficacy of vr during cancer treatment. J. Cell. Physiol. (2015)
12. Schneider, S.M., Kisby, C.K., Flint, E.P.: Effect of virtual reality on time perception in patients receiving chemotherapy. Support. Care Cancer **19**(4), 555–564 (2011)
13. Buhusi, C.V., Meck, W.H.: What makes us tick? Functional and neural mechanisms of interval timing. Nat. Rev. Neurosci. **6**(10), 755–765 (2005)
14. Burle, B., Casini, L.: Dissociation between activation and attention effects in time estimation: Implications for internal clock models. J. Exp. Psychol. Hum. Percept. Perform. **27**(1), 195–205 (2001)
15. Droit-Volet, S., Gil, S.: The time-emotion paradox. Philos. Trans. R. Soc. Lond. B Biol. Sci. **364**(1525), 1943–1953 (2009)
16. Wittmann, M., Paulus, M.P.: Decision making, impulsivity and time perception. Trends Cogn. Sci. **12**(1), 7–12 (2008)
17. Grondin, S., Plourde, M.: Judging multi-minute intervals retrospectively. Q. J. Exp. Psychol. (Hove) **60**(9), 1303–1312 (2007)
18. Schneider, S.M., Hood, L.E.: Virtual reality: a distraction intervention for chemotherapy. Oncol. Nurs. Forum **34**(1), 39–46 (2007)
19. Schneider, S.M., Prince-paul, M., Allen, M.J., Silverman, P., Talaba, D.: Women Receiving Chemotherapy. Oncol. Nurs. Forum **31**(1), 81–88 (2004)
20. Lin, M.-F., Hsieh, Y.-J., Hsu, Y.-Y., Fetzer, S., Hsu, M.-C.: A randomised controlled trial of the effect of music therapy and verbal relaxation on chemotherapy-induced anxiety. J. Clin. Nurs. **20**(8), 988–999 (2011)

Author Index

G. De Pietro et al. (eds.), *Intelligent Interactive Multimedia Systems and Services 2016*, Smart Innovation, Systems and Technologies 55,
DOI 10.1007/978-3-319-39345-2

Printed in the United States
By Bookmasters